Applied Anatomy & Physiology

A Case Study Approach

Brian R. Shmaefsky, PhD

Kingwood College
Kingwood, Texas

Senior Editor	Sonja Brown
Project Editor	Courtney Kost
Developmental Editor	Nadia Bidwell, Barking Dog Editorial
Content Expert	Melissa Curfman-Falvey
Cover and Text Designer	Leslie Anderson
Illustrator	Graphic World
Copy Editor	Colleen Duffy
Desktop Production	Leslie Anderson, Petrina Nyhan
Proofreader	Kay Savoie
Indexer	Nancy Fulton

Publishing Team—Robert Cassel, Publisher; Janice Johnson, Vice President, Marketing; Kerri Goughnour, Marketing Manager; Shelley Clubb, Electronic Design and Production Manager.

Photo Credits

Chapter 1: *Page 2* Krista Kennell/ZUMA/Corbis; *Page 4* A public domain image courtesy of Wikipedia; *Page 5* Bill Varie/CORBIS; *Page 14* Mediscan/CORBIS; **Chapter 2**: *Page 30* Chip East/Reuters/Corbis; *Page 34* USDA National Organic Program; **Chapter 3**: *Page 74* Jonathan Blair/CORBIS; *Page 131 top* CMSP, *bottom* Centers for Disease Control/Dr. W. Winn; **Chapter 4**: *Page 134* Gary Gladstone/CORBIS; *Page 137* CMSP; *Page 141* Brown/CMSP; *Page 153* courtesy of William S. Graham Foundation for Melanoma Research, Inc; *Page 156* Wood/CMSP; **Chapter 5**: *Page 170* George Steinmetz/Corbis; *Page 206* NMSB/CMSP; *Page 208 top* courtesy of National Museum of Archaeology, Anthropology, and History, Peru, *bottom* CMSP; **Chapter 6**: *Page 222* NASA TV/epa/Corbis; *Page 243* RAHMAT GUL/epa/Corbis; **Chapter 7**: *Page 258* Robert Essel/Corbis; *Page 289* CMSP; *Page 290 top* CMSP, *bottom* CMSP; **Chapter 8**: *Page 292* George Shelley/CORBIS; *Page 311* NMSB/CMSP; **Chapter 9**: *Page 326* Ruaridh Stewart/Ruaridh Stewart/ZUMA/Corbis; *Page 355* Siebert/CMSP; *Page 371 top* RUET STEPH/CORBIS SYGMA, *bottom* CMSP; **Chapter 10**: *Page 372* RNT Productions/CORBIS; *Page 389* English/CMSP; *Page 392* CMSP; **Chapter 11**: *Page 406* K. Solveig/zefa/Corbis; *Page 413* Lund/CMSP; *Page 415* Articulate Graphics/CMSP; **Chapter 12**: *Page 448* Michael Pole/CORBIS; *Page 452* Kalab/CMSP; *Page 453* Hossler/CMSP; *Page 454 counterclockwise from top left* Lester V. Bergman/CORBIS, Lester V. Bergman/CORBIS, Hossler/CMSP; *Page 461 left* Hossler/CMSP, *right* Educational Images/CMSP; *Page 474 left* Staats, MD/CMSP, *right* EpConcepts/CMSP; *Page 475* Joel Stettenheim/CORBIS; **Chapter 13**: *Page 488* Royalty-Free/Corbis; **Chapter 14**: *Page 528* David Pollack/CORBIS; *Page 533* NMSB/CMSP; **Chapter 15**: *Page 564* Philip Harvey/CORBIS; *Page 570* SIU/CMSP; *Page 571* Siebert/CMSP; *Page 572* Lester V. Bergman/CORBIS; *Page 577* Peres/CMSP; *Page 579* Morgan/CMSP; *Page 581* Rawlins/CMSP; *Page 583* Rawlins/CMSP; *Page 584* CMSP.

Softcover Edition
Text: ISBN 978-0-76382-340-5
Text + Encore CD: ISBN 978-0-76382-337-5

Hardcover Edition
Text: ISBN 978-0-76383-314-5
Text + Encore CD: ISBN 978-0-76383-313-8

Care has been taken to verify the accuracy of information presented in this book. The author, editors, and publisher, however, cannot accept any responsibility for errors or omissions or for consequences from application of the information in this book and make no warranty, expressed or implied, with respect to its content.

Trademarks
Some of the product names used in this book have been used for identification purposes only and may be trademarks or registered trademarks of their respective manufacturers.

© 2007 by Paradigm Publishing, Inc.,
   875 Montreal Way
   St. Paul, MN 55102
   (800) 535-6865
   E-mail: educate@emcp.com
   Web site: www.emcp.com

All rights reserved. Making copies of any part of this book for any purpose other than your own personal use is a violation of the United States copyright laws. No part of this book may be used or reproduced in any form or by any means, or stored in a database retrieval system, without prior written permission of Paradigm Publishing Inc.

Printed in the United States of America
16 15 14 13 12 11 10 09 08 07   2 3 4 5 6 7 8 9 10

I dedicate this book to my daughter, Kathleen, and son, Timothy, who tolerated having Dad parked in front of the computer for many nights and weekends while working on the manuscript. My mother, Adele, and my siblings also deserve to be mentioned because they were my cheerleaders throughout the writing process. Last but not least, my late father, David, deserves special mention because he nurtured my interest in the medical sciences.

Contributors—The author and publisher wish to thank the following instructors and professionals for their valuable suggestions during the development of this book.

Jerri Adler, AA, CMA, CMT
Family and Health Careers Department
Lane Community College
Eugene, Oregon

William M. Clark, MD, MBA, MS
Kingwood College
Kingwood, Texas

Robert Spears, PhD
Department of Biomedical Sciences
Baylor College of Dentistry-Texas A&M University System Health Science Center
Dallas, Texas

Melissa Curfman-Falvey
Jefferson College
Hillboro, Missouri

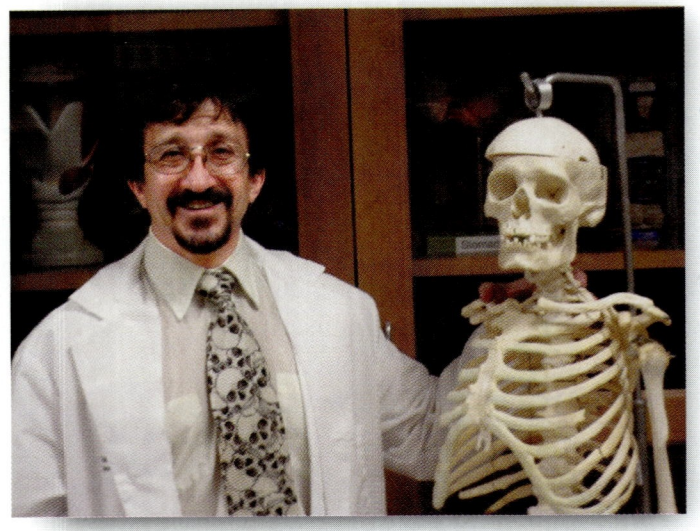

About the Author

Dr. Brian Robert Shmaefsky had a diverse career background that included three years as an industrial biochemist and a university professor before coming to Kingwood College in 1992 to establish the college's 2-year biotechnology program. His current duties at Kingwood College include biology teaching and coordinating the service-learning program. He taught one-semester and two-semester A&P courses for allied health and kinesiology majors since 1986. Dr. Shmaefsky has authored many articles on science teaching, five books on human diseases, a biotechnology book, and technical articles for industrial and science journals.

Dr. Shmaefsky carried out his undergraduate and graduate studies in biology. He then pursued doctoral studies in environmental physiology at the University of Illinois and science education at Southern Illinois University. He also carried out studies in journalism and accumulated numerous continuing education credits in pharmacology.

He received several teaching awards from his college district, Exxon Corporation, the National Association of Biology Teachers, and Teaching For Success for his innovations in college science teaching. His disease books have been recognized as recommended readings by various public health agencies.

Dr. Shmaefsky is a fervent proponent of using case studies, creative thinking, and problem-based learning in the classroom. He is currently working on grant-funded projects aimed at improving science teaching. In addition, he is applying these principles to industrial hygiene and safety training for employees in science and technology occupations.

Much of Dr. Shmaefsky's spare time is spent volunteering on environmental protection committees and serving various professional teaching organizations including the Human Anatomy and Physiology Society, National Biology Teachers Association, and National Science Teachers Association. He lives in Kingwood, Texas with his two dogs and guinea pig. His children Timothy and Kathleen live nearby.

TABLE OF CONTENTS

Preface ix

CHAPTER 1: OVERVIEW OF THE BODY . . . 2
CSI—Case Study Investigation 2
Introduction 4
Human Body Orientation 7
   Direction 8
   Positions 10
   Movement 13
Body Regions 14
   General Locations 15
   Abdominopelvic Regions and Quadrants . . . 16
Body Cavities 18
CSI Conclusion 21
Study Guide 21
   Chapter Summary 22
   Key Terms 23
   Check Your Understanding 24
   A Case Study 26
   Where Do We Go from Here? 27
   Skills Activities 27

CHAPTER 2: THE BODY'S CHEMICAL MAKEUP 30
CSI—Case Study Investigation 30
Introduction 32
   Atomic Structure and Function 35
   Properties of Molecules 38
Acids and Bases 43
Human Molecules 45
   Lipids 46
   Carbohydrates 50
   Peptides 53
   Nucleic Acids 57
Molecules and Nutrition 59
Aging of the Body's Chemistry 61
CSI Conclusion 63
Study Guide 64
   Chapter Summary 64
   Key Terms 65
   Check Your Understanding 67
   A Case Study 69
   Where Do We Go from Here? 70
   Skills Activities 71

CHAPTER 3: ORGANIZATION OF THE BODY 74
CSI—Case Study Investigation 74
Introduction 76
Hierarchy of Human Structure 76
The Human Physiological
   Environment 79
   Water 79
   Ions 81
   Enzymatic Reactions and Energy 83
   Molecular Transport 86
Cell Structure 92
   Cells of Microbes 93
   Humans Cells 94
Cell Function 97
   Metabolism 97
   Genetics 99
   Cell Cycle 102
Tissues 107
   Epithelial Tissue 108
   Connective Tissue 109
   Muscle Tissue 109
   Nervous Tissue 111
Organs and Systems 112
Wellness and Illness over the Life Span 119
   Pathology of Cells 119
   Cellular Aging 120
CSI Conclusion 1121
Study Guide 122
   Chapter Summary 122
   Key Terms 123
   Check Your Understanding 126
   A Case Study 128
   Where Do We Go from Here? 129
   Skills Activities 130

CHAPTER 4: THE SKIN AND ITS PARTS . 134
CSI—Case Study Investigation 134
Overview 136
The Integumentary System 136
   Skin Structure 138
   Skin Appendages 142
   Functions of the Integumentary System . . . 148
Wellness and Illness over the Life Span 152
   Pathology of the Integumentary System . . . 152

    Aging of the Integumentary System 156
CSI Conclusion. 159
Study Guide 160
    Chapter Summary 160
    Key Terms 161
    Check Your Understanding 163
    A Case Study 165
    Where Do We Go from Here? 166
    Skills Activities 167

Chapter 5: The Skeletal System . . . 170

CSI—Case Study Investigation 170
Overview. 172
The Human Skeletal System 174
    Axial Skeleton 176
    Appendicular Skeleton 183
Bone 187
    Bone Types 187
    Bone Structure 188
Joints 192
    Joint Structure 193
    Joint Function 194
Human Bone Charts 196
Bone Development and Healing 197
Wellness and Illness over the Life Span 203
    Pathology of the Skeletal System 203
    Aging of the Skeletal System 207
CSI Conclusion. 209
Study Guide 211
    Chapter Summary 211
    Key Terms 213
    Check Your Understanding 215
    A Case Study 217
    Where Do We Go from Here? 218
    Skills Activities 219

Chapter 6: The Muscular System . 222

CSI—Case Study Investigation 222
Muscle. 225
    Types of Muscle Tissue 226
    Muscle Cell Structure 227
    Muscle Cell Function 229
Musculature. 231
    Gross Skeletal Muscle Types 231
    Skeletal Muscle Structure 235
    Skeletal Muscle Action 236
Musculature Charts 238
Wellness and Illness over the Life Span 241
    Pathology of Musculature 241
    Aging of the Muscular System 244
CSI Conclusion. 246
Study Guide 247

    Chapter Summary 247
    Key Terms 248
    Check Your Understanding 249
    A Case Study 251
    Where Do We Go from Here? 252
    Skills Activities 253

Chapter 7: The Endocrine Glands and Hormones 258

CSI—Case Study Investigation 258
Overview. 260
    Hormone Function 264
    Endocrine Secretions 265
    Types of Hormones 266
The Endocrine Glands 268
    Pituitary Gland 268
    Pineal Gland 270
    Adrenal Glands 271
    Thyroid Gland and Parathyroid Glands . . . 273
    Pancreas 275
    Thymus Gland 276
    Gonads 277
Wellness and Illness over the Life Span 278
    Pathology of the Endocrine System 278
    Aging of the Endocrine System 280
CSI Conclusion. 281
Study Guide 282
    Chapter Summary 282
    Key Terms 284
    Check Your Understanding 285
    A Case Study 287
    Where Do We Go from Here? 288
    Skills Activities 289

Chapter 8: Function of the Nervous System 292

CSI—Case Study Investigation 292
Overview. 294
Types of Nervous System Cells. 295
    Neurons 296
    Neuroglia and Stem Cells 298
Neuron Physiology 301
Types of Neuron Communication 305
Reflexes 308
Wellness and Illness over the Life Span 310
    Pathology of Nervous System Function . . . 310
    Aging of the Nervous System 314
CSI Conclusion. 316
Study Guide 318
    Chapter Summary 318

    Key Terms 319
    Check Your Understanding 321
    A Case Study 323
    Where Do We Go from Here? 324
    Skills Activities 325

Chapter 9: Structure of the Nervous System 329

CSI—Case Study Investigation 329
Overview 330
Nerve Structure 331
Nervous System Components 334
    Central Nervous System 335
    Peripheral Nervous System 341
Human Senses 347
    Taste 347
    Smell 348
    Vision 349
    Hearing and Balance 352
Wellness and Illness over the Life Span 354
    Pathology of Nervous System Structure . . . 354
    Aging of the Nervous System Structure . . . 359
CSI Conclusion 361
Study Guide 363
    Chapter Summary 363
    Key Terms 364
    Check Your Understanding 366
    A Case Study 368
    Where Do We Go from Here? 369
    Skills Activities 369

Chapter 10: The Respiratory System 372

CSI—Case Study Investigation 372
Overview 374
Components of Human Respiratory System . 375
    Nose 375
    Nasal Cavity 376
    Paranasal Sinuses 376
    Pharynx 376
    Larynx 378
    Trachea 379
    Bronchial Tree 381
    Lungs 382
Breathing 384
    Mechanics of Breathing 385
    Gas Exchange 387
Wellness and Illness over the Life Span 388
    Pathology of the Respiratory System 388
    Developmental Diseases 389
    Infectious Diseases 391
    Aging of the Respiratory System 393
CSI Conclusion 395
Study Guide 397
    Chapter Summary 397
    Key Terms 398
    Check Your Understanding 399
    A Case Study 401
    Where Do We Go from Here? 402
    Skills Activities 403

Chapter 11: The Cardiovascular System 406

CSI—Case Study Investigation 406
Overview 408
Circulatory System Vessels 420
    Arteries and Veins 410
    Small Vessels and Capillaries 415
Structure of the Human Heart 417
    Adult Heart 420
    The Fetal Heart 423
Heart Function 425
Electrocardiography Basics 427
Wellness and Illness over the Life Span 430
    Pathology of the Cardiovascular System . . 430
    Aging of the Cardiovascular System 436
CSI Conclusion 438
Study Guide 439
    Chapter Summary 439
    Key Terms 440
    Check Your Understanding 442
    A Case Study 444
    Where Do We Go from Here? 445
    Skills Activities 445

Chapter 12: The Lymphatic System and The Blood 448

CSI—Case Study Investigation 448
Overview 450
Blood Cells 451
    Red Blood Cells (RBCs) 451
    White Blood Cells (WBCs) 453
    Platelets 454
Blood Cell Function 455
    Red Blood Cells (RBCs) 455
    White Blood Cells (WBCs) 457
    Platelets 458
Blood Cell Formation 455
Lymphatic System 462
    Structures of the Lymphatic System 462

Immune Response 465
Immunization and Vaccination **471**
Wellness and Illness over the Life Span **472**
 Pathology of the Blood and
  Lymphatic System 472
 Aging of the Blood and Lymphatic System. 475
CSI Conclusion 477
Study Guide 478
 Chapter Summary **478**
 Key Terms **479**
 Check Your Understanding **481**
 A Case Study **483**
 Where Do We Go from Here? **484**
 Skills Activities **485**

CHAPTER 13: THE DIGESTIVE SYSTEM. 488

CSI—Case Study Investigation 488
Overview **490**
Components of the Digestive Tract **491**
 Mouth & Pharynx 491
 Esophagus and Stomach 498
 Small Intestine 500
 Large Intestine and Rectum 502
Glandular Structures of Digestive System ... **504**
 Pancreas 504
 Liver and Gall Bladder 506
The Digestive Process **508**
Wellness and Illness over the Life Span **510**
 Pathology of the Digestive System 510
 Aging of the Digestive System 514
CSI Conclusion 516
Study Guide 517
 Chapter Summary **517**
 Key Terms **519**
 Check Your Understanding **520**
 A Case Study **522**
 Where Do We Go from Here? **523**
 Skills Activities **524**

CHAPTER 14: THE URINARY SYSTEM. 528

CSI—Case Study Investigation 528
Overview **530**
Gross Anatomy of the Urinary System **530**
The Kidneys – External Anatomy 531
 The Kidney - Internal Anatomy 533
 The Ureters 533
 The Urinary Bladder 534
 The Urethra 535
Urine Voiding **538**
The Nephrons **536**
Urine Formation **538**
 Filtration in the Glomerulus 540
 Reabsorption in the Proximal
  Convoluted Tubule 542
 Reabsorption in the Loop of Henle ...
 Reabsorption in the Distal Convoluted
  Tubule and Collecting Duct 543
Tubular Secretion **544**
Hormonal Regulation of Urine Formation ... **545**
Wellness and Illness over the Life Span **546**
 Pathology of the Urinary System 546
 Aging of the Urinary System 546
CSI Conclusion 553
Study Guide 554
 Chapter Summary **554**
 Key Terms **556**
 Check Your Understanding **557**
 A Case Study **559**
 Where Do We Go from Here? **560**
 Skills Activities **561**

CHAPTER 15: THE REPRODUCTIVE SYSTEMS AND HUMAN DEVELOPMENT 564

CSI—Case Study Investigation 564
Overview **566**
Female Reproductive System **567**
 Reproductive Tract 568
 Mammary Glands 573
Male Reproductive System **576**
 Testes 577
 Seminal Vessels 578
 Penis 579
Basics of Sexual Reproduction **580**
 Female Sexual Cycle 580
 Copulation 582
 Embryology and Pregnancy 583
Wellness and Illness over the Life Span **590**
 Pathology of the Human Reproductive
  System 590
 Aging of the Human Reproductive System ....
CSI Conclusion 595
Study Guide 597
 Chapter Summary **597**
 Key Terms **598**
 Check Your Understanding **600**
 A Case Study **602**
 Where Do We Go from Here? **603**
 Skills Activities **604**

Glossary **609**
Index **639**

TABLE OF CONTENTS

PREFACE

The study of human anatomy and physiology is not simply an exercise in memorizing facts about body parts and functions. A full understanding of the human body takes into consideration that anatomy and physiology varies greatly from one person to another for reasons that are not always easy to predict. Professionals who work in medicine and the human sciences are well aware that the body's function and structure are molded and continually modified by disease, environmental interactions, and lifestyle. Consequently, medical educators have learned that the human body is best studied by applying facts to situations that stretch the limits of the body's response to change.

The goal of this book is to encourage you to use the basic facts about the human body to discover the possible causes of these variations. It is also my intent to share the thrill of investigating human disease by inviting you to solve actual case studies related to the content of a chapter. I have had the pleasure of teaching anatomy and physiology to allied health professionals and pre-medical students since 1985. I also apply my knowledge of the human body to educate the public about the effects of lifestyle and pollution on human health. All of this experience has produced a perception of student learning that is incorporated into this book. I hope that this book becomes a permanent addition to your library that can be used for reviewing information that may have been forgotten and for educating others about the marvels of the human body.

Hints for Using This Book

Please do not read this book as if you were reading a newspaper or magazine. A "once through" reading of the content of this book will not provide you with a full learning experience. Every aspect of each chapter was designed to improve your retention of the multitude of facts and terms required to comprehend human anatomy and physiology. Much of the information provided in this book is a fundamental overview of the structure and function of the human body. The book also contains components that challenge you to search the Internet and conduct experiments for more information about how the body responds to disease, environmental interactions, and lifestyle. When first reading this book, it is important that you are able to recall all of the key terms defined in a section before continuing to the next section. It is also important that you use the Concept Check questions distributed throughout each chapter. These questions are intended to ensure that you have the background information necessary for effectively learning the material in the sections to come.

Take the time to read the Case Study Investigation at the beginning of each chapter and make use of the case study hints dispersed throughout the chapter. Don't get frustrated if you cannot solve the case study right away. The information in the chapter builds step-by-step so that it is possible to solve some aspect of the case study by the end of the chapter. Each case study is designed to model real-world situations encountered by health professionals and scientists. Educational researchers have ample evidence showing that memorizing facts without putting that knowledge to use does not encourage retention of information. Teaching using cases studies is a time-proven way of improving retention of anatomy and physiology.

The features of this book are designed to improve your understanding of human anatomy and physiology.

- Case Study Investigations: Each chapter begins with a CSI, presenting you with a medical mystery. Throughout the chapter, CSI Breaks offer clues to help you solve the case using the comprehensive information provided about each body system. The CSI feature asks you to

think about how the body systems work together and gives you practice developing diagnoses.
- Concept Check Questions: Located after major content sections, these questions help you test your understanding of what you have read.
- Sidebars: Sidebars in every chapter ask you to consider the world around you in light of what you're learning:
  ➤ Civic Responsibility: Emphasizes the importance of the healthcare worker's role in the community.
  ➤ Good Choice-Bad Choice: Asks you to examine issues and think about consequences.
  ➤ Cutting Edge Research: Focuses on current medical discoveries and news.
- Boxes: Each chapter includes special topic boxes that let you explore current news and fun facts related to anatomy and physiology.
- Key Terms: The essential terms needed for your understanding of anatomy and physiology are printed in bold and defined in the margins. After reading each chapter, you should be able to define these terms and use them to explain the structure and function of the human body.
- Study Guide: Use the study guide at the end of each chapter to check your comprehension of the chapter's contents. Each guide includes a chapter summary, review of key terms, multiple-choice questions, Internet research, and laboratory exercises. It is valuable resource when reviewing for any tests or board examinations related to your career.
- Glossary: The glossary provides an easy way to look up the definitions of terms you may have forgotten. It is also a useful tool for reviewing terms you may encounter on tests or board examinations.
- Ancillary Materials: A selection of ancillary materials can be used to supplement your understanding of anatomy and physiology. Used along with this book, they add up to a learning resource that can be referenced and reviewed throughout your education and career.
  ➤ Companion Workbook: A 4-color illustrated workbook includes short answer and multiple-choice questions, matching and key term exercises, practical application questions, and crossword puzzles.
  ➤ Encore CD-ROM: A multimedia CD-ROM that includes Flash animations, interactive chapter quizzes, a glossary with related images from the text, crossword puzzles, and a link to our Internet Resource Center.
  ➤ Internet Resource Center: An online hub for students and instructors featuring course planning and evaluation tools, studying and test-taking guidance, and other resources.

Acknowledgments

Almost every book that is published today is a team effort. The author is just one part of a cooperative group dedicated to producing a useful product. Many thanks go Sonja Brown and Tony Galvin of EMC Corporation for truly believing I had the ambition and talent to bring this book into being. Courtney Kost's wonderful leadership with the design of this book deserves recognition.

The book would not have been a realistic project for me to tackle without the superb editorial leadership of Nadia Bidwell of Barking Dog Editorial. Joanna Pellegrini of Van Brien & Associates deserves equal gratitude for her relentless efforts to build a high-quality art program for this book. Much gratitude goes to Melissa Curfman-Falvey, of Jefferson College in Missouri, whose keen eye and technical editing skills ensured scientific accuracy and appropriate readability of this book.

I also want to thank the reviewers and many other people who made this book a work to be proud of. They provided valuable contributions and insights in areas outside of my expertise.

Applied Anatomy & Physiology

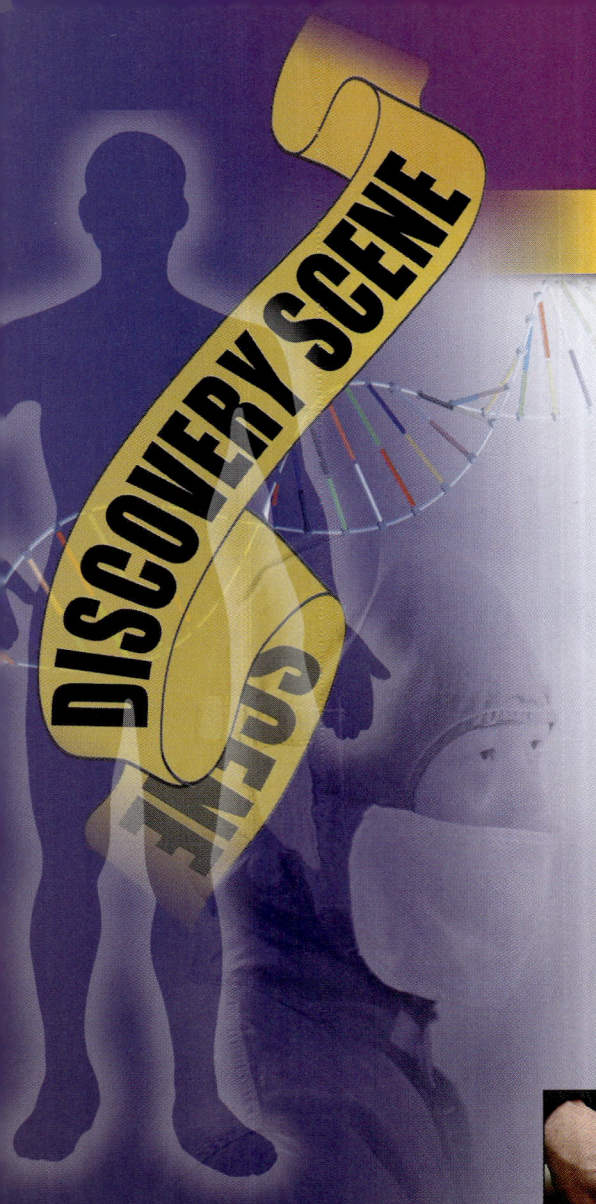

Case Study Investigation

Case Study Investigation #1

You are in an ambulance taking a young adult male to Springville Hospital Emergency Department. The patient was at a barbecue cutting some ribs when the knife slipped and cut him in the stomach. Emergency Department physicians note that the patient is complaining of chest pain and stomach pain around the knife wound. The patient is short of breath, coughing, and breathing rapidly. He is also showing redness and extreme swelling just below the waist. The ambulance crew is somewhat perplexed by the fact that the patient is showing breathing problems as a result of this wound. By reading this chapter, you will eventually conclude that the patient's conditions are due to a knife wound that has altered the body's structure. It is the job of emergency department physicians and medical laboratory personnel to determine which parts of the body are being affected. Then, they must determine what is causing the problems and whether the symptoms are all related to the knife injury. At the end of the chapter, you will be asked to determine, based on the patient's symptoms, which part(s) of the body may be damaged.

CHAPTER 1
Overview of the Body

Chapter Outline

Case Study Investigation (CSI)
Applied Learning Outcomes
Introduction
Human Body Orientation
   Direction
      Directional Orientation
      Directional Planes
   Positions
Movement
Body Regions
   General Locations
   Abdominopelvic Regions and Quadrants
Body Cavities
CSI Conclusion
Study Guide

Applied Learning Outcomes

- Learn body orientation terms that explain or describe the following:
  - body direction
  - various views of the body and body parts
  - positioning of the body for medical procedures
  - movement of major body parts
- Learn the locations of
  - the major body regions and cavities, and the structures contained within them
- Use the terminology associated with the major body regions and cavities
- Understand aging and pathology as they relate to body organization

INTRODUCTION

Key Terms: anatomy, developmental anatomy, embryology, fine (microscopic) anatomy, gross anatomy, morphology, pathology, physiology

The field of medicine is filled with loads of seemingly incomprehensible and almost unpronounceable technical terms. These terms are not meant to confuse people. Actually, they were developed to help medical practitioners and scientists to better communicate information about the body. Your friends may not understand you after telling them that their feet are inferior to their head. They may think this means the head is more valuable than the feet. However, a physician or scientist would without a doubt know that you meant that the feet are positioned below the head. People working in the health and medical fields must be able to communicate consistently, using the accepted terminology to be effective on the job.

It is first important to know and understand the terms **anatomy** and **physiology** found in the title of this book. These terms distinguish body structure from body function. Anatomy is best defined as the structural make-up of an organism such as a human. Viewing the body with your eyes or with special instruments that magnify or see into body parts is the major means of investigating anatomy. The term anatomy comes from the Greek word *anatome* or *ana-temnein*, meaning to cut (*temnein*) apart (*ana*). Anatomy in ancient times was studied by cutting up a body. The term **morphology** is commonly used in medical reports in place of the term anatomy. Morphology refers to the differences and similarities in the anatomy of individuals. It also takes into account how body structures form from birth or take on different appearances in response to damage or disease.

Anatomy The structural make-up of an organism

Physiology The functions of an organism

Morphology The structural make-up of an organism, referring to differences and similarities in anatomy

Fine (Microscopic) Anatomy The study of anatomy concerned with microscopic features of the body

Gross Anatomy The study of anatomy concerned with the features of the body visible to the naked eye

ROOTS OF ANATOMY

Modern anatomical studies involve the use of powerful microscopes and specialized imaging machines to produce detailed information about body structures. However, all of this technology is simply a follow-up of anatomical investigations carried out by the Greek scientists Alcmaeon and Empodocles in the 5th century BC. They produced detailed drawings of the human body. Hippocrates, another Greek scientist who lived from 460 to 377 BC, used anatomical studies to better explain medical conditions and treatments. However, all of their work was constrained by the fact that they were not allowed to cut open the human body. So, in the 3rd century BC, the philosopher Aristotle studied the internal structures of animals to better understand humans. It was not until 300 BC that Herophilus and Erasistratus performed the first internal studies on humans. In effect, they were truly the originators of anatomy because they cut (*temnein*) apart (*ana*) the body. Society changed its ideas

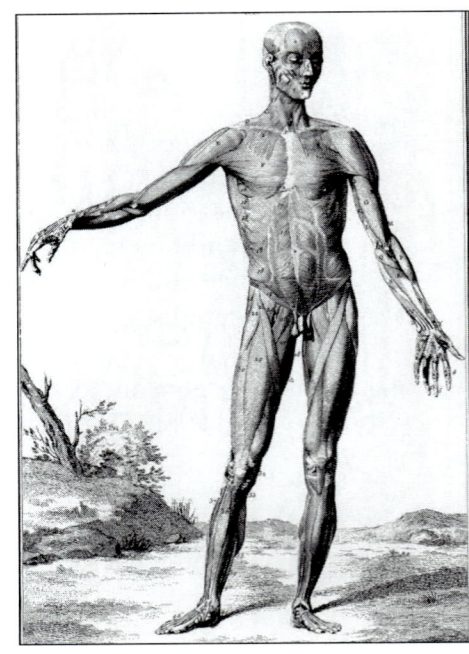

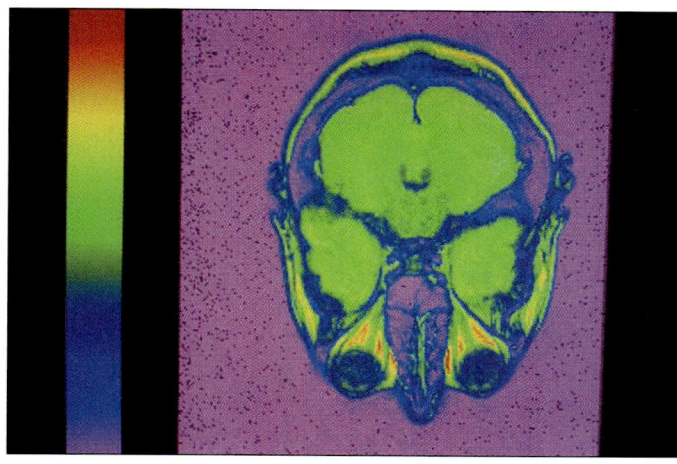

about the sacredness of the human body, giving Herophilus and Erasistratus the opportunity to do their work. It was not until the 13th century that a bulk of fine anatomy and physiology work was conducted by scientists throughout Europe. They were helped by Islamic scholars who were able to interpret ancient Greek anatomy documents that made their way to the Middle East.

Many medical practitioners and scientists divide the study of anatomy into **fine anatomy** and **gross anatomy** (Figure 1.1). Fine anatomy examines microscopic features of the body. This is usually performed with laboratory

Figure 1.1 Fine to Gross Anatomy
Gross anatomy looks at the larger anatomical structures, such as body systems and organs. Fine anatomy focuses on small components of the body, such as tissues and cells.

Cell

Tissue

Organ

Body system

Organism

OVERVIEW OF THE BODY

Civic Responsibility

HELPING OTHERS WITH YOUR KNOWLEDGE

It is valuable to use what you have learned about anatomy and physiology terms to help others better understand the world around them. It is very important to check your facts and seek further information about certain topics before discussing health and science issues. Here are some suggestions to foster a better public awareness of anatomy and physiology terminology:

1. Assist people who are not native English speakers with anatomy and physiology terms.
2. Work with sports clubs to educate players about body terminology associated with sports injuries.
3. Help elderly persons to better understand the terms used by nurses, therapists, and physicians.
4. Volunteer at a school health day to teach children body-part terms.

Embryology The study of the anatomical changes that occur during the growth of an embryo

Developmental Anatomy The study of anatomical changes that occur during the growth of a human being

Pathology The study of human diseases

instruments, such as microscopes and imaging machines (Figure 1.2). The fine anatomy of body structures is often used to investigate the cause of disease and bodily injury. Gross anatomy deals with larger parts of the body that are easily viewed with naked eye. Imaging equipment can also be used to study gross anatomy features inside the body. Gross anatomy studies provide the first indication that something may be wrong with the body. The term **embryology**, or **developmental anatomy**, is also commonly encountered in health professions. It is the investigation of the anatomical changes that take place during human growth, first as an embryo (embryology), and then after birth (developmental anatomy).

The term physiology refers to the function and role of anatomical features. It investigates the chemical reactions that make the body function. Physiology comes from the Greek words *physis* and *logos*. *Physis* means the nature of something, whereas *logos* means to study or investigate. So, physiology literally means looking at the characteristics of how body parts carry out their jobs. Physiology can be studied on fine or gross anatomical parts. Gross physiology studies, like gross anatomy, provide the first clues for the causes of disease. Likewise, fine physiology investigations reveal detailed information about the body parts affected by disease. Physiology can be investigated by observing changes to the fine and gross anatomical features. However, this method is not always accurate and does not provide many of the details needed to fully understand the body. So, specialized chemical investigations carried out in laboratory tests were developed to better understand physiology. A related term, **pathology**, is used to describe anatomical and physiological studies of human disease. Pathology comes from the Greek term *pathos*, meaning pain or suffering.

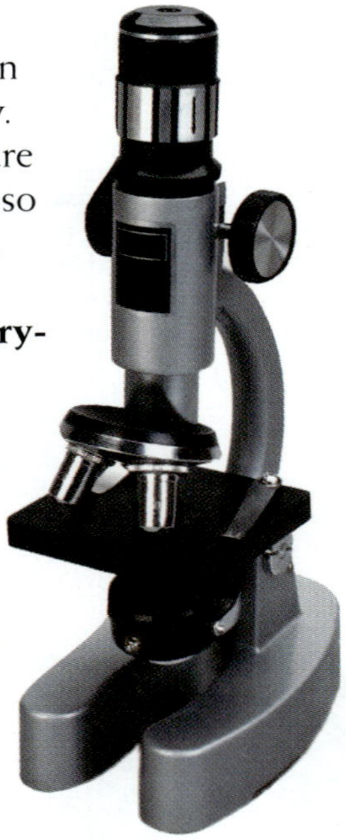

Figure 1.2 Microscope
Microscopes are the basic instruments for studying fine anatomy.

✓ Concept Check

1. Define the term anatomy.
2. What is the difference between the study of fine anatomy and gross anatomy?
3. What is the difference between the terms embryology and morphology?
4. Distinguish the difference between the terms physiology and pathology.

DISCOVERY SCENE PLEASE ENTER DISCOVERY SCENE PLEASE ENTER

What terms covered so far may help you solve the CSI? Do you see a need for any of the terms in explaining the body damage and problems associated with the injury?

CONFUSING POINTS OF REFERENCE

Veterinarians, or doctors of veterinary medicine, are skilled in the care of animals and have to learn orientation terms unique to animals. Most of the terms are similar to those used to explain human physical orientation. However, these terms do not necessarily have the same meaning when referring to animals as they do when viewing the human body. Compare the illustration of the fish shown here to the human diagram in Figure 1.4. In fish and four-legged animals, an anterior view of the animal means you are looking mostly at the face. The belly, or ventral view, cannot fully be seen. Note that in humans the anterior view shows the complete ventral view of the person as well as the face. In the same fashion, the posterior view of the fish shows only the tail, while in humans, it includes the "tail end" and the entire back, or dorsal, region. Veterinarians have to be very specific when describing the physical orientation of the animals they care for. Professionals working with people are free to interchange the term anterior with ventral, and posterior with dorsal.

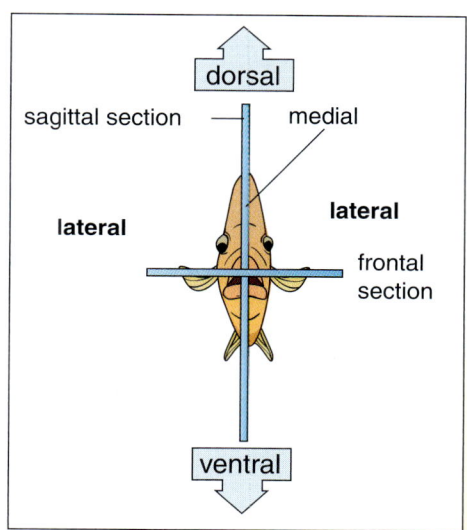

HUMAN BODY ORIENTATION

If a doctor asked you to "please lie down facing up," it is obvious what he or she is asking. However, if she says, "please get in a supine position," you might be confused. Body orientation and positioning, as well as anatomical parts, have a specialized terminology that communicates precise information in the health and medical fields. People carrying out a variety of tasks related to healthcare and medical practices need to understand the different ways patients are positioned for various medical procedures and therapeutic practices. These terms are commonly seen in instructions for client or patient care as well as in reports describing medical conditions.

Direction

Key Terms: coronal plane, cephalic, cranial, caudal, directional planes, directional orientation, distal, frontal plane, inferior, lateral, medial, midsagittal plane, proximal, sagittal plane, superior, transverse plane

Body direction is divided into two sets of terms referring to **directional orientation** and **directional planes**. Directional orientation refers to the particular view you see of a person, for example, the face or the back of the head. Directional planes include a series of terms that describe the way a body can be divided into parts for viewing surface features or internal structures.

Directional Orientation Figure 1.3a shows a front view of a person standing in the customary anatomical position. Note how his arms are placed with his palms forward. From this view, the midline of the body can be seen. It is an imaginary line that runs up and down through the center of the body. Body parts closest to the body's midline are said to have a **medial** orientation. Thus, by looking at this figure, it can be said that the nose is medial in location to the ears. Likewise, the pinky finger is medial to the thumb. **Lateral** refers to structures away from the midline. Thus, the ears are lateral to the nose, and the thumb is lateral to the little finger.

Superior is a term referring to any structure that is located above another or closer to the head. The eyes are superior to the nose as is evident in Figure 1.3a. Sometimes the term **cephalic** or **cranial** is used in place of superior. The term "superior" literally means nearest to the head. The terms cephalic and cranial mean, "pertaining to the head." In contrast, the term **inferior** refers to a body part that is below another. The nose is inferior to the eyes. The term **caudal** is regularly used in place of inferior. It means near the tail end. Caudal is the scientific term for "pertaining to the tail." Superior and inferior are often confused with the next two directional orientation terms **distal** and **proximal**. Distal specifically means any body part located far from an attachment point. By looking at the arm in Figure 1.3a, you can see that the hand is distal to the shoulder. Proximal has the opposite meaning of distal. Body parts closest to an attachment point are considered proximal. The shoulder is proximal to the hand. Similarly, the wrist is proximal to the fingers.

Now, note in Figure 1.3b that there are still four more terms to learn. When looking directly at a person's face, you are viewing his or her **anterior**. Anterior comes from the root word *ante* meaning "up front" or "coming before." The nose is anterior to the ears because it is closer to the front of the body. In human anatomy, the term **ventral** can be used in place of anterior. This holds true only for humans. Normally, the term ventral refers to the belly. Ventral comes from the term *venter* or "underside." **Posterior** refers to structures closer to the backside of the body. The term literally means, "to come after." The ears are posterior in location to the nose because they are closer to the back of the head. **Dorsal** is commonly used in place of posterior. In actuality, dorsal refers to the back, while posterior refers to the buttocks. However, in human anatomy, the terms are used interchangeably (see **Confusing Points of Reference Box**, page 7).

Directional Orientation Refers to the view one has of a person

Directional Planes A series of terms that describe the way a body can be viewed and divided

Medial Nearest to the midline of the body

Lateral Furthest from the midline of the body

Superior Nearest to the head

Cephalic Pertaining to the head

Cranial Pertaining to the head

Inferior A body part that is below another

Caudal Meaning near the tail

Distal A body part located far from an attachment point

Proximal A body part located near an attachment point

Anterior Meaning toward the front

Ventral In humans, toward the front. In other animals, refers to the belly

Posterior Toward the back

Dorsal Toward the back

Figure 1.3 Anatomical Position
This diagram represents the major human anatomy directional orientations commonly used in healthcare communication.

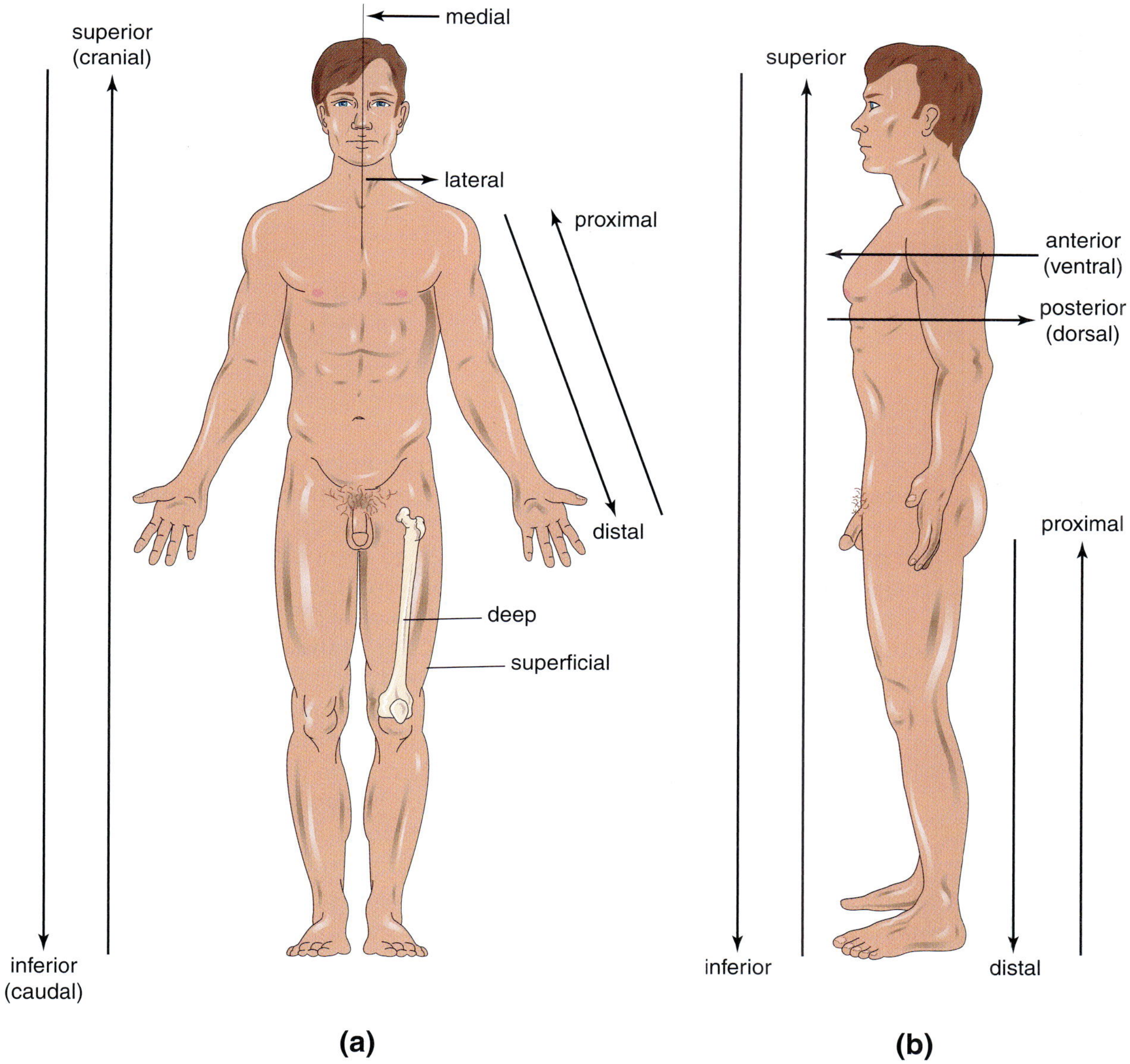

✓ Concept Check

1. Distinguish between the terms lateral and medial.
2. What is the difference between the terms inferior and distal?
3. What are alternate terms for anterior and posterior?

OVERVIEW OF THE BODY

Figure 1.4 Body Planes
This diagram represents the major human anatomy directional planes commonly used in healthcare communication.

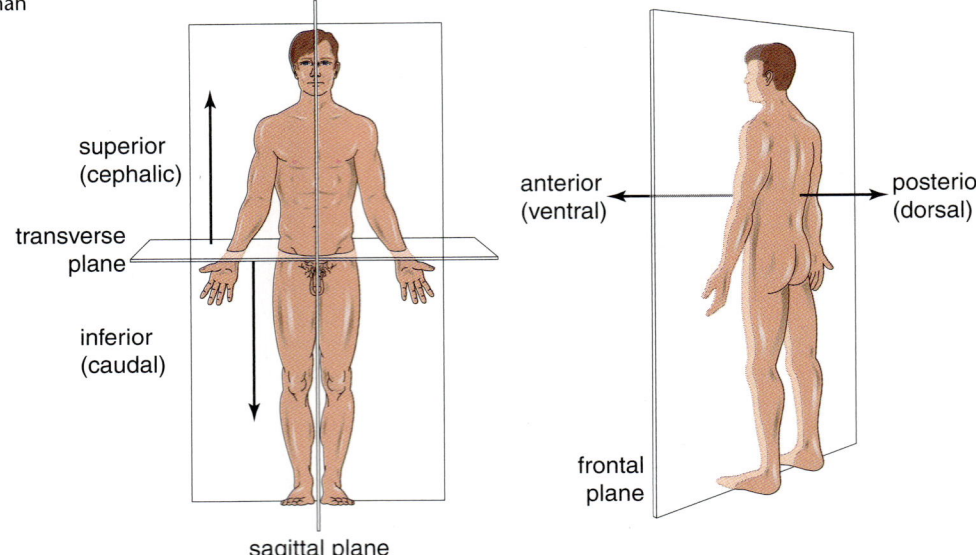

Sagittal Lengthwise planes dividing the body into right and left sections

Midsagittal The lengthwise plane that divides the body into equal halves

Frontal The plane dividing the body vertically into anterior and posterior sections

Coronal The plane dividing the body vertically into anterior and posterior sections; referring to the crown of the head

Transverse The plane dividing the body horizontally into upper and lower sections

Directional Planes The human figures shown in Figure 1.4 illustrate a person sliced three ways. These slices produce what are called directional planes for viewing the human body. Directional planes are used to describe the location of parts when imaginary or real cuts are sliced through the body. A sagittal plane cuts the body lengthwise into left and right sections. Sagittal sections can be cut into any body part. However, it is only the midsagittal plane that cuts the body into equal halves. Perpendicular to the sagittal plane is the frontal plane. The frontal plane slices the body vertically into anterior and posterior sections. Sometimes the term coronal plane is used in place of the frontal plane. Coronal refers to the fact that the section cuts through the body parallel to the crown of the head. There is no exact midsection cut for the frontal plane. A transverse plane cuts the body along a horizontal plane. It divides the body into lower and upper sections. The bottom section of the transverse plane is inferior or caudal. In contrast, the top section of the transverse plane is superior. Again, the terms cephalic or cranial can be used in place of superior.

✓ Concept Check

1. What is the difference between a sagittal plane and midsagittal plane?
2. Define the term frontal plane.
3. What does transverse plane mean?

Positions

Key Terms: dorsal recumbent position, Fowler's position, knee-chest position, left lateral position, lithotomy position, modified Trendelenburg position, prone position, Sim's position, sitting position, supine position, Trendelenburg's position

Many medical procedures and therapies require that a patient be placed into a certain position on a chair or special table. Special terms are used to describe the specific ways a patient can be situated. Since each health and medical practice has a number of terms unique to the procedures carried out in those fields, only the major positions common to many healthcare and medical practices will be mentioned. The names of these positions have a variety of origins. Some are standard names used by many scientists, while others are named after a medical procedure or the physician who developed the position.

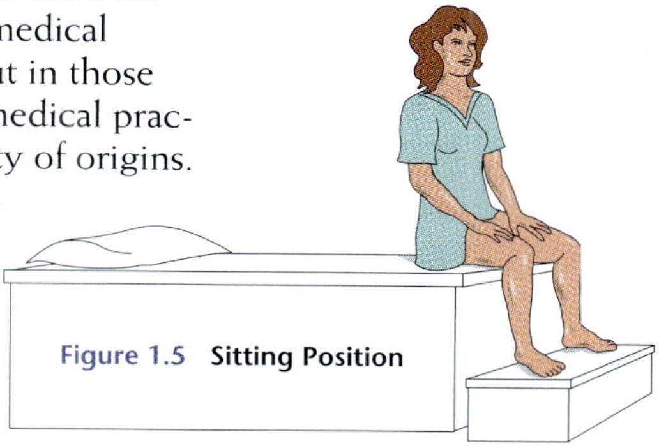

There are many ways to place a client or patient in a sitting position. The most common is the **sitting position** shown in Figure 1.5. As expected, it is the standard way people would sit in a chair or at the edge of a table. Another way to sit a person is in **Fowler's position**, demonstrated in Figure 1.6. The legs are held straight out and the back is supported by the back of a chair, a partition, or a wall. These positions are often modified so that the arms or legs are held in various ways.

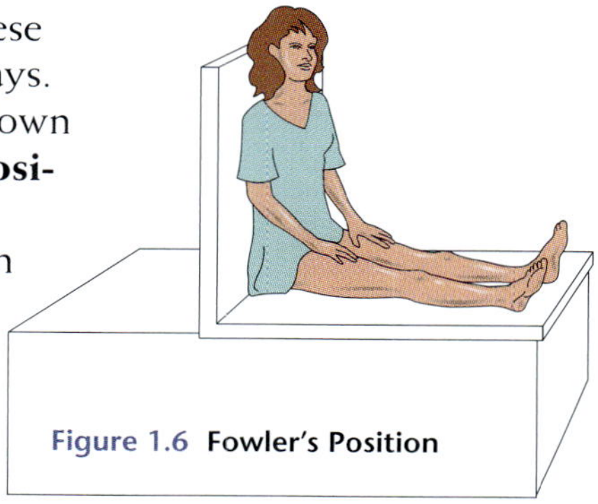

Figure 1.5 Sitting Position

The most commonly used positions require that a client or patient lie down on a table. However, there are many ways to lie on a table. The **supine position** places a person flat on his back facing up, as shown in Figure 1.7. Opposite to the supine position is the **prone position** in which the person is lying face down on the table, as shown in Figure 1.8. It is not unusual for a person to be placed in the **Trendelenburg's position**, in which the patient is supine with the body tilted so that the head is lower than the legs, shown in Figure 1.9. Trendelenburg's position can be adjusted into the **modified Trendelenburg's position**, demonstrated in Figure 1.10. The supine position can also be modified by special adjustments of the legs.

Figure 1.6 Fowler's Position

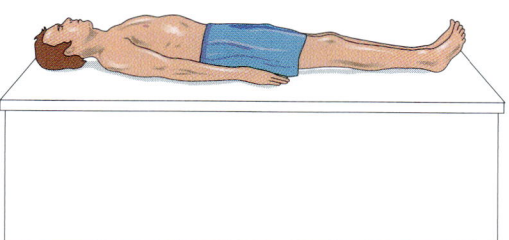

Figure 1.7 Supine Position

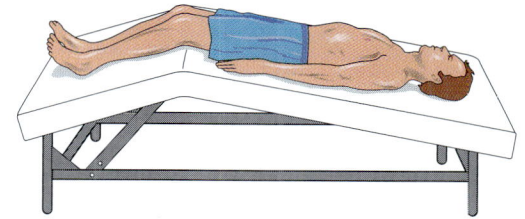

Figure 1.9 Trendelenburg's Position

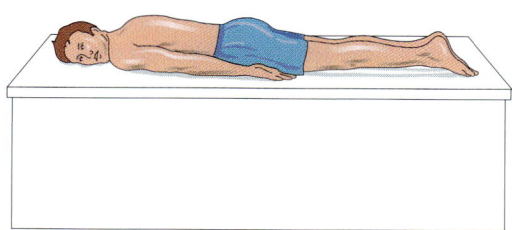

Figure 1.8 Prone Position

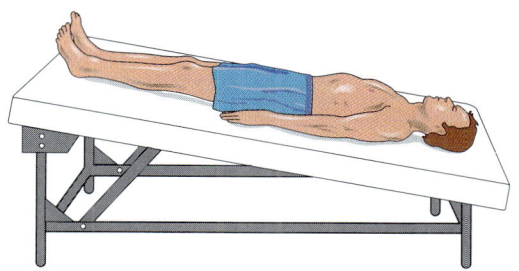

Figure 1.10 Modified Trendelenburg's Position

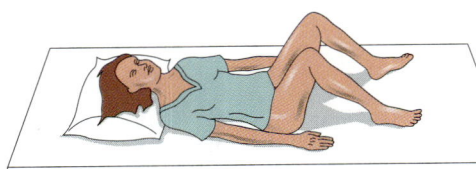

Figure 1.11 Dorsal Recumbent Position

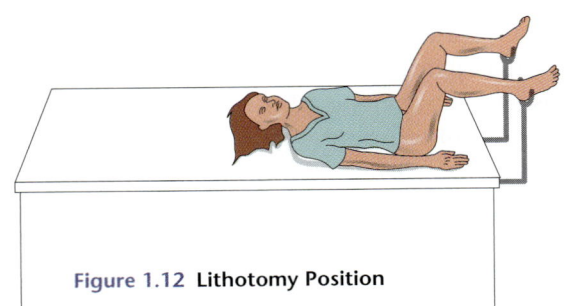

Figure 1.12 Lithotomy Position

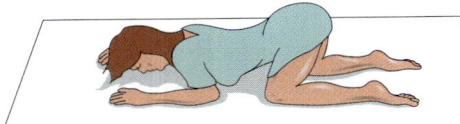

Figure 1.13 Knee-Chest Position

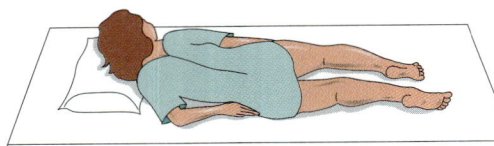

Figure 1.14 Sim's Position

A **dorsal recumbent** position places the patient supine with his or her knees bent up, as shown in Figure 1.11. A similar position, illustrated in Figure 1.12, is the **lithotomy position** in which the person's legs are placed on supports that hold the ankles and spread out the legs.

There are also two commonly modified prone positions. In the **knee-chest position**, the patient is facing down with his or her legs bent and the buttocks pushed up, as shown in Figure 1.13. In the **Sim's position**, the person lies face down on his or her left side with the left arm behind the back and the left knee bent, as demonstrated in Figure 1.14. This is also called the left lateral position.

Good Choice Bad Choice

Friends of yours enjoy bungee jumping. They are so excited about it that they want to do at least two jumps every weekend. Studies show that bungee jumping can place stress on the body that exceeds three times the force of gravity. In addition, the end of the jump involves a high-speed bounce that jolts the body up and down. What advice can you give your friends about the possible hazards of bungee jumping? What impact could bungee jumping have on the body cavities? What anatomical regions of the body would be most affected by the bouncing effect? How does the increased force of gravity affect the body? Are the risks associated with bungee jumping so great that your friends should reduce the amount of time they spend doing it?

✓ Concept Check

1. Distinguish between the supine position and the prone position.
2. What are three variations of the supine position?
3. Describe two modifications of the prone position.

Aging and Body Structure

The human body undergoes many anatomical and physiological changes as it ages. Some of these changes are part of the normal aging process, while others are due to pathology. The most obvious gross signs of aging occur in the body cavities mentioned in this chapter. It is not unusual to find fluid build-up in the thoracic and abdominopelvic cavities as a person ages. Long-term infections to the lungs or heart can cause fluid to build up in the thoracic cavity. Irritation or injury to the stomach or intestines may produce fluids in the abdominopelvic cavity. Cancer can also cause fluids to build up in these cavities. Another problem associated with aging is the effect that gravity has on the internal structures of the abdominopelvic cavity. The stomach and liver settle on top of the intestines, which causes them to compress, sag, and twist. This sometimes interferes with the functioning of the intestines. The continuous pressure of gravity also weakens the lower wall of the abdominopelvic cavity. This may produce hernias, which are protrusions of the internal structures through the cavity walls. Last, one serious problem is that the stomach may sometimes protrude into the thoracic cavity, affecting breathing and causing severe chest pain.

Movement

Key Terms: abduction, adduction, antagonistic, eversion, extension, flexion, inversion

Movement terms are critical in describing the way the arms and legs are moved in relation to the body. Note that each movement has an opposite movement, which is determined by how the part is attached to the body. The term **antagonistic** movement is regularly used to describe an opposing movement. Refer to Figure 1.15 to see illustrations of the movements as they are described. The term **flexion** means to bend a joint. This brings the distal end of the arm or leg close to the body. Bending the elbow or the knee are two

Antagonistic Opposing movements
Flexion To bend a joint

Figure 1.15 Directional Terms

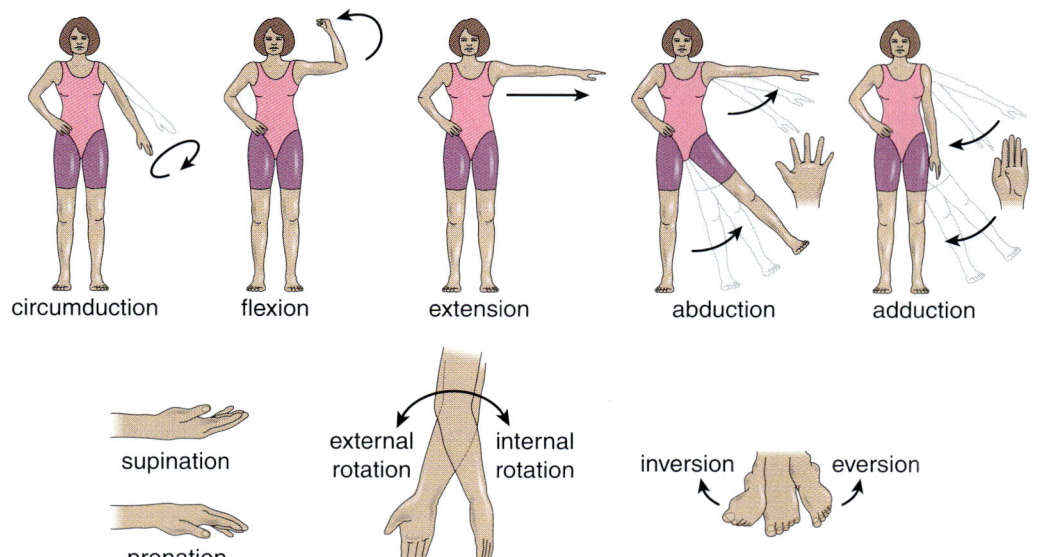

circumduction — flexion — extension — abduction — adduction

supination — external rotation — internal rotation — inversion — eversion

pronation

Overview of the Body

Extension To straighten a joint

Abduction Movement of the arm or leg away from the midline of the body

Adduction Movement of the arm or leg toward the midline of the body

Eversion Movement of the hand or foot so that the thumb or great toe moves away from the midline of the body

Inversion Movement of the hand or foot so that the thumb or great toe moves toward the midline of the body

common flexion movements. **Extension** is the opposite of flexion. It is a movement that straightens out a joint. Straightening the elbow or the knee are two common extension movements. **Abduction** means moving the whole arm or leg away from the midline of the body. The antagonistic movement is **adduction**, which means that the whole arm or leg is placed flat against the body. **Eversion** is a movement that rotates the hand or foot so that the thumb or great toe moves away from the body's midline. Its antagonistic movement is **inversion**, which turns the hand or foot inward toward the body.

✓ Concept Check

1. Define the term antagonistic in relation to body movement.
2. Distinguish between the terms flexion and extension.
3. Name and describe the antagonistic movements for flexion, abduction, and eversion.

MOVEMENT ISSUES

Under normal conditions, the joints are designed to permit a full range of motion for the body movements described in this chapter. However, aging and diseases of the joints produce a variety of conditions that reduce movement. One condition, called arthritis, or "irritation of the joints," can distort the joint so that it is not capable of its intended motions. Arthritis can cause a joint to stiffen so that the person can only make a partial movement or no movement at all.

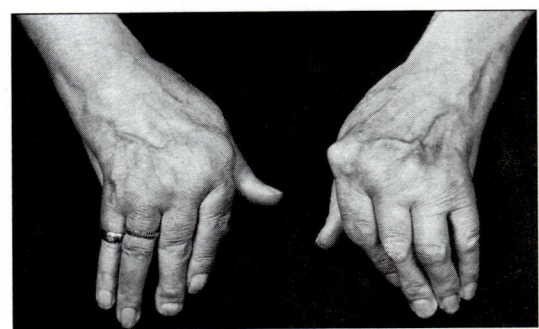

Motion can be limited because of extreme pain associated with moving the joint or because the stiffness restricts the joint surfaces. Some forms of arthritis will twist the joint into a position where it is no longer capable of its normal motion.

DISCOVERY SCENE PLEASE ENTER DISCOVERY SCENE PLEASE ENTER

Do the terms referring to body directions provide any help in communicating information about the patient with the knife wound? How would you explain where the wound is located using the terms described at this point in the chapter?

BODY REGIONS

Body region terms describe the body as if viewing a map. They describe the specific locations of body parts or regions on the surface or inside the body. These terms are universally used so that anyone hearing or reading these terms can envision the exact location on or within the body.

General Locations

Key Terms: abdominal, acromial, bilateral, brachial, carpal, cervical, clavicular, cubital, deep, geniculate, ocular, palmar, parietal, pedal, pelvic, plantar, pubic, superficial, thoracic, unilateral, visceral

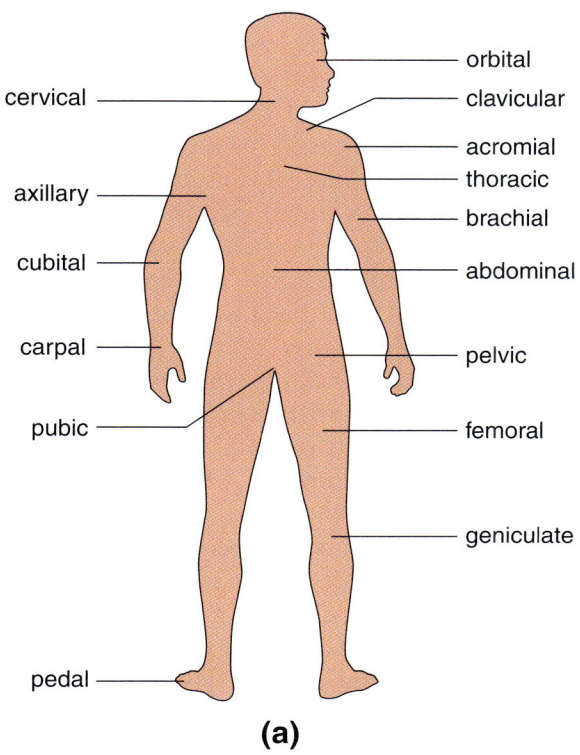

Figure 1.16a Surface Feature Coverage

Cervical Refers to the neck region
Ocular Refers to the eyes
Clavicular Refers to the region around the collar bone
Acromial Refers to the shoulder region
Brachial Refers to the arm
Cubital Refers to the elbow
Carpal Refers to the wrist
Abdominal Refers to the stomach region
Thoracic Refers to the chest region
Pelvic Refers to the region around the hip bone
Pubic Refers to the groin region
Geniculate Refers to the knee region
Pedal Refers to the region around the foot
Superficial Refers to any body part or region close to the skin
Deep Refers to any structure or region located away from the body's surface and toward the inside
Bilateral Refers to body structures located laterally on both sides of the body
Unilateral Refers to a single body part found in a lateral location
Parietal Refers to the outer wall of a hollow body part, such as the stomach. It also refers to the thin linings covering body cavities

Anatomical terminology is essential to pinpoint particular regions of the body that may need attention or that show signs of injury. The terms related to the body features shown in Figure 1.16a are equivalent to street signs. They provide a universal way to communicate major surface features and are used with other medical terms to further clarify surface locations. For example, **cervical** is specific to the neck region. **Ocular** means located near the eyes. The term **clavicular** refers to the region around the collar bone, while **acromial** refers to the shoulder. The term axillary is used in place of armpit. Even the common words arm (**brachial**), elbow (**cubital**), and wrist (**carpal**) are replaced with anatomical terms. The stomach region is called **abdominal**, and the chest region is called **thoracic**. **Pelvic** is used to refer to the hip bone region, while the term **pubic** designates the groin. The knee is called the **geniculate** area, and **pedal** is the appropriate term for foot.

The terms described here are used in the same way that a person would describe the location of objects or rooms in a house. For example, you may say that there is an attic storage space above the living room or a furnace in

OVERVIEW OF THE BODY

Figure 1.16b **Body System, Musculosketletal**

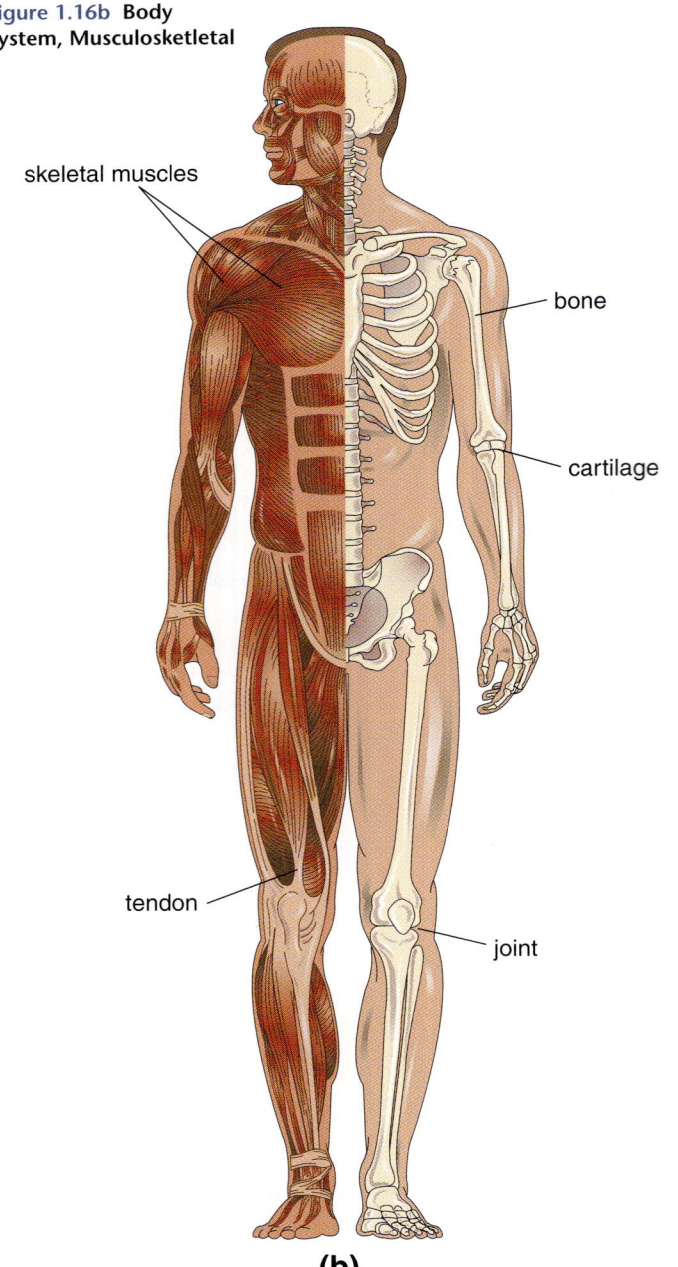
(b)

the basement below the kitchen. In this way, the terms presented here help you locate a specific area of the body. The term **superficial** describes any part or region close to the skin or outer surface of the body. In Figure 1.16b, the muscles would be superficial because they are found just beneath the skin. A **deep** structure is one that is found away from the body surface toward the inside of the body. The thigh bone is a deep structure lying beneath the muscles. **Bilateral** describes body structures that are located laterally on both sides of the body. The eyes and ears are bilateral structures. The term **unilateral** is used for a single body part found in a lateral location. For example, the heart's location is unilateral because it lies slightly left of the body's midline. The stomach is also unilateral because it is only on the left side of the body.

Certain terms that describe location can be somewhat confusing at first. This is true of the terms **parietal** and **visceral.** Parietal has two meanings: First, it can refer to the outer wall of a hollow body part. The parietal wall of the stomach refers to the outer layer of the stomach surface. Second, it also refers to the thin linings that cover whole body cavities. Parietal structures are, in effect, superficial coverings. The term visceral describes the inner wall of a body organ. It also refers to a covering found directly on a body part. Visceral structures are located deep in the body compared with parietal structures. Some location terms are specific to a particular body part. For example, the term **palmar** pertains to the palm of the hand. Similarly, **plantar** describes the sole, or lower surface, of the foot.

Visceral Refers to the inner wall of an organ. It also refers to the coverings found directly on body parts

Palmar Pertaining to the palm of the hand

Plantar Pertaining to the sole of the foot

Abdominopelvic Regions and Quadrants

Key Terms: abdominopelvic region, epigastric, hypogastric, left hypochondriac, left inguinal, left lower quadrant (LLQ), left lumbar, left upper quadrant (LUQ), right hypochondriac, quadrant, right inguinal, right lower quadrant (RLQ), right lumbar, right upper quadrant (RUQ), umbilical

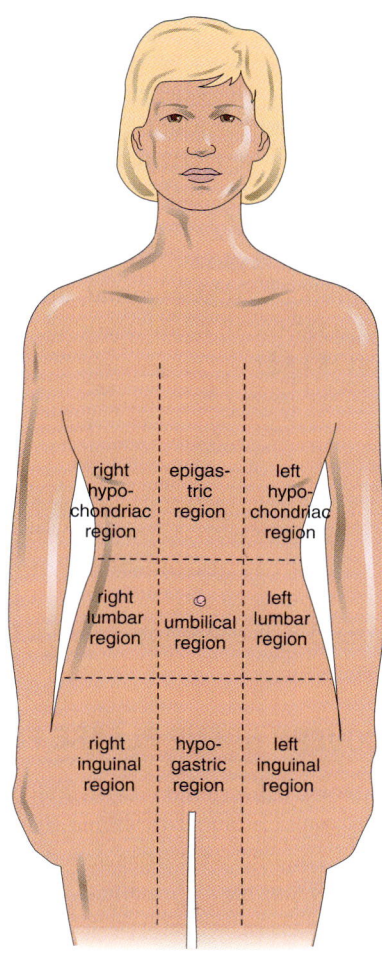

Figure 1.17 Abdominopelvic Regions

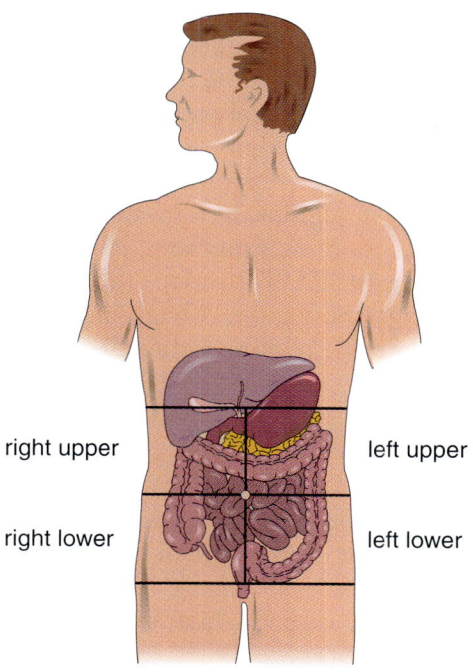

Figure 1.18 Body Quadrants

Abdominopelvic Region Refers to the region of the body found below the breasts and above the groin

Right Hypochondriac The upper right corner of the abdominopelvic region

Left Hypochondriac The upper left corner of the abdominopelvic region

Epigastric The upper middle of the abdominopelvic region

Right Lumbar The middle right corner of the abdominopelvic region

Left Lumbar The middle left corner of the abdominopelvic region

Umbilical The middle section between the right and left lumbar of the abdominopelvic region

Right Inguinal The lower right corner of the abdominopelvic region

Left Inguinal The lower left corner of the abdominopelvic region

Hypogastric The middle section between the right and left inguinal sections of the abdominopelvic region

The **abdominopelvic region** describes the part of the body lying just below the breasts and just above the groin, as shown in Figure 1.17. It is a very important region of the body because it contains almost all of the major body organs. The abdominopelvic region is divided into sections using terms that describe a particular location. In the upper right corner of the abdominopelvic region lies the **right hypochondriac**. Its name comes from the fact that it sits below the ribs. The **left hypochondriac** is in the opposite position in the body. Between these two regions is the **epigastric** region. It has this name because is sits above the stomach.

Below the right hypochondriac is the **right lumbar** region. On the other side of the body is the **left lumbar**. The term lumbar means at waist level, or at the "small of the back." Medial to these two regions is the **umbilical** region. As is evident in the name, the umbilical region contains the navel.

The **right inguinal** region makes up the area of the right lower lateral part of the abdominopelvic region. On the other side is the **left inguinal** region. These regions lie directly upon the pelvis. Inguinal is the scientific term for groin. Medial to the left and right lumber regions is the **hypogastric** region. Although it is below the navel, its name means "located below the stomach."

OVERVIEW OF THE BODY

Quadrant Refers to the abdominopelvic regions as divided into four sections

Right Upper Quadrant The quadrant containing the right hypochondriac, lumbar, epigastric, and umbilical regions

Left Upper Quadrant The quadrant containing the left hypochondriac, lumbar, epigastric, and umbilical regions

Right Lower Quadrant The quadrant containing the right inguinal, lumbar, hypogastric, and umbilical regions

Left Lower Quadrant The quadrant containing the left hypochondriac, lumbar, epigastric, and umbilical regions

Most clinicians use a simpler way to describe the body region divisions. They use what is called the **quadrant** naming system, which divides the region into four parts. These four parts are shown in Figure 1.18. The quadrant system is simple to remember. However, it provides less detail for identifying the location of pain or injury to the abdominopelvic region. The **right upper quadrant** (**RUQ**) overlaps the right hypochondriac, epigastric, right lumbar, and umbilical regions. In turn, the **left upper quadrant** (**LUQ**) overlies the left hypochondriac, epigastric, left lumbar, and umbilical regions. The **right lower quadrant** (**RLQ**) overlaps the right lumbar, umbilical, right inguinal, and hypogastric regions. Overlying the left lumbar, umbilical, right inguinal, and hypogastric regions is the **left lower quadrant** (**LLQ**).

✓ Concept Check

1. Define the abdominopelvic region.
2. What is the name of the center-most section of the abdominopelvic region?
3. How does the quadrant system differ from the abdominopelvic sections?

DISCOVERY SCENE PLEASE ENTER DISCOVERY SCENE PLEASE ENTER

How does an understanding of the terminology describing body regions help solve the CSI? Do you see how the terms would help in determining the type of damage to the person's body? How would you apply the terms deep and superficial in trying to gather more information about the knife wound?

BODY CAVITIES

Key Terms: abdominal cavity, abdominopelvic cavity, cervical region, coccyx region, cranial cavity, diaphragm, lumbar region, mediastinum, nasal cavity, oral cavity, pelvic cavity, pericardial cavity, pleural cavity, sacral region, sinuses, spinal cavity, spinal column regions, thoracic cavity, thoracic region

Abdominopelvic Cavity The body cavity containing the abdominal and pelvic cavities

Abdominal Cavity The body cavity containing the liver, gallbladder, intestines, kidneys, spleen, and stomach

Pelvic Cavity The body cavity containing the rectum, reproductive system, and urinary bladder

The human body is divided into distinct body cavities that contain particular body organs as shown in Figure 1.9. These cavities wall off the various organs with thin sheets of wet membrane called serosa, or flattened layers of muscle. An awareness of the terms describing these body cavities is important in understanding the arrangement of internal organs.

A large cavity called the **abdominopelvic cavity** forms a hollow space within the abdominopelvic region already described in this chapter. The abdominopelvic cavity is actually composed of two cavities: the **abdominal cavity** and the **pelvic cavity**. The abdominal cavity contains the liver, gall bladder, intestines, kidneys, spleen, and stomach. Inferior to the abdominal cavity is the pelvic cavity, which includes the rectum, reproductive system, and urinary bladder.

Cutting Edge Research
SMOKING AND THE THORACIC CAVITY

A story about the thoracic cavity is not the hot topic you would expect to see in the news. However, a condition called pneumothorax is becoming more prevalent among certain groups of people. *Pneumothorax* is a condition in which one or both lungs collapse. One form of this condition, called spontaneous pneumothorax, occurs in the absence of injury to the chest or lungs. Recent studies show that *spontaneous pneumothorax* is seven times more likely to occur in males than in females. Compared with nonsmokers, male smokers are 20 times more likely to develop spontaneous pneumothorax. Female smokers are only nine times more likely than female nonsmokers to experience spontaneous pneumothorax. In addition, the condition mostly occurs during the fall or winter. Smoking can be avoided, warding off the chance of developing this potentially life-threatening condition. However, current studies indicate that air pollution may have the same effect as smoking. Some air pollution comes from second-hand smoke. Much of the pollution comes from automobiles and industrial operations. People living in areas with poor air quality have to be vigilant to reduce factors that further damage the lungs. This includes staying indoors during poor-air-quality days and avoiding situations of greater exposure to polluted air. That means avoiding cigarette or other tobacco smoke, gas-powered lawn equipment, fireplace smoke, and fumes from outdoor grills.

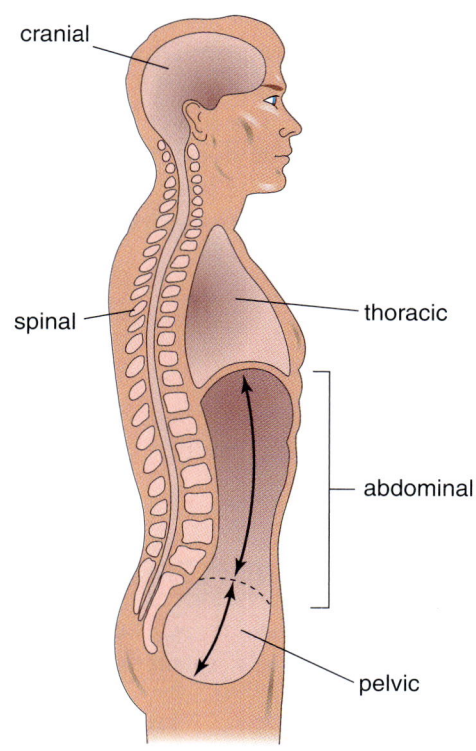

Figure 1.19 **Body Cavities**

Superior to the lumbar region is the **thoracic cavity**. It provides a covering that encases the esophagus, heart, lungs, and respiratory tree. Smaller partitions within the thoracic cavity separate the heart and each lung. For example, the **pericardial cavity** directly encases the heart, while the lungs are individually contained in the left and right **pleural cavities**. The **mediastinum** is a cavity region between the lungs. It contains the pericardial cavity and major structures, such as blood vessels passing through the region. Within the skull is the **cranial cavity**, which surrounds the brain. Anterior to the cranial cavity are smaller cavities called the **oral cavity**, which contains the mouth, and the **nasal cavity**, which lies behind the nose. Smaller cavities called **sinuses** are found in certain bones surrounding the cranial cavity. Connected to the cranial cavity and running along a medial dorsal

Thoracic Cavity The body cavity containing the esophagus, heart, lungs, and respiratory tree

Pericardial Cavity The body cavity containing the heart

Pleural Cavities The body cavities containing the left and right lungs

Mediastinum The body cavity between the lungs containing the pericardial cavity

Cranial Cavity The body cavity containing the brain

Oral Cavity The body cavity containing the mouth

Nasal Cavity The body cavity behind the nose

Sinuses Small cavities found in bones surrounding the cranial cavity

OVERVIEW OF THE BODY

Spinal Cavity The body cavity containing the spinal cord

Diaphragm A large muscular partition below the thoracic cavity

Cervical Region The part of the spinal column comprising the neck

Thoracic Region The part of the spinal column comprising the thorax, or chest

Lumbar Region The part of the spinal column comprising the dorsal section of the umbilical region

Sacral Region The part of the spinal column comprising the pelvic area

Coccyx Region The part of the spinal column comprising the tail bone

Figure 1.20 Spinal Column Regions

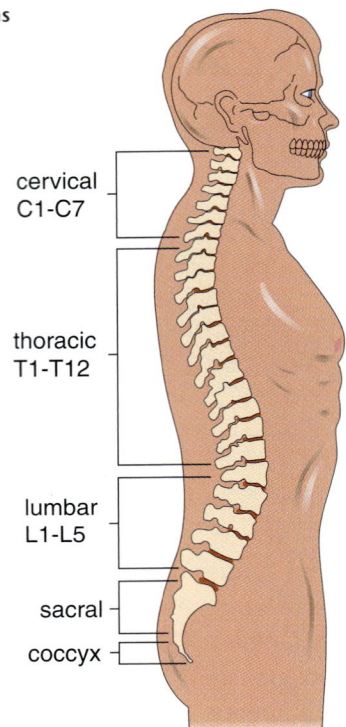

position in the body is the **spinal cavity**. As its name implies, the spinal cavity covers the spinal cord. Below the thoracic cavity is a large muscular partition called the **diaphragm**.

The spinal cavity can be divided into spinal-column regions, as shown in Figure 1.20. The uppermost section is called the **cervical region**, which makes up the neck. Inferior to the cervical region is the **thoracic region**. It is so named because it is composed of the thorax or chest area. The lower back contains the **lumbar region** of the spinal cavity. It lies along the dorsal portion of the umbilical region. The **sacral region** makes up much of the back of the pelvic bone. Below the sacral region is a tiny area called the **coccyx region**. It is also called the "tail bone" region.

✓ Concept Check

1. What structures are found in the abdominopelvic cavity?
2. Distinguish between the thoracic cavity and the spinal cavity.
3. What are the sections of the spinal cavity?

DISCOVERY SCENE PLEASE ENTER DISCOVERY SCENE PLEASE ENTER

What additional information have you gathered about the man's injuries in the CSI? How does knowledge of the body cavities help explain the problems noted by the ambulance crew?

(Hint: The knife pointed up as it entered the man's left hypochondriac region of the abdominopelvic region.)

CSI – Case Study Investigation Conclusion

What can you conclude about damage caused by the knife wound? It may seem obvious that the person would be experiencing pain in the abdominopelvic region, but why would the pain occur more toward the pelvic cavity, which is inferior to the knife wound? Also, what would explain the man's breathing problem?

Answer:

The redness and swelling noted in the pelvic region are due to the fact that the knife entered the left hypochondriac region. This region contains the stomach and the pancreas. Both of these structures produce chemicals capable of degrading the body's internal parts. These chemicals very likely leaked from the punctured structures and settled at the bottom of the abdominopelvic cavity as a pool of destructive material. But what explains the breathing difficulties? It is likely that since the knife was pointing up, it may have penetrated the left portion of the thoracic cavity. This could damage the lung coverings needed for proper lung function. A puncture wound to this covering would have created a condition in the left lung called pneumothorax. Pneumothorax means that the lung has collapsed and is not able to breathe.

This CSI was adapted from the following articles:

1. Bergman R. Abdominal wound with protruding viscera. *Virtual Hospital* at: http://www.vh.org/adult/provider/anatomy/firstaid/AbWound.html.
2. Di Nunno N, Costantinides F, Bernasconi P, Di Nunno C. Suicide by hara-kiri: A series of four cases. *Am J Forensic Med Pathol.* 2001;22(1):68-72.

Chapter Summary

The healthcare and medical fields are notorious for the numerous terms needed to accurately communicate in day-to-day duties. These terms are not meant to exclude or confuse people. They are necessary to ensure that everyone working with a client or patient is relaying and understanding the same information. At first, it may be difficult to remember all of these terms. However, they become easier to commit to memory as they are used regularly on the job.

Body organization terms are divided into sets of vocabularies that explain body direction, position, and movement. Directional terms describe the relative location of different body features and the ways the body can be viewed when sliced along imaginary lines. Position terms are used to explain how a client or patient must be situated for a particular healthcare or medical procedure. Position terms make their way into a variety of health-related fields, including police crime-scene analysis and athletic training. Movement terms are also important for helping to place a person in a particular position. They are most commonly used to explain a person's ability to move a body structure in a particular direction in relation to the center of the body.

Body region terms provide a way of accurately mapping the location of particular parts. They are also used to describe the location of pain or injury. In addition, body region terms provide a way to divide the body into sections that contain particular external or internal body structures. Assigning names to the body cavities serves a similar purpose. It gives healthcare and medical workers a common language for describing a particular region.

Introduction

- Anatomy is the study of body structure.
- Fine (microscopic) anatomy looks at tiny body structures.
- Gross anatomy looks at large body structures.
- Physiology is the study of body function.

Directional Orientation

- Directional orientation explains a particular view of a person.
- Medial refers to the middle of the body.
- Lateral refers to the sides of the body.
- Superior, or cephalic, refers to a location near the head.
- Inferior, or caudal, refers to a location near the feet.
- Anterior, or ventral, refers to the front of the human body.
- Posterior, or dorsal, refers to the back of the human body.

Directional Planes

- Directional planes are views of imaginary lines sliced through a person.
- A sagittal section slices the body vertically into left and right sections.
- A midsagittal section slices the body into equal left and right halves.
- A frontal, or coronal, section slices the body vertically into anterior and
- posterior sections.
- A transverse section slices the body horizontally into inferior and superior sections.

Positions
- Position terms indicate how a person is to be situated.

Movement
- Movement terms describe the movement of body parts in relation to the standing person.
- Each movement has an opposing movement, which is called antagonistic movement.
- Body joints can be bent or straightened in movements called flexion and extension.
- Body joints can be moved away or toward the body in movements called abduction and adduction.
- Body joints can be rotated in movements called eversion and inversion.

Body Regions
- Body regions are divided into general locations, abdominopelvic regions, and quadrants.
- Abdominopelvic regions include nine sections that divide up the abdominal and pelvic portions of the body.
- Quadrants section the abdominal region into four portions.

Body Cavities
- The human body is naturally divided into internal cavities.
- The body cavities contain specific body structures.

Key Terms

Introduction
Anatomy
Anterior
Developmental anatomy
Embryology
Fine anatomy
Gross anatomy
Morphology
Pathology
Physiology

Human Body Orientation: Direction
Cephalic
Cranial
Caudal
Coronal plane
Directional planes
Directional orientation
Distal
Dorsal

Frontal plane
Inferior
Lateral
Medial
Midsagittal plane
Posterior
Proximal
Sagittal plane
Superior
Transverse plane
Ventral

Human Body Orientation: Position
Dorsal recumbent position
Fowler's position
Knee-chest position
Left lateral position
Lithotomy position
Modified Trendelenburg's position
Prone position
Sim's position

OVERVIEW OF THE BODY

Study Guide

Sitting position
Supine position
Trendelenburg's position

Human Body Orientation Movement
Abduction
Adduction
Antagonistic
Eversion
Extension
Flexion
Inversion

Body Regions: General Locations
Abdominal
Acromial
Bilateral
Brachial
Carpal
Cervical
Clavicular
Cubital
Deep
Geniculate
Ocular
Palmar
Parietal
Pedal
Pelvic
Plantar
Pubic
Superficial
Thoracic
Unilateral
Visceral

Body Regions: Abdominopelvic Regions and Quadrants
Abdominopelvic region
Epigastric
Hypogastric
Left hypochondriac
Left inguinal
Left lower quadrant (LLQ)
Left lumbar
Left upper quadrant (LUQ)
Right hypochondriac
Right inguinal
Right lower quadrant (RLQ)
Right lumbar
Right upper quadrant (RUQ)
Umbilical

Body Cavities
Abdominal cavity
Abdominopelvic cavity
Cervical region
Coccyx region
Cranial cavity
Diaphragm
Lumbar region
Mediastinum
Nasal cavity
Oral cavity
Pelvic cavity
Pericardial cavity
Pleural cavity
Sacral region
Sinuses
Spinal cavity
Spinal column regions
Thoracic cavity
Thoracic region

Check Your Understanding

1. Anatomy is the study of:
   a. body function
   b. body structure
   c. human development
   d. disease

2. The term pathology refers to the study of:
   a. pain and suffering
   b. disease
   c. dead people
   d. genetics

3. Which body structure is visible from a dorsal view?
   a. nose
   b. back
   c. eyes
   d. toes

4. The eyes are located _____ to the nose.
   a. anterior
   b. inferior
   c. lateral
   d. medial

5. The fingers are located _____ to the wrist.
   a. dorsal
   b. ventral
   c. proximal
   d. distal

6. The midsagittal plane cuts the body into:
   a. upper and lower sections
   b. two equal left and right sections
   c. two equal front and back sections
   d. unequal left and right sections

7. Which position requires a patient to sit?
   a. Fowler's
   b. supine
   c. Trendelenburg's
   d. dorsal lateral recumbent

8. The person is lying prone in this position:
   a. supine position
   b. knee-chest position
   c. Trendelenburg position
   d. lithotomy position

9. Moving the whole leg away from the center of the body is an example of:
   a. extension
   b. adduction
   c. flexion
   d. abduction

10. Feet twisted toward the body are in the _____ position.
    a. extension
    b. adduction
    c. eversion
    d. inversion

11. Which structure is considered unilateral?
    a. lungs
    b. eyes
    c. kidneys
    d. heart

OVERVIEW OF THE BODY

Study Guide

12. The _____ layer forms the outer wall of the stomach.
    a. inferior
    b. superficial
    c. parietal
    d. visceral

13. The heart is mostly located in the:
    a. LLQ
    b. RLQ
    c. LUQ
    d. RUQ

14. The thoracic cavity contains the following structures:
    a. heart and lungs
    b. lungs and stomach
    c. stomach and intestines
    d. reproductive system and urinary bladder

15. The inferior part of the spinal cavity is called the _____ region.
    a. cervical
    b. thoracic
    c. sacral
    d. coccyx

A Case Study

Urine in the Lungs

Imagine discovering that something produced in one body cavity ended up in another cavity. An unusual condition called *urothorax* results in urine entering the lungs. Urine is a waste product made by the kidneys, which are buried deep in the abdominopelvic cavity. Normally, urine does not enter the thoracic cavity. Rather, it is expelled from the body by the urinary bladder. Also, the lungs are not at all associated with structures of the abdominopelvic cavity; they are isolated in the thoracic cavity. So, how is it possible for urine to enter the lungs?

Use the information in this chapter and in the following Web sites to explain the most probable ways that urine could enter in the lungs:

1. Biology On-line.com
   http://www.biology-online.org/dictionary/urothorax

2. National Institutes of Health
   http://kidney.niddk.nih.gov/kudiseases/pubs/yoururinary/

3. WebMD
   http://my.webmd.com/webmd_today/home/default

4. American Lung Association
   http://www.lungusa.org/site/pp.asp?c=dvLUK9OoE&b=22576

Where Do We Go from Here?

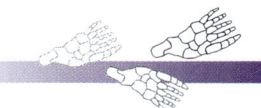

Study Guide

People in health fields can use their knowledge of anatomy and physiology to solve everyday problems. You may wish to use other resources, such as the suggested Web sites, in addition to your textbook to investigate a way to solve each of the following situations:

1. A friend is complaining of extreme chest pain and wants to go to the hospital. What terminology do you need to know to communicate information about your friend to an Emergency Department attendant?
   www.biologydaily.com/biology/terms_for_anatomical_location

2. You are volunteering at a summer science camp. One of the counselors asks you to create a song for children that would help them remember the directional terms of the human body.
   www.emedicinehealth.com/includes/basic_search.asp

3. You are watching a movie with some friends in which a "bad guy" is shot in the left lumber region. Explain to your friends the possible damage caused by the bullet. www.emcp.net/biologydaily
   www.biologydaily.com/biology/terms_for_anatomical_location

4. You are assisting at an animal shelter for abused cats and dogs. How would your knowledge of human anatomy help communicate information about the injured animals to veterinarians and caretakers?
   www.nlm.nih.gov/medlineplus/medlineplus.html

5. You witnessed a traffic accident in which people were thrown from their cars. An ambulance took the accident victims away before the police could assess the scene. How would your knowledge of anatomy and physiology terms assist the police with reconstructing the accident?
   www.webmd.com

Skills Activities

1 Drawing the Abdominopelvic Regions & Quadrants

Materials
- Blank overhead transparency sheet
- Washable black marker
- Washable red marker

It is important to be able to visualize the abdominopelvic regions of the body without having to reference a book. This activity will give you practice imagining the precise locations of these regions on an illustration of the human body.

Study Guide

Place the clear transparency sheet over the diagram provided here. First, use the black marker to draw the lines representing the abdominopelvic regions. Then, label the diagram. Next, draw the quadrant lines using the red marker. Add the quadrant labels. Compare your drawing and labels to the information provided in this chapter.

Figure 1.21 Directional Terms

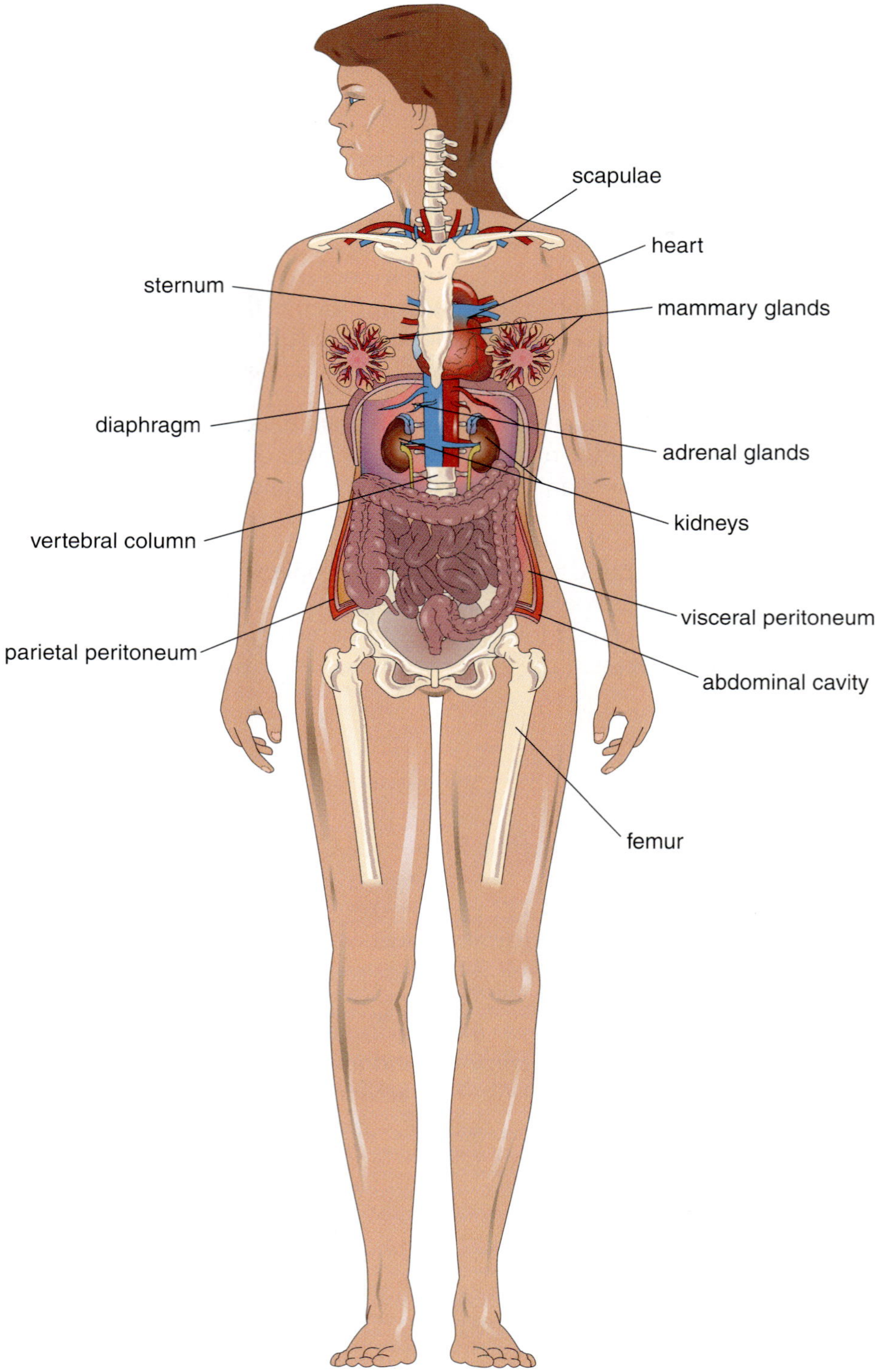

Skills Activities

2 Drawing the Body Cavities

Materials
- Blank overhead transparency sheet
- Washable black marker
- Blank sheet of paper

Study Guide

Place a clear transparency sheet over the diagram provided here. Use the marker to draw on the transparency sheets outlines of the body cavities. Then label the cavities. Next, on the blank piece of paper, make a list of the major structures found in each body cavity. Compare your drawings and labels with the information provided in the chapter.

Figure 1.22 Anatomical Position

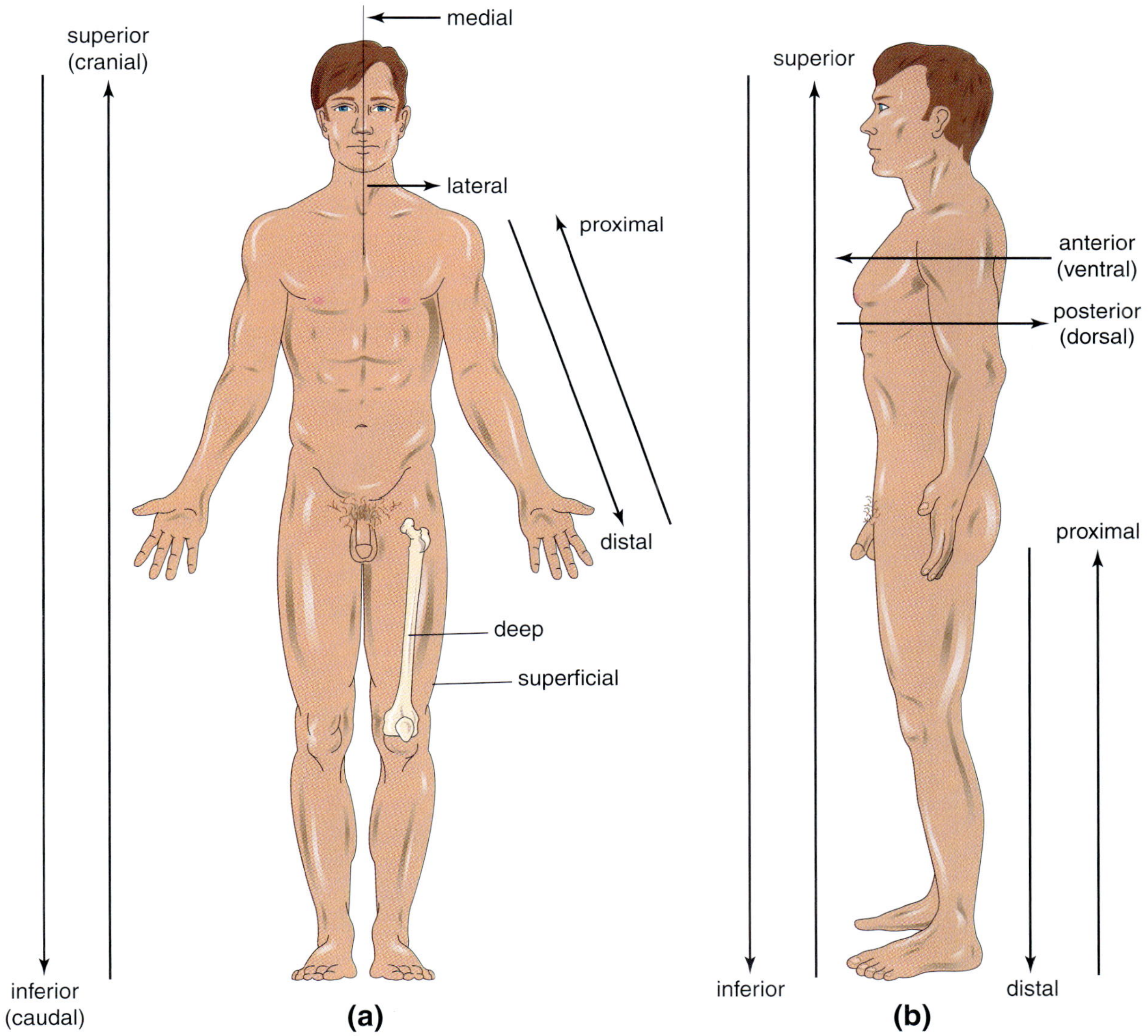

(a) Anterior view labels: superior (cranial), medial, lateral, proximal, distal, deep, superficial, inferior (caudal)

(b) Lateral view labels: superior, anterior (ventral), posterior (dorsal), proximal, distal, inferior

OVERVIEW OF THE BODY

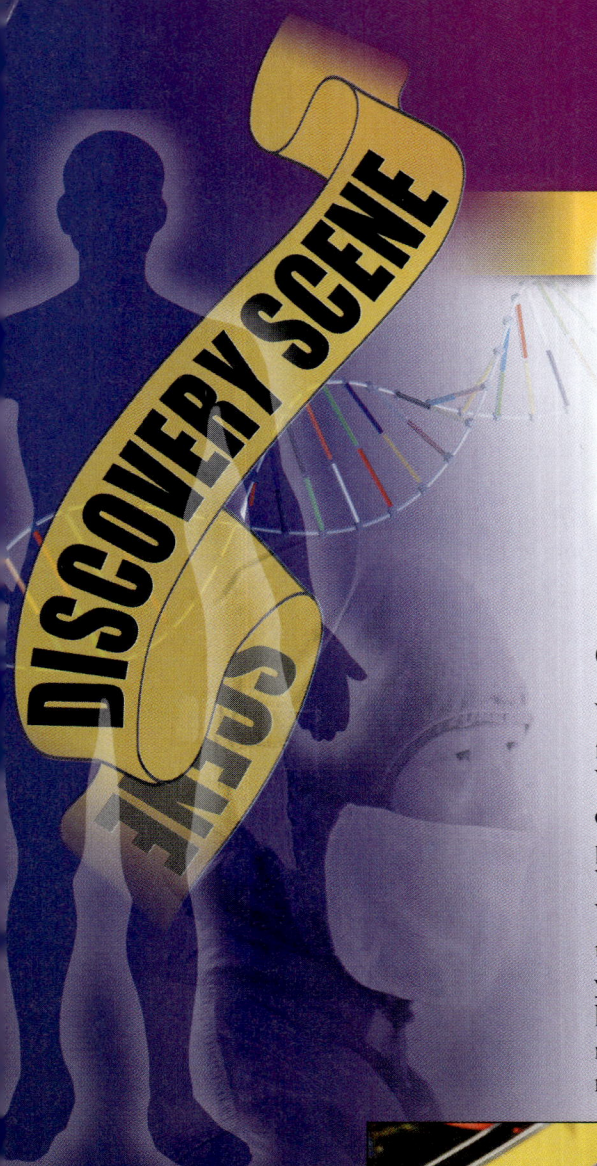

Case Study Investigation

Case Study Investigation #2

You have a friend who is preparing for a charity marathon, but he feels that he is too out of shape to compete with seasoned runners. You go with him to shop for energy drinks that would give him a competitive edge. He finally finds a drink called "Endurance" that promises to boost an athlete's energy during strenuous competition. Your friend drinks several bottles of "Endurance" before the race. Your friend collapses after 2 hours of running, only 10 miles into the 26-mile race. The medical team on the scene discovers that your friend is dizzy, vomiting, and breathing rapidly. In addition, his muscles are quickly alternating between very tense and overly relaxed. An on-the-spot examination shows that his heartbeat is rapid, yet his blood pressure is low. It is the job of the medical team to determine the chemical factors causing your friend's illness. Then, they must determine how to remedy the problem to prevent your friend from becoming more ill or even dying. At the end of the chapter, you will be asked to determine the possible body chemistry problems causing this set of conditions.

CHAPTER 2

The Body's Chemical Makeup

Chapter Outline

Case Study Investigation (CSI)
Applied Learning Outcomes
Introduction
Atoms and Molecules
   Atomic Structure and Function
   Properties of Molecules
      Types of Bonds
      Parts of a Molecule
Acids and Bases
Human Molecules
   Lipids
      Glycerides
      Sterols
      Other Lipids
Carbohydrates
   Monosaccharides
   Disaccharides
   Polysaccharides
Peptides
   Amino Acids
   Proteins
Nucleic Acids
Molecules and Nutrition
Wellness and Illness over the Life Span
   Aging of the Body's Chemistry
CSI Conclusion
Study Guide

Applied Learning Outcomes

- Use the terminology associated with the body's chemical makeup
- Learn about:
  - atomic structure and bonding
  - molecular structure
  - characteristics of the biochemical groups composing the human body
  - the chemical environment in which human biochemicals function
- Understand the aging and pathology of the body's chemical makeup

NATURAL VERSUS SYNTHETIC DRUGS

To most biochemists, a drug is a drug is a drug, no matter its source. Drugs are compounds that somehow alter the body's chemistry to gain some desirable effect. Many people feel that there is a difference between drugs that are made "naturally" and those that are produced "synthetically." Many people assert that organic, or naturally occurring drugs, differ in their chemistry from those made in the laboratory. However, strict chemical manufacturing processes ensure that the chemistry of the synthetic drug is identical to the one obtained from nature. The body uses the chemical no matter where it comes from.

INTRODUCTION

Key Terms: biochemistry, carbon, chemistry, energy, heat, mass, matter, molecular biology, organic chemistry

Biochemistry or Molecular Biology The chemistry of the body's structures and functions

Chemistry The branch of natural sciences dealing with the composition of substances, their properties, and reactions

Energy The ability of chemical systems to do work or carry out change

Matter Material that has mass and occupies space

Heat A form of energy

Organic Chemistry The field of chemistry that studies matter composed of carbon

Carbon An element found in all living organisms

An e-mail circulating among biologists and chemists states: "Organic chemistry is the chemistry of carbon compounds. **Biochemistry** is the study of carbon compounds that crawl." – Mike Adams." It simply summarizes the fact that any study of the human body requires knowledge of the chemistry of the body's structure and function. **Chemistry** is best defined as a branch of the natural sciences dealing with the composition of substances, and their properties and reactions. The body is composed of two substances: **energy** and **matter**. Energy is usually explained as the ability for chemical systems to do work or carry out some type of change. Much of the energy taken in by the human body through food ingestion is used to operate chemical pumps that make nerves and muscles work. Most of the body's work is converted into a form of energy known as **heat**. Heat is a measure of how fast the particles making up a substance are vibrating. The body uses many types of energy to operate. (This energy is found in two forms that will be described in the next chapter.)

Matter is the substance that will be covered in this chapter. It is defined as a material that has mass and occupies space. Mass is the property of a material that causes it to have weight in the presence of gravity as found on Earth. The basic unit of body structure is the matter making up human chemistry (Figure 2.1). The composition and amount of matter making up the human body significantly impacts its energy and matter needs. Scientists who study human anatomy and physiology must be aware of two fields of chemistry that investigate particular types of matter. **Organic chemistry**, as mentioned in the quote above, describes the study of chemicals made up of a type of matter called **carbon**. Carbon's role in organic chemistry will be described later.

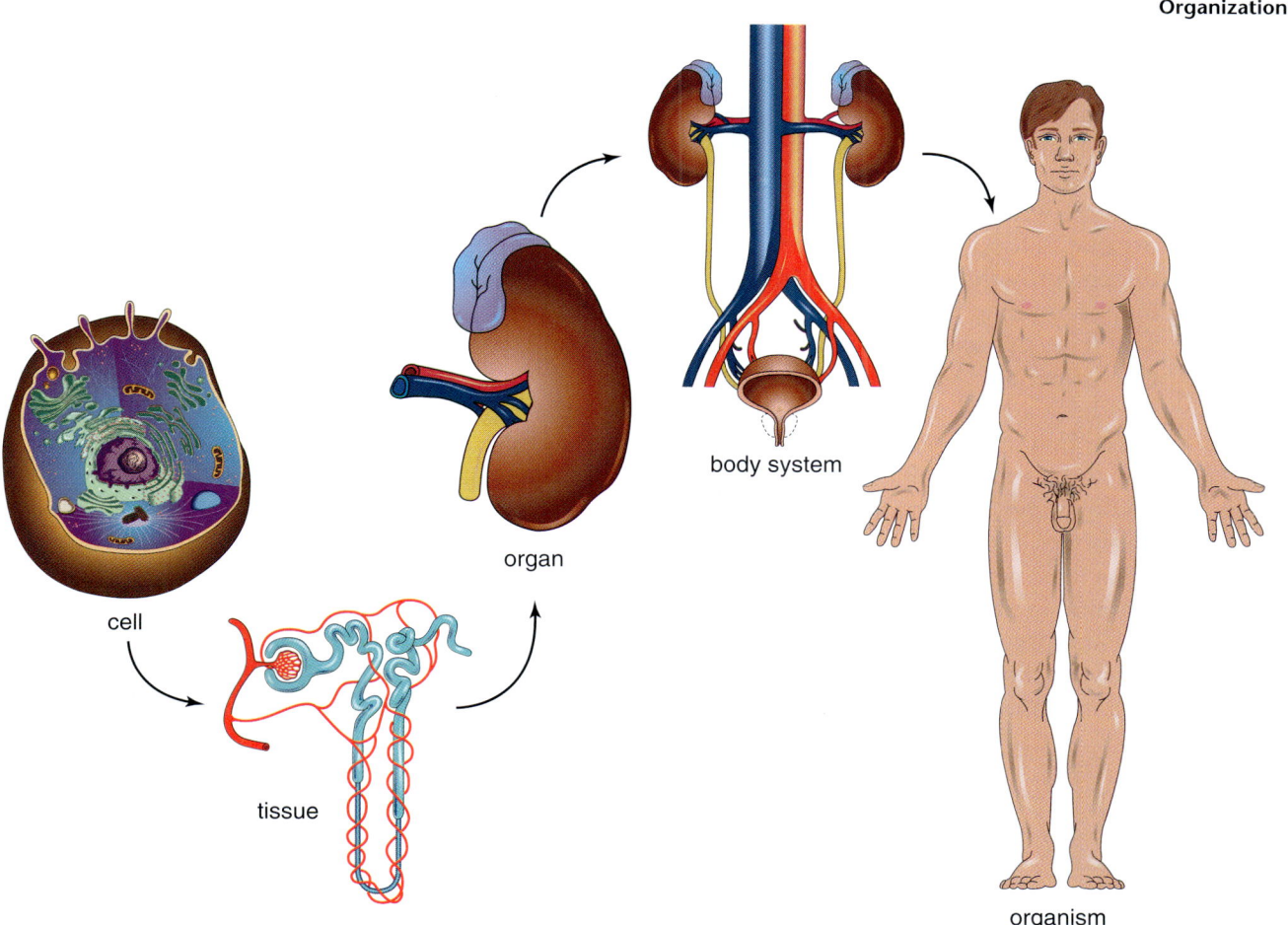

Figure 2.1 Levels of Organization

The word "organic" in organic chemistry has nothing to do with the concept of "organically grown." It merely describes chemicals that commonly compose all organisms. Biochemistry looks at how organic chemicals work together to build and run an organism. The CSI in this chapter will challenge a medical team to think about the biochemistry causing the runner's apparent illness.

✓ Concept Check

1. Define the term chemistry.
2. Distinguish between the terms energy and matter.
3. What is the difference between organic chemistry and biochemistry?

What Does Organic Mean?

Technically, all foods could be given an "organic" label, since they are made of organic chemicals. After all, scientists consider anything that contains at least one atom of carbon to be organic. However, many nutritionists and health conscious people would disagree with this sweeping definition. They tend to define organic as food that is produced without pesticides, chemical fertilizers, growth hormones, antibiotics, artificial additives, food coloring, or ionizing radiation, and has not been genetically modified in any way. This description is merely the "popular" definition of organic. The United States Department of Agriculture (USDA) and related governmental organizations in other countries have a legal definition of organic. In 2002, the USDA defined organic as food produced by farmers who emphasize the use of renewable resources and the conservation of soil and water to enhance environmental quality for future generations. Organic meat, poultry, eggs, and dairy products come from animals that are given no antibiotics or growth hormones. Organic food is produced without using the following: most conventional pesticides; petroleum-based fertilizers or sewage sludge-based fertilizers; bioengineering; or ionizing radiation. A government-approved certifier must inspect a food before it can be labeled "organic." Researchers have no solid evidence that organic foods are healthier, safer, or tastier than conventionally grown foods. Current food safety laws strictly control food quality, ensuring that no foods should cause undue harm when consumed.

Atoms and Molecules

Key Terms: atom, element, periodic table, subatomic particles

Atom The smallest portion of an element that still retains its properties

Subatomic Particles The parts of an atom

Element A substance composed of atoms having identical numbers of subatomic parts that cannot be broken down into simpler substances by normal chemical means

Periodic Table A chart of all known elements arranged according to chemical properties

All the matter found on the Earth, including what makes up human chemistry, is composed of **atoms**. The atom is usually defined as the smallest undividable unit of matter. However, it will be shown later that atoms can be divided into parts called **subatomic particles**. The subatomic parts give different types of atoms unique characteristics. Each different type of atom is called an **element**. Most scientists define elements as a substance composed of atoms with identical numbers of subatomic parts. In addition, elements are described as unable to be broken down into simpler substances by normal chemical means. There are 92 different types of naturally occurring elements making up the Earth. Scientists have created many other types in laboratory experiments.

Scientists keep track of the chemical properties of each element by organizing them into a diagram called the **periodic table** (Figure 2.2). The periodic table is a chart of the elements arranged according to chemical properties that reappear periodically as their complexity increases. Russian scientist, Dmitri I. Mendeleev, developed this chart in 1871. The periodic table is arranged in columns and rows (Figure 2.2). The 18 columns provide information about how the elements in a particular column interact with other elements. Each row represents an increasing complexity to the overall structure of the elements. Larger and heavier elements, having many subatomic parts, are found toward the bottom of the table. The periodic table is not a mysterious piece of information

Figure 2.2 Periodic Table A modern periodic table of the elements.

Periodic Table of the Elements

	IA	IIA	IIIB	IVB	VB	VIB	VIIB		VIII		IB	IIB	IIIA	IVA	VA	VIA	VIIA	0
1	1 H 1.008																	2 He 4.003
2	3 Li 6.939	4 Be 9.0122											5 B 10.811	6 C 12.011	7 N 14.007	8 O 15.999	9 F 18.998	10 Ne 20.183
3	11 Na 22.99	12 Mg 24.312											13 Al 26.982	14 Si 28.086	15 P 30.974	16 S 32.064	17 Cl 35.453	18 Ar 39.948
4	19 K 39.102	20 Ca 40.08	21 Sc 44.956	22 Ti 47.9	23 V 50.942	24 Cr 51.996	25 Mn 54.938	26 Fe 55.847	27 Co 58.933	28 Ni 58.71	29 Cu 63.546	30 Zn 65.37	31 Ga 69.72	32 Ge 72.59	33 As 74.922	34 Se 78.96	35 Br 79.904	36 Kr 83.8
5	37 Rb 85.47	38 Sr 87.62	39 Y 88.905	40 Zr 91.22	41 Nb 92.906	42 Mo 95.94	43 Tc (97)	44 Ru 101.07	45 Rh 102.91	46 Pd 106.4	47 Ag 107.87	48 Cd 112.4	49 In 114.82	50 Sn 118.69	51 Sb 121.75	52 Te 127.6	53 I 126.9	54 Xe 131.3
6	55 Cs 132.91	56 Ba 137.34	57 *La 138.91	72 Hf 178.49	73 Ta 180.95	74 W 183.85	75 Re 186.2	76 Os 190.2	77 Ir 192.2	78 Pt 195.09	79 Au 196.97	80 Hg 200.59	81 Tl 204.37	82 Pb 207.19	83 Bi 208.98	84 Po 210	85 At 210	86 Rn 222
7	87 Fr 215	88 Ra 226.03	89 +Ac 227.03	104 Rf (261)	105 Db (262)	106 Sg (266)	107 Bh (264)	108 Hs (269)	109 Mt (268)	110 Ds (271)	111 Rg (272)							

• Lanthanide series

58 Ce 140.12	59 Pr 140.91	60 Nd 144.24	61 Pm 145	62 Sm 150.35	63 Eu 151.96	64 Gd 157.25	65 Tb 158.92	66 Dy 162.5	67 Ho 164.93	68 Er 167.26	69 Tm 168.93	70 Yb 173.04	71 Lu 174.97

+ Actinide series

90 Th 232.04	91 Pa 231	92 U 238.03	93 Np 237.05	94 Pu 239.05	95 Am 241.06	96 Cm 244.06	97 Bk 249.08	98 Cf 252.08	99 Es 252.08	100 Fm 257.1	101 Md 258.1	102 No 259.1	103 Lr 262.11

only useful for scientists. Medical personnel and drug developers use the periodic table to better understand the effects of different elements on the body. Sometimes it is possible for one element to interfere with the function of another element in the body.

Atomic Structure and Function

Key Terms: atomic structure, atomic mass, atomic nucleus, atomic number, atomic orbitals (shells), atomic structure, electrons, ion, isotope, nuclear decay, neutron, physical properties, proton, radioactive, radiocarbon dating

As discussed earlier, atoms are the simplest material composing all matter. The **atomic structure** that makes up each element determines how the body uses a particular element. The atomic structure of each element is composed of two fundamental components: the **atomic nucleus** and the **atomic orbitals**, or **shells** (Figure 2.3).

A small core of dense material forms the atomic nucleus. The atomic nucleus provides each element with its mass and certain **physical properties**. A physical property

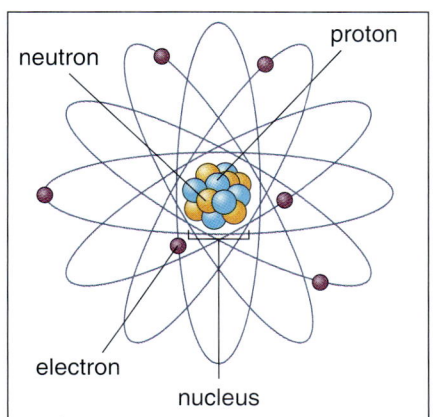

Figure 2.3 Atomic Structure Atoms are composed of a central nucleus surrounded by atomic orbitals.

Atomic Structure The structure of atoms composing each element, consisting of an atomic nucleus and atomic orbitals

Atomic Nucleus A core of dense material providing an element with its mass and physical properties

Atomic Orbitals (Shells) The regions around the nucleus where electrons are located

Physical Properties Any characteristic that can be detected by the five human senses or devices that are extensions of the senses

Proton A positively charged subatomic particle found in the nucleus

Atomic Number The number of protons in an atom

Atomic Mass The sum of protons and neutrons in an atom's nucleus

Isotope A variation of an element having the same number of protons, but a different number of neutrons

Nuclear Decay When the nucleus of an atom breaks down

Radioactive Describes a substance that gives off energy due to the decay of its unstable atoms

Radiocarbon Dating A method for dating organic remains based on their content of carbon-14

includes any characteristic that can be detected using the five human senses or devices that are extensions of the senses, such as a microscope. The nucleus influences physical properties, such as boiling point, color, density, hardness, heat conduction, electrical properties, melting point, and rigidity.

The nucleus itself is composed of two major subatomic particles. One particle is called the **proton**. Protons carry a positive electrical charge. The number of protons in an atom is used as a means of identifying each element. This identifying number is called the **atomic number**. The periodic table shown in Figure 2.2 arranges elements by their atomic number, which is represented by the number in the upper left corner of the box containing the element's abbreviated name.

Neutrons are the other subatomic particles found in the nucleus. They get their name from the fact that they have no electrical charge. The state of having no charge is called neutral, hence, the name neutron. Neutrons have almost the same mass as protons and contribute to the **atomic mass** of an element. Atomic mass is defined as the sum of protons and neutrons in an element's nucleus. For most elements, the number of neutrons is equivalent to the number of protons. In the periodic table, the atomic mass is the number found in the lower part of the box containing the element's abbreviated name. Note that the atomic mass is typically not the exact number you might expect after looking at the number of protons. This is because some elements contain more or fewer neutrons than the expected number. Variations of elements having the same number of protons, but different numbers, of neutrons are called **isotopes**. The result is a difference in atomic masses between two isotopes of the same element. For example, all carbon isotopes have six protons. However, carbon-12 has six neutrons, carbon-13 has seven neutrons, and carbon-14 has eight neutrons. The periodic table represents the average mass of an element including the relative proportion of naturally occurring isotopes. This is equivalent to taking the average weight of 99 people weighing 150 pounds along with one person weighing 152 pounds. The average weight would be 150.02 pounds.

Isotopes are very important to medical-care workers and to scientists who study biochemistry. An element's isotopes have very similar chemical and physical properties, but greatly different nuclear properties. One nuclear property has to do with the stability of the element. Unlike traditional elements, an isotope is very likely to change the composition of its nucleus. Some transformations cause the nucleus to break down, resulting in what is called **nuclear decay**. These violent changes can cause alterations to the body's biochemistry, including destructive changes that may result in cancer. This damaging potential of **radioactive** isotopes is, ironically, a very important strategy for killing cancer cells. For example, an isotope of gold (Au-198) is used to treat brain, ovarian, and prostate cancer. Scientists also use radioactive isotopes for tracing the path of chemicals through the body. An isotope of the element iodine (I-131) helps researchers trace the biochemical processes of the kidneys, liver, and thyroid gland. Isotopes also play a role in telling how long an organism has been dead in a procedure called **Radiocarbon dating**. The box on page 37 describes how an isotope of carbon is used to investigate the age of unusual human remains.

DATING A MUMMY?

Most people in health vocations work to keep people alive and vigorous. There is a group of researchers interested in studying people who have been dead for thousands of years. Physical anthropologists and archeologists study the diseases and lives of ancient people by studying excavated remains. Dr. Mike Parker Pearson of Sheffield University in northern England made an unusual discovery in the nearby island of South Uist. The archeologist found two mummified bodies buried under a prehistoric house. It appeared to Parker that the bodies dated back to a period of time called the Bronze Age, which means the people lived in that area over 4000 years ago. Parker was surprised to learn that people living in England at that time had the knowledge to mummify bodies in the same fashion as the ancient Egyptians. Therefore, he had to use a precise means of proving that the bodies dated back to that period of time or were mummified much later as people gathered more knowledge about the process. The team carried out a procedure called radiocarbon dating to calculate the age of the two bodies. Radiocarbon dating relies on the fact that Earth's upper atmosphere is bombarded by radiation, the unstable isotope carbon-14. Carbon-14 first makes its way into the molecules of plants and then into the human body when the plants are eaten. When a person dies, the amount of carbon-14 within the body begins to slowly decrease as the element decays. The speed of decay is 1/2 the quantity at death every 5730 years. Parker found that the carbon-14 content of the bodies he found was consistent with the Bronze Age. Most unusual was the age of the materials used to mummify the bodies. The materials were much younger than the bodies, meaning that the bodies were buried 300 to 600 years after the people died and were subsequently mummified. It appears that the mummies were kept above ground in some type of shrine, or even in a house. The mystery of who buried the mummies and why they were buried requires further investigation.

Surrounding the nucleus of each element is the second major component of the atom called the atomic orbitals (Figure 2.3). Atomic orbitals, or shells, are regions encasing the nucleus where rapidly moving particles, called **electrons**, can be found. Electrons are tiny negatively charged particles. Their negative charge is equal in strength to the positive charge of the proton. However, protons have almost 2000 times the mass of an electron. Elements usually have an equal number of electrons and protons. This sets up the atom so that it has no net electrical charge. Many elements are readily capable of losing or gaining electrons. They exchange these electrons with other elements. This ability to swap electrons is due to the stability of the atomic orbitals. An element that has gained or lost electrons is called an **ion**. Ions are defined as atoms that carry positive or negative charges due to the loss or gain of an electron. An element that loses an electron gains a positive charge unit for each electron lost. So, the element calcium, which is Ca on the periodic table, would have a positive charge value of 2 if it lost two electrons and would be written as Ca^{+2}. In contrast, the element chlorine, Cl on the periodic table, gains a negative charge value of 1 if it acquires an electron from another element and is written as Cl^{-1}. Ions are very important to the human body because they are involved in electrical responses associated with nerve and muscle function. In addition, certain numbers of ions are essential for maintaining the chemical reactions needed for everyday life.

Electron A negatively charged particle that orbits the nucleus

Ion An element that has gained or lost an electron

THE BODY'S CHEMICAL MAKEUP

✓ Concept Check

1. Describe the two major components of an atom.
2. What is the difference between the terms atomic mass and atomic number?
3. Distinguish between the terms ion and isotope.

Properties of Molecules

Key Terms: adhesiveness, amino group, bioactive molecules, biochemical, carbon skeleton, carbonyl group, carboxyl group, chemical bonds, chirality, cohesiveness, compounds, covalent bond, functional group, hydroxyl group, hydroxyl ions, hydrogen bond, ionic bond, isomer, molecular formula, phosphate group, sulfhydryl group, pure molecules, structural formula, structural molecules

> **Chemical Bond** The way atoms are attached to each other
>
> **Pure Molecules** Identical elements bonded together
>
> **Compound** Molecules of two or more different elements bonded together

Molecules are the true building blocks of the human body. A molecule is defined as two or more atoms joined together by **chemical bonds**. A person eats for the sole purpose of taking in molecules to maintain the body. Many of these molecules provide the raw materials needed to replace worn out molecules. Other molecules are broken down to provide energy to run body processes. Chemical bonds are forms of attachment between atoms. **Pure molecules** are composed of identical elements bonded together. Oxygen gas is an example of a pure molecule because the bonding of two oxygen atoms forms one molecule of oxygen gas. **Compounds** are molecules formed by different elements bonded to each other. Water is an example of a compound. One molecule of water contains two hydrogen atoms joined with one atom of oxygen. Another important feature of molecules is that they are the smallest particle of a substance that retains the properties of the substance. This piece of information is important in understanding why certain molecules lose their function in the body when altered or broken down. It is important not to confuse the roles of elements with the roles of molecules in the human body. Their different responsibilities will be discussed throughout the rest of this chapter.

Civic Responsibility

CIVIC RESPONSIBILITY: HELPING OTHERS WITH YOUR KNOWLEDGE

It is valuable to use what you learned about biochemicals to help others better understand the world around them. It is very important to check your facts and seek further information about certain topics before discussing health and science issues. Here are some suggestions to foster a better public awareness of the body's chemical makeup:

1. Speak to schoolchildren about the roles of different molecules in their diet.
2. Work with sports clubs educating the players about proper nutrition.
3. Help the elderly to better understand the basis of antiaging strategies.
4. Volunteer at a school health day to teach children about their body's chemistry.

CHAPTER 2

Types of Bonds Three major types of bonds make up the molecules of the human body. An element's placement on the periodic table predicts the types of bonds it can form with other elements. For example, elements found in the first two columns and the next-to-the last column of the periodic table effortlessly form **ionic bonds**. The ionic bond is a force that holds together two electrically charged elements. Remember, ion is the name given to electrically charged particles. This is the origin of the name ionic bond. Ionic bonds form only between ions of opposite charge, as shown in Figure 2.4. Elements in the first column of the periodic table form ionic bonds because they readily lose an electron giving them a single positive charge. Similarly, elements in the next-to-last column easily gain electrons from other elements, giving them a single negative charge. Any of these oppositely charged ions can come together to form ionic bonds.

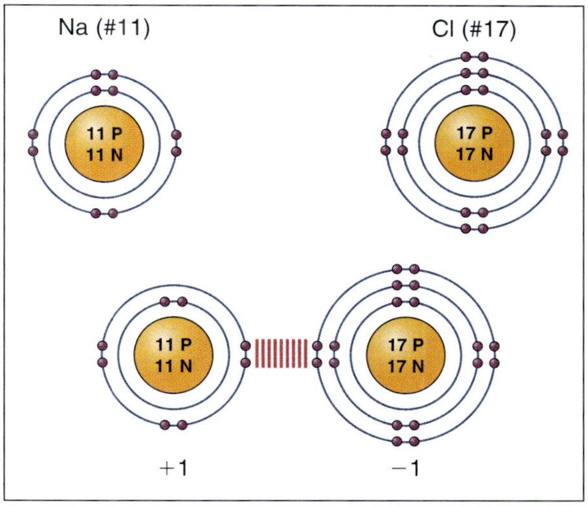

Ionic bonds usually form so that the molecule ends up with a balance of positive and negative charges. For example, sodium ions (Na^{+1}) will bond to chlorine atoms (Cl^{-1}) forming the molecule sodium chloride, written NaCl. Calcium loses two electrons to form a calcium ion (Ca^{+2}). It forms the molecule calcium chloride, which contains one calcium ion for every two chlorine ions. Scientists represent calcium chloride as $CaCl_2$. Note that the number 2 is written as a subscript following the element. Ionic bonds are strong and rigid in a solid state, and most of the molecules they form take on a crystalline structure similar to table salt. The body takes advantage of such structure when the ionically bonded molecules form the hard "nonliving" matrix of bone tissue. In water, however, ionic bonds easily come apart and, therefore, do not form most of the molecules used to build and run the body. Ionic bonding is vital in the electrical processes involved in nerve and muscle activity as will be discussed in later chapters.

Covalent bonds are more characteristic of the **biochemicals** making up the human body than are ionic bonds. They are found typically in organic molecules and involve only a few types of elements. Carbon (C), hydrogen (H), nitrogen (N), oxygen (O), phosphorus (P), and sulfur (S) form covalent bonds with each other. Carbon can also form covalent bonds with many elements that usually form ionic bonds. Covalent bonds are characterized by the sharing of one or more electrons between two or more elements (Figure 2.5). This is what holds the molecule together. Covalent bonds are similar in strength to ionic bonds. However, these bonds do not weaken in water, as do ionic bonds. The properties of organic materials, such as hair, rubber, plastic, skin, and vegetable oil, are imparted by covalent bonds. Covalent bonds are an important source of energy for all living things. Most organisms break down the covalent bonds of molecules in their foods to supply the energy needed to carry out a variety of body functions. Covalent bonds stored in human fat are broken down to supply energy when food intake is not adequate to meet these needs.

Figure 2.4 Properties of Molecules (Ionic Bond) Ionic bonds are formed between ions with opposing charges.

Ionic Bond A bond between two electrically charged elements

Biochemicals Organic molecules produced by the chemical reactions of living organisms

Covalent Bond A bond between two elements sharing electrons

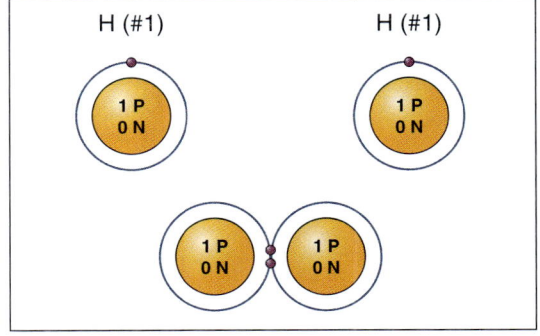

Figure 2.5 Properties of Molecules (Covalent Bond) Covalent bonds involve a sharing of orbitals.

THE BODY'S CHEMICAL MAKEUP

Hydrogen Bond A temporary weak bond between a partial positive hydrogen atom and a partial negative oxygen, nitrogen, or fluorine atom

Adhesiveness A glue-like property, as with water

Cohesiveness A substance's tendency to stick to itself

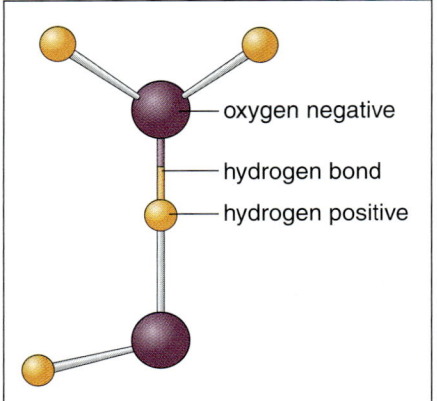

Figure 2.6 **Hydrogen Bond** Hydrogen bonds are weak electrical attractions.

A third group of bonds is identified by a type of attractive force that takes place between two partial, opposite electrical charges. These typically temporary and weak bonds are called **hydrogen bonds**. Hydrogen bonds form when a hydrogen atom carrying a partial positive charge is near oxygen, nitrogen, or fluorine atoms that are carrying an excess negative charge (Figure 2.6). Basically, hydrogen bonds form under precise conditions, usually attaching one molecule to another. Water is the most common molecule in the body that forms hydrogen bonds with both itself and other molecules. This gives water its glue-like property called **adhesiveness**. Many people take advantage of water's adhesive features when they use moisture to help a suction cup attach to a surface. The ability of water to form hydrogen bonds with itself is called **cohesiveness**. The great degree of water's cohesiveness explains why water evaporates slowly and has a high boiling point. In humans, hydrogen bonds are essential for carrying out most of the chemical reactions that run the body. They also play important roles in holding together body structures, and they permit the body to recognize and fight disease.

✓ Concept Check

1. Define the term molecule.
2. Distinguish between covalent, hydrogen, and ionic bonds.
3. What is the difference between a compound and a pure molecule?

Bioactive Molecules Molecules that promote chemical reactions

Structural Molecules Molecules that compose body parts

Carbon Skeleton Carbon atoms held together with covalent bonds

Functional Group One or more elements attached to a carbon skeleton, responsible for the chemical activities of a biochemical

Parts of a Molecule Biochemicals are organic molecules produced by chemical reactions carried out by living organisms. Many of the biochemicals needed for human existence are synthesized in plants. People must eat plants, or animals that eat plants, to obtain many of the biochemicals essential for survival. Other biochemicals can be manufactured in the body. Biochemicals can be classified into two major categories: **bioactive molecules** and **structural molecules**. Bioactive molecules, as the name implies, carry out or promote chemical reactions in the body. Caffeine is an example of a bioactive molecule. Many people are familiar with the way caffeine stimulates chemical reactions that can keep a person awake and jittery all night. Structural molecules are used as components of body parts. Nails and tendons are made of structural biochemicals. Some structural biochemicals, such as fat, take on a storage role. This means that they are stored in the body for later use. Certain biochemicals have both roles in the body. The abundant simple sugars found in many candy bars can take on bioactive or structural tasks.

Almost all biochemicals have two fundamental components that permit them to carry out their particular functions in the body. One component is the **carbon skeleton** and the other is the **functional group**. The carbon skeleton is composed of carbon atoms held together with covalent bonds. Biochemicals are categorized initially by the arrangement of carbons making up their carbon skeleton. Carbons can be arranged into straight chains, branched chains, single rings, or a collection of fused rings. The shape of the carbon skeleton affects the utility of the molecule in the body in the same way that the shape of a tool handle affects its utility. For example, screwdrivers have handles that permit users to twist the tool, and wrenches have flat handles that better fit a grip for pulling the tool as a lever.

Attached to various regions of the carbon skeleton are one or more functional groups. A functional group is defined as one or more elements attached to the carbon skeleton that is responsible for the chemical activities of a biochemical. Functional groups also encourage the formation of covalent bonds between two or more biochemicals. To fully understand the nature of the human body, it is essential to understand a few functional groups. One of the most common functional groups is called the **hydroxyl**, or alcohol, **group**. The hydroxyl group, comprised of **hydroxyl ions**, contains oxygen, which helps molecules dissolve well in water. In addition, the group attaches to other functional groups to form covalent bonds. Related to the hydroxyl group is the **carbonyl group**. Carbonyl groups are involved in special chemical reactions that transfer electrons between molecules. They also attach to other functional groups for bond formation. Hydroxyl and carbonyl groups are commonly associated with hydrogen bonding. The **carboxyl**, or organic acid, **group** should not be confused with carbonyl. This group plays an important role in exchanging both hydrogen ions and electrons with other molecules. The **amino group** possesses nitrogen and has a role similar to that of the carboxyl group. Their roles are complementary: the carboxyl group usually gives off hydrogen ions, while the amino group collects them. Energy transfer is the major role of the **phosphate group**. This group contains phosphorus, and is able to capture and release energy needed to run body processes. Last is the **sulfhydryl group**. As its name indicates, this functional group contains sulfur. The sulfhydryl group behaves much like the hydroxyl group. It readily bonds to other sulfhydryl groups, helping it to provide unique shapes to molecules containing several of these functional groups.

Many biochemicals have identical **molecular formulas**, but different **structural formulas**. Molecular formula refers to the number of atoms of each element present in a molecule. Structural formula describes the bonding and arrangement of elements making up a compound. Some molecules have the same molecular formula, but different arrangements of elements. These molecules are called **isomers** and, in turn, may have different physical and chemical properties. Ethanol, or drinking alcohol, is a simple molecule with no isomers. Its molecular formula is C_2OH, indicating that there are two carbon atoms:

Hydroxyl Ions Negatively charged oxygen particles bonded to hydrogen

Hydroxyl Group The functional group containing oxygen; helps molecules to dissolve in water

Carbonyl Group The functional group involved in the transfer of electrons between molecules

Carboxyl Group The functional group involved with the exchange of hydrogen ions and electrons with other molecules

Amino Group The functional group containing nitrogen, involved with the exchange of hydrogen ions and electrons with other molecules

Phosphate Group The functional group containing phosphorus, involved with the capture and release of energy

Sulfhydryl Group The functional group containing sulfur, involved with creating the structure of molecules

Molecular Formula The number of atoms of an element present in a molecule

Structural Formula The arrangement and bonding of elements comprising a compound

Isomer One of two or more molecules with the same molecular formula, but different structural formulas

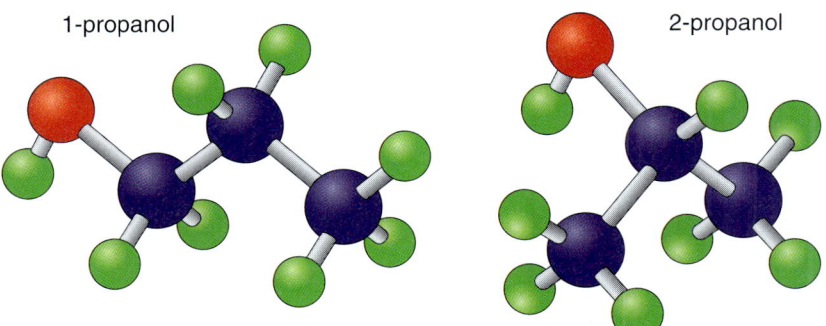

Figure 2.7 Propanol Isomer Propyl alcohol, or propanol, is an example of a molecule with isomer forms.

one oxygen and one hydrogen. Only one structural formula, C-C–OH, exists for ethanol. The "C" represents the carbon, "OH" is the hydroxyl group, and the "-" is a covalent bond. In contrast, an alcohol called propanol has two isomers. Propanol has a molecular formula of three carbons and one hydroxyl group, or C_3OH. One isomer is called 1-propanol and can be written as C-C-C-OH. The hydroxyl group is attached to one end or the other. It is a common component in perfumes. The other isomer is 2-propanol or isopropanol. It has the hydroxyl group attached to the middle carbon. This isomer is sold in stores as rubbing, or isopropyl, alcohol (Figure 2.7).

Chirality Describes isomers that form mirror image molecules

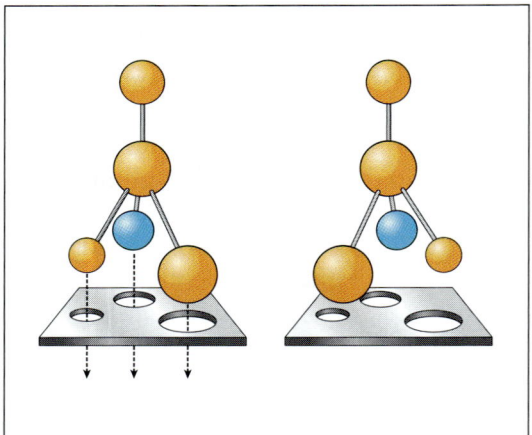

Figure 2.8 Isomers and Chirality
Isomers are variant forms of a molecule.

A special group of isomers form mirror image molecules. They posses a property called **chirality** (Figure 2.8). Chirality is similar to the structure of a person's hands or feet. One is a mirror image of the other and cannot be superimposed. Chiral molecules are often described as being left-handed or right-handed. Chirality does not usually affect the chemical or physical properties of a molecule. Chiral molecules are usually identified using special chemical analysis techniques. One technique, called optical rotation, looks at the way a beam of light is altered as it passes through a solution of biochemicals. Left-handed chiral molecules deflect the beam of light toward the left, whereas right-handed chiral molecules deflect the light to the right. The body responds very differently to molecules with chiral forms. The simple sugar called glucose, used as a food sweetener, has right and left forms. Its right-handed form is called *d*-glucose. The "d" stands for dextrorotary or "twists to the right." This form of glucose tastes sweet and is edible. In contrast, the left-handed form called *l*-glucose has no taste and can poison the body. Just as people need appropriately designed gloves to fit left or right hands, the human body needs specific chiral forms to build body structures and carry out chemical reactions.

✓ Concept Check

1. Distinguish between bioactive molecules and structural molecules.
2. Describe the functions of the six common functional groups found in biochemicals.
3. Define the term chiral isomer and describe how this molecule can affect the human body.

Chiral Drugs?

Most people would not expect to see front-page national news coverage about chirality. However, the 1957 release of a new sleeping pill and "morning sickness" drug called thalidomide hit the news after a dangerous side effect was discovered. The company manufacturing thalidomide did not take into account that it had chiral forms. It was unfortunately discovered that one chiral form caused severe birth defects in the children of mothers who had taken thalidomide to relieve the nausea associated with pregnancy. The birth defects, known as "thalidomide baby syndrome," produced defective limbs leaving the children profoundly handicapped. It was estimated that 15,000 babies worldwide were affected by thalidomide. Government tests on mice in 1961 showed that only one isomer caused the defects, while the other possessed the desired therapeutic activity. Later tests on rabbits indicated that both forms were capable of causing birth defects, so thalidomide was banned for human use. Recently, the pharmaceutical company Celgene Corporation received permission by the United States Food and Drug Administration to market a thalidomide drug called Thalidomide®. The drug was initially developed as a treatment for a condition associated with the infectious disease leprosy. It was later discovered that thalidomide could inhibit the growth of human immunodeficiency virus (HIV) in test-tube studies. HIV is the virus that causes acquired immunodeficiency syndrome (AIDS). The reestablishment of thalidomide was met with much protest because of fears that it might once again produce a spate of birth defects. It is probable that thalidomide could once again be prescribed for pregnant women. Currently, thalidomide drugs carry a warning label that explains the risks for birth defects.

DISCOVERY SCENE PLEASE ENTER DISCOVERY SCENE PLEASE ENTER

What information about elements would help solve the CSI? Is there any additional information about elements that is needed to determine possible causes of your friend's illness?

Acids and Bases

 Key Terms: acid, alkaline, base, buffer, denaturation, dissociate, electrolyte, hydrogen ion acceptors, hydrogen ion donors, hydroxyl ions, neutral, pH, pH scale

Before continuing with a description of biochemicals, it is important to briefly discuss the chemical environment in which they work. All human biochemicals

pH "Potential for hydrogen atoms," the measure of hydrogen ion concentration in water

Acid Water containing large amounts of hydrogen ions

Base or Alkaline Solution Water with a low concentration of hydrogen ions

pH Scale A scale measuring the concentration of hydrogen ions

Neutral Neither basic nor acidic

Dissociate Break-down of molecules in a compound into simpler molecules, atoms, or ions

Electrolytes Ions capable of conducting electricity in water

Hydrogen Ion Donor Acidic molecules that dissociate in water, releasing hydrogen ions

Hydrogen Ion Acceptors Basic molecules that dissociate in water, absorbing hydrogen ions

Buffer Molecules that act as either hydrogen ion acceptors or donors

work in water, a substance that makes up a significant proportion of the body. One property of water that has an important impact on the function and structure of molecules is called **pH**. The term pH stands for the "potential of hydrogen atoms and is always written small case "p" followed by a capital "H." It is a measure of the hydrogen ion concentration in water. Water containing large amounts of hydrogen ions is called **acid** or an acidic solution. Many foods, such as cheese, lemons, vinegar, yogurt, and wine, get their sour or sharp flavor from naturally occurring acids. A **base** or basic solution is also known as **alkaline solution**. It has much lower concentrations of hydrogen ions in solution. Bases normally feel slimy to touch; bases give soap its characteristic feel. The pH values are represented using the **pH scale** (Figure 2.9). The pH scale is a 0-to-14 scale measuring the concentration of hydrogen ions. The lower pH range of 1 through 6 indicates acidic conditions; 7 is **neutral**; and numbers greater than 7 are basic. Each decrease on the pH scale represents a 10-fold increase in the concentration of hydrogen ions. This association sounds backward, but that is the nature of the way pH is mathematically calculated. Thus, a pH 5 solution has 10 times more hydrogen ions than a pH 6 solution.

Water has a natural tendency to break down or **dissociate** into hydrogen ions and **hydroxyl ions**. Hydroxyl ions are negatively charged particles of oxygen bonded to hydrogen. It is usually written as OH^-. Hydrogen ions are represented with the symbol H^+. Pure water dissociates in equivalent amounts of hydrogen and hydroxyl ions. This is the condition called neutral pH (pH 7) previously described. Many molecules held together by ionic bonds dissociate in water to produce a variety of ions called **electrolytes**. Electrolytes are ions that, in a solution of water, become capable of conducting electricity. Acidic molecules dissociate to release many hydrogen ions into the water. They are called **hydrogen ion donors**, which is a typical role of acids in the body. Basic molecules dissociate to absorb hydrogen ions from water or contribute to the hydroxyl ion concentration. These are called **hydrogen ion acceptors**. Molecules called **buffers** can act as either hydrogen ion donors or acceptors. The chemical make up of some buffers allows them to neutralize acids or bases in solution. Buffers are very important components of blood and body fluids. They prevent the body's pH from fluctuating into dangerous ranges.

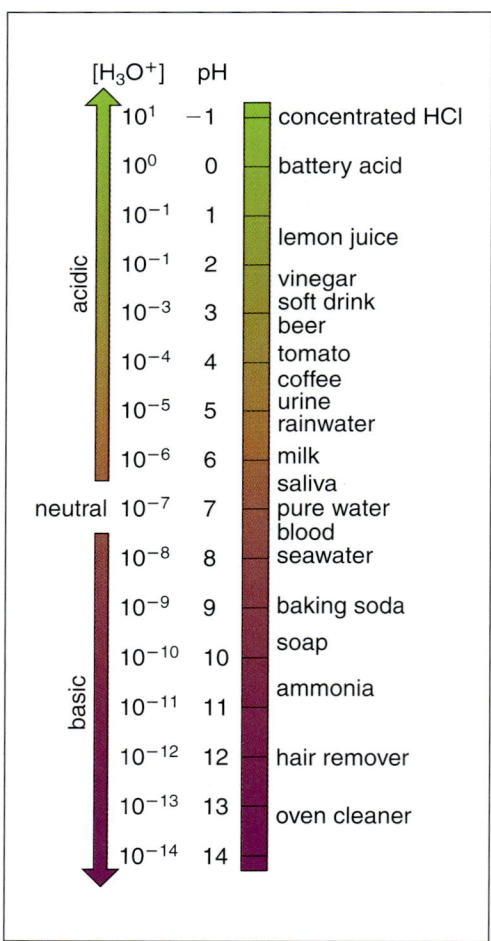

Figure 2.9 Acids and Bases
A pH scale.

It is important to have an understanding of pH, as you will be finding out how the pH within certain body systems can be adversely altered and how the body responds.

Not all organisms thrive under the same pH conditions. Bacteria prefer basic conditions, while fungi do best in acidic conditions. Fluctuations of pH in a person's saliva, sweat, or vaginal fluids can encourage the conditions for bacterial or fungal infections. Some parts of the body have extreme pH values. The stomach maintains a highly acidic pH to help in breaking down proteins in the food. In contrast, the upper part of the small intestines is basic to help with the digestion of fats. Highly acidic or basic pH values can create a variety of problems for the body. Most significant is that biochemicals can alter their shape if the pH is too high or too low for that particular molecule. **Denaturation** is a term used to describe the alteration of a biochemical's shape. In addition, large amounts of hydrogen ions are capable of breaking down many types of covalent bonds, rendering many molecules useless. Too few hydrogen ions in solution can disable some molecules that are responsible for transferring energy needed to run body functions.

Denaturation The process of altering a biochemical's structure

✓ Concept Check

1. Define pH.
2. Distinguish between a hydrogen ion acceptor and a hydrogen ion donor.
3. What is the role of a buffer in the human body, and how can it prevent denaturation.

DISCOVERY SCENE PLEASE ENTER DISCOVERY SCENE PLEASE ENTER

What have you learned so far about molecules that may help you solve the CSI? Does information about molecular structure assist in determining possible causes of your friend's illness? What role could a particular molecule play in causing disease or illness?

HUMAN MOLECULES

 Key Terms: carbohydrate, lipids, monomer, nonpolar, nucleic acid, peptide, polar, polymer, water insoluble, water soluble

Thousands of molecules are needed to keep the human body intact and running. A majority of the molecules are organic molecules that build body structure and regulate physiology. Figure 2.10 shows a few of the many diverse organic molecules making up a human's biochemicals. Biochemicals can be categorized in four organic chemical groups: **lipids**, **carbohydrates**, **peptides**, and **nucleic acids**. Each group has a unique chemistry that determines its role in the body, which will be discussed in more detail below.

Carbohydrates, peptides, and nucleic acids are predominantly **water soluble** and **polar**. Polarity is characteristic of molecules that have a greater electron density at one end than the other. This gives one end of the molecule a positive charge and the other end a negative charge.

Lipids or Fats Simple molecules that provide the body with chemical signals, insulation, protective padding, and stored energy

Carbohydrates Compound molecules that provide the body with energy

Peptides Linear polymers of amino acids

Nucleic Acid Molecules involved in converting food to energy and an essential component of genetic material

Water Soluble Able to dissolve in water

Polar having a stronger negative or positive charge concentrated on one side or region

THE BODY'S CHEMICAL MAKEUP

Figure 2.10 Human Molecules
The human body contains thousands of simple to complex biological molecules.

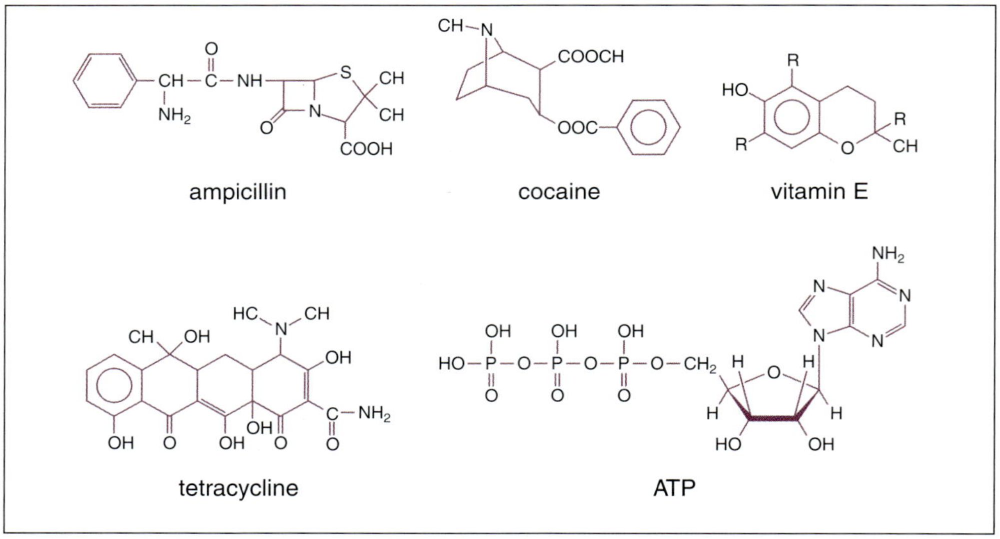

ampicillin

cocaine

vitamin E

tetracycline

ATP

> **Water Insoluble** Not able to dissolve in water
>
> **Nonpolar** Describes molecules lacking the electrically charged functional groups that add in water solubility
>
> **Monomer** The simplest form of a biochemical
>
> **Polymer** A complex biochemical

Water soluble means that they dissolve in water. Lipids differ from the other groups because a majority of lipids do not dissolve in water. They are called **water insoluble** for this reason. Lipids are also nonpolar molecules because they possess no polar charges. The term nonpolar refers to the fact that these molecules lack the electrically charged functional groups that help them to dissolve in water. Carbohydrates, peptides, and nucleic acids also come in simple and complex forms. The simplest form of a biochemical is called the **monomer**. Monomer means "one part" and refers to a single identifiable unit of a molecule. Complex biochemicals are called **polymers**. Polymers are chains or groupings of monomers held together by covalent bonds. Lipids do not have true monomer and polymer forms. The basic and simplest component of a lipid is a chain of carbons, which many chemists refer to as polymers of carbon.

You will see that monomers and polymers play different roles in the body.

✓ Concept Check

1. What are the four groups of molecules that make up the human body?
2. Distinguish between the terms polar and nonpolar.
3. What is the difference between a monomer and a polymer?

Lipids

Key Terms: bioactive, cholesterol, detergent, diglyceride, emulsion, estrogen, fat, fat soluble, fatty acid, glyceride, glycerol, hydrophilic, hydrocarbon, hydrogenate, hydrophobic, linear, lipophilic, monoglyceride, phospholipid, progesterone, saturated, steroid, sterol, surfactant, suspension, testosterone, terpenoid, triglyceride, unsaturated, vitamin

Lipids, or fats, are probably the most talked about and, undoubtedly, the most infamous of the four biochemical groups. The media is loaded with stories about "good fats" and "bad fats." Fats are essential molecules that provide the body with chemical signals, insulation, protective padding, and stored energy. They are needed for the proper functioning of the nervous system, and, in females, help ensure fertility. Chemically, the lipids are very simple molecules mostly composed of a carbon skeleton. Fats carry only a few functional groups, making them predominantly nonpolar. Most lipids are repelled by water.

Many of the lipids used by the human body possess soap-like or **surfactant**, or **detergent** properties. Surfactants or detergents are defined as organic chemicals that have a **hydrophilic** chemical structure at one end and a **hydrophobic**, or **lipophilic**, component at the other end. Hydrophilic, literally meaning "water-loving," refers to a polar substance that dissolves or mixes with water. Its electrical charge combines to the slight electrical charge of the water molecule. Hydrophobic or lipophilic molecules are nonpolar and, therefore, not attracted to water. The term lipophilic literally means, "likes fats." Lipophilic molecules are also called **fat soluble**. Unlike salt, which readily dissolves and disappears in water, soaps form a hazy **emulsion** or **suspension**. Emulsion is defined as a mixture of two or more liquids that do not readily combine, as when oil is mixed with vinegar.

Many scientists categorize fats into three groups based on carbon skeleton structure: glycerides, sterols, and terpenoids. The most abundant lipid in the human body is **glyceride**. The name comes from the fact that glycerides contain a molecule called glycerol. **Sterols** are complex lipids composed of a pattern of carbon rings. There is a wide-ranging category of lipids that will be covered under Other Lipids (see page 49). Their carbon skeletons have a great diversity of shapes.

Surfactant or Detergent Organic chemicals with a hydrophilic functional group at one end and a hydrophobic component at the other end

Hydrophilic A polar substance that dissolves in water

Hydrophobic or Lipophilic A nonpolar substance that does not mix with water

Fat Soluble Able to dissolve in fat

Emulsion or Suspension A mixture of two or more liquids that do not readily combine

Glyceride The most abundant lipid in the body

Sterols Complex lipids composed of carbon rings

✓ Concept Check

1. Distinguish between the terms hydrophobic and hydrophilic.
2. What is the relationship between the terms fat soluble and emulsion?
3. Describe how scientists categorize lipids.

Glycerides The basic unit of all glycerides is the **fatty acid**, or **hydrocarbon**. A fatty acid is simply a **linear** chain of carbons that end in a carboxyl functional group (Figure 2.11). Linear refers to polymers made up of one long, continuous chain without branching. Fatty acids are also called hydrocarbons because the chain contains hydrogen atoms that stabilize the structure of the molecule. The hydrogen prevents other atoms from attaching to the carbon atoms. The length of a fatty acid chain ranges from 2 to 80 carbons. The fatty acids used in humans generally range from 12 up to 24.

There is a correlation between carbon number and the nature of the fatty acid. Fatty acids with small numbers of carbons are oils, while those with a large carbon number are greases or waxes.

Fatty Acid or Hydrocarbon A linear chain of carbons ending in a carboxyl functional group

Linear Refers to chains of polymers without branching

THE BODY'S CHEMICAL MAKEUP

Figure 2.11a and 2.11b Lipids and Glycerides
Glycerides are composed of carbon chains called fatty acids attached to a glycerol base molecule.

(a) lipids

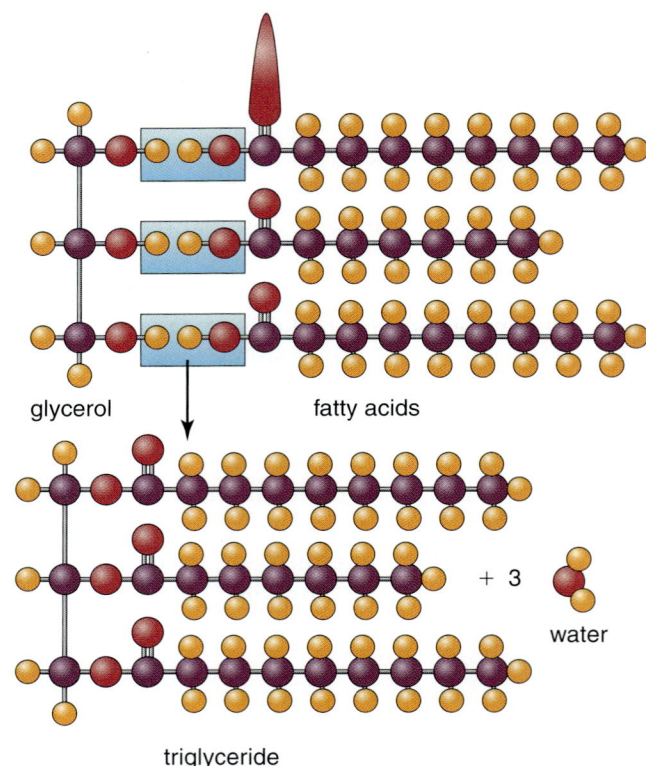

(b) glycerides

Saturated Refers to fatty acids with one covalent bond

Unsaturated Refers to fatty acids that have double bonds and a reduced number of hydrogen atoms

Hydrogenate To add hydrogen to unsaturated fats

Monoglycerides A form of glyceride with one molecule of fatty acid

Diglycerides A form of glyceride with two molecules of fatty acid

Triglycerides A form of glyceride with three molecules of fatty acid

Fatty acids come in two bonding patterns: **saturated** and **unsaturated**. Saturated fats are shown at the upper part of Figure 2.11a. Each carbon in a saturated fatty acid forms only one covalent bond with the carbon next to it. The carbon bonds are filled up, or saturated, with hydrogen. Unsaturated fatty acids lack hydrogen atoms and have double bonds between some carbons as shown in the bottom of Figure 2.11. Polyunsaturated is a nutritional term that means that the fats have many double bonds in the chain.

The chemistries of saturated and unsaturated fats differ greatly. Saturated fats are chemically stable, which means they do not spoil and are not easily oxidized. Unsaturated fats spoil easily and are subject to breakdown by oxidation. The double bonds of unsaturated fats readily bond to other molecules, causing them to change their chemical properties. This is why butter, cheeses, and milk sometimes take on the flavor of other things stored in the refrigerator. Many food odor molecules covalently bond to the unsaturated fats. The food industry **hydrogenates** the unsaturated fats in certain foods to prevent the fats from spoiling and to create a thicker consistency.

Figure 2.11b shows the role of the glycerol molecule in glyceride fats. It is an attachment point for one, two, or three fatty acids resulting in the formation of **monoglycerides**, **diglycerides**, and **triglycerides**, respectively. Monoglycerides are natural surfactants that play important roles in digestion and lung function. These functions will be discussed later in the book. The body transports fats in the form of monoglycerides. Physicians monitor the amount of monoglycerides in the blood, which indicates how much fat a person is consuming. Some people have naturally high levels of monoglycerides, which can worsen heart disease and blood vessel conditions.

Diglycerides are the most common lipid found in the body. These are the fats that build body structures. Diglycerides can have two saturated, two unsaturated, or a combination of saturated and unsaturated fatty acids. The glycerol of many diglycerides is attached to a phosphate group and, therefore, is called a **phospholipid**. Phospholipids are the fats that build body structures. They are an essential component of the diet, and are needed for body growth and maintenance. Children in particular need phospholipids for proper brain development. Triglycerides are the primary storage fats of the human body. They can contain various combinations of a total of three saturated and unsaturated fatty acids. Large amounts of triglycerides in the blood can aggravate heart and blood vessel ailments.

Sterols Sterols have a complex carbon skeleton. The body makes sterols by converting fatty acids into intricate multiringed shapes (Figure 2.12). Sterols belong to a group of fats called **steroids**. The steroids include a diverse group of molecules that have a variety of roles in animals and plants. **Cholesterol** is the most common of the sterol group. It is found throughout the body and is essential in body structure. Cholesterol also plays a role in fat digestion and is needed for brain development. You will later learn how a body can make cholesterol if it is not taken in through the diet. Other sterols include the sex hormones, **estrogen**, **progesterone**, and **testosterone**, which will be discussed in detail later in the book.

Other Lipids The third group of lipids is commonly represented by polymers of short-chain fats called **terpenoids**. Terpenoids play an important role in fighting off disease and include **bioactive** molecules called **vitamins** (Figure 2.13). Vitamins A, D, E, and K are fat soluble. These vitamins tend to be stored in the body's fat deposits and can build up to toxic levels if regularly eaten in

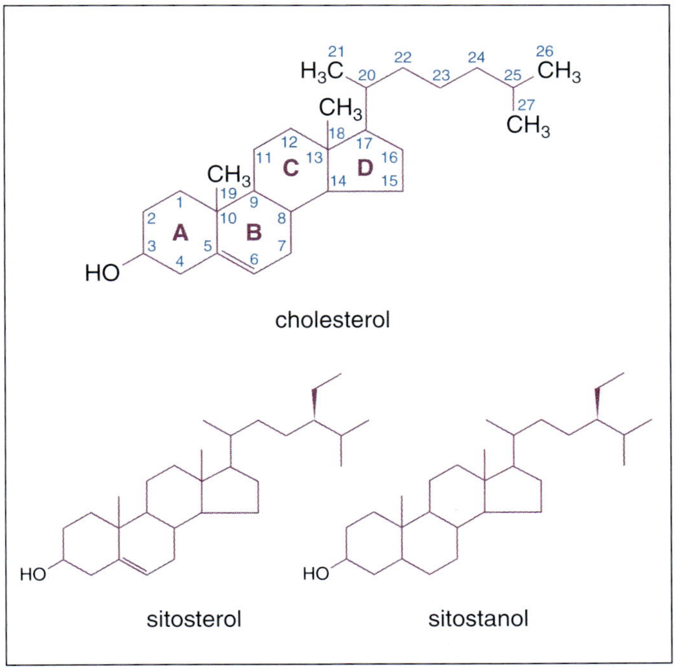

Phospholipid A lipid with a phosphate group attached to the glycerol of a diglyceride

Steroids Long-chain fatty acids converted into ring shapes with a variety of functions

Cholesterol The most common sterol, essential to body structure and brain development

Estrogen A sex hormone, part of the sterol group

Progesterone A sex hormone, part of the sterol group

Testosterone A sex hormone, part of the sterol group

Terpenoids Short-chain fatty acids that help fight disease

Bioactive Describes a substance that has an effect on living organisms

Vitamin Bioactive molecules essential to the health of an organism

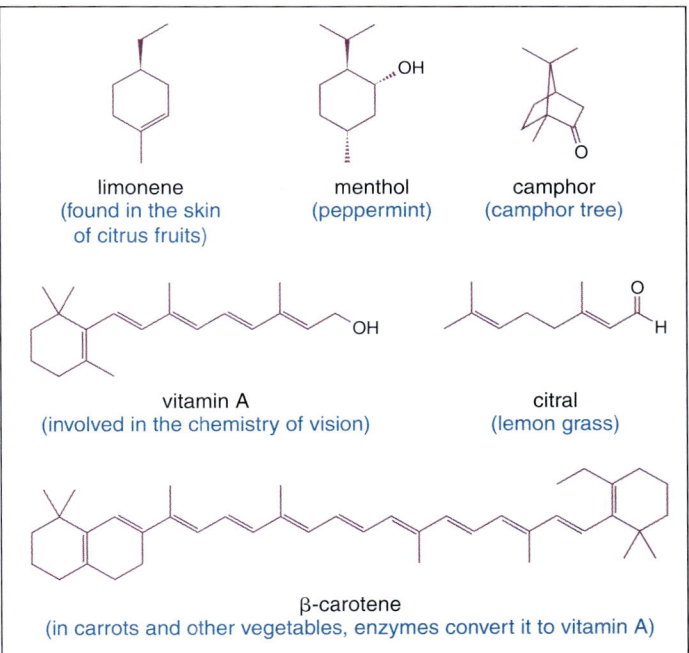

Figure 2.12 Sterols
Sterols are comlex lipids that play a variety of roles in the body.

Figure 2.13 Other Lipids
Vitamins and many immune system chemicals are modified forms of lipids.

THE BODY'S CHEMICAL MAKEUP

large amounts. The body loses water-soluble vitamins readily through the sweat and urine. They do not become toxic unless large amounts are taken over a short period of time. The vitamins B and C make up the water-soluble vitamins. Vitamin B is actually not one vitamin, but a group of vitamins called the vitamin B complex. Biotin, choline, folic acid, inositol, para-aminobenzoic acid (PABA), and the six numbered B vitamins—vitamin B-1 (thiamin), B-2 (riboflavin), B-3 (niacin), B-5 (pantothenic acid), B-6 (pyridoxine), and B-12 (cobalamin)—make up the B complex group.

Monosaccharide A simple sugar consisting of a single sugar molecule that cannot be further decomposed. They are commonly used as a source of energy in organisms.

Disaccharide Two covalently bonded monosaccharides

Glycosidic Linkage A type of covalent bond

Polysaccharide A chain of monosaccharides

Oligosaccharide A polysaccharide consisting of three to 10 monosaccharides

✓ Concept Check

1. Describe the structure of the three types of glycerides.
2. What are the roles of sterols in the body?
3. Why are terpenoids important for normal body function?

Carbohydrates

Key Terms: alpha isomer, amino-sugar, amylose, arabinose, beta isomer, cellobiose, cellulose, dextrose, disaccharide, fructose, galactose, glucose, glycogen, glycosidic linkage, hexose, inulin, lactose, lactose intolerance, levulose, maltose, mannose, monosaccharide, mucopolysaccharide, oligosaccharide, pentose, polysaccharide, starch, sucrose, trisaccharide

Carbohydrates are defined as rings of carbon atoms attached to hydroxyl functional groups. The large amount of hydrogen and oxygen in their molecular structure is due to the large number of hydroxyl groups found in carbohydrates. Carbohydrates can exist as single units called **monosaccharides** (Figure 2.14). They are primarily known as a source of energy for the body. As evident in the name, **disaccharides** are composed of two monosaccharides attached by a type of covalent bond called a **glycosidic linkage**. Several types of glycosidic linkages can be formed between monosaccharides. The type of glycosidic linkage determines the ability of an organism to use the disaccharide as a dietary sugar. **Polysaccharides** are composed of linear or branched chains of either one type of monosaccharide or a combination of different ones. Polysaccharides containing three to 10 monosaccharides are called **oligosaccharides**.

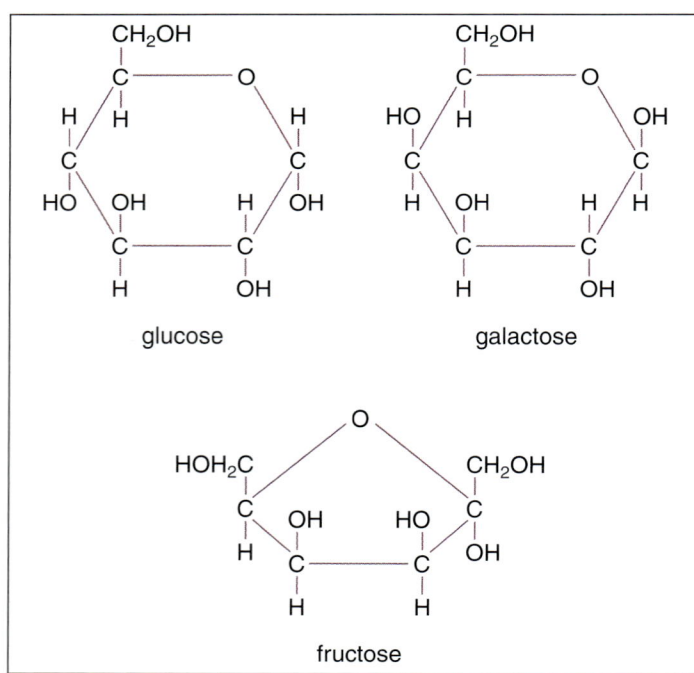

Figure 2.14 **Monosaccharides**
Monosaccharides are the simplest unit of carbohydrates.

Good Choice Bad Choice

An elderly relative heard that eating pure carbohydrate diets for 2 months at a time cleanses the body and reduces the amounts of free radicals in the blood. Should she be encouraged to take part in this practice? What are the truths and fallacies of the claim?

Monosaccharides Monosaccharides are almost exclusively used as a source of energy in all organisms. Plants produce almost all of the hundreds of monosaccharides found in nature. Scientists have developed a multitude of artificial monosaccharides called sweeteners. Most sweeteners provide the pleasant taste of sugar without contributing to energy intake. The body cannot utilize many types of monosaccharides. Consequently, many of them pass out of the body without being taken into the blood and used for energy. Humans commonly use what are called **hexose** monosaccharides in everyday life. Hexose means that the carbohydrate contains six carbons in its molecular makeup. These hexose monosaccharides are **glucose**, **fructose**, **galactose**, and **mannose**.

Glucose is the major energy source for most bodily functions. It exists as d- and l-isomers. The body is primarily able to use the d-glucose isomer, also called **dextrose**, to distinguish it from the nondietary form. The l-glucose isomer is not common in nature and can cause ailments if large amounts are taken into the body. Scientists use l-glucose to study various diseases, and it is an ingredient in certain cancer-fighting drugs. Fructose, or **levulose**, is a fruit sugar that can be used in place of glucose as an energy source. It is the source of carbohydrate energy preferred by nutritionists who work with children and diabetics. It does not cause the blood sugar problems that glucose produces, and it requires less water when taken up by the body. Fructose, however, is not as readily taken into the body as is glucose, and much of it can be lost as a waste product during a high-fructose meal.

The next most common monosaccharide is galactose. It is not as prevalent in the diet, and is mostly available in milk products and sugar beets. Human mammary glands are able to convert glucose into galactose. The body can use galactose as an optional source of energy. Like fructose, it is not readily taken up by the body and can cause dehydration if eaten in large amounts. Mannose is taken in as a nutrient and can be used in place of glucose. Mannose plays an important role in helping the body fight infectious diseases. It is important to mention one other hexose monosaccharide: gulose, not to be confused with glucose, is a little known hexose monosaccharide found in foods made from algae carbohydrates. It is becoming more common in the human diet because of the common use of algae as a food thickener and an ingredient in Japanese sushi. It is needed in certain organisms as raw material for making vitamin C. Unfortunately, humans cannot produce vitamin C. Certain **pentose**, or five-carbon, monosaccharides are taken in the diet and used by the body.

Hexose Monosaccharide A monosaccharide with six carbon molecules

Glucose A hexose monosaccharide that is the major energy source for most body functions

Fructose or Levulose A hexose monosaccharide that can be used as an energy source; fruit sugar

Galactose A hexose monosaccharide found in milk products and sugar beets

Mannose A hexose monosaccharide that can be used as an energy source

Dextrose The dietary form of glucose isomers

Pentose A monosaccharide with five carbon molecules

Figure 2.15 Disaccharides
Disaccharides are composed of two monosaccharides attached by a covalent bond.

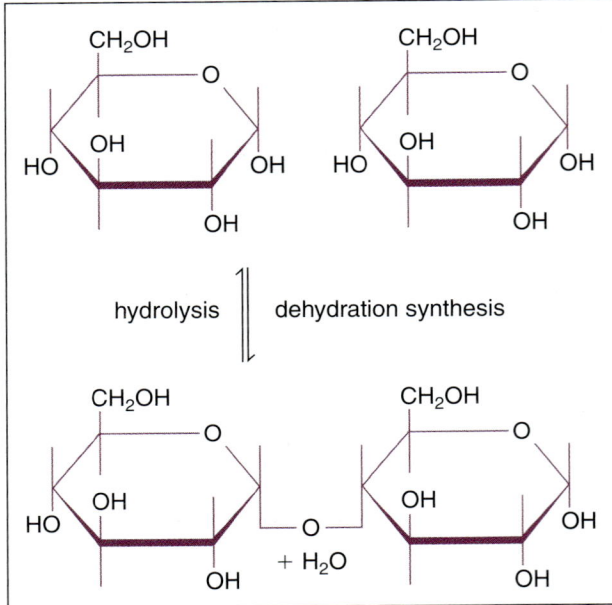

Arabinose A pentose monosaccharide found in vegetables and nutritional drinks

Sucrose A disaccharide; a combination of glucose and fructose

Lactose A disaccharide; a combination of glucose and galactose

Maltose A disaccharide; a combination of two glucose molecules

Cellobiose An indigestible disaccharide; a combination of two glucose molecules]

Lactose Intolerance The inability to digest lactose

Glycogen A highly branched polysaccharide of alpha-bonded glucose

Cellulose A fibrous polysaccharide

Starch or Amylose An alpha-bonded glucose found in plants

Inulin A polysaccharide containing large amounts of glucose

One of these, **arabinose**, is commonly found in vegetables and nutritional drinks. Many nutritionists believe it is a better source of energy than glucose because it does not raise blood sugar levels.

Disaccharides As mentioned earlier, disaccharides are composed of two similar or different monosaccharides, as shown in Figure 2.15. **Sucrose**, **lactose**, **maltose**, and **cellobiose** are the four common disaccharides in the human diet. Sucrose is the carbohydrate found in table sugar. It is a combination of glucose bonded to fructose. Sucrose is preferred for flavoring foods because it is one of the sweetest natural sugars. Plus, it does not become thick and syrupy when moderate amounts are added to water. Sucrose is a nutrient for humans because the body can break it down into glucose and fructose. Lactose, or milk sugar, is a combination of glucose and galactose. The amount of lactose varies depending on the type of milk. Cow and goat milk have almost half the amount of lactose as human milk. Infant diets benefit by lactose because it is taken up more slowly than glucose, providing a gradual supply of energy. However, many adults lose the ability to break down lactose and develop a condition called **lactose intolerance**. Undigested lactose can cause cramping, gas buildup, and extreme pain. Maltose is the least common disaccharide in the diet. This disaccharide is composed of two glucose molecules, and it is primarily found in starchy foods and beer. Cellobiose is a common thickening agent in food, and it adds to the food's fiber content.

Polysaccharides The polysaccharides are a very diverse group of molecules (Figure 2.16). However, only a few are commonly encountered in the diet. The four most common polysaccharides are **glycogen**, **starch**, **cellulose**, and **inulin**. Glycogen is a highly branched polysaccharide of alpha-bonded glucose stored in the human liver and muscles. Starch, or **amylose**, is a major source of glucose in the diet. It is an alpha-bonded glucose polymer produced by plants, and it is a common component of many fruits and vegetables. The bodies of plants are composed of a beta-bonded glucose polymer called cellulose. Paper and wood products are primarily cellulose mixed with other polymers. Humans lack the ability to digest cellulose. Cellulose makes up the fiber component of the diet. Inulin is a little known polysaccharide that contains large amounts of fructose. Jerusalem artichokes and many tropical roots are rich in inulin. Many nutritionists prefer inulin to starch because it does not cause extreme fluctuations in blood sugar. However, there is evidence that many people have inulin allergies.

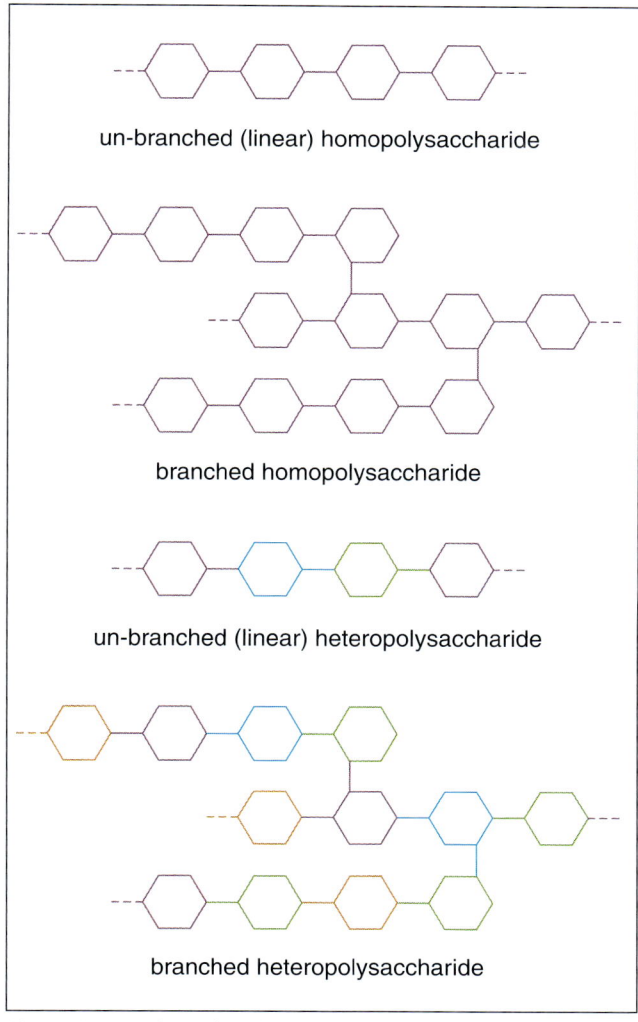

Figure 2.16 **Polysaccharides**
Polysaccharides are composed of chains of covalently bonded monosaccharides.

Other polysaccharides include mucopolysaccharides. They are polymers made of specialized carbohydrates called **amino-sugars**. Amino-sugars have an amino functional group. These gummy polymers have adhesive properties. They form body secretions and help hold together many body structures. Various **oligosaccharides** are also important to the body. Oligosaccharides are carbohydrates containing three to 10 monosaccharides. The A, B, and O blood types are due to oligosaccharides found on blood cell membranes.

Raffinose is a **trisaccharide** found in asparagus, beans, cabbage, and whole grains. It is broken down into sucrose and galactose. Too much raffinose in the diet can cause gas because it is difficult to digest. People with lactose intolerance may have difficulty breaking down raffinose.

Amino-Sugars A mucopolysaccharide with an amino functional group

Trisaccharide A three-bonded monosaccharide

✓ Concept Check

1. Describe the three groups of carbohydrates.
2. Describe four common monosaccharides encountered in the diet.
3. Distinguish between the body's use of monosaccharides and polysaccharides.

Peptides

Key Terms: amino acids, enzyme, functional proteins, helix, oligopeptides, peptide bond, polypeptides, primary structure, proteins, quaternary structure, R group, secondary structure, sheet, side chain, structural proteins, and tertiary structure

THE BODY'S CHEMICAL MAKEUP

Figure 2.17 **Amino Acids**
The human body uses about twenty different types of amino acids.

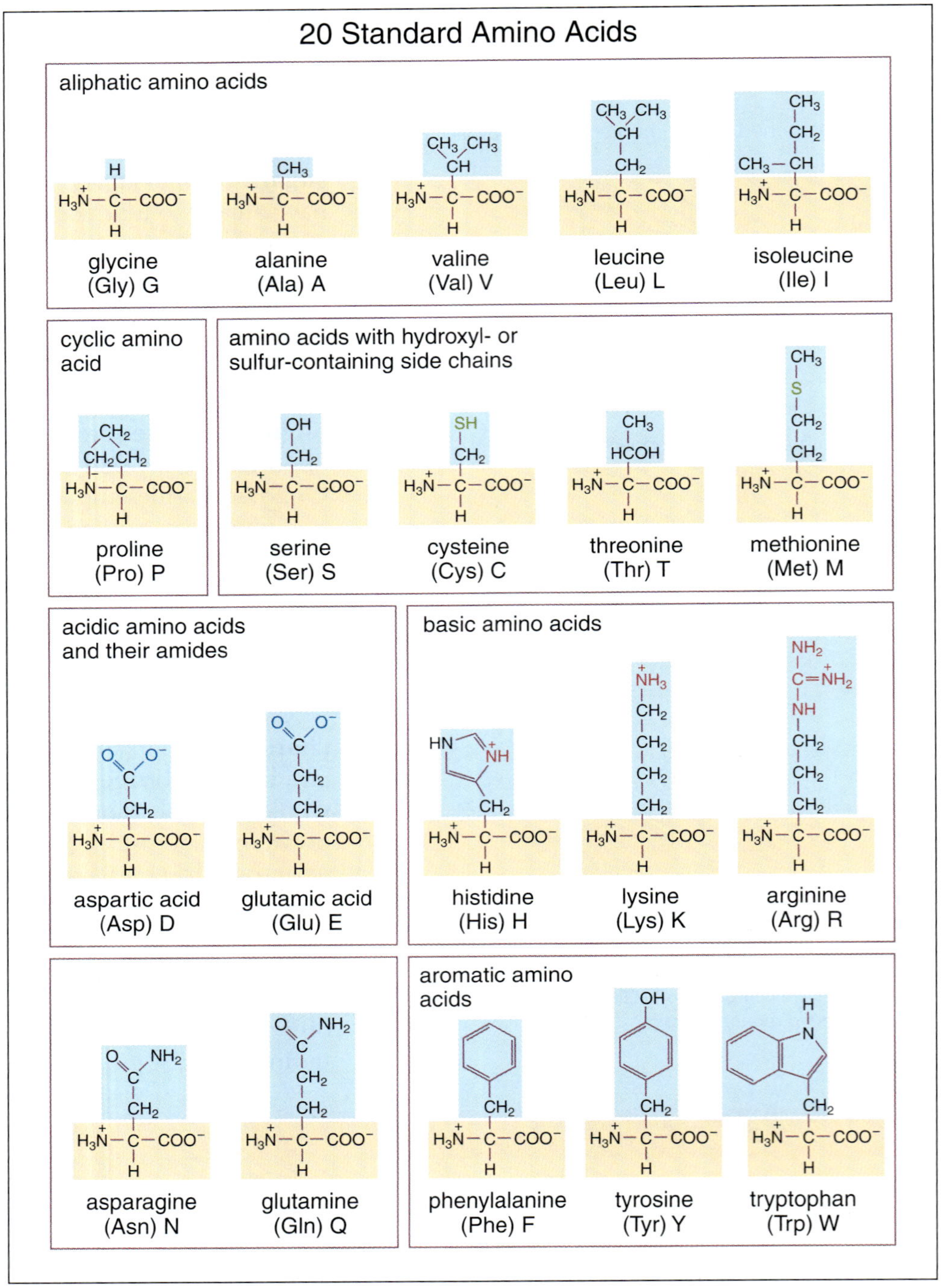

20 Standard Amino Acids

aliphatic amino acids

glycine (Gly) G   alanine (Ala) A   valine (Val) V   leucine (Leu) L   isoleucine (Ile) I

cyclic amino acid

proline (Pro) P

amino acids with hydroxyl- or sulfur-containing side chains

serine (Ser) S   cysteine (Cys) C   threonine (Thr) T   methionine (Met) M

acidic amino acids and their amides

aspartic acid (Asp) D   glutamic acid (Glu) E

basic amino acids

histidine (His) H   lysine (Lys) K   arginine (Arg) R

asparagine (Asn) N   glutamine (Gln) Q

aromatic amino acids

phenylalanine (Phe) F   tyrosine (Tyr) Y   tryptophan (Trp) W

Amino Acids Structural units of proteins

Oligopeptides Small chains of amino acids

Polypeptides or Proteins Large chains of amino acids

Structural Proteins Proteins that help build body structures

Functional Proteins Proteins that carry out functions that run the body

Peptides are linear polymers of molecules called **amino acids**. Amino acids are a group of nitrogen-containing compounds that combine to form chains. Small chains of amino acids are called **oligopeptides**. They play an important role in the body, serving as chemical signals that regulate body function. Oligopeptides are also important in fighting disease. Larger collections of amino acids are called **polypeptides**, or **proteins**. They are categorized into two major categories: **functional proteins** and **structural proteins**. Functional

CHAPTER 2

proteins carry out a variety of jobs that run the body. One type of functional protein is called an **enzyme**. Enzymes perform chemical reactions needed for body maintenance. Structural proteins help build body structures, such as tendons. Other proteins are a means for the body to store amino acids. These are called storage proteins, which are commonly found in the blood.

Amino Acids The amino acids that make up peptides have an amino group, a carboxylic acid group, and one of many types of organic molecules attached to a single carbon. The amino and carboxyl groups are on opposing sides of the carbon (Figure 2.17). The group of atoms attached to the center-carbon characterizes the amino acid. This is often referred to as the **side chain**, or **R group**. Humans and most other organisms use 20 different types of amino acids. The R groups vary from a simple hydrogen molecule to a complex array of functional groups. Some R groups are polar, while others are nonpolar. R groups can add acid or base properties to the polypeptide as well as impart electric charges. There are two types of R groups that form attachments to other R groups. They control the shape and stability of many polypeptides. It is important to understand that amino acids are what the body "really wants" out of proteins in the diet. Humans use specific amino acids from the diet as the building blocks of peptides. Much of the human genetic material serves as a blueprint for the peptides that the body must build. Diets have to be balanced to provide the appropriate amino acid constituents needed to maintain the body's peptide composition.

Proteins Proteins are not merely arbitrary strands of amino acids. They develop their various functions based on a unique arrangement and composition of amino acids. Proteins get their start when amino acids in the diet attach to each other with **peptide bonds**. Peptide bonds are formed when the amino functional group of one amino acid is covalently bonded to the carboxyl functional group of a neighboring amino acid. The R groups are all aligned on one side as the amino acids attach to each other (Figure 2.18).

> **Enzyme** A functional protein
> **Side Chain or R Group** Used to characterize amino acids by the types of organic molecules they carry
> **Peptide Bond** A bond created when the functional group of one amino acid covalently bonds to the carboxyl functional group of a neighboring amino acid

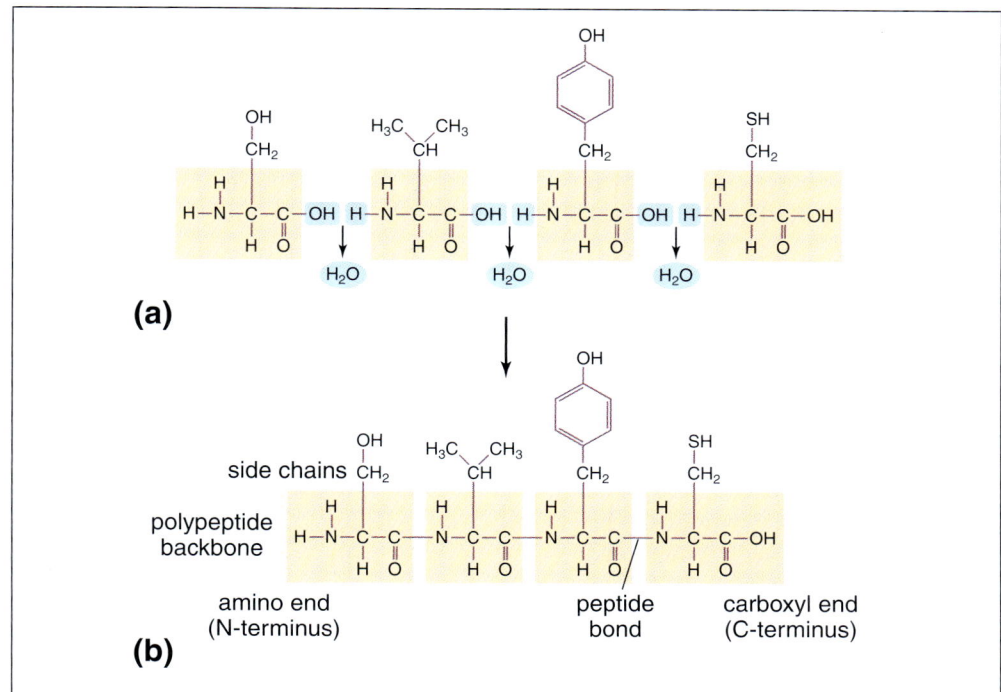

Figure 2.18 Peptide Bond The peptide bond is a covalent bond formed between two amino acids.

THE BODY'S CHEMICAL MAKEUP

Figure 2.19 **Protein Structure**
Protein levels of organization.

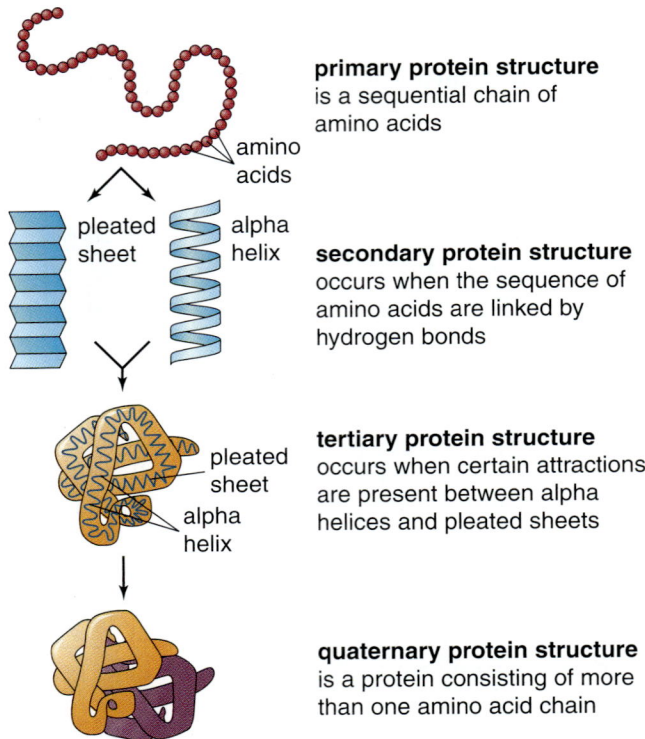

Primary Structure The initial arrangement of amino acids in a peptide bond, with the R group aligned on one side

Secondary Structure The characteristic shape of peptide chains

Helix A type of secondary structure with a spring-like shape

Sheet A type of secondary structure shaped similarly to a pleated sheet of paper

Tertiary Structure The three-dimensional structure of a polypeptide chain

Quaternary Structure Two or more polypeptide chains combined to form a larger and more complex molecule

This initial arrangement of amino acids is called the protein's **primary structure** (Figure 2.19). The primary structure is an important clue to genetic information. Many genetic diseases cause illness because they alter the arrangement of critical peptides. Most peptide chains take on a characteristic shape called the **secondary structure**. There are two major types of secondary structures: the **helix** and the **sheet**. Helix is the name given to a spring-like shape. The helix takes on the shape of a somewhat rigid, but straight, tube. Sheets form when the amino acids arrange themselves into a linear structure similar to a pleated strip of paper. Most proteins are a combination of helix and sheet shapes. The primary structure determines the distribution of secondary structures in a protein.

Proteins develop another degree of complexity called the **tertiary structure**. It is defined as the three-dimensional structure of the entire polypeptide chain. The function of most proteins depends on its tertiary structure. Enzymes in particular have a precise tertiary structure that determines their ability to carry out a specific chemical reaction. The tertiary structure is very sensitive to electrolyte concentration, pH, and temperature, as alterations in these factors can easily break the bonds creating this structure level. This means that many proteins can only carry out their function under precise conditions. Some proteins develop **quaternary structure**. A quaternary structure is two or more polypeptide chains combined to form a larger and more complex protein molecule. These proteins carry out complex jobs in the body and sometimes bind to other elements. Hemoglobin, the molecule that carries oxygen in the blood, has a quaternary structure.

Cutting Edge Research

MOLECULES THAT MAKE COWS MAD

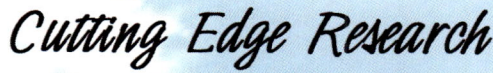

It is a standard practice to look for some type of life form when a physician is diagnosing an infectious disease. Even the most difficult-to-find organisms can ultimately be distinguished from the body parts they are infecting. However, the rise in the incidence of mad cow disease is bringing into question the meaning of "life form." Mad cow disease is not a typical infectious disease. An aberrant molecule called a prion (pronounced *pree-yon*) is believed to cause it. Prions push the limit of what it means to be an organism. They are no more than a protein particle capable of causing a variety of nervous-system diseases, such as Creutzfeldt-Jakob (*croits-feld yah-cob*) disease in humans and scrapie disease in sheep. The fact that a protein can be infectious contradicts much of what scientists understand about reproduction. Prions have no genetic material and, therefore, should be incapable of reproduction. This did not stop Dr. Stanley B. Prusiner of the University of California at San Francisco from hypothesizing a mode of replication for prions. Dr. Prusiner discovered prions in 1982 and was awarded the Nobel Prize in physiology and medicine in 1997 for his discovery. He believes that prions are somehow capable of converting body proteins into replicas of the invading prion. This produces disease as the nerves dedicate time and resources making a continuous stream of prions. Not all scientists support Prusiner's ideas. Some believe that a type of unidentified infectious organism is making the prions that harm the body. In spite of the controversy, food safety inspectors are vigilant in testing for prions in all agricultural animals capable of carrying the disease.

✓ Concept Check

1. Distinguish between the terms amino acid and peptide.
2. What are the four types of structures that can make up a particular protein?
3. Describe why body conditions, such as pH and temperature, are important to protein function

Nucleic Acids

Key Terms: adenine, adenosine triphosphate (ATP), deoxyribonucleic acid (DNA), deoxyribose, nitrogen base, guanine, nucleotide, purine, pyrimidine, ribonucleic acid (RNA), ribose

Nucleic acid refers to a group of molecules involved in two major functions of the body. First, as monomers, they are critical for transferring energy from food to body functions. Second, as polymers, they are an essential component of genetic material. The building blocks of nucleic acids are molecules called **nucleotides**. Nucleotides are small molecules composed of three parts: a

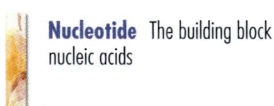

Nucleotide The building block of nucleic acids

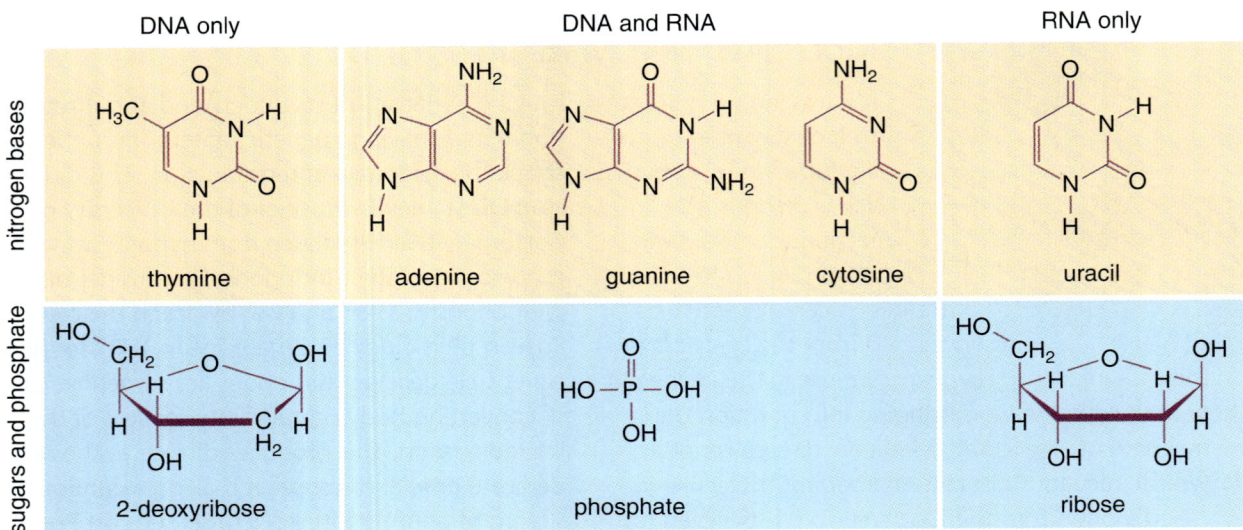

Figure 2.20 Nucleic Acids Nucleic acids are composed of a monosaccharide bonded to a nitrogen base.

Purines Nitrogen bases composed of adenine or guanine

Pyrimidines Nitrogen bases composed of cytosine, thymine, or uracil

Ribose A pentose sugar involved in the composition of nucleotides

Deoxyribose A pentose sugar involved in the composition of nucleotides

Deoxyribonucleic Acid (DNA) The molecule inside cells that contains the genetic information and passes it from one generation to the next

Ribonucleic Acid or RNA Nucleotides containing ribose

Adenosine Triphosphate (ATP) The most common energy transfer molecule; contains three phosphate functional groups

nitrogen base, a pentose sugar, and a phosphate functional group (Figure 2.20). The nitrogen base and the phosphate functional group are attached to the pentose sugar. A nitrogen base is defined as a nitrogen-containing molecule with a basic pH. There are two categories of nitrogen bases: **purines** and **pyrimidines**. Purines are single-ring compounds. Almost all organisms build their nucleotides out two types of purines: adenine and guanine. Pyrimidines are composed of three types of nitrogen bases: cytosine, thymine, and uracil. There are two types of pentose sugars that make up nucleotides: **ribose** and **deoxyribose**. Nucleotides containing ribose are called **ribonucleic acids**, or **RNA**. RNA is usually limited to using adenine, cytosine, guanine, and uracil as its nitrogen bases. Nucleotides containing deoxyribose use adenine, cytosine, guanine, and thymine, and are called **deoxyribonucleic acids**, or **DNA**.

Monomer nucleotides are primarily used for energy transfer. The most common energy transfer molecule is **adenosine triphosphate** (**ATP**). Triphosphate refers to the molecule with three phosphate functional groups attached as a linear strand on its ribose sugar. Other monomer nucleotides are necessary for fueling specific chemical reactions in the body. They do this by assisting with the function of certain enzymes. The nucleotides that make up the genetic material function as polymers. Polymers of RNA are essential for the production or synthesis of proteins from amino acids. DNA functions as two side-by-side polymers to comprise the genetic material. Figure 2.21 shows how the two DNA strands line up next to each other. The strands are attached to each other by hydrogen bonds between matched pairs of purines and pyrimidines. Adenine is normally paired to thymine by two hydrogen bonds, and cytosine is matched to guanine by three hydrogen bonds. It is the arrangement of nucleotides in a strand that permits DNA to store the genetic information.

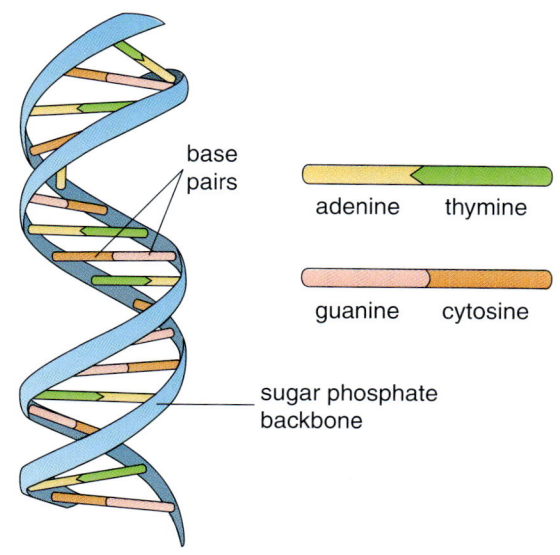

Figure 2.21 **DNA Structure**
DNA is a polymer of nucleic acids.

✓ Concept Check

1. Describe the structure of a nucleotide.
2. How does DNA differ from RNA?
3. What are the roles of nucleic acids in the human body?

DISCOVERY SCENE PLEASE ENTER DISCOVERY SCENE PLEASE ENTER

So, what have you learned so far that may help you solve the CSI? Do you see how a particular molecule from one of the four biochemical groups could be associated with your friend's condition? What molecular information would the medical team have to test for to draw any conclusions about the cause of the illness?

MOLECULES AND NUTRITION

Key Terms: "carbo loading", craving, glycemic index, homeostasis, malnutrition, undernutrition

Many people equate eating with enjoying good food rather than as a method for balancing the molecular **homeostasis** of the body. In the health fields, homeostasis implies the natural tendency of a person to maintain physiological and psychological stability. At the molecular level of the body, it means taking in a constant supply of energy and raw materials needed to keep each body component in a continuous state of well-being. The body's physiology continually monitors its needs for energy and raw materials. It does this through a variety of methods, including chemical messengers that coordinate all of the

Homeostasis The natural tendency of a person to maintain psychological and physiological stability

Craving The powerful and often uncontrollable desire for a substance

Malnutrition The condition resulting from a diet lacking a balanced molecular composition

Undernutrition The condition due to a diet lacking sufficient molecules

Glycemic Index A measurement indicating the amount of glucose available in a particular food

"Carbo Loading" Eating large amounts of glucose before an endurance event to maintain adequate glucose levels during activity

body's systems. Unfortunately, humans have little intuitive sense of what the body needs to stay healthy. Many people experiment with fad diets that compromise the body's homeostasis. Other people feel that their bodies will develop **cravings** that tell them what to eat. The medical definition of craving is a powerful and often uncontrollable desire for a substance. There is evidence that many people develop cravings for certain foods that fulfill a nutritional role. For example, there are many stories about pregnant women have cravings for ice cream, which contains calcium and other essential elements. Research now shows that the craving is merely a sign of stress registered by the body. The craving does not direct the person to seek a particular substance, such as milk, to supplement its calcium needs. People learn by trial and error that certain substances satisfy the cravings, while others do not. Diets lacking a balanced molecular composition create a condition called **malnutrition**. **Undernutrition** describes a condition when the diet lacks all of the needed molecules for energy and raw materials. Malnutrition and undernutrition will be discussed in detail in Chapter 13.

Carbohydrates and fats in the diet usually play the role of providing the body with energy. The body monitors the monosaccharide glucose in particular as a way of registering the body's energy intake. A measurement called the **glycemic index** indicates the amount of glucose readily available in a given food. The glycemic index is traditionally represented as a ranking of foods based on their immediate effect on blood glucose levels. Most people use the term blood sugar in place of blood glucose. The glycemic index of a particular food is measured by monitoring how much the blood glucose increases over a period of 2 or 3 hours after eating that food. Foods rich in carbohydrates, which break down quickly during digestion, have a high glycemic index. The glycemic index of a meal should be balanced with the body's energy needs. Taking in large amounts of glucose to boost a person's energy does not work. The excess glucose is not used for energy producing reactions; rather, it is stored in the body as glycogen in the liver and muscles. An excess of glucose is converted to fat. High-performance athletes learn to regulate glucose intake to keep the body at peak performance. Many athletes ensure ample glucose for energy by taking part in a strategy called "**carbo loading**." A day or two before an endurance event an athlete maintains a low activity level while eating large amounts of glucose. This encourages glycogen storage as well as some fat accumulation. The glycogen then provides adequate glucose to the muscles if the blood glucose should drop too low to keep the body going.

The health benefits and risks of fats in the diet are highly debated. Fats are an essential nutrient for the body and needed as a fuel for energy when blood glucose runs low. They are also needed as raw material for building body parts. Children in particular need ample amounts of fats for normal growth and development. However, there are concerns that some diets take in too many fats and that certain fats can cause a variety of medical problems ranging from cancer to heart disease. It is known that excess fat in the diet is stored as fat despite the body's energy needs. In most people, fats signal the body that the appetite is sated. However, some people have a physiology that does not respond to that signal. They tend to take in too much food, leading to the excessive accumulation of body fat. Dietary proteins are mainly broken down to amino acids, which are used as building blocks for creating other body proteins and nucleic acids. This should be their primary role in the diet. However, it was discovered that diets high in protein could contribute to weight loss. Many athletes follow high-protein diets to accelerate muscle growth. The philosophy behind using proteins in this way is correct, but there are dangers to using proteins as the primary food source. Diets that are low in carbohydrates and high in proteins use the proteins as an energy source. Proteins do not "burn clean" as do carbohydrates. The breakdown of proteins into energy produces a variety of waste products that can harm the heart, kidneys, and liver. This will be discussed further in Chapter 3.

✓ Concept Check

1. Define homeostasis.
2. What is the role of diet in maintaining homeostasis?
3. Distinguish between the terms malnutrition and undernutrition.

DISCOVERY SCENE PLEASE ENTER DISCOVERY SCENE PLEASE ENTER

Have you come closer to solving the CSI? What nutritional factors may be associated with the problems noted in your friend? How are these related to the race in which he was competing?

WELLNESS AND ILLNESS OVER THE LIFE SPAN
Aging of the Body's Chemistry

Key Terms: aging, antioxidant, electromagnetic radiation, electromagnetic spectrum, entropy, free radical, oxidation, senescence, ultraviolet (UV) light, wavelength

Television, radio, and the Web are blasted by advertisements for nutritional supplements that slow down **aging**, or make a person look and feel younger. Most of the claims are based on true scientific studies. But this does not mean that the supplements will work to slow body aging. Aging is not a simple process. It has to do with various aspects of the body ranging from chemical-level changes to the way body systems interact with each other. Each chapter of this book will identify aspects of aging as they relate to the topic of the chapter. The discussion of aging here addresses the molecular changes that occur as people age. Medical practitioners and scientists use the term **senescence** to describe the aging process in organisms.

Aging is a general term used to describe the gradual deterioration of a material or an object. There is a law of nature called **entropy**, which is defined as a measure of energy dispersal and disorganization in the universe. In simple terms, it means that all things decay to a simpler form unless energy is put forth to repair the damage in some manner. The human body is continuously decaying at the chemical level. Functional and structural biochemicals are always being broken down. Functional molecules, such as enzymes, break down as a result of the work they carry out. They wear out just like machines do after

Aging The gradual deterioration of a material or an object

Senescence The aging process in organisms

Entropy The measure of energy dispersal and disorganization in the universe

Free Radicals Aggressive chemicals that readily react with biochemicals

Oxidant A substance that causes or hastens oxidation

Oxidation The process of joining oxygen with another molecule; a chemical change in which an atom loses electrons

Antioxidants Chemicals that protect the body by oxidizing before oxidizing agents can hurt the body

Ultraviolet (UV) Light The invisible rays of the electromagnetic spectrum

Electromagnetic Radiation A form of energy that travels in waves

Wavelength The length of a wave

Electromagnetic Spectrum The full breadth of wavelengths

years of use. Structural molecules decay of their own accord, or are destroyed by injury or disease. Covalent bonds will break down over time simply due to the natural stress and strain placed on bonds in a molecule. The environment of the body is a major cause of molecular aging. **Free radical** waste products formed by various body processes alter many essential body chemicals. Scientists define free radicals as aggressive chemicals that readily react with biochemicals. They can modify or degrade biochemicals in such a way that prevents them from functioning in the body. Free radical oxygen is a waste product of the body and a component of the atmosphere. Certain types of pollutants, such as ozone, are rich in free radical oxygen activity. Free radical oxygen is an **oxidant** and, therefore, causes a chemical modification called **oxidation**. Oxidation is explained as the process of joining oxygen with some other molecule. It is also defined as a chemical change in which an atom loses electrons.

Oxidation is considered a major aging factor by many researchers who study senescence. Its least damaging effects involve the conversion of hydroxyl to carbonyl groups. This can prevent many biochemicals from carrying out their normal jobs. Proteins and nucleic acids are very sensitive to oxidation damage. Unfortunately, oxidation is a naturally occurring event that cannot be avoided. Its major effects on aging take place when the body cannot keep up with the molecular damage. This inability to keep up with damage can be due to age-related problems that slow the body's ability to take in molecules or build new molecules. Air pollution, smoking, and many diets high in meat contribute to the oxidative damage of the body. Chemicals called **antioxidants** can slow the oxidation process. They do this by becoming oxidized themselves, thereby protecting the body's chemicals from being affected by the oxidizing agents. Vitamins A, C, and E have antioxidant properties. Certain plants are also high in antioxidant chemicals. Broccoli, cabbage, garlic, red grapes, soybeans, and tomatoes are a few food plants with large amounts of antioxidants. Research shows that antioxidants do reduce chemical damage in test tubes. However, it has not been confirmed that antioxidants have the same effects on the body when eaten or applied to body surfaces.

Ultraviolet (UV) light is another contributing factor to the molecular decay related to senescence. Light is a form of energy called **electromagnetic radiation**. It travels as a wave and is measured by the length of the wave, or the **wavelength**. The full breadth of wave sizes or wavelengths is called the **electromagnetic spectrum**. UV light is distinguished as being the invisible rays of the spectrum. It is composed of high-energy light with small wavelengths. UV light is capable of modifying or damaging many types of molecules. It can cause inappropriate covalent bonds to form between two molecules. It can also break down the covalent bonds holding molecules together. It is a primary factor in breaking down critical polymers in the body. Proteins and nucleic acids are readily damaged by UV light. This is why the adage "ultraviolet light will age your skin" is actually true. It can damage the skin faster than the skin can repair itself. In spite of some claims, antioxidants do not reduce the molecular damage caused by UV light.

✓ Concept Check

1. Define the term senescence.
2. Describe three factors that can cause molecular aging.
3. Explain one strategy for reducing molecular aging.

DISCOVERY SCENE PLEASE ENTER DISCOVERY SCENE PLEASE ENTER

Have you solved the CSI yet? Did the information about molecular aging provide any more clues about your friend's condition?

CSI – Case Study Investigation Conclusion

What can you conclude about what happened to your friend? This is not a simple case to figure out. What could cause him to be dizzy and vomit? Why the rapid heart rate? What would cause his muscles to have bouts of tension and relaxation? Was it something in the sports drink? Could it have been due solely to the activity? Was it a combination of the energy drink and the overexertion that created his problems?

Answer:

Your friend's energy drink was loaded with caffeine and sugar. He did get an initial boost of energy from the sugar; however, this ran out rather quickly, leaving him in a weakened state. His dizziness was probably due to the fact that the brain was not getting any glucose for energy because it was all going to muscle activity. In addition, it is very likely he was subjected to too much caffeine. The following are indications of caffeine overdose:

difficulty sleeping	vomiting
muscle twitching	diarrhea
confusion	irregular heartbeat
increased urination	rapid heart beat
increased thirst	hallucinations
fever	dizziness
difficulty breathing	convulsions

This CSI was adapted from the following articles:

1. Greenleaf JE. Problem: thirst, drinking behavior, and involuntary dehydration. *Med Sci Sports Exerc.* 1992;24:645-656.
2. Nightingale SL, Flamm WG. Caffeine and health. Current status. In: Weiniger J, Briggs GM, eds. Nutrition Update. Vol 1. New York, NY: John Wiley & Sons; 1983: 3-19.

Chapter Summary

Matter and energy work together to build and run the human body. Elements are the raw materials of the body, but they do not work in isolation. They are bonded together to form a wide array of structural and functional molecules. Energy is constantly needed to ensure that the molecules can carry out their roles. Humans are primarily composed of organic molecules called biochemicals. The four categories of human biochemicals are carbohydrates, lipids, peptides, and nucleic acids. The body cannot manufacture most of the biochemicals found in humans. They have to be taken in through the diet. It is not unusual to discover that many human ailments are caused by diets that do not balance the body's need for particular biochemicals. The essence of life is to take in the appropriate proportions of biochemicals needed to replace those that are used up or decay in the process of running the body.

Atoms and Molecules

- All organisms are composed of energy and matter.
- All matter is composed of atoms.
- Atoms have a central core called the nucleus, which is composed of protons and neutrons that determine the atomic mass.
- The proton number or atomic number determines the properties of an atom.
- Atoms with a particular proton number are called elements.
- Elements that vary in neutron number are called isotopes.
- Ions are elements that have more or fewer electrons than protons and carry an electric charge.
- Elements can form ionic, covalent, and hydrogen bonds.
- An isomer is a molecule that has the same elemental makeup, but a different elemental arrangement.
- Organisms are primarily composed of organic molecules called biochemicals.
- All biochemicals are composed of a carbon skeleton and a functional group.

Acids and Bases

- pH is a measure of the hydrogen ion concentration in a solution.
- Acid solutions have a high hydrogen ion concentration.
- Base or alkaline solutions have a low hydrogen ion concentration.
- pH is represented by the pH scale, which ranges from pH 1 through 14.
- pH numbers 1 through 6 are acids, 7 is neutral, and bases are pH 8 through 14.
- Buffers are used in the body to prevent pH fluctuations.

Human Molecules

- The human body is composed of lipids, carbohydrates, peptides, and nucleic acids.
- Lipids are classified as glycerides, sterols, and an assorted group containing vitamins.
- Fatty acids can be saturated or unsaturated.
- Carbohydrates are classified as monosaccharides, disaccharides, oligosaccharides, and polysaccharides.

- Glucose, fructose, galactose, and mannose are common monosaccharides.
- Sucrose, lactose, maltose, and cellobiose are common disaccharides.
- Polysaccharides usually have structural or storage roles in organisms.
- Glycogen, starch, cellulose, and mucopolysaccharides are common polysaccharides.

Molecules and Nutrition

- Biochemicals must be eaten in the right amounts to maintain homeostasis.
- Homeostasis is the ability to maintain body function and structure.
- Malnutrition means the body is not getting the proper proportion of biochemicals.
- Undernutrition means the body is not getting all the biochemicals it needs to maintain homeostasis.

Wellness and Illness over the Life Span

- Molecular aging is due to destruction of essential biochemicals.
- Molecular aging is also called senescence.
- Other factors, such as oxidation and ultraviolet light, cause molecular decay.
- Antioxidants are chemicals that prevent or slow molecule decay due to oxidation.

Key Terms

Introduction
Biochemistry
Carbon
Chemistry
Energy
Heat
Mass
Matter
Molecular biology
Organic chemistry

Atoms and Molecules
Atom
Element
Periodic table
Subatomic particle

Atoms and Molecules: Atomic Structure and Function
Atomic mass
Atomic nucleus
Atomic number
Atomic orbitals
Atomic structure
Electron
Ion
Isotope
Nuclear decay
Neutron
Physical properties
Proton
Radioactive
Radioisotope dating
Shells

Atoms and Molecules: Properties of Molecules
Adhesiveness
Chemical bonds
Cohesiveness
Compounds
Pure molecules

Atoms and Molecules: Types of Bonds
Covalent bond

Study Guide

Hydrogen bond
Ionic bond

Atoms and Molecules: Parts of a Molecule
Amino group
Biochemical
Bioactive molecule
Carbon skeleton
Carbonyl group
Carboxyl group
Chirality
Functional group
Hydroxyl group
Isomer
Molecular structure
Phosphate group
Sulfhydryl group
Structural formula
Structural molecule

Acids and Bases
Acid
Alkaline
Base
Buffer
Denaturation
Dissociate
Electrolyte
Hydrogen ion acceptor
Hydrogen ion donor
Hydroxyl ion
Neutral
pH
pH scale

Human Molecules
Carbohydrate
Lipid
Monomer
Nonpolar
Nucleic acid
Peptide
Polar
Polymer
Water soluble

Human Molecules: Lipids
Bioactive
Cholesterol
Detergent
Diglyceride
Emulsion
Estrogen
Fat
Fat soluble
Fatty acid
Glyceride
Glycerol
Hydrophilic
Hydrophobic
Hydrocarbon
Hydrogenate
Linear
Lipophilic
Monoglyceride
Phospholipid
Polyunsaturated
Progesterone
Saturated
Short chain
Steroid
Sterol
Surfactant
Suspension
Terpenoid
Testosterone
Triglyceride
Unsaturated
Vitamin

Human Molecules: Carbohydrates
Alpha isomer
Amino-sugar
Amylose
Arabinose
Beta isomer
Cellobiose
Cellulose
Dextrose
Disaccharide
Fructose
Glucose
Glycogen
Glycosidic linkage
Hexose
Inulin
Lactose
Lactose intolerance
Levulose
Linkage
Maltose
Mannose

Monosaccharide
Mucopolysaccharide
Oligosaccharide
Pentose
Polysaccharide
Starch
Sucrose
Trisaccharide

Human Molecules: Peptides
Amino acid
Enzyme
Functional protein
Helix
Oligopeptide
Peptide bond
Polypeptide
Primary structure
Protein
Quaternary structure
R group
Secondary structure
Sheet
Side chain
Structural protein
Tertiary structure

Human Molecules: Nucleic Acids
Adenine
Adenosine triphosphate (ATP)
Deoxyribonucleic acid (DNA)
Deoxyribose
Nitrogen base
Guanine
Nucleotide
Purine
Pyrimidine
Ribonucleic acid (RNA)
Ribose

Molecules and Nutrition
"Carbo loading"
Craving
Glycemic index
Homeostasis
Malnutrition
Undernutrition

Wellness and Illness over the Life Span
Aging
Antioxidant
Electromagnetic radiation
Electromagnetic spectrum
Entropy
Free radical
Oxidation
Senescence
Ultraviolet (UV) light
Wavelength

Check Your Understanding

1. All matter is composed of:
   a. mass
   b. atoms
   c. peptides
   d. isotopes

2. The atomic number of an atom is determined by:
   a. multiplying the mass by two
   b. counting the number of electrons
   c. counting the number of protons
   d. the number of protons added to the number of neutrons

THE BODY'S CHEMICAL MAKEUP

Study Guide

3. The bonding properties of an element are primarily determined by the:
   a. protons
   b. neutrons
   c. electrons
   d. atomic mass

4. Which bond is common to all biochemicals?
   a. ionic
   b. sulfhydryl
   c. hydrogen
   d. covalent

5. A molecule with an abundance of this functional group would readily dissolve in water:
   a. hydroxyl
   b. sulfhydryl
   c. amino
   d. neutrino

6. A chemical capable of neutralizing an acid or a base is called a(n):
   a. indicator
   b. moderator
   c. buffer
   d. catalyst

7. pH is a measure of the _____ concentration of a solution.
   a. hydrogen ion
   b. hydroxyl ion
   c. electrolyte
   d. electron

8. Which molecule is a monosaccharide or simple sugar?
   a. starch
   b. glucose
   c. table sugar
   d. glycogen

9. Human muscle stores sugars in the form of:
   a. glycogen
   b. fatty acid
   c. starch
   d. sucrose

10. Which of these lipids act as chemical messengers in the body?
    a. triglycerides
    b. fatty acids
    c. sterols
    d. monoglycerides

11. Fats are needed in the diet to help provide these important structural lipids:
    a. diglycerides
    b. monoglycerides
    c. saturates
    d. steroids

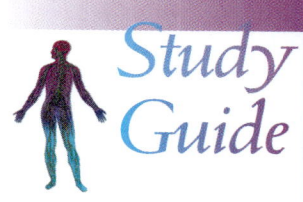

Study Guide

12. Which essential nutrients are converted into proteins in humans?
    a. nucleic acids
    b. amino acids
    c. carbonic acids
    d. inorganic acids

13. Peptides that carry out chemical work for the body are called?
    a. nucleosomes
    b. aminozymes
    c. enzymes
    d. albumins

14. Monomers of nucleic acids are essential for the following role:
    a. converting into structural proteins
    b. long-term storage of body energy
    c. transmitting genetic information
    d. energy transfer

15. Which of the following is not a common cause of molecular aging?
    a. ultraviolet (UV) light
    b. oxidation
    c. antioxidants
    d. stress on covalent bonds

A Case Study

Breast-Feeding

Studies of infant and childhood nutrition saturate the news, and whenever someone is expecting a baby, babies become the primary topic of conversation. It would seem rational that the best way to provide a baby with the molecules it needs is through breast, or mothers', milk. Some nutritional studies show that cow's milk and infant formulas do not provide all the nutrients needed for normal human development. In contrast, some physicians say that mother's milk is inferior to formula. In addition, a group of environmental researchers argue that breast milk can be harmful because it can contain a variety of pollutants that affect the infant's development.

Use the information in this chapter and the web sites below to resolve the issue of whether mothers' milk is the safest and most nutritious way to feed a baby. Answer the following questions to assist in this query:

- What are the "real" safety concerns of mothers' milk?
- How does formula differ from milk?
- Is there adequate information about the nutritional value of mother's milk compared with cow's milk and formula?
- What does the research show about the health of infants raised on mothers' milk versus other forms of nutrition?
- Are there any government guidelines or some type of standards related to infant nutrition?
- How are studies on infant nutrition carried out, and what types of data are collected?

Study Guide

1. Breast Feeding Basics
   http://www.breastfeedingbasics.org

2. Breastfeeding
   http://www.4woman.gov/Breastfeeding/

3. La Leche League
   http://www.lalecheleague.org/

4. National Institutes of Health
   http://ehp.niehs.nih.gov

5. Institute of Food Science & Technology
   http://www.ifst.org/

Where Do We Go from Here?

People in health fields can use their knowledge of the body's chemical makeup to solve everyday problems. You may wish to use other resources, such as the suggested web sites, in addition to your book to investigate the answers to each of the following situations:

1. Ozone air pollution causes oxidation. How would you explain to a person the effects of ozone on the body?
   http://science.howstuffworks.com/ozone-pollution2.htm

2. An athlete wants to know more about carbo loading. What would you tell him or her about the practice?
   http://www.physicalmag.com/index.php?p=9&a=926&pfv

3. Neighbors want to put their infant on a low-fat diet. Explain to them the possible risks of doing this to the child. www.emcp.net/personalmd
   http://www.personalmd.com/news/n0824083337.shtml

4. Neighbors want to know the value of knowing the glycemic index of their foods? www.emcp.net/carbs
   http://www.carbs-information.com/glycemic-index.htm

5. A child asks you to explain where proteins come from and what they do in the body. www.emcp.net/contexo
   http://www.contexo.info/

Skills Activities

1 Sources of Glucose

Materials
- three glucose test strips for urine sugar testing
- one pair of sharp scissors
- one marker
- two eyedroppers
- small container of water
- six small plastic cups
- three metal teaspoons
- one slice of potato
- one grape
- one small bottle of colorless soda
- one small piece of ground beef
- one small bottle of orange juice

As mentioned in this chapter, simple sugars, such as glucose, are a primary source of quick energy for the body. However, large amounts of simple sugars taken in the diet at one time can elevate blood sugar. This, in turn, can create immediate problems for the body. The body must find a way of removing the sugars from the blood before the sugars damage blood vessels and the kidneys. This excess glucose is stored away as glycogen. Excessive consumption of glucose in the diet can lead to the conversion of glucose to fat. People with diseases such as diabetes have trouble removing glucose from the blood. Therefore, they have to be very careful about the glucose content of the foods they eat. This activity provides a simple way to look for glucose in various foods and beverages. It uses the chemical reaction in which the carbonyl functional group of glucose interacts with an indicator solution in a urine test strip.

Use the marker to label the six cups as follows: 1) control; 2) potato; 3) grape; 4) soda; 5) meat; and 6) juice. Then cut each of the urine test strips in half lengthwise so that there are six narrow strips containing the colored test patch on the bottom. Use the marker to label each of the six strips as follows: 1) "C" for control; 2) "P" for potato; 3) "G" for grape; 4) "S" for soda; 5) "M" for meat; and 6) "J" for juice. Now, set up the experiment as follows:

Setup
- Add a dropper full of water to the "C" cup.
- Use a clean spoon to grind up the piece of potato in the "P" cup.
- Use a clean spoon to grind up the grape in the "G" cup.
- Use a clean eyedropper to add a dropper full of soda to the "S" cup.
- Use a clean spoon to grind up the meat in the "M" cup.
- Use a clean eyedropper to add a dropper full of juice to the "J" cup.

THE BODY'S CHEMICAL MAKEUP

Study Guide

Experiment

Place the appropriately labeled test strips next to each cup. Then, dip the colored portion of the test strip into the liquid of its corresponding cup. Compare any changes to the urine test strip with the color chart on the test-strip container. Note how much glucose is available in each substance. What were your expectations about the glucose content of each food? How could you use the information you obtained to help people make dietary choices?

Skills Activities

2 Factors That Affect the Glycemic Index of Food

Materials
- one bottle of medicinal iodine solution (potassium iodide)
- one marker
- one eyedropper
- a small container of water
- seven 3-inch by 3-inch squares of aluminum foil
- one slice of fresh potato
- one slice of baked potato
- one slice fresh yellow banana
- one slice of artichoke heart
- one slice of cauliflower
- one slice of Jerusalem artichoke or taro

As mentioned in this chapter, the glycemic index is a measure of how fast a particular food elevates the blood sugar level after eating. Foods with a high glycemic index are loaded with simple sugars, such as glucose. Starch has a somewhat lower glycemic index than glucose, but it is still a major source of "hidden" simple sugars in the diet. In this activity, you will use the iodine test to indicate the presence of starch in certain foods. The iodine test indicates the presence of starch by turning from an amber or brown color to a dark blue or black color.

Use the marker to label the seven squares of aluminum foil as follows: 1) control; 2) uncooked potato; 3) cooked potato; 4) banana; 5) artichoke; 6) cauliflower; and 7) Jerusalem artichoke or taro. Now, set up the experiment as follows:

Setup
a. Add a dropper full of water to the "control" square.
b. Add the slice of uncooked potato in the "uncooked potato" square.
c. Add the slice of cooked potato in the "cooked potato" square.
d. Add the slice of banana in the "banana" square.
e. Add the slice of artichoke in the "artichoke" square.
f. Add the slice of cauliflower the "cauliflower" square.
g. Add the slice of Jerusalem artichoke or taro to the appropriately labeled square.

Experiment

Add one drop of the iodine solution to the samples on each square. Make sure the iodine solution is mixed with the water and soaked into each sample of food. Look for the presence of iodine by watching for the change in color of the iodine solution. The control should not change color. As mentioned above, foods with starch will turn the iodine solution blue or black. Which of the foods contain starch? How should these findings influence dietary choices about the intake of foods with a low glycemic index? Some people believe that baking starchy foods lowers the glycemic index. Was this confirmed or refuted by your findings?

Case Study Investigation

Case Study Investigation #3

Your job takes you to a public health office in a small Texas community bordering Mexico. An indigent family comes by to get advice about their ill infant. The child was admitted to the hospital one month after birth because of severe dehydration. The hospital staff concluded that the dehydration may have been due to periodic episodes of diarrhea and vomiting. They treated the child until she appeared normal and sent her home with the parents. The child has grown slowly from its original 7.2 lb birth weight. Now, she is beginning to lose weight again at six months old. You notice that the child has a faint blue tinge to her skin and seems very inactive

or lethargic for a baby. The parents explain that the child has a low body temperature, does not feed well, and still has mild bouts of vomiting and diarrhea. It is the job of the public health office to determine whether the child needs more assistance for her apparent illness. The office must first figure out what type of physician or medical care is needed. At the end of the chapter you will be asked to determine the possible cellular problems causing this set of conditions.

CHAPTER 3
Organization of the Body

Chapter Outline

Case Study Investigation (CSI)
Applied Learning Outcomes
Introduction
Hierarchy of Human Structure
The Human Physiological
  Environment
    Water
    Ions
    Enzymatic Reactions and Energy
    Molecular transport
      Diffusion
      Passive Transport
      Osmosis
      Active Transport
Cell Structure
  Cells of Microbes
  Human Cells
Cell Function
  Metabolism
  Genetics
  Cell Cycle
    Asexual Reproduction
    Sexual Reproduction
Tissues
  Epithelial Tissue
  Connective Tissue
  Muscle Tissue
  Nervous Tissue
Organs and Systems
Wellness and Illness over the Life
  Span
  Pathology of Cells
  Cellular Aging
CSI Conclusion
Study Guide

Applied Learning Outcomes

- Use the terminology associated with cell structure and function.
- Learn about:
  - Body hierarchy
  - Cell organization
  - Cell physiology
  - Cell life cycle
  - The chemical environment in which human biochemicals function
  - Tissue organization
  - Tissue form and function
- Understand the cellular basis of aging and pathology.

Introduction

 Key Terms: hierarchies, levels of organization

 Levels of Organization or Hierarchies A series of ordered groupings within a system

Chapter 2 showed how knowing the body's chemical makeup is important for a comprehensive understanding of body structure and function. In this chapter the molecules will be placed into higher **levels of organization**, or **hierarchies**. Hierarchy is defined as a series of ordered groupings within a system. Each higher category or a hierarchy relies on properties of the lower levels. Level of organization specifically refers to a biological concept describing the hierarchical interactions of an organism's components. It also looks at how these components respond with other organisms and the environment. Molecules do not work in isolation to keep the body running. They work together in structural and physiological units that cooperate to maintain the body's homeostasis. Many simple organisms, such as bacteria, have fewer levels of organization than complex organisms like humans. Eventually, six levels of organization for the human body will be described in this book. Health care professionals are particularly attentive to the impacts of environmental factors and disease on hierarchical functions of the body.

Hierarchy of Human Structure

Key Terms: cell, cellular level of organization, differentiation, envirome, molecular level of organization, multicellular, organ, organ system, organ system level of organization, organismic level of organization, society, tissue, tissue level of organization, unicellular

Cell The basic structural and functional unit of the human body
Unicellular Consisting of only one cell
Multicellular Consisting of many cells
Differentiation A process by which cells mature in order to carry out specific physiological tasks
Molecular Level of Organization Groups of atoms making up molecules

According to many biologists, the **cell** is the basic structural and functional unit of the human body. It is also considered the simplest characteristic of all organisms. Many microscopic organisms, such as bacteria and yeast, are called **unicellular**. This means that their "body" comprises only one cell. The cell of a unicellular organism must be able to carry out all the jobs needed for survival. It is in continuous communication with the environment and must have the capability to contend for resources with competing unicellular organisms. Other organisms, such as humans, are multicellular, meaning their bodies are composed of many cells. It is estimated that humans are made up of over 100 trillion cells. The cells of most **multicellular** organisms cooperate with one another to carry out specific jobs. This job specificity is called **differentiation**. Differentiation is defined as the process by which cells mature in order to carry out the physiological tasks they were meant to do in the body. Cooperation between differentiated cells is what forms the basis of an organism's physical hierarchy.

Cells make up the cellular level of organization and are one step above the **molecular level of organization** (Figure 3.1). Chapter 2 discussed how molecules do not work in isolation. They are organized into cellular structures.

Groups of cells that perform similar specific function form the **tissue level of organization**. The term tissue is usually defined as an organized assembly of cells that have a similar structure and perform a special function. Tissues are able to function as one unit because the cells communicate and interact with each other as they respond to the environment surrounding the cells. Groups of tissues are arranged into functional components called organs. An **organ** is best defined as a structure composed of more than one tissue that is specialized for some particular function. Organs form the **organ level of organization**. The types of tissues that make up an organ determine its function. The brain, heart, kidneys, liver, lungs, skin, and stomach are examples of organs.

At the next level of hierarchy, organs are arranged into **organ systems** making up the **organ system level of organization**. An organ system is a collection of organs having related roles in an organism's function. Each organ system has a specific set of distinct tasks it performs to maintain homeostasis in the body. For example, the digestive system is primarily for the breakdown and uptake of food. The circulatory system in turn takes the food and moves it throughout the body.

Above the organ system level is the **organism**. There are a variety of definitions for organism. In its simplest definition, it means any living thing. The most comprehensive description defines an organism as an individual biological unit capable of reproduction. In truth, organisms are not the highest level of organization. Many organisms interact with other organisms to form a

Tissue Level of Organization Groups of cells that perform similar specific functions

Tissue An organized assembly of cells that have similar structures and perform a specific funtion

Organ A structure composed of more than one tissue that is specialized for a particular function

Organ Level of Organization Groups of tissues that perform similar specific functions

Organ Systems A collection of organs having related roles in the body's function

Organ System Level of Organization Groups of organs that perform similar specific functions

Organism An individual biological unit capable of reproduction

Figure 3.1 **Levels of Organization**

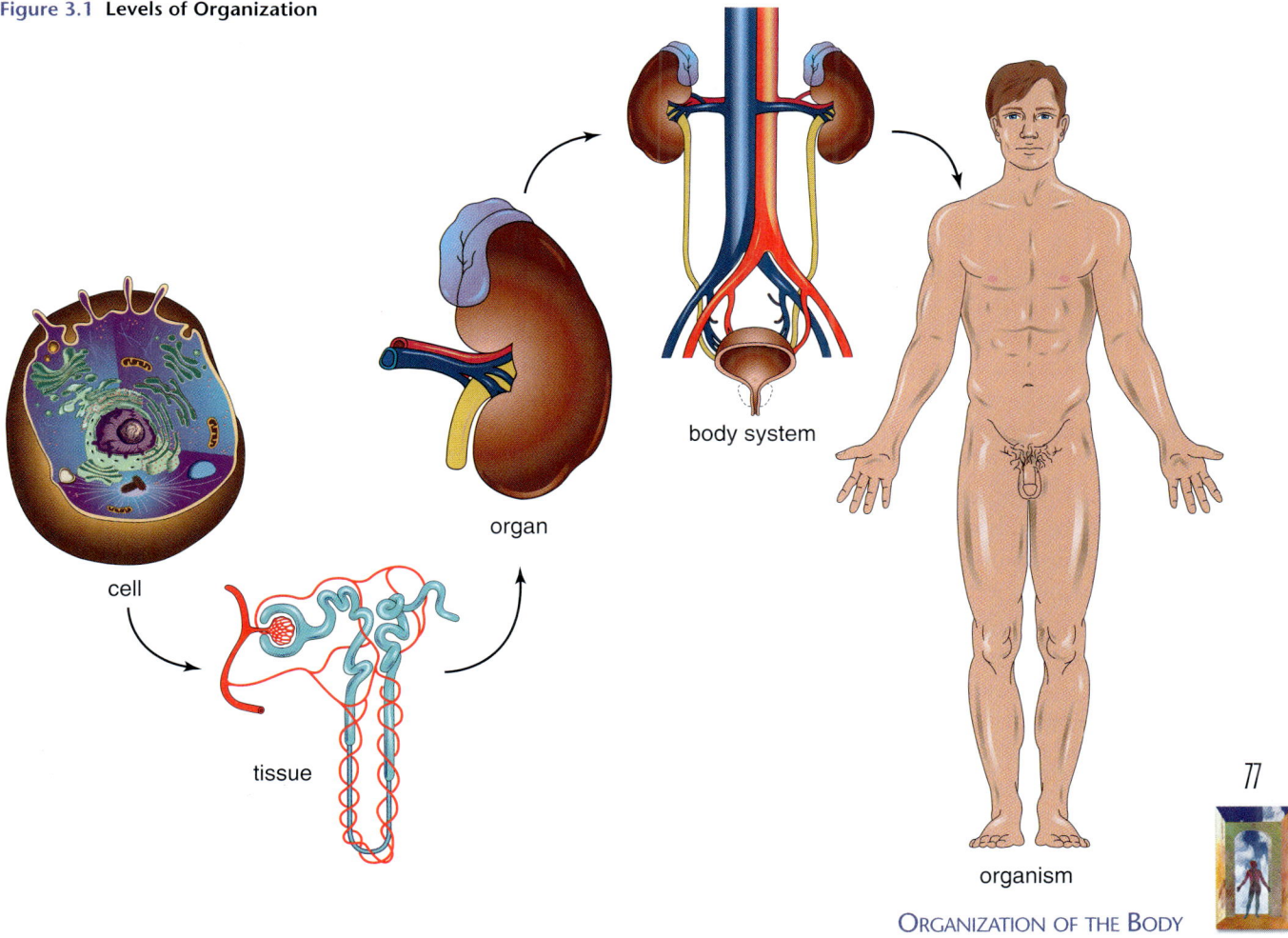

cell

tissue

organ

body system

organism

ORGANIZATION OF THE BODY

Society Groups of organisms interacting with each other

Envirome All of the environmental factors that affect the survival of an organism or society of organisms

society. Human society is defined as a group of individuals living and working together in a particular region. Society has major implications on the homeostasis of the human body. Scientists know that the outcomes of human interactions in a society affect the homeostasis of individuals. Society contributes to the highest level of organization called the **envirome**. This term is used commonly in genetics and psychological studies. It refers to all the environmental factors that affect the survival of an organism or society of organisms. Enviromics, or the study of the envirome, investigates the balance of environmental conditions and processes that affect an organism's or society's successful adaptations needed for survival.

✓ Concept Check

1. Define the term cell and explain how cells contribute to an organism's survival.
2. What is the difference between the cells of a unicellular organism and a multicellular organism?
3. Explain the role of enviromics in human survival.

DISCOVERY SCENE PLEASE ENTER DISCOVERY SCENE PLEASE ENTER

What have you learned so far that may help you solve the Case Study Investigation? Does information about hierarchy or levels of organization help you understand the condition? Is there a role of the envirome that might determine the cause of this illness?

TRAINING STEM CELLS

The 2005 Biotechnology Industry Organization's international conference held in Philadelphia hosted an out of the ordinary unannounced keynote speaker. She was not a business analyst, politician, scientist, or anyone with a professional connection to biotechnology-based medicines. Yet her talk about stem cell research captured the audience's imagination more than the other speakers who had Ph.D.s and medical degrees. Her child and husband, both with genetic disorders, would be the direct recipients of the stem cell research discussed by the other speakers.

There is much ethical and moral opposition to stem cell research. Regardless of this opposition, many people accept the fact that stem cells may one day be used to cure and prevent many illnesses and injuries. Advances in biotechnology make it possible to understand the mechanisms of "training" stem cells to carry out a variety of healing jobs. Genetic engineering may even pave the way to make "metabolically engineered" cells produce drugs based on bodily needs. However, realizing the promise of stem cell therapies requires painstaking attention to the details of cell differentiation. Scientists have devised a way for neural stem cells to differentiate into other types of tissue cells. This was not an easy task because differentiation requires a subtle and incredibly complex mixture of growth factors, mitogens, and environmental conditions. The United States National Institutes of Health provides frequently updated information on their website at *http://stemcells.nih.gov/index.asp*.

The Human Physiological Environment

Key Terms: environment, external environment, internal environment physiological environment

The body's levels of organization would be ineffective in maintaining homeostasis if the cells were not surrounded by a suitable **physiological environment**. The physiological environment includes all the internal conditions that optimize individual cell function and body organization. Environment simply means conditions within the cell (**internal environment**) or outside of the cell (**external environment**) affecting cell function. It takes the back and forth collaboration of the body's hierarchy to maintain the environment inside the cells and the environment bathing the tissues. The physiological environment takes into account that both internal and external environments of cells rely on water. This watery, or aqueous, environment in turn must have precise conditions that ensure homeostasis. These conditions are determined by pH, ions, chemical reactions, and the transport of molecules between internal and external environments.

Physiological Environment All the internal conditions that optimize individual cell function and body organization

Internal Environment Conditions within the cell

External Environment Conditions outside the cell

✓ Concept Check

1. Define the term physiological environment.
2. Distinguish between the terms internal environment and external environment.
3. What aqueous factors contribute to the body's physiological environment?

Water

Key Terms: dehydrated, overhydration, solute, solution, solvent, specific heat, water excess, water intoxication

Life on Earth would not be possible without an abundant supply of water. It makes up a small percentage of all the molecules on Earth, yet it is the most abundant molecule in the body. Scientists estimate that 55% to 65% of a person's body weight is made up water. This water is not just sloshing around underneath the skin. It includes all the water contained in the cells, the blood, body cavities, organs systems, and in the fluids bathing the tissues. When the water content of the body falls below a certain level the body is said to be **dehydrated**. Dehydration is usually defined as a state in which the body tissues are deprived of water. It commonly results from inadequate water intake or excessive water loss. Diarrhea, sweating, and vomiting are frequent causes of water loss. Indicators of dehydration include extreme thirst, nausea and exhaustion. Surprisingly, it is also possible to have too much water in the body. **Overhydration** is a condition in which the body contains too much water. This condition occurs when the body takes in more water than it loses over a period of time. Some health professionals call overhydration **water excess** or **water intoxication**. This condition is harmful because it disrupts the physiological environment.

Dehydrated The state in which the body tissues are deprived of water

Overhydration The state in which the body contains too much water

Water Excess or Water Intoxication The state in which the body contains too much water

ORGANIZATION OF THE BODY

Civic Responsibility

HELPING OTHERS WITH YOUR KNOWLEDGE

It is valuable to use what you learned about the organization of the body to help others better understand the world around them. It is very important to check your facts and seek further information about certain topics before discussing health and science issues. Here are some suggestions to foster a better public awareness of the body's organization:

1. Speak to schoolchildren about the health of particular organ systems.
2. Work with sports clubs educating the players about metabolism enhancing products.
3. Help the elderly better understand the limitations of anti-aging supplements to reduce cellular aging.
4. Volunteer at a school health day to teach children the basics of human body organization.

Solvent A substance that dissolves other chemicals

Solution A uniform mixture of two or more substances

Solute Any particle that dissolves in a solvent

Water's atomic structure provides it with all of the properties that make it useful for the body's physiological environment. First, its atomic arrangement (Figure 3.2) induces polarity on the water molecule. The oxygen end of the molecule takes on a slight negative charge while the part with two hydrogen atoms develops a slight positive charge. Polarity permits water to dissolve most of the biochemicals needed for human survival. Many scientists call water the universal **solvent** for this reason. A solvent is a substance that dissolves other chemicals and in so doing forms a **solution**. A solution is a uniform mixture of two or more substances. Any particle that dissolves in a solvent is called a **solute**. Water is such a good biochemical solvent because it readily forms hydrogen bonds with the hydroxy groups of carbohydrates and attracts the electrical charges of R groups on peptides. Enzymes in particular require an aqueous polar environment to facilitate chemical reactions needed for homeostasis. Water is also a natural solvent for ions, including the electrolytes needed for many cell functions.

The polarity of the water molecule also gives it the adhesive and cohesive properties needed to keep it from evaporating quickly and thereby causing dehydration. Adhesive properties give water the ability to latch on to other molecules reducing evaporation. The ability of water to be electrically attracted to other water molecules is its cohesive property. Both of these properties give

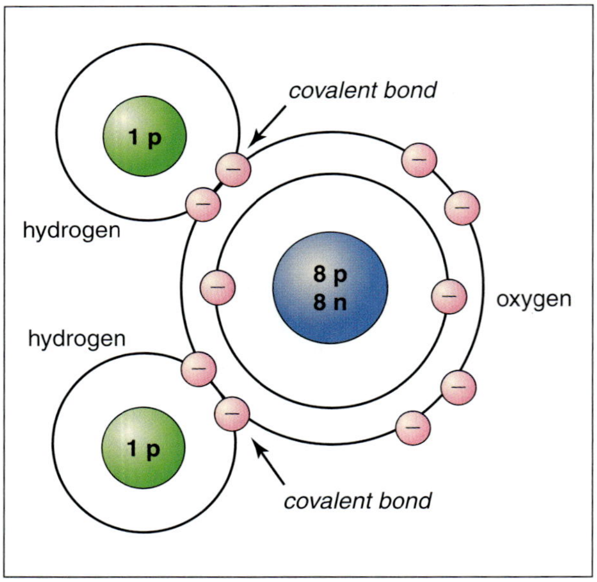

Figure 3.2 Hydrogen Bond
Molecular structure of water

water a high **specific heat**. Specific heat is the of heat energy required to raise the temperature of a substance. Water's high specific heat prevents the body from heating or cooling too quickly. As mentioned in Chapter 2, biochemicals can only perform their intended jobs within a certain temperature range. Water provides the ideal temperature range and narrow temperature fluctuations necessary for homeostasis.

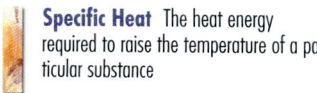 **Specific Heat** The heat energy required to raise the temperature of a particular substance

✓ Concept Check

1. Distinguish between dehydration and overhydration.
2. How do the polar properties of water contribute to the body's physiological environment?
3. What are the similarities and differences of waters adhesive and cohesive properties?

Ions

 Key Terms: anion, cation, metal, mineral

Ions play a variety of roles in maintaining the physiological environment for the body. As mentioned in Chapter 2, ions are electrically charged particles that disperse in water. Some ions are categorized as elements called **minerals**. Minerals are nutrients needed by the body in small amounts. Ions called **metals** are any of several elements that can form shiny solids that conduct heat and electricity. Most ions are composed of a positively or negatively charged element. Some are electrically charged molecules. Ions having a positive charge are called **cations** while those with a negative charge are called **anions** (Figure 3.3). All ions are water soluble, meaning that they are easily lost by the body whenever fluids are eliminated. Most of the body's ions are lost by the kidneys as they produce urine. Sweating is the second major ion loss mechanism. Athletes must pay close attention to replenishing ions lost during profuse or prolonged periods of sweating. Ions can be lost by the digestive system as feces accumulate water before being removed. Vomiting is also a way the body loses ions. Health care workers must be alert to excessive ion loss in people exhibiting diseases that produce diarrhea and vomiting. The major ions related to human health are:

 **Minerals** Nutrients needed by the body
Metal Any of several elements that conduct heat and electricity
Cations Positively charged ions
Anions Negatively charged ions

- **Bicarbonate (HCO3-)** is a major body fluid buffer that makes acidic conditions neutral to slightly basic. The bicarbonate concentration of the body fluids is regulated by the kidneys and the lungs.
- **Calcium (Ca2+)** is found in two forms in the body: diffusible calcium and nondiffusible calcium. Diffusible calcium is found in salts mostly in bone, whereas nondiffusible calcium is bound to blood and cell proteins.
- **Chloride (Cl-)** - is usually found in association with potassium and sodium. It is commonly taken in the body as the salts potassium chloride (KCl) and sodium chloride (NaCl).
- **Copper (Cu,** which stands for Cuprum), **iodine (I)**, and **iron (Fe,** meaning Ferrum) are common biologically important metals that help carry out chemical reactions in the cell.
- **Magnesium (Mg^{2+})** carries out many of the jobs done by calcium. In

ORGANIZATION OF THE BODY

addition, it is critical for energy production and proper nerve function. Is also supports muscle relaxation and helps regulate blood sugar.
- **Phosphate** (PO_4^{2-}) is another body fluid buffer that is regulated by the kidney. Over 80% of the phosphate used in the body bonds to calcium to assist with bone hardening. Much of the remaining phosphate is associated with body energy.
- **Potassium** (K^+) - The "K" stands for its Latin name kalium. It is the most abundant mineral inside of cells. Its movement in and out of cells is essential to control muscle and heart contractions.
- **Sodium** (Na^+, the Na is Latin for natrium) is the most common mineral found outside of the cell being abundant in the blood and body fluids. One of its major roles is controlling the amount of water retained by the body.
- **Sulfate** (SO_4^{2-}) is usually found as an anion attached to other biologically important molecules. Sulfate ions are acidic and lower the pH of body fluids.

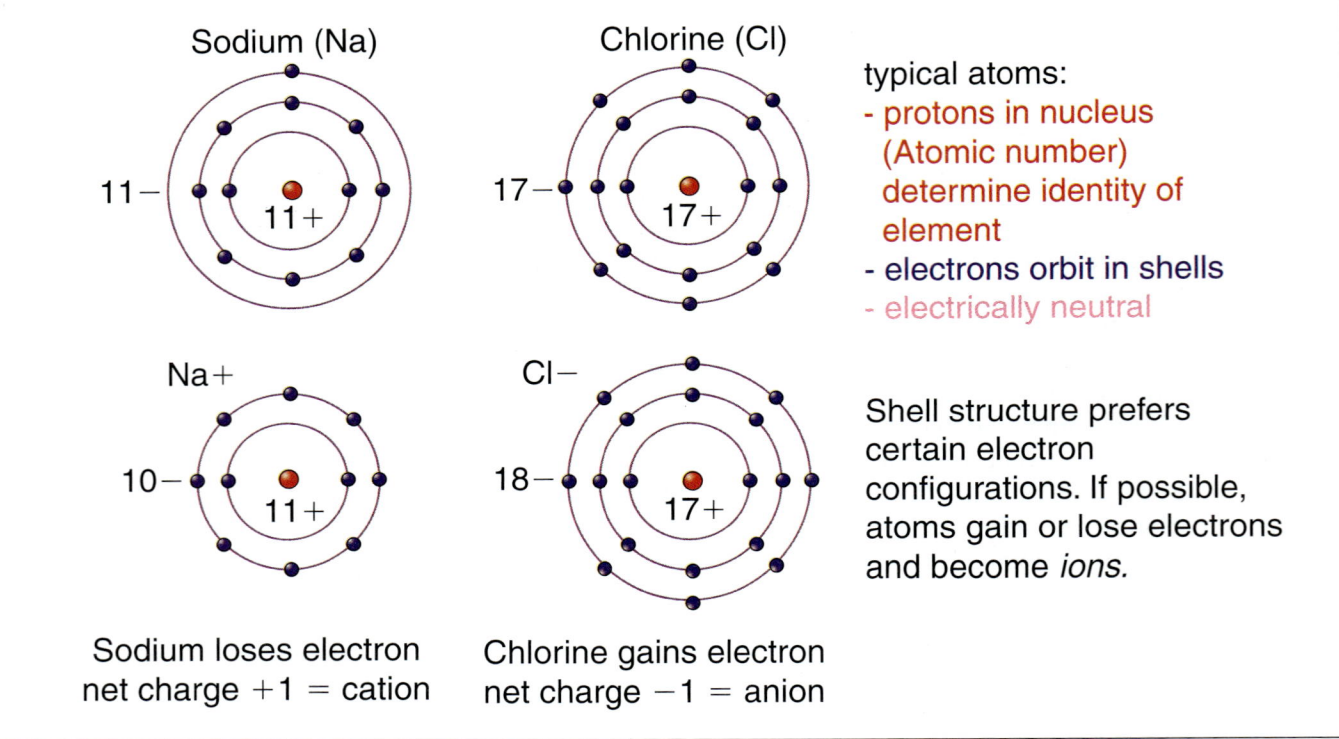

Figure 3.3 Ions
Ionic molecules disassociate in water to form independent charged particles

✓ Concept Check

1. What are some ways that ions are lost from the body?
2. Distinguish between the terms minerals and metals.
3. What the differences and similarities of anions and cations?

Enzymatic Reactions and Energy

Key Terms: active site, calorie, catalyst, chemical energy, coenzyme, cofactor, electrical energy, endergonic, exergonic, hydrolysis, kilocalorie, kinetic energy, mechanical energy, potential energy, product, reduced, substrate, synthesis, thermal energy

The body's physiological environment would not exist without the chemical reactions that permit cells, tissues, organs, and organ systems to carry out their jobs. Enzymes facilitate most of these chemical reactions. Chemicals that start chemical reactions are called **catalysts**. In order to the run the body certain molecules have to be broken down in reactions called **hydrolysis**. Hydrolysis means to break (lysis) with water (hydro). Water is needed for a hydrolysis reaction to occur. Other molecules have to be built, or **synthesized** by cells and body fluids. Then some molecules are modified by being oxidized or **reduced**. An oxidized molecule loses an electron or a hydrogen ion from its molecular structure. An oxygen atom can also be added to a molecule as it is being oxidized. Reduced molecules gain an electron or a hydrogen ion to its structure. Individual elements can be oxidized or reduced thereby giving an extra positive or negative charge to the atom.

Scientists have many ways to categorize chemical reactions. One important set of reactions looks at the energy use status of a chemical reaction. Most of the chemical reactions in the human body involve some form of energy transfer. The chemical reactions of primary concern in medicine transfer energy from food to some type of cell function. It is important to know which reactions require cell energy and which provide the cell with energy. **Endergonic** chemical reactions require energy to set off and carry out the reaction. In contrast, **exergonic** chemical reactions release energy. They generate energy that can be used to build cell structures or carry out physiological functions.

Energy is usually defined as the capacity for doing work. This can be measured as the ability to do work (**potential energy**) or the conversion of this ability to so some type of action (**kinetic energy**). Think of food as potential energy and muscle action as kinetic energy. The body uses four types of energy to maintain homeostasis: **chemical**, **electrical**, **mechanical**, and **thermal**. The conversion of a chemical into another form is called chemical energy. Electrical energy occurs in the body in two ways: the oxidation and reduction of elements and molecules, or the transport of ions from one location to another. Mechanical energy is the movement of a molecule or a large structure. Thermal energy is the production of heat. In the human body, the chemical energy of the glucose in a candy bar is converted into electrical energy that drives the mechanical energy of a moving muscle. The moving muscle then produces thermal energy which people feel as body heat. People in the health care field measure body energy using the **calorie**. A calorie is technically defined as a standard unit of measurement equal to the amount of heat required to raise the temperature of one gram of water one degree Celsius. Food calories are actually a measure called the **kilocalorie**, which is 1,000 times larger than the standard calorie.

Catalyst Chemicals that start chemical reactions

Hydrolysis The chemical process of breaking down with water

Synthesized Refers to the process of synthesis; built

Reduced The process of a molecule gaining an electron or a hydrogen atom to its structure

Endergonic Chemical reactions that require cell energy

Exergonic Chemical reactions that release cell energy

Potential Energy The ability to do work

Kinetic Energy The energy associated with motion or action

Chemical Energy The conversion of a chemical into another form

Electrical Energy The energy associated with the movement of electrons to produce a current

Mechanical Energy The energy of motion or movement used to perform work

Thermal Energy The production of heat

Calorie A standard unit of measurement equal to the amount of heat required to raise the temperature of one gram of water one degree Celsius

Kilocalorie A standard unit of measurement, 1,000 times larger than a calorie

Good Choice Bad Choice

A young woman is feeling tired and irritable after entering the second month of her pregnancy. She is hesitant about taking prescription drugs or other medications out of concern of harming the developing baby. She hears that the natural herb ginseng provides energy "safely" when consumed as tea. What advice should be given to her about the effectiveness and safety of ginseng? What concerns must be considered before drinking ginseng when pregnant? In what situations would it be appropriate and inappropriate to use ginseng?

Now, where do enzymes fit into the concept of body energy? Enzymes promote the exergonic reactions needed to obtain cell energy. In addition, they carry out endergonic reactions that build body structures or are used in cell work. Enzymes do this by modifying molecules in patterns that release or store energy. The amino acid sequence of an enzyme provides them with a tertiary shape that permits them to carry out reactions. Enzymes have a region of the protein called an **active site**. The active site provides the attachment area

 Active Site An area of a protein providing an attachment area and energy

Figure 3.4 Enzymatic Reactions Enzymes lock on to a substrate molecule converting it into another molecule called the product.

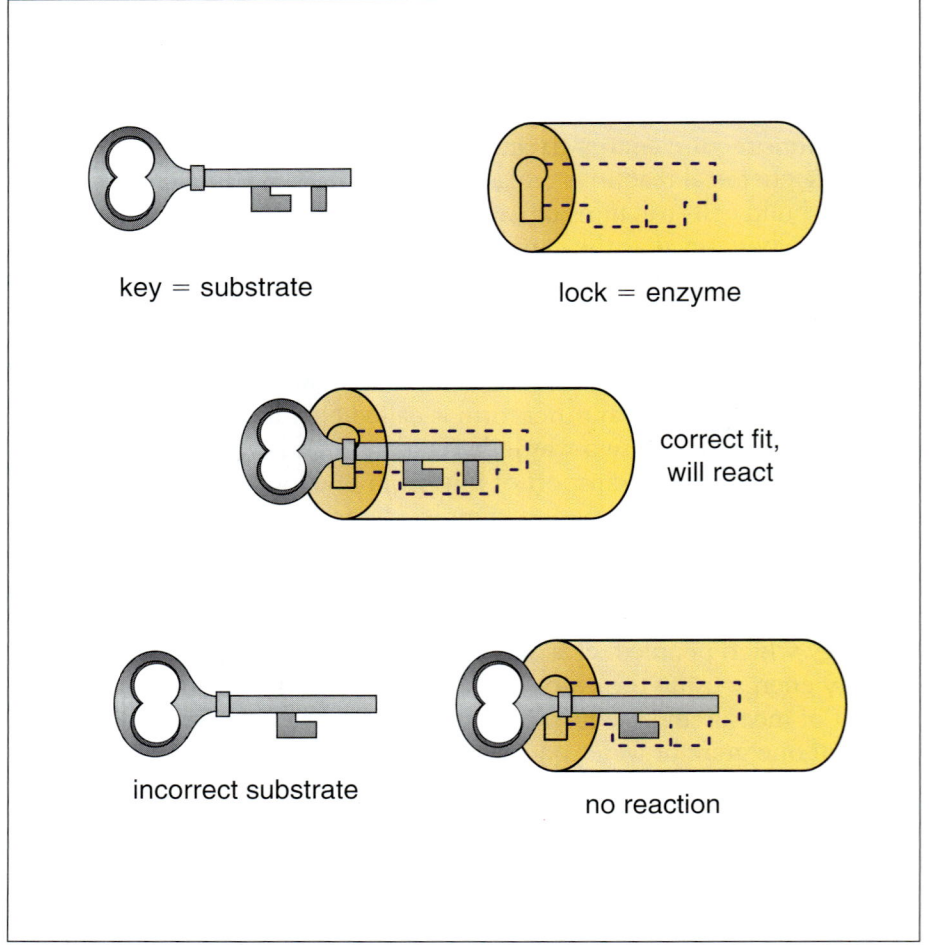

84

CHAPTER 3

and energy needed to modify a molecule. Hydrogen bonds in the active site are generally used to hold onto the molecule. Each active site fits a particular molecule in the way a lock fits a specific key. **Substrate** is the name given to a chemical that an enzyme modifies in the active site. The substrate in turn is converted into the enzyme **product** (Figure 3.4). The active site of many enzymes can be controlled by a process called **allostery**. An enzyme exhibits allostery if its activity or efficiency is changed by the binding of a molecule to a region of the enzyme called an **allosteric site**. The nucleic acid ATP mentioned in Chapter 2 binds to the active sites of many enzymes causing them to slow down or inhibit their function. Oxygen has an allosteric role in the binding of other oxygen molecules to the hemoglobin protein of blood.

Other factors that affect enzyme function are competitive inhibition, feedback inhibition, pH, electrolyte concentration, temperature, heavy metals, **coenzymes**, and **cofactors**. Competitive inhibition occurs when a chemical resembling the substrate binds to the active site. This in effect reduces the enzyme's ability to perform chemical modification of the substrate. Many foods such as raw eggs and uncooked dough, have competitive inhibitors that block digestive enzymes. This is one reason why eggs and wheat products are best eaten when adequately cooked. Feedback inhibition takes place when an enzyme's product is not removed. As a product builds up it can go back to active site and block it from binding to substrate molecules. In some situations it is possible for the enzyme to convert the product back into substrate. The body devises a variety of ways to prevent this from happening. Some chemical reactions in the cell have developed feedback inhibition as a way to regulate itself.

Non-competitive inhibition is caused by the elements, molecules, or environmental factors that alter the enzymes structure or electrical charges. Out of the ordinary pH values, temperatures, and electrolyte concentrations can alter the shape of an enzyme. They could also prevent hydrogen bonds in the active site from binding to substrate molecules. Heavy metals have the same effect on many enzymes. Coenzymes are organic molecules that activate certain enzymes. They do this by binding to the enzyme or by assisting with the function of the active site. Nucleic acids and vitamins are common coenzymes. Cofactors are elements or ions that facilitate enzyme activity. Many scientists have stopped using the term coenzyme and use the term cofactor as any molecule that assists or completes the function of an enzyme. Unprocessed foods are rich in coenzymes and cofactors. Many nutritionists worry that processed foods and food supplements may cause illness because they usually lack these essential nutrients.

> **Substrate** A chemical that will be modified in an enzyme's active site
>
> **Product** The chemical produced as a result of an enzymatic reaction
>
> **Allostery** Modification of an enzyme by a chemical so that the enzyme works faster or slower
>
> **Allosteric Site** The region of an enzyme where a chemical binds, causing the enzyme to work faster or slower
>
> **Coenzymes** Organic molecules that activate certain enzymes
>
> **Cofactor** Elements or ions that facilitate enzyme activity

✓ Concept Check

1. Define the term energy and describe the types of work energy produces in a cell.
2. What are the roles of enzymes in the body?
3. Explain how diet and environmental factors can affect the functions of enzymes.

ORGANIZATION OF THE BODY

Molecular Transport

Key Terms: active transport, bulk active transport, bulk mechanical transport, carrier protein, channel, diffusion, diffusion gradient, endocytosis, excretion, exocytosis, facilitated diffusion, filtration, hyperosmotic, hypertonic, hypoosmotic, hypotonic, isoosmotic, isotonic, osmolarity, osmosis, membrane, membrane diffusion, passive transport, phagocytosis, pinocytosis, receptor-mediated endocytosis, secretion, selectively permeable membrane, voltage-gated

Membrane A sheet-like structure that surrounds a cell or its organelles and keep its internal environment contained

Selectively Permeable Membrane A membrane allowing only certain molecules to pass through it

Diffusion The mixing of liquid or gas molecules as a result of random thermal stirring

Passive Transport Diffusion across a membrane that requires no cell energy

Active Transport Diffusion across a membrane that requires cell energy

Osmosis Diffusion of water across a membrane

Bulk Mechanical Transport The movement of large volumes of molecules and ions from one location to another

A cell's internal environment is separated from the external environment by a covering called a **membrane**. A membrane is a sheetlike structure that surrounds a cell and keeps the internal environment contained. Cells contain what scientists call a **selectively permeable membrane**. A selectively permeable membrane allows certain molecules to pass through it, in what is called molecular transport. There are five methods of molecular transport: **diffusion**, **passive transport**, **active transport**, **osmosis**, and **bulk mechanical transport**.

Diffusion Scientists define diffusion as the mixing of molecules in gases and liquids as a result of thermal stirring (Figure 3.5). Thermal stirring refers to the fact that all molecules are vibrating as the result of heat in the environment. The amount of heat or temperature determines the speed and intensity of these vibrations. Diffusion is evident when a container of something with an objectionable odor is opened. The odor molecules mix with the atmosphere and are distributed around the room by collisions with the gas molecules in the air. The warmer the room the faster the smell travels. Smells travel even faster in humid air. The vibrating water molecules in the humid air act as additional participants in the diffusion frenzy. Diffusion is valuable for cell survival. It is the only way that certain molecules make their way into the cell. Without diffusion, cells would have no way of making contact with essential molecules. Diffusion also provides a means of transporting wastes away from the cell.

Figure 3.5 Diffusion
Diffusion of dye in a container of water

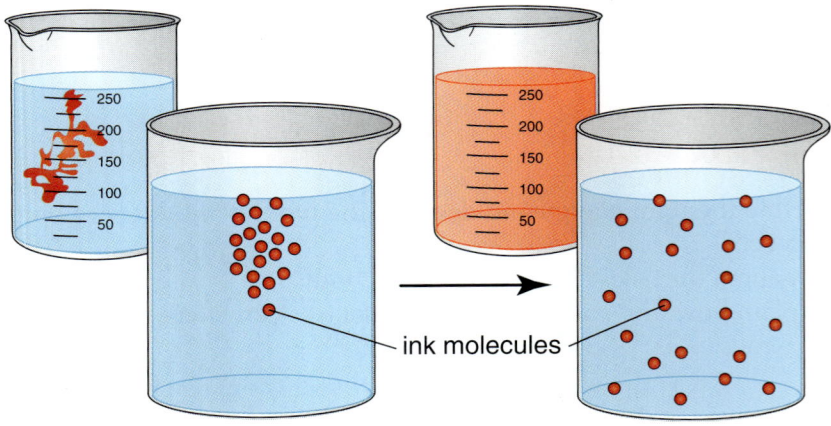

Diffusion gives cells a perplexing environmental problem to deal with. Cells must prevent their contents from diffusing into their environment while at the same time diffusing certain materials in and out of the cell. The cell's selectively permeable membrane helps it tackle this problem through a process called passive transport. Passive transport allows certain molecules to pass in and out of cell based on their natural flow. Molecules that cannot be transported this way are moved using active transport mechanisms.

Passive Transport Passive transport is defined as diffusion across a membrane in which the cell requires no energy. The cell relies on a **diffusion gradient** as the driving force for moving the molecules. A gradient is a variation in some quantity with respect to another. In cells a diffusion gradient is set up when internal and external environments have unequal quantities of particles in its solution. For example, the cell might have more molecules in its solution than the fluids surrounding the cell. This situation sets up a gradient. According to the principles of diffusion, molecules having the ability to pass through the membrane will travel from a region where they are in a higher quantity to a region where their quantity is lower. There are three primary types of passive transport: **membrane diffusion**, **facilitated diffusion**, and **filtration**.

Membrane diffusion relies on the principle that certain particles can pass through a membrane. The membrane covering a typical cell is composed of lipids with gaps and protein gateways distributed on the surface (Figure 3.6). Molecules that dissolve in lipids usually have no trouble passing back and forth across the membrane. After all, they just need to dissolve into the lipid covering to enter the cell. Polar molecules cannot dissolve through lipids. However, certain small molecules can squeeze through minute protein gateways in the membrane. Carbon dioxide, oxygen, and certain cell wastes move in and out of the cell this way. A cell's survival is determined by the gradient being favorable for the cell's needs. There are three terms that refer to the direction of a gradient found between a cell and the environment: **hypertonic**, **hypotonic**, and **isotonic**.

Diffusion Gradient The condition when a cell's internal and external environments have unequal quantities of particles

Membrane Diffusion A type of passive transport in which certain particles pass through a membrane

Facilitated Diffusion A type of passive transport in which carrier proteins are used to move particles through a membrane

Filtration A passive transport in which particles are removed from water by passage through a porous membrane

Hypertonic Refers to an environment with a greater quantity of a particular molecule than exists in the cell

Hypotonic Refers to an environment with a lesser quantity of a particular molecule than exists in the cell

Isotonic Refers to an environment with an equal quantity of a particular molecule in the solution and the cell

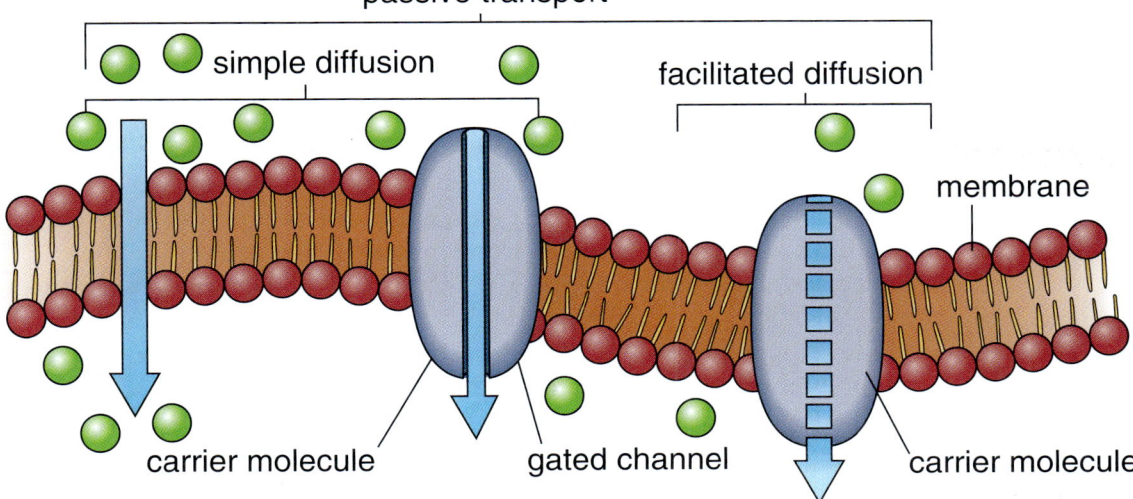

Figure 3.6 Passive Transport
Passive transport across a cell surface

ORGANIZATION OF THE BODY

A hypertonic (hyper means "more than") environment has a greater quantity of molecules of a particular substance than the cell. This situation would cause these molecules to enter the cell through the membrane. Oxygen in the atmosphere is hypertonic and as a result, oxygen molecules move into the cells. In contrast, hypotonic (hypo means "less than") environments are those with fewer molecules of a substance in the solution than the cell. Consequently, cells in a hypotonic environment lose the higher quantity molecules in their interior to the environment as they pass through the membrane. Carbon dioxide in the atmosphere is hypotonic and thereby moves out of the cell under normal circumstances. An isotonic environment has an equal (iso means "equal to") concentration of molecules as the cell. Molecules in this situation move back and forth across the membrane in equal numbers. Therefore, there is no appreciable change in the molecule concentration on either side of the membrane. Nitrogen gas in the atmosphere is isotonic. It is the most abundant gas in the air and only poses problems when people are subjected to rapid altitude changes.

It's a Diffusion Issue

You decide to give mountain climbing a try and you learn that most people need oxygen assistance at altitudes greater than 18,000 feet (6300 meters). On the flight to the mountain, you find out the plane is pressurized to a value called 1 atmosphere. The bus that picked you up at the airport for the drive to the mountain has all of its windows open. You were told this was to prevent an exhaust leak building up carbon dioxide in the passenger area. What do all of these situations have to do with diffusion?

Carrier Protein or Channel A protein that moves molecules through a membrane as part of facilitated diffusion

Facilitated diffusion is a form of passive transport which uses a specific **carrier protein**, or **channel**, to move ions and molecules through a membrane. It is reserved for molecules that normally do not pass across the membrane. A slight amount of energy is used in this process. However, the energy does not come from the cell performing the transport. The energy is provided when the carrier protein changes its shape as it bonds to the ion or molecule it is transporting. Dietary amino acids and monosaccharides enter the body by facilitated diffusion. This can create problems when diets are very high in these molecules. The amount of material absorbed by facilitated transport is limited by the number of carrier proteins available, so it is possible for these molecules to pass out of the body if eaten in excess. This can lead to digestive system problems that include cramping and diarrhea. There are various types of carrier proteins used in facilitated transport.

FISH, FROGS, AND VOLTAGE-GATES

From 1974 through 1983 there were 646 reported cases of poisonings from a fish called "fugu" in Japan. Fugu, or pufferfish, is a dangerous luxury ingredient that became popular in Japan during that time period. Fugu poisoning was recognized by the ancient Chinese as early as 2800 BC. Egyptian carvings from 2700 BC warned about eating pufferfish. Yet, in spite of this information people continue to get ill after eating this potentially deadly fish. Pufferfish contain an interesting toxin that is also present in many tropical insects and a group of frogs called the poison arrow frogs. The toxic substance in these animals is a chemical called Tetrodotoxin or TTX. It is one of the most toxic of the natural poisons.

Tetrodotoxin has a very specific effect on the body. It targets the molecular functioning of the cell membrane. It blocks cell membrane proteins called sodium channels. Sodium channels affect the operation of voltage-gated active transport mechanisms needed for nerve function. In effect, it stops communication between nerve cells. Muscle cells are also affected but to a lesser degree. Indigenous people of South America extract the poison from frogs to poison hunting arrows and darts. They have learned that an animal becomes paralyzed once the poison sets in. In people, the poison sets in around 14 hours after the meal. Stomach pain, paralysis of extremities, loss of muscle coordination, respiratory distress, and a drop in blood pressure characterize fugu poisoning. Despite its lethal effects in humans, International Wex Technologies, a Vancouver-based company, sees a benefit in administering tetrodotoxin to people. They foresee its use as a painkiller for people with cancer. It may also benefit drug addicts going through withdrawal programs

Osmosis Osmosis is best described as the diffusion of water across a selectively permeable membrane. Because there is more water in a region of low solute concentration, water molecules move to a region of high solute concentration which has relatively less water (Figure 3.7). Water has a diluting effect and always moves toward the more concentrated solution regardless of the type of molecule dissolved in the solution. Osmosis is carried out until the concentration of dissolved particles is equalized on both sides of the membrane. In effect, osmosis counteracts diffusion of particles into and out of a cell. Osmosis uses slightly different terminology than diffusion; **osmolarity** is the potential of water to move across a membrane. A **hyperosmotic** solution contains a higher concentration of solute than a **hypoosmotic** solution. **Isoosmotic** solutions are equal in solute concentrations.

Osmolarity is very important to your understanding of fluid flow into and out of a cell. Certain osmotic situations can cause the body to swell with water, as in overhydration, or lose water, as in dehydration. For example, a cell with a hyperosmotic internal environment has a hypoosmotic external environment. This is a hypotonic situation because the environment outside of the cell is more dilute. The diffusion gradient causes certain molecules to pass out of the cell as water moves into the cell. As a result, the cell swells with water and can even burst under extreme gradients. In contrast, a cell with a hypoosmotic internal environment has a hyperosmotic external environment making it a hypertonic situation. As you would expect, certain molecules enter the cell as water leaves and the cell shrivels.

Osmolarity Refers to water's potential to move across a membrane

Hyperosmotic Refers to a solution with a greater concentration of solute than exists in the cell

Isoosmotic Refers to a solution with an equal concentration of solute in the solution and the cell

Hypoosmotic Refers to a solution with a lower concentration of solute than exists in the cell

Figure 3.7 **Osmosis**
Osmosis and its affect on cells

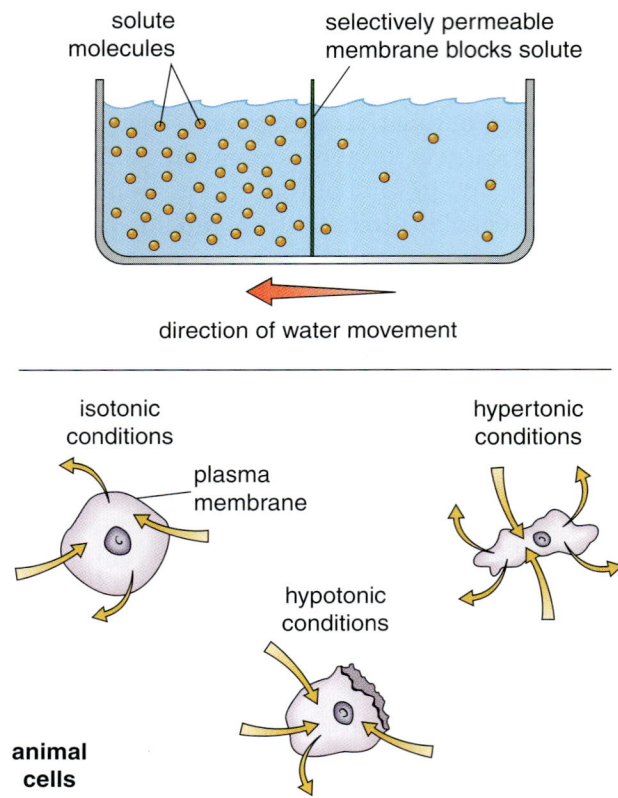

Figure 3.8 **Active Transport Pumping**
Active transport protein on the surface of a cell

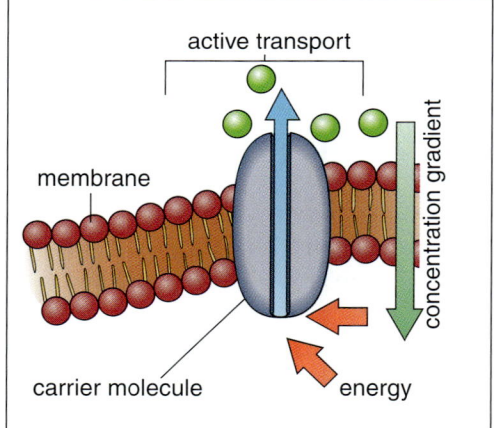

Active Transport Pumping The movement of particles across a membrane by protein "pumps"

Bulk Active Transport The movement of large amounts of particles across a membrane due to membrane movement

Active Transport Active transport differs greatly from passive transport. It is an energy-expending process by which a cell moves ions and molecules across its membrane. Active transport is useful for moving materials from an area of lower concentration to an area of higher concentration. This is in contrast to the direction of diffusion and works against the diffusion gradient. There are two types of active transport: **active transport pumping** and **bulk active transport**. Active transport pumping uses proteins called pumps to move several ions or molecules at a time across the membrane (Figure 3.8). These pumps require an expenditure of cell energy to carry out their jobs. Active transport pumps consume approximately 70% of the body's energy needs. You will be learning more about this type of transport in your study of the nervous system.

Bulk active transport moves large amounts of particles into and out of the cell. Large polymers and even cell components can be taken in or expelled this way. Bulk active transport relies on membrane movement and modification to do its job (Figure 3.9). **Endocytosis** is the process of moving particles into the cell. A special type of endocytosis called **receptor-mediated endocytosis** uses a special protein on the cell membrane called a receptor to bind specific types of molecules. Receptor-mediated endocytosis is used to take in large amounts of a particular molecule. One type of receptor-mediated endocytosis maintains cholesterol levels in the blood. Defects of this mechanism can cause heart disease and other problems.

Exocytosis is a bulk active transport method that removes large amounts of molecules or materials from the cell. It works in opposition to endocytosis; a large sack of material is gradually pushed out of the cell. **Excretion** is the removal of waste using exocytosis. Worn out cell components are regularly removed from the cell by exocytosis. **Secretion** is the transport of important molecules by exocytosis. Digestive juices and mucus are released using this method. Nerves communicate to each other and to muscles using secretions.

> **Endocytosis** A form of bulk active transport; the process of moving particles into a cell
>
> **Receptor-Mediated Endocytosis** A form of endocytosis using proteins to bind molecules
>
> **Exocytosis** A form of bulk active transport; The process of moving particles out of a cell
>
> **Excretion** The removal of waste from a cell using exocytosis
>
> **Secretion** The transport of molecules using exocytosis

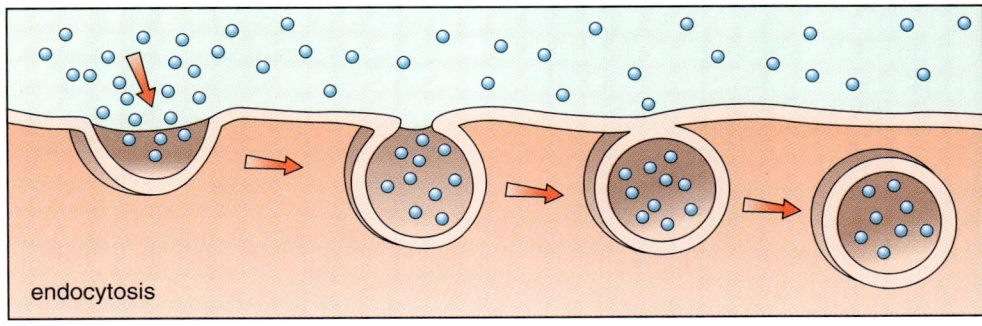
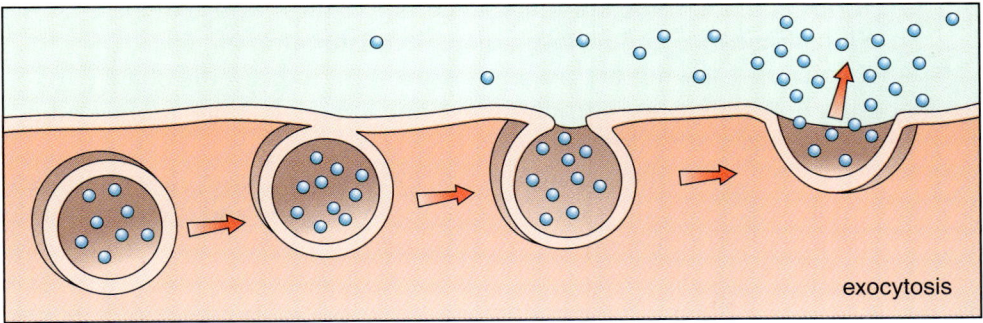

Figure 3.9 Other Types of Transport
Endocytosis and exocytosis

✓ Concept Check

1. Describe the term selectively permeable membrane and explain what it means to the survival of a cell.
2. Distinguish between passive transport and active transport and provide examples of each.
3. How does endocytosis differ from exocytosis?

ORGANIZATION OF THE BODY

DISCOVERY SCENE PLEASE ENTER DISCOVERY SCENE PLEASE ENTER

CSI Break

So, what information about the physiological environment is useful for resolving the Case Study Investigation? Could the infant's condition have something to do with water or ions? How could further knowledge about enzymes help understand the disease? Does the information on molecular transport provide any more clues for solving the case? If so, what information does it provide? How does the information help explain the infant's condition?

Cutting Edge Research
FIGHTING THE DOWNHILL BATTLE: ANTIAGING RESEARCH

The news media abound with stories about aging research. Much of that research has to do with lifestyle risk factors that accelerate aging. But lately there is growing public interest in the "genetics of aging." Scientists recently reported that they have discovered what are believed to be antiaging genes. The genes of interest are four sequences found in humans and other organisms.

- SIR2/SIRT – These genes have been shown to increase lifespan when expressed excessively in yeast and flies. They play a role in lipid metabolism.
- Insulin/insulin-like growth factor receptor – This gene increases lifespan in worms and mice when disabled in certain tissues. The gene plays a role in cell maintenance and may stimulate the maturation of cells.
- AMP kinase – Slowing of this gene's expression reduces the chances of obesity. In certain cells, nutrient intake and energy balance in the body and individual tissues are associated with this gene.
- INDY – This cell surface transporter is known to increase lifespan in flies when mutated. It is called the "I'm not dead yet "gene and is critical for providing cells with energy for homeostasis

No one is sure if modifying the expression of these genes can make people live longer. A certain amount of animal research evidence supports the idea that human life can be prolonged. However, these studies are neglecting the dozens of other genes and enviromic factors that contribute to longevity.

CELL STRUCTURE

Key Terms: cell doctrine, cell theory, cytoplasm, genome, prion, viroid, virus

Before the invention of the microscope in the 1500's AD little was known about the fine anatomy of the human body. The first cells were not seen until

CHAPTER 3

the 1700's AD. Anthony Leeuwenhoek (1632-1723) of the Netherlands and Robert Hooke (1635-1703) of England were the first to recognize the microscopic attributes of the body. Their findings contributed to the modern understanding of life called **cell theory**. Cell theory, which is also called **cell doctrine**, asserts that all organisms are composed of similar units of organization called cells. The concept was first proposed in 1839 by Matthias Schleiden and Theodor Schwann of Germany. Cell theory remains the foundation of modern medicine. Another important principal of cell theory was supported by the research of Louis Pasteur (1822-1895). He showed that all the essential functions of an organism occur within cells and that the cells use biochemicals to carry out these functions. All of these findings paved the way for modern drugs that treat many diseases and disorders.

Scientists divide cells into three structural components: cell membrane, **cytoplasm**, and **genome**. The cell membrane is the covering around the cell that is involved in transport of material into and out of the cell. Cytoplasm refers to the cell contents within the cell membrane. The term genome refers to the complete genetic material that is passed down from one generation to the next. Cells vary greatly in the complexity and use of these parts. Today it is accepted that not all living organisms have a cell as the basic unit of structure. **Viruses**, **viriods**, and **prions** are disease-causing agents that do not use cells as a basic unit of structure. Viruses are infectious agents composed of just a genome in a protein coat. Viroids are merely short pieces of RNA. Prions are the most puzzling organism because they are no more than a piece of protein resembling abnormal proteins found in other organisms. There is still much to learn about the way prions cause disease and reproduce.

Cell Theory or Cell Doctrine The assertion that all organisms are composed of cells

Cytoplasm The cell contents with the cell membrane

Genome The genetic material within the cell passed from generation to generation

Virus An infectious agent composed of a genome in a protein coat

Viriod An infectious agent composed of pieces of RNA

Prion An infectious agent composed of a protein

Cells of Microbes

Key Terms: bacteria, cell wall, filamentous, flagella, fungi, microbe, microorganism, nucleoid, nucleus, organelles, prokaryote, protista

Microbes includes any of a diverse group of simple organisms that must be viewed with a microscope. Microbe is another term for **microorganism**. **Bacteria**, **fungi**, prions, **protista**, viroids, and viruses are all categorized as microorganisms. Bacteria are the most common microorganism of the human body. A majority of bacteria help the body and do not cause disease. They are defined as single-celled organisms that have a very simple cell structure and a circular genome composed of DNA (Figure 3.10). The cells of bacteria are called **prokaryotic** cells. Prokaryotes (organisms that have prokaryotic cells) are distinct in that their genome is located in the cytoplasm in a region called the **nucleoid**. Another prokaryotic characteristic is that their cytoplasm has no specialized compartments. Prokaryotic cells are usually a thousand times smaller than those found in the human body. This makes it very difficult to view the fine details of bacteria. Many bacteria possess swimming appendages called **flagella**. Almost all bacteria have a peptidoglycan structure called a **cell wall** covering the cell membrane. The characteristics of the cell wall determine the way bacteria produce certain ill effects in humans. Bacterial secretions also establish how they invade the body and cause disease.

Microbe or Microorganism Any simple organism that can only be seen with a microscope

Bacteria The most common microorganism of the body

Fungi A microorganism found on the body, both single and multicellular

Protista A group of eukaryotes associated with disease

Prokaryote A cell with its genome located in the nucleoid and without specialized compartments in the cytoplasm

Nucleoid A region of the cytoplasm containing the genome

Flagella Swimming appendages found on some types of microorganisms

Cell Wall The covering of the cell membrane

ORGANIZATION OF THE BODY

Figure 3.10 Microbe Cell Structure
Typical bacterial cell

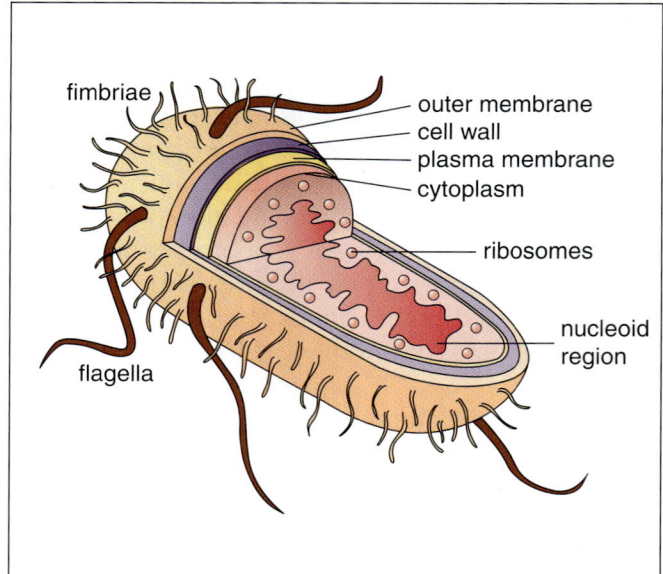

A fungus is another cellular organism likely to be found on the human body. Fungi are defined as a diverse group of organisms ranging in form from a single cell to a body mass of branched elongated and stringy cells. Single cell fungi are usually called yeast. **Filamentous fungi** are characterized by a mass of stringy cells. The cells of fungi are very similar to human cells and are categorized as **eukaryotes**. Eukaryotes have their DNA contained in a structure called the **nucleus**. In addition, their cytoplasm is compartmentalized into specialized functional units called **organelles**. Most fungi produce specialized reproductive structures, such as mushrooms. The striking colors noted on spoiled foods are mostly likely the reproductive structures of fungi. Several types of fungi are commonly found living harmlessly on the skin and in the digestive system. However, under certain conditions these fungi can cause mild to fatal diseases. Protista, another group of eukaryotes, are primarily associated with diseases such as malaria and sleeping sickness.

Filamentous Fungi Fungi characterized by masses of stringy cells

Eukaryotes Cells characterized by containing their DNA inside a nucleus and having compartmentalized cytoplasm

Nucleus The central structure of a cell containing the DNA

Organelle Specialized functional units of compartmentalized cytoplasm

✓ Concept Check

1. Define the term microbe and provide some examples with the definition.
2. What is the difference between bacteria and fungi?
3. Why is it important to know the types of microorganisms found on a person?

Human Cells

Key Terms: antigen, apoptosis, centriole, cilia, cytoskeleton, cytosol, endoplasmic reticulum, ER, fluid mosaic model, genetic expression, Golgi apparatus, Golgi body, lysosome, mitochondria, nuclear envelope, programmed cell death, receptor, rough endoplasmic reticulum, RER, ribosome, SER, smooth endoplasmic reticulum, transport vesicle

As discussed earlier, the human body is composed of eukaryotic cells (Figure 3.11). The human body has a diversity of cells. Each type carries out a particular function that contributes to the body's homeostasis. This diversity is achieved by the way the cell's genetic material adapts the cell membrane and organelles to carry out specialized jobs. The cell membrane is a continuous double layer of phospholipids stabilized by cholesterol molecules. It encloses the contents of the cell while at the same time acting as a two-way selectively permeable transport system. Floating around the membrane, embedded in the lipid layer, are cell membrane proteins. This "ocean" of proteins and lipids is called the **fluid mosaic model**. Fluid describes the motion of the proteins in the membrane. Mosaic refers to the fact that the membrane is composed of a variety of molecules. The proteins involved in cell transport were described earlier in this chapter.

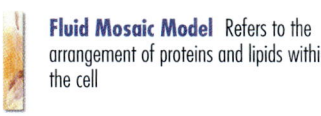

Fluid Mosaic Model Refers to the arrangement of proteins and lipids within the cell

Figure 3.11 **A Human Cell**

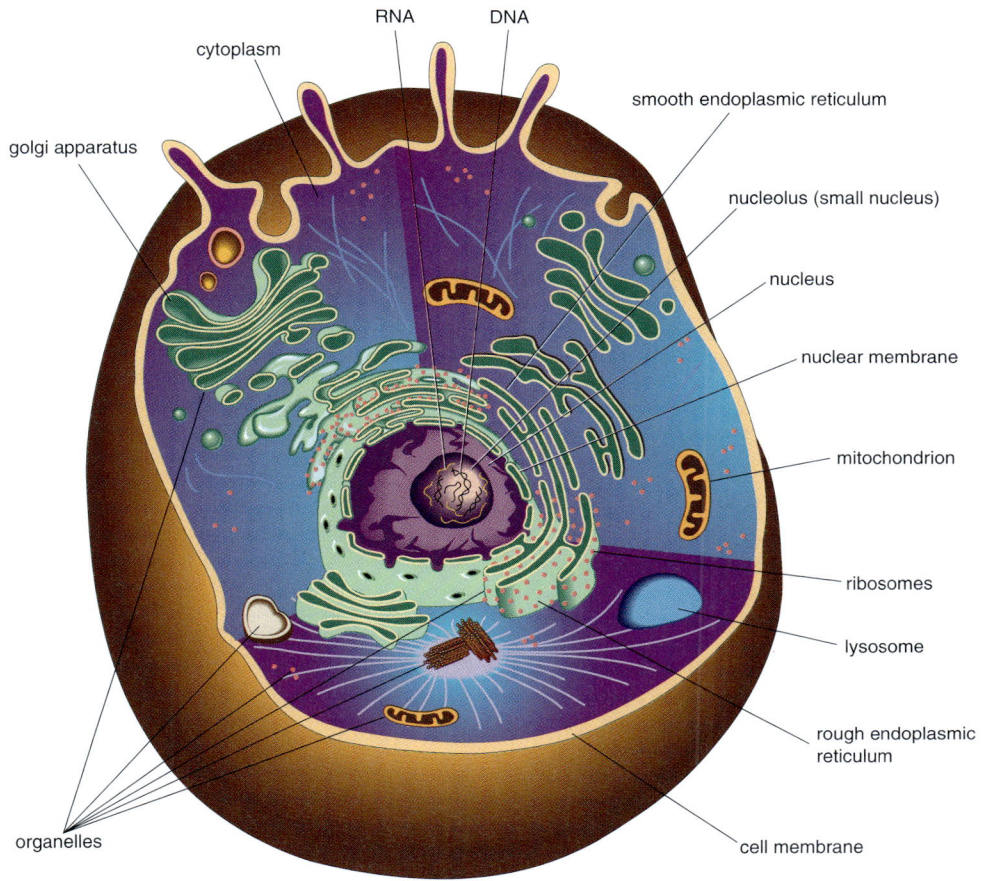

It is within the cytoplasm that the cell carries out the chemical reactions for homeostasis. The cytoplasm is divided into the **cytosol** and the organelles. Cytosol is a gel-like fluid composing about 54% of the cells total volume. It contains thousands of enzymes that conduct a variety of cell functions mostly associated with obtaining cell energy. Most of the chemical reactions in the cytosol are regulated by chemical information from the genetic material and the cell membrane. The organelles in the cytosol perform specialized cell functions.

A group of five organelles form a succession of membrane-bound structures involved in the manufacture and movement of molecules and cells parts: the **nuclear envelope, endoplasmic reticulum, Golgi bodies**, vesicles, and cell membrane. Components of this group transfer materials to each other through direct contact and through the use of **transport vesicles**. The nuclear envelope is responsible for transmitting genetic information. It also permits the inward

Cytosol A gel-like fluid component of the cytoplasm

Nuclear Envelope An organelle responsible for transmitting genetic information

Endoplasmic Reticulum or ER An organelle responsible for the production of most of a cell's protein and lipid components

Golgi Body or Golgi Apparatus An organelle responsible for modifying, storing, and shipping certain products from the ER

Transport Vesicles Organelles responsible for transporting products within the cell

Rough Endoplasmic Reticulum or RER A region of the ER responsible for manufacturing proteins

Ribosome A structure found in the RER responsible for the manufacture of proteins

Smooth Endoplasmic Reticulum or SER A region of the ER responsible for carbohydrate and lipid production

Lysosome A vesicle responsible for recycling cell components

Programmed Cell Death or Apoptosis The process by which cells program their own death

Mitochondria An organelle responsible for producing much of a cell's energy

Cytoskeleton A meshwork of protein filaments in the cytoplasm giving the cell its shape and capacity for movement

Centriole An organelle that assists the cell with reproduction

Cilia A hair-like organelle that helps move fluids over the surface of the cell

Genetic Expression A process by which the genetic information in a cell is used to produce cell structures and carry out cell functions

passage of chemicals that control genetic material function and formation. The endoplasmic reticulum, or **ER**, is an extensive network of membrane tubes derived from the nuclear membrane and connecting to the cell membrane. It is responsible for the production of the protein and lipid components of most of the cell's organelles. A region of the ER called the **rough endoplasmic reticulum** (**RER**) usually lies closest to the nuclear membrane and is responsible for manufacturing proteins. Complex structures called **ribosomes** carry out this job for the RER. Ribosomes are composed of nucleic acids and proteins. Most of the proteins made in the RER are secreted from the cell. The **smooth endoplasmic reticulum** (**SER**) has a variety of functions including carbohydrate and lipid production.

Next to the SER is a structure called the Golgi body or **Golgi apparatus**. There can be many Golgi bodies depending on the cell's function. It is responsible for modifying, storing, and shipping certain cell products from the ER. Transport vesicles move the products from the ER to the Golgi body. Cells that specialize in producing secretions usually have a large number of Golgi bodies. The Golgi body also produces vesicles that carry out specific chemical reactions. A lack of some of these vesicles is the basis of many human diseases. Another specialized vesicle called the **lysosome** contains enzymes capable of digesting the cell from inside out. These organelles recycle cell components and can be activated to cause cell death if needed. Cells can program their own death using a strategy called **programmed cell death** or **apoptosis**. Vacuoles are related to vesicles except that they are produced by the cell membrane. They are mostly for storing materials taken in by endocytosis.

Another group of organelles originates from the egg's cytoplasm. This means that every person gets these organelles from his or her mother. They contain genetic material and work in cooperation with the cell's genome. The health of a cell is monitored by information transmitted between the mitochondria and the cell's nucleus. **Mitochondria** take oxygen and simple molecules from the cell to produce much of the energy needed for cell function. Human cells can have hundreds of mitochondria. Mitochondria will take on different appearances and jobs depending on the type of cell in which they are located. The **cytoskeleton** is a meshwork of protein filaments and tubules in the cytoplasm that gives the cell shape and capacity for movement. Additionally, it coordinates the function of **centrioles**, **cilia**, and flagella. Centrioles assist the cell with reproduction. Cilia are hairlike processes on the cell membrane and are capable of rhythmic motion. This motion helps to move body fluids on the surface of the cell including the lining of mucus inside the respiratory system. Flagella are found only on sperm and give them the ability to swim in body fluids.

The nucleus is sometimes called the "brain of the cell." This interpretation is not quite accurate. The genetic material housed within the nucleus is more like an instruction manual than a brain. With a few exceptions, every cell of the body contains a nucleus carrying an identical set of genomic information. The nucleus's main role in the cell is **genetic expression**. This is a process by which the genetic material's coded information is used to produce cells structures and carry out cell physiology. Physicians use the shape and size of the nucleus as an indirect way of monitoring cell activity and health. Within the nucleus is a bundle of DNA looped around special proteins called histones. The nucleolus is responsible for building ribosomes needed for protein synthesis. Most cells have one nucleolus. However, the number of nucleoli can range from zero to several depending on the activity of the cell. A healthy cell can manufacture 10,000 ribosomes a minute.

✓ Concept Check

1. Describe the function of Golgi bodies.
2. What is the difference between cilia and flagella?
3. What is the role of mitochondria?

DISCOVERY SCENE PLEASE ENTER DISCOVERY SCENE PLEASE ENTER

What information about cell structure helps you to better understand the condition of the infant in the Case Study Investigation? Is it possible that a particular organelle is responsible for some of the characteristics of the illness?

CELL FUNCTION

Key Terms: cell cycle, metabolism, trait

The microscopic study of cell anatomy provides important clues to cell function. However, it is impossible to view the chemical reactions carried out by each organelle. Over the past 60 years scientists devised a variety of methods for analyzing the step-by-step activities of the cell. The cellular activities covered in this chapter are **metabolism**, genetics, and **cell cycle**. Metabolism is usually defined as the sum of all chemical reactions in the body that maintain homeostasis. Genetics is the study of genomic function and heredity. Genome function examines the ways genetic information is converted into cell and body characteristics. Heredity is the study of how particular qualities or **traits** of an organism are transmitted from parents to offspring. A cell cycle is the events a cell goes through to carry out daily functions and to divide.

Metabolism The sum of all chemical reactions in the body that maintain homeostasis

Cell Cycle The events a cell goes through to carry out daily functions and the steps it takes to reproduce

Trait A particular characteristic that distinguishes one person from another

Metabolism

Key Terms: acetyl coenzyme A, acetyl CoA, adenosine triphospate (ATP), aerobic respiration, anaerobic respiration, anabolism, catabolism, cellular respiration, electron transport chain, ETC, fermentation, glycolysis, Krebs cycle, lactic acid, oxidative phosphorylation, protein synthesis, pyruvic acid, tricarboxylic acid cycle, TCA, urea

The full scope of a cell's metabolism is mind boggling in its complexity. Hundreds of enzymes are working together to carry out a staggering number of chemical conversions. Many of these are daily reactions, others only occur when environmental conditions are not favorable. They help the body to compensate for unusual diets and extreme fluctuations in the environment. The metabolic reactions covered in this chapter are limited to **anabolism** and **catabolism**. Anabolism includes metabolic reactions that use cell energy and result in the production of body and cell components. Catabolism is defined as the metabolic breakdown of molecules to provide the cell with energy and raw materials for anabolism.

Anabolism A metabolic reaction that uses cell energy and results in the production of cell components

Catabolism A metabolic reaction that breaks down molecules to provide the cell with energy and materials to perform anabolism

Cellular Respiration The extraction of energy from the chemical breakdown of stored food molecules

Aerobic Respiration Cellular respiration that requires oxygen

Anaerobic Respiration or Glycolysis Cellular respiration that does not require oxygen

Fermentation A form of anaerobic respiration that produces lactic acid

Lactic Acid A byproduct of fermentation that can cause soreness if built up in the muscles

Cellular respiration is the extraction of energy for the cell using the chemical breakdown of stored food molecules (Figure 13.12). Human cells primarily carry out a type of cellular respiration called **aerobic respiration**. This type of respiration that requires oxygen to release food energy in a sequence of steps that take place in the cytoplasm and mitochondria. Muscle cells are able to carry out another type of cellular respiration called **anaerobic respiration or glycolysis**. It is defined as the oxidation of molecules to produce energy in the absence of oxygen. The oxidation reaction performed in aerobic respiration combines oxygen with food molecules to cause a chemical change in which atoms lose electrons. Anaerobic respiration in human cells is linked to another metabolic pathway called **fermentation**. Fermentation is an energy-capturing process that produces **lactic acid** wastes in muscle. A buildup of lactic acid can cause muscle soreness after heavy exercise or prolonged high-endurance activity.

Figure 3.12 **Metabolism**
Cellular respiration

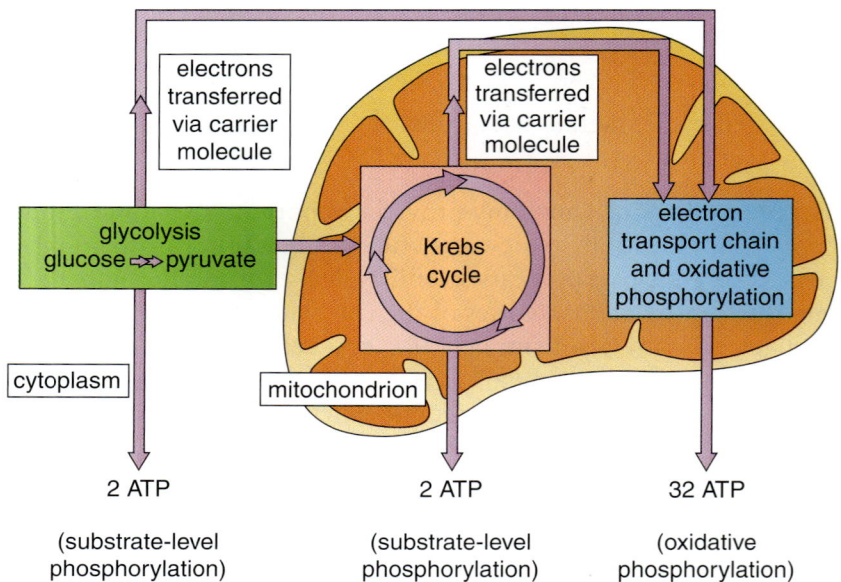

Aerobic respiration typically uses oxygen and glucose as input for energy production. The waste products are usually carbon dioxide and water. This process occurs in four distinct stages. Stages 1 and 2 are collectively called glycolysis. These reactions take place in the cytoplasm. The aerobic events take place during stages 3 and 4 in the mitochondria.

Stage 1 – Glucose is broken down to **pyruvic acid** in the cytoplasm with the release of four hydrogen ions. Some cell energy is needed to run the enzymes for this reaction.

Pyruvic Acid A product of glucose produced during glycolysis

Acetyl Coenzyme A (Acetyl CoA) A metabolic compound related to acetic acid and vinegar

Stage 2 – If oxygen is present, pyruvic acid is oxidized to **acetyl coenzyme A** (**acetyl CoA**), with the release of four more hydrogen ions and electrons. Acetyl CoA is an important metabolic compound related to acetic acid or vinegar. Carbon dioxide is also given off at this point. In the absence of oxygen the pyruvic acid is directed into the fermentation pathway where it is converted into lactic acid.

Stage 3 – Acetyl CoA enters the mitochondrion where a series of chemical reactions called the **Krebs cycle** releases 16 hydrogen ions and electrons from the molecule. This results in the conversion of acetyl CoA into carbon dioxide.

The Krebs cycle is sometimes called the **tricarboxylic acid cycle** or **TCA**. Stage 4 – The hydrogen ions and electrons produced in all the stages enter the **electron transport chain** (**ETC**) to form a high-energy molecule called **adenosine triphospate** (**ATP**). Oxygen then comes along to remove the electrons and hydrogen ions. They combine to form water. The electron transport chain produces ATP using a process called **oxidative phosphorylation**.

Other metabolic pathways exist when glucose is not available for the process. Diets high in protein bypass glycolysis. The amino acids of the protein are converted into small molecules used by the Krebs cycle. This results in the accumulation of amino functional groups that are removed from the body as **urea**. High proteins diets can dehydrate the body due to the increased water needed to eliminate the urea. High-fat diets have a complex fate. As discussed in Chapter 2, triglycerides are composed of glycerol and three fatty acid components. When metabolized, the glycerol enters glycolysis while the fatty acids are fed into the Krebs cycle. Too many calories in the diet redirects the Krebs cycle to convert any metabolized molecules into fat which is then stored in the body. Diets too low in calories force the body to convert stored molecules into energy. Carbohydrates and lipids are removed first. If these are depleted, a process called wasting occurs in which the body consumes proteins from muscles and other organs.

> **Krebs Cycle or Tricarboxylic Acid Cycle (TCA)** A series of chemical reactions whereby energy is obtained from the oxidation of certain molecules
>
> **Electron Transport Chain (ETC)** A chain of proteins on the mitochondrial membrane that transfers electrons for ATP production
>
> **Adenosine Triphosphate (ATP)** A high-energy molecule formed during aerobic respiration
>
> **Oxidative Phosphorylation** The process by which the electron transport chain produces ATP
>
> **Urea** A waste product consisting of amino functional groups

✓ Concept Check

1. Distinguish between aerobic and anaerobic respiration.
2. Describe the stages of aerobic respiration.
3. How does the body compensate for diets high in protein or in lipids?

Genetics

Key Terms: anticodon, antisense strand, chromosome, codon, gene, gene expression, genetic code, gene regulatory network, GRN, messenger RNA, mRNA, pre-mRNA, protein synthesis, regulatory DNA, sense strand, structural DNA, transfer RNA (tRNA), transcription, translation,

Protein synthesis or **gene expression** is the characteristic activity of genetic function. It is defined as the process by which cells build amino acids into proteins according to genetic information contained within that cell's genome. Many proteins build the structural features of a person while hundreds of enzymes give humans their metabolic characteristics. The **genetic code** is the basis of DNA information. It is comprised of information units called genes. A **gene** can be defined many ways. It is usually interpreted as a functional unit of heredity consisting of a segment of DNA located in a specific site of the genome. In humans, the genome is divided into 23 pairs of **chromosomes** totaling 46 chromosomes. Chromosomes are threadlike collections of genes and other DNA in the nucleus of a cell. There are two major types of code in DNA programming: **regulatory DNA** and **structural DNA**. Regulatory DNA is DNA segments and whole genes that function to regulate the expression of other genes. Structural genes carry the code for structural polypeptides and enzymes that build other structural components of a cell.

> **Protein Synthesis or Gene Expression** The process by which cells turn amino acids into proteins according to the genetic information contained within the genome
>
> **Genetic Code** The basis of DNA information, the specific order of DNA and RNA
>
> **Gene** A functional unit of heredity, one of many segments of DNA
>
> **Chromosome** A threadlike collection of genes and other DNA found in the nucleus
>
> **Regulatory DNA** Genes and DNA segments that regulate the expression of other genes
>
> **Structural DNA** Genes that carry the instructions for building structural components of a cell

ORGANIZATION OF THE BODY

Transcription The first phase of gene expression

Codon A unit of genetic code consisting of a set of three consecutive nucleotides

Messenger RNA or mRNA A nucleic acid derived from a copied segment of DNA during transcription

Translation The second stage of gene expression

Gene expression is composed of two stages. (Figure 3.13). The first stage, which takes place in the nucleus, is called **transcription**. This stage copies a particular sequence of DNA into RNA to fulfill a cell's needs. Every three sequential nucleotide bases in the DNA molecule form a "code" to match a specific amino acid, thus each "trio," or triplet, of bases is known as a **codon**. For example, the DNA code ACC programs for the UGG codon. This codon is the information for the amino acid tryptophan. You will soon see that it is the order of codons in DNA that determines the amino acid sequence in a protein. The copied segment of DNA derived through transcription forms a nucleic acid known as **mRNA**, or **messenger RNA**. The next stage of gene expression is called **translation**. It takes place on ribosomes located either in the cytoplasm or the endoplasmic reticulum. Translation is the process by which the mRNA directs the synthesis of specific proteins from amino acids.

Figure 3.13 **Genetics** DNA function

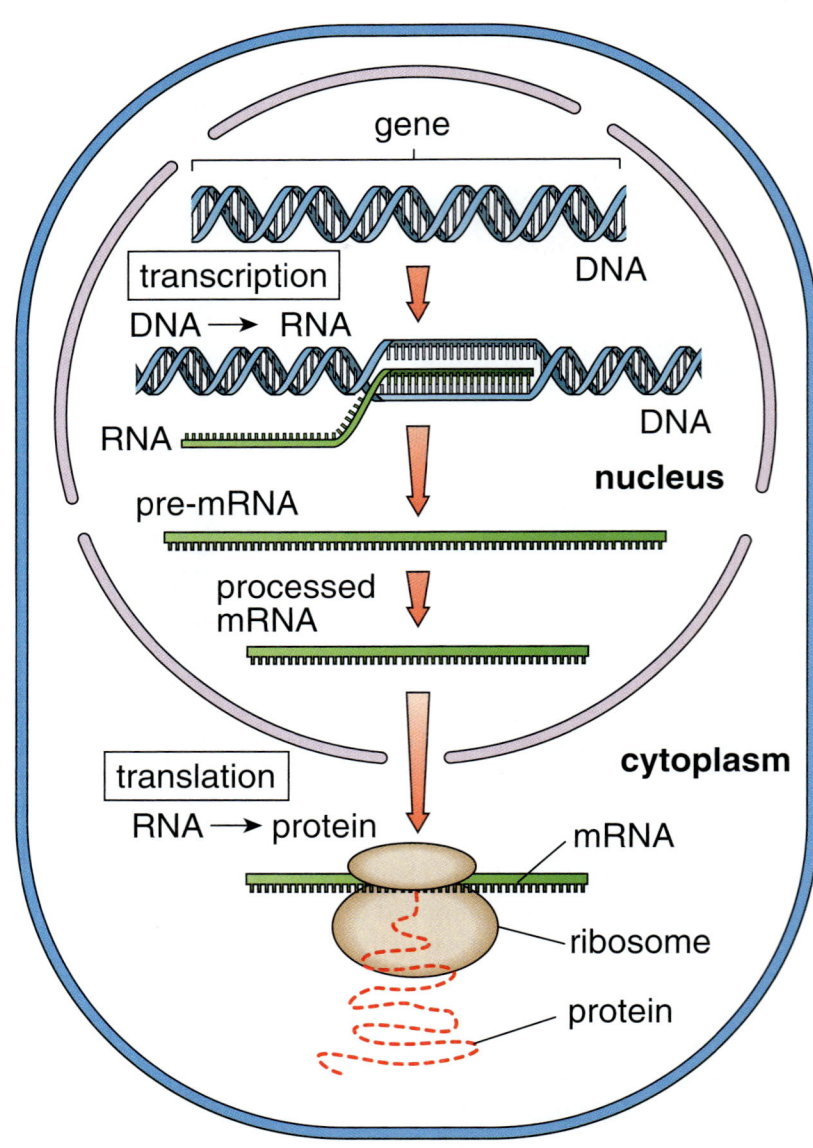

Gene expression begins when information from the environment or from within the cell communicates the need for a gene product. Information from the environment is either detected by the cell membrane or communication proteins inside the cytoplasm. Regulatory proteins or transcription factors are usually produced in response. These proteins locate pieces of DNA called **gene regulatory networks** (**GRNs**) that are the on and off switches of genes (Figure 3.14). The double helix of the DNA is unraveled to expose the genetic code, which in humans is located on only one strand of the DNA, the **sense strand**. **Antisense** refers to the strand that does not code for gene information. It carries the complementary sequence to the sense strand and serves as a blueprint for reducing genetic errors when DNA is somehow damaged.

Gene Regulatory Network or GRN Pieces of DNA that function as on and off switches for genes

Sense Strand The strand of DNA carrying the genetic code

Antisense Refers to the DNA strand that does not carry the genetic code

Figure 3.14 Gene Process
Gene regulation

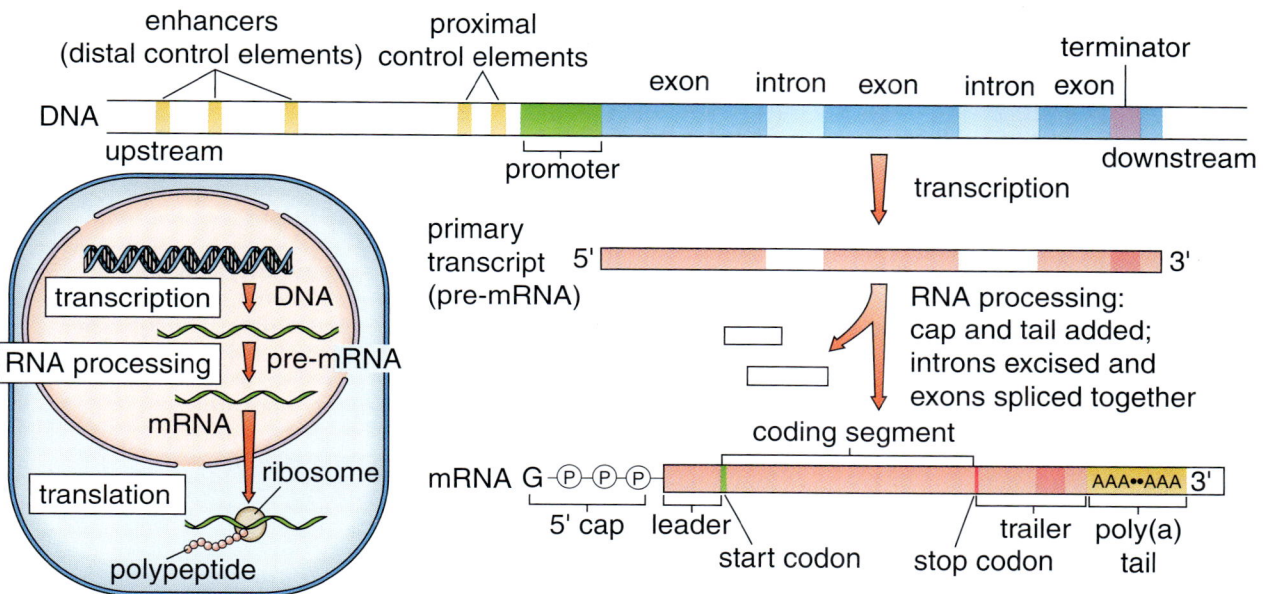

Once the sense strand is exposed, a host of proteins help carry out transcription. Transcription, as indicated above, involves the synthesis of mRNA using DNA as the blueprint. Transcribed mRNA is really in a form called **pre-mRNA**. Pre-mRNA contains genetic information called introns and exons. Introns are noncoding sequences of DNA interspersed among the protein-coding sequences in a gene. They are removed from the mRNA sequence before translations occurs. Various diseases can result from errors in this deletion process. Exons are the protein-coding DNA segments of a gene which remain following removal of introns. They are joined together while still in the nucleus to form the resulting mRNA which is then sent out across the nuclear envelope to ribosomes either in the rough endoplasmic reticulum or in the cytoplasm.

The mRNA molecule now enters the translation stage. In this stage the mRNA binds with a ribosome and a host of molecules called **transfer RNA**, or **tRNA**. Transfer RNA has structures with three nucleotide sequences that are complementary to the codon sequences of mRNA. These sequences are called

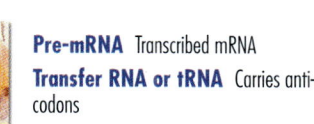

Pre-mRNA Transcribed mRNA

Transfer RNA or tRNA Carries anticodons

ORGANIZATION OF THE BODY

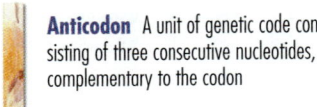

Anticodon A unit of genetic code consisting of three consecutive nucleotides, complementary to the codon

anticodons. Their job is to bond with specific amino acids and transfer them to the respective codons on the mRNA. This "matching" of codon and anticodon occurs on the ribosomes and allows the protein's amino acid sequence to be assembled according to the genetic code of the DNA. Many proteins can be made this way using one mRNA molecule. The resultant proteins are then modified and carried to particular regions of the cell. Proteins meant for secretion are synthesized in the rough endoplasm reticulum and sent to the Golgi body for packaging and transport out of the cell. Bacteria carry out protein synthesis in a manner similar to humans. Table 3.1 presents the variations in the way certain steps are conducted. Fortunately, the mechanisms of many antibiotics utilize these differences to combat bacteria without harming humans.

Table 3.1 Differences in Gene Expression between Prokaryotes and Eukaryotes

Prokaryotes	Eukaryotes
All RNA species are synthesized by a single RNA polymerase.	Three different RNA polymerases are responsible for the different classes of RNA molecules.
mRNA is translated during transcription.	mRNA is processed before transport to the cytoplasm, where it is translated. Caps and tails are added, and internal portions of the transcript are removed.
Genes are contiguous segments of SNA that are co-linear with the mRNA that is translated into a protein.	Genes are often split. They are not contiguous segments of coding sequences; the coding sequences are interrupted by intervening sequences.
mRNAs are often polycistronic.	mRNAs are monocistronic.

✓ Concept Check

1. Define the term gene expression.
2. Distinguish between regulatory DNA and structural DNA.
3. Describe the steps of transcription and translation.

Cell Cycle

Key Terms: asexual, chromatid, diploid, division, dormancy, egg, G_0, gamete, gap 1 (G_1), gap 2 (G_2), haploid, interphase, meiosis, mitogen, mitosis, S, sexual, sperm, synthesis, anaphase, centromere, cytokinesis, equatorial plane, karyokinesis, M phase, metaphase, prophase, somatic cell, spindle fiber, telophase, anaphase I, anaphase II, crossing over, homologous, meiosis I, meiosis II, metaphase I, metaphase II, prophase I, telophase I

Dormancy The suspension of active cell growth and development
Division The initiation of cell replication

In simple terms, the cell cycle is the sequence of stages that a cell passes through between one division and the next. It also includes stages of cell activity not involved in cell division. Cells can carry out one of three major jobs in the body: **dormancy**, differentiation, and **division**. Very few body cells in an

adult are in dormancy. For cells dormancy is the suspension of active growth and development. Certain stem cells are dormant cells. They carry out no cell function until called upon to divide and differentiate into functional cells. Differentiation, or the **G_0** stage, as discussed earlier, is when the cell modifies its organelles to carry out a specific job for the body. Division takes place when the cell initiates the process of replicating itself. **Asexual division** is a type of reproduction in which two new cells develop from a single cell. **Sexual** division carries out chromosome replication and cell divisions that result in the formation of cells called **gametes** or **haploid** cells. Haploid cells contain half of the chromosomes of a regular, or **diploid**, body cell. **Egg** and **sperm** are gametes.

A variety of factors control the cell cycle. Differentiation is due to chemical messages sent by the body and the cell's placement in the body. Most cells remain in G_0 as long as they are in contact with neighboring cells. Division occurs if a nearby cell dies and leaves a space. Loosely attached cell membrane proteins regulate this process. Certain complex cells such as fat, muscle, and nerve cells are not able to replicate. A loss of those cells means that there are no replacements unless nearby stem cells can generate new ones. The rate of division is controlled by environmental factors, nutrition, and chemical information from other body parts. Factors that direct a cell to undergo division are called **mitogens**. Cancer cells undergo uncontrolled cell division because they secrete mitogens that stimulate their own division. In addition, they readily divide when in contact with each other and other cells. Some bacteria produce mitogens and thereby cause abnormal cell development when they invade the body.

The cell cycle oscillates between a division stage and a stage called **interphase**. There are two types of division stages: asexual division (**mitosis**) and sexual division (**meiosis**). Interphase is the nondivisional stage of the cell cycle. It is during interphase that the DNA is replicated in the nucleus in preparation for division stages. Human cells can spend up to 20 hours in interphase. An average cell cycle in humans takes about 24 hours. Interphase is divided into a sequence of events called **G_1 (gap 1)**, **S (synthesis)**, and **G_2 (gap 2)** as shown in Figure 3.15. The cell produces the cytoplasmic components and enzymes needed for cell division while in the G_1 stage. During the S stage, the cell doubles its DNA content as an outcome of chromosome replication. This doubled DNA is called the chromosome. Each half of the doubled chromosome is called a **chromatid** and is an exact copy of the other. The G_2 phase carries out final preparations for the cell division phases.

G_0 Differentiation stage of interphase

Asexual Division A type of division where two new cells develop from a single cell

Sexual Division A type of division where gametes are created

Gamete or Haploid Cells that contain half the number of chromosomes of regular cells

Diploid Cells that contain the normal amount of chromosomes

Egg The female reproductive cell

Sperm The male reproductive cell

Mitogen A factor that directs a cell to undergo division

Mitosis Asexual division

Meiosis Sexual division

Interphase The non-divisional stage of the cell cycle

G_1 or Gap 1 The first phase of interphase

S or Synthesis The second phase of interphase

G_2 or Gap 2 The last phase of interphase

Chromatid One half of a doubled chromosome

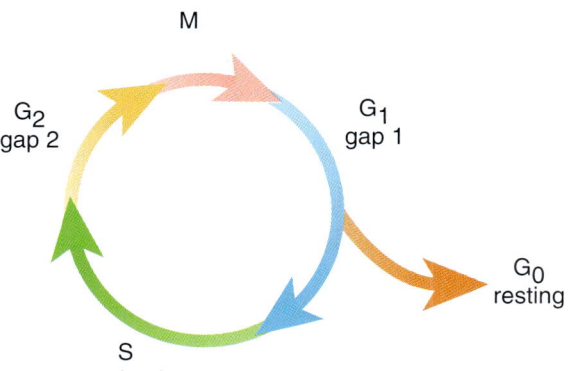

Figure 3.15 **Cell Cycle**
Eukaryote cell cycle

ORGANIZATION OF THE BODY

✓ Concept Check

1. Define the term cell cycle.
2. Distinguish between the cell division stage and interphase.
3. What are the roles of mitogens in regulating cell division?

Somatic Cells Any cells in the body other than reproductive cells

M Phase Mitosis

Prophase The first phase of mitosis

Metaphase The second phase of mitosis

Anaphase The third phase of mitosis

Telophase The last phase of mitosis

Asexual Reproduction Mitosis is the most probable type of division carried out by **somatic cells**. Somatic cells are any of the cells in the human body other than reproductive cells. Sets of genes direct the cell to undergo mitosis after the G_2 is completed. At least another 60 genes control the outcomes from interphase through mitosis. Cancers are usually caused by genetic mutations of these genes. Mutations in one of the most studied of these genes, called p53 gene, are found in a diversity of human cancer and tumor cells. Tumors are an abnormal mass of cells resulting from excessive cell division. They may be benign (not cancerous) or malignant (cancerous). As mentioned earlier, the aim of mitosis or the **M phase** is to produce two genetically and structurally similar cells from the original cell. In the simplest explanation, mitosis is divided into four stages: **prophase**, **metaphase**, **anaphase**, and **telophase** (Figure 3.16).

Figure 3.16 Asexual Reproduction – Mitosis

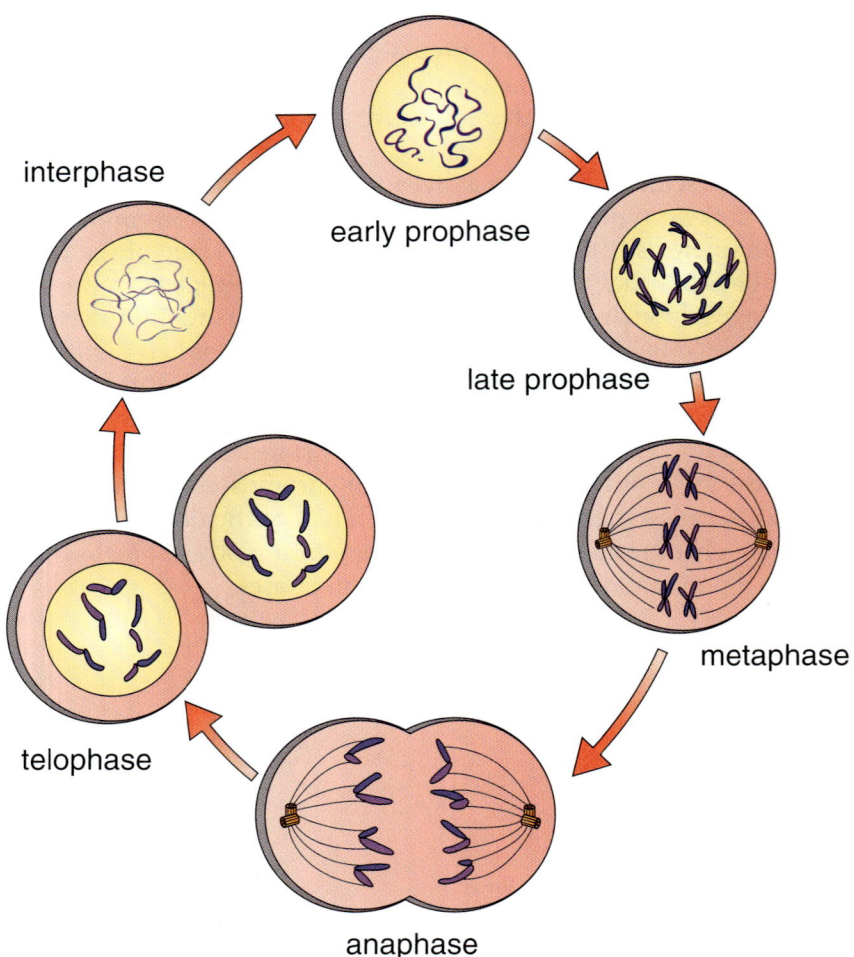

During prophase the doubled chromatids are attached to one another at a region called the **centromere**. This makes up the structure called the chromosome. The chromosome now contracts into a compact tightly coiled structure, followed by the break down of nuclear envelope. Proteins called **spindle fibers** begin to form and attach to the centrioles. The centrioles then start to separate and move apart in opposite directions. Metaphase follows prophase. In metaphase, the chromosomes are pulled into a flat line midway between the two centrioles which are now at opposite ends, or poles, of the cell. This midline is called the **equatorial plane** and represents the region where the whole cell will divides into two. The chromatids now attach the spindle fibers to the centromeres. Mitochondria are also attached to spindle fibers.

Anaphase starts to progress at the end of metaphase. During anaphase the two chromatids of each chromosome begin to separate, moving to opposite ends of the cell. They are pulled along the spindle fibers by the centromeres. This is immediately followed by telophase. In telophase, a new nuclear envelope forms around the separated DNA at each end of the cell. Now the spindle fibers disappear as the chromosomes uncoil. The separation of the DNA into different nuclei is called **karyokinesis**. The result of this process could be described as a double-nucleated cell. In order to actually produce two separate cells, a process called **cytokinesis** has to occur. Cytokinesis is the division of the cytoplasm after karyokinesis has occurred. Cells having completed these M phase stages can either re-enter dormancy, differentiate, or undergo another round of division.

> **Centromere** A region of the chromosome where doubled chromatids attach to one another
>
> **Spindle Fibers** Proteins attached to the centrioles during prohase
>
> **Equatorial Plane** The midline of the cell along which the cell will divide
>
> **Karyokinesis** The separation of the DNA into different nuclei
>
> **Cytokinesis** The division of cytoplasm after karyokinesis

✓ Concept Check

1. What is the goal of mitosis in human cells?
2. Describe the stages of mitosis.
3. What would happen to a cell if karyokinesis occurred without being followed by cytokinesis?

Sexual Reproduction The term meiosis, or reduction division, was derived from the Greek word "decrease." Scientists viewing what they thought was mitosis noticed a strange sequence in which the amount of DNA halved after two cell divisions. This type of division occurred only in gamete producing cells. Therefore, it was hypothesized that the cell division being viewed was a method of decreasing the DNA content for the formation of gametes. Meiosis consists of two stages of nuclear division, called **meiosis I** and **meiosis II** (Figure 3.17). Cells begin in the same state whether going through meiosis or mitosis. However, special mitogens turn on genes that direct the cell to undergo meiosis.

Similarly to mitosis, meiosis I is divided into four stages: **prophase I**, **metaphase I**, **anaphase I**, and **telophase I**. Prophase I is almost identical to the prophase stage of mitosis. The main difference is that during prophase I the chromosomes arrange into **homologous** pairs. Homologous chromosome pairs have the same lengths, the same centromere positions and in most cases, the same number of genes arranged in similar linear order. It is possible at this time for the maternal and paternal chromosomes to swap segments of DNA in a process called **crossing over**. In metaphase I the centrioles attach spindles to only one set of the chromosomes. The spindle fiber of one pole is attached to the maternal chromosome, while the spindle at the other pole attaches to the

> **Meiosis I** The first stage of meiosis
>
> **Meiosis II** The second stage of meiosis
>
> **Prophase I** The first phase of meiosis I
>
> **Metaphase I** The second phase of meiosis I
>
> **Anaphase I** The third phase of meiosis I
>
> **Telophase I** The last phase of meiosis I
>
> **Homologous** Referring to chromosomes that are derived from a maternal egg or paternal sperm
>
> **Crossing Over** The process of maternal and paternal chromosomes swapping DNA segments

Figure 3.17 Sexual Reproduction by Meiosis

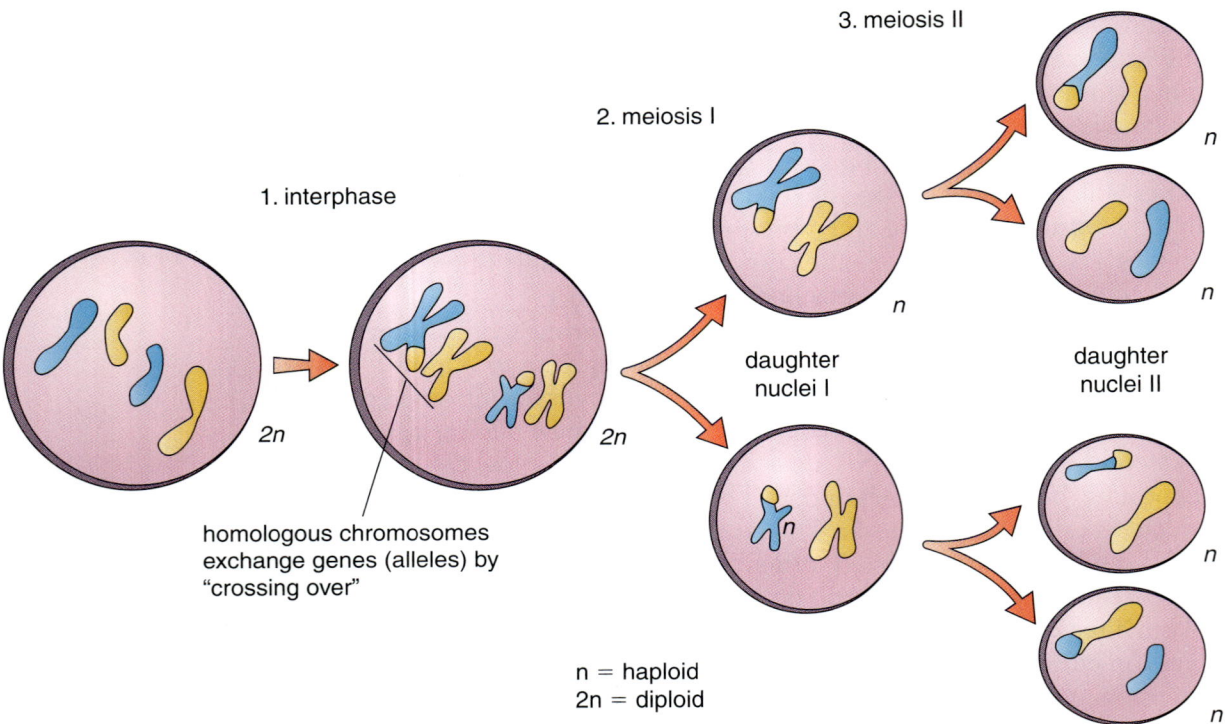

paternal chromosome. Metaphase I lines up the homologous chromosomes to ready them for separation during anaphase I. Anaphase I then separates the maternal and paternal pairs to opposite poles of the cell. The cells at the end of telophase I each have half the number of chromosomes but each chromosome consists of a pair of chromatids. Meiosis II then jumps into metaphase II and anaphase II, which line up and separate the chromatids. Metaphase II is essentially the same as mitosis in that chromatids of each chromosome are being separated. By the end of telophase II, four gametes are formed.

✓ Concept Check

1. Define reduction division.
2. Describe how mitosis differs from meiosis.
3. How does meiosis I differ from meiosis II?

DISCOVERY SCENE PLEASE ENTER DISCOVERY SCENE PLEASE ENTER

Could there be a link to a problem with the way aerobic respiration is carried out in the child? What role could genetics and cell reproduction play?

Tissues

Key Terms: connective tissue, ectoderm, embryological germ layer, endoderm, epithelium, mesoderm, muscle, nervous tissue, stem cell

All of the body's cells originate from the fertilized egg cell. The rapid mitosis involved in growing the fetus is interspersed with periods of cell differentiation. As the fetus develops, groups of cells are directed to form into functional units called tissues. The first three tissues that form are called the embryological germ layers. An outer layer called the **ectoderm** forms the skin and brain. **Mesoderm** is a middle layer that builds bone and muscle. The innermost layer is **endoderm** which produces the digestive organs. Certain cells called **stem cells** retain their ability to undergo cell division (Figure 3.18). They assist further body development and healing later in life.

Each germ layer is responsible for laying down the four human tissue types into the hierarchy that eventually forms a human. The four human tissue types are **epithelial**, **connective**, **muscle**, and **nervous tissue** (Figure 3.19). Epithelial tissue forms layers of cells that line body cavities creating coverings over external and internal body surfaces. Connective tissue forms the supporting framework of the organs and the body. Muscle is a contractile tissue that provides the body with movement. Nervous tissue is made of highly specialized cell types capable of conducting and coordinating body information.

Ectoderm The outer embryological germ layer forming the skin and brain
Mesoderm The middle embryological germ layer forming bone and muscle
Endoderm The inner embryological germ layer forming the digestive organs
Stem Cells Cells that retain their ability to undergo division
Epithelial Tissue Tissue that covers external and internal body surfaces
Connective Tissue Tissue forming the supporting framework of the organs and the body
Muscle Tissue Tissue providing the body with movement and support
Nervous Tissue Tissue that conducts and coordinates body information

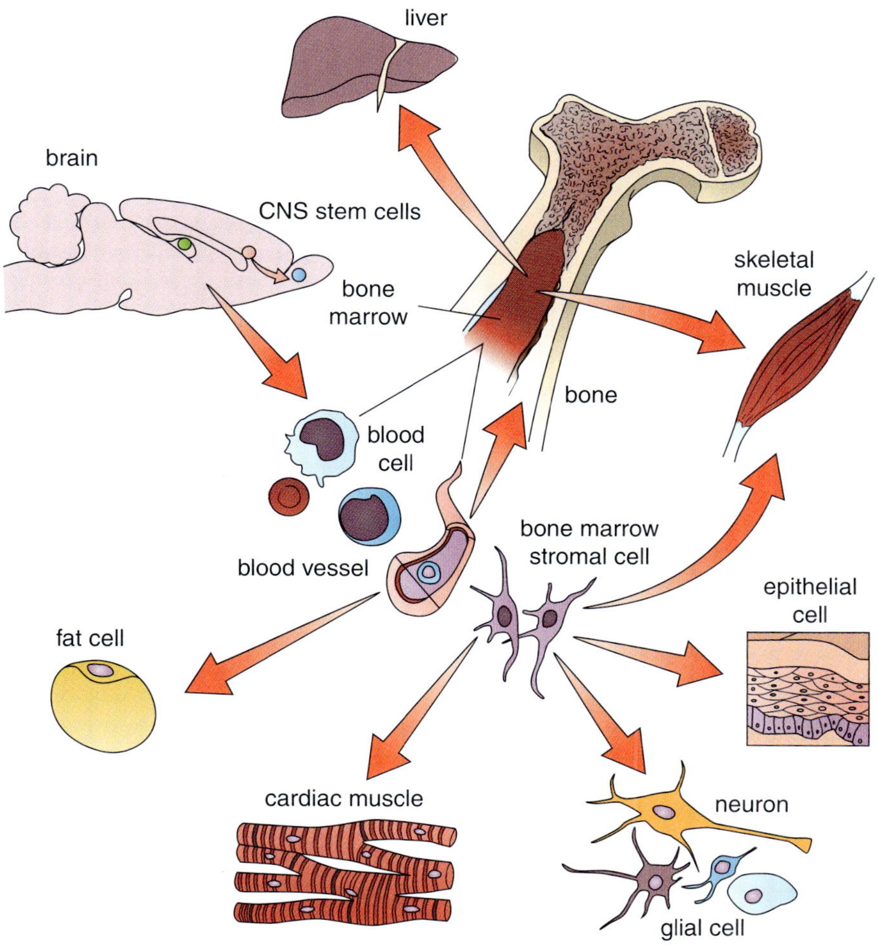

Figure 3.19 Tissue Types
Stem cells from bone can produce a variety of human tissues. Current research indicates that brain stem cells can develop in many types of cells.

Figure 3.19 **Four Types of Tissue**
The types of human tissues

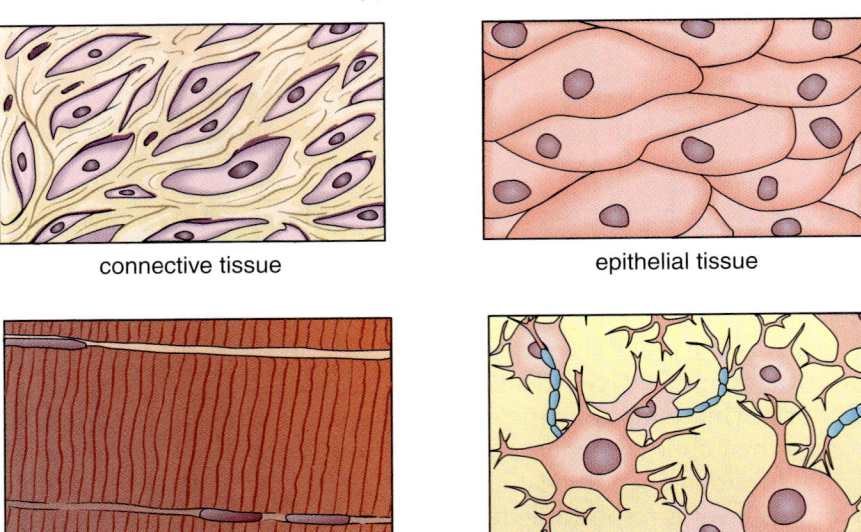

Epithelial Tissue

Key Terms: ciliated, columnar, cuboidal, pseudostratified, simple, squamous, stratified, transitional

Figure 3.20 **Epithelium**
Epithelium types

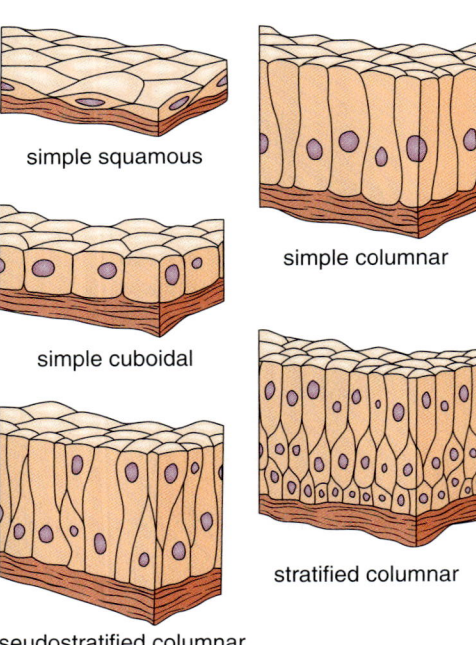

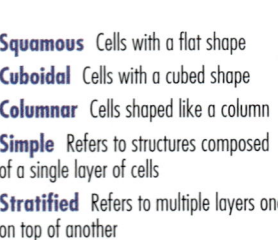

Squamous Cells with a flat shape
Cuboidal Cells with a cubed shape
Columnar Cells shaped like a column
Simple Refers to structures composed of a single layer of cells
Stratified Refers to multiple layers one on top of another
Pseudostratified Appearing to be stratified but consisting of only one layer

The cells making up the epithelium come in a variety of shapes and arrangements determined by their function in the body (Figure 3.20). Thin coverings of epithelium are commonly composed of flat or **squamous** cells. **Cuboidal** or cube-shaped epithelial cells are usually associated with structures that produce secretions. Tall column-shaped or **columnar** cells are involved in secretion and the uptake of materials into the body. Epithelial cells may be laid down in a single layer of cells or in a multi-layered stacking of cells. **Simple** epithelium refers to structures composed of a single layer of cells. **Stratified** epithelium refers to multi-layered arrangements found in areas where the cells are likely to be subjected to wear and tear. Skin is a stratified layer of squamous cells. An unusual type of arrangement occurs with **pseudostratified** epithelium as it actually has only a single layer of cells but the arrangement gives the impression that it is stratified. This type is found in the tubes leading to the lungs. **Transitional** epithelium occurs in areas where cells change shape from columnar to squamous for stretch and expansion. Some types of epithelial cells such as columnar or pseudostratified may be **ciliated**.

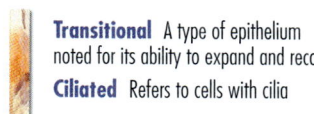

Transitional A type of epithelium noted for its ability to expand and recoil

Ciliated Refers to cells with cilia

Connective Tissue

Key Terms: collagen, dense, elastin, loose, matrix, reticulum

Matrix Intercellular material found in connective tissue

Collagen A type of connective tissue that provides strength

Elastin A type of connective tissue that provides flexibility

Reticulum A type of connective tissue that provides support

Loose Refers to connective tissue that provides attachment, stabilization, structure, and support for other tissues

Dense Refers to connective tissue that provides strength, storage, and flexibility

Connective tissue is composed of cells dispersed throughout a **matrix** of gel, liquid, protein fibers, or salts. Common proteins found in the matrix are **collagen** (for strength), **elastin** (for flexibility), and **reticulum** (for support). They are classified according to the composition of the matrix and the specialization of the cells. Connective tissue is classified as being **loose** or **dense** connective tissue (Figure 3.21). Loose connective tissue is the most abundant type found in the body. It provides attachment, stabilization, structure, and support for other tissues. Blood is a specialized type of loose connective tissue that connects the many physiological systems of the body. Dense connective tissue provides strength (bone and cartilage), storage (adipose), and flexibility (ligaments and tendons).

Figure 3.21 Connective Tissue
Connective tissue types

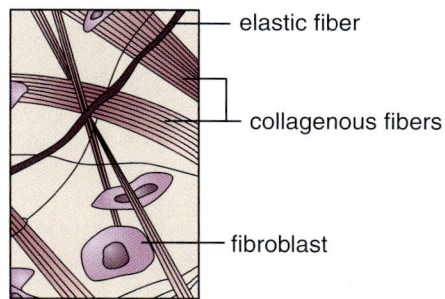

Type: loose

Common locations:
under skin, most epithelia

Function: support, elasticity

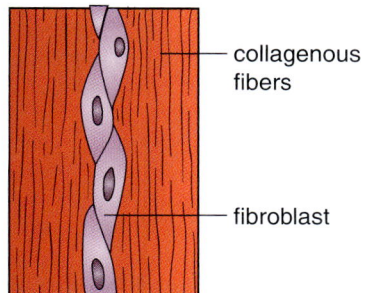

Type: dense, regular

Common locations:
tendons, skin, kidney capsule

Function: support, elasticity

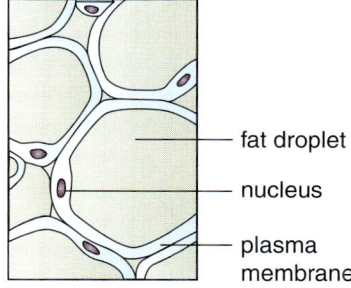

Type: adipose

Common locations:
under skin, around kidneys, heart

Function: energy reserve, insulation, padding

Muscle Tissue

Key Terms: cardiac muscle, intercalcated disks, involuntary muscle, skeletal muscle, smooth muscle, striations, voluntary muscle

Smooth Muscle Muscle found in the linings of blood vessels and tubular organs

Cardiac Muscle The muscle of the heart

Skeletal Muscle Muscle attached to bone

Muscle tissue consists of cells with cytoskeleton fibers organized into bands or bundles that contract to cause body movement. Three types of muscle tissue are found in the human body: **smooth muscle**, **cardiac muscle**, and **skeletal muscle** (Figure 3.22). Smooth muscle is made of spindle or teardrop shaped cells in which the fibers are not visible. They provide the body with

Figure 3.22 **Muscle Tissue**
Muscle tissue types

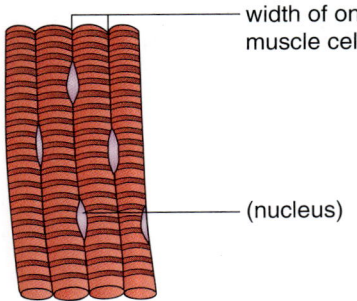

Type: skeletal muscle
Description:
   long, striated cells with multiple nuclei
Common locations: in skeletal muscles
Function: contraction for voluntary movements

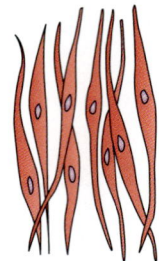

Type: smooth muscle
Description:
   long, spindle-shaped cells, each with a single nucleus
Common locations: in hollow organs (e.g., stomach)
Function: propulsion of substances along internal passageways

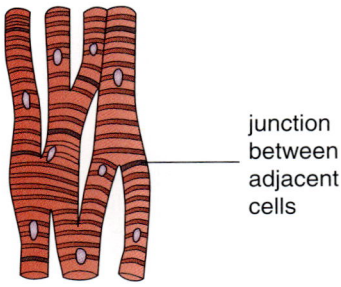

Type: cardiac muscle
Description:
   branching, striated cells fused at plasma membranes
Common locations: wall of heart
Function: pumping of blood in the circulatory system

Striations Muscle fibers that are grouped as visible bands

Intercalated Disks Structures that connect cardiac muscles cells to each other

Voluntary Muscle Muscles that are under conscious control

Involuntary Muscle Muscles that work without conscious effort

weak contractions that can last for long periods of time and are often found in the linings of blood vessels and tubular organs. Cardiac muscle makes up the heart. The fibers are lined up as visible bands called **striations**. Cardiac muscle cells are usually connected to each other by **intercalated disks**. Skeletal muscle is composed of large cells with distinct striations which provide strong directional contractions. They attach to bones and joints in a manner that produces body movement. Most skeletal muscles are under conscious control. This is why they are sometimes called **voluntary muscle**. Smooth and cardiac muscle work without conscious effort and are therefore considered to be types of **involuntary muscle**. Both skeletal and cardiac muscle are referred to as striated while smooth muscle is non-striated.

Nervous Tissue

Key Terms: astrocytes, ependymal cells, microglia, myelin, neuroglia, neurons, oligodendrites

Nervous tissue is composed of two highly specialized cell types called **neurons** and **neuroglia**. Neurons, or nerve cells, are made up of a nerve cell body and various extensions from the cell body that conducts electrical impulses from and to other nerves and muscles. As shown in Figure 3.23, there are a variety of types of nerve cells that transmit information to one or several other structures at a time. These three major types of neurons and their anatomical structure will be described in detail with the nervous system. Neuroglia are not involved in the conduction of impulses. They assist neuron function in a variety of ways (Figure 3.24). **Astrocytes** provide organization and support for the nervous system. Certain **oligodendrites** form sheaths called **myelin** around neuron fibers. The myelin speeds up the transmission of nerve impulses. **Ependymal cells** secrete fluids that protect the brain and **microglia** are believed to maintain the ion balance needed for nerve cell function.

Neurons Nerve cells that transmit impulse throughout the body
Neuroglia Nerve cells that assist neurons
Astrocyte A type of neuroglia providing organization and support for the nervous system
Oligodendrite A type of neuroglia that produces myelin
Myelin A sheath protecting neurons
Ependymal Cells A type of neuroglia that secretes fluids to protect the brain
Microglia A type of neuroglia that maintains ion balances

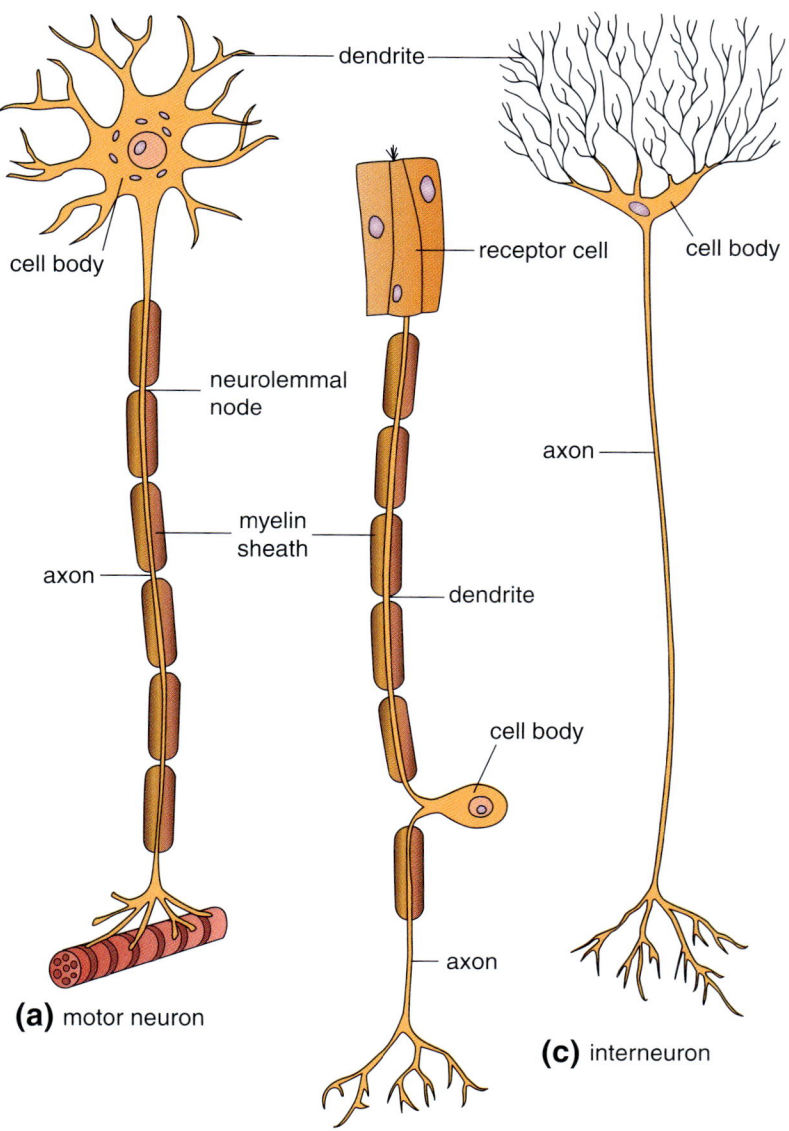

Figure 3.23 Nervous Tissue - Neurons
Types of neurons

ORGANIZATION OF THE BODY

Figure 3.24 **Nervous Tissue - Neuroglia**
Types of glial cells

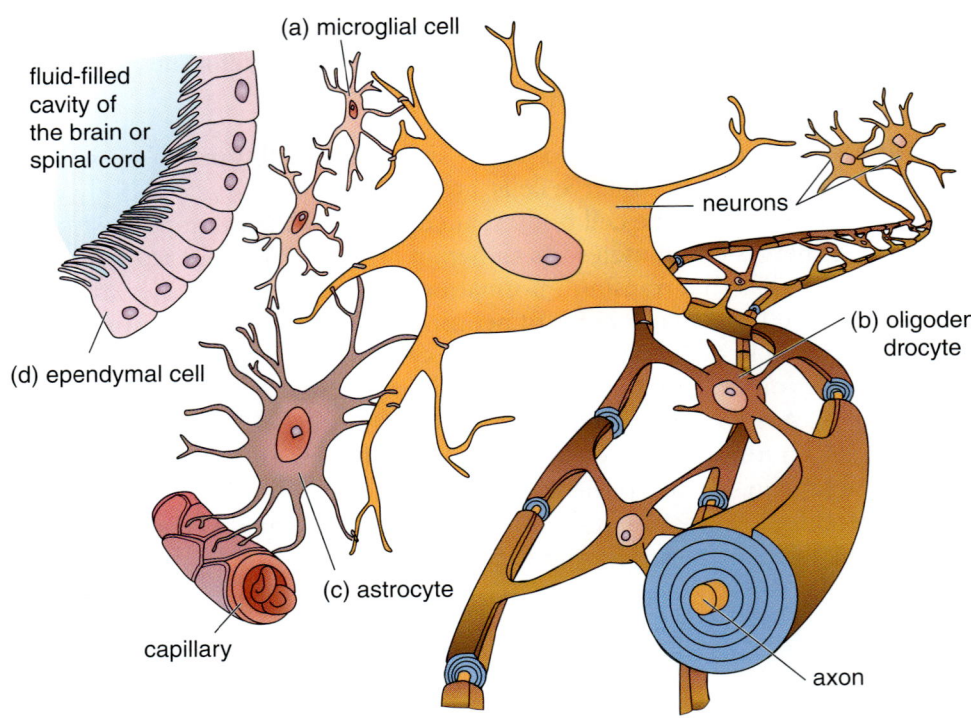

✓ Concept Check

1. Distinguish between pluripotential and totipotential stem cells.
2. Name the four tissues types and provide an example of each.
3. Describe the way that scientists classify exocrine glands.

ORGANS AND SYSTEMS

Key Terms: cardiovascular system, digestive system, endocrine system, integumentary system, lymphatic system, muscular system, nervous system, reproductive system, respiratory system, skeletal system, urinary system

Organs are defined as a group of tissues which perform some function within an organ system. The body's organ systems are made of related organs working together for a specific purpose that contributes to overall homeostasis. No one organ system works alone. They communicate with each other to ensure fully integrated responses to body demands and environmental changes. The remaining chapters of this book describe each organ system in detail. It is important to remember that the information covered in Chapters 1 through 3 is necessary for understanding the explanations of organ system function and structure. The organ systems of humans are the **cardiovascular**, **digestive**, **integumentary**, **lymphatic**, **muscular**, **nervous**, **reproductive**, **respiratory**, **skeletal**, and **urinary** systems (Figure 3.25).

Cardiovascular System Body system that regulates blood flow

Digestive System Body system that regulates nutrition**Endocrine System** Body system that regulates body function and development

Integumentary System Body system that provides protection

Lymphatic System Body system that regulates body fluids and helps fight disease

Muscular System Body system that provides structure and movement

Nervous System Body system that regulates the flow of information

Reproductive System Body system that regulates sexual function

Respiratory System Body system that regulates atmospheric gases and certain bodily wastes

Skeletal System Body system that provides support and movement

Urinary System Body system that regulates production, storage, and removal of urine

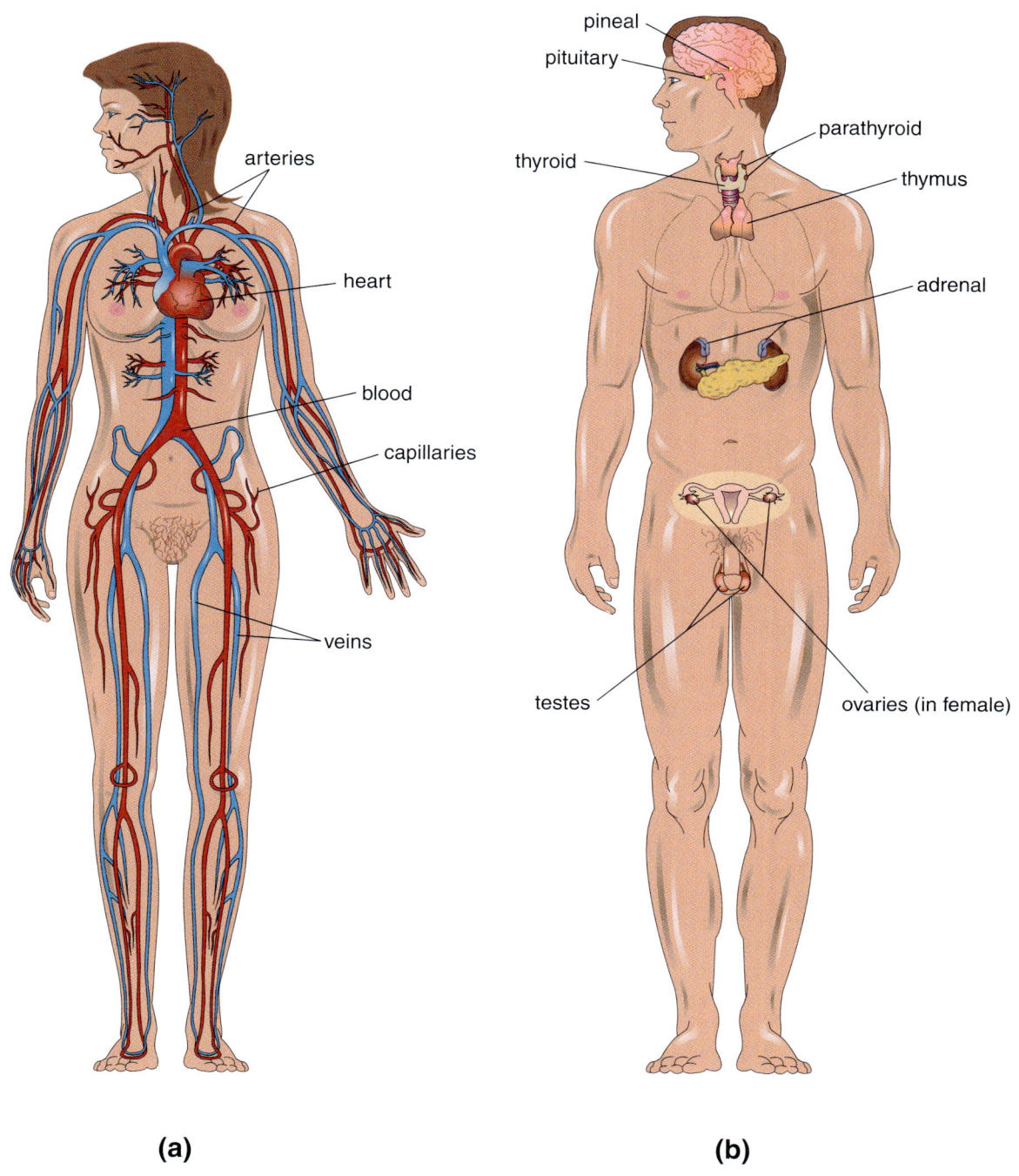

Figure 3.25 Organs and Systems
(a) Cardiovascular or Circulatory System
(b) Endocrine System

ORGANIZATION OF THE BODY

Figure 3.25 Organs and Systems
(c) Gastrointestinal or Digestive System
(d) Immune or Lymphatic System

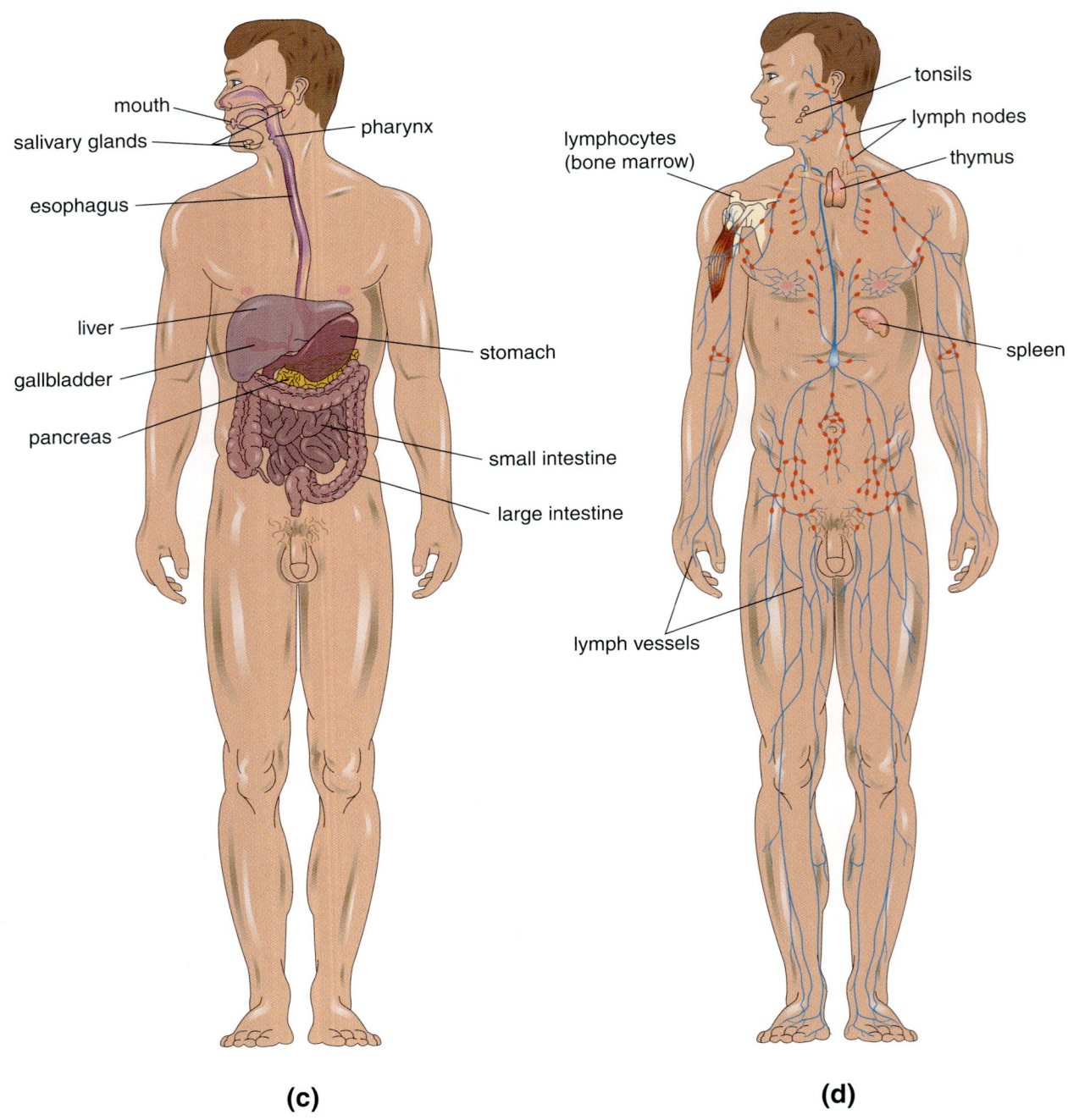

(c)

(d)

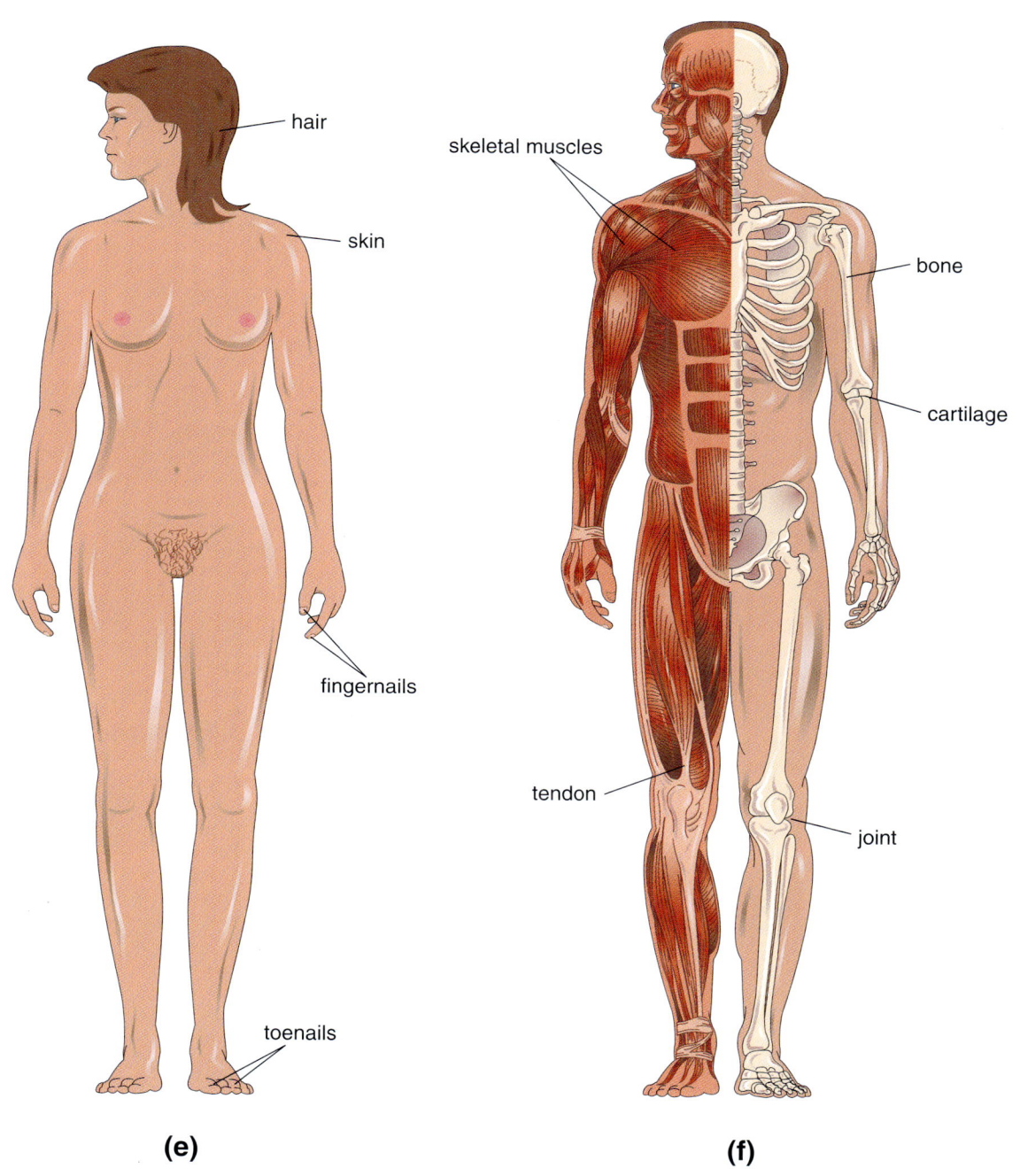

Figure 3.25 Organs and Systems
(e) Integumentary System
(f) Musculosketetal System

ORGANIZATION OF THE BODY

Figure 3.25 **Organs and Systems**
(g) Nervous System
(h) Respiratory System

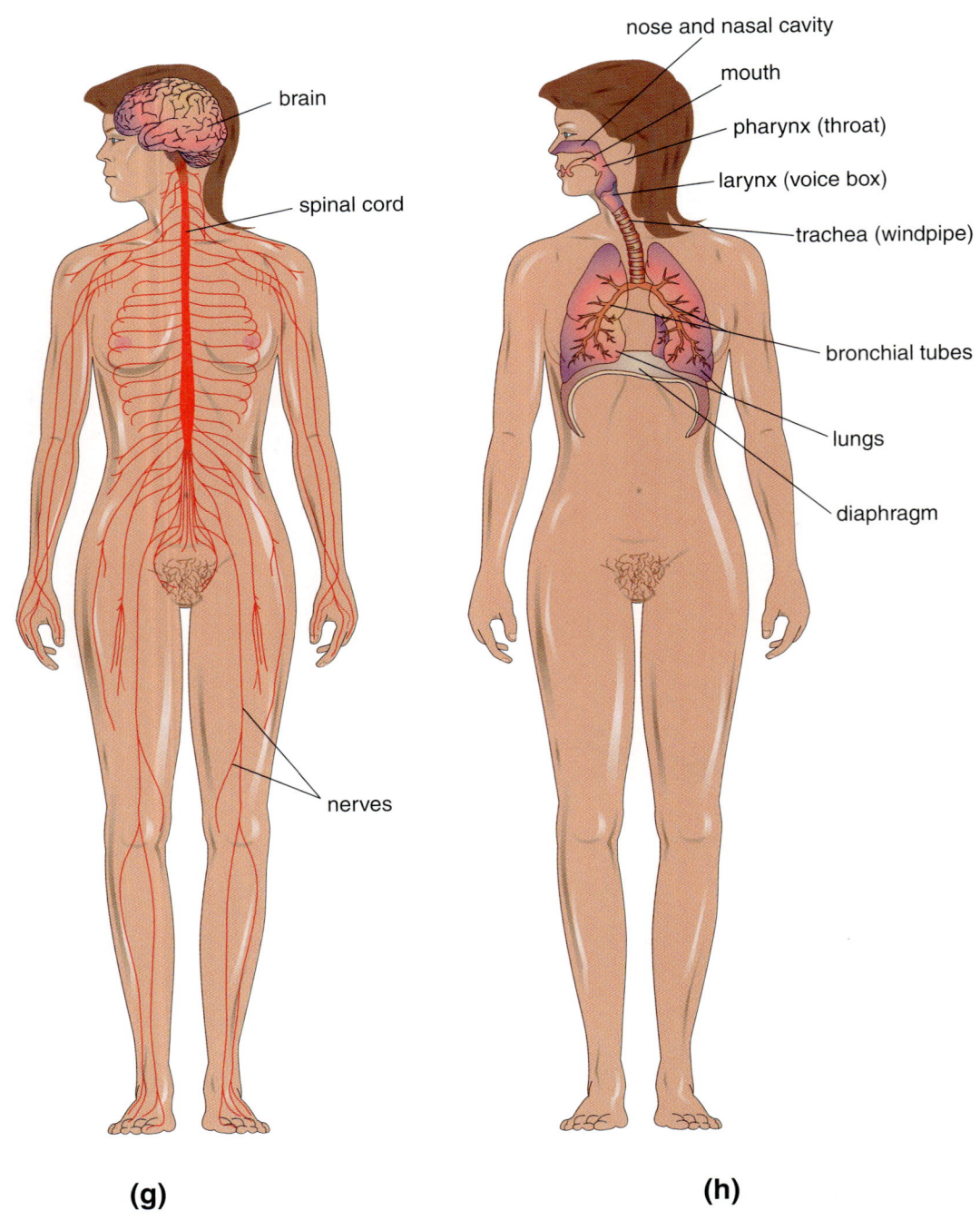

(g) (h)

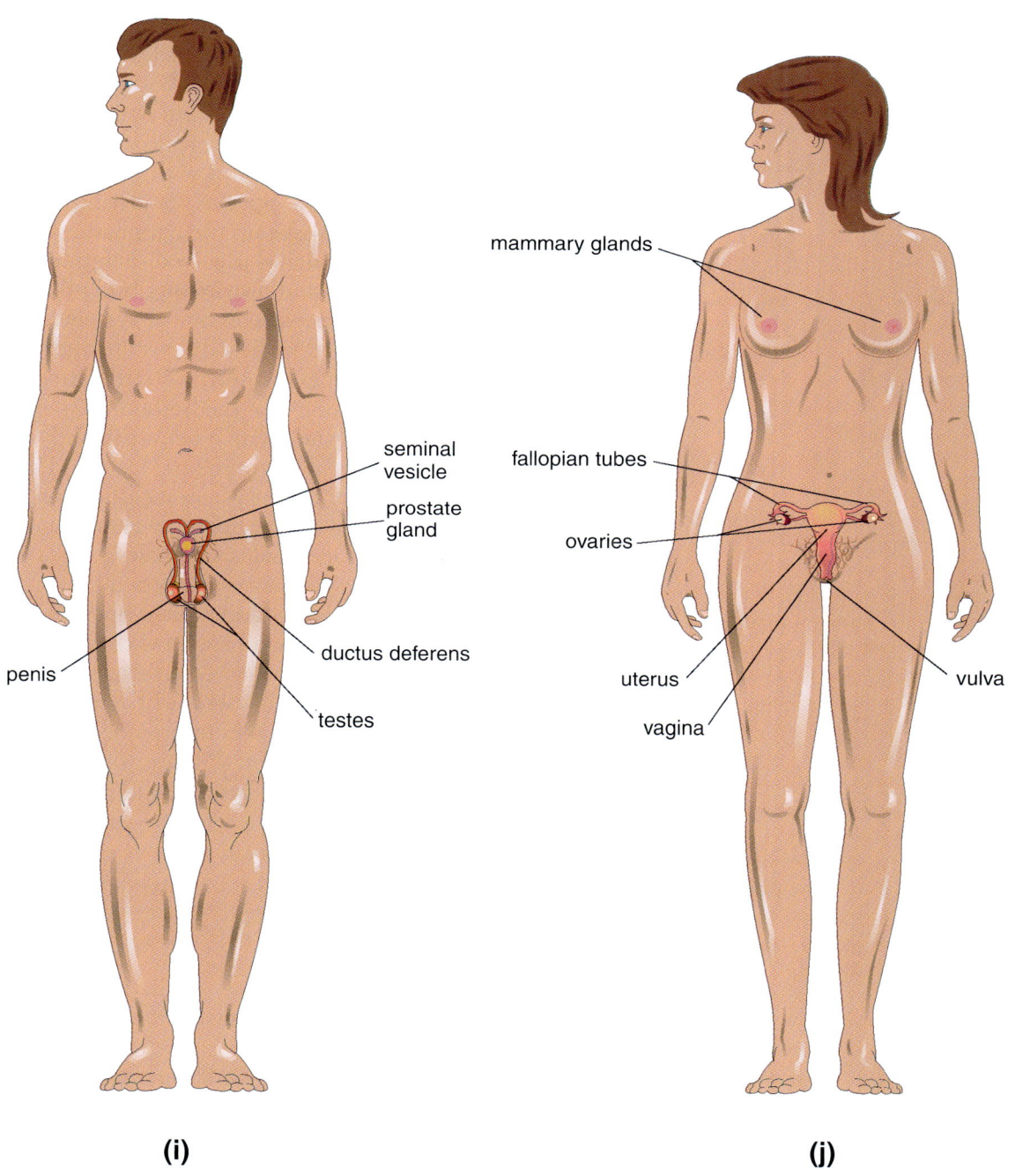

Figure 3.25 Organs and Systems
(i) Reproductive System (male)
(j) Reproductive System (female)

ORGANIZATION OF THE BODY

Figure 3.25 Organs and Systems
(k) Urinary System

- kidneys
- ureters
- urinary bladder
- urethra

(k)

The cardiovascular system distributes the blood flow needed for maintaining the proper cell environment for each organ system. Nutrients needed for each cell are provided by the activities of the digestive system. The endocrine system helps coordinate body function and plays an important role in development and aging. Covering the body is the integumentary system which provides protection for the underlying systems and helps with temperature regulation. Special vessels and glands make up the lymphatic system, which is involved in fighting disease and distributing body fluids. The extensive muscular system is involved with body posture and movement. Another coordinating component of the body is the nervous system which helps the body register environmental factors. The reproductive system includes specific organs that regulate all sexual functions in females and males. Atmospheric gases needed for homeostasis and certain bodily wastes are exchanged with the environment by the respiratory system. The skeletal system works with the muscular system to provide support and movement. Processes involved in producing, storing and removing urine from the body are carried out by the urinary system.

✓ Concept Check

1. What is the relationship between tissues and organ system function?
2. Name the human organ systems.
3. Give two general examples of how organ systems work together to maintain the body's homeostasis.

DISCOVERY SCENE PLEASE ENTER DISCOVERY SCENE PLEASE ENTER

Have you come closer to solving the Case Study Investigation? What other information have you gathered after reading the organization of body tissues? What other information about tissues and organ system function is needed to solve the case study?

Wellness and Illness over the Life Span

Pathology of Cells

Key Terms: amyloid, amyloid deposition, amyloidosis, atrophy, biopsy, dysfunction, dysplasia, dystrophy, fatty change, hyperplasia, hypertrophy, metaplasia, metastasis, and necrosis

Cell pathology is the basis of understanding **dysfunction** of the body's hierarchy. Dysfunction is defined by biologists as abnormal, impaired, or incomplete functioning of an organism, organ system, organ, tissue or cell. Almost all gross diseases are due to the dysfunction of one or more tissues in an organ system. Cell pathology is studied at the microscopic level by examining cells visually or by measuring cell metabolism using some type of chemical or genetic testing. **Biopsy** is the removal of diseased cells for study (Figure 3.26). The major types of cell pathology lead to the conditions described below.

Amyloid deposition involves the accumulation of **amyloids** in a cell. An amyloid is a proteinlike material that can collect in cells and tissues. There is great disagreement as to whether amyloids cause disease or are simply the result of a disease. Scientist do know that amyloid accumulation in a cell, called **amyloidosis**, can occur in a variety of tissues and organs. It is believed that amyloids are intended to help the cell but in turn cause harm when they build up in the cytoplasm. Amyloid proteins are important indicators of cell damage. The cell uses this information to try to correct the cell injury. Cell death is usually the result of amyloidosis. Alzheimer's disease is an example of amyloid deposition. **Atrophy** is the wasting or decrease in size of a cell, tissue, or organ. Undernutrition and blood flow problems are common causes of atrophy. Muscle atrophy occurs during nerve damage or lack of muscle use. **Dysplasia** is disorderly growth pattern in a tissue or organ. It is not cancerous but can have a significant impact on the functioning of the body structure. **Dystrophy** means "ill growth." It involves progressive changes in a tissue that is almost always due to long-term malnutrition, undernutrition, or decreased blood flow. **Fatty change** is the accumulation of lipids in the cell in response to cellular injury. Excessive alcohol intake can lead to fat accumulation in the liver cells. Fat can keep building up in the cell's organelles until it disrupts normal cell and tissue function. **Hyperplasia** is the abnormal multiplication in the number of normal cells in a tissue. Usually the normal cell arrangement and distribution of the tissue is not affected in this condition, but it

Dysfunction Abnormal, impaired, or incomplete functioning of an organism, organ system, organ, tissue, or cell

Biopsy The surgical removal of diseased cells for study

Amyloid Deposition or Amyloidosis The accumulation of amyloids in a cell

Amyloid A protein-like material

Atrophy The wasting of a cell, tissue, or organ

Dysplasia A disorderly growth pattern in a tissue or organ

Dystrophy A progressive change in tissue due to a loss of nutrition or decreased blood flow

Fatty Change The accumulation of lipids in the cell in response to cellular injury

Hyperplasia The abnormal multiplication of normal cells

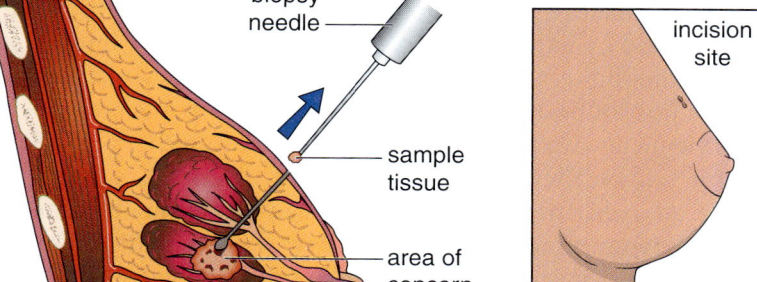

Figure 3.26 **Pathology of the Cell** Biopsy

Hypertrophy The abnormal increase in cell size

Metaplasia An abnormal change in cell and tissue function

Metastasis The movement of diseased cells away from their original location to establish themselves in a new area of the body

Necrosis Localized tissue death

increases the risks of certain cancers and distorts the function of the tissue or organ. A separate condition, called **hypertrophy**, can also distort tissue or organ function and throw off the balance of homeostasis but should not be confused with hyperplasia. Hypertrophy is the enlargement of a tissue or organ due to an increase in cell size, not cell number, as is indicative of hyperplasia.

A change in cell and tissue function from normal to abnormal is called **metaplasia**. It can be a reversible change in which one cell type is replaced by another cell type. Such substitution can produce inappropriate functioning of the tissue or organ. A variety of factors can cause this including DNA damage or exposure to certain hazardous chemicals. Some long-term metaplasias can lead to cancer due to a breakdown in cell communication and tissue organization. **Metastasis** is usually restricted to cancerous or highly abnormal cells. It describes a condition in which the diseased cells break away from the original location and establish themselves in new areas of the body. They can then carry out their abnormal functions in the new location. **Necrosis** is described as localized tissue death. This can result from blood flow decrease, burns, chemical damage, infections, injury, and a variety of other factors. Obviously, it results in diminished functioning of the tissue, organ, and organ system.

✓ Concept Check

1. What does the term dysfunction mean in respect to cell pathology?
2. What is the relationship between cell pathology and the body's structural hierarchy?
3. Describe the major types of cell pathology.

Cellular Aging

Key Terms: accumulated cell damage, cellular aging, telomere, telomere shortening

Cellular Aging The accumulation of molecular decay

Accumulated Cell Damage The sum of years of accumulated damage to a cell

Cellular aging is an accumulation of the molecular decay described in Chapter 2. As mentioned earlier in the discussion of mitosis, fat cells, skeletal muscle, and nervous tissue cannot undergo mitosis. Mitosis is not only for the replication of cells; it is also when cells carry out major cellular repairs including the correction of minor DNA damage. Cells accumulate years of damage in the cytoplasm over the lifetime of an individual. This phenomenon is called **accumulated cell damage**. Eventually these cells will fail at their tasks. This negatively affects other tissues and ultimately all of the organ systems. Damage of this type is worsened by exposure to hazardous chemicals, pollution, smoking, radiation, ultraviolet light, and viruses. Even bodily stress can cause cell aging. Stress releases a host of body chemicals that somehow contribute to cellular decay. Some chemicals can bind to the DNA, increasing the likelihood that gene expression will be altered in harmful ways. Cells that accumulate too much chemical damage can die prematurely from metabolic disturbances or can undergo programmed cell death (apoptosis). This type of cell death is a normal cellular process involving a genetically programmed series of events leading to the death of a cell without harming nearby cells.

Cells that regularly replicate accumulate a different type of damage. Their DNA can become damaged every time they pass through the S stage of mitosis. Deleterious mutations are likely to occur during the DNA replication. After several hundred rounds of mitosis, cells of the digestive, respiratory, and

CHAPTER 3

integumentary systems may perform abnormally due to the accumulation of mutations. Some of these metabolic changes lead to cancer, particularly when certain genes related to mitosis are damaged. In addition, every round of mitosis shortens the ends of the chromosomes. These ends are called **telomeres**. Telomeres do not carry any genes; however, too much **telomere shortening** can make the chromosome structure abnormal, possibly causing cells to malfunction or undergo programmed cell death. Some scientists believe that telomere shortening acts like a molecular clock for the body, tracking the time that a cell will die. Cancer cells are one of a few cells that do not go through telomere shortening; they remain immortal.

Telomere The end of a chromosome

Telomere Shortening The process of telomeres becoming shorter after each round of mitosis

✓ Concept Check

1. How is cellular aging related to molecular aging?
2. Describe how cells can accumulate cytoplasmic and DNA damage as a person gets older.
3. What is the role of telomere shortening on aging?

DISCOVERY SCENE PLEASE ENTER DISCOVERY SCENE PLEASE ENTER

Have you solved the Case Study Investigation yet? How did the information in this chapter contribute to your understanding of the infant's condition? Is the condition due the function of a particular cell or is it due to the malfunctioning of a whole tissue?

CSI – Case Study Investigation Conclusion

What did you conclude about the infant's condition? Which level of body organization was creating the pathology? What type of pathology did it appear to be? Could the dehydration and vomiting have been due to cell transport problems? Was the blue coloration and inactivity the result of metabolic errors? Is it one disease or several?

Answer:

This condition comprises a very rare group of diseases that affect the mitochondria. The disease group is called mitochondrial myopathy, encephalopathy, lactic acidosis, and stroke syndrome (MELAS). It develops when mitochondria undergo their own replication during regular cell division. Like our DNA, their genetic material can be damaged by mutations as they replicate. This condition causes few problems when passed along through somatic cells. It is deadly, however, to an infant when the condition is transmitted through the egg cell. Sperm do not contribute mitochondria to the child. So, the disease is only passed to a child by the mother. MELAS and related diseases prevent mitochondria from carrying out adequate aerobic respiration. Its greatest effect is on organs and organ systems that require an abundance of cell energy to function.

This CSI was adapted from the following articles:

1. Holt, IJ, Harding, AE, Petty, RK, Morgan-Hughes, JA. 1990. A new mitochondrial disease associated with mitochondrial DNA heteroplasty. American Journal of Human Genetics. 46(3):428-433.
2. Singh, B, Low, PS, Yeo, JF. 2004. MELAS: a case report. Annals of the Academy of Medicine - Singapore. 33(4 Suppl):69-71.

Chapter Summary

The body's organization is not understood by merely memorizing a long list of its parts. Each part is one piece of a complex organism arranged in a hierarchical fashion. The molecular level alone does not describe the body. Molecules are organized into cells and in turn cells are arranged into patterns of tissues, organs, and organ systems. Each level of complexity is dependent on the proper functioning of the levels below them. Components of each level must work in unison to maintain the overall homeostasis of the body.

Hierarchy of Human Structure
- Molecules make up the lowest level of the body hierarchy.
- Cells compose tissues, tissues form organs, organs are combined into organ systems, and the organ systems make up the human organism.
- Organisms interact with other organisms (society) and with the environment (envirome).

The Human Physiological Environment
- The human physiological environment is aqueous.
- Water is a solvent for most biological molecules or solutes.
- Ions are charged particles that dissolve in water.
- Anions carry a negative charge and cations carry a positive charge.
- Enzymes are proteins that carry out chemical reactions for the body.
- Exergonic chemical reactions give off energy.
- Endergonic chemical reactions absorb energy.
- Energy can be chemical, electrical, mechanical, or thermal.
- Forms of molecular transport are passive transport, active transport, and bulk mechanical transport.
- Endocytosis moves large materials into a cell, while exocytosis moves large materials out the cell.
- Osmosis is the diffusion of water.

Cell Structure
- Cell theory or cell doctrine explains that all life has a cell structure.
- All cells have three components: cell membrane, cytoplasm, and genome.
- Bacteria are prokaryotic cells which have no organelles or nucleus.
- Human cells are eukaryotic meaning they have organelles and contain the genome in a nucleus.

Cell Function
- Cell function is carried out by metabolism.
- ATP is a molecule that transfers energy from respiration to the cell.
- Aerobic respiration uses oxygen to break down glucose for energy.
- Aerobic respiration is composed of three states: glycolysis, Krebs cycle, and the electron transport chain.
- Anaerobic respiration is carried out by muscles as fermentation.
- A gene is the genetic information programming or DNA sequence for a characteristic.
- Gene expression involves transcription and translation.
- The cell cycle describes what a cell is doing at a particular time during its differentiation or division stages. Some cells are dormant and carry out no function.

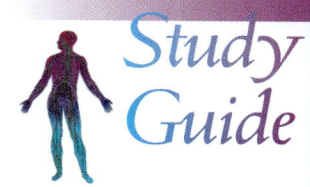

- Asexual reproduction uses mitosis to make two similar diploid cells from one cell. Somatic cells in humans are diploid.
- Sexual reproduction uses meiosis to make four haploid cells carrying one of the set of homologous chromosomes. Haploid cells in humans are gametes.

Tissues
- Tissues are made up of similar cells carrying out a specific function.
- The human body is composed of four tissue types: epithelial tissue, connective tissue, muscle tissue, nerve tissue.
- Epithelium forms coverings, linings, or glands.
- Connective tissue is composed of cells surrounded by a secreted matrix.
- Muscle is a contractile tissue. There are three muscle tissue types: smooth, cardiac, and skeletal.
- Nerve tissue comprises neurons and neuroglia.

Organs and Systems
- Tissues form into organs and organs are connected through organ systems.
- The organ systems of humans are the integumentary, skeletal, muscular, endocrine, nervous, cardiovascular, lymphatic, respiratory, digestive, urinary, and reproductive system.

Wellness and Illness over the Life Span
- Cell pathology causes hierarchy dysfunctions of the body.
- Cell aging is due to an accumulation of molecular decay.
- Mutations cause DNA damage during the S stage of interphase.
- Some aging is due to telomere shortening.

Key Terms

Introduction
Hierarchies
Levels of organization

Hierarchy of Human Structure
Cell
Cellular level of organization
Differentiation
Envirome
Molecular level of organization
Multicellular
Organ
Organ system
Organ system level of organization
Organism
Organismic level of organization

Society
Tissue
Tissue level of organization
Unicellular

The Human Physiological Environment
Environment
External environment
Internal environment
Physiological environment.

Water
Dehydrated
Overhydration
Solute
Solution

Study Guide

Solvent
Specific heat
Water excess
Water intoxication

Ions
Anion
Cation
Mineral

Enzymatic Reactions and Energy
Active site
Calorie
Catalyst
Chemical energy
Coenzyme, cofactor
Electrical energy
Endergonic
Exergonic
Hydrolysis
Kilocalorie
Kinetic energy
Mechanical energy
Potential energy
Product
Reduced
Substrate
Synthesis
Thermal energy.

Molecular Transport
Active transport
Bulk active transport
Bulk mechanical transport
Carrier protein
Diffusion channel
Diffusion gradient
Endocytosis
Excretion
Exocytosis
Facilitated diffusion
Filtration
Hyperosmotic
Hypertonic
Hypoosmotic
Hypotonic
Isoosmotic
Isotonic
Osmolarity
Osmosis
Membrane
Membrane diffusion

Passive transport
Phagocytosis
Pinocytosis
Receptor-mediated endocytosis
Secretion
Selectively permeable membrane
Voltage-gated

Cell Structure
Cell doctrine
Cell theory
Cytoplasm
Genome
Prion
Viroid
Virus

Cells of Microbes
Bacteria
Cell wall
Filamentous
Flagella
Fungi
Microbe
Microorganism
Nucleoid
Nucleus
Organelles
Prokaryote
Protista

Human Cells
Antigen
Apoptosis
Centriole
Cilia
Cytoskeleton
Cytosol
Endoplasmic reticulum (ER)
Fluid mosaic model
Genetic expression
Golgi apparatus
Golgi body
Lysosome
Mitochondria
Nuclear envelope
Programmed cell death
Receptor
Ribosome
Rough endoplasmic reticulum (RER)
Smooth endoplasmic reticulum (SER)
Transport vesicle

Study Guide

Cell Function
Cell cycle
Metabolism
Trait

Metabolism
Acetyl coenzyme A
Acetyl CoA
Adenosine triphospate (ATP)
Aerobic respiration
Anaerobic respiration
Anabolism
Catabolism
Cellular respiration
Electron transport chain (ETC)
Fermentation
Glycolysis
Krebs cycle
Lactic acid
Oxidative phosphorylation
Protein synthesis
Pyruvic acid
Tricarboxylic acid cycle (TCA)
Urea

Genetics
Anticodon
Antisense strand
Chromosome
Codon
Gene
Gene expression
Genetic code
Gene regulatory network (GRN)
Messenger RNA (mRNA)
Pre-mRNA
Protein synthesis
Regulatory DNA
Sense strand
Structural DNA
Transfer RNA (tRNA)
Transcription
Translation

Cell Cycle
Asexual
Chromatid
Diploid
Division
Dormancy
Egg
G_0
G_1 (Gap 1)
G_2 (Gap 2)
Gamete
Haploid
Interphase
Meiosis
Mitogen
Mitosis
S
Sexual
Sperm
Synthesis

Cell Cycle: Asexual Reproduction
Anaphase
Centromere
Cytokinesis
Equatorial plane
Karyokinesis
M phase
Metaphase
Prophase
Somatic cell
Spindle fiber
Telophase

Cell Cycle: Sexual Reproduction
Anaphase I
Anaphase II
Crossing over
Homologous
Meiosis I
Meiosis II
Metaphase I
Metaphase II
Prophase I
Telophase I

Tissues
Connective tissue
Ectoderm
Embryological germ layer
Endoderm
Epithelium
Mesoderm
Muscle
Nervous tissue
Stem cell

Epitheilal Tissue
Ciliated

ORGANIZATION OF THE BODY

Study Guide

Columnar
Cuboidal
Psuedostratified
Serous
Simple
Squamous
Stratified
Transitional

Connective Tissue
Collagen
Dense
Elastin
Loose
Matrix
Reticulin

Muscle Tissue
Cardiac muscle
Intercalated disk
Involuntary muscle
Skeletal muscle
Smooth muscle
Striations
Voluntary muscle

Nervous Tissue
Astrocyte
Ependymal cell
Microglia
Myelin
Neuroglia
Neuron
Oligodentrite

Organs and Systems
Cardiovascular system
Digestive system
Endocrine system
Integumentary system
Lymphatic system
Muscular system
Nervous system
Reproductive system
Respiratory system
Skeletal system
Urinary system

Wellness and Illness over the Life Span

Pathology of the Cell
Amyloid
Amyloid deposition
Amyloidosis
Atrophy
Biopsy
Dysfunction
Dysplasia
Dystrophy
Fatty change
Hyperplasia
Hypertrophy
Metaplasia
Metastasis
Necrosis

Aging at the Cellular Level
Accumulated cell damage
Cellular Aging
Telomere
Telomere shortening

Check Your Understanding

1. The ability of the body to maintain its internal environment is called:
   a. Chemotaxis
   b. Homeostasis
   c. Catabolism
   d. Enzymology

Study Guide

2. Which of the following is directly above the cell level of organization?
   a. Molecular level
   b. Organ level
   c. Tissue level
   d. Metabolic level

3. Many of the life-giving properties of water are due to this characteristic of water:
   a. It is composed of covalent bonds
   b. It forms ionic bonds
   c. It is made of low atomic weight elements
   d. It has polarity

4. Enzymes convert _____ to _____ in the act of carrying out chemical reactions
   a. product   substrate
   b. molecules   elements
   c. substrate   product
   d. excretions   secretions

5. The movement of water across the cell membrane is called:
   a. Dehydration
   b. Hydrolysis
   c. Osmosis
   d. Active transport

6. Depriving a cell of ATP would affect the following cell transport method:
   a. Diffusion
   b. Endocytosis
   c. Passive transport
   d. Osmosis

7. A scientist notices that a molecule moved into a cell from an area of low concentration of the molecule to an area of high concentration. Which transport mechanism was observed?
   a. Osmosis
   b. Membrane diffusion
   c. Facilitated diffusion
   d. Active transport

8. Which structure is common to bacterial and human cells?
   a. Nucleus
   b. Genome
   c. Cell wall
   d. Organelles

9. Nicotine slows down the movement of mucus in the respiratory system. Given this information, which organelle is most likely directly affected by nicotine?
   a. Flagella
   b. Vesicles
   c. Cilia
   d. Golgi body

Study Guide

10. A defect of SER would directly affect the following cell function:
    a. Lipid production
    b. Energy release from food
    c. Water balance
    d. Genetic expression

11. Which stage of respiration requires oxygen to operate?
    a. Glycolysis
    b. Electron transport chain
    c. Krebs cycle
    d. Anaerobic respiration

12. Sperm and eggs are produced by this type of cell division:
    a. Mitosis
    b. Meiosis
    c. Asexual reproduction
    d. Clonal selection

13. Which stage of the cell cycle occurs during interphase?
    a. Anaphase
    b. Metaphase
    c. G1
    d. Differentiation

14. Cuboidal cells are found in this type of tissue
    a. Epithelium
    b. Muscle
    c. Nerve
    d. Connective

15. Which organ system removes a bulk of the body's nitrogen wastes?
    a. Endocrine system
    b. Respiratory system
    c. Digestive system
    d. Urinary system

A Case Study

Cytoplasm

Recall that the cytoplasm takes information from the cell membrane and transmits it to the nucleus so a cell knows how to respond to the environment. In addition, the nucleus uses the cytoplasm as a place for carrying out the metabolic processes that give a cell its characteristics. Many parts of the world have serious air and water pollution problems. A large proportion of these pollutants are chemicals that affect cytoplasmic function. Public health officials and scientists are noting increasing numbers of birth defects in areas containing these pollutants. Most of the chemicals thought to be causing these problems do not directly damage the DNA. So explanations are needed to explain how an unborn child can be affected by these chemicals. Recently, the government of Mexico and the state of Texas are investigating an unusual number of birth

Study Guide

defects called neural tube defects. Some people believe that this condition is caused by pollutants that make their way into the soil and the water. Others feel that it has more to due with the diets of the Hispanic women giving birth to these children. Think about the type of information that must be collected to explain the problem.

Use the information in this chapter and the following websites to resolve the issue of toxins and birth defects. Answering the following questions will assist you:

- What is the nature of the chemicals that cause birth defects?
- How are birth defects passed along to offspring?
- Can cytoplasmic damage to an adult be passed on to an unborn child?
- Should companies or the government compensate people for exposure to chemicals that caused cytoplasmic damage?
- Who should be responsible for monitoring chemicals that effect cytoplasmic function?

1. Birth Defect Research for Children
   http://www.birthdefects.org/

2. e.Hormone
   http://e.hormone.tulane.edu/

3. March of Dimes
   http://www.marchofdimes.com/

4. Centers for Disease Control
   http://www.cdc.gov/ncbddd/

5. National Birth Defects Prevention Network
   http://www.nbdpn.org/

6. Thalidomide Victims Association of Canada
   http://www.thalidomide.ca/en/index.html

Where Do We Go from Here?

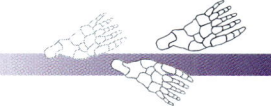

People in health fields can use their knowledge of the body's levels of structure to answer everyday problems. You may wish to use other resources, such as the following Web sites, in addition to your book to investigate the answers to each of the following situations.

1. A person who always feels tired asks you if she should buy ATP from a nutritional supplement store. What would you tell her?
   http://www.biologyinmotion.com

2. An aerobics instructor comes up to you and asks you about a drink that improves oxygen transport into the cells. What would you tell him about the probable effectiveness of the drink?
   http://www.nutriwatch.org/04Foods/ff.html

Study Guide

3. A personal trainer tells you that it is impossible to take in too much water during exercise. What would you explain to her about the effects of water on the body?
   http://www.emedicine.com/emerg/topic275.htm

4. An athletic performance-enhancing product claims to have amino acids needed to "fuel the cell cycle" needed for body growth. How would you explain the validity of this claim to a group of young athletes?
   http://www.cellsalive.com/

5. An elderly person asks you if it is possible to slow down cell aging. What would you explain to her?
   http://www.gslc.genetics.utah.edu/units/basics

Skills Activities

1 Identifying Cell Structure and Function

Materials
- Blank overhead transparency sheet
- Microscope with high power capability (400X)
- Prepared slide of skin
- Prepared slide of Trachea

Early histologists looked at microscopic preparations of body structures to determine the function of a particular tissue. They suggest a purported function by observing the composition of cell types in the specimen. But, it takes a trained eye to recognize the different cell types in a tissue or organ specimen. Much of the human body has been microscopically analyzed. Yet sometimes new structures are discovered. The following activities practice looking for cells in tissue and organ samples.

Place a slide of the skin under the microscopic. Start with the low magnification and move to the higher magnification to see details and individual cells. Can you find layers of squamous cells composing the outer portion of the skin? How do the cells in the outermost layer of skin differ from those below it? Is it possible to see a transition in tissue structure as you view inner regions of the skin? What happens to the cellular composition of the innermost layers of skin? What does this mean about the function of the innermost layers compared with that of the outer layers?

Next, look at a prepared slide of the trachea under the microscope. Can you identify cells in this specimen that were not present in the skin sample? What is the probable function of this structure based on the cell types present?

Now, look at the photographs below and answer the questions for each image.

Study Guide

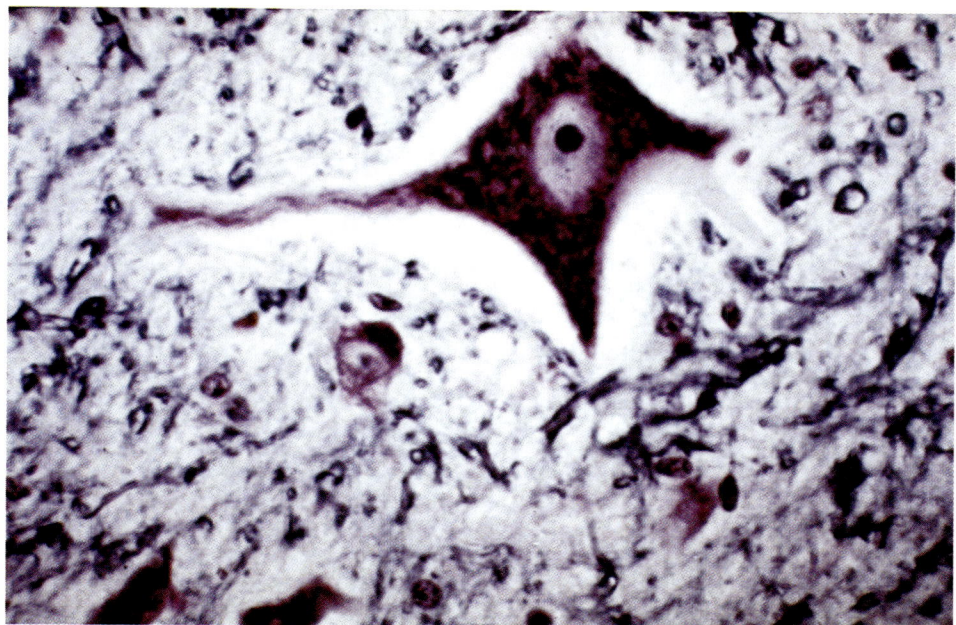

Image 1

Image 1:

What types of cells can be seen in Image 1? What is the appearance of the matrix surrounding the cells? What body structure is most likely the origin of this specimen?

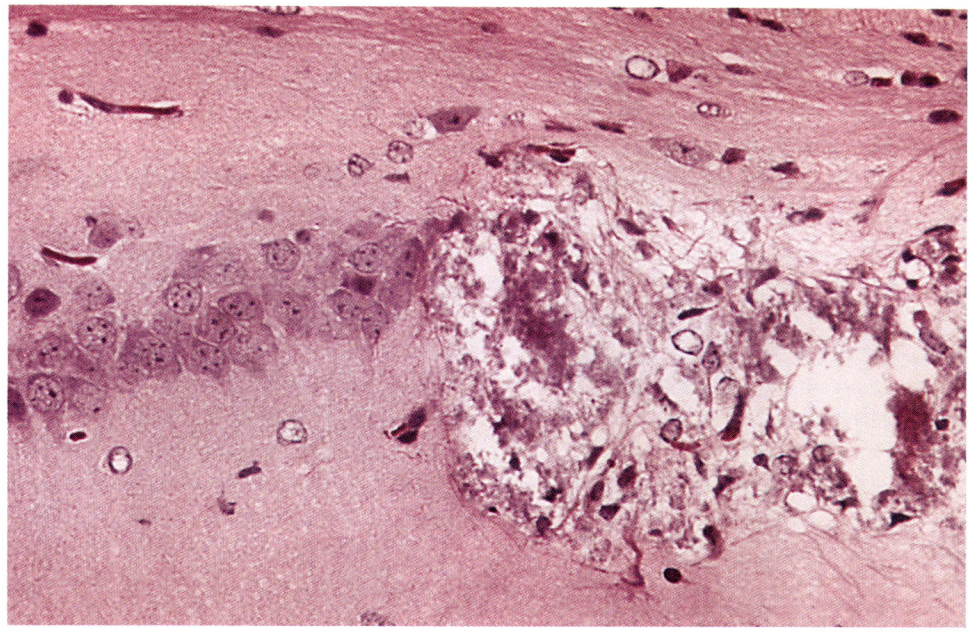

Image 2

Image 2:

How does Image 2 compare with Image 1? What are the similarities and differences? What could be the probable cause of the differences noted in the two specimens?

ORGANIZATION OF THE BODY

Study Guide

Skills Activities

2 Effects of Aspirin on Cell Function

Materials
- Elodea or American waterweed kept in a container of clean fresh water
- Forceps for removing elodea leaf
- 3 aspirin tablets ground up and soaked in 1 tablespoon of water
- 1 teabag soaked in ½ cup of cool water
- Microscope
- 2 droppers for collecting samples aspirin and tea solutions
- 2 clean microscope slides
- Surgical gloves (optional)

Aspirin and caffeine are bioactive molecules that affect cell function. Many plants produce aspirin or acetylsalicylic acid to control cellular metabolism. Caffeine is another plant substance used to regulate some aspect of the plant's metabolism. Many of these plant chemicals are produced to alter the metabolism of animals that eat the plants. Laboratory studies on cells show that both of these compounds have specific effects on metabolism. This laboratory activity looks at the effects aspirin and caffeine on a model cell. The cell model being investigated in this activity is the cytoskeleton of an aquatic plant called elodea. Plant cells can be used in certain human investigations because they have many similar metabolic processes as human cells. Usually, cytoskeleton physiology varies very little between different types of cells.

Remove two leaves of elodea and place them on separate clean microscope slides. Observe the cells of one specimen under the microscope at high power and look for the movement of small green spots. These spots are plant organelles called chloroplasts. They are attached to the cell's cytoskeleton. Plant cells use the cytoskeleton as a conveyor belt to move chloroplasts and other cell components around the cytoplasm. Make sure you keep adding water to the slides to keep them from drying out and dying. Keep track of the speed and direction at which the chloroplasts are moving around the cell.

First, add one drop of the aspirin solution to the leaf on the slide. While observing the leaf, record anything that happens to the movement of the chloroplasts. It may take one or two minutes for the aspirin to enter the cells. Add another drop or two of aspirin and see if this has any other effect. Next, look at the second elodea specimen. This time add one drop of caffeine to the slide after observing the chloroplast movement. Again, record what happens to the movement of the chloroplasts. Add another drop or two of caffeine to see if it has any different effect.

- How do aspirin and caffeine affect the cytoplasmic function you observed?
- What is the effect of adding an additional drop of aspirin or caffeine?
- How do aspirin differ from that of caffeine?
- How could these data be interpreted to understand the effects of aspirin or caffeine on human cells?
- What information is available on the Internet to help investigate the effects of aspirin and caffeine on cell function?

Precautions

It is recommended to wear surgical gloves when handing the aspirin and caffeine solutions. All of the solutions can be stored in labeled bottles in the refrigerator for two or three weeks. Excess should be flushed down a drain.

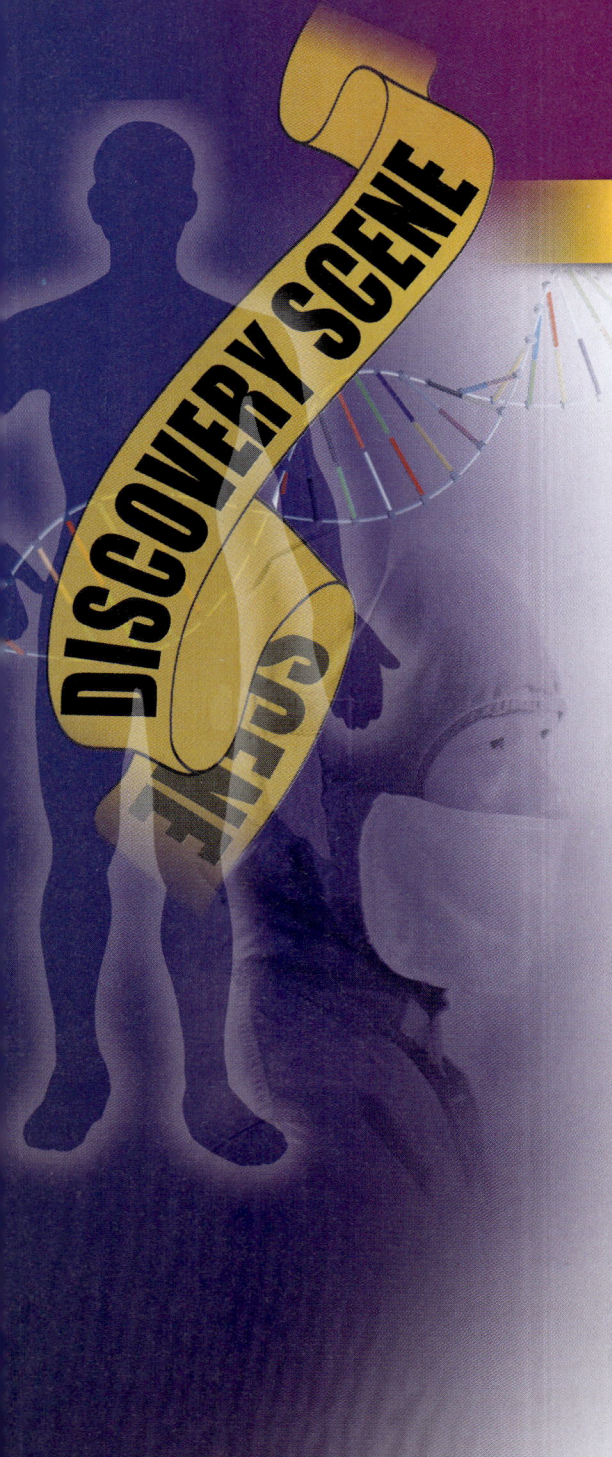

Case Study Investigation

Case Study Investigation #4

You are observing an elderly woman who is seeing a personal trainer as part of her rehabilitation therapy. The woman was hospitalized for problems associated with obesity and high blood sugar. She is now going through an exercise program to help bring her back to health. While working out, one of her socks rolled down exposing the lower part of her left leg. A glance at the side of her left leg revealed a large, shiny, deep, red sore. The sore had a dark margin, like tanned skin. Parts of it looked as if you could see right through to the muscle. The woman saw your face reacting to the sore and kindly said, "Do not worry about that, it doesn't cause me any pain." Part of the personal trainer's responsibility is to pay attention to any pathology that can be worsened by the patient's rehabilitation. How would you use your observation to assist the personal trainer in judging the possible physical limitations of this patient? What is the most likely cause of this woman's sore, and how could it affect any exercise or rehabilitation programs? At the end of the chapter, you will be asked to determine the possible problems causing this skin condition.

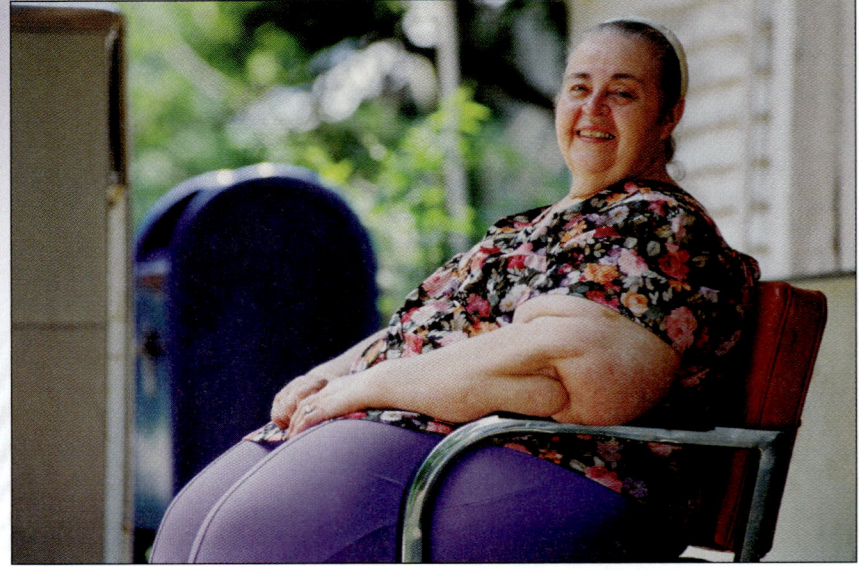

CHAPTER 4

The Skin and Its Parts

Chapter Outline

Case Study Investigation (CSI)
Applied Learning Outcomes
Overview
The Integumentary System
   Skin Structure
   Skin Appendages
Glands
   Nerves
   Nails
   Hair
Functions of the Integumentary System

Wellness and Illness over the Life Span
   Pathology of the Integumentary System
      Degenerative Skin Disorders
      Genetic Skin Disorders
      Infectious Skin Disorders
   Aging of the Integumentary System
CSI Conclusion
Study Guide

Applied Learning Outcomes

- Use the terminology associated with the integumentary system.
- Learn about skin structure, function, appendages, glands, and care.
- Understand the aging and pathology of the integumentary system.

OVERVIEW

Key Terms: adaptive, inherent, integument, integumentary system

Integument Skin

Integumentary System A system of tissues and cells associated with the skin

Inherent Refers to qualities a person or animal is born with

Adaptive Refers to the ability of an organism's genes to respond to environmental changes

After reading this chapter, it will become evident that skin, or **integument**, is more than a covering that holds the guts in place. It is not merely a body ornament as declared in the phrase "beauty is only skin deep." The **integumentary system** comprises a complex association of tissues and cells that play critical roles in maintaining the body's homeostasis. It is the largest organ system of the human body and is composed of blood vessels, connective tissue structures, glands, hair, nails, nerves, and skin. Much of the skin's structure is organized according to the genetic programming of human development. These are the skin's **inherent**, or inborn, features. For example, the genes for making nails only turn on in the upper tips of the fingers and toes. However, there is some flexibility designed into the genetics of skin cells. These are called **adaptive**, or enviromic, features.

Each organism has an integumentary system with genes that adapt the creature to its environment. This is also true for humans. Calluses on the feet are produced in response to walking barefoot on rough surfaces. The callus protects the underlying tissues and bones from damage. For many people, the skin darkens with regular exposure to the sun. This prevents the sunlight from damaging sensitive cells underneath the skin. Skin stretches as the body grows to keep it from tearing and exposing the internal organs to infection and injury. It also shrinks back, as happens after a woman gives birth.

Human skin does best under certain environmental conditions. It prefers temperatures well above freezing, but it is not comfortable with too much heat. The body temporarily compensates for extreme temperatures by adjusting its blood flow and glandular activities. The skin uses sunlight to carry out some of its functions; however, too much sunlight can cause damage, including sunburns and premature aging of the skin. There is even an optimum amount of moisture for skin to remain healthy. Too much moisture makes the skin subject to infection. Too little moisture weakens the skin's protective-barrier role.

✓ Concept Check

1. Define the integumentary system.
2. Distinguish between the adaptive and inherent factors that make different areas of the integumentary system dissimilar.
3. What role does the environment play in skin function?

THE INTEGUMENTARY SYSTEM

Key Terms: angiogenic factor, fibroblast, lanugo, melanoblast, melanocyte, mesenchyme, mucous membrane, pigmentation

Mucous Membrane Lubricated inner linings that secrete mucus

The integumentary system is a dynamic continuous body covering that includes the skin with its associated structures and the **mucous membranes**

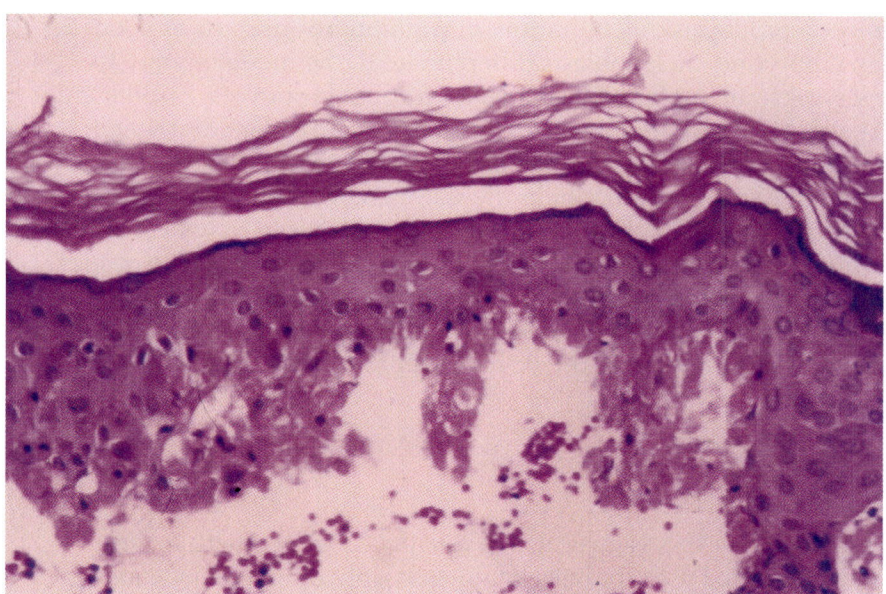

Figure 4.1 Cross-Section of the Skin

adjacent to the anus, mouth, nose, and reproductive tract. Mucous membranes are lubricated inner linings that secrete mucus. These membranes have many of the same functions as skin; however, they carry out their tasks in different ways. A microscopic image of the integumentary system (Figure 4.1) shows a layered structure that is formed through embryonic development. The integumentary system has a very complex developmental anatomy that starts 4 to 5 weeks after an egg is fertilized by the sperm. In humans, it develops from two embryological germ layers: the ectoderm and the mesoderm. The ectoderm forms the outermost layer of skin from a simple squamous tissue that becomes stratified as the embryo develops. In some areas, the ectoderm forms nervous tissue that later becomes integrated into the skin structure. This ectodermal layer does not develop any further until the inner layers of skin form.

Deeper parts of the skin start forming from the mesoderm embryonic tissue after 6 to 7 weeks. At about 8 to 9 weeks, a group of stem cells called **mesenchyme** starts maturing. Mesenchyme is an embryonic connective tissue composed of star-shaped cells in the gel-like matrix or extracellular material. Within the matrix, **fibroblasts** develop from certain mesenchyme cells. They start secreting collagen fibers that strengthen this delicate tissue. At about the same time, small dents begin forming in the upper squamous layer. These will slowly develop into body and head hair until about 24 weeks. Babies are usually born with a temporary, fine body hair called **lanugo**. At 10 weeks, small ridges start to form between the outer and inner layers of skin. These ridges ensure a large surface area of contact between these layers. This prevents separation of the layers when the skin is stretched or rubbed. Glandular structures in the skin start forming after 20 weeks and may not be functional until after birth. They are inward growths of the outermost squamous layer. These glands will eventually produce oils and sweat. Small nails at the tips of the fingers and toes start growing after 11 weeks. This is followed by the development of special blood vessels in the inner layers of skin. The growth of these vessels is initiated by secretions from the newly developing skin, called **angiogenic factors**. Angiogenesis is a term describing the formation of blood vessels.

Mesenchyme Embryonic connective tissue

Fibroblast A cell that secretes proteins that form collagen and elastin fibers

Lanugo Temporary, fine body hair found on babies

Angiogenic Factor A secretion that helps develop blood vessels

THE SKIN AND ITS PARTS

Pigmentation Skin coloration
Melanoblast A cell that will develop into a melanocyte and produce pigment
Melanocyte A cell that produces pigment

The ability to produce skin coloration, or **pigmentation**, starts at about 25 weeks when embryonic melanoblasts differentiate in the mesenchyme. In another few weeks, the **melanoblasts** mature into **melanocytes**, or pigment-producing cells. Pigments are chemicals that emit a particular color when exposed to light. These cells usually do not carry out their job until after birth. The nervous-tissue structures that help the skin perceive pain, temperature, and touch also form at this late period of development. It should now be evident that skin is not a simple structure, and many genes are involved in its proper development. There are many errors that can occur during skin development and lead to a variety of problems. These errors can be due to genetic defects or environmental factors that affect the development of particular cells.

HABITS AND HAIR LOSS

Many people accept the fact that hair loss is genetic and usually associated with males. However, the genetic condition called male-pattern baldness is only one of many causes of hair loss. In Canada, Northern Europe, and the United States, physical stress is a major cause of hair loss. Hair loss in women is particularly associated with physical stress due to childbirth, illness, injury, and pregnancy. Too much exercise can also place the body under physical stress. Hair loss is actually a slowing down of hair growth and replacement. When the body is using its energy to cope with physical stress, hair growth is placed at a lower priority for cell resources. Malnutrition and undernutrition also lead to hair loss. This is especially true for diets low in essential amino acids. Too much exercise can exacerbate the bodily effects of an inadequate diet. Another cause of hair loss includes rising hormone levels in males. Some dermatologists have discovered that certain hair styles, such as tight braids, pull off hair faster than it is able to be replaced. Excessive hair combing and styling does the same thing. Most physicians agree that hair growth remedies available without a prescription most likely have no effect on hair growth or replacement. Surprisingly, emotional stress seems to accelerate hair growth.

✓ Concept Check

1. What two embryonic cell layers contribute to the integumentary system?
2. Describe how blood vessels form in the early integumentary system.
3. What is a melanocyte, and how does it form in the skin?

Skin Structure

Key Terms: areolar connective tissue, capillary, dermal papilla, dermis, desquamation, epidermis, fascia, fasciitis, hemidesmosome, hypodermis, keratin, keratocytes, Malpighian layer, melanin, melanosome, stratum basale, stratum compactum, stratum corneum, stratum germinativum, stratum granulosum, stratum lucidum, stratum spinosum, subcutaneous layer

Epidermis The outermost layer of skin

At the time of birth, the skin matures into three distinct tissue layers that are loosely attached to underlying tissues covering bones and muscles (Figure 4.2). Making up the outermost layer of skin is the **epidermis**, which means "upon

Figure 4.2 Skin Structure

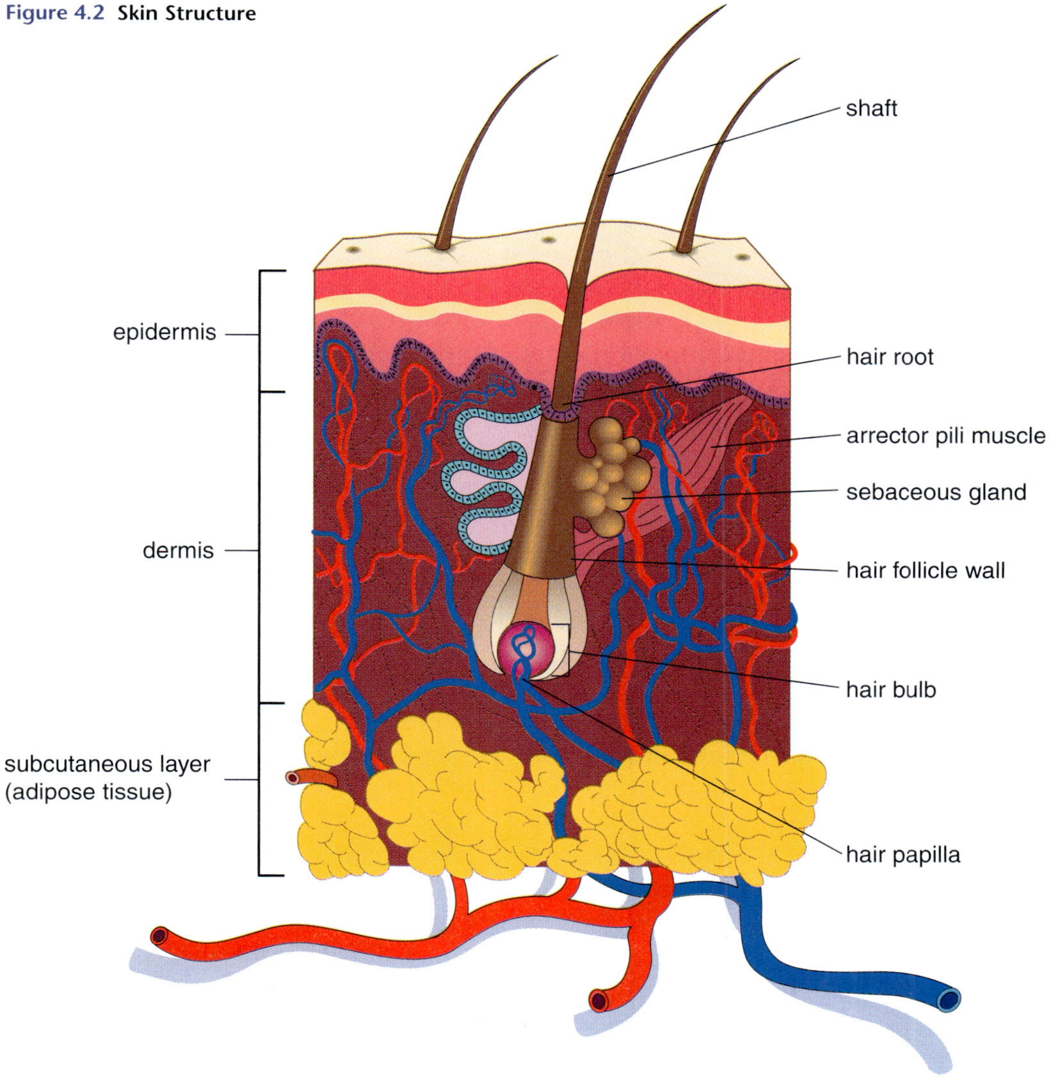

(epi) the skin (dermis)." It is composed of stratified squamous epithelium that continuously regenerates itself from the inside out. This is the layer that is wholly derived from the embryonic ectoderm layer. Underneath the epidermis lies the **dermis**. This is the "true skin" layer. The dermis sits upon the innermost layer of skin, called the **hypodermis** or **subcutaneous layer**. This layer is the last to form during development and results from the infiltration of cells derived from other body regions. Each layer of skin has its own distinct anatomical and physiological features that contribute to skin function.

Epidermis   The epidermis is composed of layers of squamous epithelium cells stacked sequentially in different levels of maturity. Older cells are on the outside, while the innermost layer is made up of metabolically active cells that are usually undergoing mitosis. This lower layer is called the **stratum basale**, or **stratum germinativum**. These terms mean the same thing. Most anatomists call the innermost layer of dividing epidermal cells the stratum basale. The stratum basale and any newly divided cells immediately above it are collectively called the stratum germinativum. The term "stratum" means layer; it is used consistently in epidermis terminology. The term basale means the innermost portion, while germinativum means the part that produces the

Dermis The middle layer of skin, formed from mesenchyme cells

Hypodermis or Subcutaneous Layer The innermost layer of skin

Stratum Basale or Stratum Germinativum The innermost layer of the epidermis

THE SKIN AND ITS PARTS

Dermal Papilla A ridged layer of the dermis that is bound tightly to the stratum germinativum

Malpighian Layer The layer of epidermis containing melanocytes

Melanin A black- or brown-colored chemical that gives color to skin

Melanosome The organelle in which melanin is produced

Stratum Spinosum The second innermost layer of the epidermis

Stratum Granulosum The middle layer of the epidermis

Keratin A yellow sulfur-rich protein that gives skin its strength

Keratocytes Cells that contain keratin

Stratum Compactum The second outermost layer of the epidermis

Stratum Lucidum A breakable layer of epidermis found in areas of thick skin

Stratum Corneum The outermost layer of the epidermis

Desquamation The process of shedding dead skin cells

Hemidesmosome A specialized junction between an epithelial cell and the basement membrane

Areolar Connective Tissue A connective tissue that binds blood vessels, membranes, muscles, nerves, and skin to other structures

---

other epidermis layers. This layer of epidermis is responsible for generating the layers above it. The cells start out cuboidal or columnar, but they become flattened as they migrate to the skin surface. It takes about 60 to 75 days for cells in this layer to reach the outermost surface. Mitosis in the stratum germinativum can occur in two planes. The horizontal plane divides the cells into an outer and inner cell. This division occurs continuously and produces new skin layers. It is stimulated by certain chemicals related to vitamin A and accelerates when exposed to continuous pressure or rubbing. The vertical plane helps the skin to grow longer. It does this in response to growth and stretching.

Cells of the stratum germinativum sit on ridges called the **dermal papillae** These fingerlike extensions of the underlying dermis are bound tightly to the epidermal cells above by a thin layer of fibrous connective tissue known as the basement membrane. Interspersed in and above the stratum germinativum are melanocytes in what is called the **Malpighian layer**. Melanocytes secrete a black- to brown-colored chemical called **melanin**. Melanin is made from an amino acid building block in special organelles called **melanosomes**. The melanin is deposited into the cells in this layer. Some people take the amino acid tyrosine as a supplement hoping to enhance melanin production for tanning; however there is no evidence supporting the effectiveness of this practice. Cells of the stratum germinativum move up into the next layer of epidermis called the **stratum spinosum**, or prickly layer. These cells develop many desmosomes on their outer surface, which give them their characteristic spiny appearance when they shrink during tissue-slide preparation. Mixed in this layer are immune system cells called Langerhans cells. They are important in fighting skin infections and healing injured skin.

As the stratum spinosum is pushed up, it develops into the next epidermal layer, called the **stratum granulosum**. This layer gets its name from the fact that the cells have granules within the cytoplasm. The granules reflect an accumulation of yellowish sulfur-rich protein, called **keratin**, in the cells, which can now be called **keratocytes**. Sulfur is what gives the characteristic smell to burning fur, hair, and leather. Keratin is a tough protein that gives the skin its strength. The keratocytes also produce a glycolipid substance that acts as a waterproofing material that prevents the free flow of water into and out of the skin. The cells of the stratum granulosum eventually flatten into a single layer of waterproof cells called the **stratum compactum**. This layer is a sheet of dying cells that are filled with keratin. In areas of thick skin, the stratum compactum forms a breakable layer called the **stratum lucidum**. It is found on the palms of the hands and soles of the feet. The uppermost layer of skin is the **stratum corneum**. It is composed of flattened dead cells that regularly shed in a process called **desquamation**.

Dermis Dermis is a thick layer of connective tissue attached to the stratum germinativum by **hemidesmosomes** along the dermal papilla. Making up a bulk of the dermis is a dense, irregular connective tissue, and also contains loose connective tissue called **areolar connective tissue**. Areolar connective tissue is used extensively throughout the body for binding blood vessels, membranes, muscles, nerves, and skin to other structures. The tissue consists of an extensive meshwork of protein fibers secreted by fibroblasts. Fibroblasts are involved primarily in body growth, connective tissue maintenance, and wound healing. For years, researchers have been able to grow fibroblasts in cell cultures to produce proteins and artificial tissues for transplantation.

Civic Responsibility

CIVIC RESPONSIBILITY: HELPING OTHERS WITH YOUR KNOWLEDGE

It is valuable to use what you have learned about the integumentary system to help others to better understand the world around them. It is very important to check your facts and seek further information about certain topics before discussing health and science issues. Here are some suggestions to foster a better public awareness of the integumentary system:

1. Speak to schoolchildren about the benefits of skin care.
2. Speak to a civic group about the effects on the skin of working or doing other activities outdoors.
3. Help the elderly to better understand the limitations of antiaging skin creams and therapies purported to make you look younger.
4. Volunteer at a school health day to teach children about the structure and function of their skin.

Subcutaneous Layer The subcutaneous layer makes up the third and innermost layer of the skin. Its thickness and composition vary throughout the body and from one person to another. The subcutaneous layer is absent or very thin in the eyelids, penis, scrotum, and nipples. After puberty, the subcutaneous layer is responsible for most of the increase in size of female breasts and hips. It differs from the dermis in many ways. First, it is primarily composed of loosely arranged elastic fibers that anchor the skin to the underlying **fascia**. Fascia is a sheet of fibrous tissue that covers muscles, skull bones, and some organs. In some areas of the body, this layer is directly attached to bone and joint surfaces. A second distinction is the large amount of adipose tissue, or fat cells. This adipose tissue is often a subject of interest to people involved in areas of human health. The thickness of the adipose tissue is measured as an indirect indicator of body fat (Figure 4.3). In addition, the subcutaneous layer has a wealth of large blood vessels that branch out in smaller networks of **capillaries** upward into the dermis. Another vessel system, called the lymphatic system, is found alongside the blood vessels. You will learn later in the book how the lymphatic system regulates body fluids, and facilitates immune responses to disease and injury. The immune system presence does not fully protect the subcutaneous layer from infection. Some microbes can enter and irritate the subcutaneous layer causing a disease called **fasciitis**. Fasciitis can also result from injuries that leave fragments of bone or other material in skin that is subject to pressure and stretching. Deep nerves are also located in the subcutaneous layer.

Fascia Fibrous tissue covering muscles, the skull, and some organs
Capillary A narrow thin-walled blood vessel
Fasciitis Inflammation of the fascia

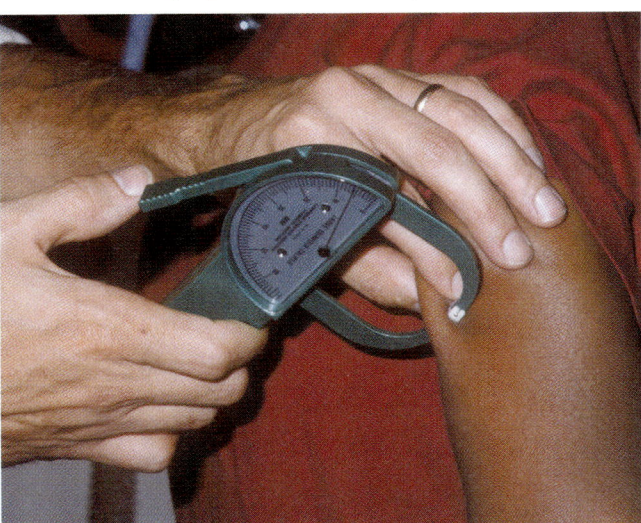

Figure 4.3 **Skinfold Calipers**

✓ Concept Check

1. Describe the three layers making up the skin.
2. What are the functions of the fibers making up the dermis?
3. Describe the structure of the subcutaneous layer.

Skin Appendages

Key Terms: apocrine, apocrine sweat gland, arrector pili muscle, cerumen, ceruminous gland, cuticle, eccrine sweat gland, free nerve ending, hair bulb, hair cortex, hair cuticle, hair cycle, hair follicle, hair medulla, hair papilla, hair shaft, holocrine gland, Krause end bulb, lunula, Meissner's corpuscle, Merkel cell, nail body, nail matrix, nail plate, nail root, Pacinian corpuscle, pheromone, pubic hair, Ruffini receptor, sebaceous gland, sebum, sensory receptor, skin appendages, skin-nail fold, sweat gland, tactile corpuscle, terminal hair, vellus hair

Skin Appendages Complex structures that assist skin function

Skin appendages are any complex structures that assist the skin with its functions. Some of these appendages are expelled cells of the epidermis, while others are modifications of the epidermis. Most of the appendages found in the dermis and subcutaneous layer develop from special cells that migrate to those layers during embryological development. These appendages include glands, nerves, nails, and hair. Figure 4.4 shows the locations of these structures.

Figure 4.4 Skinfold Appendages

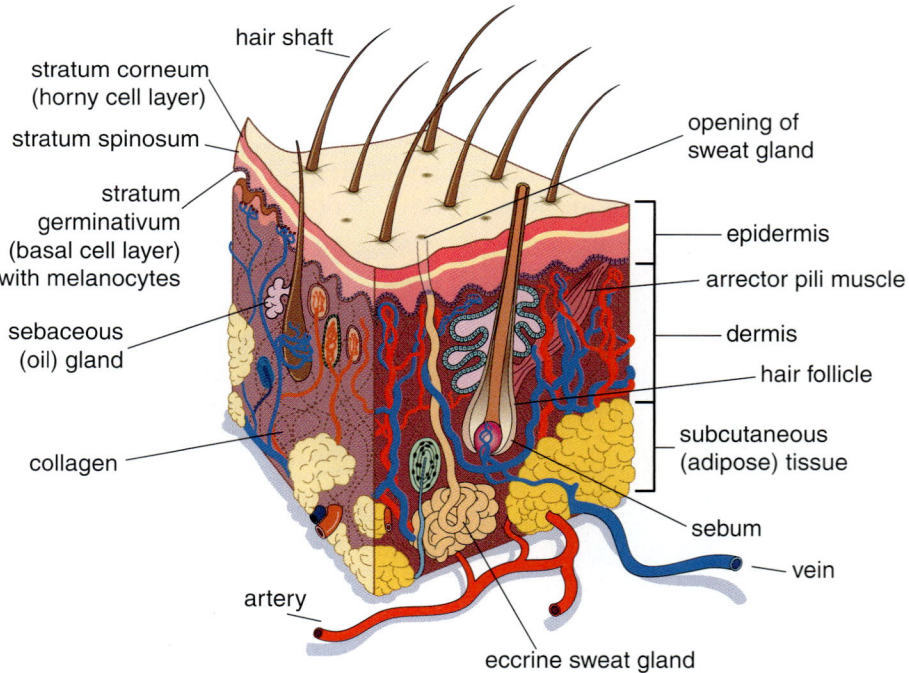

Glands Most physicians identify three types of glands commonly found in the skin: **ceruminous glands**, **sebaceous glands**, and **sweat glands** (Figure 4.5). Ceruminous glands produce a waxy secretion called **cerumen**. These large glands are found in the thin skin lining the ear canal, and their openings are usually surrounded by small hairs. They produce an **apocrine** secretion called ear wax. Sebaceous glands are **holocrine glands**, meaning that they secrete whole dead cells. The cells secreted by sebaceous glands produce and store an abundance of fat along with their membranous organelles. Once these cells are secreted in the gland ducts they burst open and die. This releases the fats onto the surface of the skin in an oily secretion called **sebum**. Sebaceous glands secrete sebum into **hair follicles** or **hair bulb**, (described below). The sebum then moves to the surface of the skin.

Ceruminous Gland The large gland found in the skin lining the ear canal]

Sebaceous Gland A gland that secretes sebum into the hair follicle

Sweat Gland A gland that produces sweat

Cerumen A waxy secretion produced by the ceruminous gland

Apocrine Refers to secretions consisting of pinched-off cytoplasm, including some plasma membrane

Holocrine Gland Secretions composed of dead cells that swell with fat and then rupture

Sebum An oily secretion produced by the sebaceous glands

Hair Follicle or Hair Bulb An inward protrusion of the epidermis

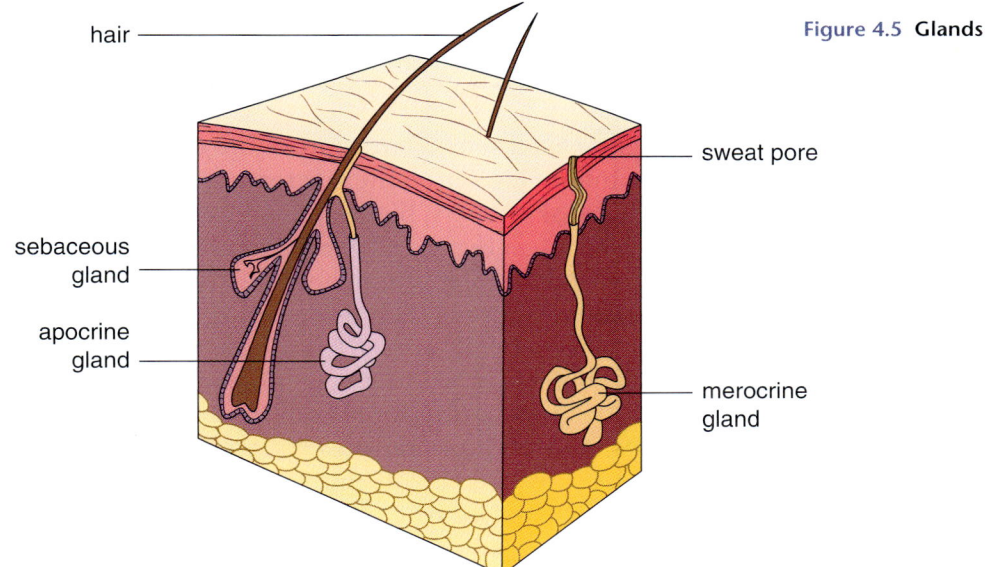

Figure 4.5 Glands

Sweat glands are categorized into **apocrine sweat glands** and **eccrine sweat glands**. Apocrine sweat glands secrete an odorous, sweatlike material into the hair follicles of the armpits, navel, groin region, and areolae. They are inactive until puberty and usually taper off in activity in elderly people. The secretion contains chemicals called **pheromones**, believed to play a role in courtship and social behavior. Dogs are notorious for their embarrassing habit of sniffing these secretions on people. Another problem that these secretions pose is that they are readily broken down by bacteria, which creates body odors. Eccrine sweat glands are found mostly on the skin of the armpits, forehead, palms, and soles. There are considerable differences in how much one individual sweats in these body areas compared with another individual, as eccrine sweat gland activity, concentration, and distribution are genetically determined. Human eccrine sweat is composed primarily of water with various concentrations of salts, organic compounds, and wastes, including urea. Many of the chemicals causing food odors leak out of the body through eccrine sweat. Microbes also feed off the nutrients in eccrine sweat, again contributing to body odors. Antiperspirants can be used to reduce the amount of sweat produced by both of these glands, thereby reducing body odors and dampness. It is an interesting note that most antiperspirants work due to an electrical repulsion between a negatively charged ingredient and the negatively charged sweat molecules.

Apocrine Sweat Glands Sweat glands that secrete sweat into hair follicles

Eccrine Sweat Glands Sweat glands that secrete mostly water and salts

Pheromone A chemical secreted by apocrine sweat glands, believed to play a role in courtship and social behavior

THE SKIN AND ITS PARTS

Good Choice Bad Choice

A pregnant friend starts developing a severe case of acne due to the pregnancy. She goes online and gets a prescription for a drug called isotretinoin, commonly sold as Accutane. It is chemically related to vitamin A and is taken as a pill for 15 to 20 weeks to rid one of acne. At the same time, she sees her physician to request a prescription for vitamin treatments as a precaution for her pregnancy. What problems can arise from taking both the isotretinoin and prenatal vitamin treatments? What are the implications of not seeking a physician's advice for self-treatment of what appear to be mild medical conditions?

Sensory Receptor A nerve cell found in all skin layers

Free Nerve Endings Pain-sensing nerves found in the lower part of the epidermis

Merkel Cells Nerves sensitive to gentle physical sensations

Tactile or Meissner's Corpuscles Nerves found in the upper region of the dermis that responds to pressure

Nerves Specialized nerve cells, called **sensory receptors**, are critical for the skin's ability to communicate information from the environment to the body. Sensory receptors are found in all skin layers, but they are mostly located in the innermost regions and in the fascia (Figure 4.6). **Free nerve ending**s are numerous pain-sensing structures distributed throughout the inner part of the epidermis. They detect chemicals associated with tissue damage and bleeding, registering a response as pain. **Merkel cells** are sensitive to gentle physical sensations and are found in small numbers in the stratum germinativum. They are most numerous in regions of the body having special sensitivity, such as the fingertips. **Tactile corpuscles**, or **Meissner's corpuscles**, are nerve receptors surrounded by an elongated, club-shaped pile of connective tissues. They are found in the upper region of the dermis in the dermal papilla where they respond to touch. Pressure placed on tactile corpuscles compresses the connec-

Figure 4.6 **Nerves**

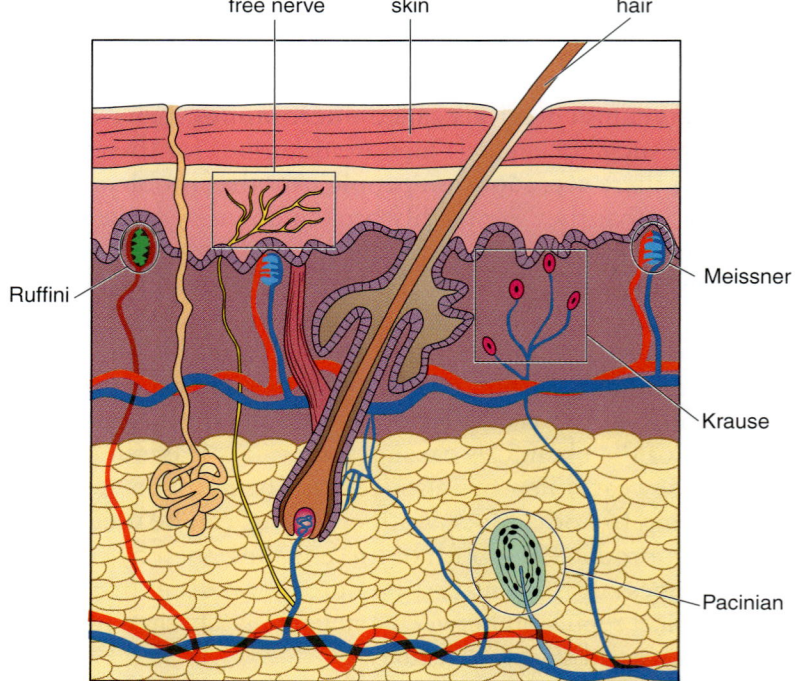

tive tissue covering, which causes the nerve to respond. **Lamellated corpuscles**, or **Pacinian corpuscles**, look like onions and are located in deeper parts of the hypodermis. They respond to hard pressure, including vibrations. **Ruffini receptors** are another type of nerve cell capable of providing the sensory experience of pressure or constant touch. **Krause end bulbs** are sensitive touch receptors that are mostly found in the mucous membrane of the mouth.

Nails Fingernails and toenails are nothing more than a keratin secretion. It is the same keratin that fills the cells of the stratum corneum. Each nail grows forward from a **nail root** that lies beneath an area called the **skin-nail fold** (Figure 4.7). Nails will grow back if removed as long as the nail root and skin-nail fold are not severely damaged. Keratocytes at the base of the nail root gradually move up to the surface of the skin. The cells closest to the surface die, flatten, press tightly together, and disintegrate to form the **nail body**, or **matrix**. As these cells accumulate, they continuously push the nail forward. The whitish area at the base of nail body is called the **lunula**. The pinkish portion of the nail underlying the nail body is called the **nail plate**. The **cuticle** is merely an outgrowth formed by the upper part of the skin-nail fold. Fingernails of a healthy person typically grow about an eighth of an inch per month, which is slightly faster than toenails.

> **Lamellated or Pacinian Corpuscles** Nerves that respond to hard pressure
>
> **Ruffini Receptors** Nerves that respond to pressure or constant touch
>
> **Krause End Bulbs** Nerves found in the mouth that respond to touch
>
> **Nail Root** The area of the nail where growth takes place
>
> **Skin-Nail Fold** A fold of hardened skin overlapping the base a fingernail or toenail
>
> **Nail Body or Nail Matrix** The nail
>
> **Lunula** The white, half moon shape at the base of the nail body
>
> **Nail Plate** The pink area underneath the nail body
>
> **Cuticle** An outgrowth of the skin-nail fold

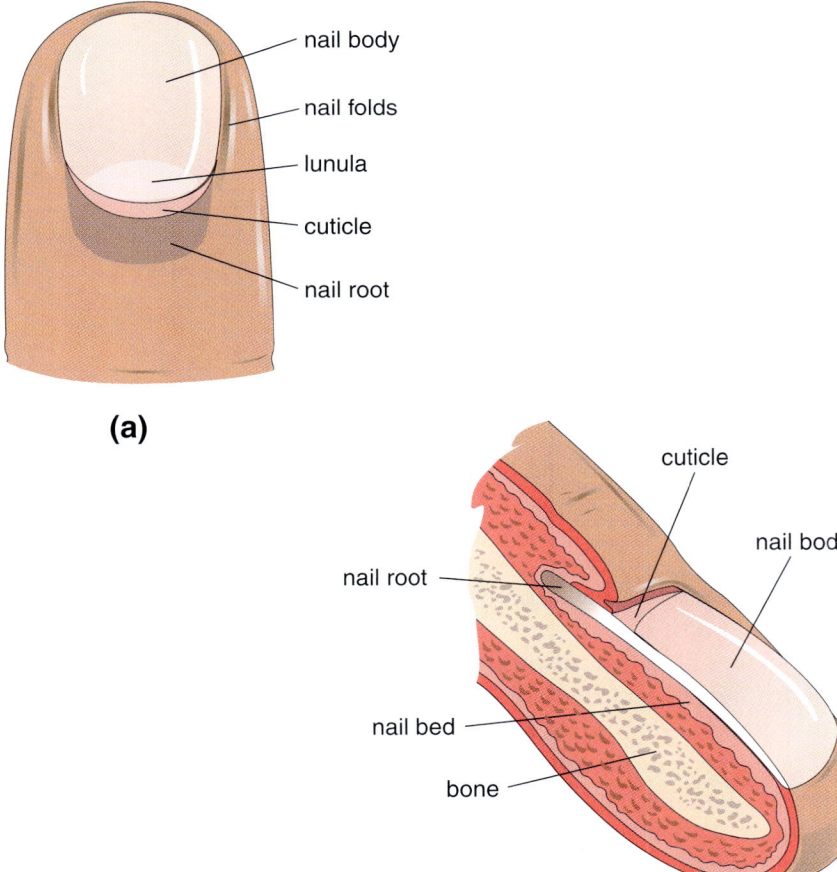

Figure 4.7 **Nails**

Hair Papilla The base of the hair follicle

Hair Shaft The main part of the hair structure

Hair Cortex The principle layer of the hair shaft

Hair Cuticle The outer layer of the hair shaft

Hair Medulla The inner layer of the hair shaft

Hair Hair is a modified stratum corneum formed by an inward protrusion of epidermis called the hair follicle, or hair bulb. Each strand of hair grows from an individual follicle buried in the subcutaneous layer (Figure 4.8). The base of the follicle has a structure called the **hair papilla**, which is associated with small blood vessels. A nerve also serves the hair papilla. The dead, hardened cells protruding from the skin form the main part of the hair, called the **hair shaft**. Most hair shafts are composed of two distinctive cylindrical regions. An outer **hair cortex** is made of dead, densely packed, keratinized cells that are similar to the stratum corneum cells. The dried surface found on larger hairs is called the **hair cuticle**. Inside of the cortex is the **hair medulla**, which is composed of more loosely arranged cells. Keratin gives a yellowish color to hair. Melanocytes at the base of the hair secrete the red, brown, and black pigmentations seen in hair. The differences in hair color are due to the amount and location of melanin deposited in the cortex and medulla, which is genetically controlled.

Figure 4.8 **Hair**

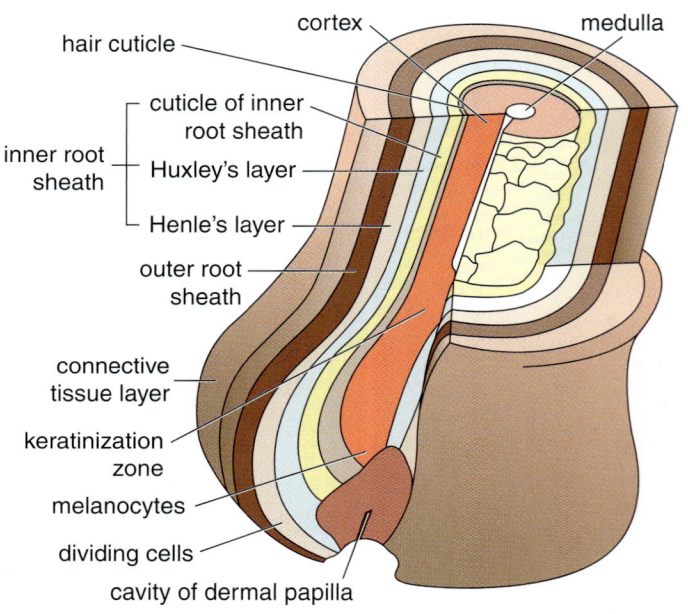

Hair Cycle The mitosis cycle of the hair follicle

Arrector Pili Muscle A smooth band of muscle that holds the hair erect

Vellus Hair Fine body hair

Terminal Hair The hair of the head

Pubic Hair The hair around the genitals

The **hair cycle** describes the mitosis cycles that the hair follicle undergoes to produce a length of hair (Figure 4.9). Newly formed hair cells move up the follicle as newer cells are forming beneath them. As in the skin, the cells dry out, fill with keratin, harden, and die. Hairs have two accessory structures: a sebaceous gland and an **arrector pili muscle**, which is a band of smooth muscle. The arrector pili muscle attaches to the outside of the follicle and functions to hold the hair erect. It may also work with sensory structures, helping the hair to act like a sensitive touch receptor. This is mostly associated with the fine body hairs called **vellus hair**. Head hair is called **terminal hair**, while hair around the genitals is called **pubic hair**. Hair shaft cells play an important role in police work and pathology. The cells of hair capture many types of chemicals that pass through the blood. Any drug or poison that entered a person's body can be found in the hair (Figure 4.10). Current

Figure 4.9 New Hair Development
New hair developing: old hair separates from the papilla in the early stage of shedding. New hair will grow from the same healthy papilla soon after.

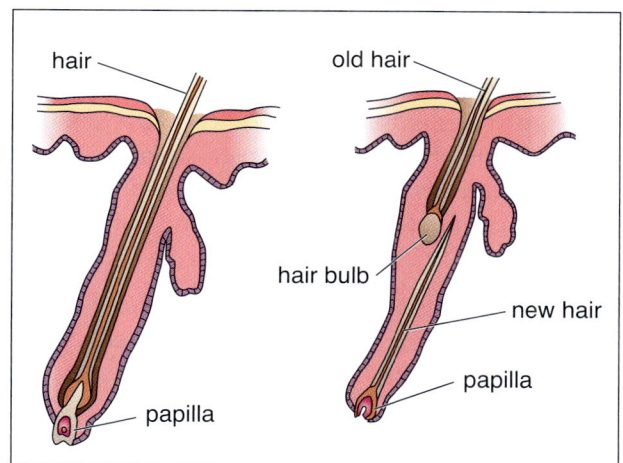

Figure 4.10 Hair Attached to a Vein and Artery

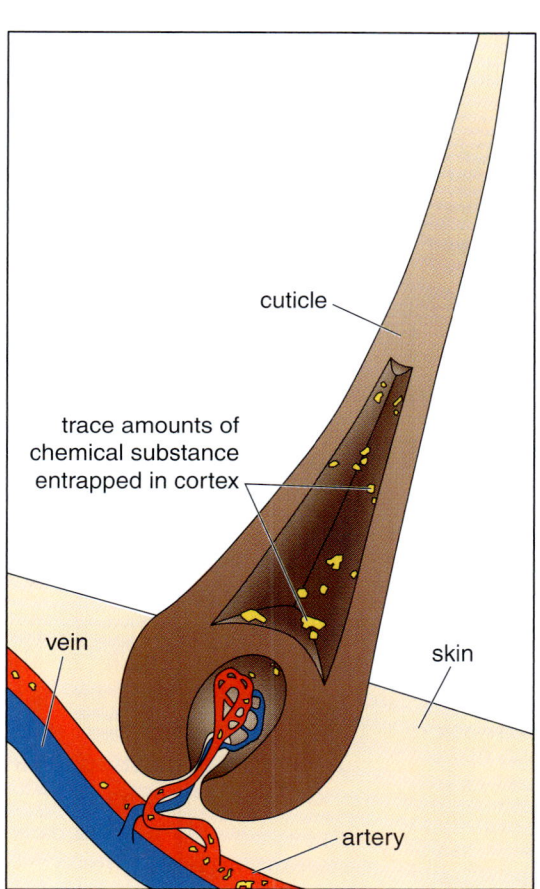

drug use can be determined by testing drug-breakdown products in the living cells at the base of the hair follicle. Scientists are able to detect the use of morphine as a painkiller by studying the hair of ancient mummies. Drug-breakdown residues can also be detected in older cells of the shaft. The DNA and mitochondria can also be removed from the hair cells for genetic testing. Scientists are able to conduct accurate genetic testing on the hair of bodies buried for hundreds of years.

✓ Concept Check

1. Describe the different types of skin glands.
2. What are the functions of the different skin nerves?
3. Distinguish between nails and hair.

THE SKIN AND ITS PARTS

Skin Trivia: Just for Fun

The true intellect of any scientist lies in his or her ability to use scientific information to answer complex questions or solve various problems. However, sometimes it is helpful to have a repertoire of science fun facts to entertain friends. Sharing trivia is also a good way to get people interested in a particular topic. The following bits of skin trivia serve the purposes of being both educational and entertaining:

- The largest human organ is the skin, which has a surface area averaging 25 square feet (1.5 square meters).
- A human being loses an average of 40 to 100 strands of hair each day.
- An average human scalp has 100,000 hairs.
- The average human sheds over 40 pounds of skin in a lifetime.
- The body loses 30,000 to 40,000 dead skin cells every minute.
- There are 45 miles of nerves in the skin of a human being.
- 1 square inch (6.5 square centimeters) of skin contains up to 15 feet (4.5 meters) of blood vessels.
- Blondes have more hair than dark-haired people.
- Fingernails grow faster than toenails.
- Fingerprints provide traction for the fingers to grasp things.
- Almost 50% of the body's heat can be lost through the skin of the head as a result of its extensive blood vessel network.
- Based on the selling price of cowhide, which is approximately $.25 per square foot, the value of an average person's skin is about $6.25.

Functions of the Integumentary System

Key Terms: commensal, first-degree burn, second-degree burn, third-degree burn, transducer

All the skin components mentioned above work together to give the skin its four major roles as an organ system. Some of the skin's jobs help it carry out specific functions of the integumentary system. Other activities of the skin assist other organ systems to maintain the body's homeostasis. These four functions, in order of magnitude, are as follows: protection, heat regulation, sensation, and waste excretion.

The integumentary system acts as a barrier against three types of environmental damage: **chemical**, **mechanical**, and **microbial**. Chemical damage is due to any chemical, including water, that can break down the connections between cells or completely disintegrate the cells. Sweat dilutes potentially hazardous chemicals, plus it contains buffers that neutralize small amounts of acidic and basic chemicals. Cerumen and sebum act as oily barriers that repel excessive water and any potentially dangerous chemicals dissolved in water. This oily layer also prevents water from escaping the body through the skin, which would cause dehydration. Mechanical damage is any type of force that can compress, erode, stretch, or tear the skin. The flexibility imparted by the skin's loose connective tissue helps it resist stretching and tearing. The shedding of the stratum corneum reduces skin erosion. The "shock-absorbing" capability of adipose tissue, reticular fibers, and callus (thickened layers of stratum corneum) minimizes compression damage. Microorganisms can damage the

Chemical Damage Damage caused by any chemical that breaks down cells or the connections between cells

Mechanical Damage Damage caused by any force that compresses, erodes, tears, or stretches the skin

Microbial Damage Damage caused microorganisms on the skin

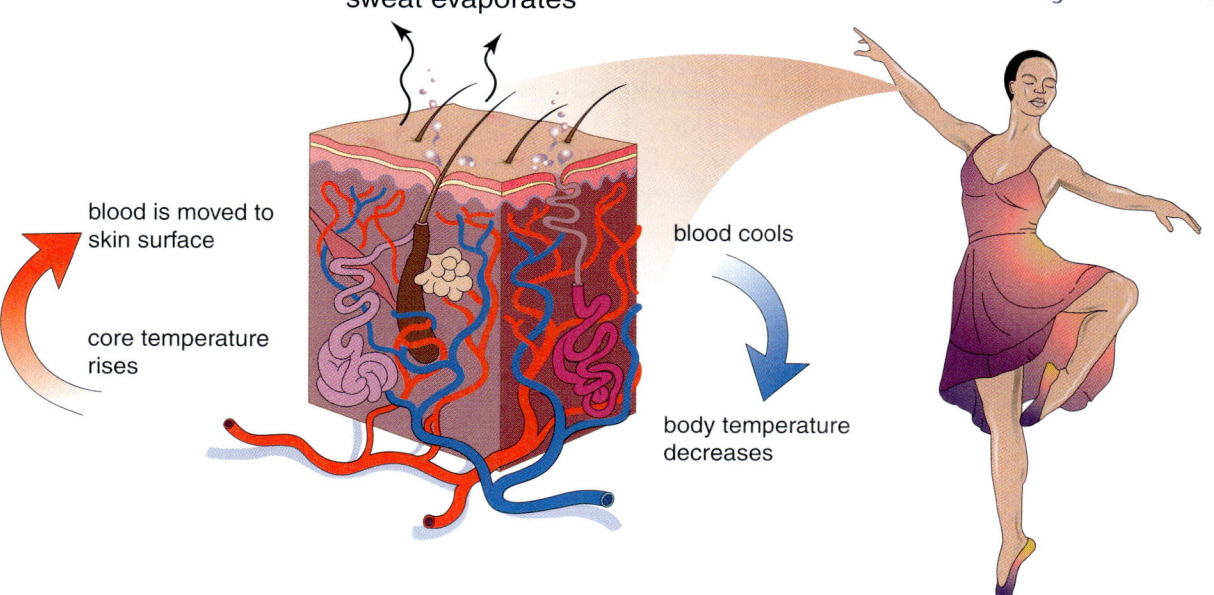

Figure 4.11 Functions of the Skin

skin by producing destructive secretions in the sebum and sweat. These secretions are kept in check by chemicals in the sweat that assist beneficial bacteria called **commensals**. Commensal bacteria and yeast reduce the chance that harmful microorganisms will survive on the skin. In addition, the shedding of the stratum corneum removes microorganisms from the skin. Skin is an important first line of defense for keeping microorganisms out of the body.

Heat regulation is the body's ability to maintain a constant internal temperature. The integumentary system assists with this need in several ways (Figure 4.11). A special network of blood vessels in the skin expands or contracts according to the internal temperature of the body. These blood vessels work like the cooling coils in an air conditioner, transferring the heat carried in the blood to the environment (Figure 4.12). Blood flow to the skin vessels increases when the body is warm, encouraging the loss of body heat through the skin. When the body is cold, the supply of blood to these vessels diminishes, retaining body heat. The evaporation of the sweat on skin lowers the body's temperature. Adipose tissue in the subcutaneous layer is a natural blanket that keeps the body from losing too much heat.

Commensals Beneficial bacteria found on the skin

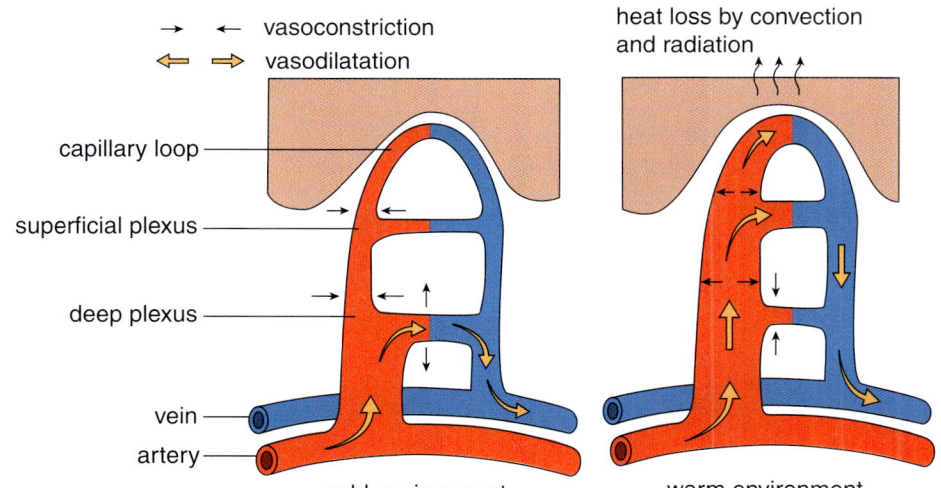

Figure 4.12 Blood Supply to the Skin
Variations in the blood

THE SKIN AND ITS PARTS

Transducer A nerve cell that converts environmental stimuli into body signals

Sensation is best described as received stimuli from the environment, which is interpreted in a way that the brain can comprehend. Sensory nerves in the skin convert the stimuli into signals that tell the brain about environmental conditions and any hazards. These sensory structures permit the skin to detect cold, heat, injury, pressure, stretching, and touch. The name **transducer** is given to nerve cells that convert various environmental messages into body signals. Waste excretion is handled primarily by the eccrine sweat glands. They excrete waste products, such as urea, organic chemicals, and excess salts. However, the skin is not nearly as efficient as the excretory system at removing body wastes. Skin performs other tasks, too, including the production of vitamin D when exposed to sunlight. Vitamin D production is prevalent in lighter-skinned people. Physicians can use the skin as an indicator of a person's health. Special lasers can "see through" the skin to monitor the chemistry and flow of blood (Figure 4.13). Some lasers can detect glucose, fats, toxins, and various ions without having to take a blood sample.

Burns are one of the common ways the skin loses its ability to maintain homeostasis locally and for the whole body. Exposure to sunlight and cooking accidents make up the majority of skin burns. The number of people who are burned this way is not fully known because few people report these incidents to physicians. Acids, bases, corrosive chemicals, electricity, fires, and steam rooms are other sources of burns. Different types of burns produce various degrees of damage to the skin. The severity of a burn can range from minor to fatal. Physicians classify the severity of a burn based on the extent of skin damage (Figure 4.14), as follows:

Figure 4.13 **Laser to the Skin**

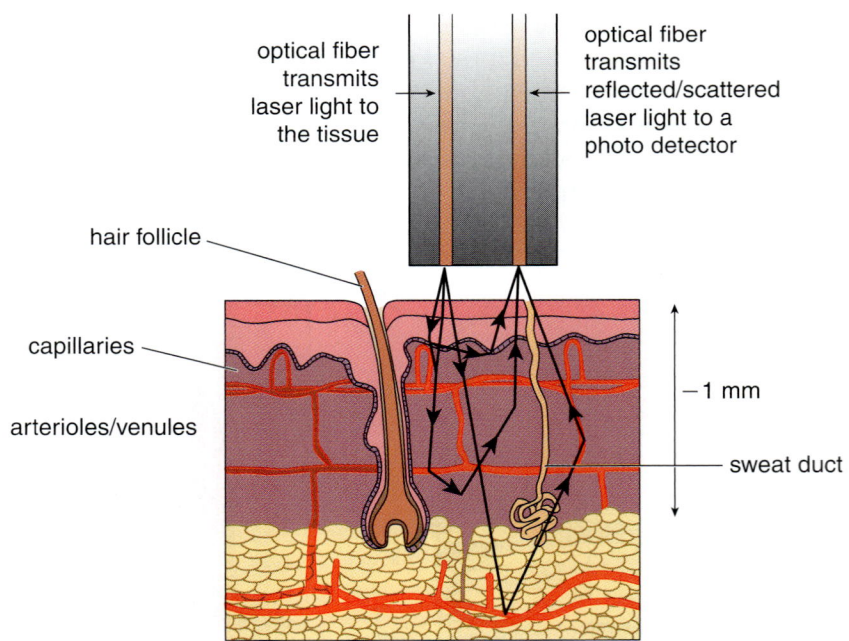

Figure 4.14 Burns

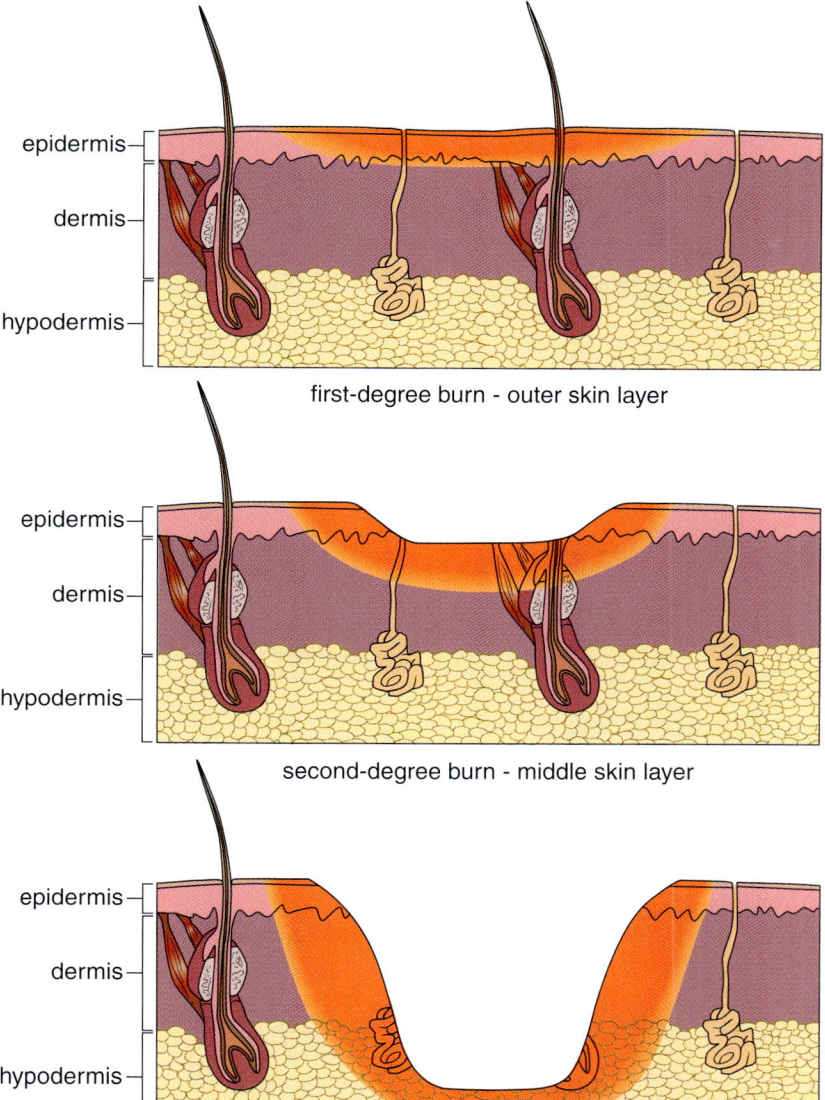

1. **First-degree burn:** These burns involve superficial damage that causes reddening and swelling of the skin. Usually only the outer layers of the epidermis are damaged. Certain chemicals, steam, heat from a flame, and mild sun exposure can cause this type of burn.
2. **Second-degree burn:** This category refers to damage to the stratum spinosum and stratum germinativum. Blisters, reddening, swelling, and fluid buildup under the epidermis are indicators of second-degree burns. These burns are very painful and take a longer time to heal than first-degree burns. Second-degree burns are caused by longer exposure to the same factors that cause first-degree burns. Minor second-degree burns are no larger than 2 to 3 inches in diameter. Larger burns and those found on the skin of the buttocks, face, feet, groin, hands, and major joints should receive medical attention.
3. **Third-degree burn:** In this type of burn the entire epidermis is charred or missing. This means that the stratum germativum is not available for healing the open wound. There is usually damage to the dermis that can

First-Degree Burn A burn involving superficial damage

Second-Degree Burn A burn involving the stratums spinosum and germinativum

Third-Degree Burn A burn that affects the entire epidermis

cause permanent nerve-cell loss. This means that tissue damaged directly cannot feel pain or other sensations. However, the nearby skin registers the pain as histamine leaks over from the damaged skin. Immune system cells release histamine in response to tissue damage. Glands can also be damaged past the point of repair. In severe burns, adipose cells, muscle, and bone may be affected. Direct flame and prolonged exposure to corrosive chemicals and electricity create these burns. This degree of burn destroys the skin's ability to act as a barrier. Therefore, people with third-degree burns are susceptible to dehydration, loss of body heat, and infection.

✓ Concept Check

1. What are the three types of barriers the skin provides?
2. What is the role of commensal microorganisms on the skin?
3. Describe the three categories of skin burns.

DISCOVERY SCENE PLEASE ENTER DISCOVERY SCENE PLEASE ENTER

So, what have you learned so far that may help you solve the CSI? What general information about skin structure sheds light on the nature of the wound? Which skin layers are associated with the tissue damage seen? Does the sore have damage affecting any of the skin appendages? Which skin and body functions are being affected by the sore?

WELLNESS AND ILLNESS OVER THE LIFE SPAN

Pathology of the Integumentary System

Key Terms: acne, albinism, arthropod, candida albicans, cyst, degenerative, demodex, dermatitis, dermatophytes, follicle mite, folliculitis, furuncle, human papilloma virus, hypopigmentation, impetigo, lice, lipoma, melasma, mole, monilia, nodule, port wine stain, precancerous gene, psoriasis, ringworm, sebaceous hyperplasia, seborrhoeic keratosis, skin cancer, skin tags, solar lentigene, spider vein, staphylococcal scalded skin syndrome (SSSS), Staphylococcus aureus (S. aureus), Streptococcus pyogenes (S. pyogenes), strawberry hemangioma, syringoma, tinea, vitiligo

Degenerative Refers to diseases that progressively deteriorate tissues

The integumentary system, like any of the other organ systems, has its share of disorders. It is a host for **degenerative**, genetic, and infectious diseases. Degenerative diseases are usually due to progressive deterioration of tissue caused by continuous injury from environmental and/or physical stresses. Genetic disorders are the result of mutations that diminish skin function and structure. Infectious diseases are caused by microorganisms that damage tissues and organs. The most common of the many human skin diseases are mentioned in this section.

Degenerative Skin Disorders Freckles are probably viewed more as a facial "decoration" than as a degenerative skin disorder. Most facial freckling is genetic and not considered a skin disorder. However, sun exposure can darken faint freckles on light-skinned people. Another type of freckling found all over the body is called **solar lentigene**. These freckles usually appear on people in their 30s who overexpose their skin to sunlight or tanning beds. Strong sunscreens can prevent solar lentigines, but they can also be removed with laser treatments. It may surprise many people to learn that use of cosmetics is the most common cause of skin degeneration. Much of this damage is due to the mild **dermatitis** that many of these products cause. Dermatitis is an inflammation of the skin caused by an allergic reaction or contact with an irritant. Itching, redness, and some swelling characterize this condition. Extensive use of many types of facial cleansers and toners can produce scaling skin and accelerate skin aging.

Skin cancer is a degenerative disorder with an underlying genetic component, which scientists call **precancerous genes**. These genes, if damaged by sunlight or certain chemicals, produce abnormal divisions of the affected skin cells. The common type of skin cancer induced by sun exposure starts out as irregularly shaped brown or black spots that can develop into open sores. Other skin cancers are due to injury deep within the skin. These cancers first appear as a variety of discolored blisters or sores. They then erode the skin and readily spread to other parts of the body. Skin cancers can be fatal if not treated. Treatment varies with the type of cancer and the extent to which it has spread throughout the body.

The skin develops many types of tumors that are not associated with cancer. **Moles** are flat squamous-cell tumors that are heavily pigmented by melanocytes (Figure 4.15). Their origin is unknown, but some may be induced by genetic damage to the skin. These are likely to lead to cancer. **Skin tags** are soft, colored, knob-shaped tumors that grow out of the skin. They usually occur on the neck, armpits, and body, and are easily removed by minor surgery. **Seborrhoeic keratosis** is a black to brown growth on the face or body. It creates a greasy, rough appearance to the skin and generally is removed for cosmetic reasons. Oil glands can also develop cell masses called **sebaceous hyperplasia**. Remember, the term hyperplasia means an overgrowth of cells. The glands appear as small, yellow bumps with an opening in the center. They are removed by simple surgery. Sweat-gland ducts can form tumors called **syringomas**. These painless growths mostly appear as small lumps on the cheeks and eyelids. Sometimes fat cells underneath the skin form raised tumors called **lipomas**. Usually, they do not go away and mainly cause problems only when they impede movement or occur around the mouth.

Solar Lentigene A type of freckling

Dermatitis A skin inflammation caused by an allergic reaction or contact with an irritant

Skin Cancer A degenerative disorder with an underlying genetic component

Precancerous Genes Genes that have the potential to become cancerous under certain conditions

Moles Heavily pigmented squamous cell tumors

Skin Tag Soft tumors

Seborrhoeic Keratosis A black or brown growth on the face or body

Sebaceous Hyperplasia A disorder that affects oil glands

Syringomas Tumors that form in sweat glands

Lipomas Tumors formed in fat cells

LOOK FOR DANGER SIGNS IN PIGMENTED LESIONS OF THE SKIN

A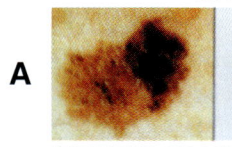
Asymmetry—one half unlike the other half.

B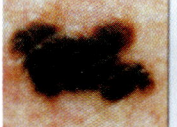
Border irregular—scalloped or poorly circumscribed border.

C
Colored Varied from one area to another: shades of tan and brown; black; sometimes white, red or blue.

D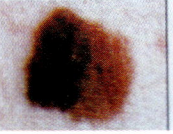
Diameter larger than 6mm as a rule (diameter of pencil eraser).

Figure 4.15 Moles
The ABCDs of moles: The signatures of skin cancer include Asymmetry (one half unlike the other), Border irregularity (scalloped edges), Color variation (shades of tan, brown, and black), and Diameter (larger than a pencil eraser). Images from NASA.

Acne A common skin condition resulting in the overproduction of sebum

Cyst or Nodule A sack-like structure filled with a liquid or semisolid substance

Furuncle or Boil An inflammation of a hair follicle

Psoriasis An inflammation of the skin accompanied by increased skin cell production

Port Wine Stain A red birthmark

Spider Veins A birthmark caused by enlarged veins, appears as a spider's web

Strawberry Hemangiomas A birthmark caused by enlarged blood vessels

Vitiligo A skin disorder resulting in white patches on the skin

Hypopigmentation The decrease in melanin production

Genetic Skin Disorders Acne is the most common skin condition that likely has a strong genetic component. It is believed to be stimulated by hormonal changes that cause the overproduction of sebum around the hair follicles. It is most common in teenagers following puberty, and in women undergoing hormone changes due to maturity or pregnancy. The disorder consists of blackheads, pimples, red spots, whiteheads, and, sometimes, lesions called **cysts**, or **nodules**, which are found deep in the skin. A boil, or **furuncle**, is the inflammation of hair follicles resulting in the buildup of dead cells and blood components. The terms cyst and nodule refer to any sacklike structures swollen with a liquid or semisolid substance. Skin bacteria that feed on the sebum and decaying cells aggravate acne. These bacteria produce irritating waste products that are the result of feeding on the acne pimple. The blackhead, pimple, and whitehead simply refer to small skin swellings that may be colored by fluids contained within the cyst. Most people outgrow the acne. However, the intensity of mild cases can be reduced with soaps that open the cysts, remove some of the bacteria, and remove the excess sebum. Severe acne is treated with a variety of drugs, including antibiotics that kill the bacteria and isotretinoin, which helps shed the sebum-filled cysts.

Another commonly encountered disease is **psoriasis**. It is an inflammatory skin disease accompanied by an increased amount of skin cell production. This causes a buildup of thick scales appearing on the skin. The inflammatory component refers to the fact that the skin in these areas is painful, red, swollen, and warm. These features are a consequence of the body trying to protect affected skin from disease or injury. The affected skin is dry, itchy, and unsightly. Hair loss sometimes occurs if the psoriasis appears on hairy parts of the body. Scientists do not know the exact cause of psoriasis. However, they do know that there is a gene that is activated by certain environmental conditions and infectious diseases. There is no cure for psoriasis, but it can be treated with drugs that reduce the swelling and slow the skin production.

A variety of disorders called birthmarks can affect the skin. **Port wine stains**, **spider veins**, and **strawberry hemangiomas** are three common birthmarks. A port wine stain birthmark looks like a spot of red wine spilled on the skin, hence the name. It is due to an abnormality of skin blood vessels, and it affects less than 1% of the population. The birthmark, which can appear anyplace on the body, begins at birth and usually grows larger and darkens slowly through the years. In some cases, this condition may cause disfigurement of the skin. Some larger port wine stains can bleed as the skin above the vessels cracks. Strawberry hemangiomas are caused by enlarged blood vessels. These birthmarks grow rapidly after birth. However, in many people, the birthmark lightens in color by age 6 years. They produce no complications unless they bleed excessively, or affect skin near the eyes and mouth. The name "spider veins" describes another type of enlarged blood vessel condition of the skin. This birthmark consists of a central blood vessel with smaller vessels branching out from the center, like a spider web. This type of birthmark causes no problems, except that many people find them unattractive.

Vitiligo is the skin disorder made famous by singer Michael Jackson. It is not known whether he truly has the disease, which is found in 1% of the population. The disease appears as white spots, called **hypopigmentation**, and patches on the skin, which are due to a regular, localized decrease in melanin production. This loss of melanin results from the destruction of

Cutting Edge Research
USING ROBOTIC HANDS TO SCREEN BREAST CANCER

Imagine a robotic hand with a covering that works the same way as skin. Scientists at Michigan State University recently developed a remote-controlled robotic hand that can feel for breast tumors. The robotic hand is controlled by a medically trained operator wearing a motion-sensing glove. It is able to detect breast lumps using ultrasound sensors and video cameras, which provide more information than the standard touch examination. The goal is to use the machine to examine people in remote areas of the United States and in countries that do not have enough doctors. One physician can remotely control several machines from a distant location. In addition, it will allow specialists to give a second opinion without the patient or the physician having to travel. There are also robots with the dexterity and senses to carry out surgery. Surgeons in Australia are testing new robots that perform operations on patients. The robots give the surgeons improved precision and reduce complications associated with traditional surgery. Much of the success of these robots lies in their ability to respond better than skin to particular sensations.

---

melanocytes in the skin. Currently, the cause of vitiligo is unknown, although it is thought to be caused by an immune system attack on the melanocytes. The disease is treated with therapeutic vitamins and drugs that are activated by ultraviolet light. Vitiligo should not be confused with **albinism**. Albinism is a genetic disease in which there is no melanin production in the eyes, hair, or skin. **Melasma** consists of brown patches on both sides of the face, especially on the cheeks, upper lips, nose, and chin. The distribution is usually symmetrical. Melasma is a darkened area of skin that usually shows up in pregnant women. Sunlight exposure usually darkens the melasma. Melasma is believed to be a consequence of female hormonal changes, as it also appears in women who take oral contraceptives. The coloration tends to fade with age, but it usually does not completely disappear.

Infectious Skin Disorders Arthropods, bacteria, fungi, protista, and viruses all have relatives that cause many skin diseases. As mentioned earlier, skin is the home to many beneficial bacteria and fungi. However, it is possible for "safe" bacteria and fungi to cause skin infections. In most instances, vitiligo and microbial infections are mild and treatable, unless they breach the skin and enter the deeper tissues or blood. It is possible to alter the population of microorganisms on the skin, making way for disease-causing bacteria. This commonly occurs when the skin is washed frequently or remains wet for long periods of time. The most common bacterial skin infections are caused by **Staphylococcus aureus** (**S. aureus**). This species of bacteria produces destructive secretions that erode and inflame the skin. The four most common conditions are as follows: **folliculitis**, an inflammation of the hair follicles; **impetigo**, an easily spread rash in children; boils; and **staphylococcal scalded skin syndrome** (**SSSS**), which is a potentially fatal shedding and swelling of the skin.

A common fungal infection of the skin is caused by yeast cells that live beneficially in the digestive system and the female reproductive system. **Candida albicans** (**C. albicans**), also called **monilia**, can spread to skin

---

Albinism A genetic disease characterized by the lack of melanin production in the eyes, hair, and skin

Melasma A disorder resulting in brown patches on the face

Staphylococcus aureus A species of bacteria that causes skin infections

Folliculitis An inflammation of the hair follicles

Impetigo A rash occurring in childhood

Staphylococcal Scalded Skin Syndrome (SSSS) A potentially fatal disorder resulting in shedding and swelling of the skin

Candida albicans or Monilia A fungal infection caused by yeast

Dermatophytes Fungi that eat hair, nails, and outer layers of the epidermis

Ringworm or Tinea A type of dermatophyte infection

Human Papilloma Virus A group of viruses that cause warts in humans

Arthropod A class of animals that have exoskeletons, jointed limbs, and segmented bodies

Follicle Mite or Demodex An arthropod that causes an inflammation of the eyelash follicles

Lice Arthropods that irritate the skin and spread disease

and nails when the immune system is weakened. The fungi degrade the tissues causing mild to severe inflammation. Other fungi called **dermatophytes** specialize in eating keratin-rich materials, such as hair, nails, and outer layers of the epidermis. At most, this readily spread condition can cause itching, hair loss, and deformation of the nails. **Ringworm**, or **tinea**, is a common type of dermatophyte infection. It is usually contracted from furry pets, or it can spread from one person to another by contact with clothing and skin. It is related to "athlete's foot" and "jock itch."

Warts are one of the many viral diseases that can affect the skin. There are about 60 different types of wart viruses that affect skin and other tissues. The common skin wart is caused by the readily spread **human papilloma virus** (HPV), a group of viruses that cause various types of warts in humans. Warts are incurable and are treated with some type of removal strategy. Protista are mostly involved in exotic tropical diseases that are spread by insect bites. Many of these diseases spread to internal organs and cause severe bodily damage.

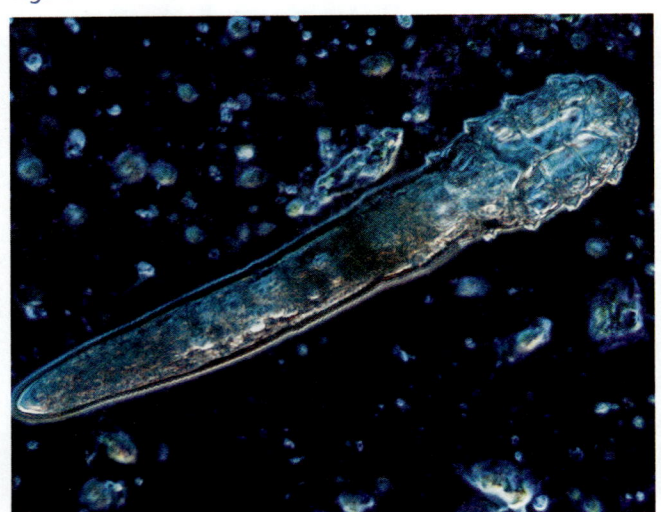

Figure 4.16 Parasites

There is also a little-known skin infection caused by an **arthropod**. Arthropods are a class of animals with hard, outside skeletons, segmented bodies, and jointed limbs. Insects, mites, spiders, and ticks are common arthropods. Inflammation of the eyelash follicles can be caused by an arthropod called the **follicle mite**, or **demodex** (Figure 4.16). Follicle mites usually live undetected in most people until they accidentally contribute to infections of the eyelash follicles. The condition is harmless to people, but it can be life-threatening in animals that carry similar mites. **Lice** are blood-sucking insects that can irritate the skin and spread infection as they feed. Lice can spread a variety of diseases. Both of these arthropods are treated with special types of pesticides.

✓ Concept Check

1. Explain the three causes of skin disorders.
2. Describe two types of degenerative skin conditions.
3. Why is it important to know the type of microorganism causing an infectious skin condition?

Aging of the Integumentary System

Extrinsic Aging Factor or External Aging Aging caused by environmental factors

Intrinsic Aging Factor Aging caused by the natural decline of cells

Key Terms: external aging, extrinsic aging factor, intrinsic aging factor, skin needling

Most physicians and scientists characterize skin as **extrinsic aging factors** or **intrinsic aging factors**. Extrinsic, or **external**, **aging** is caused by environmental factors, such as disease, pollution, and sun exposure. Aging brought about by genetic and natural maturation factors is called intrinsic, or internal, aging. Stress placed on skin by the pathology of other organ systems or unusual envi-

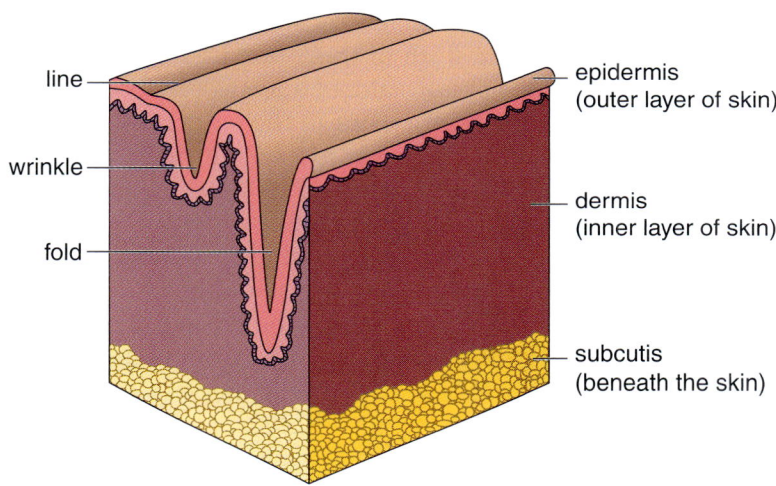

Figure 4.17 **Aging of the Integumentary System**
As we age, two components of our skin—collagen and elastin—degenerate, setting the stage for the appearance of wrinkles, creases, folds, and furrows. The breakdown of these components, accelerated by sun exposure and gravity, results in the sagging skin of old age.

romic interactions can accelerate intrinsic aging. Skin aging is typified by the following gross anatomical changes (Figure 4.17):

- loss of head hair
- graying and whitening of hair
- excessive growth of body and facial hair
- drying of the skin due to diminished oil secretion
- decreased ability to sweat
- thinning of the skin
- loss of melanin leading to transparent skin
- loss of subcutaneous fat
- wrinkling
- skin sagging due to muscle atrophy
- skin stretching due to connective-tissue changes in the dermis
- skin stretching due to gravity pulling on skin
- thinning of the nail plate
- increased probability of skin tumors
- prolonged wound healing due to diminished blood flow and suppressed immune system function
- regular irritation due to changes in microbial populations on the skin

Intrinsic aging is impossible to stop despite the many antiaging-remedy claims seen in advertisements and infomercials. The natural decline of cells in connective tissue and muscle is uncontrollable, but it can be slowed by living in a climate that lacks the physical abuse of extremely cold or hot environments. The connective tissues of the dermis and subcutaneous regions naturally reduce elastin production, and become dominated by collagen and reticular fibers. This makes the skin thinner, less flexible, and more brittle. There are no known creams, potions, or salves that can reverse or undo this tissue transformation. A decrease in blood flow to the skin slows down the stratum germativum thinning the epidermis and making it more difficult to repair skin damage. People with diseases that affect the blood vessels exhibit premature aging of the skin. Melanocyte decline is also natural, as is the loss of many nerves in the skin. This condition makes the skin more susceptible to environmental damage. Older skin is less able to protect itself from sunlight and has difficulty registering injury.

THE SKIN AND ITS PARTS

Skin Needling An antiaging treatment in which needles are inserted into the skin to promote skin growth and swelling

Skin is very susceptible to DNA damage because it is constantly exposed to oxidizing chemicals and sunlight. Both of these factors are known to cause DNA damage that can produce abnormalities, such as cancers and tumors. In addition, the high rate of mitosis carried out in the epidermis makes the skin prone to DNA damage that is naturally introduced during the S stage of interphase. Telomere shortening is pronounced in the skin. Most dermatologists prescribe strategies for reducing skin aging that control extrinsic aging factors. A proper diet, reducing exposure to sunlight, avoiding skin irritation, protecting the skin from air pollutants, and not smoking all reduce skin aging. The two biggest factors that can slow aging are avoiding sun exposure by covering the skin with clothing or sunscreen on bright days, and not smoking. Smoking reduces blood flow to the skin, making it more difficult for the skin to carry out growth and repair. This decrease in blood flow also affects the skin's temperature regulation, so it is more susceptible to the damaging effects of extreme hot and cold weather. Oxidizing chemicals are also introduced into the skin during smoking. Some people seek a treatment called **skin needling** (Figure 4.18). This procedure promotes growth and swelling in wrinkled depression areas, making the skin smooth for a while. Skin aging can be slowed if a person starts taking care of his or her skin at a young age. It is very difficult to reverse skin aging once the damage is done.

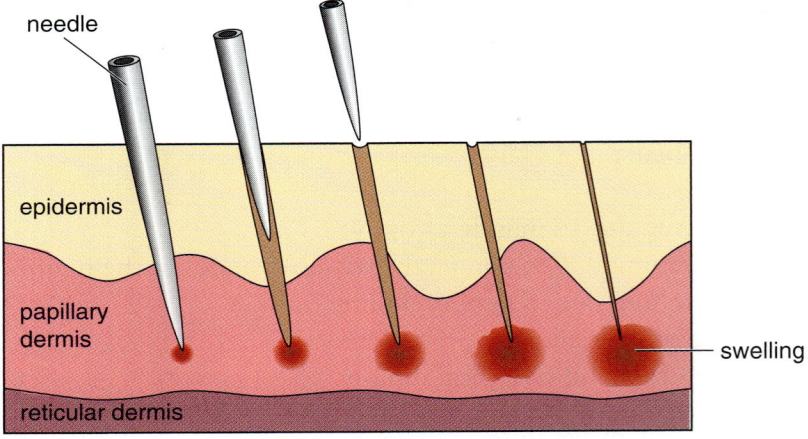

Figure 4.18 Skin Needling Skin needling used to treat wrinkled upper lips and difficult scars—an alternative to laser resurfacing

✓ Concept Check

1. What are the two major categories of skin aging?
2. Describe how internal aging occurs in the skin.
3. Explain the roles of smoking and sunlight exposure on skin aging.

DISCOVERY SCENE PLEASE ENTER DISCOVERY SCENE PLEASE ENTER

Have you come closer to solving the CSI? What information about skin disease and aging is useful to better understand the nature of the sore?

CSI – Case Study Investigation Conclusion

What can you conclude about the sore from the information provided in this chapter? Was the sore a result of pathology or injury? Why was the dermis exposed? The sore also appeared to be old because of the dark scarring around it. Why would such an "ugly" sore not cause any pain sensation? Could it be an old burn that never healed because of the woman's ill health? Would rigorous exercise cause this sore to get better or worse?

Answer:

The sore was due to the decline of blood vessel integrity associated with prolonged or untreated diabetes. Obesity and high blood sugar are good indicators of this condition. Continuous high blood-sugar levels cause smaller blood vessels in the skin to break down. This significantly affects the skin's ability to produce the epidermis needed to replace the stratum corneum that naturally sheds and is removed by physical damage. It is very typical for the skin of diabetes patients to form long-lasting sores. The woman's exercise program should include routines that reduce weight and increase blood flow. However, the inflexible skin over the sore could cause some movement restrictions. Elderly people are much more likely to develop sores like this due to the natural decrease in blood flow that occurs with aging. Increasing the amount of protein in the diet may help to keep the sore from becoming worse. The extra protein assists the immune system and provides the stratum germinativum with the amino acids needed for an adequate rate of mitosis. Strenuous activities should be avoided because stress on the body reduces the immune response and would slow healing of the sore.

This CSI was adapted from the following articles:

1. The Expert Committee on the Diagnosis and Classification of Diabetes Mellitus, American Diabetes Association. Report of the expert committee on the diagnosis and classification of diabetes mellitus. *Diabetes Care.* 1997;Vol 20:1183-1197.
2. Gilgor RS, Lazarus GS. Skin manifestations of diabetes mellitus. In: Rifkin H, Raskin P, eds. *Diabetes Mellitus.* Vol 5. Bowie, MD: Robert J. Brady Company.

Chapter Summary

The integumentary system is a remarkable structure that is fundamentally integrated with the other organ systems. It is the largest human organ system. It contains blood vessels, connective tissue, glands, hair, nails, nerves, and skin. The integumentary system is an accumulation of embryonic structures that must come together with precision timing to ensure a fully functioning organ system. Developmental defects of the integumentary system compromise the ability of other organ systems to maintain the body's homeostasis. The three distinct layers of skin carry out most of the functions of the integumentary system. They protect the body from damage by chemicals, physical damage, and microorganisms. The skin detects many of the sensations that cue the body to react to environmental changes. It is also able to adapt the body to certain environmental conditions. The skin's ability to lose or retain body heat plays a major role in regulating body temperature. Skin damage and disease compromise the skin's ability to do its work. Burns are the most likely type of damage that significantly impair skin function. Physicians categorize burns according to the degree of damage to the skin layers. Skin diseases vary greatly and can be due to degenerative changes, genetic damage, and skin infections. The cells of the skin mature and age, which causes the skin to change over time. Aging of the skin is not only an indicator of health, but it also accelerates the aging of the entire body as it slowly loses its homeostatic capabilities.

Overview

- The integumentary system is composed of the following:
  - blood vessels
  - connective tissue structures
  - glands
  - hair
  - nails
  - skin

Integumentary System

- a continuous body covering
- formed by ectoderm and mesoderm

Skin Structure

- Skin is composed of three distinct layers:
  - epidermis
  - dermis
  - subcutaneous layer or hypodermis
- The epidermis is composed of stratified squamous epithelium and is divided into five or six discernable layers.
- The upper layers of skin fill with keratin and die.
- The dermis is primarily dense irregular connective tissue.
- Three types of fibers are found in the dermis:
  - collagen
  - elastin
  - reticulin
- The dermis contains nerves and blood vessels.
- The subcutaneous layer contains adipose tissue.

Skin Appendages
- The skin contains several types of glands.
- Nerves provide the skin with sensation.
- Nails are sheets of dried, flattened, keratinized cells.
- Hair is composed of cylinders of keratinized cells.

Functions of the Integumentary System
- Skin provides four functions for the body:
  - protection
  - heat regulation
  - sensation
  - waste excretion
- Burns are categorized according to severity of skin damage:
  - first-degree burn
  - second-degree burn
  - third-degree burn

Pathology of the Integumentary System
- Diseases of the integumentary system come in three categories:
  - degenerative
  - genetic
  - infectious

Aging of the Integumentary System
- Some of skin aging is due to intrinsic factors.
- Some of skin aging is due to extrinsic, or environmental, factors.
- Lifestyle may accelerate skin aging.

 **Key Terms**

Overview
Adaptive
Inherent
Integument
Integumentary system

The Integumentary System
Angiogenic factor
Fibroblast
Lanugo
Melanoblast
Melanocyte
Mesenchyme
Mucous membrane
Pigmentation

Skin Structure
Areolar connective tissue
Capillary
Dermal papilla
Dermis
Desquamation
Epidermis
Fascia
Fasciitis
Hemidesmosome
Hypodermis
Keratin
Keratocytes
Malpighian layer
Melanin
Melanosome
Stratum basale
Stratum compactum
Stratum corneum
Stratum germinativum

THE SKIN AND ITS PARTS

Study Guide

Stratum granulosum
Stratum lucidum
Stratum spinosum
Subcutaneous layer

Skin Appendages
Apocrine
Apocrine sweat gland
Arrector pili muscle
Cerumen
Ceruminous gland
Cuticle
Eccrine sweat gland
Free nerve ending
Hair bulb
Hair cortex
Hair cuticle
Hair cycle
Hair follicle
Hair medulla
Hair papilla
Hair shaft
Holocrine gland
Krause end bulb
Lunula
Meissner's corpuscle
Merkel cell
Nail body
Nail matrix
Nail plate
Nail root
Pacinian corpuscle
Pheromone
Pubic hair
Ruffini receptor
Sebaceous gland
Sebum
Sensory receptor
Skin appendages
Skin-nail fold
Sweat gland
Tactile corpuscle
Terminal hair
Vellus hair

Functions of the Integumentary System
Commensal
First-degree burn
Second-degree burn
Third-degree burn
Transducer

Pathology of Integumentary System
Acne
Albinism
Arthropod
Candida albicans
Cyst
Degenerative
Demodex
Dermatitis
Dermatophytes
Follicle mite
Folliculitis
Furuncle
Human papilloma virus
Hypopigmentation
Impetigo
Lice
Lipoma
Melasma
Mole
Monilia
Nodule
Port wine stain
Precancerous gene
Psoriasis
Ringworm
Sebaceous hyperplasia
Seborrhoeic keratosis
Skin cancer
Skin tags
Solar lentigene
Spider vein
Staphylococcal scalded skin syndrome, SSSS
Staphylococcus aureus
Strawberry haemangioma
Syringoma
Tinea
Tumor
Vitiligo

Aging of the Integumentary System
External aging
Extrinsic aging factor
Intrinsic aging factor
Skin needling

Check Your Understanding

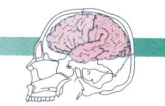 

Study Guide

1. The outermost layer of skin is:
   a. Hypodermis
   b. Dermis
   c. Epidermis
   d. Fascia

2. Which of the following is *not* part of the integumentary system?
   a. Melanocyte
   b. Mucous membrane of mouth
   c. Nail
   d. Iris of eye

3. Which statement is true about the dermis?
   a. It protects underlying layers from sunlight
   b. It provides structural strength for the skin
   c. It contains large amounts of adipose tissue
   d. It produces epidermis

4. Which skin or associated structure lacks blood vessels?
   a. Epidermis
   b. Dermis
   c. Subcutaneous layer
   d. Fascia

5. Which layer of epidermis is composed of dead cells?
   a. Stratum corneum
   b. Stratum granulosum
   c. Stratum spinosum
   d. Stratum basale

6. Where does keratinization occur?
   a. In fascia
   b. During the formation of ectoderm
   c. As epidermis cells mature
   d. As fibroblasts secrete proteins

7. Eccrine sweat glands produce the following type of secretion(s):
   a. Oil
   b. Wax
   c. Water and salts
   d. Pheromones

8. Earwax is produced by:
   a. Sebaceous glands
   b. Lacrimal glands
   c. Cerumenous glands
   d. Eccrine sweat glands

THE SKIN AND ITS PARTS

Study Guide

9. Hair is produced by the:
    a. Hair cortex
    b. Hair follicle
    c. Hair bulb
    d. Hair medulla

10. Melanin in the integumentary system provides the following function:
    a. Skin coloration
    b. Protection from ultraviolet light
    c. Hair coloration
    d. All of the above

11. Which of the following is *not* a function of blood vessels in the skin:
    a. Provides nutrients for cell division
    b. Regulates skin temperature
    c. Carries wastes to the sweat glands
    d. All of the above

12. Nails are produced by the:
    a. Nail matrix
    b. Nail root
    c. Nail gland
    d. Keratin gland

13. Which burn is defined by damage to the dermis:
    a. Steam burn
    b. First-degree burn
    c. Third-degree burn
    d. Rope burn

14. Birthmarks are best defined as:
    a. Skin cancers
    b. Benign tumors
    c. Glandular tumors
    d. Abnormal blood vessels of the skin

15. The gradual loss of skin pigmentation due to melanocyte abnormalities is called:
    a. Hypermelanism
    b. Albinism
    c. Lanugo
    d. Vitiligo

A Case Study

Study Guide

TREATING WOUNDS BY "EATING THEM"

Imagine visiting a physician whose office has shelves of bugs and worms instead of the typical cartons of gauze and medications. No, this physician is not merely a collector of curious things. Rather, the critters in his office are a growing strategy for treating skin ailments. This physician practices an ancient medical treatment, which today is called larval therapy. Larval therapy takes advantage of the fact that the larvae, or maggots, of certain flies can eat dead flesh without harming living tissues. The larvae are supposed to be effective for wound healing in several ways. They prevent infection by eating microorganisms that aggravate the wound and cause fatal illnesses. They also secrete chemicals with antibacterial and wound-healing action. They dissolve and devour rotting tissue, and promote healing.

The body's immune response does not respond to the larvae, so there is no inflammation. Fly larvae are raised sterilely so that they do not introduce diseases into the wound. They are removed from the wound after about three to five days. Many physicians believe that larval therapy is more cost-effective than the standard treatment of scraping the wound and dressing it with hydrogel. Hydrogel dressings are carbohydrate polymers that are usually shaped into sheets. They provide the wound with a moist, sterile environment. Increasing the moisture around the wounds helps remove dying tissues. Hydrogels do not stick to the wound and can be removed without causing further damage. Various federal and state public health agencies have conducted research showing that larval therapy is effective in promoting healing in large, festering wounds.

Use the information in this chapter and the following web sites to resolve the issue of larval therapy. Answering the following questions will assist you:

- Give two ways in which larval therapy is effective.
- In what situations should larval therapy be used?
- Should physicians be required to use larval therapy if it proves more effective than traditional medicine? Explain your view.
- Should physicians be required to use larval therapy to reduce medical costs? Explain your view.
- Should patients have a right to request or refuse larval therapy? Explain your view.
- What precautions need to be taken before using larval therapy as a regular practice?

1. Worldwide Wounds
   http://www.worldwidewounds.com/1997/october/LarvalTherapy/Larval-CaseStudy.html

2. National Institutes of Health
   http://www.nih.gov/

THE SKIN AND ITS PARTS

3. National Center for Complementary and Alternative Medicine
   http://nccam.nih.gov/

4. Maggot Therapy
   http://www.larvaltherapy.com/

Where Do We Go from Here?

People in health fields can use their knowledge of the integumentary system to solve everyday problems. You may wish to use other resources, such as the following web sites, in addition to your textbook to investigate the answers to each of the following situations:

1. A sales representative for a shampoo company tells you their product "feeds" your hair cells protein. What arguments could you use to debate the validity of their claim?
   www.ehealthmd.com

2. A woman tells you she is considering botulinum injections to remove wrinkles. What would you tell her about the procedure?
   www.nsc.gov.sc

3. A relative asks your advice about a skin cream that reduces oxidation of the lower skin layers and, thereby, reduces skin aging. Explain to her the effectiveness of the skin cream based on that claim.
   www.aad.org/public

4. A friend says to you that he was told to go to a tanning booth to get a "healthy" tan. He asks you to explain the difference between a "healthy" tan and an "unhealthy" tan. How would you respond?
   www.skincancer.org

5. You are talking to the family of a boy who suffered third-degree burns on over 25% of his body. They want you to tell them why he is being given fluids directly into his bloodstream even though he appears healthy and able to drink?
   www.webmd.com

Skills Activities

Study Guide

1 Histology of the Integumentary System

Materials
- microscope with high-power capability (400X)
- prepared slide of human skin – normal or thin section
- prepared slide of human skin – with hairs
- prepared slide of human skin – thick skin or callus
- computer with access to the Internet
- browser bookmark at Virtual Hospital Medical Atlas: http://www.vh.org/adult/provider/anatomy/Microscopic Anatomy/MicroscopicAnatomy.html

As a fetus develops, the skin forms particular structures in specific areas of the body. This differentiation is what gives the hair on the head a different appearance than the hair on the rest of the body. The skin also adapts itself to conditions imposed on a particular region. For example, the skin on the sole of the foot is subject to different conditions than the skin on the back of the arm. So, these skin areas become slightly modified in response to environmental exposure, as well as pressure placed on them during day-to-day living. Scientists and physicians are able to look at the skin structure under a microscope to see if it is carrying out its normal function in a particular region. However, to do this they must have a database of microscopic skin images for comparison. Activities such as the one provided below are used to train physicians and technicians to use image databases to help understand skin disorders.

On a computer, load the Virtual Hospital Medical Atlas Website. Then, navigate to Section 7, which provides information about the human integument. Click on the various plates to microscopic images of skin from different body regions.

Normal Skin Now, place a microscopic slide of the normal skin under the microscope. Start with the low magnification, and move to the higher magnification to see details and individual cells. Can you find the three layers of skin by comparing the slide with the photographs shown in Section 7 of the Virtual Hospital Medical Atlas? Draw a simple picture showing the relative thickness of each skin layer. Are any other skin structures or appendages visible in the atlas images that are not evident in the slide?

Skin with Hair Look at a specimen of the skin with hair under the microscope. How does it compare with the slide of the normal skin? Besides the hair, are there any structures in this skin section that are not present in normal skin. Use the Virtual Hospital Medical Atlas to find the part of the body where this skin section is most likely found.

Thick Skin Place the thick skin specimen on the microscope and view the various regions of the skin. How do these areas compare with those on the slide of the normal skin and skin with hair sections? Again, use the Virtual Hospital Medical Atlas to find the part of the body where this skin section is

THE SKIN AND ITS PARTS

Study Guide

most likely found? What environmental factors or body conditions cause this skin to differ from the normal skin? What causes the skin to grow the thick portion?

Skills Activities

2 Effectiveness of Sunscreen at Blocking Ultraviolet Light

Materials
- ultraviolet light or "black light"
- black Sharpie® marker
- bright yellow highlighter marker
- black sheet of paper 8 ½ in × 11in
- five clean microscope slides
- 10 small cords
- one small bottle of sunscreen rated SPF 10
- one small bottle of sunscreen rated SPF 60
- one small bottle of hand lotion
- one small bottle of shampoo
- digital camera (optional)
- surgical gloves (optional)

Sunscreen is meant to protect the skin from ultraviolet light damage. As mentioned in this chapter, the skin secretes melanin into the epidermis in an attempt to block some ultraviolet light from reaching the lower layers of skin. However, this is most effective in people with dark skin or after the skin has tanned in people with lighter skin. Melanin does not stop all the ultraviolet light from penetrating the inner layers of skin. So, people are urged to use sunscreen to prevent ultraviolet light damage of nontanned and lightly tanned skin. Not all substances are equally effective at blocking ultraviolet light. This activity provides a way of testing the effectiveness of sunscreens and chemicals claiming to be sunscreen.

Use the yellow highlighter marker to draw an 11-inch line through the middle of the black paper. Turn on the ultraviolet or black light, and hold it over the center of the black piece of paper. What happens to the marker line? The line should glow brightly in response to the ultraviolet light. This glow is called fluorescence. It is due to the interaction of the ultraviolet light with light-absorbing chemicals in the highlighter ink. The glow is due to the light energy given off by electrons. Now, use the black marker to label the slides one through five. Next, prop the five slides on the corks so that they each sit over the yellow line, as shown in the figure opposite. Make sure they are arranged in numbered order from one through five. Keep little or no space between the slides.

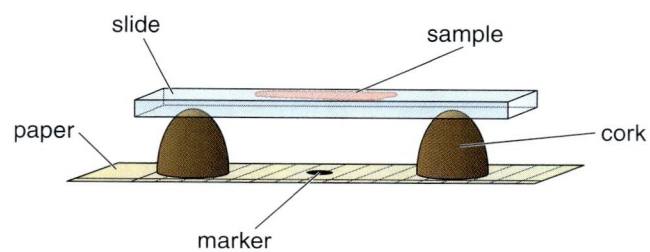

Gently smear slide number 2 with the SPF 10 sunscreen. To slide 3, add the SPF 60 sunscreen. Repeat this procedure adding hand lotion to slide 4, and the shampoo to slide 5. Next, shine the ultraviolet light or black light over the slides to evaluate whether the highlighter glows. The brighter glow means that more ultraviolet light is passing through the substance to the highlighter on the paper. What effect does glass have on blocking ultraviolet light? Is it possible to see the benefits of using strong sunscreen in this activity? What is the sun-blocking value of hand creams and shampoos? How could this activity be used to show people the value or effectiveness of different sunscreen substances?

Precautions It is recommended that surgical gloves are worn when handling any of the chemicals, in case of allergies or sensitivities to a particular product. Never aim the ultraviolet light at the eyes or the skin. The black light should not be stared into for more than a few seconds. Slides should be cleaned with warm, soapy water and then reused.

Case Study Investigation

Case Study Investigation #5

There is a volunteer neighborhood cleanup day going on in your community. A young boy who is helping out by dragging a heavy bag of litter all of a sudden lets out a loud cry of pain. You are thinking that maybe the boy pulled a muscle. To your surprise, you notice that his hand is dangling as if he broke his forearm. You rush the child to his parents, and he is hurried off to the hospital. Later that day, you see the boy's father, and he tells you how the boy is doing. He laments that his son should really be more careful about his activities. His son has broken both his legs twice, as well as that same arm, on the padded school playground. He then explains that it takes a long time for the boy's bones to heal. The father mentions that two years ago the boy lost his hearing in one ear after getting too close to an exploding fire cracker. The father ends the story saying, "Otherwise, he is your typical kid. He hardly has a sick day off from school, and he is growing like a weed." Why is this boy so accident-prone? How is it that his bones break so easily? At the end of the chapter, you will be asked to determine the possible reasons why this boy repeatedly hurts himself so severely.

CHAPTER 5

THE SKELETAL SYSTEM

Chapter Outline

Case Study Investigation (CSI)
Applied Learning Outcomes
Overview
Human Skeletal System
    Axial Skeleton
        Skull
        Vertebral Column and Rib Cage
    Appendicular Skeleton
Bone
    Bone Types
    Bone Structure
        External Features
        Internal Features
Joints
    Joint Structure
    Joint Function
Human Bone Charts
Bone Development and Healing
Wellness and Illness over the Life Span
    Pathology of the Skeletal System
    Aging of the Skeletal System
CSI Conclusion
Study Guide

Applied Learning Outcomes

- Use the terminology associated with the skeletal system.
- Learn about the following:
  - skeleton structure
  - bone structure and types
  - bone function
  - bone tissue
  - bone development and growth
  - bone physiology
  - bone articulations
- Understand the aging and pathology of the skeletal system.

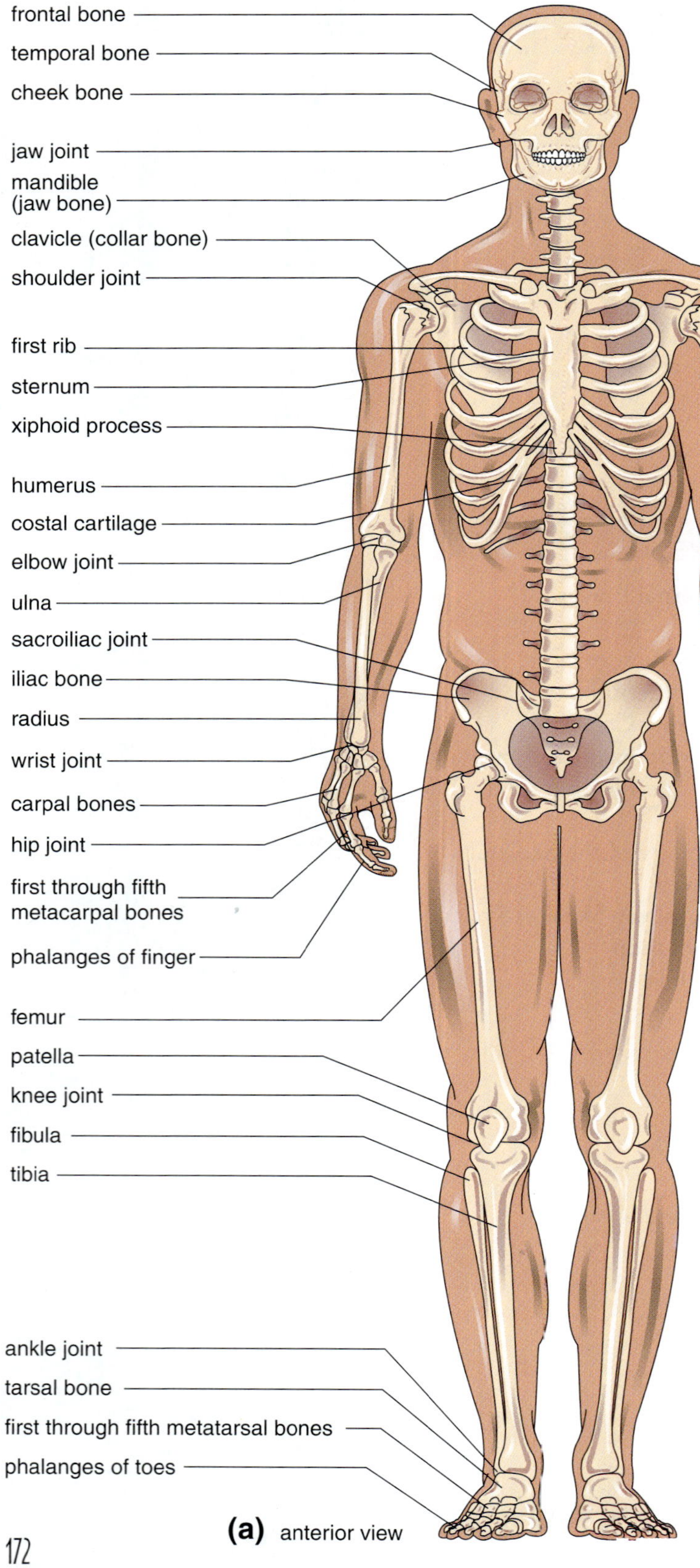

Labels (anterior view):
- frontal bone
- temporal bone
- cheek bone
- jaw joint
- mandible (jaw bone)
- clavicle (collar bone)
- shoulder joint
- first rib
- sternum
- xiphoid process
- humerus
- costal cartilage
- elbow joint
- ulna
- sacroiliac joint
- iliac bone
- radius
- wrist joint
- carpal bones
- hip joint
- first through fifth metacarpal bones
- phalanges of finger
- femur
- patella
- knee joint
- fibula
- tibia
- ankle joint
- tarsal bone
- first through fifth metatarsal bones
- phalanges of toes

(a) anterior view

OVERVIEW

Key Terms: bone, cartilage, endoskeleton, ligament, tendon

Essential to any animal that walks on land is a strong, but lightweight, skeleton. Humans possess what is called an **endoskeleton**, or internal skeleton, that develops from mesenchyme cells of the embryonic mesoderm. The skeletal system is an organ system composed of **bones**, **cartilage**, **ligaments**, and **tendons**. The human skeleton is composed of over 200 bones of various shapes and sizes. Bones are complex organs associated with blood vessels, nerves, and stem cells. The stem cells are capable of producing a variety of body cells. The human skeleton is effective at providing movement, protection, shape, and support for the human body. Muscles alone do not give the body the capacity for movement. The human skeletal system provides the scaffolding needed for efficient movement. The smallest fall could severely damage the brain and internal organs if there were no skeletal system to protect them. Enclosures made of bone protect the brain and many internal organs. The various shapes and arrangements of bones in the skeletal regions form almost all of the contours of the face and body. The posture needed to hold the body upright would not be possible without the skeletal system. It supports the body in any position, during any type of movement.

Figure 5.1 Human Skeletal System – Anterior (a) and Posterior (b)

CHAPTER 5

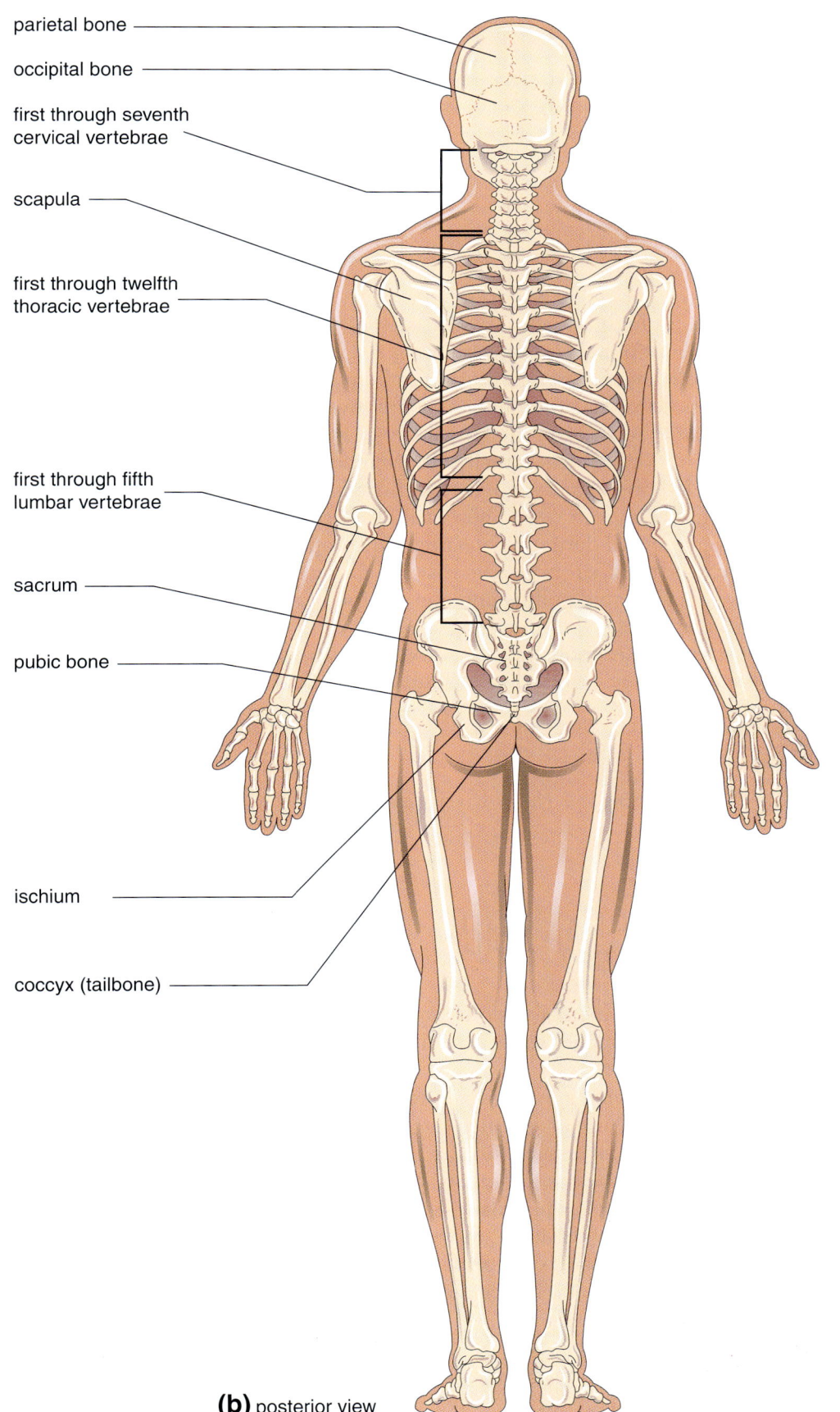

(b) posterior view

THE SKELETAL SYSTEM

✓ Concept Check

1. Explain why the skeletal system is an organ system.
2. Describe the components of the skeletal system.
3. Describe the four structural roles of the skeletal system.

STRETCHING: THE TRUTH

Most personal trainers would recommend a short stretching routine before and after any physical activity. This is true for high-impact activites, such as weightlifting, or low-impact aerobic exercises. Their rationale is that stretching reduces the risk for joint injury. This generally accepted practice is now being scrutinized by medical researchers. Certain studies of sports injury are showing that stretching may not lower the risk for injury. Exercise physiologist, Dr. David A. Lally, at the University of Hawaii–Manoa, discovered that preworkout stretching in 1543 runners actually contributed to the number of running injuries. His research paralleled other studies covered in Austalia. It is estimated that 65% of all runners sustain an injury in a given year. These data show that 47% of male runners who stretched regularly were injured during a 1-year period compared with 33% of male runners who did not stretch. Some doctors believe that stretching before exercise does not condition the tendons for the rapid pulling and contraction of physical activity. Researchers cannot find a reason why the stretching contributed to more injuries; however, studies do show that stretching is good *after* a workout. It is shown to be a good way to reduce tension on tendons and bone by relaxing the pull of tightened muscles.

Lally, D. 2002. Should Static Stretching Be Used During a Warm-Up for Strength and Power Activities? Strength and Conditioning Journal. Vol. 24(6), pp. 33-37.

Lally, D. 1994. New Study Links Stretching with Higher Injury Rates. Running Research News. Vol. 10(3), pp. 5-6.

THE HUMAN SKELETAL SYSTEM

Key Terms: appendicular skeleton, articulations, axial skeleton, lower appendage, surface features, upper appendage

Axial Skeleton The part of the skeleton composed of the spine, rib cage, and skull

Appendicular Skeleton The part of the skeleton composed of the upper and lower appendages, and the bones that girdle them to the axial skeleton

Upper Appendages The shoulders, arms, wrists, and hands

Lower Appendages The hips, legs, knees, ankles, and feet

The human skeletal system is divided into two anatomically distinct skeletal regions (Figure 5.1). Supporting the core of the body is the **axial skeleton**, which is composed of the spine, rib cage, hyoid bone, and skull (Figure 5.2). The axial skeleton is the basis of much age-related pathology because it functions nonstop throughout life, balancing and absorbing a large amount of the stress associated with lying down, moving, sitting, and standing. Attached laterally to the axial skeleton, is the **appendicular skeleton** (Figure 5.7). It is composed of the **upper** and **lower appendages**, and the bones that connect them to the axial skeleton. The lower appendages include the bones of the feet, ankles, legs, kneecap, and hips. Bones of the hands, wrists, arms, and shoulders comprise the upper appendages. The appendages take on the strains of movement and share some of the axial skeleton's work. Pathology mostly

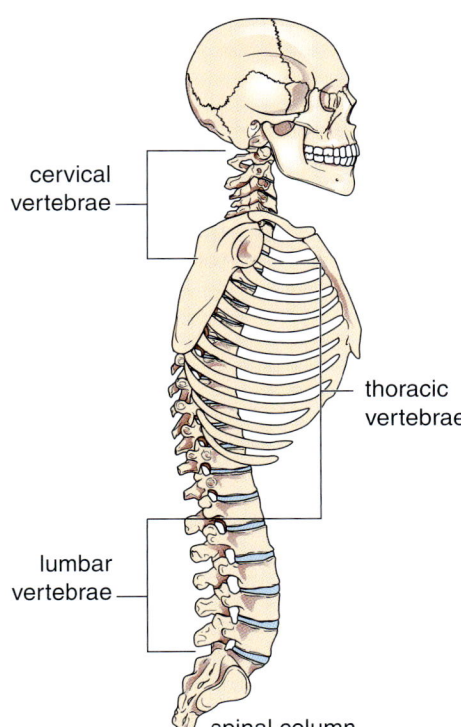

takes its toll on the joints of the appendages. **Articulations** connect each bone in the skeletal system directly to another bone. The articulations are areas where two bones meet to form a joint. Joints vary greatly in tissue composition and function (they will be described in detail later in this chapter).

Each bone has characteristic **surface features** that result from its attachments to ligaments and tendons. Ligaments and tendons pull on the bone, forming a variety of protrusions and edges. Additional bone characteristics are formed where bones meet at joints, and where blood vessels and nerves come in contact with bone. The more common bone features are listed in Table 5.1.

Articulation A junction between two or more bones

Surface Feature Protrusions and edges on bone formed by the pull of ligaments and tendons

Figure 5.2 The Axial Skeleton

Table 5.1 Characteristics of Bone

articulation	The region where adjacent bones form a joint
articular process	A bulge that contacts an adjacent bone near a joint
canal	A wide foramen that is usually a passage for larger blood vessels and nerves
condyle	A large and rounded articular process, which is usually covered with cartilage
cornu	A small, horn-shaped protrusion that attaches to ligament or tendon
crest	A large ridge where muscles attach
diaphysis or shaft	The long, main body of a bone
eminence	A small bump where muscles attach
epicondyle	A bulge near a condyle where ligaments and tendons attach
facet	A small, smooth, articular surface usually covered with cartilage
foramen	An opening through a bone where a small blood vessel or nerve passes into the bone
fossa	A wide and shallow indentation formed by a muscle or nerve pressed against the bone
hamulus	A small, hooked protrusion that attaches to ligament or tendon
head	The proximal or distal articular end of a bone, usually covered by cartilage
line or ridge	A long, thin bulge, usually having a rough surface that attaches to muscle
malleolus	A hammer-shaped knob on the end of a bone that attaches to ligament and tendon
meatus	A short canal with a variety of purposes
neck	A region of bone between the head and the shaft
process	A large bulge that serves as a muscle attachment
sinus	A cavity within a cranial or facial bone
spine	A long, narrow projection for muscle attachment
sulcus	A long furrow formed by a muscle or nerve pressed against the bone
suture	A ridged articular surface attached to other bones of the skull by fibrous connective tissue
tuberosity	A large, irregularly shaped bump that attaches to ligament and tendon
tubercle	A small, irregularly shaped bump that is generally smaller than a tuberosity and attaches to ligament and tendon
trochanter	A large, ridged tuberosity that attaches to ligaments and tendons

✓ Concept Check

1. Describe the two divisions of the human skeleton.
2. Which body parts make up the axial and appendicular skeletons?
3. Describe the different origins and purposes of bone-surface features.

DISCOVERY SCENE PLEASE ENTER DISCOVERY SCENE PLEASE ENTER

Which terms and concepts covered so far can help you to better understand the CSI? What would you need to ask to understand the nature of the bone damage?

Axial Skeleton

Key Terms: alveolus, atlas, axis, calvaria, cervical vertebrae, coccygeal vertebrae, coccyx, coronal suture, cranial base, cranial bone, cranium, dentin layer, enamel, ethmoid bone, facial bone, false vertebrae, fixed vertebrae, foramen, frontal bone, gladiolus, hyoid, inferior nasal concha, inferior orbital foramen, lacrimal bone, lumbar vertebrae, mandible, manubrium, mastoid process, maxillary bone, mental foramen, movable vertebrae, nasal bone, nasal septum, occipital bone, occipital foramen, orbit, orbital ridge, palatine bone, parietal bone, ribs, sacral vertebrae, sacrum, sagittal suture, sphenoid bone, squamousal suture, sternum, styloid process, temporal bone, thoracic vertebrae, thorax, true vertebrae, tympanic region, vertebrae, vertebral column, vomer, xiphoid process, zygomatic bone, zygomatic process

The axial skeleton is composed of all the bones that are located along the vertical axis of the body. The skull is the superior-most component of the axial skeleton. The vertebral column is the medial-most portion of the axial skeleton. Ribs and the other bones making up the rib cage are considered part of the axial skeleton because they are functionally linked to the vertebral column.

Civic Responsibility

CIVIC RESPONSIBILITY: HELPING OTHERS WITH YOUR KNOWLEDGE

It is valuable to use what you have learned about the skeletal system to help others to better understand the world around them. It is very important to check your facts and seek further information about certain topics before discussing health and science issues. Here are some suggestions to foster a better public awareness of the skeletal system:

1. Speak to schoolchildren about eating right to maintain healthy bones.
2. Work with sports clubs to educate the players about joint care.
3. Help the elderly to better understand the effects of aging on bones and the skeleton.
4. Volunteer at a school health day to teach children the basics of their skeletal system.

Figure 5.3 **Skull**

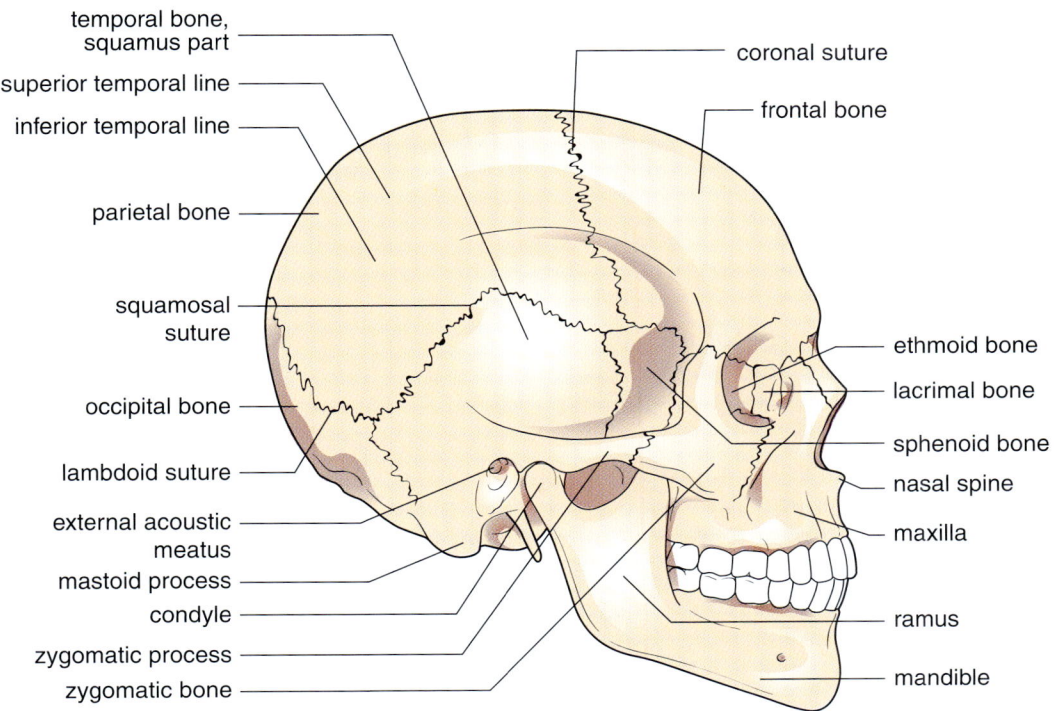

Cranium The skull
Cranial Bones Eight bones of the cranium that protect the brain
Calvaria The dome-shaped superior portion of the cranium
Cranial Base The portion of the cranium composed of the ethmoid and sphenoid bones
Frontal Bone Bone forming the front part of the skull, shaping the forehead
Occipital Bone Bone forming the back part of the skull
Parietal Bones The two bones on either side of the skull that form the sides and roof of the skull
Temporal Bones The two bones that form the sides and base of the skull
Ethmoid Bone The bone forming the roof of the nasal cavity and inner wall of the eye socket
Sphenoid Bone The bone forming the anterior base of the cranium and the posterior orbit
Facial Bones The 15 bones forming the face
Inferior Nasal Conchae Bones forming an inferior protrusion in the nasal cavity
Lacrimal Bones Bones forming the medial region of the orbits
Mandible The lower jawbone
Maxillary Bones Bones forming the upper jawbone
Nasal Bones Bones forming the bridge of the nose
Palatine Bone The bone forming the walls of the nasal cavity and the posterior roof of the mouth
Vomer The bone forming the inferior part of the nasal septum
Zygomatic Bones Bones forming the cheeks
Orbits Eye sockets
Orbital Ridge The thickened area of frontal bone above the orbits

Skull The skull is composed of several bones that come together in the embryo to form one functional structure called the **cranium** (Figure 5.3). **Cranial bones** protect the brain, ears, and eyes from physical damage. Some of the cranial bones are attachment points for muscles that move the head, jaw, and neck. Muscles that produce facial features are also attached to cranial bones. Many physicians divide the cranial bones into two categories: the **calvaria** and the **cranial base**. The calvaria is the dome-shaped superior portion of the cranium. It is composed of the **frontal**, **occipital**, and **parietal bones**, and the flat portion of the **temporal bones**. The cranial base is composed of the two remaining cranial bones, the **ethmoid bone** and the **sphenoid bone**. Fourteen **facial bones** form the other components of the skull. The facial bones are composed of the **inferior nasal conchae**, **lacrimal bones**, **mandible**, **maxillary bones**, **nasal bones**, **palatine bones**, **vomer**, and **zygomatic bones**. Several of the skull bones contain sinuses. Sinuses are air cavities lined with pseudostratified, ciliated, columnar epithelia containing goblet cells that secrete mucus. Sinuses carry out many functions depending on the bone in which they are contained.

The frontal bone makes up the forehead and eyebrow ridges. It is actually two bones that fuse into one at birth. It consists of a vertical portion that forms the forehead and an orbital region that forms the upper parts of the eye sockets, or **orbits**. A thickened area of the frontal bone above each orbit is called the **orbital ridge**. The orbital ridge underlies the eyebrows and varies among people. Males have a thicker orbital ridge than females. In addition, the male's eyebrows sit lower on the orbital ridge. Posterior to these ridges, embedded in

Coronal Suture The suture joining the frontal and parietal bones

Sagittal Suture The suture joining the parietal bones

Squamousal Suture The suture joining the parietal and temporal bones

Mastoid Process The attachment for neck muscles

Styloid Process The slender process that projects from the temporal bone

Tympanic Region The region containing the ear bones

Zygomatic Process Articulates with the zygomatic bone to form the cheek

Foramen Magnum or Occipital Foramen The opening formed by the occipital bones through which the spinal cord enters the brain

Inferior Orbital Foramen A passageway to the eye for blood vessels and nerves

the frontal bone, are the left and right frontal sinuses. It is surprising to learn that the frontal sinuses are not present at birth. They start developing at 2 to 5 years of age and reach their final size during the late teens. It is believed that these sinuses lighten the weight of the skull and warm the air taken in for breathing. The posterior portion of the frontal bone is joined to the two parietal bones along a joint called the **coronal suture**. Parietal bones form the sides and roof of the cranium. They are fused to each other medially at the **sagittal suture**.

The two temporal bones are located at the sides and base of the skull. They are fused to the parietal bones by the **squamousal suture**. The temporal bones have a complex shape with several distinct anatomical features: the **mastoid process**, **styloid process**, **tympanic region**, and **zygomatic process**. The mastoid process is on the posterior part of the bone and is an important attachment for neck muscles. It contains a complex network of cavities called the mastoid sinuses. These sinuses are connected to the inner parts of the ear; they can be invaded by bacteria, causing ear and throat infections. Little is known about the function of the mastoid sinuses. It is believed they may help the inner ear adjust to pressure changes. The styloid process is a slender, pointed protrusion that projects downward and forward from the inferior surface of the temporal bone. It varies in length among people, and is an important attachment point for muscles and tendons of the neck and throat. The tympanic region houses the internal components of the ears, including three small ear bones, the malleus, incus, and stapes, which develop in the embryo from facial bones.

The occipital bone forms the posterior base of the cranial vault. It is actually a series of bones that come together in the embryo, leaving an opening called the **foramen magnum**, or **occipital foramen**, which encircles the part of the spinal cord that enters the brain. Lateral to the occipital foramen are the occipital condyles. These very important processes attach the head to the vertebral column, and allow head movement. The sphenoid bone is situated anterior to the temporal bones. It makes contact with almost all of the other skull bones and helps form the facial features. The sphenoid is a very important bone for many reasons. First, it contains the sphenoidal sinuses, which are connected to the inner part of the nasal cavities. They appear after birth and are formed completely within a year after puberty. Their function is debated; they may be related to the sphenoid's role in forming part of the nasal cavities. Another important feature of the sphenoid is that major blood vessels, and nerves of the eyes and face pass through this bone. In addition, the superior surface of the sphenoid contains a structure called the sella turcica. The sella turcica encases the pituitary gland, which is a part of the brain and endocrine system. Anterior to the sphenoid bone is the small, spongy ethmoid bone. Its upper surface contains the cribriform plate, which is the passageway for the nerves that detect smell. The ethmoid bone also contains the ethmoid labyrinth, which forms the upper part of the nasal cavity. It is believed to clean and moisten the air that enters the nose.

The mandible, maxillary bone, and zygomatic bones form most of a person's facial features. The two maxillary bones form the upper jaw, and is of one of the larger facial bones. They form the areas around the nose (the lateral surfaces of the nasal cavity) and underneath the eyes (the medial border of the orbits). Underneath the orbit is the **inferior orbital foramen**, which is a passageway for major blood vessels and nerves. Within the maxillary bones are two large sinuses (the maxillary sinuses) that, again, have speculative functions.

On the lower surface are special structures in which the teeth develop. The zygomatic process of the temporal bone connects with the zygomatic bone to form the cheek. The zygomatic and temporal bones form the zygomatic arch. Muscles that assist with chewing attach to the zygomatic arch. In females, the zygomatic arch sits higher and further forward than in males.

The mandible, or lower jaw, is the largest of the facial bones. In males, the mandible is usually thicker and larger than in females. This complex bone is divided into the alveolus, body, condyle, and ramus. The body contains the teeth, and the condyle articulates with the temporal bones to form the joint that moves the jaw. A tooth is actually a bone that grows out of an **alveolus** in the mandible, or maxilla (Figure 5.4). Each tooth has a crown and a root. The crown projects above the gum while the root is embedded in the alveolus. A hard material called **enamel** covers the crown of each tooth. It is a hard, thin, transparent layer of calcium and protein that protects the dentin layer. **Dentin** is the layer of the tooth that protects the pulp. Teeth will be described in more detail in Chapter 13, which describes the digestive system. **Mental foramina**, holes that serve as conduits for blood vessels and nerves to the skin of the chin and lower lip, are also present on the mandibular body. The mandible, along with the ear bones and **hyoid bone**, are the only skull bones that are not fused to other bones of the skull. The tiny nasal bones are fused to the frontal and maxillary bones, forming the bridge of the nose. Facial injuries can dislocate these fragile bones. Contrary to many stories, the nasal bones cannot be smashed into the brain by a hard, upward hit to the nose. They will merely dislocate or shatter as they press against the thick frontal bone. The smallest and most fragile of the facial bones is the lacrimal bone. It is a sliver of bone on the anterior region of each orbit. The lacrimal bone helps form the lacrimal groove, which directs tears into the nasal cavity. The horseshoe-shaped hyoid bone is suspended below the mandible by muscles. It anchors anterior muscles of the neck and attaches to the muscles of the tongue.

Alveolus A socket in the jawbone out of which a tooth grows

Enamel A hard layer that protects the dentin

Dentin The bone-like substance of the tooth

Mental Foramen Passageways to the teeth for blood vessels and nerves

Hyoid Bone The U-shaped bone in the neck supporting the muscles of the tongue, larynx, and pharynx

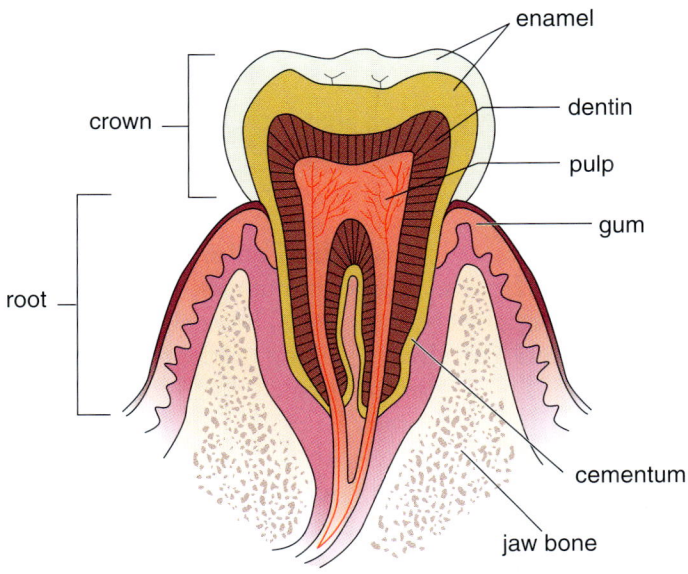

Figure 5.4 **The Tooth**

Good Choice Bad Choice

A young woman is very concerned about developing osteoporosis as she ages. She decides to visit a health-food store after reading an advertisement for a calcium-enriched functional drink. Functional drinks are juices and mixtures enriched with minerals, nutrients, and vitamins. The functional-drink sales market is larger than the market for sport and energy drinks. The companies selling these products claim that the drinks apply scientific evidence to effectively respond to customer needs and concerns. Do functional drinks provide any real benefits? Should they be taken without consulting a physician? What is the harm of taking these drinks to rebuild the body after athletic performance or to ward off disease?

Nasal Septum A bony plate that divides the nasal cavity

Vertebral Column A flexible column formed by the vertebrae

Vertebrae Bones that form the vertebral column

Cervical Vertebrae Seven vertebrae of the cervical region

Thoracic Vertebrae Twelve vertebrae of the thoracic region

Lumbar Vertebrae Five vertebrae of the lumbar region

Sacral Vertebrae Five vertebrae of the sacral region

Coccygeal Vertebrae Three to five vertebrae of the coccyx region

True (Movable) Vertebrae Vertebrae of the cervical, thoracic, and lumbar regions

False (Fixed) Vertebrae Vertebrae of the sacral and coccyx regions

The palatine bones are located at the back part of the nasal cavity, between the maxillary and sphenoid bones. They help form the walls of the nasal cavity, the roof of the mouth, and the bottom of the orbits. This bone pair grows in height as the facial features mature from birth through adolescence. The single vomer sits medially in the nasal cavity, fused to the ethmoid and sphenoid bones. This bone starts out as a sliver of cartilage until it completely hardens into bone after puberty. It is covered with mucous membranes and helps form the **nasal septum** in the nasal cavity. The nasal septum is a wall dividing the left and right nostrils. It is formed by the ethmoid bone, vomer, and a strip of cartilage. The inferior nasal conchae are located on the lateral sides of the nasal cavity; they are covered with mucous membranes that clean, warm, and moisturize air.

✓ Concept Check

1. Describe the two groupings of bones that form the skull.
2. Which bones make up the cranial bones?
3. Which bones make up the facial bones?

Vertebral Column and Rib Cage The **vertebral column** is a flexible column formed of a series of bones called **vertebrae**. In humans, the vertebral column is composed of five distinct regions (Figure 5.5). The neck, or cervical region, is composed of seven **cervical vertebrae**. Making up the thoracic region are twelve **thoracic vertebrae**. This region is articulated with the rib cage. In the lower back are the five **lumbar vertebrae**. The sacral region is composed of five fused **sacral vertebrae** that articulate with the hipbones of the appendicular skeleton. A series of three to five fused **coccygeal vertebrae** form the tail end of the vertebral column. The number of vertebrae present in a region can vary from one individual to the next, however, the number of cervical vertebrae rarely varies. Vertebrae in the three upper regions of the vertebral column are called **true**, or **movable**, **vertebrae**. The sacral and coccygeal vertebrae are called **false**, or **fixed**, **vertebrae** because they are fused to one another.

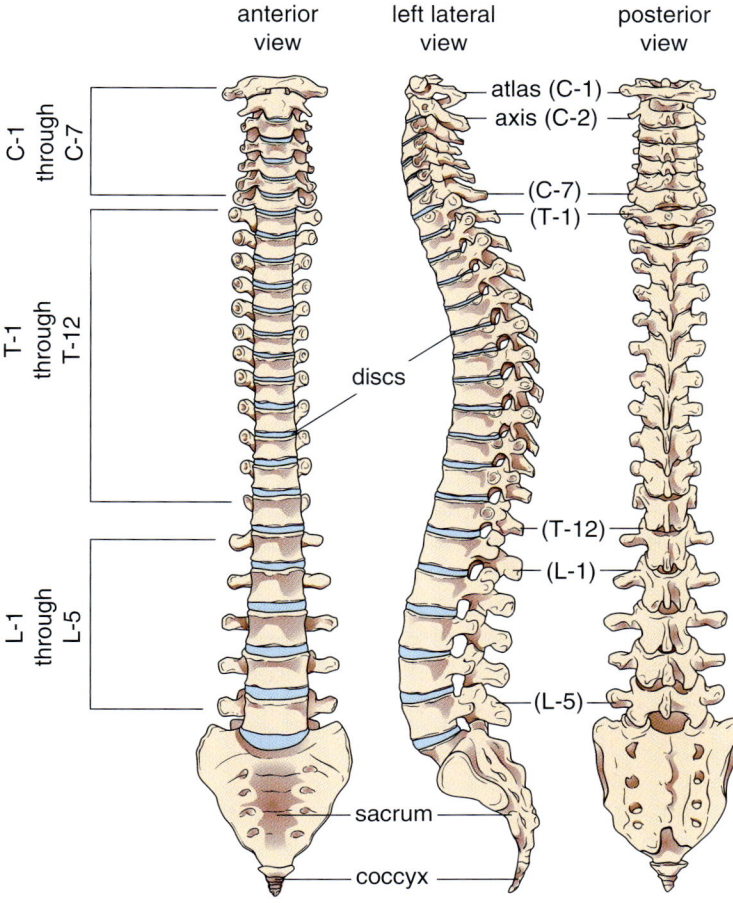

Figure 5.5 **The Vertebral Column**

A typical vertebral bone is composed of the vertebral body, vertebral arch, and the vertebral foramen. The vertebral body supports the weight of the body. Dorsal to the body is the vertebral arch, which is composed of a pair of pedicles, four articular processes, two laterally projecting, transverse processes, and a spinous process. These processes are important muscle attachment points that allow movement of the appendages, head, neck, and vertebral column, while the articular processes act to articulate the vertebrae with each other. The vertebral foramen is the passageway for the spinal cord and is continuous with the foramen magnum of the occipital bone. Cervical vertebrae have a smaller body than the other vertebrae. They also have small transverse processes and a short, bifurcated spinous process. Cervical vertebrae differ from other vertebrae in that they have two transverse foramina that serve as passageways for major blood vessels into the skull. The first and second cervical vertebrae, the atlas and the axis, respectively, differ from the others because of their specialized jobs. The **atlas** supports the skull; it lacks a body, but has a set of large articular surfaces that attach to the occipital bone. Below the atlas is the **axis**, upon which the atlas and head rotate. It is characterized by a large protrusion called the dens, or odontoid process. The dens articulates with the atlas to allow free movement of the skull.

Thoracic vertebrae have a larger body than the cervical vertebrae. Their spinous process is usually long and narrow. Unique to the thoracic vertebrae are two articulation points with the ribs. Each transverse process has an articular facet for one rib. Another rib facet is located on the dorsal portion of the

Atlas The first cervical vertebra, which supports the head

Axis The second cervical vertebra; the atlas and head rotate upon it

THE SKELETAL SYSTEM

Sacrum Five fused sacral vertebrae
Coccyx Fused coccygeal vertebrae
Thorax The rib cage and chest cavity
Rib One of the 12 bones that form the rib cage
Sternum The breastbone
Manubrium The upper region of the sternum
Gladiolus The body of the sternum

body. These facets form an articulation that permits the ribs to flex for breathing. The lumbar vertebrae are larger and thicker than the other vertebrae because they support most of the body's weight. They also work with the back muscles to balance the body while moving, sitting, and standing. The spinous process is short and stubby, allowing the vertebral column to bend backward easily. The five sacral vertebrae are tightly fused to form the **sacrum**. The lateral portions of the sacrum have a large articular surface that fuses with the hipbones. In the female, the sacrum is shorter, wider, and less curved than in the male. In most people, the vertebral arch disappears from the lower two sacral vertebra, exposing the sacral canal, which houses the spinal cavity. The spinal cord, interestingly, ends in the lumbar region. From a clinical perspective, this affords the opportunity to collect spinal fluid from this area with little risk for damage to the nerves. The coccygeal vertebrae fuse to form the **coccyx**, commonly known as the tailbone. It has no vertebral foramen and serves as an attachment point for muscles of the upper leg.

The rib cage, or **thorax**, is composed of the costal cartilage, **ribs**, and **sternum** (Figure 5.6). It is a protective structure that also assists with breathing. The ribs are large arches of bone that articulate with the thoracic vertebrae and sternum. Normally, there are 12 ribs; however, some people are born with small cervical and lumbar ribs. Scientists are discovering that each vertebra has the genetic potential to become a small rib. The seven upper ribs are called true ribs because they attach directly to the sternum via flexible costal cartilages. The eighth through tenth ribs are attached to the costal cartilage of the rib above and are called false ribs. The two bottom ribs are not attached to costal cartilage and are given the term floating, or vertebral, ribs. Lying medially and ventrally in the rib cage is the flattened sternum. The sternum is divided into three regions. An upper region, called the **manubrium**, articulates with the collarbones and ribs 1-2, and serves as an attachment point for chest and shoulder muscles. The **gladiolus**, or body, of the sternum articulates with the

Figure 5.6 The Thoracic Cage

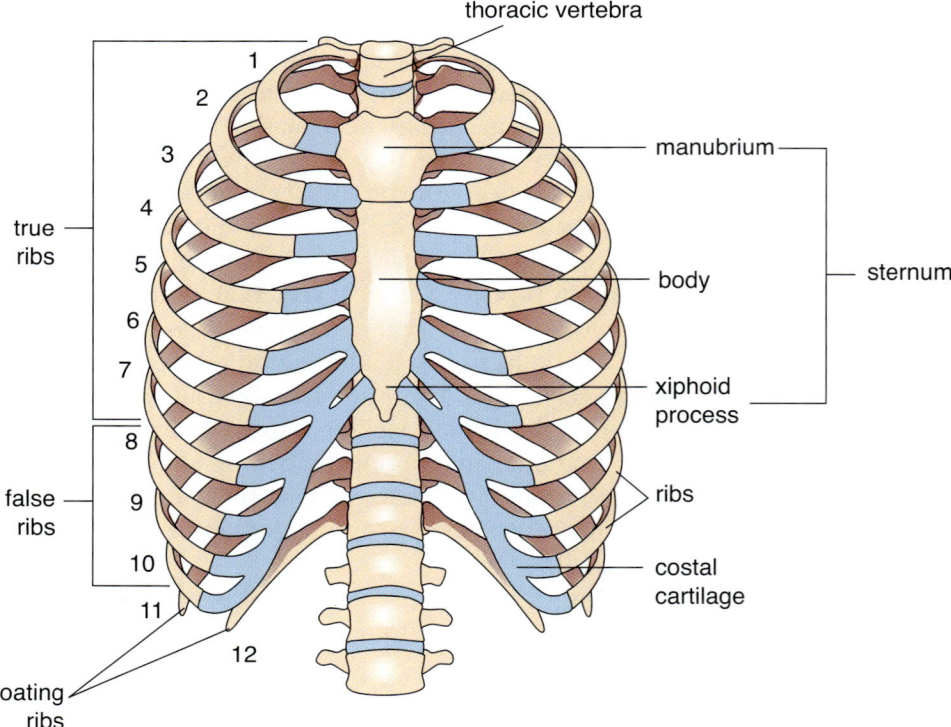

second through seventh ribs. It is also a chest-muscle attachment point. The lower portion of the sternum is the **xiphoid process**. It is a thin, elongated bone that is formed of cartilaginous material in children and turns to bone in adults. The xiphoid process is an attachment point for the stomach muscles.

Xiphoid Process The tip of the sternum

Extremities The upper and lower limbs

✓ Concept Check

1. Describe the four regions of the vertebral column.
2. What are the parts of a vertebral bone?
3. Describe the bones of the rib cage.

Appendicular Skeleton

Key Terms: acetabulum, acromion process, calcaneus, capitate, carpal, clavical, coronoid fossa, cuboid, cuneiform, distal phalanx, extremity, femur, fibula, hamate, humerus, ilium, ischium, lunate, medial phalanx, metatarsal, navicular, obturator foramen, olecranon fossa, olecranon process, patella, phalanges, pisiform, proximal phalanx, pubic bone, pubic symphysis, pubis, radius, scaphoid, scapula, shoulder blade, shoulder girdle, sternal extremity, talus, tarsal, tibia, trapezium, trapezoid, triangular, triquetral, ulna

The appendicular skeleton contains the various bones that make up the **extremities**. The extremities are defined as the upper and lower limbs (Figure 5.7). The upper limbs, or superior extremities, also known as the pectoral appendages, are attached to the rib cage by ligaments in a region called the shoulder, or pectoral, girdle. Bones of the lower limbs, or inferior extremities, are often referred to as the pelvic appendages. They are fused to the sacrum in a region called the pelvic girdle. The pectoral girdle bones are lightweight and flexible compared with the pelvic girdle. The two primary functions of the pelvic girdle are to support the weight of the body and to maintain an upright posture. These jobs require thick bones and a rigid attachment to the body. Each of the upper extremities is made up of two shoulder bones, three arm bones, eight wrist bones, and 19 hand bones. Each lower extremity has three fused hipbones, four leg bones, seven ankle bones, and 19 foot bones.

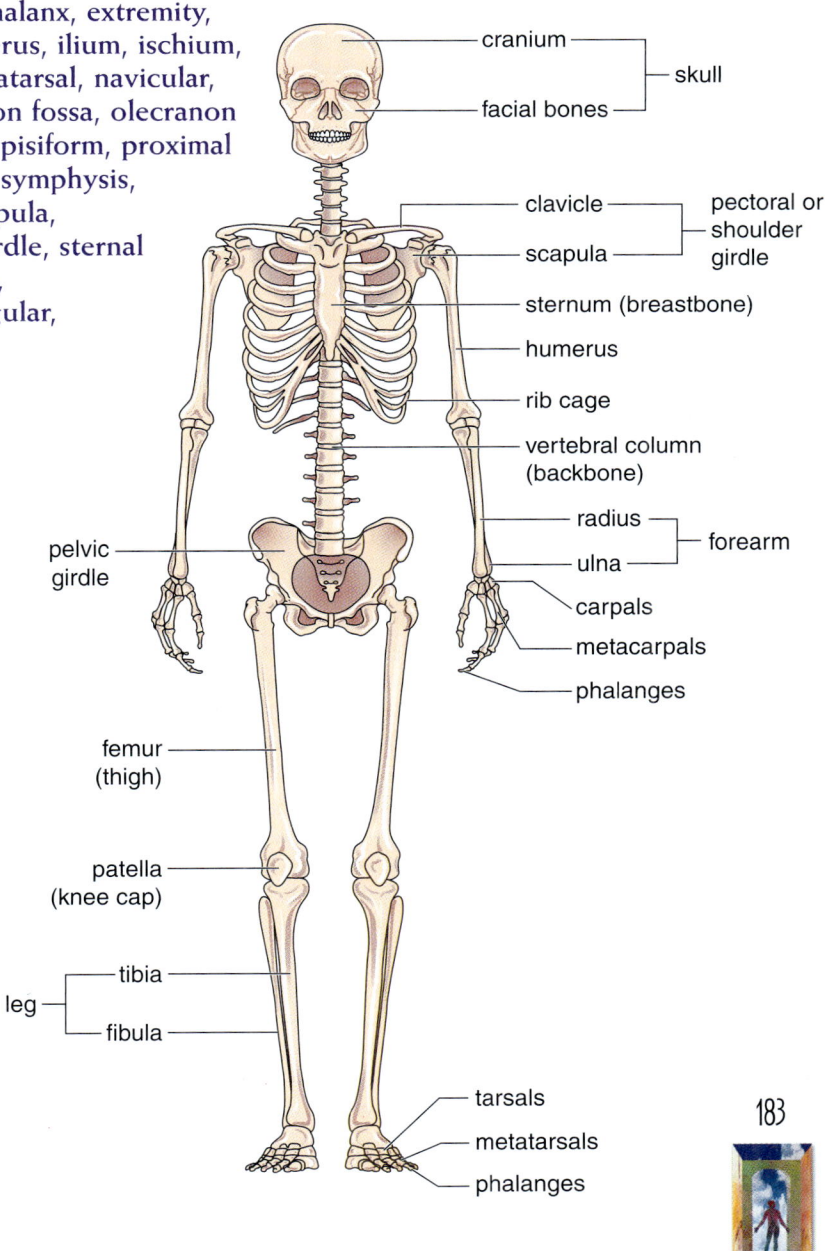

Figure 5.7 The Appendicuar Skeleton

THE SKELETAL SYSTEM

Clavicle or Collarbone The long bone that runs parallel to the first rib

Scapula or Shoulder Blade The flat bone parallel to the vertebral column

Humerus The longest bone of the upper arm

Radius The shorter of the two forearm bones

Ulna The longer of the two forearm bones

Trochlear Notch A crescent-shaped notch on the proximal end of the ulna that articulates with the humerus

Olecranon Process The proximal extremity of the ulna; "funny bone"

Carpal Refers to the wrist

Navicular or Lunate Bone A proximal carpal bone

Triangular or Triquetral Bone A proximal carpal bone

Pisiform Bone A proximal carpal bone

Trapezium The lateral distal carpal bone

Trapezoid A distal carpal bone

Capitate A large distal carpal bone

Hamate A distal carpal bone

Phalanges The bones of the fingers and toes

Ilium The upper bone of the pelvic girdle

Ischium The lower bone of the pelvic girdle

Pubis or Pubic Bones The lower front bones of the pelvic girdle

Acetabulum A deep-socket depression on the pelvic girdle that forms the point of articulation with the femur

Forming the shoulder girdle are the ventrally located **clavicle** and the dorsally located **scapula**. The clavicle, or collarbone, is a long, curved bone that runs parallel to the first rib. Its medial end, or sternal extremity, is attached to the sternum by ligaments. The lateral end is called the acromial extremity; it forms a complex joint with the scapula. Muscles of the arms, back, chest, and neck attach to the clavicle. This is why a broken clavicle severely restricts upper-head and upper-extremity movement. The scapula, or shoulder blade, is a flat, triangular bone with a blade-like medial surface that is parallel to the vertebral column. Its lateral end contains the glenoid cavity, which articulates with the humerus to form the shoulder joint. The scapula is a major attachment point for muscles that move the shoulders and arms. On the scapula's dorsal and lateral edges is the acromion process, a large, curved outgrowth that articulates with the acromial extremity of the clavicle.

The three arm bones are the **humerus**, **radius**, and **ulna**. As mentioned earlier, the humerus articulates with the glenoid cavity of the scapula. Significant features of the humerus are tubercles located near the head and the deltoid tuberosity in the lateral edge. They are important muscle attachment points for arm movement. On the distal portion of the humerus is a very specialized articular surface composed of two structures, the trochlea and the capitulum. Together with the radius and ulna they form the elbow joint. Just above the elbow joint are the coronoid fossa on the ventral side and the olecranon fossa on the dorsal side. Working as a unit, this joint and the fossae (plural for fossa) permit the characteristic hinge-like movement of the elbow. The radius and ulna form the forearm. When the hand is facing palm up, the radius lies lateral and parallel to the ulna. This bone forms a large articulation with the wrist and is responsible for rotating the hand. On the proximal portion of the ulna is the arch-shaped **trochlear notch**, which directly articulates with the trochlea of the humerus. The most proximal area of the humerus, the **olecranon process**, forms the hard, pointy part of the elbow. It is an important muscle attachment point for lower-arm movement.

A double row of eight articulated bones forms the **carpal**, or wrist. Articulated with the radius and ulna is the proximal row of carpal bones. It is composed of the **navicular** (or **scaphoid**), **lunate**, **triangular** (or **triquetral**), and **pisiform** bones. The navicular bone is immediately distal to the radius, and the lunate articulates with the ulna. A small, hard bump, which can be felt at the base of the wrist under the pinky finger, is the pisiform. The distal row of carpals contains the **trapezium**, **trapezoid**, **capitate**, and **hamate**. This row of carpals articulates with the five metacarpals that form the palm of the hand. The trapezium articulates with the first metacarpal, which moves the thumb. Each metacarpal is attached to a series of bones called the **phalanges** (singular is phalanx), which make up the finger bones. With the exception of the thumb, each finger has three phalanges: the proximal, middle, and distal. The thumb has no middle phalanx. Each distal phalanx has a flattened distal end that helps form the fingertip.

The pelvic bones brace the body to the lower extremities (Figure 5.8). The three bones making up the pelvic girdle are the **ilium**, **ischium**, and **pubis** (**pubic bones**). These bones are separate from each other at birth, but they become fused in adulthood to form an os coxae, or coxal bone. The **acetabulum**, a cup-shaped articular surface formed at the lateral point of the pubic bones' fusion to one another, forms the hip joint with the **femur**. Each set of pelvic bones meets at an articulation on the pubic bones called the

CHAPTER 5

Figure 5.8 Pelvic Bones – Male and Female

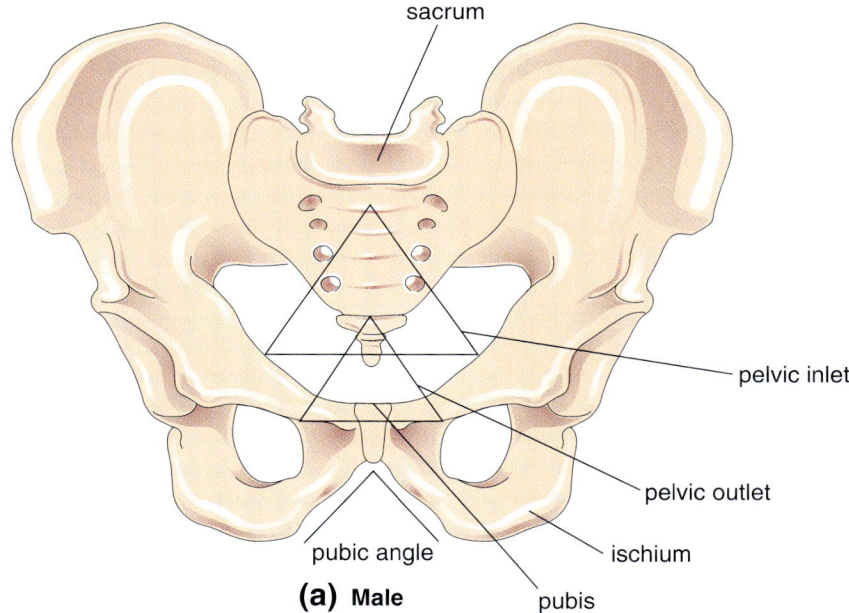

(a) Male

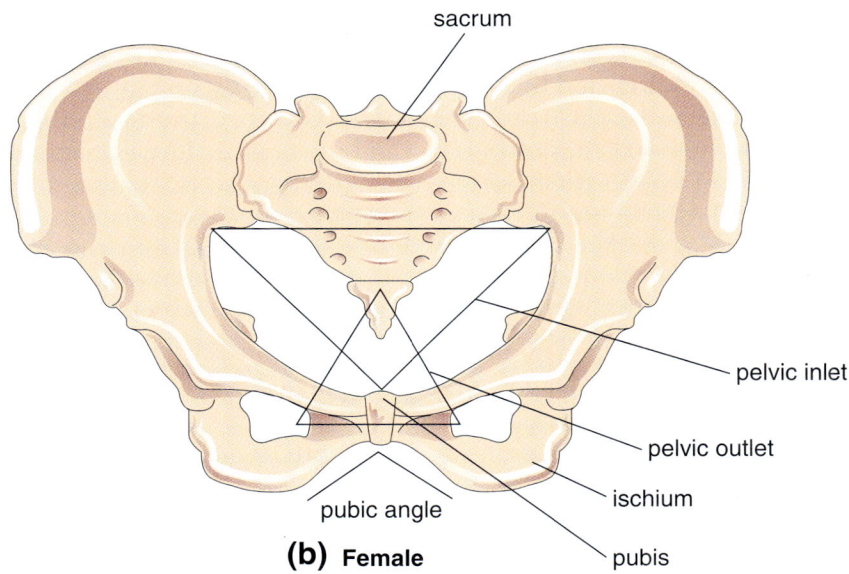

(b) Female

pubic symphysis. Many large muscles that permit back, hip, and leg movements are attached to the various crests and processes of the hipbones. Two large openings are present on the floor of the pelvic girdle between the ischium and pubis. Each one is called an **obturator foramen**, and forms a passageway for major blood vessels and nerves. In males, the obturator foramen is large and oval, while in females it is smaller and somewhat triangular. The angle formed by the dorsal union of the pubic bones is known as the pubic arch. Its size, as well as the position of the iliac bones, also differs in males and females. Males have a sharp angle between the pubic bones, and the iliac bones lie more parallel to the body. Female pubic bones have a wide pubic arch, and the iliac bones flair outward. The large cavity formed by the sacrum and pelvic bones is called the pelvic aperture. In females, the upper

Pubic Symphysis The articulation of the pubic bones

Obturator Foramen A passage for major blood vessels and nerves of the pelvis

THE SKELETAL SYSTEM

Tibia The medial bone of the lower leg
Fibula The lateral bone of the lower leg
Patella The kneecap
Tarsals The group of ankle bones
Metatarsals The long bones of the foot
Calcaneus One of seven tarsal bones, located at the lower and dorsal region of the foot forming the heel
Cuboid A large tarsal bone that articulates with the calcaneus
Navicular One of the tarsal bones located near the heel
Talus One of seven tarsal bones
Cuneiforms Three of seven tarsal bones forming the middle region of the foot

part of the pelvic aperture, called the pelvic inlet, is larger and more oval-shaped, while in males it is more circular, or heart-shaped.

Making up the leg bones are the femur, **tibia**, **fibula**, and **patella**. The femur is the longest and strongest bone in the body. It supports much of the body's standing weight. The proximal end of the femur is characterized by the large, rounded head that projects into the acetabulum to form the ball-and-socket joint of the pelvic girdle. The large process around the neck of the bone is made up of the greater trochanter and lesser trochanter. Powerful muscles involved in leg movement attach to the trochanters. The head and neck of the femur are specially arched to take the stress and strain of carrying the weight of the body. Breaks to this region require a long time to heal because the bone has to be completely restored before it can bear the weight of the body again. At the distal end of the femur is a specialized articular surface that forms the knee joint with the tibia and the patella. The patella, commonly referred to as the kneecap, is a flat, triangular bone located on the front of the knee joint. It protects the front of the joint and increases the leverage for knee joint movement. Making up the medial side of the lower leg is the tibia. It is the larger of the lower leg bones and articulates with the fibula on its lateral side. The fibula is the thinner of the two bones and is more subject to breakage from leg injuries. At the distal ends of each bone is a process called a malleolus, which articulates with the foot.

The foot is composed of the three groupings of bones called the **tarsals**, **metatarsals**, and phalanges. Seven bones make up the tarsals: **calcaneus**, **cuboid**, **navicular**, **talus**, and the first, second, and third **cuneiforms**. The talus plays a pivotal role in the ankle by forming the only articulation with the leg. It nestles between the distal ends of the tibia and the fibula. The calcaneus, the largest of the tarsals, forms the heel and part of the foot arch. Forming the middle portion of the foot are the cuboid, the navicular, the three cuneiform bones, and the five metatarsals. The arch of the foot depends on the integrity of the ligaments and muscles holding these bones in place. Defects in this arrangement can result in the condition commonly referred to as "flat-feet." The phalanges of the foot match up in number and organization with the phalanges of the hand. There are two phalanges, a proximal and a distal, in the big toe and three in each of the other toes, as they each have the additional middle phalanx. The phalanges of the toes are flatter and shorter than those in the hand.

✓ Concept Check

1. Describe the differences between the upper and lower extremities.
2. Name the parts and the bones of the upper extremities.
3. Name the parts and the bones of the lower extremities.

DISCOVERY SCENE PLEASE ENTER DISCOVERY SCENE PLEASE ENTER

CSI Break

Is there some type of connection that can be made between the function and structure of the axial skeleton and the boy's tendency to break bones? Why is this child's appendicular skeleton taking the brunt of the bone breakage? Is it due to the nature of the bones or merely the result of the types of accidents that cause the breakage?

Bone

Key Term: osseus

Osseus Bone tissue
Long Bone A bone having an elongated shape

As discussed in the overview, a bone is a living organ made up of a complex arrangement of tissues. A bone is composed of bone, or **osseus**, tissue, blood vessels, nervous tissue, ligaments, tendons, and a specialized connective tissue capable of producing stem cells. Bones can be categorized by their general shape, the relative composition of compact versus spongy bone, or the way that they develop in the embryo. A bone's shape is key to the role it plays in the body. Bone is adaptable to the various stresses and strains placed on the body, so bone structure is somewhat flexible. It can be modified by the way gravity presses on the body and the way muscles pull on attachment surfaces. Nutrition and exposure to various diseases during bone growth and development can also affect bone shape in adulthood. The variety of bone modifications becomes evident after examining the bones of many individuals. Anthropologists and pathologists use the variations in bone shape as indicators of a person's lifestyle or as evidence of certain diseases they may have experienced.

Bone Types

Key Terms: alveolar bone, dermal bone, endoskeletal bone, flat bone, irregular bone, long bone, sesamoid bone, short bone, sutural bone, wormian bone

The most common way to categorize bones is by their overall shape or dimensions (Figure 5.9). Bones are generally categorized into four shapes: **long**,

Figure 5.9 Bone Types

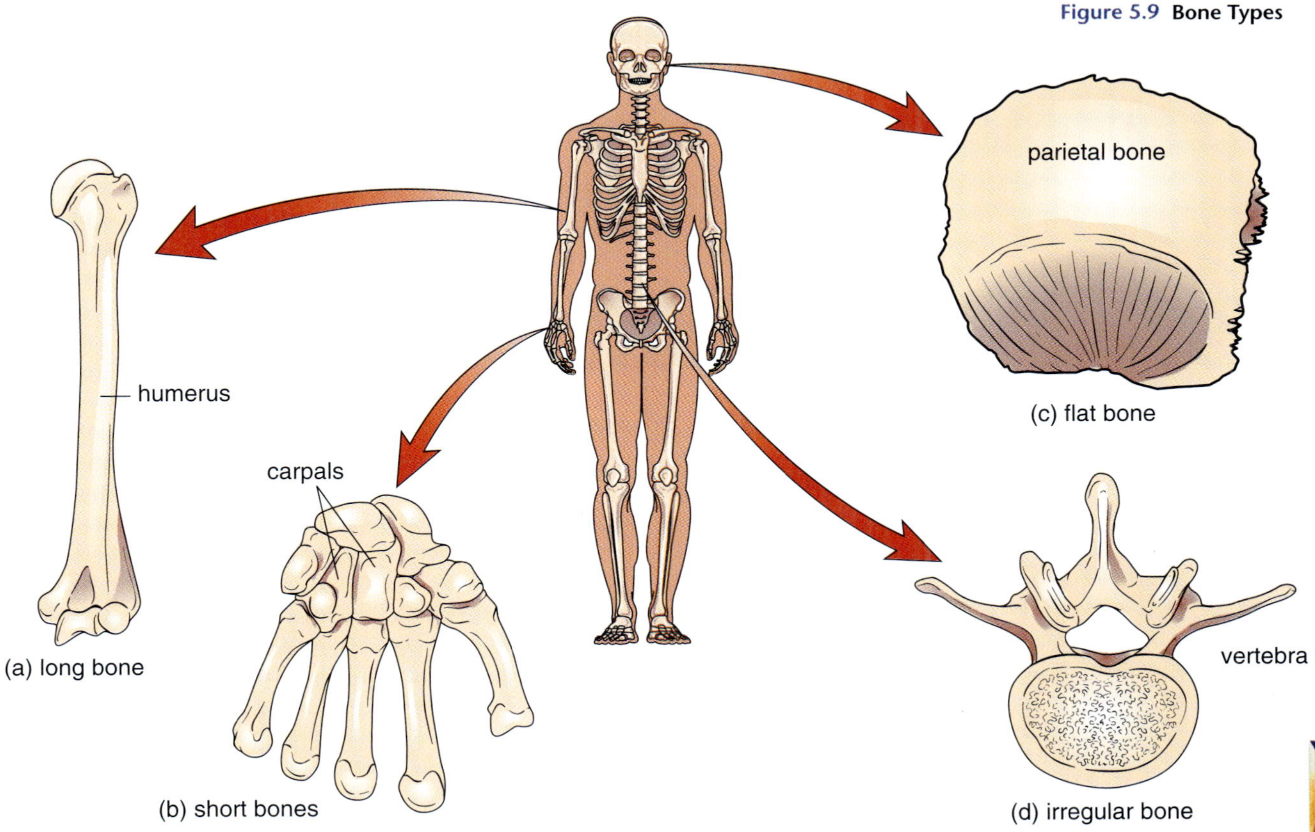

(a) long bone
(b) short bones
(c) flat bone — parietal bone
(d) irregular bone — vertebra

humerus
carpals

THE SKELETAL SYSTEM

Short Bone A bone having a square-like shape; one of four bone shapes

Irregular Bone A bone having a unique, often complicated, shape that is not geometrically describable

Flat Bone A bone that is thin and flattened, and often slightly curved

Endochondral Bones Bones developed from embryonic cartilage

Dermal (Intramembranous) Bones Bones that develop from embryonic connective tissue

Alveolar Bones Bones that develop from special cells found only in the jaw bones

Sesamoid Bones Bones that develop within tendons

Wormian or Sutural Bones Bones that develop within flat bones of the skull

short, **irregular**, and **flat**. Long bones are longer than they are wide. They can be a variety of sizes. Their design enables them to work as levers for the appendages. The femur, humerus, metacarpals, phalanges, and ulna are long bones. Short bones are usually small four-sided bones that form articulations with limited movement. The carpals and tarsals are short bones. Irregular bones are highly sculptured and usually have many processes and articular surfaces. Their shapes are determined by the role they play in the body. Vertebrae are irregular bones that permit bending and rotation of the vertebral column. Most of the facial bones are characterized as irregular bones. Flat bones have sheet-like surfaces for encasing structures or fastening to broad muscle attachments. The cranial bones, ileum, nasal bones, patella, ribs, scapula, and vomer are flat bones.

Bones can also be categorized according to the way they develop in the fetus. **Endochondral bones** represent most of the bones that make up the face and appendages. They develop from pieces of cartilage deposited as sheets of tissue in the developing embryo. **Dermal (intramembranous) bones** comprise the clavicle, scapula, and the flat cranial bones. They are formed from sheets of fibrous connective tissue. Teeth are formed within the **alveolar bone** by specialized groups of bone cells in the mandible and maxilla. Most humans produce two sets of teeth over their lifetime: deciduous teeth and permanent teeth. The permanent teeth progressively grow and push out the deciduous teeth. Permanent teeth cannot be restored once they are damaged or lost.

The patella, pisiform, and small rounded bones scattered in the metacarpals and metatarsals are categorized as **sesamoid bones**. Sesamoid bones form within tendons, where they remain embedded. Many sesamoid bones develop in response to excessive stress on a tendon. **Wormian** (or **Sutural**) bones are small, irregular, isolated bones that develop within the sutures of the skull. It is not fully understood why these bones form so haphazardly. Usually only two or three are found in an individual. However, certain skull formation disorders are characterized by dozens of wormian bones.

Bone Structure

Key Terms: bone marrow, bone proper, canaliculi, cancellous bone, compact bone, cortical bone, diaphysis, epiphyseal line, epiphyseal plate, epiphysis, Haversian canal, hyaline cartilage, lacunae, medullary cavity, osteocalcin, osteocyte, osteon, periosteum, red bone marrow, spongy bone, trabecular bone, Volkmann's canal, yellow bone marrow

Bone is a connective tissue produced early in embryonic development. As the bones mature, blood vessels, muscles, and nervous tissues come together with the bone to produce its final structure (Figure 5.10). It is important to remember that bone is a dynamic organ—it responds to the body's health and any stresses placed on it.

Periosteum A connective-tissue membrane that covers bones

External Features Covering much of the surface of almost all bones is a sheet of fibrous connective tissue called the **periosteum**. The term means, "to surround the bone." It contains blood vessels and nerves, which provide the bone with nutrition and sensation. The periosteum is also important in bone growth, maintenance, and healing. Attached to the periosteum are ligaments and tendons. Ligaments are bands of fibrous connective tissue that connect one bone

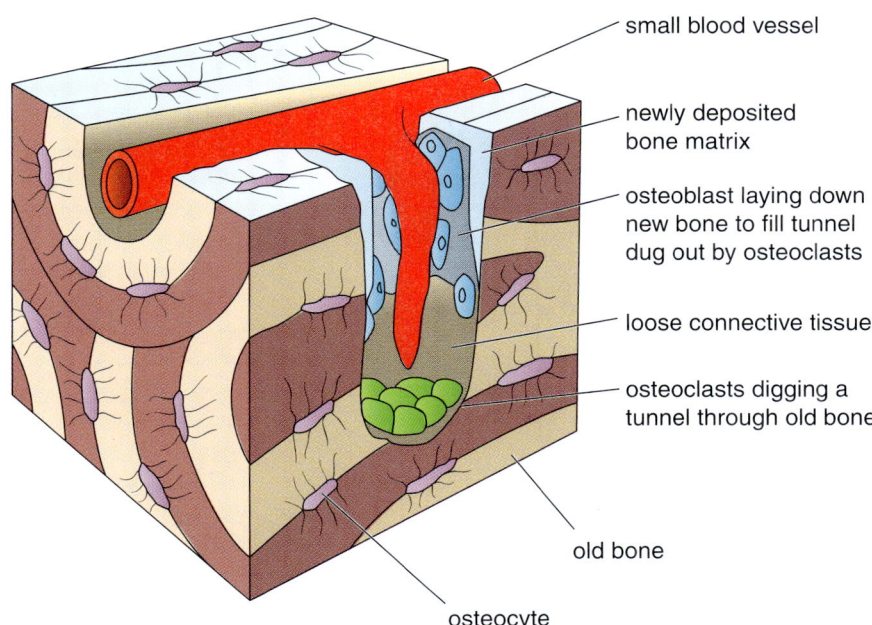

Figure 5.10 Bone Structue

to another, or attach bone to cartilage. Tendons are structurally similar to ligaments. They differ in that they connect muscle to the periosteum. Underneath the periosteum is the bone proper.

A typical long bone has various external features that provide physicians clues to normal bone function and pathology (Figure 5.11). These features vary as a person matures, providing an indicator of a person's age and developmental progression. The main body of a long bone is called the **diaphysis**, or shaft. Its name means "to grow apart" because the tips of the diaphysis grow in opposite directions to elongate the bone. The diaphysis usually contains a variety of surface features as discussed earlier in the chapter. Each end of a long bone terminates in a region called the **epiphysis**. It is the end of the bone that makes up a joint. The epiphysis is also called a secondary center for bone formation. Epiphyses (plural for epiphysis) develop separately and mature later than the rest of the bone structure. Underneath each epiphysis of mature bone is the **epiphyseal line**. The epiphyseal line is a thin strip of bone marking the fusion of the epiphysis to the diaphysis. Bones that are still maturing have an actively growing area called the **epiphyseal (growth) plate**. It is replaced by the epiphyseal line when the bone stops growing. The proximal and distal ends of the long bones are capped with **hyaline cartilage** on their articular surface. These cartilage coverings form a protective covering for joint function.

The short, irregular, and flat bones do not have a recognizable diaphysis. Their anatomy is defined by the dominant surface features characteristic of each bone. They also have various types of growth regions that are unlike the epiphyseal plates of the long bones. Some short bones, such as the talus and the proximal row of carpals, have hyaline cartilage on their articular surfaces.

Diaphysis The main body of a long bone

Epiphysis The end of a long bone that makes up a joint

Epiphyseal Line A thin strip of bone marking the fusion of epiphyses to diaphysis

Epiphyseal Plate An actively growing area of bone

Hyaline Cartilage A smooth cartilage covering the articular surfaces bones

THE SKELETAL SYSTEM

Figure 5.11 **Structure of Long Bones**

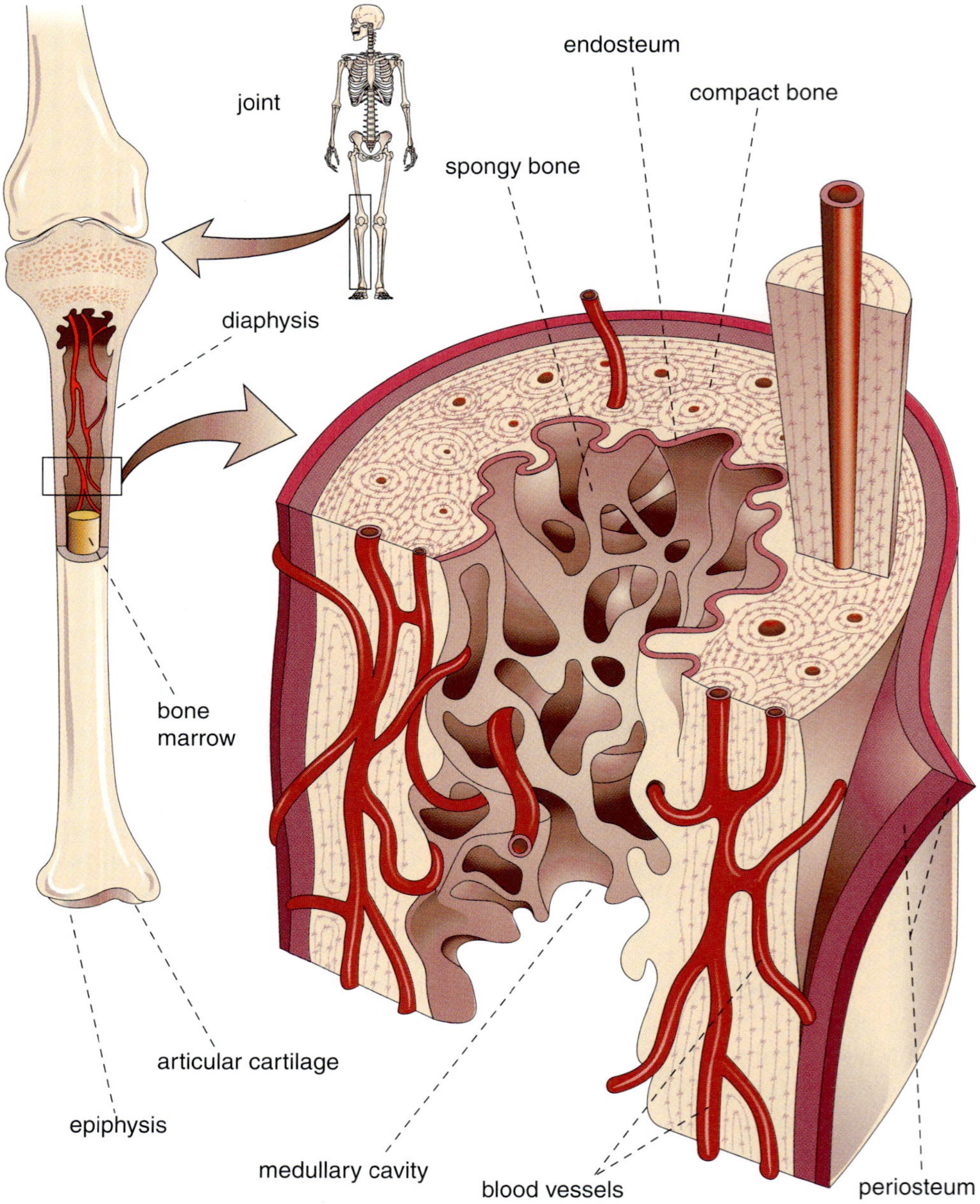

Compact or Cortical Bone Rigid outer shell of the bone

Osteon The structural unit of compact bone tissue

Haversian Canal The internal structure of the osteon

Osteocytes The cells of bone tissue

Lacunae Cavities that store osteocytes

CHAPTER 5

Internal Features Human bones are composed of two types of bone tissue: compact and spongy (trabecular) bone (Figure 5.12). **Compact**, or **cortical**, **bone** tissue makes up almost 20% of the skeletal structure, although it represents about 80% of the skeleton's weight. It forms a rigid, supportive outer shell around each bone. This dense tissue is composed of mineral deposits embedded in a collagen matrix. The structural unit of compact bone is called an **osteon**. It forms an arrangement of concentric circles of hollow tubes of bone matrix around a central **Haversian canal**. Blood vessels, lymphatic vessels, and nerves pass through this opening. Embedded in the bone-matrix rings that surround each Haversian canal are bone cells, or **osteocytes**. The cavities that the osteocytes sit in are called **lacunae**. The osteocytes secrete bone matrix. Bone matrix is composed of collagen fibers and a protein called

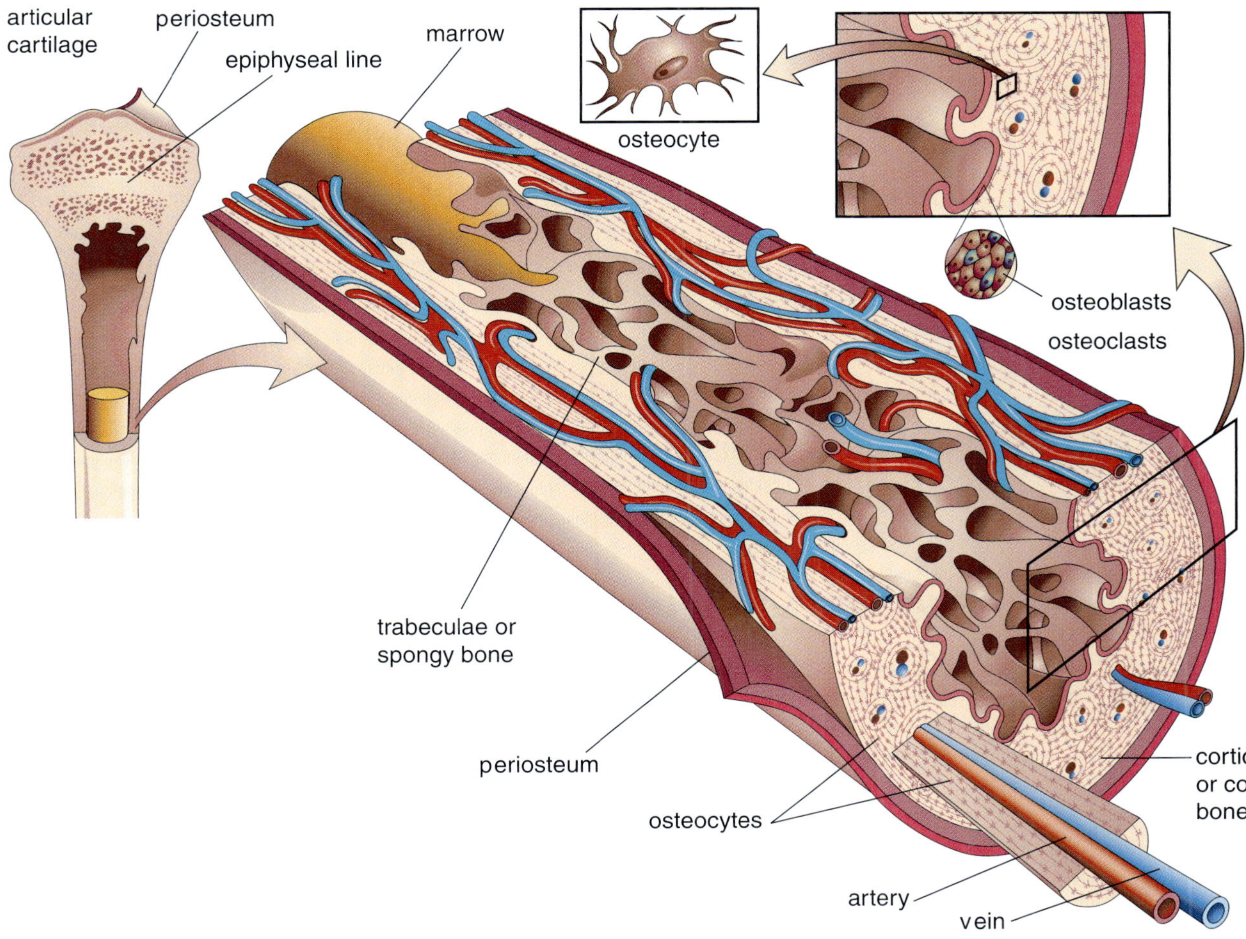

Figure 5.12 Internal Features of Bone

osteocalcin, which is impregnated with calcium salts, or hydroxylapatite. Osteocytes connect to each other and to blood vessels through small channels called **canaliculi**. **Volkmann's canals** lie at right angles to the long axis of the bone. They carry blood vessels and nerves of the periosteum to the central Haversian canals.

Cancellous, or **trabecular**, **bone** tissue represents about 80% of the bone structure, but contributes only a small proportion of bone weight. It also is known as **spongy bone** because it has many "open spaces" as does a sponge. The mineral and collagen making up spongy bone are arranged into a system of archlike braces called trabeculae. Spongy bone is less dense and more flexible than compact bone. The osteocytes in spongy bone are distributed throughout the tissue. The blood vessels of spongy bone are scattered throughout the trabeculae and do not form the uniform patterns seen in compact bone.

The center of long bones contains a hollow space called the **medullary (marrow) cavity**. This cavity is bordered by spongy bone and lined with a thin layer of connective tissue called endosteum. Endosteum is capable of generating new bone cells. The medullary cavity is usually filled with a soft

Osteocalcin Protein comprising bone matrix

Canaliculi Small channels that connect osteocytes

Volkmann's Canals Canals in the osteon through which nerves and blood vessels pass from the periosteum to the Haversian canals

Cancellous or Trabecular Bone A type of bone that forms the ends of the long bones and the center of other bones

Spongy Bone A synonym for cancellous bone; bone composed of a honeycomb-like network bony struts

Medullary (Marrow) Cavity The hollow center of long bones

THE SKELETAL SYSTEM

Bone Marrow Soft tissue within the medullary cavity

Yellow Marrow A food reserve for bone cells

Red Marrow Produces red blood cells

tissue called **bone marrow**. Bone marrow is also found in the spongy bone of certain flat and irregular bones. There are two types of bone marrow: yellow and red. **Yellow marrow** consists predominantly of fat cells and is found in most bones. It is used as a food reserve for bone cells. A rich blood supply ensures that a continuous source of dietary fats is available for the replenishment of yellow marrow. **Red marrow** is mainly found in spongy bone and in the ends of long bones. It is made up of stem cells that form blood cell components. In infants, the skull contains approximately 30% of the body's red marrow compared with only 7% in adults. Most of the red marrow in adults is found in the pelvic girdle, ribs, sternum, and the long bones of the lower limbs. Yellow marrow can be converted to red marrow when the body suffers severe blood loss.

✓ Concept Check

1. Describe the different ways that bone can be categorized.
2. Explain the two different types of bone tissue.
3. What is the function of bone marrow?

DISCOVERY SCENE PLEASE ENTER DISCOVERY SCENE PLEASE ENTER

What information about bone structure helps explain the easily broken bones in this child? What do you need to know about bone structure to better understand the probable cause of the breakage? Could the bone tissue itself be the factor? Or could it be due to other anatomical features of the bone?

JOINTS

Key Terms: bursa, bursitis, functional classification, structural classification, synovial fluid

Joints, or articulations, provide two functions for the skeletal system. Joints attach bones, affording support and protection. This is the primary purpose of joints in the cranial bones. Other joints allow muscles to reposition two or more articulated bones to produce body movement. Joints are categorized using two different criteria: **structural classification** and **functional classification**. The structural classification scheme categorizes joints based on their tissue composition and structural complexity. A functional classification system focuses on the type of movement that a joint permits. The type of movement a joint is capable of is determined by several factors:

Structural Classification A system of joint classification based on tissue composition and structural complexity

Functional Classification A system of joint classification based on the way joints move

Bursa A fibrous sack filled with synovial fluid

Synovial Fluid Clear fluid that lubricates joint linings

- the manner in which the bones fit together
- the tightness of fit between the articular surfaces
- the tension of the tissues forming the articulation
- the position of the ligaments, muscles, and tendons associated with the joint

Certain moveable joints are accompanied by a structure called a **bursa**. A bursa is a sac-like space composed of fibrous tissue. The sac contains a thick, lubricating fluid called **synovial fluid**. Synovial fluid is a clear, oily fluid that lubricates, nourishes, and protects a joint and its surface. It is secreted by epithelium cells on the inner surface of the bursa. Bursae (plural of bursa) are found in articular areas where rubbing between skin, muscle, ligaments, or bones could occur. They provide a cushion against the rubbing. However, bursae are likely to become inflamed if they are damaged, causing a condition called **bursitis**.

Bursitis The inflammation of a bursa

Joint Structure

Key Terms: cartilaginous joint, fibrous joint, synovial capsule, synovial joint

Structurally, the joints are classified into the following types:

- **Cartilaginous** – Cartilage covers the articulating bone surfaces. The articulation between the pubic bones is cartilaginous.
- **Fibrous** – The articulated bones are attached by fibrous connective tissue. Fibrous joints attach the radius and ulna to each other.
- **Synovial** – The articulating suface is covered with a fluid-filled, fibrous, connective-tissue sack called the **synovial capsule** (Figure 5.13). Synovial joints are found in the elbows and knees. The fluid in the synovial capsule is similar in composition and function to the fluid found in bursae.

Cartilaginous Joint A joint formed of cartilage

Fibrous Joint A joint formed of fibrous connective tissue

Synovial Joint A joint formed by a synovial capsule

Synovial Capsule A fibrous, fluid-filled sack

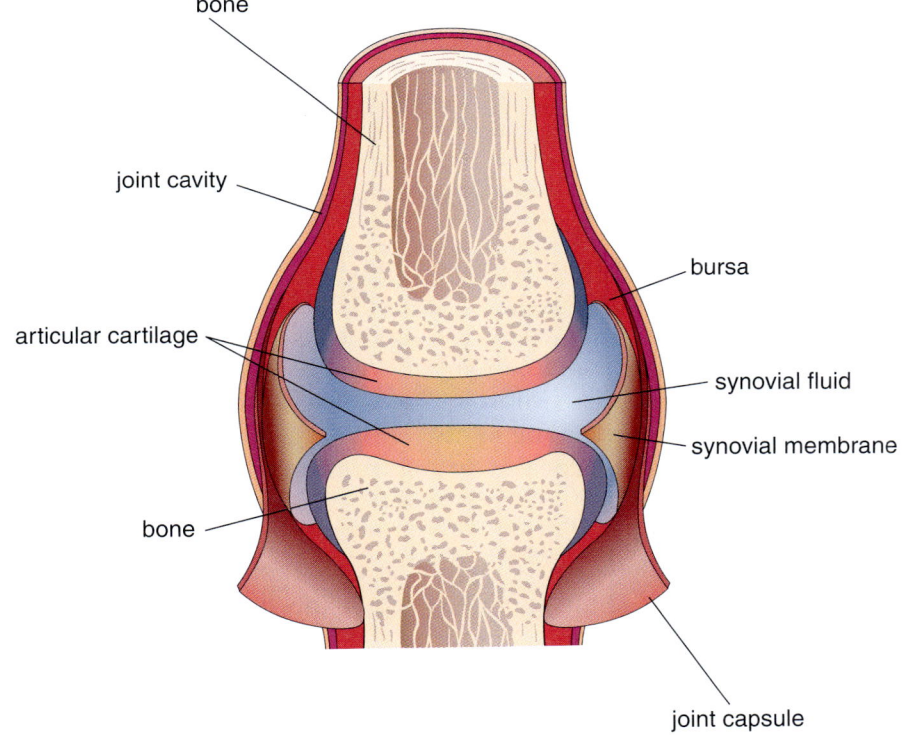

Figure 5.13 Joint Structure

Joint Function

Key Terms: amphiarthrosis, ball-and-socket joint, condyloid joint, diarthrosis, ellipsoid joint, gliding joint, gomphosis, hinge joint, pivot joint, symphysis, synarthrosis, synchondrosis, syndesmosis, synostosis

Functionally, joints can be classified as follows (Figure 5.14):

- **Synarthrosis** – These joints permit no movement, which is typical of certain fibrous joints.
- **Amphiarthrosis** – A slight degree of movement can occur. Some cartilaginous and fibrous joints allow this type of movement.
- **Diarthrosis** – These joints provide a variety of movements. Synovial joints permit this degree of movement.

Synarthroses are subdivided into four categories: **gomphosis**, **syndesmosis**, **synchondrosis**, and **synostosis**. The gomphosis is formed by a conical process that is held in a socket with ligament. Teeth form a gomphosis with the mandible and maxilla. Syndesmoses are formed when two bones are joined by one or more ligaments. A syndesmosis holds together the tibia and fibula in the ankle region. Synchondroses, or **symphysis**, are joints where the two bones are joined by a piece of cartilage. The epiphyses are attached to the shaft by this type of joint. Another example is found in the attachment between pubis bones. Synostoses, or sutures, are formed by the fusion of two bones. Some fuse so tightly that they are, in effect, one bone. The cranial bones of adults contain sutures.

Synovial joints are further categorized by the type of motion they permit. The shape and complexity of the synovial capsule determine the degree and direction of their movement. The major functional categories of synovial joints are as follows:

- **Ball-and-socket joint** – They produce a wide array of movements. The hip forms this type of joint.
- **Condyloid**, or **ellipsoid**, **joint** – The condyloid joint occurs where a ball-like articular surface rests against the curve-shaped end of another articular surface. This arrangement permits a circular or oval pattern of motion. The knee is a condyloid joint between radius and carpals.
- **Gliding joint** – These joints allow a wide variety of side-to-side movements. The carpals form gliding joints.
- **Hinge joint** - These joints permit an angular motion along one plane, which is similar to the opening and closing of a door. The elbow is an example of a hinge joint.
- **Pivot joint** – This allows a rotation similar to the turning of a dial.
- **Saddle joint** – These joints resemble a saddle in which one articular surface rocks back and forth upon another.

✓ Concept Check

1. Describe the structural classifications of joints.
2. Describe the functional classifications of joints.
3. What are the different types of motions permitted by synovial joints?

Synarthrosis Joint A joint that does not permit movement

Amphiarthrosis Joint A joint that permits only slight movement

Diarthrosis Joint A joint that permits a variety of movements

Gomphosis A synarthrodial joint formed by a conical process; it is held in a socket by a ligament

Syndesmosis A synarthrodial joint formed by two or more ligaments

Synchondrosis or Symphysis A synarthrodial joint formed by cartilage

Synostosis or Suture A synarthrodial joint formed by the fusion of two bones

Ball-and-Socket Joint A synovial joint that permits a variety of movements

Condyloid or Ellipsoid Joint A synovial joint that permits a circular or oval pattern of movement

Gliding Joint A synovial joint that permits a variety of side-to-side movements

Hinge Joint A synovial joint that permits an angular movement along a plane

Pivot Joint A synovial joint that permits a rotation

Saddle Joint A synovial joint that permits a variety of movements, primarily a rocking movement in two planes

DISCOVERY SCENE PLEASE ENTER DISCOVERY SCENE PLEASE ENTER

CSI Break

What information about the joints will help you understand bone breakage? Could the boy's tendency to break bones be due to problems with the articulations? How would you determine the role of joint function as a factor in bone breakage?

Figure 5.14 **Joint Function**

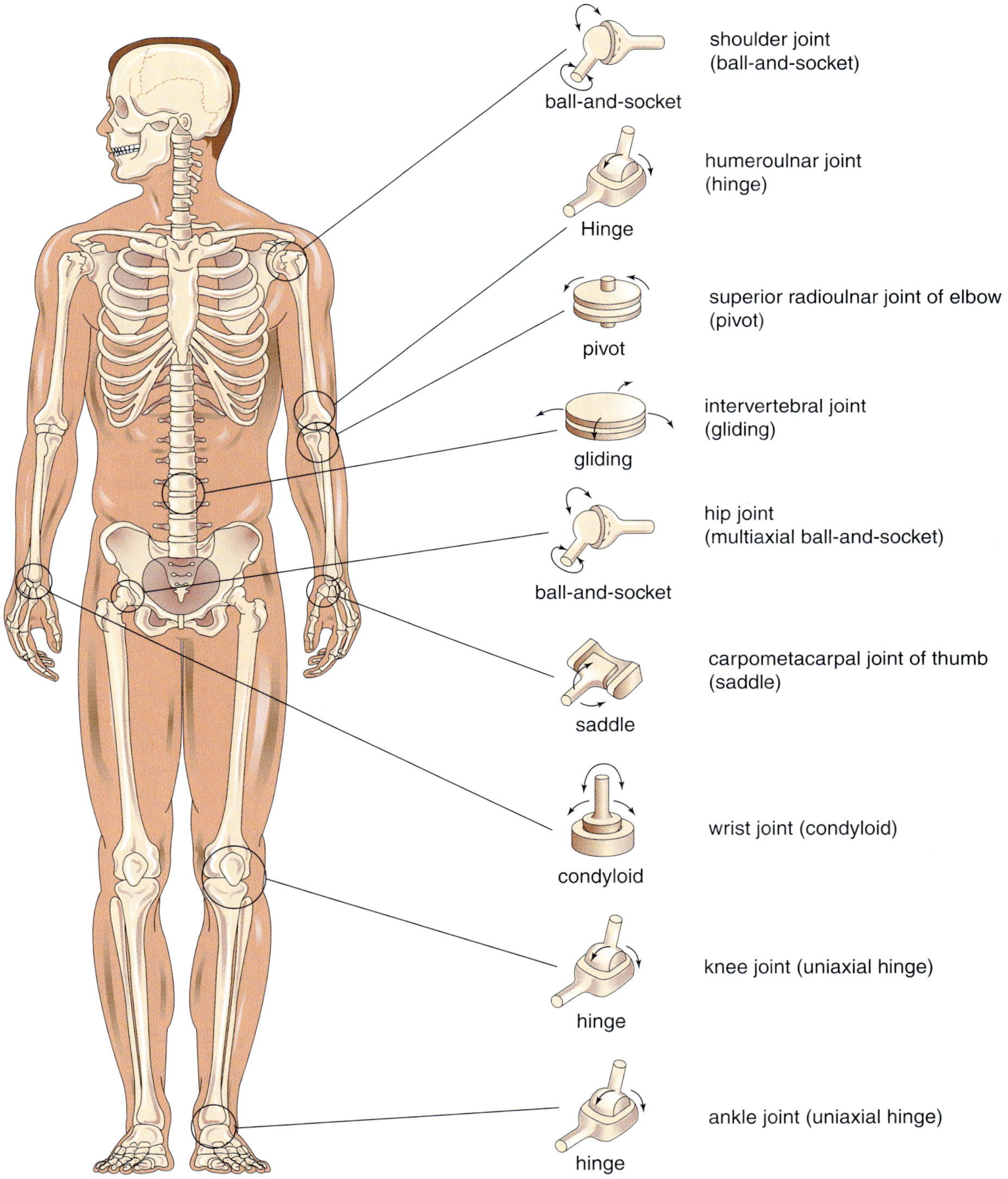

- shoulder joint (ball-and-socket) — ball-and-socket
- humeroulnar joint (hinge) — Hinge
- superior radioulnar joint of elbow (pivot) — pivot
- intervertebral joint (gliding) — gliding
- hip joint (multiaxial ball-and-socket) — ball-and-socket
- carpometacarpal joint of thumb (saddle) — saddle
- wrist joint (condyloid) — condyloid
- knee joint (uniaxial hinge) — hinge
- ankle joint (uniaxial hinge) — hinge

THE SKELETAL SYSTEM

Human Bone Charts

Use Figures 5.15 and 5.16 to review the anatomy and function of the skeletal system. Pay attention to the orientation of each bone, and any clues for recognizing distal and proximal ends, and lateral and medial sides. Note which bones are articulated and what component of the skeletal system they comprise. Look for evidence of the types of articulations formed between the different bones. Also, think about the stresses placed on the different bones when the body is stationary or mobile. Think how the stresses change when the body is placed in different positions, such as lying down or sitting.

Figure 5.15 Human Bone Chart–Anterior View

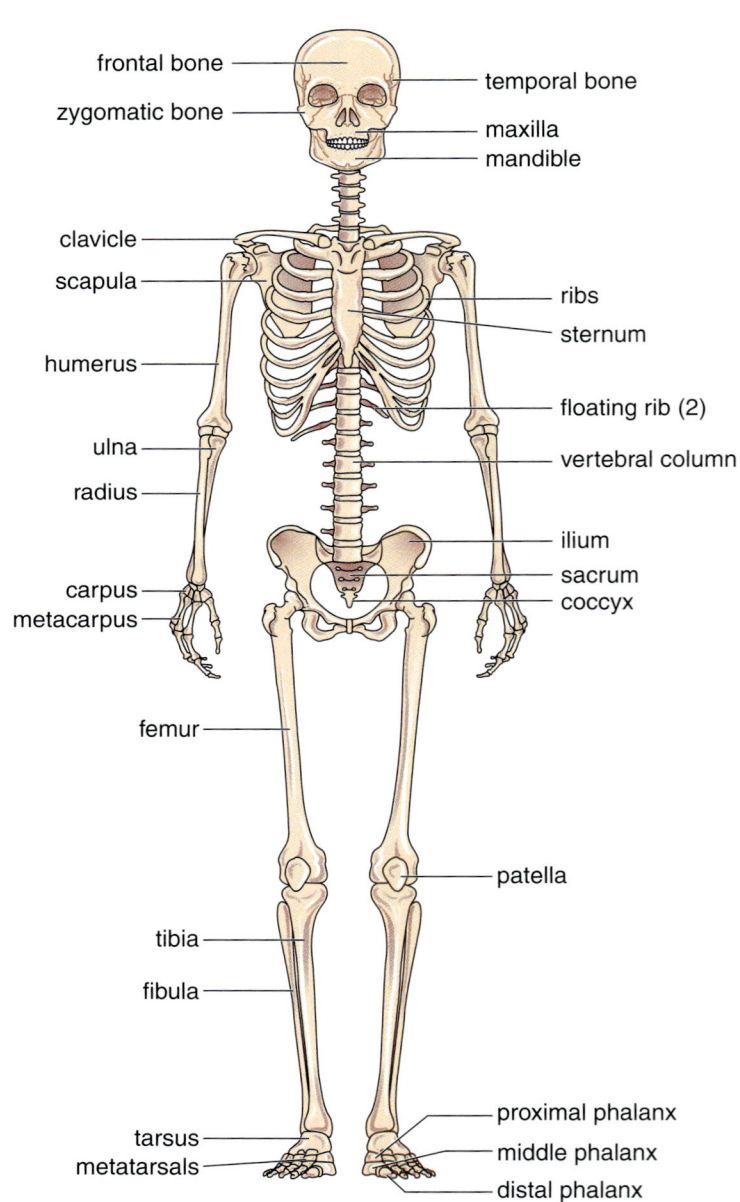

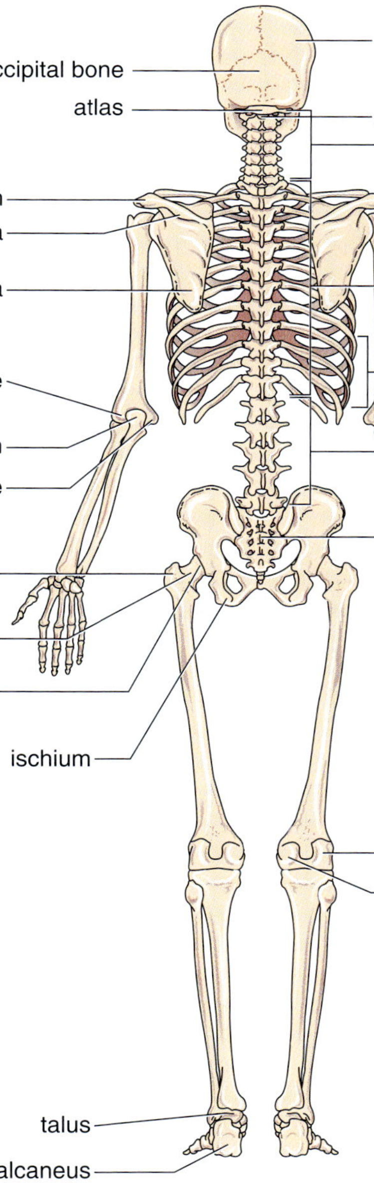

Figure 5.16 Human Bone Chart–Posterior View

Bone Development and Healing

Key Terms: angulated, bone collar, calcification, commuted fracture, compound fracture, endochondral ossification, fontanelles, fracture, greenstick fracture, intramembranous ossification, oblique, open fracture, ossification, osteoblast, osteoclast, simple fracture, spiral, transverse, white blood cell

Human bones develop at different rates and times as a person progresses through the developmental stages of growth. The cells that permit human bone formation are present after the first three weeks of development. They form from mesoderm cells, which are responsible for all of the body's connective tissues and muscles. The conversion of embryonic tissues into recognizable bone

THE SKELETAL SYSTEM

Ossification The process of bone formation

Endochondral Ossification Bone formation that begins within a cartilage

Intramembranous Ossification The formation of bone from connective-tissue membranes

Calcification The process by which bones harden

Osteoblasts Cells that build bone tissue

Osteoclasts Cells that break down bone and cartilage

Bone Collar A layer of bone on the surface of the diaphysis

is called **ossification**. Scientists have identified two primary ways that bones form in the embryo: **endochondral ossification** and **intramembranous ossification**. Certain bones form as a combination of the two types of ossification. Long bones are usually formed by endochondral ossification. Intramembranous ossification usually forms most flats bones. Irregular bones are usually formed by a combination of the two types of ossification.

Endochondral ossification is bone formation that begins within (endo) a cartilage (chondral). It starts after the fifth week of fetal development when special mesoderm cells form cartilaginous centers in locations where bones will ultimately form (see Figure 5.17) These cartilaginous centers grow and develop into cartilage pegs that resemble a miniature bone. As bone develops, these cartilage pegs are restructured and filled with hydroxylapatite in process called **calcification**. The cartilage pegs are first formed and molded by fibroblasts. Bone-forming cells called **osteoblasts** and **osteoclasts** then move into the cartilage peg. Osteoblasts build bone tissue, while osteoclasts break down bone and cartilage. Both work together to sculpt growing bones. Diet plays an important role in the function of these cells. Calcium and vitamin D stimulate bone growth and maintenance. Specific parts of the endocrine system produce chemicals that regulate the rate of bone growth.

Three distinct growth events take place during endochondral bone formation: primary ossification, bone collar formation, and secondary ossification. Primary ossification begins when osteoclasts carve out a cavity in the center of the cartilage peg that corresponds to the long-bone diaphysis. Blood vessels begin to grow into the bone entering the cavity. These blood vessels provide nutrients for bone growth and maintenance. Osteoblasts enter and start secreting bone tissue to replace the cartilage. Compact bone is deposited first, and then it is converted into spongy bone as the bone cavity enlarges. As this is going on, osteocytes in the periosteum are producing a dense, sheetlike layer of bone on the surface of the diaphysis called the **bone collar**. This step is the bone-collar formation stage. Both of these stages work together to elongate and widen the bone. Bone elongation takes place at the epiphyseal plate. Growth of the plate pushes the epiphyses in opposite directions. Primary ossification and bone-collar formation continue until the bone is fully mature, by the late teens or early twenties. The rate of long-bone growth is an important determinant of overall body height.

Secondary ossification occurs late in the development of the fetus. Sometimes it begins just before birth. Osteoclasts form a small cavity in the epiphyses. As with primary ossification, blood vessels and osteoblasts enter the cavity and convert the cartilage matrix into bone. Secondary ossification progresses much more slowly than the primary ossification carried out in the diaphysis, and a full bone collar does not form on the epiphyses. In some bones, hyaline cartilage is deposited over the areas forming the articular surfaces. Secondary ossification carries on through adolescence. It ceases when the cartilage growth slows, and at last stops by the early twenties. At this point, the epiphyseal plate fuses to the diaphysis forming the epiphyseal line. Bone elongation is not possible once the epiphyseal line forms. Bone thickening and sculpting can continue throughout a person's lifetime. The osteoblasts become osteocytes in the mature bone tissue. Osteocytes are more involved in bone maintenance than in bone formation.

Intramembranous, or direct, ossification is much simpler than endochondral bone development. In addition, it does not require cartilage for bone formation. It involves the replacement of sheetlike, fibrous connective-tissue mem-

Figure 5.17 **Bone Growth**

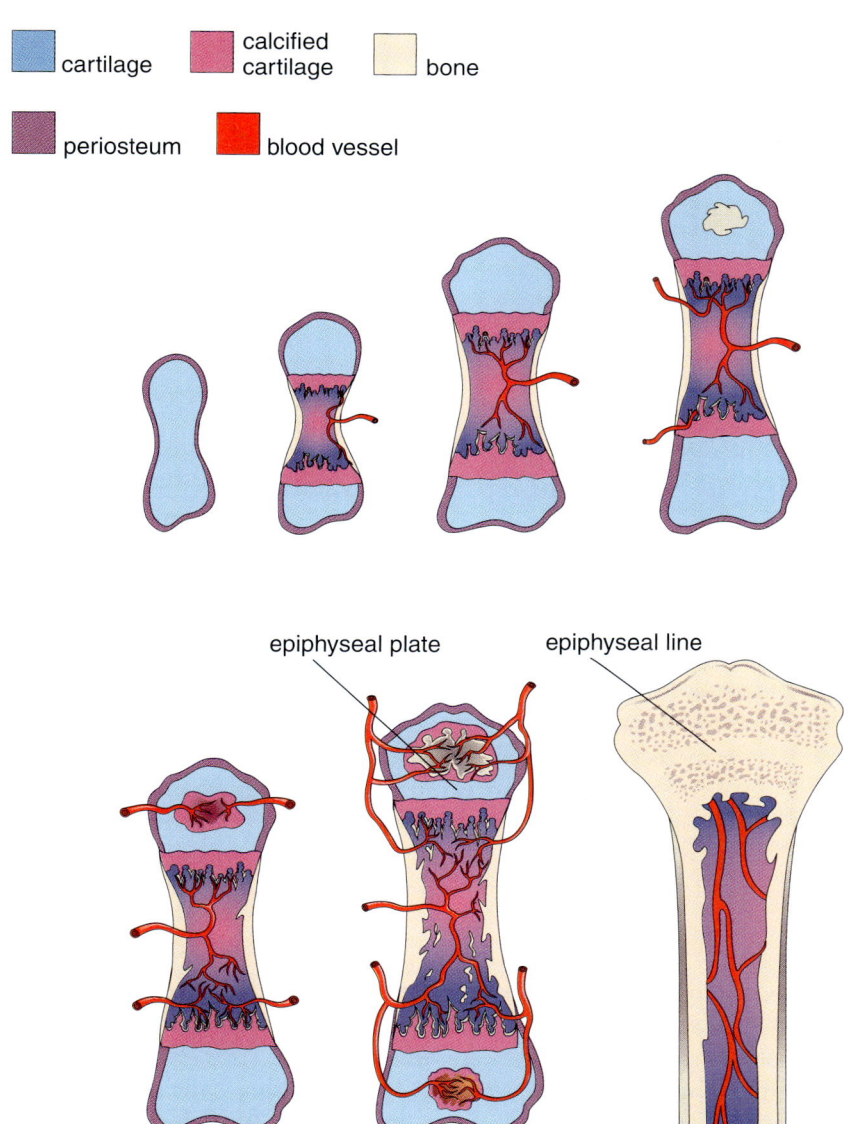

branes with bone tissue. This process begins with the laying down of the connective-tissue membranes where the future bone will form. Osteoblasts then settle into the membranes and deposit a spongy bone matrix. Blood vessels and osteoclasts also move into the bone to assist with further growth. A thin cap of dense bone is deposited on the surface of most intramembranous bones. This cap is secreted by osteoblasts in the periosteum. The shape of the intramembranous sheet is an important factor in determining the contours of the intramembranous bone. It is possible to find skeletal birth defects caused by abnormally formed or missing connective-tissue membranes. Intramembranous development is evident in a newborn baby's skull (Figure 5.18).

The large sutures and the structures called **fontanelles**, or soft spots, are regions of flat bone that have not completed intramembranous development. The skull will fully ossify in the teenage years.

 **Fontanelle** A soft spot on an infant's skull where intramembranous development has not been completed

THE SKELETAL SYSTEM

Figure 5.18 A Fetal Skull

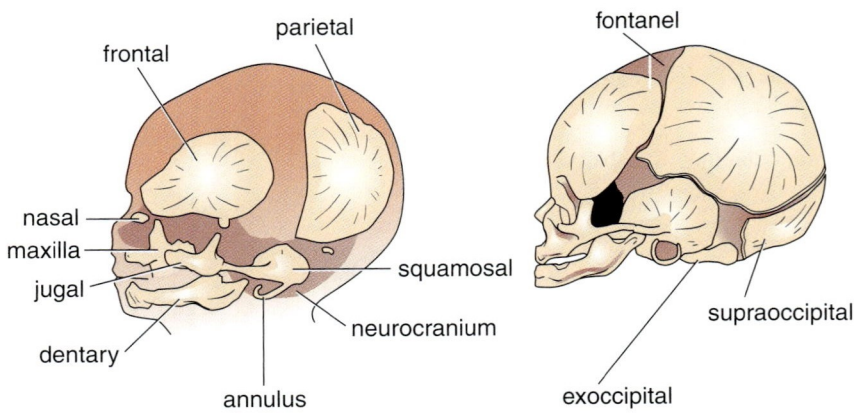

Fracture A crack or splinter in a bone
Simple Fracture A crack in the bone structure
Greenstick Fracture A fracture in which one side of the bone is broken and the other side is bent
Comminuted or Compound Fracture A fracture in which one or more areas of bone are displaced
Open Fracture A fracture in which the skin is pierced by the bone
Transverse Fracture Refers to a fracture occurring horizontally on the bone
Oblique Fracture Refers to a fracture occurring at angles on the bone
Spiral Fracture Refers to a fracture occurring in a twisting pattern on the bone
Angulation A twisted change in the original shape of a bone as a result of damage

Bone growth not only takes place as a part of growth and development. Sometimes it is needed for bone repair. Bone **fractures** are the most common type of bone damage (see Figure 5.19). A fracture is a medical condition in which a bone becomes cracked or splintered as a result of physical injury. Fractures are categorized by the severity of the break or the angle at which the break occurs in a bone. A **simple fracture** is the least severe and involves a crack in the bone structures. Simple fractures can be so small that they are not readily noticeable. Some are large and involve bleeding, pain, and swelling. Children commonly have **greenstick fractures**. In these fractures, one side of the bone is frayed from the fracture, while the other side is twisted, but not broken. **Comminuted**, or **compound**, **fractures** are large fractures in which one or more areas of bone are displaced or shattered. The overall bone structure is altered and involves bleeding and swelling. **Open fractures** involve tearing of the skin and are easily infected. The angle of breakage does not determine the severity of damage. However, it has much impact on how the bone heals and how it is medically corrected (Figure 5.19). Breaks can occur horizontally (**transverse**), at angles (**oblique**), and in twisting patterns (**spiral**). Some compression causes **angulation**, which means the bone changes its overall shape. Angulation causes the bone to take on a bent, or angular, shape. Bone healing involves the same cells that carry out bone growth and development (Figure 5.20). For bone healing to occur, there usually has to be enough damage to cause some type of blood accumulation in the injured bone tissue. It is possible for bone fractures to occur without any restoration of the bone by the body.

Figure 5.19 Types of Bone Fractures

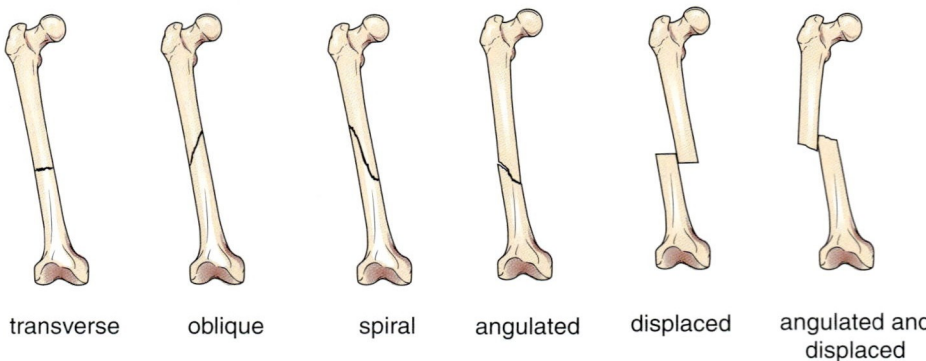

Figure 5.20 Bone Fracture Repair

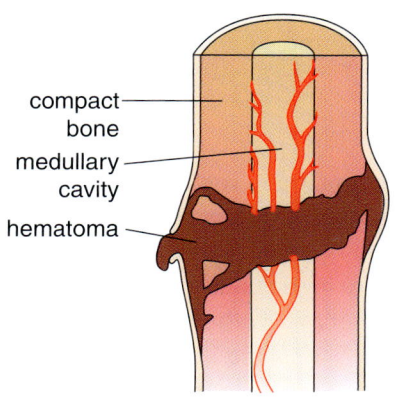
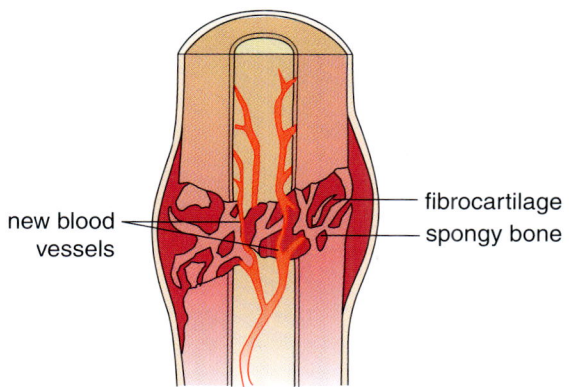

(a) blood escapes from ruptured blood vessels and forms a hematoma

(b) spongy bone forms in regions close to developing blood vessels, and fibrocartilage forms in more distant regions

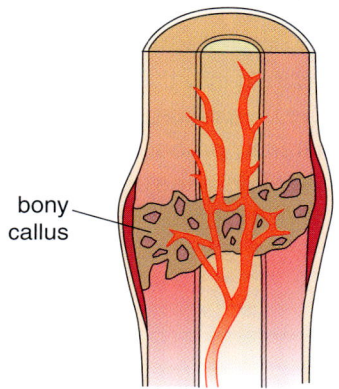
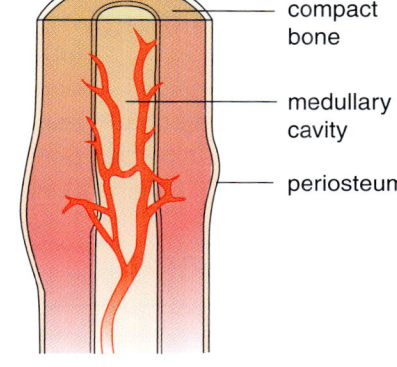

(c) a bony callus replaces fibrocartilage

(d) osteoclasts remove excess bony tissue, restoring new bone structure much like the original

Bone healing involves the following distinct stages: fracture, granulation, callus, lamellar bone, and normal contour. Some people lump these stages together into the reactive phase (fracture and granulation), reparative phase (callus and lamellar bone), and the restorative phase (normal contour). The fracture stage starts the body's reaction to something that needs to be repaired. Bleeding and tissue damage with the subsequent formation of a blood clot is required to elicit a healing response. This stage occurs within hours or a couple of days after the initial damage. The body's reaction to bone damage is completed when fibroblasts, **white blood cells**, and new blood vessels enter the injured area. Chemicals called cytokines leak out of the damaged tissues and clot, stimulating this step. The bone tissue now takes on a granular appearance, which may persist for weeks. This is due to the action of the white blood cells breaking down and digesting the damaged tissues and clotted blood.

Bone repair begins when the fibroblasts secrete fibrous connective tissue and cartilage into the damaged area. A mass of tissue called a callus is then formed. The callus is the mold that osteoblasts use to secrete new bone tissue. In a large fracture, osteoblasts from the periosteum start repairing any

White Blood Cell A large blood cell that helps fight infection

THE SKELETAL SYSTEM

Cutting Edge Research
ARTIFICIAL BONES

Many people take it for granted that a broken bone will heal and be as good as new. However, this is not always true for the elderly people or those with severe breaks involving bone loss. Dr. Molly Shoichet, at the University of Toronto, is developing a strategy for bone replacement to be used when a bone will not heal properly. She is researching a polymer that can be sculpted to fit into a damaged bone. Her 2003 study, published in the Journal of Biomedical Materials Research, investigates whether the healthy bone near the polymer would accept the implant as if it were the person's own bone matrix. Not just any material can be used as a bone replacement. Bone cells are finicky about the type of matrix they will inhabit. Past investigations by other scientists discovered that bone cells had to be "enticed" to grow into artificial materials. Even if the bone cells did enter the matrix, there was always the chance that the immune system would attack the matrix.

Dr. Shoichet's preliminary investigations show that the new bone grows naturally without the addition of chemical growth stimulants. The polymer acts as a scaffold, which provides a framework for the growing tissue. In addition, there is no evidence that the immune system will reject and, consequently, degrade the new matrix. The artificial bone is so similar to the natural bone that it is almost impossible to distinguish between the two at the histological level. This spongy polymer makes it very easy to grow back new bone. It is simply cut into a desired shape and inserted into the damaged bone. Bone-marrow cells are then placed in the new matrix. Osteoblasts start naturally producing new bone in the matrix and sculpt it into the original bone. Eventually, the polymer is replaced by the person's collagen.

damaged bone surface. This new bone tissue does not have the original bone-tissue structure. It is less organized and does not respond in the same way to stresses placed on the bone. This stage can take weeks to months to complete. The final stage of bone healing is a slow period of bone remodeling. Osteoclasts and osteoblasts make an attempt to return the bone to its original structure or contour. Small injuries are usually completely repaired. However, continuous bone damage and large fractures rarely return to normal. Physicians can see old fractures on x-rays because the tissue density and organization of a repaired bone differ slightly from that of the original bone.

Recent studies show that bone healing changes the nutritional needs of a person. Studies have indicated that athletes who damage a large bone, such as the femur, can increase their body's metabolic needs by 20% for a few weeks. It is also important to recognize that bone healing requires higher intake of calcium and protein in the diet than is usual. Undernutrition and malnutrition can slow or even stop bone healing. Physicians can monitor bone healing in laboratory tests that measure osteoblast activity. The most widely used test measures the presence of an enzyme called alkaline phosphatase. Other tests measure the byproducts of bone restructuring. Physicians can use these tests to determine potential bone diseases as well as bone healing.

✓ Concept Check

1. Explain the steps of endochondral bone formation.
2. Explain the steps of intramembranous bone formation.
3. Describe the stages of bone healing.

THE PHYSICS OF BOUNCING JOINTS

Biomechanics is the use of physics and engineering principles to describe the motion of body parts. Joints are of particular interest because of the forces they must endure during body movement. When people walk or run, they bounce along the ground relying on ligaments and tendons to store and return elastic energy. These stresses can cause joint damage if they reach a certain point. Scientists are now able to understand the stress and strain on joints by using a mathematical model that measures the mechanics of springs. The study model consists of a central weight that represents the body pressing on springs, which act like the joints. This model accurately predicts the mechanics of all hopping, jogging, and running. Current research studies are investigating the effects of various degrees of leg stiffness on joint stress. Now it is possible for scientists to calculate the types of the movements that may cause joint injuries. Ultimately, this information may be simple enough to use as a regular part of athletic training and physical therapy.

DISCOVERY SCENE PLEASE ENTER DISCOVERY SCENE PLEASE ENTER

CSI Break

Could the child's condition be due to bone development problems? Could it be due to bone healing problems? Or is it a combination of both bone growth and healing problems?

WELLNESS AND ILLNESS OVER THE LIFE SPAN
Pathology of the Skeletal System

Key Terms: ankylosing spondylitis, arthritis, bone density loss, cavity, fibromyalgia, gout, juvenile arthritis, lupus, myeloma, osteoarthritis, osteomyelitis, osteonecrosis, osteoporosis, rheumatoid arthritis, scleroderma, shin splint, stress fracture, systemic lupus erythematosus, tooth decay

The most common bone and joint pathologies are related to atypical stress and strain placed on bones during various activities. Intensive athletic exertion, heavy lifting, and continuous repetitive motions commonly lead to premature wear and tear on joints, tendons, and ligaments. A common pathology is called a **shin splint**. It is a mildly to severely painful condition that develops on the medial side of the tibia or shin. Overuse or high-impact use of the ankle joint can lead to abnormal stretching of the ligaments and tendons. It may also

Shin Splint A painful condition of the anterior lower leg that develops from overuse of the ankle joint

Stress Fracture Thin breaks in a bone that may be too small to detect and, generally, do not heal

Arthritis A condition causing swelling and stiffness in the joints

Osteoarthritis The deterioration of the articular cartilage covering the ends of bones

Rheumatoid Arthritis A condition in which the immune system attacks connective tissues

Ankylosing Spondylitis Arthritis affecting the spine

Juvenile Arthritis Arthritis that affects children

Gout A metabolic disorder that causes severe inflammation of the joints

Systemic Lupus Erythematosus or Lupus An autoimmune disorder that causes inflammation of connective tissues throughout the body

Scleroderma A connective-tissue disorder that causes thickening of the skin

Fibromyalgia A disorder that causes widespread joint pain

Osteoporosis A degenerative bone disorder

Bone Density Loss A less-severe form of osteoporosis

inflame the fascia over the ankle and lower leg. Shin splits are not the same as a **stress fracture**. Stress fractures are thin breaks that may cause pain, but sometimes are too small to detect and, generally, do not heal.

Arthritis is also a common bone and joint ailment, which simply means "swelling of the joint." The incidence of arthritis has unexplainably doubled since 1985. Data collected in 2005 indicate that it affects one in six American adults. The disease is generally categorized into five or more types. **Osteoarthritis** is deterioration of the articular cartilage covering the ends of bones. This common form of arthritis causes pain and loss of movement due to rubbing of the rough bone surfaces. **Rheumatoid arthritis** develops when the immune system attacks connective tissues of a joint. It is sometimes caused by bacteria or viruses that enter the bloodstream in common types of infections. Rheumatoid arthritis is probably the most serious and disabling form of arthritis, and it is most common in females. An arthritis specific to the articulation of the spine is called **ankylosing spondylitis**. It is common for the vertebrae to fuse after long-term inflammation. This limits mobility, and causes muscle and tendon pain along the back. **Juvenile arthritis** is joint inflammation in children. Most likely, it is a form of rheumatoid arthritis, though it also may be caused by genetic factors.

Other organ-system diseases cause inflammation of the bones and joints. **Gout** is a metabolic disorder that causes the body to produce a waste product called oxalic acid. Oxalic acid forms crystals that are difficult to remove from the body. These crystals enter certain joints, where they cause severe inflammation. Gout most often affects males, and usually occurs in the small joints of the foot and in the vertebral column. **Systemic lupus erythematosus**, or **lupus**, is an inflammation of connective tissues throughout the body. It also affects joints, causing arthritis-like conditions. This disease is caused by the immune system accidentally attacking the body's own proteins in the connective tissue. It is most common in women. Another general connective-tissue disease is **scleroderma**. The ailment causes thickening and hardening of the skin, usually in localized patches, but, in rare cases, it can become widespread and affect other organs. Little is known about the cause of scleroderma, although it is likely an immune-system disorder. **Fibromyalgia** is a disorder that causes widespread joint pain that affects the tendons. It affects mostly women, and its cause is unknown.

Osteoporosis is a degenerative bone disorder that mostly affects women. Approximately 4% of the North American population has been diagnosed with osteoporosis. Of this group, 68% are women, and it is believed that another 12% have a mild form of the disease known as **bone density loss**. The decrease in bone density is due to a gradual loss of osteons in the compact bone and fewer supportive structures in the spongy bone (Figure 5.21) Osteoporosis has a variety of causes and can occur at any age. Most research studies relate osteoporosis to decreased levels of sex hormones. Malnutrition and undernutrition are other causes. Some research indicates that diets deficient in calcium and vitamin D lead to osteoporosis. Smoking, excessive alcohol intake, and certain anti-inflammatory medications can predispose a person to osteoporosis. Data from the National Institutes of Health show that osteoporosis is responsible for more than 1.5 million bone fractures in adults annually in North America. This includes approximately 700,000 vertebral fractures, 300,000 hip fractures, and 250,000 wrist fractures. Over 80% of osteoporosis related injuries occur in women. Osteoporosis is related to

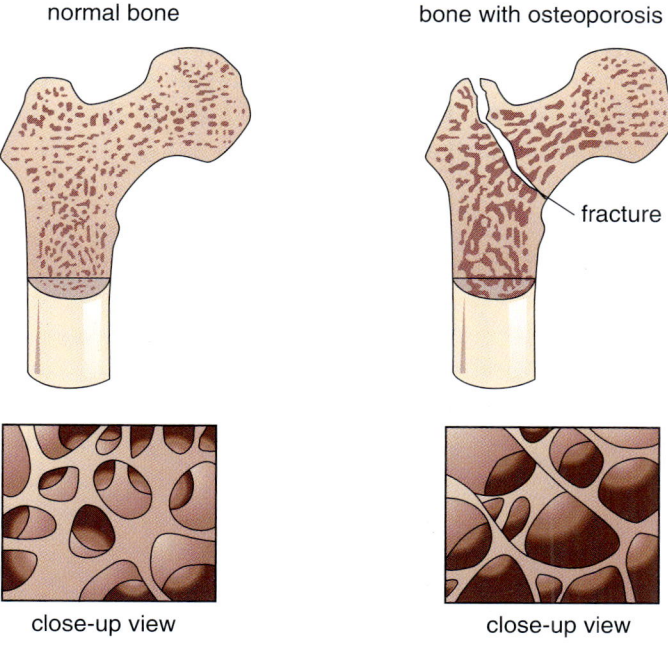

Figure 5.21 Osteoporosis Close Up

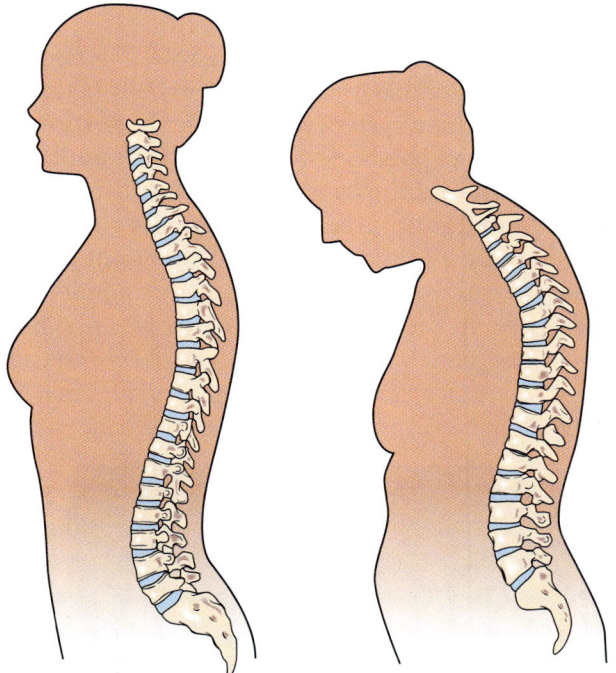

Figure 5.22 Normal and Osteoporotic Vertebrae

osteomalacia and other types of degenerative bone diseases caused by genetic disorders or exposure to toxic substances (Figure 5.22).

Tooth decay is considered a type of skeletal disease. It is the process of tooth destruction caused by mouth bacteria (Figure 5.23). Food that is not brushed or flossed off the teeth is a rich source of nutrients for mouth bacteria. In the process of digesting food, the bacteria produce acids and enzymes. The acids remove calcium from the teeth, and the enzymes degrade the remaining matrix. Both activities weaken the tooth's structure. Enough of this decay can

Tooth Decay Tooth destruction caused by bacteria

THE SKELETAL SYSTEM

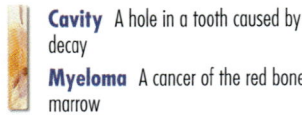
Cavity A hole in a tooth caused by decay

Myeloma A cancer of the red bone marrow

create a **cavity** in the tooth. These enzymes also break down the collagen fibers that hold the teeth in place, making the teeth loose. Tooth decay is not as common today as it was years ago. However, it still is a problem for certain people. Advanced tooth decay can be painful and lead to the loss of teeth and the underlying bony structures of the jaw.

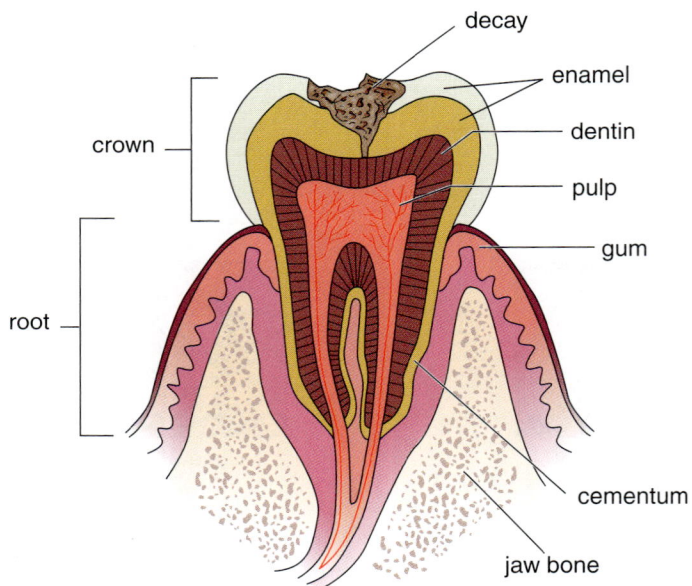

Figure 5.23 **Tooth Decay**

Myeloma is a deadly disease of the red bone marrow (Figure 5.24). It starts out when the stem cells that produce immune-system cells become cancerous. Initially, this weakens the immune system and disables a person's ability to fight disease. It also causes damage to the bone tissues. It then rapidly spreads throughout the body, distributing the cancer cells to almost every major organ. Body chemicals produced to fight off the cancerous cells end up damaging the body organs. They eventually weaken the body to the point of causing death. Myeloma is traditionally treated with chemotherapeutic agents including the notorious drug thalidomide. In the 1960's thalidomide was first used for treating morning sickness in pregnant women until it was shown to cause birth defects. It later showed value as a cancer treatment. Today, scientists are investigating the use of stem cells to displace and replace the cancerous bone cells.

Figure 5.24 **Myeloma**

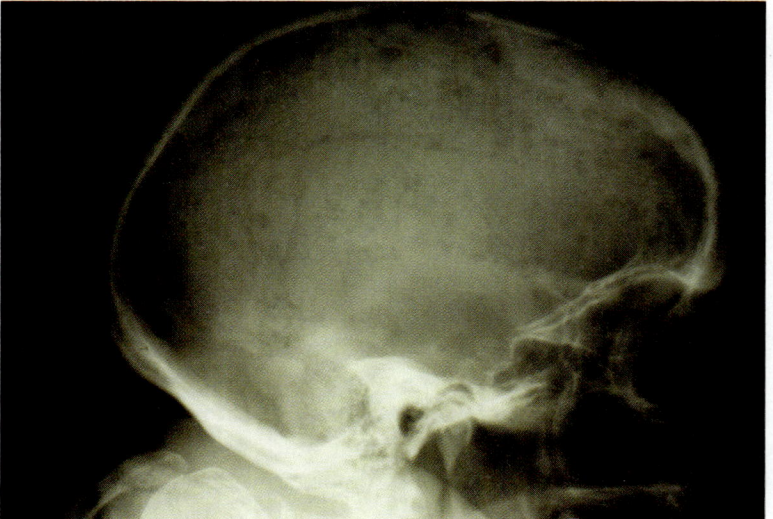

Two other bone conditions are **osteomyelitis** and **osteonecrosis**. Osteomyelitis is inflammation of the bone caused by bacterial infections of the blood. It can affect bone marrow, the compact-bone osteons, regions of the spongy bone, and the periosteum. It is located in one bone or spread throughout many bones. Osteonecrosis is caused by the death of osteocytes due to the obstruction of blood flow to the Haversian canals. It can be caused by a variety of conditions that cut off blood flow. This condition can occur in deep-sea divers and fighter pilots who develop bubbles of gas in their blood. These bubbles can block small blood vessels.

Osteomyelitis Inflammation of the bone caused by bacterial blood infections

Osteonecrosis A condition caused by osteocyte death due to the obstruction of blood flow

✓ Concept Check

1. Name five different types of skeletal-system diseases.
2. What are the different types of arthritis?
3. Describe the possible causes of osteoporosis.

Aging of the Skeletal System

Key Term: plaque

Lifestyle and aging of the other organ systems contribute to the overall wellness of the skeletal system as a person matures. The most prevalent condition of aging is the deterioration of articular surfaces. Some of this may be due to pathology rather than aging. However, cartilage has no direct blood supply and lacks nerves. This means that cartilage can unknowingly wear out without the body being able to repair it. So, over time, any joint surface becomes subject to arthritis. Compression of cartilage over a lifetime of standing and walking will produce changes in muscles and tendons that stress the joints. As a person matures, it is very likely that they have encountered several bacterial infections that contribute to joint decay.

Throughout a person's life, bone is regularly degraded as new bone is added. The rate of bone production is greater in young people, causing their bones to become denser, heavier, and larger. As a person ages, the bone degradation activities of the osteoclasts outpace the bone rebuilding action of the osteoblasts. This process usually begins around the age of 30 years due to a variety of factors. Most often, it is due to a decline in sex hormones needed for bone maintenance. Second, it can be due to poor nutrition or individual genetic factors. This natural decomposition of bone can lead to osteoporosis in 50 % of women and 25% of men over age 50 years. Inactivity associated with disability and maturity promotes weakened bones. Bones can only retain their density when placed under the stresses of muscle action. An active lifestyle can slow this form of body aging. However, it must be noted that too much activity can wear out the joints prematurely. Another factor causing bone density loss is the inability of the digestive system to take up nutrients needed for bone growth.

Bones, including those of the skull, deform in response to continuous muscle action. The tendons pull on articular surfaces making them rough. This, along with the decline of bone density, exaggerates many bone features. Tight clothing and certain cultural practices can affect the shapes of flat bones over the life span (Figure 5.25). For example, wearing restrictive bands over the

Plaque A film of bacteria and bacterial waste that forms on teeth

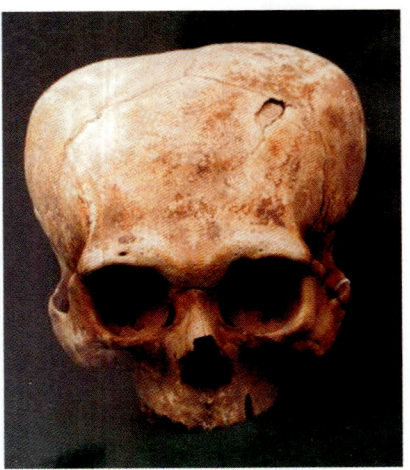
Figure 5.25 **Cranial Deformation**

head or tight shoes can permanently deform adult bones. Certain measures can be taken to correct age-related decay of bone. It is common today for people to have worn joints replaced (Figure 5.26). Some physicians find it necessary to replace the major components of the bones that form articulations. For example, a hip replacement can involve replacing the top of the femur and the acetabulum.

Changes in the teeth are also part of life span changes (Figure 5.27). It is common for the alveolar surfaces of the mandible and maxilla to lose bone density, leading to tooth loss. Plaque-forming bacteria take their toll on the teeth. A persistent film of bacteria that grows on the teeth causes **plaque**. The film collects bacterial wastes that contribute to tooth and alveolar decay. Gravity and posture contribute to overall skeletal deformation. Softening bones distort in response to the pull of gravity. It also makes them bend in response to muscle tugging. The long-term effects on the body can cause severe postural changes (Figure 5.28). Good standing posture is important for the even distribution of weight on the spine. A distorted posture associated with age-related skeletal changes can cause back pain and predispose a person to hip fractures. It also can change the shape of the rib cage, making it more difficult to breathe.

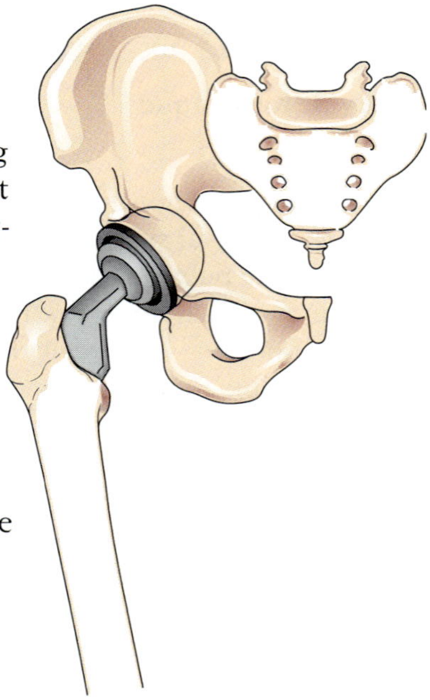

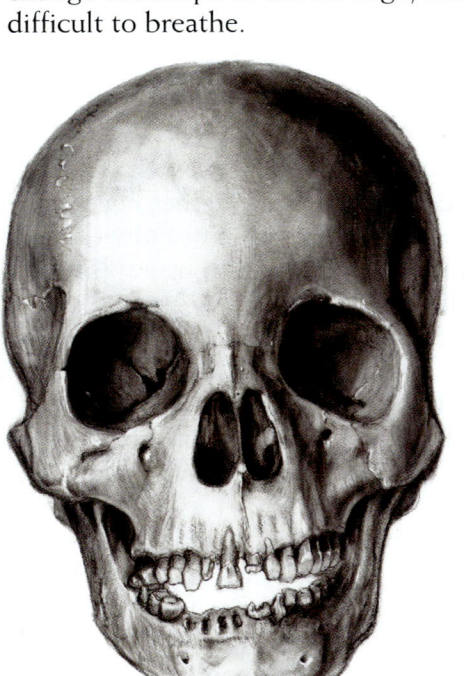

Figure 5.26 **Hip Replacement (Ball and Socket)**

Figure 5.27 **Aging of the Skull**

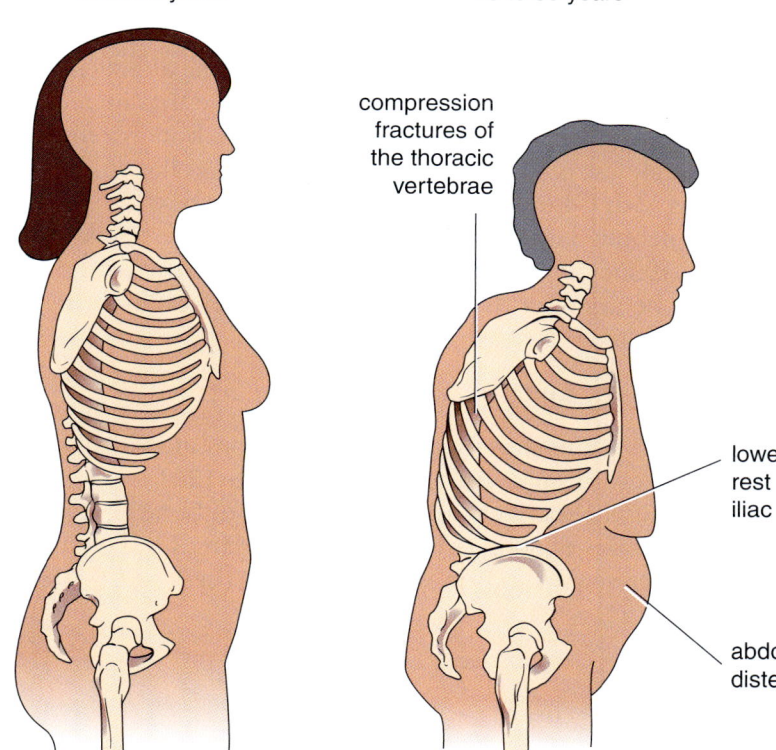

Figure 5.28 Skeletal Aging in the Thorax

✓ Concept Check

1. Describe two factors that contribute to skeletal-system aging.
2. What contributes to bone density loss?
3. What are the effects of bone density changes on posture?

DISCOVERY SCENE PLEASE ENTER DISCOVERY SCENE PLEASE ENTER

Is the boy's problem related to one of the diseases mentioned above? Did the information about changes to the skeletal system over the life span provide any more clues about the boy's bone-breakage condition?

CSI – Case Study Investigation Conclusion

What can you conclude about the ease with which the child's bones break? Is it normal for a child to be this accident-prone? Should activities such as dragging a heavy object cause bone breakage?

Answer:

This boy has a genetic condition called osteogenesis imperfecta (OI). It is the most common cause of childhood fractures. The defective gene affects how collagen is laid down in the bone matrix. Many people believe that increased dietary calcium helps the condition; however, calcium has no bone strengthening effects. The bone weakness is solely due to defects in the collagen. It is even possible for infants with OI to have a bone break during birth. Approximately 50% of people with OI develop hearing loss. The fragile ear bones eventually break, reducing the transmission of sounds to the brain. Unfortunately, little can be done for the child except to protect him or her from activities that cause bone breakage.

This CSI was adapted from the following articles:

1. Chamberlain JR, et al. Gene targeting in stem cells from individuals with osteogenesis imperfecta. *Science*. 2005;303(5661):1198-1201.
2. Shapiro J. Osteogenesis imperfecta and other defects of bone development as occasional causes of adult osteoporosis. In: Marcus R, Feldman D, Kelsey J, eds. *Osteoporosis*. Burlington, MA; Academic Press; 1996:703-13.

Chapter Summary

The skeletal system works together with muscles, ligaments, tendons, and articulations to provide the body with support and movement.

Bone development begins early in human development, but it is not complete until a person is in his or her twenties. Bones are classified according to their shape or origins in the developing individual. Each bone develops at a different rate, which is determined by its role in the body. Most bone diseases are due to some condition that reduces bone or joint structure. The wear and tear of everyday life is the most common contributing factor to skeletal-system disorders and aging. Bones can lose mass due to genetic factors related to DNA defects or genetic changes associated with aging. Inadequate nutrient intake can reduce the levels of calcium and protein needed for bone maintenance. Infections will also decay teeth and joints, causing a loss of skeletal function. It is important to remember that bone is a living tissue that responds to the daily activities of a person. A certain amount of activity is needed to maintain the integrity of the skeletal system. However, too much activity will prematurely wear out its components.

Overview
- Humans have an endoskeleton.
- The skeletal system is composed of bones, cartilage, ligaments, and tendons.
- The skeletal system provides movement, protection, and shape.

Human Skeletal System
- The axial skeleton is composed of the spine, rib cage, and skull.
- The appendicular skeleton is composed of the bones of the arms, hips, legs, and shoulders.
- Bones have surface features that function as articulations, muscle attachment areas, openings for blood vessels and nerves, and regions where blood vessels, muscles, and nerves press against bone.

Axial Skeleton
- The skull is composed of the cranial and facial bones.
- The cranial bones are the ethmoid, frontal, occipital, parietal, sphenoid, and temporal bones.
- The facial bones are the inferior nasal conchae, lacrimal, mandible, maxillary, nasal, palatine, vomer, and zygomatic bones.
- The vertebrae are divided into the cervical, thoracic, lumbar, sacral, and coccygeal bones.
- The rib cage is composed of the ribs and sternum.
- The hyoid bone of the neck is also part of the axial skeleton.

Appendicular Skeleton
- The appendicular skeleton is composed of the upper and lower extremities.
- The upper extremities are composed of the following bones: the carpals, clavicle, humerus, metacarpals, phalanges, radius, scapula, and ulna.
- The lower extremities are composed of the following bones: the femur, fibula, ilium, ischium, metatarsals, phalanges, pubis, tibia, and tarsals.

THE SKELETAL SYSTEM

Study Guide

Bone Types
- Bones can be categorized by shape: flat, irregular, long, and short.
- Bones are also classified by their origin in the embryo.

Bone Structure
- Long bones are composed of a diaphysis, two epiphyses, and two epiphyseal lines.
- The marrow cavity can contain yellow or red bone marrow.
- Bone cells are called osteocytes.
- Bone matrix is made of collagen, osteocalcin, and the calcium mineral, hydroxyapatite.
- Bone is composed of compact (cortical) and cancellous (trabecular) bone.

Joints
- Joints can be categorized by their structural components. The categories are cartilaginous, fibrous, and synovial.
- Joints can also be categorized by their function. The categories are amphiarthrosis, diarthrosis, and synarthrosis.
- Synarthroses allow movement. The following are the different types of synarthroses: ball and socket, condyloid or ellipsoidal, gliding, hinge, pivot, and saddle.

Bone Development and Healing
- Endochondral bones form from cartilage pegs in the embryo. They usually produce long bones, and parts of irregular and short bones.
- Endochondral bones have primary and secondary ossification centers, and a region that produces the bone collar. The growth of the diaphysis determines bone length.
- Dermal bones form in subcutaneous membranes. They are mostly composed of cancellous bone with a covering of bony plates. They usually produce flat bones and parts of irregular bones.
- Bone fractures can be simple, commuted or compound, and open.
- Fractures can also be categorized by the shape or displacement of the fracture line.
- Bone healing involves four stages: fracture, granulation, callus, and normal contour. Sometimes these stages are classified as three phases: reactive, reparative, and restorative.

Pathology of the Skeletal System
- Arthritis is a common skeletal disease that affects the joints. The types of arthritis include osteoarthritis, rheumatoid arthritis, ankylosing spondylitis, and juvenile arthritis.
- Other diseases that affect joints are gout, lupus, scleroderma, and fibromyalgia.
- Osteoporosis is one of the diseases of bone density loss.
- Myeloma is a bone marrow cancer.
- Osteomyelitis is a bone inflammation caused by bacteria.
- Osteonecrosis is death of bone tissue caused by loss of blood flow to the osteons.

Aging of the Skeletal System
- During a person's lifetime, bone is constantly degraded and replaced.
- Some bone aging is due to wear and tear on the bones and joints.
- Some bone aging is due to a slowing of bone tissue replacement.
- Lifestyle can affect bone and joint aging.
- Changes in sex-hormone production contribute to skeletal aging.
- The teeth are degraded by bacteria that form plaque.
- Skeletal aging is affected by gravity and posture.

Key Terms

Overview
Bone
Cartilage
Endoskeleton
Ligament
Tendon

Human Skeletal System
Appendicular skeleton
Articulation
Axial skeleton
Lower appendage
Surface features
Upper appendage

Axial Skeleton
Alveolus
Atlas
Axis
Calvaria
Cervical vertebrae
Coccygeal vertebrae
Coccyx
Coronal suture
Cranial base
Cranial bone
Cranium
Dentin layer
Enamel
Ethmoid bone
Facial bone
False vertebrae
Fixed vertebrae
Foramen
Frontal bone
Gladiolus
Hyoid
Inferior nasal concha

Inferior orbital foramen
Lacrimal bone
Lumbar vertebrae
Mandible
Manubrium
Mastoid process
Maxillary bone
Mental foramen
Moveable vertebrae
Nasal bone
Nasal septum
Occipital bone
Occipital foramen
Orbit
Orbital ridge
Palatine bone
Parietal bone
Ribs
Sacral vertebrae
Sacrum
Sagittal suture
Sphenoid bone
Squamosal suture
Sternum
Styloid process
Temporal bone
Thoracic vertebrae
Thorax
True vertebrae
Tympanic region
Vertebrae
Vertebral column
Vomer
Xiphoid process
Zygomatic bone
Zygomatic process

THE HUMAN SKELETON

Study Guide

Appendicular Skeleton
Acetabulum
Acromion process
Calcaneus
Capitate
Carpal
Clavical
Coronoid fossa
Cuboid
Cuneiform
Distal phalanx
Extremity
Femur
Fibula
Hamate
Humerus
Ilium
Ischium
Lunate
Medial phalanx
Metatarsal
Navicular
Obturator foramen
Olecranon fossa
Olecranon process
Patella
Phalanges
Pisiform
Proximal phalanx
Pubic bone
Pubic symphysis
Pubis
Radius
Scaphoid
Scapula
Shoulder blade
Shoulder girdle
Sternal extremity
Talus
Tarsal
Tibia
Trapezium
Trapezoid
Triangular
Triquetral
Ulna

Bone Types
Alveolar bone
Dermal bone
Endoskeletal bone
Flat bone
Irregular bone
Long bone
Sesamoid bone
Short bone
Sutural bone
Wormian bone

Bone Structure
Bone marrow
Bone proper
Canaliculi
Cancellous bone
Compact bone
Cortical bone
Diaphysis
Epiphyseal line
Epiphyseal plate
Epiphysis
Haversian canal
Hyaline cartilage
Lacunae
Medullary cavity
Osteocalcin
Osteocyte
Osteon
Periosteum
Red bone marrow
Spongy bone
Trabecular bone
Volkmann's canal
Yellow bone marrow

Joints
Bursa
Bursitis
Functional classification
Structural classification
Synovial fluid

Joint Structure
Cartilaginous joint
Fibrous joint
Synovial capsule
Synovial joint

Joint Function
Amphiarthrosis
Ball-and-socket joint
Condyloid joint
Diarthrosis
Ellipsoid joint
Gliding joint

Study Guide

Gomphosis
Hinge joint
Pivot joint
Symphysis
Synarthrosis
Synchondrosis
Syndesmosis
Synostosis

Bone Development and Healing
Angulated
Bone collar
Calcification
Commuted fracture
Compound fracture
Endochondral ossification
Fontanelle
Fracture
Greenstick fracture
Intramembranous ossification
Oblique
Open fracture
Ossification
Osteoblast
Osteoclast
Simple fracture

Spiral
Transverse
White blood cell

Pathology of the Skeletal System
Ankylosing spondylitis
Arthritis
Bone density loss
Cavity
Fibromyalgia
Gout
Juvenile arthritis
Lupus
Myeloma
Osteoarthritis
Osteomyelitis
Osteonecrosis
Osteoporosis
Rheumatoid arthritis
Scleroderma
Shin splint
Stress fracture
Systemic lupus erythematosus
Tooth decay

Aging of the Skeletal System
Plaque

Check Your Understanding

1. Which bone is part of the appendicular skeleton?
   a. Sphenoid
   b. Clavicle
   c. Sternum
   d. Rib

2. Carpals are found in this region:
   a. Lower appendages
   b. Upper appendages
   c. Skull
   d. Rib cage

3. Which is the most superior portion of the spinal column?
   a. Lumbar
   b. Cervical
   c. Thoracic
   d. Sacral

4. Which is not a cranial bone?
   a. Maxilla
   b. Ethmoid
   c. Occipital
   d. Temporal

THE HUMAN SKELETON

Study Guide

5. Which bones would be directly damaged by a blow to the cheek?
   a. Occipital and sphenoid
   b. Zygomatic and ethmoid
   c. Maxilla and zygomatic
   d. Temporal and parietal

6. Where would you look for wormian bones?
   a. Cranium
   b. Pelvic girdle
   c. Shoulder girdle
   d. Facial bones

7. Fusion of this type of joint would have the greatest effect on body movement:
   a. Synovial
   b. Suture
   c. Synarthrosis
   d. Diarthrosis

8. Damage to this bone structure could restrict blood flow into a bone:
   a. Sulcus
   b. Foramen
   c. Crest
   d. Trochanter

9. A reduction of these cells would slow bone growth and hardening in adults:
   a. Osteoblasts
   b. Fibrocytes
   c. Osteoclasts
   d. Melanocytes

10. Which of the following bones carries more of the body's weight than the other bones?
    a. Malleolus
    b. Calcaneus
    c. Fibula
    d. Atlas

11. Intramembraneous bone formation is most important for the development of the:
    a. Upper extremities
    b. Lower extremities
    c. Skull
    d. Lower vertebrae

12. Bone density is likely to be lost by overactivity of these cells:
    a. Osteoclasts
    b. Osteocytes
    c. Osteoblasts
    d. Chondrocytes

13. Which condition is less likely to convert yellow marrow into red marrow?
    a. Excessive blood from a major injury
    b. The intake of toxins that damage blood forming cells
    c. Intake of too many fats in the diet
    d. Starvation

14. Which of the following is a bone cell cancer:
    a. Caries
    b. Myeloma
    c. NICO
    d. Lupus

15. Aging is usually accompanied by the following event:
    a. Increased bone density
    b. Increased blood flow to bones
    c. A loss of bone density
    d. Rapid replacement of the joint cartilage surfaces

A Case Study

NICO: A Hidden Epidemic?

A growing group of dentists and physicians are discovering that what was once thought to be a rare condition might be very common. Many medical professionals however, believe that this problem is overplayed and might even be the basis of fraudulent medical practices. This controversial condition is called neuralgia inducing cavitational osteonecrosis (NICO). In simple terms, cavitational osteonecrosis refers to holes, or cavities, that form in bone as a result of bone cell death. Neuralgia means pain felt along a particular nerve. So, NICO refers to a consistent pain located along a nerve that serves a bone with osteonecrosis. Apparently, the benefits provided by better access to healthcare had a downside: a large body of research shows that frequent dental work can introduce bacteria into the blood and facial bones, causing localized and widespread cavitational osteonecrosis. Cavitational osteonecrosis could also be caused by blood-flow problems and other degenerative changes to the body. Many medical practitioners feel that anyone could be a candidate for NICO. They believe that the disease can go undetected until it starts causing intense pain. By then, it may be too late to adequately treat the condition. This is giving them the justification to do routine NICO testing on patients who come for dental and medical visits.

Use the information in this chapter and the following Web sites to resolve the issue of NICO. Answering the following questions will assist you:

- After weighing the evidence from different sources, is it evident that NICO is a serious concern worthy of more investigation?
- Should every person be tested for NICO even though it may not be as prevalent as believed?
- Should dentists and physicians be required to have patients seek second and third opinions when diagnosed with NICO?
- Should the government put money into investigating the extent of NICO?

THE HUMAN SKELETON

Study Guide

- Should insurance companies have the right to investigate the validity of a NICO diagnosis and treatment before deciding to pay out benefits?

1. Quackwatch
   http://www.quackwatch.org/01QuackeryRelatedTopics/cavitation.html

2. Facial Neuralgia Resources
   http://facial-neuralgia.org/conditions/nico.html

3. National Institutes of Health – NINDS Trigeminal Neuralgia Information Page
   http://www.ninds.nih.gov/disorders/trigeminal_neuralgia/trigeminal_neuralgia.htm

4. Healthy People 2010
   http://www.healthypeople.gov/document/html/volume2/21oral.htm

Where Do We Go from Here?

People in health fields can use their knowledge of the skeletal system to solve everyday problems. You may wish to use other resources, such as the following Web sites, in addition to your book to investigate the answers to each of the following situations:

1. You are told that a neighbor's young boy has bone cancer. Explain to the parents why the cancer must be treated with chemicals that spread throughout the body rather than with the surgical removal of the tumor.
   www.patient.cancerconsultants.com

2. You are advising a 12-year-old boy who wants to start lifting heavy weights to bulk up for junior high school football. What advice would you give him related to the benefits and problems of weightlifting at his age?
   www.kidshealth.org

3. A young woman read about a dietary supplement that reduces osteoporosis. How would you determine the effectiveness and safety of this supplement?
   www.healthfinder.gov

4. You were asked to help a woman plan an exercise routine after recovering from a prolonged illness. Her physician said she should avoid anything that can cause stress fractures. What activities should she avoid?
   www.nlm.nih.gov

5. A friend was told to keep his food intake slightly higher and to take in more proteins after breaking a leg. Why did the physician give this advice?
   www.innerbody

Skills Activities

1 Bone Studies

Study Guide

Materials
- lab partner
- entire articulated human skeleton
- male and female pelvis models
- prepared slide of thyroid gland
- rulers in centimeters
- measuring tape in centimeters
- meter stick
- calculators
- black board or large paper chart to record data
- computer with internet access linked to:
  http://medstat.med.utah.edu/kw/osteo/index2.html

Bones can tell much about persons, even after they are long dead. Forensic anthropologists rely on the fact that the everyday life of a person can be discovered by studying the wear and tear found on bones and joints. An anthropologist is a scientist who studies human origins. They do this by studying human behavior, physical features, social structures, and cultural development. Forensic anthropology is the application of various sciences to answer questions based on the examination and comparison of evidence that gives some information about human origins. Some forensic anthropologists use these strategies and human remains to solve crimes. The remains are what the courts call biological evidence. In this activity, you will use actual human skeletons or casts of real bones to determine certain features of a person using forensic anthropological methods.

Load up the Forensic Anthropology Web site. First, go to the Human Osteology link to learn how scientists orient and measure human bones for anthropological studies. Is the information consistent with what you learned in this chapter? Next, go to the Forensic Anthropology link to see how bone measurement is important in determining features such as age, gender, pathology, race, stature, and trauma. Now find a lab partner, and take the following measurements in centimeters on your lab partner and on the human skeleton:

- total height
- length of the foot
- length of the ankle to the hip
- length of the knee to the ankle or tibia
- length of the thigh or femur
- length of the arm from wrist to shoulder
- length of the forearm or radius from the wrist to the elbow
- length of the upper arm or humerus from the elbow to the shoulder
- length of the index finger
- circumference of the wrist
- circumference of the neck
- circumference of the leg just above the knee
- distance of an outstretched arm span from finger tip to finger tip
- width of the back from shoulder to shoulder

THE HUMAN SKELETON

Study Guide

Take each measure and divide it by the total height of the person. This is called the bone-to-height ratio. Record your bone-to-height data on the board for class data comparison.

Gender Differences Now, use this information to see if there are any skeletal patterns that can determine gender. Gender can be determined in several ways from different parts of a skeleton. For example, males have a sloping forehead, while females have a straighter forehead. Plus, the hipbones show obvious differences as mentioned earlier in the chapter. Males and females may also have differences in arm and leg bone-to-height ratios. Use this information to answer the following questions:

- Do all males have the same ratios?
- Is the same true for females?
- What other factors can affect bone ratios?
- How do ratios of the skeleton compare with the people in class?

Determining Height Forensic anthropologists are able to accurately determine the height of a person. For example, it is possible to measure the height of a female just by knowing the length of a femur. The customary way to calculate female height from femur length is to multiply the length of the femur (measured in centimeters) by 1.945, and then add 72.84, as shown in the tables below. Try to see if this is true for your partner. Now, use the information to estimate the height of the human skeleton, and see if its measurements are consistent for the estimated gender.

Base Equation for Females	Factor for Wet Bone (with cartilage)	Factor for Dry Bone (without cartilage)
1.945 x length of femur in cm +	71.163	72.844
2.754 x length of humerus in cm +	70.046	71.475
2.352 x length of tibia in cm +	73.369	74.774
3.343 x length of radius in cm +	80.189	81.224
(1.117 x length of femur in cm) + (1.125 x length of the tibia in cm) +	67.939	69.561
(1.339 x length of femur in cm) + (1.027 x length of humerus in cm) +	65.763	67.435
Base Equation for Males	**Factor for Wet Bone (with cartilage)**	**Factor for Dry Bone (without cartilage)**
1.88 x length of femur in cm +	79.971	81.306
2.894 x length of humerus in cm +	69.454	70.641
2.376 x length of tibia in cm +	77.547	78.664
3.271 x length of radius in cm +	85.205	85.925
(1.22 x length of femur in cm) + (1.080 x length of the tibia in cm) +	70.069	71.443
(1.03 x length of femur in cm) + (1.557 x length of humerus in cm) +	67.027	68.397

Skills Activities

2 Comparative Anatomy of Primates

Materials
- computer with Internet access linked to:
  http://www.eskeletons.org/
- disarticulated human skeleton with skull

Study Guide

It might be surprising to learn that physicians are now using comparative anatomy to better understand dental and skeletal abnormalities and birth defects in humans. Primates are particularly important in these studies because of their close kinship to humans. Traditionally, scientists compared bones by having the actual bones in front of them and recorded any differences or similarities. Some physicians today do the same with bone defects due to mutations in bone-development genes. They try to determine whether any of the mutations match up with growth patterns found in primates and other animals. It is hoped that they can get a better understanding of how the mutation occurs by seeing whether it causes changes that resemble the normal growth of other animals.

Load up the eSkeletons Project Web site on the computer. First, go to the Select the Taxon drop-down menu, and review the different bones of the human and the primates. Note that the different body regions are colored to indicate the similar structures. Then, go to the Comparative Anatomy link. Select the various bones to compare side-by-side for humans and primates. Use what you observed to answer the following questions:

- What are the major differences in dentition between humans and the other primates?
- What are three main differences between the human skull and the skulls of the other primates?
- What explains the differences between the human and gorilla tibia?
- How do the calcaneus bones differ in each of the specimens, and what about their way of life explains the differences?
- How do the thoracic vertebrae differ between each? What human conditions can explain how vertebrae resemble those of the other primates?

THE HUMAN SKELETON

Case Study Investigation

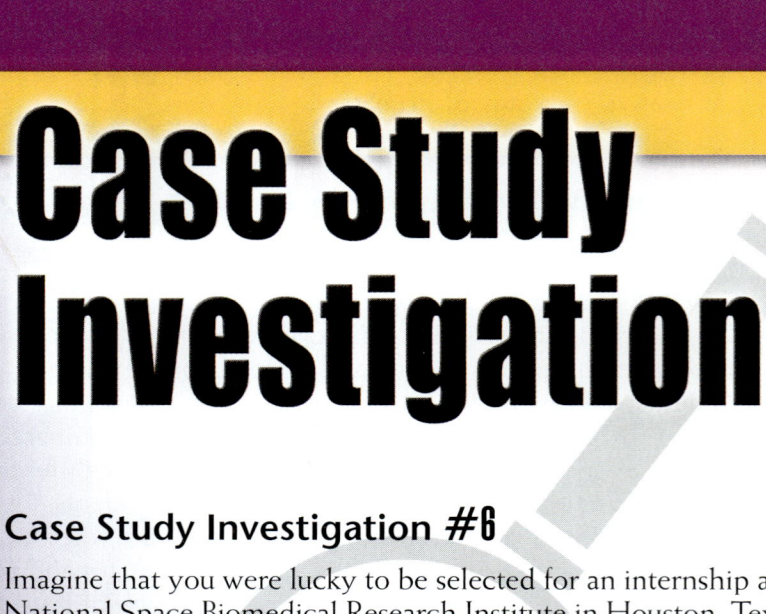

Case Study Investigation #6

Imagine that you were lucky to be selected for an internship at the National Space Biomedical Research Institute in Houston, Texas. Part of your job is to monitor the health of the astronauts for the National Aeronautics and Space Administration (NASA) before, during, and after a space mission. By chance, your time in the lab corresponds with the return of an American astronaut from the International Space Station. She just spent 171 days in the space station and is now getting back to her research studies on earth. Unfortunately, it was discovered that she would have to go through at least 3 weeks of physical therapy. You learned from reading her physical health report that she lost 30% of her skeletal muscle mass. Most of it occurred within the last 2 months of the mission in space. The normal muscle loss during a space mission is less than 20%. A majority of muscle atrophy results from disuse, yet she followed all the exercise programs NASA requires. Your job is to assist the research team in investigating her problem. At the end of the chapter, you will be asked to determine the most likely cause of her accelerated loss of muscle mass.

CHAPTER 6

THE MUSCULAR SYSTEM

Chapter Outline

Case Study Investigation (CSI)
Applied Learning Outcomes
Overview
Muscle
   Types of Muscle Tissues
   Muscle Cell Structure
   Muscle Cell Function
Musculature
   Gross Skeletal Muscle Types

Skeletal Muscle Structure
   Skeletal Muscle Action
Musculature Charts
Wellness and Illness over the
  Life Span
   Pathology of the Musculature
   Aging of the Muscular System
Case Study Conclusion
Study Guide

Applied Learning Outcomes

- Lean to use the terminology associated with the musculature system.
- Learn about the following:
  - different types of muscle cells
  - muscle tissue development
  - gross and fine muscle structure
  - gross muscle function
  - muscle cell physiology
  - muscle types and actions
  - muscle development and growth.
- Understand the aging and pathology of the musculature.

Overview

Key Terms: body mass index (BMI), calcium, contractile cell, electrolytes

Contractile Cell A cell with a specialized membrane and cytoskeleton that permits it to change shape

Contraction Muscle cell shortening

Electrolytes An ion capable of conducting an electrical current in a solution

Body Mass Index (BMI) An indirect measure of body density

Even the smallest external or internal body movement would not be possible without muscle tissue. Human muscle tissue does a variety of jobs, including moving the skeletal system, passing food through the digestive system, and pumping blood throughout the body. Muscles are able to do their jobs because they are composed of arrangements of **contractile cells**. Contractile cells have a specialized cell membrane and cytoskeleton that permit them to change their shape. The cytoskeleton is arranged in such a way that muscle cells change their shape by shortening along one or more planes. **Contraction** is the term used to describe the shortening motion of a muscle cell. Muscle cells are usually laid out as sheets of muscle tissue that produce coordinated contractions.

Over half of the body's mass is composed of muscle tissue, and over 90% of this muscle tissue is involved in skeletal movement. Contractile cells have high energy needs, so it is common to see them associated with an ample blood supply. The blood provides the cells with much needed glucose and oxygen while removing the large amounts of metabolic wastes. **Electrolytes**, which are transported by the blood, are also essential components of muscle cell contractions. Along with nervous tissue, muscle consumes almost 70% of the food energy taken into the body every day. Plus, muscle is as intensive a consumer of calcium as is the skeletal system. Much of the calcium stored in bones is made available for the muscles' needs. Muscle makes up a large component of the measurement called **body mass index (BMI)**. Body mass index, which is an indirect measure of body density, is becoming a popular indicator of health. People with a lean body have higher amounts of muscle mass compared with their body fat composition. A certain degree of leanness is known to reduce heart disease and metabolic disorders.

✓ Concept Check

1. What is a contractile cell?
2. Describe three bodily needs of muscle cells.
3. Define the term, body mass index, or BMI.

Civic Responsibility

CIVIC RESPONSIBILITY: HELPING OTHERS WITH YOUR KNOWLEDGE

It is valuable to use what you have learned about the muscular system to help others to better understand the world around them. It is very important to check your facts and seek further information about certain topics before discussing health and science issues. Here are some suggestions to foster a better public awareness of the muscular system:

1. Speak to schoolchildren about avoiding muscle strain during play.
2. Work with sports clubs to educate players about muscle enhancers.
3. Help elderly persons to better understand the limitations of taking dietary supplements or products that claim to reduce muscle aging.
4. Volunteer at a school health day to teach children the basics of human musculature.

Muscle

 Key Terms: contractile proteins

Human muscle cells are categorized several ways. The first scheme evaluates their microscopic appearance. Using this method, muscles are grouped as nonstriated and striated. The striped appearance of striations results from the uniform arrangement of the cytoskeleton proteins responsible for muscle cell contraction. The proteins of the cytoskeleton involved in contraction are called **contractile proteins**. Nonstriated muscle cells have a randomized pattern of contractile proteins that is indiscernible under a microscope. Nonstriated muscle cells usually provide weaker contractions than do striated muscle cells. The type of control a person has over the function of a type of muscle cell makes up the second categorization scheme. Muscle cells are characterized as involuntary and voluntary. Involuntary muscle cells contract without conscious control. They carry out jobs that must be done automatically or in coordination with other organ systems. People have a large degree of control over voluntary muscle cells. Some voluntary muscles function unconsciously, such as the muscles used for breathing. However, they can be contracted intentionally when a person consciously thinks about controlling the muscles. The third way to classify muscle is based on its location. This method places muscle tissue into the three categories that were briefly discussed in Chapter 3 under tissue types, which are cardiac, skeletal, and smooth (Figure 6.1).

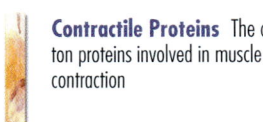 **Contractile Proteins** The cytoskeleton proteins involved in muscle contraction

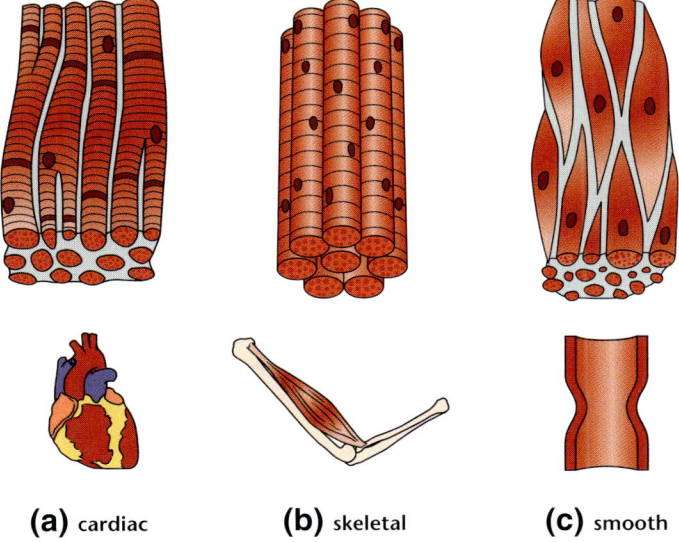

(a) cardiac **(b)** skeletal **(c)** smooth

Figure 6.1 Types of Muscle Tissue

✓ Concept Check

1. What are the three ways to characterize muscle?
2. Distinguish between nonstriated and striated muscle cells.
3. Distinguish between involuntary and voluntary muscle cells.

The Muscular System

Types of Muscle Tissue

Key Terms: growth factor, intrinsic beat, muscle fiber, myoblast, myogenesis, motor nerve cell, peristalsis

Myogenesis The process by which embryonic mesoderm cells become muscle tissue

Myoblast A stem cell that forms muscle tissue

Growth Factor Chemicals that act as signals to initiate cell division and differentiation

Intrinsic Beat A natural contraction cycle of cardiac muscles

Peristalsis Weak, pulsating contractions that move food and waste through the digestive system

Muscle Fiber A muscle tissue cell

Motor Nerve Cells Nerves that control skeletal muscle fibers

All muscle tissue develops from mesoderm cells of the embryo in a process called **myogenesis**. Myogenesis starts when a group of stems cells in the mesoderm form into **myoblasts**. Myoblasts move about other developing tissues and organs to form the three muscle types. Muscle **growth factors** produced by the other cells provide the direction for differentiation of muscle cell types. Over a dozen genes are involved in muscle cell development. It is not unusual to find genetic defects in the complex mechanisms that consequently produce a variety of problems associated with muscle activity.

Cardiac muscle forms around large blood vessels that eventually develop into the heart. Cardiac muscle is an involuntary striated muscle, which means it provides strong contractions that are not under conscious control. Unlike most cells, cardiac muscle cells have two nuclei. These branched cells communicate with one another by special cell junctions called intercalated disks. Cardiac muscle cells have a natural contraction cycle called the **intrinsic beat**. Intercalated disks help synchronize the intrinsic beat so that all the cardiac muscles act in unison.

Smooth muscle is found in many organ systems. It is part of the lining of blood vessels, digestive organs, urinary system, and parts of the respiratory system. Smooth muscle gets its name from the fact that it is nonstriated and, therefore, appears "smooth" under the microscope. These spindle-shaped cells produce weak involuntary contractions that can last for long periods of time and assist with dilation (widening) and constriction (narrowing) of the blood vessels and tubular structures in the respiratory system. Smooth muscles of the digestive system are laid in sheets that produce weak, pulsating contractions called **peristalsis**. Peristalsis moves food and wastes through the digestive system.

Skeletal muscle is the focus of this chapter. It is the muscle type that provides movement of the bones and joints. This large voluntary muscle cell has distinct striations, indicating its powerful contractile capabilities. A skeletal muscle cell is actually composed of several myoblasts that fuse together into what is called a **muscle fiber**. This is why skeletal muscle cells have many nuclei that are readily visible under the microscope. Each muscle fiber is stimulated to contract by a nerve cell that controls several muscle fibers at once. Nerves that control skeletal muscle fibers are called **motor nerve cells**.

✓ Concept Check

1. Describe the three types of muscle tissues.
2. Define the terms constriction and dilation.
3. What type of nerve controls skeletal muscle?

Artificial Muscle?

Physicians and biomedical engineers are well aware of the limitations brought about by the currently available artificial appendages. The robotic machinery of these appendages does not provide the same type of control and movement afforded by muscle. Engineers at various universities and the United States Department of Energy are investigating the feasibility of certain materials to make machines operate with the same capabilities of human muscle. These artificial muscle materials are called hydrogel polymers. Hydrogel polymers are special protein mixtures that contract in a manner similar to muscle. They can be induced to contract in a variety of directions using targeted electrical impulses or changes in the pH of particular regions of the hydrogel. It is possible to organize the hydrogels into fascicle patterns that resemble human muscle. These fascicles can be used to control movement and contraction strength very much like real muscle. So far, hydrogel muscles have been tested in simple machines to control switches and levers. Scientists are hoping to outfit artificial appendages with hydrogel in an attempt to provide people more lifelike prosthetic devices.

Dikovsky D, Bianco-Peled H, Seliktar D. 2006. The effect of structural alterations of PEG-fibrinogens hydrogel scaffolds on 3-D cellular morphology and cellular migration. *Biomaterials.* 27(8):1496-506.

Muscle Cell Structure

Key Terms: 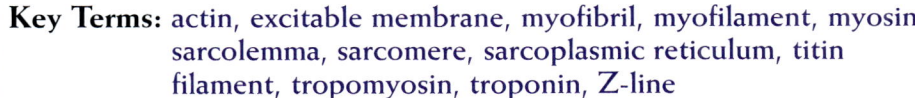 actin, excitable membrane, myofibril, myofilament, myosin, sarcolemma, sarcomere, sarcoplasmic reticulum, titin filament, tropomyosin, troponin, Z-line

Skeletal muscle cells are long, cylindrical cells covered with an **excitable membrane** and filled with a specialized cytoskeleton. The membrane of muscle cells is called the **sarcolemma**, meaning a covering (lemma) of the flesh (sarco). Flesh is an older term used for muscle. Excitable membranes contain membrane proteins that respond to signals from other cells and from the environment. Signals from nerve cells cause the sarcolemma to transmit information to the cytoskeleton and other organelles that carry out contraction of the muscle cell. The cytoskeleton is composed of bands of proteins called **myofilaments**. There are two types of myofilaments: thick and thin. A third type, called a **titin filament** runs vertically. Thick myofilaments, the first type, are composed of a protein called **myosin**. Thin myofilaments, the second type, are made up of three proteins. A major portion of the thin myofilament is composed of a band of **actin** wrapped around a length of **tropomyosin**. Speckled on the coils of the actin are small proteins called **troponin**. Tropomyosin blocks the binding of myosin to actin, which will be discussed later. The third type of myofilament is composed of a protein called titin, which is considered an elastic myofilament.

Excitable Membrane A membrane that responds to signals from other cells and the environment

Sarcolemma The membrane of muscle cells

Myofilaments Bands of proteins that compose the cytoskeleton of muscle cells

Titin Filament A protein filament that supports the myofilaments

Myosin A protein of the thick myofilaments of muscle cells

Actin One of the three proteins that form the thin myofilaments of muscle fibers; it forms the core of the fiber

Tropomyosin One of the three proteins that form the thin myofilaments of muscle fibers; it reinforces the actin core

Troponin One of the three proteins that form the thin myofilaments of muscle fibers

THE MUSCULAR SYSTEM

Sarcomere The contractile unit of a muscle cell

Myofibril Long cords of myofilaments that form parallel bundles that comprise most of a muscle cell's interior.

Z-Line A line across striated muscle fibers that marks the boundaries between each sarcomere

Sarcoplasmic Reticulum A system of tubes that stores and transports the calcium needed for muscle contraction

The contractile unit of a muscle cell is called a **sarcomere**. Many thousands of sarcomeres run the length of a muscle cell. These chains of sarcomeres form **myofibrils**. Each muscle fiber is made of many bundled myofibrils that run parallel to one another for the length of the cell. The thick and thin myofilaments arrange to form an overlapping pattern within a sarcomere (Figure 6.2). This overlapping configuration is what carries out the muscle cell's contraction and produces the striated pattern seen under the microscope. Thin myofilaments of the sarcomere are attached to a protein structure called the **Z-line**. Myofilaments run up the length of the Z-line like the rungs of a ladder. Inserted between the rungs of thin myofilaments are thick myofilaments that seem to be floating in the cell. However, they are held in place by the nearly invisible titin filament band that attaches them to the Z-Line. This protein band emerges at its end to secure it to the Z-line. Its function is to keep the thick and thin filaments aligned, and to help control the stretch and recoil limits in a muscle. Microscopically, the Z-line produces the anatomical dark striation lines of muscle tissue. Functionally, the Z-line serves a very important role in muscle contraction. It anchors the sarcomeres to the sarcolemma. Consequently, any movement of the Z-lines changes the length of the muscle cell. Surrounding each sarcomere is a modified organelle called the **sarcoplasmic reticulum**. The sarcoplasmic reticulum, a system of inner membrane tubes, stores and transports large amounts of the calcium needed for muscle contraction.

Figure 6.2 Muscle Cell Structure

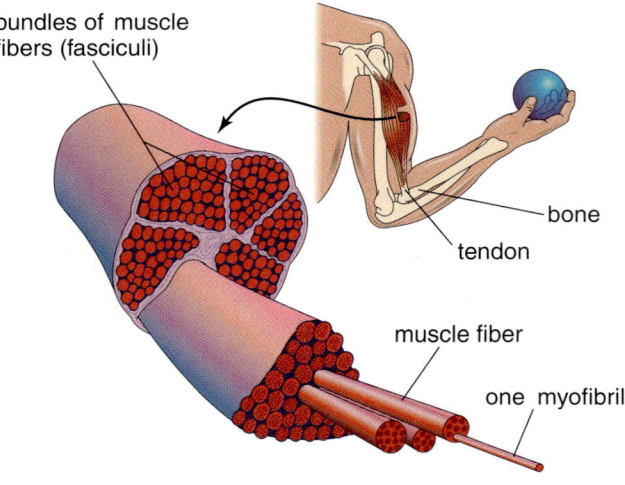

✓ Concept Check

1. What is the special name given to the muscle cell membrane?
2. Explain the protein composition of skeletal muscle myofilaments
3. Describe the structure of a sarcomere.

Muscle Cell Function

Key Terms: acetylcholine, acetylcholine receptors, creatine phosphate, glycogen, ion channels, muscle cell contraction, muscle cell relaxation, myoglobin, neural stimulation, neuromuscular junction, neurotransmitter, rigor mortis, sodium/potassium pumps

Muscle cell contraction is achieved by the simultaneous shortening of all the sarcomeres within a cell. Sarcomere shortening is visible under the microscope as a decrease in distance between the Z-lines. It is also possible to see more of an overlap between the thick and thin myofilaments as a muscle cell contracts. This is an amazing process that takes place in the following stages: **neural stimulation**, **muscle cell contraction**, and **muscle cell relaxation**.

The neural stimulation stage takes place at the **neuromuscular junction** (Figure 6.3). Nerve cells communicate information from the body to a muscle cell through the neuromuscular junction. A contraction is initiated when the end of a nerve cell releases a chemical called a **neurotransmitter**. Neurotransmitters are chemicals used for cell-to-cell communication. There are many types of neurotransmitters. Muscle cells respond to a specific neurotransmitter called **acetylcholine**. Acetylcholine binds to proteins called **acetylcholine receptors**, which are located on the sarcolemma. The binding of acetylcholine to the receptor produces changes in transport proteins embedded in the sarcolemma. These changes consequently affect the transport of ions across the sarcolemma.

In a resting muscle, the concentration of sodium ions is normally higher in the fluid outside the muscle cell, while the concentration of potassium ions is higher inside the cell. **Sodium/potassium pumps**, or **ion channels**, in the sarcolemma maintain these unequal ion concentrations. This imbalance produces an unstable, or excitable, condition. When stimulated by a chemical or environmental factor, an excitable membrane loses its ability to maintain the imbalanced sodium and potassium concentrations. Disturbance of the balance stimulates a variety of chemical reactions within the cell. Acetylcholine stimulates the sarcolemma of a muscle cell when it binds to an acetylcholine receptor.

Neural Stimulation Stage The first stage of muscle cell contraction

Muscle Cell Contraction Stage The second stage of muscle cell contraction

Muscle Cell Relaxation Stage The last stage of muscle cell contraction

Neuromuscular Junction The space between a nerve cell and a sarcolemma

Neurotransmitter A chemical used for cell communication

Acetylcholine A neurotransmitter that communicates with muscle cells

Acetylcholine Receptor A protein located on the sarcolemma

Sodium/Potassium Pump or Ion Channel A membrane protein channel that controls the ionic distribution of sodium and potassium inside and outside of the cell

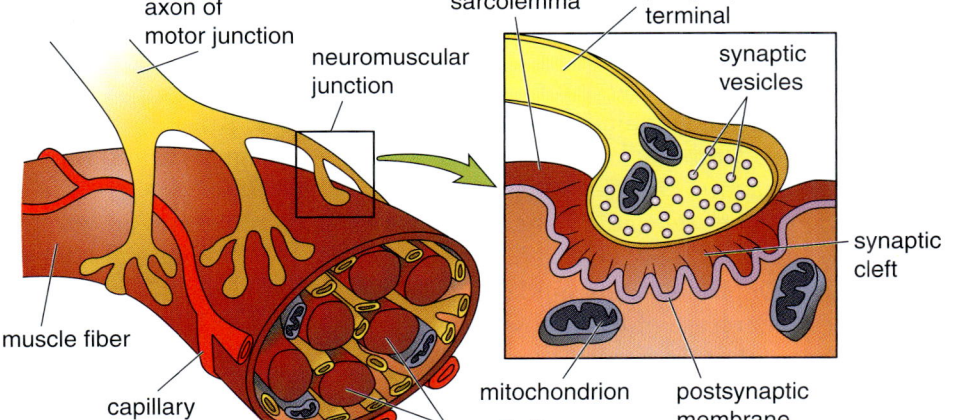

Figure 6.3 The Neuromuscular Junction

THE MUSCULAR SYSTEM

Figure 6.4 **Myofibril Contraction**

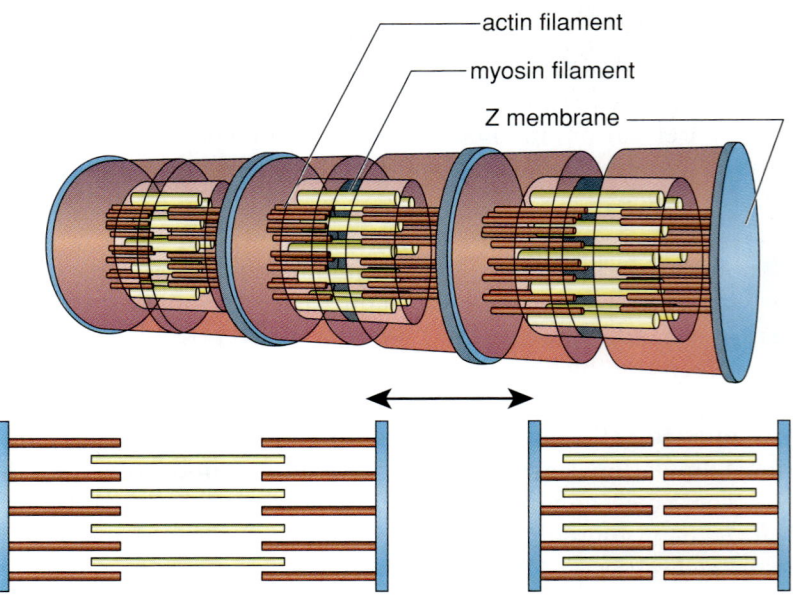

contraction

This stimulation opens the cell membrane's proteins, called **ion channels**. The opened ion channels permit the free flow of sodium into the muscle cell and potassium out of the cell. This, in turn, causes the chemical reactions that release the calcium stored in the sarcoplasmic reticulum. The calcium travels from the sarcoplasmic reticulum to the sarcomere, which initiates the muscle contraction phase (Figure 6.4).

The muscle contraction phase begins when the calcium released by the sarcoplasmic reticulum binds to the troponin on the thin myofibrils. Troponin sits on tropomyosin on the region where actin binds to myosin. Calcium moves the troponin off of the binding site, permitting myosin to attach to actin through its structures called myosin heads. This temporary bond provides the muscle cell with a rigid tension that keeps the filaments in place. Troponin is also responsible for transmitting information that activates ATP synthesis around the myosin. The ATP provides energy for the myosin head to swivel in a manner that pulls the thick myofilament across the thin myofilament. The swivel motion brings the two Z-lines together, shortening the sarcomere. While all of this is going on, the sarcoplasmic reticulum rapidly takes up the calcium. It takes another neural stimulation to continue another cycle of muscle contraction. The complete contraction of a muscle cell requires several cycles of neural stimulation and contraction phases. Any neural activity that takes place after a full contraction is used to keep the myosin tightly bonded to the actin.

A muscle reaches relaxation when no more neural stimulations are exciting the sarcolemma. The sodium and potassium ion levels are completely recovered. In addition, the sarcoplasmic reticulum has retrieved most of the calcium. This causes the release of the myosin heads from the actin. The sarcomere now has no tension on it. However, there is no mechanism within the muscle cell for lengthening the sarcomere. So, the muscle cell remains in a contracted, but pliable, state. The muscle is fully recovered when a particular body movement causes the sarcomeres within the muscle tissue to stretch. These mechanisms of muscle recovery will be discussed later in the section on skeletal muscle action.

Calcium leakage out of the sarcoplasmic reticulum into the sarcomere is common after death. This causes a muscle tension called **rigor mortis**. Eventually, the muscle cell structures start decaying, causing the muscles to become soft and loose, unless the body becomes dehydrated.

Other factors are found in muscle fibers that ensure adequate muscle contractions. **Creatine phosphate** is a molecule that stores energy in muscle cells. It collects the energy from ATP and is capable of storing this energy for long periods. Creatine phosphate then transfers the energy back to the ATP when muscle contractions require energy for contraction. Glycogen, a stored form of glucose, is an important source of energy reserve for muscle action. A continuous supply of glucose is needed for muscle cells to produce ATP. This glycogen comes in handy, as it can quickly break down into glucose when the blood supply cannot provide an adequate amount of this sugar. **Myoglobin** is a red-colored chemical that stores oxygen for certain muscle cells. This oxygen permits muscle cells to provide large amounts of ATP during continuous or heavy work.

Rigor Mortis Muscle stiffness due to calcium leakage after death
Creatine Phosphate A molecule that stores energy in muscle cells
Myoglobin A chemical that stores oxygen for muscle cells

✓ Concept Check

1. What are the three stages of muscle cell contraction?
2. What is the function of the neuromuscular junction?
3. Describe the role of calcium in muscle contraction.

DISCOVERY SCENE PLEASE ENTER DISCOVERY SCENE PLEASE ENTER

Is it important to know that a particular type of muscle cell was affected by the astronaut's space travel? What information about muscle cell's structure is useful in understanding the plight of the astronaut? Which information is most important in coming up with a research direction for solving the CSI?

MUSCULATURE

Key Terms: insertion, morphology, origin

The direction and force of movement a muscle contraction produces can often be determined by looking at the overall or gross shape of a muscle. Muscle **morphology** has to do with the arrangement of the muscle fibers and the overall muscle shape. The way muscles are attached to particular body structures also determines their role in the body. Most muscles are attached to two or more bones. Some muscles connect to skin. Each muscle has a stable or immovable attachment point known as its **origin**. A second attachment point, which connects a muscle to the body part it moves, is called the **insertion**.

Morphology The shape of the muscle
Origin The immovable attachment point of a muscle to a bone
Insertion The movable attachment point of a muscle to a bone

Gross Skeletal Muscle Types

Key Terms: biceps, brevis, deltoid, longus, maximus, minimus, parallel, pinnate, quadriceps, rhomboideus, serratus, trapezoid, trapezius, triangular, triceps

THE MUSCULAR SYSTEM

Figure 6.5a Human Musculature–Anterior

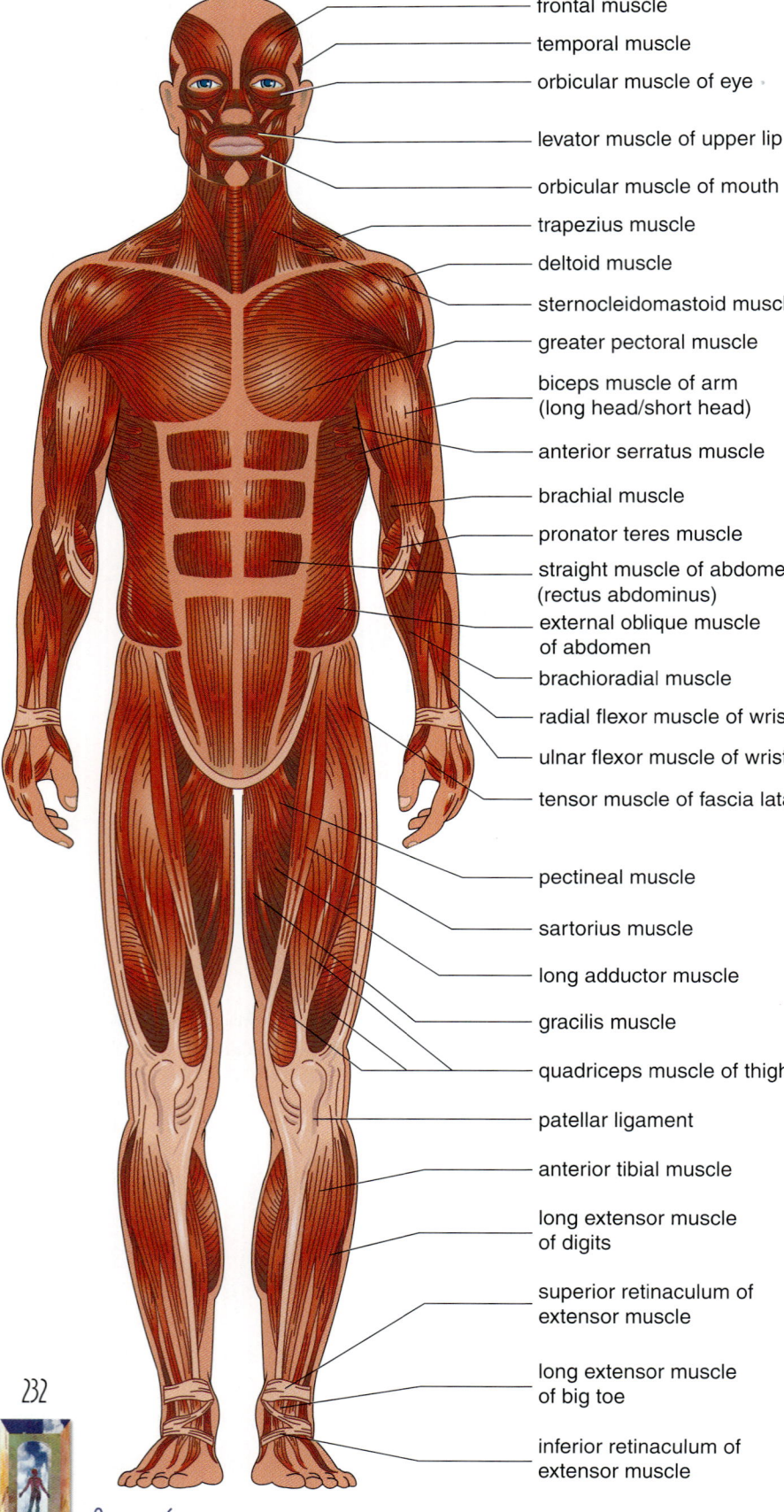

Most muscles are formed of muscle cells laid out in one of two patterns: **parallel** or **pinnate** (Figures 6.5 and 6.6). Many of the body's general-purpose muscles have a parallel structure. Parallel muscles are sheets of muscle cells that run in the same direction. They provide contractions for moving light loads over a long distance. The latissimus dorsi muscles of the back (Figure 6.5) are parallel muscles that move the arms backwards and upwards. Pinnate, or feather-patterned, muscles provide great strength for moving heavy loads over a short distance. The muscle cells are laid out in various directions to provide greater strength to the movement. Muscles that provide strong movements for the arms and legs are pinnate. Note the pinnate patterns of the gastrocnemius of the posterior lower leg and the rectus femoris of the anterior upper leg (Figures 6.5 and 6.6).

These two muscle patterns can be woven into a variety of different gross muscle shapes (Figure 6.6.). **Deltoid**, or **triangular**, muscles have a broad origin, and focus to a narrow insertion point. These muscles, such as the deltoid and latissimus dorsi, provide much pulling power. **Trapezius**, or **trapezoid**, muscles are similarly shaped and have the same role. **Rhomboideus**, or diamond-shaped, muscles, such as the levator scapulae in the shoulder provide holding power for the scapulae.

Figure 6.5b **Human Musculature–Posterior**

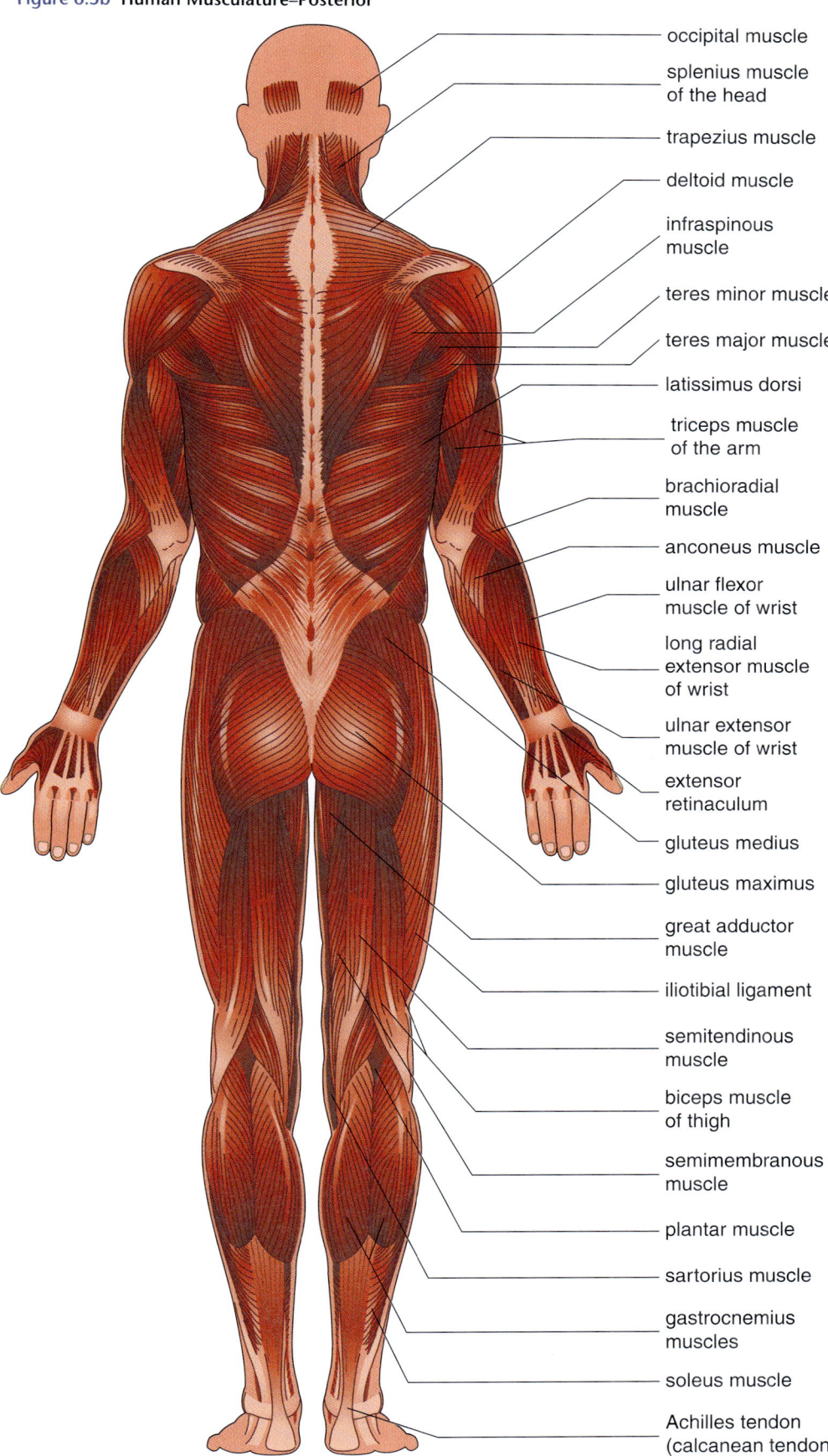

Parallel Muscle Cells Muscle cells that run in the same direction

Pinnate Muscle Cells Muscle cells that run in various directions

Deltoid, or Triangular, Muscles Muscles that have a broad origin, and focus to a narrow insertion point

Trapezius, or Trapezoid, Muscles Muscles that have a broad origin, and focus to a narrow insertion point

Rhomboideus Muscles Diamond-shaped muscles

THE MUSCULAR SYSTEM

Figure 6.6 **Skeletal Muscle Shape**

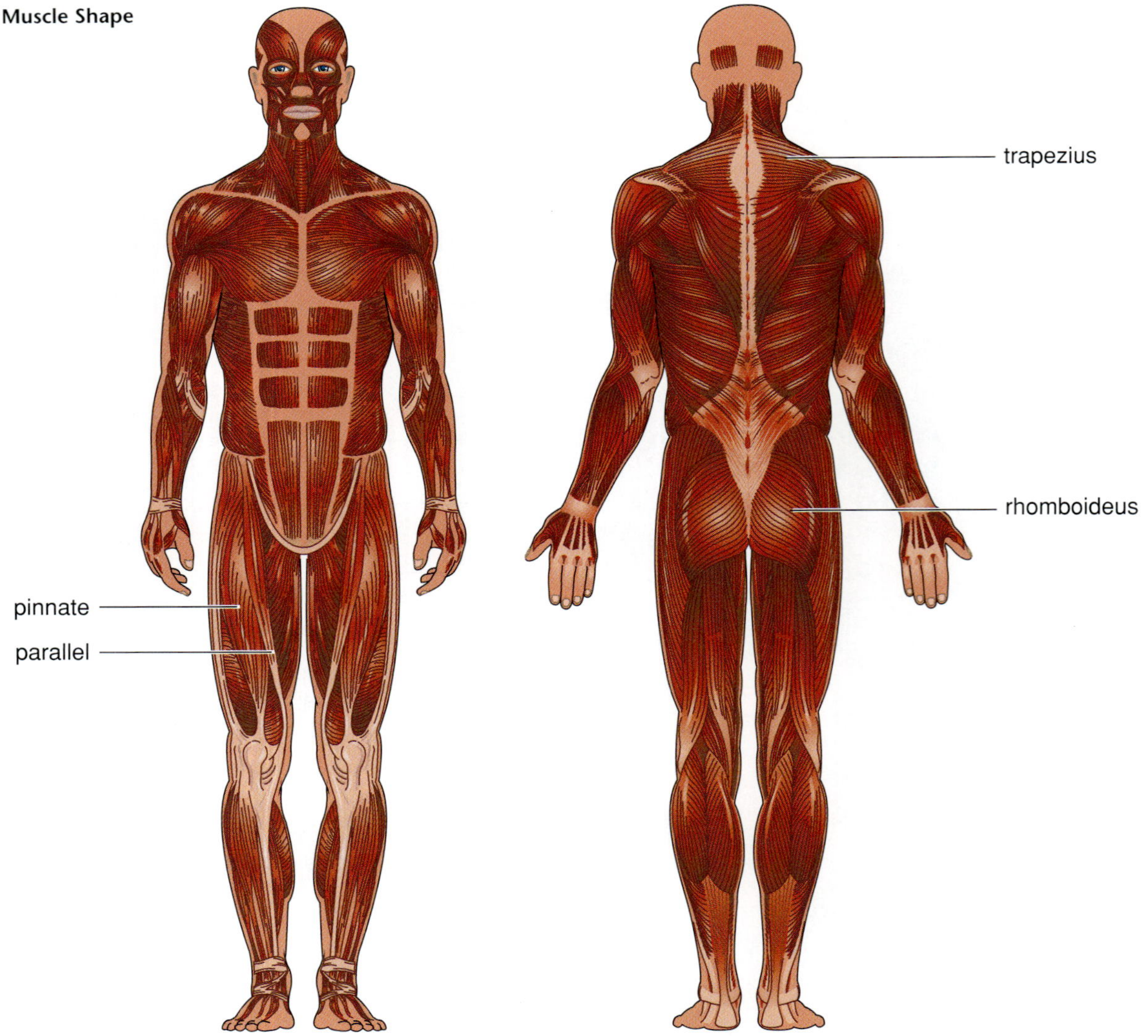

labels: trapezius, rhomboideus, pinnate, parallel

Serratus Muscles Saw-tooth-shaped muscles

Biceps Muscle A muscle with two origins

Triceps Muscle A muscle with three origins

Quadriceps Muscle A muscle with four origins

Maximus Muscle The largest muscle of a muscle group

Minimus Muscle The smallest muscle of a muscle group

Longus Muscle The longest muscle of a muscle group

Serratus, or saw-toothed, muscles are involved in short movements of the arms, rib cage, and shoulders. These are usually deep muscles that are not visible under the skin. The serratus anterior of some people is visible just below the pectoralis major muscles of the chest.

A muscle with two origins has a **biceps** shape. Biceps means "two (bi) heads"(ceps). The biceps brachii of the upper arm has two origin attachment points on the scapula. Just posterior to the biceps is the triceps brachii muscle. This muscle has three origins, or a **triceps** shape. It has two attachments on the humerus and one on the scapula. A four-headed origin muscle is called **quadriceps**. The quadriceps muscle group of the upper legs is composed of the rectus femoris, vastus lateralis, vastus intermedius, and vastus medialis. This group has one origin on the ilium and three on the femur.

Other terms refer to the relative size of a muscle. Larger and longer muscles provide greater pulling power. The largest muscle of a muscle group is called **maximus**. **Minimus** refers to a smaller muscle. These terms are evident when viewing the gluteus maximus and gluteus minimus muscles on the posterior hips. **Longus** refers to the longest of a muscle group, while

234

CHAPTER 6

brevis describes the shortest muscle. The extensors of the forearms and legs have longus and brevis groups.

Skeletal Muscle Structure

 Key Terms: endomysium, epimysium, fascicle, perimysium

A skeletal muscle is a complex organ composed of both the striated muscle tissue previously discussed and connective tissues. Nerve cells and blood vessels are also integral components of muscle structure. Motor nerves stimulate contraction. Muscles also have nerves that sense pain and stretching. Each muscle has levels of complexity in these components. The most basic level is that of the muscle fiber, or cell. Each muscle fiber is covered with a connective layer called the **endomysium**. Much of the role of the endomysium is to maintain the chemical environment that muscle cells need to contract. Bundles of muscle cells called **fascicles** (fasciculi) comprise the second level of gross muscle structure (Figure 6.7). Each fascicle is under independent control of a motor neuron. Surrounding each fascicle is a thin connective tissue covering called the **perimysium**. A fibrous connective tissue called the **epimysium** covers gross muscle, which is the third and highest organ level of the skeletal muscle structure. It is a protective covering that holds the fascicles in place, and provides a lubricating surface for muscles

Brevis Muscle The shortest muscle of a muscle group
Endomysium The connective tissue covering each muscle fiber
Fascicle A bundle of muscle cells, or fibers.
Perimysium The thin connective tissue covering each fascicle
Epimysium The fibrous connective tissue covering gross muscle

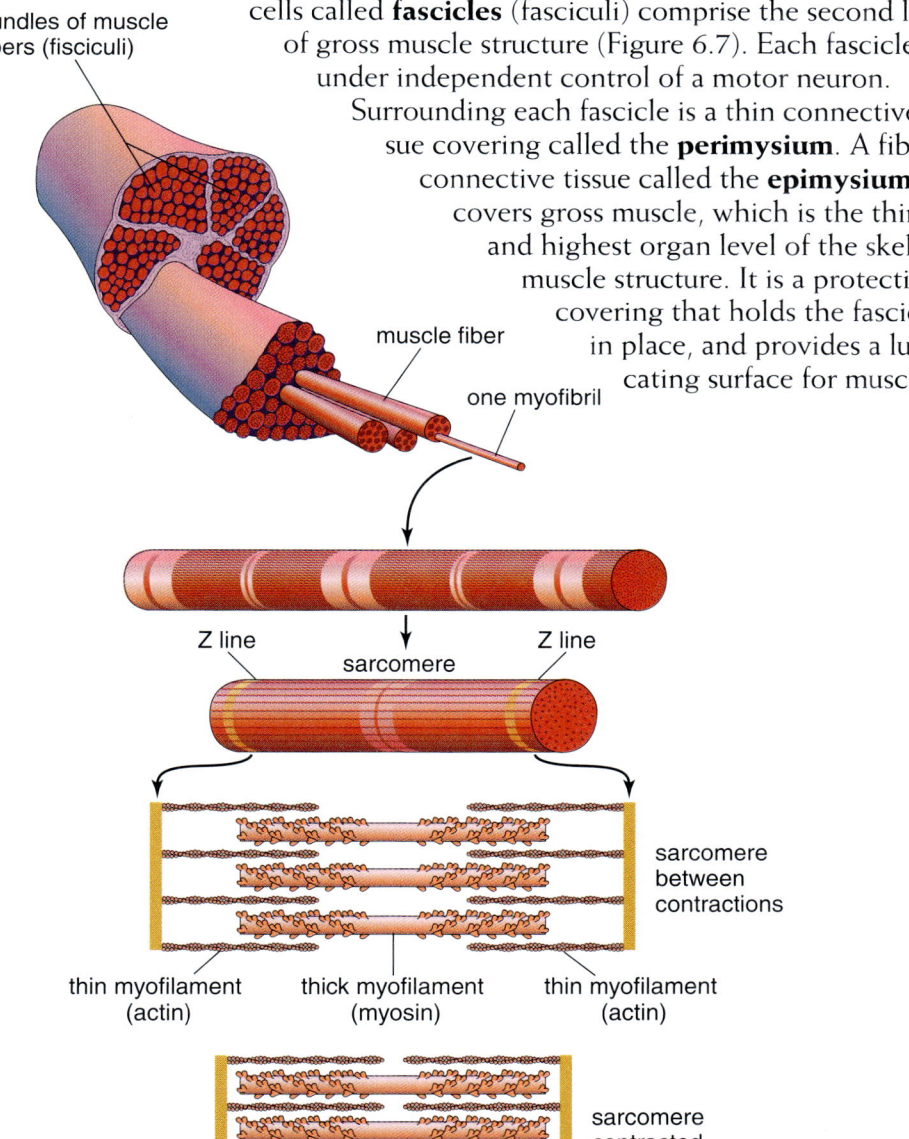

Figure 6.7 Skeletal Muscle Structure

THE MUSCULAR SYSTEM

as they rub against bone, skin, and other muscles. The epimysium also covers the tendons that attach muscle to bone and skin. The fascia underneath the skin makes contact with the epimysium of superficial muscles.

Skeletal muscle structure responds to the amount of work it must do over a period of time, and even over the lifetime. Regular disuse of a muscle will cause muscle atrophy. When a muscle is not used regularly, its cells lose sarcomere proteins, which eventually causes muscle shrinkage. At first, this results in a loss of contraction strength, which is followed by an overall decrease in muscle size. Muscles will also atrophy when there is a lack of neural stimulation. In contrast, regular use of a muscle will produce muscle cell hypertrophy. The increased blood flow to a muscle during regular use will enlarge the muscle cells. This growth in muscle diameter produces more muscle strength. Continuous, heavy muscle use will build up the sarcomere structure. In some people, this will significantly increase muscle diameter. However, some people increase sarcomere density and strength without a significant increase in overall muscle size. This difference has to do with genetic differences in muscle physiology and variations in blood flow to the muscles.

Skeletal Muscle Action

Key Terms: abductor, adductor, antagonistic, concentric, depressor, eccentric, extensor, flexor, graded effect, isometric, levator, origin, pronator, rotator, sphincter, supinator, synergistic, tensor, threshold

Muscle action is based on the fact that a tendon fixes the muscle **origin** to a firm foundation of bone (Figure 6.8). A muscle's origin is the point of attachment of a muscle that remains fixed during contraction. A muscle must also be inserted onto the surface of a movable component at its other end. Shortening of the muscle brings the insertion closer to the origin. All muscle fibers contract with a particular strength when given a **threshold** neural stimulation. A graph of a muscle contraction is represented in Figure 6.9. The height of the curve can be equated to the power of the contraction. So, how is it possible for a muscle to perform at different pulling powers? The ability to achieve different pulling forces, or **graded effects**, can be accomplished by contracting more fascicles or by working muscles together. Strength is achieved by stimulating more fascicles. Endurance can be accomplished by producing contracting and relaxing groups of fascicles. Muscles can also work together to strengthen or weaken a muscle contraction. An **antagonistic** effect occurs when one muscle opposes or resists the action of another muscle. This effect weakens muscle strength. Gravity can have an antagonistic effect if the muscle is working against the pull of gravity. Antagonistic actions are essential for pulling relaxed muscles back to their original length. Specialized insertions, such as cartilage, will antagonize a mus-

Origin The point of attachment of a muscle that remains fixed during contraction

Threshold The level of stimulation needed to induce a muscle cell to contract

Graded Effects Different pulling forces

Antagonistic Refers to one muscle opposing or resisting the action of another muscle

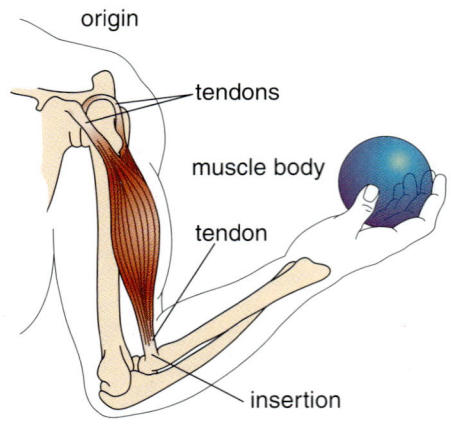

Figure 6.8 Skeletal Muscle Action

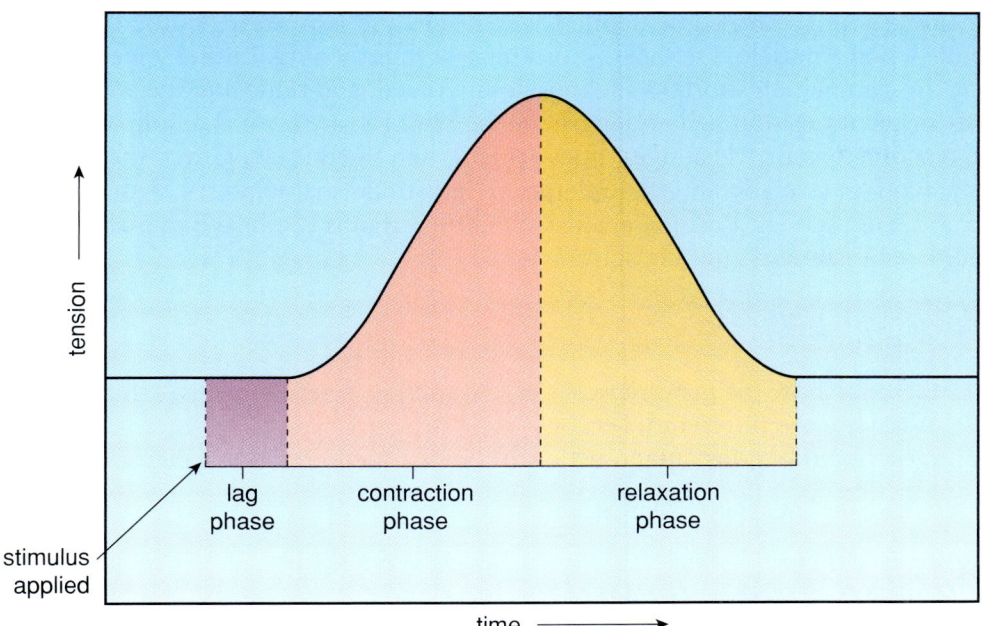

Figure 6.9 Muscle Contraction

cle. This cartilage is found in the rib cage and is an important factor during breathing. Muscles can work in a **synergistic** way when they contract to produce a common effect. The muscles of the forearm work synergistically with the muscles of the fingers to produce a fist.

Muscle can be categorized by the effect it has on joint motion. Many of the actions of one category antagonize the actions of another category. The major categories of muscle action are as follows:

- **Abductor** – These muscles move a bone away from the midline of the body. Abductor muscles antagonize adductor muscles.
- **Adductor** – These muscles antagonize abductor muscles by moving a bone closer to the body's midline.
- **Depressor** – Muscles that produce a downward movement and antagonize a levator muscle.
- **Extensor** – These muscles increase the angle of a joint and antagonize flexor muscles.
- **Flexor** – This type of muscle antagonizes an extensor muscle by decreasing the angle of a joint.
- **Levator** – These muscles provide an upward movement and are antagonistic to depressor muscles.
- **Pronator** – These muscles turn the palm downward and antagonize the effect of a supinator.
- **Rotator** – These muscles move a bone around its longitudinal axis in a circular direction.
- **Sphincter** – This specialized group of muscles decreases the size of an opening. Sphincter muscles are usually attached to skin or connective tissue.
- **Supinator** – These muscles antagonize pronator muscles by turning the palms upward.
- **Tensor** – These are important posture and positioning muscles that make a body part more rigid or tense. A variety of muscles antagonize the tensor muscles.

Synergistic Refers to muscles contracting together to produce a common effect

Abductor Muscles Muscles that move a bone away from the body's midline

Adductor Muscles Muscles that move a bone closer to the body's midline

Depressor Muscles Muscles that produce a downward movement

Extensor Muscles Muscles that increase the angle of a joint

Flexor Muscles Muscles that decrease the angle of a joint

Levator Muscles Muscles that produce an upward movement

Pronator Muscles Muscles that turn the palm downward

Rotator Muscles Muscles that move a bone around its longitudinal axis in a circular direction

Sphincter Muscles Muscles that decrease the size of an opening

Supinator Muscles Muscles that turn the palms upward

THE MUSCULAR SYSTEM

Tensor Muscles Muscles that make a body part more rigid or tense

Isotonic Refers to a muscle that is actively shortening or lengthening

Isometric Refers to a muscle that is not lengthening or shortening

Muscle action can also be defined as **isotonic** and **isometric**. Isotonic action occurs when a muscle is actively shortening or lengthening. Lifting a weight with the arm, which shortens the muscles involved, and returning the weight to its original position, which lengthens the muscles, are both examples of isotonic activity. Isometric activity does *not* lengthen the muscle; rather, the muscle remains at a steady length, undergoing indistinguishable pulses of shortening and lengthening. Pushing against something that is too heavy to move would elicit isometric muscle activity.

✓ Concept Check

1. Describe how the gross muscle shape is used to determine muscle function.
2. What are the different levels of muscle structure?
3. Describe the terms antagonistic and synergistic in relation to muscle function.

Cutting Edge Research
PAVING THE WAY FOR MUSCLE REGENERATION?

Damaging a muscle is not like fracturing a bone or injuring the skin. Muscle cells are so complex that it is almost impossible for them to undergo mitosis. Therefore, they cannot replicate to replace muscle cells killed by injury. This is a major reason why humans cannot grow back a finger or other parts that have a good deal of muscle structure. Scientists at Stanford University in California and Joslin Diabetes Center in Boston are on the way to finding stem cells that regenerate skeletal muscle tissue. Many scientists search red bone marrow as a source of adult stem cells. However, Dr. Amy J. Wagers, of Joslin, and Dr. Irving L. Weissman, of Stanford, identified cells in mature muscle that, under the right conditions, could be capable of replacing damaged muscle tissue. They discovered these cells while looking for cell membrane proteins unique to stem cells. Success at getting these cells to replicate new muscle has incredible potential for reversing the deadly effects of degenerative muscle diseases.

SOURCE: Isolation of adult mouse myogenic progenitors: functional heterogeneity of cells within and engrafting skeletal muscle. Cell. 2004 Nov 12;119(4):543-54.

DISCOVERY SCENE PLEASE ENTER DISCOVERY SCENE PLEASE ENTER

CSI Break

Did the information about muscle cell function narrow down a probable reason for the astronaut's rapid muscle loss? Does her muscle loss have anything to do with a particular category of skeletal muscle that is affected?

MUSCULATURE CHARTS

Use the muscle diagrams in Figure 6.10 to learn the locations of the major muscles. It is critical to know which are the superficial muscles, ie, the muscles lying closest to the skin. Superficial muscles are landmarks that indicate bone and joint locations. Plus, they are the muscles most often discussed when people have muscle problems. However, it is also important to know the location of the deep muscles of the back and rib cage because they are associated with life-threatening muscle pathologies.

Figure 6.10a Human Musculature Charts

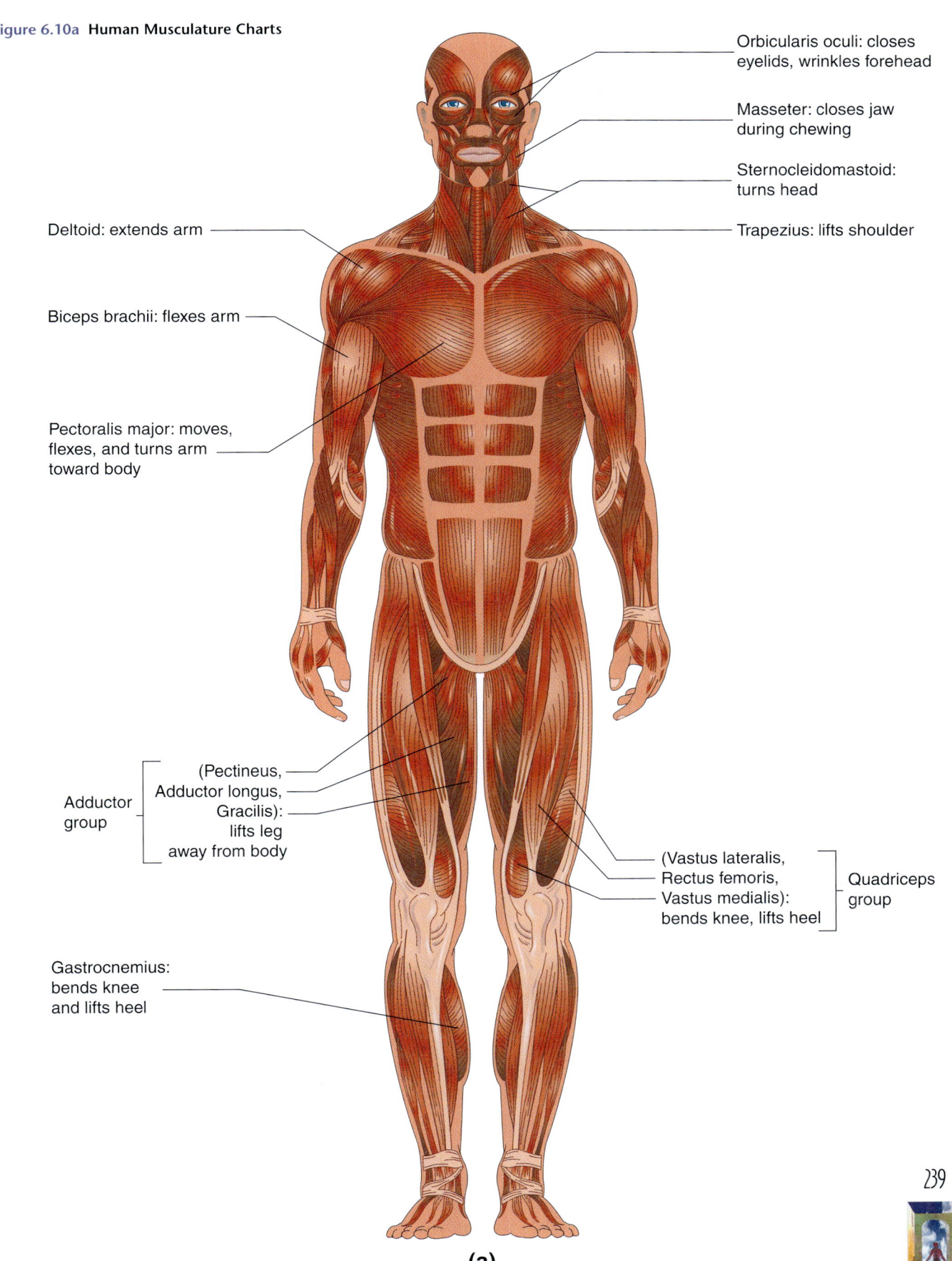

(a)

Figure 6.10b **Human Musculature Charts**

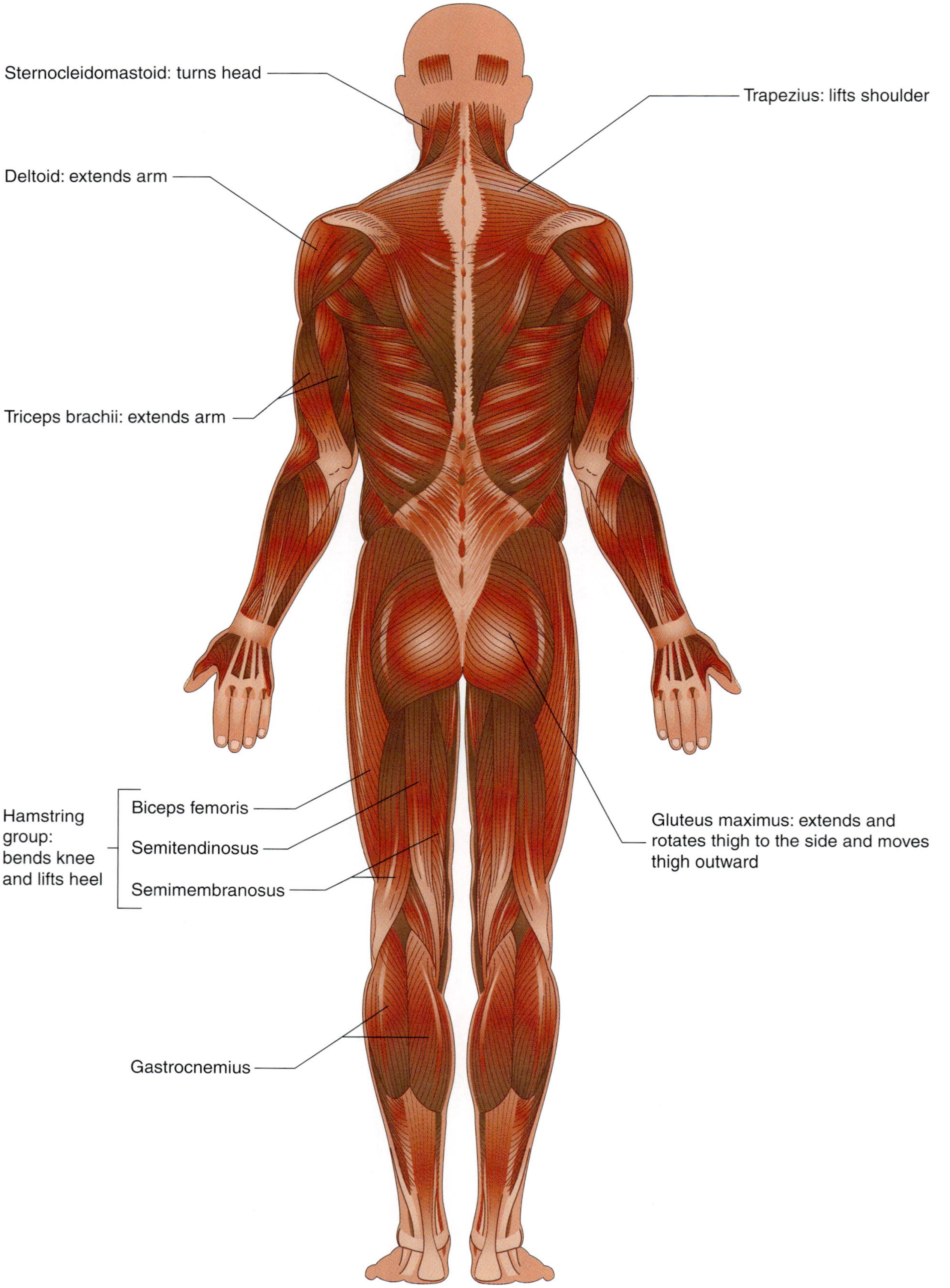

- Sternocleidomastoid: turns head
- Trapezius: lifts shoulder
- Deltoid: extends arm
- Triceps brachii: extends arm
- Hamstring group: bends knee and lifts heel
  - Biceps femoris
  - Semitendinosus
  - Semimembranosus
- Gluteus maximus: extends and rotates thigh to the side and moves thigh outward
- Gastrocnemius

(b)

ANATOMY OF COSMETIC SURGERY

Elective cosmetic or plastic surgery is becoming commonplace despite the great expense and potential risks. A typical "face lift" can average $8000 in the United States. Muscle adjustments and modifications make up a significant component of many types of cosmetic surgery. Probably the simplest cosmetic procedure performed on the face is blepharoplasty, or eyelid lift. In this procedure, excess fat, muscle, and skin are surgically removed from the upper and lower eyelids. A more complex procedure is autologous tissue breast reconstruction in female breast cancer patients. In this procedure, the transverse rectus abdominus muscle, the most superficial of the abdominal muscles, is removed and used to reshape a new breast after a mastectomy. Plastic surgeons use high-tech medical imaging and computer modeling before performing complex reconstructive procedures. Usually, a computed tomography (CT) scan develops a three-dimensional image of the tissues being modified. It is able to show fine details of blood vessels, bones, fat, fibrous connective tissues, muscle, and skin. This information can then be fed into computers that show what the person will look like after the surgery. Many cosmetic surgeons are using virtual-reality modeling language 3D facial-reconstruction software to come up with the "perfect face" for a client. It uses the client's existing features to project the possible types of new faces that can be created. The software is designed to produce a face that is as beautifully proportioned as possible following data from a body of literature on facial symmetry. It projects the face into a grid, measuring ideal proportions of the nose, cheeks, eye sockets, and forehead.

WELLNESS AND ILLNESS OVER THE LIFE SPAN

Pathology of Musculature

Key Terms: botulism, contusion, cramp, dermatomyositis, familial periodic paralysis, flaccid paralysis, glycogen storage disease, mitochondrial myopathy, muscle sensitivity, muscular dystrophy, myoglobinuria, myopathy, myositis ossificans, neuromuscular, neuromyotonia, rigid paralysis, spasm, sprain, stiff-man syndrome, strain, tetanus, tetany

It is interesting to note that many disorders of the musculature are due to interactions with the skeletal and nervous systems. Muscle **strains** due to overworking the muscle's force on joints and tendons are the most common muscle ailment. They result in pain and swelling of the fascia, joints, ligaments, and tendons. Nerves in these structures signal pain when stretched or swollen. **Sprains** should not be confused with a strain. Sprains are more severe than strains. They usually involve sudden or violent stress on a joint or muscle that causes tearing of a ligament, muscle, or tendon. Sprains are often accompanied by damage to nearby blood vessels. Treating sprains and strains is not straightforward. Strains are usually quick to heal when the muscle is kept warm and relaxed. Muscle stiffness usually accompanies a strain. Sprains require time for the muscle tissue to replace and restore the muscle cell proteins. With both injuries, it is important to keep swelling down at first by applying cold. The cold application must be followed by continuous warmth to the injured area to speed healing.

Strain An injury due to overworking the muscle's force on the joints

Sprain An injury resulting from sudden or violent stress on a joint or muscle

Contusion An injury caused by a direct hit or repeated battering of a muscle

Spasm An involuntary, abnormal muscle contraction

Cramp The painful contraction of a muscle

Muscle Sensitivity Continuous muscle pain due to tissue damage or disease

Rigid Paralysis Loss of muscle function due to excessive muscle stiffness

Flaccid Paralysis Loss of muscle function due to a lack of muscle contraction

Tetanus An infectious disease that causes rigid paralysis

Myopathy, or Neuromuscular Disorders Diseases characterized by the nervous system's inability to communicate with the muscular system

Dermatomyositis Inflammation of the muscle and skin

Familial Periodic Paralysis Periodic weakness in the arms and legs; a genetic disorder

Glycogen Storage Diseases Diseases that cause muscle weakness due to a diminished ability to use glucose

Mitochondrial Myopathies Genetic mitochondrial abnormalities that prevent muscles from producing energy

Muscular Dystrophies Diseases characterized by progressive weakness of voluntary muscles

Myoglobinurias Disorders that affect how myoglobin provides oxygen to muscles

Myositis Ossificans Bone growing within muscle tissue

Contusions are related to muscle sprains. They are caused by direct hits or repeated battering to a muscle. A contusion is a common injury caused by falls and impacts. Muscle **spasms**, which are involuntary, are abnormal contractions of a muscle or muscle group. A spasm can be due to a wide variety of medical conditions. Many spasms are associated with muscle pain. **Cramps** are defined as the painful contraction of a muscle. They are sometimes confused with muscle stiffness. Extreme muscle exertion is probably the most common cause of cramps. Cramps can develop in response to working in a cold environment. Certain poisons and bacterial infections also cause muscle cramping. **Muscle sensitivity** is a continuous muscle pain due to muscle tissue damage or inflammatory diseases of the surrounding tissues. Paralysis is the complete failure of muscle function caused by excessive muscle stiffness (**rigid paralysis**) or a lack of muscle contraction (**flaccid paralysis**). There are many causes of paralysis. Bacterial poisoning contributes to two types of paralysis: tetanus and toxic food-poisoning bacteria. **Tetanus** is caused by soil bacteria that produce poisons and cause rigid paralysis. A type of flaccid paralysis is caused by toxic secretions from a food-poisoning bacterium. Both conditions can cause death by paralyzing the breathing muscles.

The inability of the nervous system to communicate properly to muscles causes a group of diseases called **myopathy**, or **neuromuscular** disorders. Myopathies include a wide array of disorders varying from paralysis to spasms. The most common myopathies include the following:

- **Dermatomyositis** —This condition involves inflammation of the muscle and overlying skin. The cause is unknown; however, it can be treated with drugs that reduce inflammation and sun avoidance.
- **Familial periodic paralysis** —This genetic disorder causes periods of weakness in the arms and legs. Muscle weakness occurs when large amounts of potassium from the bloodstream unexplainably enter the muscle.
- **Glycogen storage diseases** —These are genetic disorders of enzymes that control the conversion of muscle glycogen to glucose. The muscles have diminished ability to use glucose, causing muscle weakness. Plus, the glycogen can accumulate to high levels and cause muscle damage.
- **Mitochondrial myopathies** —These are genetic abnormalities of the mitochondria. They are passed along from the egg or they develop during embryological development. These conditions prevent the muscles and other cells from producing energy derived from food. So, the muscles become easily cramped from carrying out a large degree of anaerobic respiration.
- **Muscular dystrophies** —This includes a group of conditions that involve progressive weakness in the voluntary muscles. They are usually due to the inability of the nervous system to stimulate muscle action. This eventually results in muscle atrophy and wasting (Figure 6.11).
- **Myoglobinurias** —This large group of disorders affects how myoglobin provides oxygen for muscle function. There are many causes, including electrocution, electrolyte imbalance, excessive exercise, and exposure to certain poisons or toxins.
- **Myositis ossificans** —This condition is most likely caused by damage to soft tissues near a muscle. It is distinguished by bone growing within muscle tissue. This produces pain associated with muscle contraction and difficulty moving the affected body part.

CHAPTER 6

- **Myotonia** —This nervous system condition causes the muscles to relax slowly after contractions. It usually takes repeated efforts to relax the muscles. Warming up the muscles before taking part in an activity can reduce the severity of the condition.
- **Neuromyotonia** —This is a rare nerve disorder that causes alternating bouts of muscle twitching and stiffness. The condition makes it difficult to carry out movements and to maintain posture.
- **Stiff-man syndrome** —This rare condition causes rigidity, and spasms of the spine and lower-extremity muscles. The disease becomes progressively worse with age.
- **Tetany** —A calcium imbalance disease that causes extended periods of spasms in the arm and leg muscles. It should not be confused with the bacterial disease tetanus.

Myotonia The slow relaxation of muscles after contraction

Neuromyotonia A nerve disorder characterized by bouts of muscle twitching and stiffness

Stiff-Man Syndrome Rigidity, and spasms of the spine and lower-extremity muscles

Tetany Periods of arm and leg muscle spasms caused by calcium imbalances

Figure 6.11 **Pathology of the Musculature**

✓ Concept Check

1. What are the differences between strains, sprains, and contusions?
2. Explain the difference between flaccid and rigid paralysis.
3. Describe four types of myopathies.

Good Choice Bad Choice

A person at work has smoked for 20 years. He recently found out from his physician that he has low blood oxygen because of lung disease caused by smoking. This was not a surprise because he was aware that he had trouble breathing while jogging. He heard about an oxygen-enhancement product for sale through a body-building Web site, which claimed to boost oxygen flow to the muscles. He hopes this will prevent him from cramping after jogging. What should he know about this product before taking it?

DISCOVERY SCENE PLEASE ENTER DISCOVERY SCENE PLEASE ENTER

What information about muscle pathology is helpful to better understand the astronaut's excessive muscle loss?

Aging of the Muscular System

Key Terms: aromatase, cachexia, insulin-like growth factor-1 (IGF-1), protein turnover

Cachexia Muscle loss

Protein Turnover The rate at which a cell replaces damaged proteins

Cachexia is normally a type of muscle loss associated with diseases such as AIDS and cancer. It is also found in starvation and is a common consequence of anorexia and bulimia. A slower form of cachexia is a normal consequence of aging because the body reduces its ability to rebuild muscle structure (Figure 6.12). Much of this muscle loss is due to sedentary life styles brought about by other age-related conditions or changes in daily routines. Neural stimulation is important in maintaining muscle upkeep. Muscles preserve their protein metabolism each time they receive a signal from a nerve. This is called muscle upregulation. Contractions further contribute to the maintenance of the muscle cell myofibrils. Muscle loss can be reduced and even reversed with appropriate exercise regimens. Physicians in geriatric medicine (doctors who specialize in caring for elderly patients) know to apply a variety of strategies to reduce this type of muscle atrophy. One new strategy is to artificially upregulate the muscle by applying gentle electrical pulses to it. A technique in physical therapy uses special muscle massages that induce mild muscle contractions.

Of the conditions that contribute to muscle atrophy related to aging, nutritional factors probably rank second only to cachexia. Muscles require a high **protein turnover** to maintain the muscle cell fibrils. Protein turnover is the rate at which a cell replaces damaged proteins. Malnutrition and undernutrition greatly affect a muscle's protein turnover. Malnutrition can be caused by diets that are deficient in protein or have the wrong types of protein. A lack of appetite or lower income can contribute to an insufficient intake of certain foods. In addition, maturation of the digestive system reduces the uptake of

Figure 6.12 Progression of Muscle Wasting

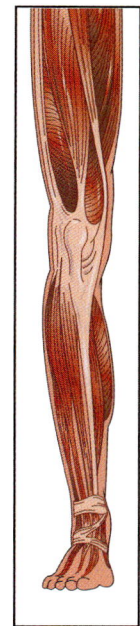
(a) normal muscle

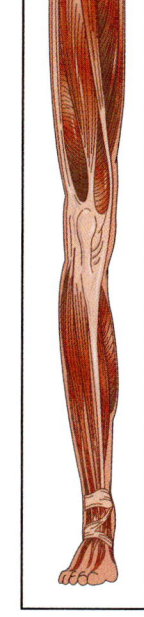
(b) early signs of muscle loss

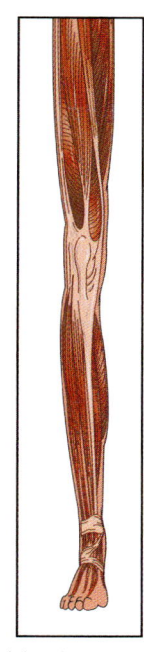
(c) advanced muscle loss

certain amino acids. The essential amino acids, arginine, isoleucine, leucine, lysine, phenylalanine, and tryptophan, are particularly important because they cannot be synthesized in cells. They are important for building muscle-cell fibril proteins. Undernutrition can reduce the intake of carbohydrates needed to maintain muscle activity, again leading to atrophy. Reduced levels of vitamin D also contribute to dietary muscle loss.

Loss of muscle mass can be compounded by a decline in sex hormones and other chemical messages needed for muscle cell maintenance. A chemical called **insulin-like growth factor-1 (IGF-1)** is needed for muscle cell growth, maintenance, and repair. IGF-1 levels are known to lower with maturity. Immune system chemicals called cytokines also cause muscle atrophy. Their levels are higher in the elderly population due to the increased incidence of disease associated with maturity. Aging also changes the body's BMI. Body fat composition increases as the proportion of muscle decreases. This stimulates the production of an enzyme called **aromatase** in fat cells. Aromatase reduces the levels of sex hormones needed for muscle mass and strength. There is no consistent scientific evidence supporting the use of therapies targeting these chemical factors leading to muscle atrophy. There are genetic differences in people that accelerate or reduce muscle atrophy. Again, at this point in time they cannot be altered to affect muscle aging.

Insulin-Like Growth Factor-1 or IGF-1 A chemical needed for muscle cell growth, maintenance and repair

Aromatase An enzyme that reduces the level of sex hormones needed for muscle mass and strength

✓ Concept Check

1. Define cachexia.
2. Describe four factors that contribute to muscle aging.
3. What is the relationship between BMI and muscle aging?

DISCOVERY SCENE PLEASE ENTER DISCOVERY SCENE PLEASE ENTER

Did the information about aging of musculature provide any more clues about the astronaut's condition? What are the likely conditions found in outer space that may lead to muscle disease? Could the conditions of outer space contribute to premature muscle aging?

THE MUSCULAR SYSTEM

CSI — Case Study Investigation Conclusion

What did you conclude about the astronaut's muscle atrophy? Why did she lose 10% more muscle mass than other astronauts? Is there one explanation that stood out amongst the other possible causes of her condition? What other information would need to be known to discover the cause of her problem?

Answer:

Everyday activities such as lifting, sitting, standing, and walking require the continuous stimulation of skeletal muscles. Much of the muscles' actions go into counteracting the affects of gravity on the body. Muscles must work harder in gravity environments. This regular resistance produces muscle mass and strength. Fighter pilots do special strength training to work under the conditions of increased gravity sustained during certain plane maneuvers. However, the weightlessness of space flight reduces the muscles need to work against gravity. So, after a prolonged period of weightlessness, the muscles lose their mass and strength, and they atrophy. This particularly weakens posture support muscles such as those in the back, calf, and thigh. Astronauts regularly perform weight-training exercises that simulate the gravity of Earth to reduce the rate of muscle atrophy.

There are other factors that maintain muscle mass and strength. Protein metabolism is important in maintaining the muscle fibrils needed for strong contractions. In particular, the essential amino acids are needed in large amounts. Sex hormones and other blood factors are needed to maintain muscle mass. Both of these factors are affected by space travel. The astronaut's muscle loss was caused by an unusual drop in sex hormones due to the stress of working in space for six months. Luckily, a few weeks of therapy will bring her back to normal.

This CSI was adapted from the following articles:

1. Keller TS, Strauss AM, and Szpalski M. Prevention of bone loss and muscle atrophy during manned space flight. *Microgravity.* 1992;2:89-102.
2. Ball, JR and Evans, CH, Jr. (ed). *Safe Passage: Astronaut Care for Exploration Missions.* Washington, DC: National Academy Press; 2001:
3. Vandenburgh H, Chromiak J, Shansky J, Del Tatto M, Lemaire J. Space travel directly induces skeletal muscle atrophy. *Fed Am Soc Exp Biol J.* 1999;13(9):1031-1038.

Chapter Summary

Muscle cells or fibers have the capabilities of being contractile and excitable. By being contractile, muscle cells can be used to provide various types of movement for the body. Strong striated muscle contractions generate skeletal system movement and pump blood throughout the body. Non-striated muscle contractions assist with blood flow, breathing, and digestion. The excitable nature of muscle cells permits them to be controlled voluntarily and involuntarily by the body. Voluntary control is mostly involved in skeletal system movements. Involuntary control moves blood throughout the body and passes food through the digestive system.

Skeletal muscle is the most common muscle type making up over half of the body's weight. Muscles can do work by shortening, lengthening, or remaining stationary. Muscle needs regular activity and proper nutrition to retain their size and functions. Immobility and malnutrition as a person matures could cause muscle atrophy. A variety of genetic diseases can also cause the loss of muscle function.

Overview
- Muscle tissue is composed of contractile cells.
- Contractile cells have specialized cell membranes and cytoskeletons.
- Muscle cells consume a large amount of the body's energy.
- Body mass index (BMI) is an indicator the body's muscle composition.

Muscle
- All muscle cells originate from myoblasts in the embryo.
- Muscle cells can be categorized by appearance as nonstriated or striated
- Muscle cells can be categorized according to their location and function in the body: cardiac, skeletal, and smooth.
- Muscle cells are covered with an excitable membrane called a sarcolemma.
- The cytoskeleton of muscle cells is modified into contractile protein units called sarcomeres.
- A muscle contraction occurs in three stages: neural stimulation, muscle cell contraction, muscle cell relaxation.

Musculature
- Muscles can be categorized by the pattern of muscle cell organization: parallel and pinnate.
- Muscle cells are laid down in sheets that produce different shapes of muscles. The shape determines the direction and strength of a muscle contraction.
- Skeletal muscles are organs containing blood vessels, connective tissues, and nerves.
- Skeletal muscles are composed of bundles, or fascicles, of muscle cells.
- Fascicles work independently to control the endurance and strength of a muscle contraction.
- Muscles can counteract (antagonistic) or assist (synergistic) other muscles.
- Muscle contractions can be categorized as isotonic or isometric.

Study Guide

Pathology of the Musculature
- Overuse of muscles can cause strains, stiffness, and sprains.
- Muscle damage will produce muscle pathology, such as contusions, cramps, paralysis, and sensitivity.
- Some muscle diseases are genetic or developmental. This includes a large group of muscle disease called myopathies.

Aging of the Muscular System
- Maturity is usually accompanied by cachexia, or gradual muscle loss.
- Some aging of the musculature is due to wear and tear.
- Diet can affect muscle aging.
- Changes in hormone production contribute to muscle aging.

Key Terms

Overview
Body mass index (BMI)
Calcium
Contractile cell
Electrolyte

Human Muscle
Contractile proteins

Types of Muscle Tissue
Growth factor
Intrinsic beat
Muscle fiber
Myoblast
Myogenesis
Motor nerve cell
Peristalsis

Muscle Cell Structure
Actin
Excitable Membrane
Myofibril
Myofilament
Myosin
Sarcolemma
Sarcomere
Sarcoplasmic reticulum
Titin filament
Tropomyosin
Troponin
Z-line

Muscle Cell Function
Acetylcholine
Acetylcholine receptors
Creatine phosphate
Glycogen
Ion channels
Muscle cell contraction
Muscle cell relaxation
Myoglobin
Neural stimulation
Neuromuscular junction
Neurotransmitter
Rigor mortis
Sodium/potassium pumps

Musculature
Insertion
Morphology
Origin

Gross Skeletal Muscle Types
Biceps
Brevis
Deltoid
Longus
Maximus
Minimus
Parallel
Pinnate
Quadriceps
Rhomboideus
Serratus
Trapezoid
Trapezius
Triangular
Triceps

Skeletal Muscle Structure
Endomysium
Epimysium
Fascicle
Perimysium

Skeletal Muscle Action
Abductor
Adductor
Antagonistic
Concentric
Depressor
Eccentric
Extensor
Flexor
Graded effect
Isometric
Levator
Origin
Pronator
Rotator
Sphincter
Supinator
Synergistic
Tensor
Threshold

Pathology of the Musculature
Botulism
Contusion
Cramp
Dermatomyositis
Familial periodic paralysis
Flaccid paralysis
Glycogen storage disease
Mitochondrial myopathy
Muscle sensitivity
Muscular dystrophy
Myoglobinuria
Myopathy
Myositis ossificans
Neuromuscular
Neuromyotonia
Rigid paralysis
Spasm
Sprain
Stiff-man syndrome
Strain
Tetanus
Tetany

Aging of the Muscular System
Aromatase
Cachexia
Insulin-like growth factor-1 (IGF-1)
Protein turnover

Check Your Understanding

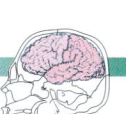

1. Muscle cells differ from most other body cells because they have this characteristic:
   a. contractibility
   b. more than one nucleus
   c. a cytoskeleton
   d. late formation in the embryo.

2. Striations are responsible for the following muscle cell function:
   a. excitability
   b. multicellularity
   c. contraction
   d. homeostasis

3. Smooth muscle has the following characteristic:
   a. striated
   b. involuntary action
   c. strong contractions
   d. multinucleated

Study Guide

4. All muscle cells develop from:
   a. fibroblasts
   b. ectoderm
   c. endodermal derivatives
   d. myoblasts

5. Which type of muscle is responsible for peristalsis?
   a. cardiac
   b. smooth
   c. striated
   d. skeletal

6. Which ion stimulates the binding of actin to myosin?
   a. potassium
   b. calcium
   c. iron
   d. sodium

7. A muscle returns to its resting state following contraction when the following occurs:
   a. Z-lines lengthen
   b. Z-lines come together
   c. actin binds to the myosin heads
   d. calcium escapes the sarcoplasmic reticulum

8. Which is *not* a property of skeletal muscle?
   a. mostly involuntary
   b. striated
   c. multicellular
   d. makes up over half of the body's weight

9. A muscle is best described as the following:
   a. complex tissue
   b. organ
   c. not part of any organ system
   d. has no distinct organ structure

10. Which of the following is the role of ATP energy in a muscle contraction?
    a. binds actin to myosin
    b. permits inward movement of the Z-lines
    c. stimulates the neurotransmitter receptor
    d. blocks the action of calcium

11. Which neurotransmitter initiates muscle contraction?
    a. adrenaline
    b. acetylcholine
    c. dopamine
    d. serotonin

12. Muscles carrying out anaerobic metabolism produce the following waste product:
    a. glycogen
    b. oxygen
    c. methane
    d. lactic acid

Chapter 6

13. Which muscle shape is noted for having a broad origin?
    a. deltoid
    b. serratus
    c. biceps
    d. pinnate

14. Which statement best represents an isometric muscle action?
    a. active lengthening
    b. passive lengthening
    c. no perceivable change in length
    d. active shortening

15. Which muscle type produces the most pull for its size?
    a. serratus
    b. triangular
    c. triceps
    d. pinnate

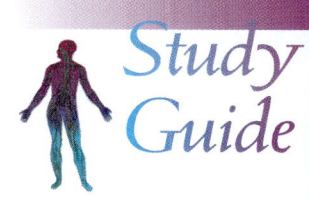

A Case Study

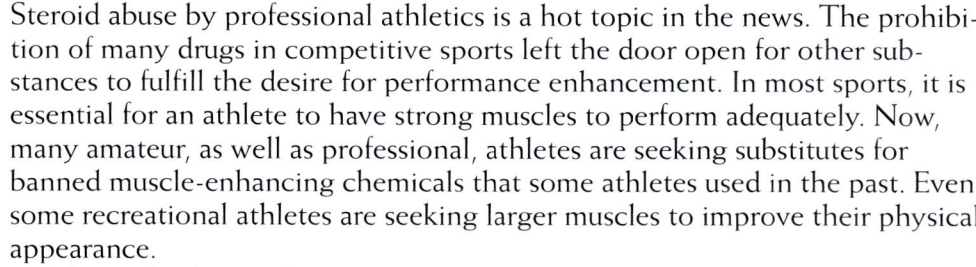

CREATINE: WHAT'S IT ALL ABOUT?

Steroid abuse by professional athletics is a hot topic in the news. The prohibition of many drugs in competitive sports left the door open for other substances to fulfill the desire for performance enhancement. In most sports, it is essential for an athlete to have strong muscles to perform adequately. Now, many amateur, as well as professional, athletes are seeking substitutes for banned muscle-enhancing chemicals that some athletes used in the past. Even some recreational athletes are seeking larger muscles to improve their physical appearance.

   A host of safe muscle-enhancement alternatives are available in nutrition stores and on health-supplement web sites. Creatine, or creatine monohydrate, is probably the most commonly sold muscle enhancer. It can be found in a variety of nutritional supplements and even in smoothies. Smoothies are cold beverages made by mixing fruit and juice and sometimes ice in a blender. Many athletes testify to creatine's effectiveness, claiming temporary gains in muscle size and strength.

   Creatine is an amino acid that converts to phosphocreatine in muscles. Phosphocreatine stores energy for muscle contraction. It is effective in providing greater energy for muscle activities that require quick bursts of energy. Phosphocreatine also diverts water to muscle cells, which causes temporary muscle hypertrophy. This increase in muscle cell size contributes to muscle strength.

Study Guide

There is a great debate about the fairness of using any substance for sports enhancement, including creatine. More important, many research studies indicate that creatine is not as safe as its manufacturers claim.

Use the information in this chapter and the web sites listed below to answer the following questions about creatine use:

- What are the possible risks of using creatine? Do the benefits outweigh the potential risks?
- What is the role of government in determining the safety of strength enhancers? Should creatine be made available to anybody who wants to use it, or should it be restricted to particular people?
- Should people have the right to use creatine for recreational purposes if they are aware of the potential risks?

1. Creatine Information Center
   http://www.creatinemonohydrate.net/

2. Muscular Dystrophy Association
   http://www.mdausa.org/research/creatine.html

3. National Institutes of Health – Creatine Information
   http://www.ninds.nih.gov/funding/research/parkinsonsweb/drug_summaries/creatine.htm

4. Quackwatch
   http://www.quackwatch.org/01QuackeryRelatedTopics/DSH/creatine.html

Where Do We Go from Here?

People in health fields can use their knowledge of human musculature to solve everyday problems. You may wish to use other resources, such as the web sites below, in addition to your textbook to investigate the answers to each of the following situations:

1. A neighbor wants your advice about some medical treatment he felt went bad. He admits that he received an injection around the scalp line to reduce his forehead wrinkles. Now he cannot raise his eyebrows. He wants to know if this is normal or he should run to an attorney. What information would you give him?
   http://www.webmd.com/

2. You are volunteering to help an elderly person recover from a stroke. The person has been confined to bed for 4 months. He confides in you that he is inconvenienced by the physical therapist pulling and bending his arms and legs. He wants to know if this routine is truly needed for his recovery. What would you explain?
   http://www.ptcentral.com/muscles/

3. A friend told you about a cousin born with shortened quadriceps muscles. She asked you how this would affect the child and whether there is a way to correct the condition. What would you tell her?
   http://www.innerbody.com/image/musfov.html

Study Guide

4. A friend's chiropractor told him that he needs treatment for a rotator-cuff impingement. Your friend wants your opinion before considering the treatment. He asks you if it is a legitimate ailment, and, if so, he wants to know what could have caused it.
   http://www.nlm.nih.gov/medlineplus/aboutmedlineplus.html

5. You overhear a personal trainer telling a person to balance her weight-lifting routine for the biceps and triceps. The trainer said it would cause problems later if these muscles were worked unevenly. This person asks your opinion about the trainer's advice.
   http://coachesinfo.com/

Skills Activities

1 Effect of Ambient Temperature on Muscle Action

Materials
- one subject and one researcher
- a clock with a second hand or a stopwatch
- two small buckets
- two thermometers
- access to warm water (40°C or 104°F)
- access to cold water to be chilled with ice to 3°C or 5°C
- access to ice
- one red pen or pencil
- one blue pen or pencil

The body's organ systems work together to ensure a constant internal environment that resists changes in the surrounding environment. However, superficial organs and tissues, such as the smaller surface muscles, cannot always adapt accordingly. Sometimes their cellular environments fluctuate with changes in outside or ambient conditions. People who work outdoors, or in extremely cold or hot environments, must be aware of the limitations these conditions impose on the body. In this activity, you will investigate the effects of ambient temperature on superficial muscle action.

Follow the steps below to investigate the effects of temperature on muscle action:

1. Select a person to be the experimental subject. Make sure the subject is able to make a fist repeatedly without feeling any pain or any resistance to his or her hand movement.
2. Set out the buckets of ice water and warm water.
3. Record the room temperature.
4. Be prepared to start timing the subject while he or she is making repeated fists with one hand.
5. First, count the number of times the subject can make a fist in 20 seconds. Start timing when the hand is completely outstretched, then making a tight fist as fast as possible each time. Record the count in the table below. Repeat this step five times. Calculate the average, and record it in the table.

THE MUSCULAR SYSTEM

Study Guide

6. Next, have the subject submerge a hand in the bucket of warm water. Keep the hand in the water for 2 minutes.
7. Now, have the subject remove his or her hand and immediately start making fists repeatedly. Count how many fists were made in 20 seconds. Record the data in the table below. Repeat this step five times. Calculate the average, and record it in the table.
8. Have the subject submerge a hand in the bucket of cold water chilled with ice. Keep the hand in the water for 2 minutes.
9. Now, have the subject remove the hand and immediately start making fists repeatedly. Count how many fists can be made in 20 seconds. Record the data in the table below. Repeat this step five times. Calculate the average, and record it in the table.

Temperature	Number of Fists/ 20 sec					Average
room temperature (___°C or ___°F)						
warm water						
cold water						

1. What effect did temperature have on the subject's muscle activity?
2. How can any differences in the rate of fist clenching be explained?
3. What other information is available to understand how temperature affects muscle cell physiology?
4. How can these findings be useful for people who exercise or work outdoors?

Skills Activities

2 Effect of Fatigue on Grip Muscle Action

Materials
- two subjects and one researcher
- two tennis balls
- a clock with a second hand or a stopwatch
- one measuring tape

Muscle fatigue occurs after extended or strong muscle contractions. Research shows that short-term muscle fatigue is associated with a lack of oxygen and subsequent build-up of lactic acid in the muscle. It is also known that long-term muscle fatigue is related to the rate of glycogen reduction. As discussed in this chapter, muscle cell fatigue is based on their metabolism. Some muscles fatigue after short bursts of activity, while others fatigue only after prolonged action. Each person has a unique muscle composition and structure that determine the ability of a particular muscle group to fatigue for particular activities. In this activity, you will investigate whether there are any differences in how muscle fatigue affects the muscle action of two people.

Follow the steps below to investigate the effects of fatigue on muscle action:
1. Using the measuring tape, measure the length, width, and circumference of the hand that each subject will use to squeeze the tennis ball. Also, measure the length of the arm from the elbow to the wrist, as well as the diameter of the wrist. Record the data in Table 6.1 below.
2. For Subject 1, count how many times he or she can tightly squeeze a rubber ball in 30 seconds. Record the number in Table 6.2.
3. Immediately, repeat this nine more times, and record the data in Table 6.2. The subject should not rest between each trial.
4. Immediately after the trials, use the measuring tape to measure the length, width, and circumference of the hand that each subject will use to squeeze the tennis ball. Again, measure the length of the arm from elbow to wrist, as well as the diameter of the wrist. Record the data in Table 1.
5. Repeat the procedure for Subject 2.
6. After completing the trials with both subjects, record the grip-fatigue data in the graph below. Use the red pen for Subject 1 and the blue pen for Subject 2.

Table 6.1 Subject's Dimensions

Measurement	Subject 1		Subject 2	
	Before	After	Before	After
hand length				
hand width				
hand circumference				
forearm length				
wrist circumference				

THE MUSCULAR SYSTEM

Study Guide

Table 6.2 Grip Fatigue Chart

Trial	Number of Squeezes per 30 Seconds	
	Subject 1	Subject 2
1.		
2.		
3.		
4.		
5.		
6.		
7.		
8.		
9.		
10.		

1. What muscle changes did you observe while the muscle was working (contracted)?

2. In Figure 3, make a line graph using the results of the fatigue experiment. Be sure to fill in the values on the vertical axis.

Figure 3 Effect of Fatigue on Grip Muscle Action Graph

Number of Attempts

1. What happened to the number of grips per 30 seconds in each trial?

2. Where on the graph does it appear that fatigue set in for each subject on grip action? Explain why you picked that point.

3. Were there any differences between the graphs of the two subjects? If so, what could explain these differences?

4. Is there a relationship between the arm and hand measurements and the point of fatigue?

THE MUSCULAR SYSTEM

Case Study Investigation

Case Study Investigation #7

Imagine that you are in a clinic examining a 12-year-old female patient. Her parents are concerned about her excessive facial hair and the deepening of her voice. Further conversation with the parents brings out the fact that the girl started developing pubic hair at age 7 years. Examination of the patient confirmed the parents' observations, plus showed that the girl was well above average height for her age. She also has unusually large amounts of acne on her face and the back of her neck. A vaginal tract examination indicates swelling of her external genitals. The girl started losing weight rapidly last year, and leg and thigh muscles became well developed. This was attributed to the girl starting intensive gymnastics practices for an upcoming competition. Everything else about the girl appears normal. By reading this chapter, you will eventually conclude that this girl's conditions are due to an abnormality of the endocrine system. It is the job of the examiner to determine what part of the endocrine system is being affected. Then, she must figure out what is causing the abnormality. At the end of the chapter, you will be asked to determine the possible problem that is causing this set of conditions.

CHAPTER 7

THE ENDOCRINE GLANDS AND HORMONES

Chapter Outline

Case Study Investigation (CSI)
Applied Learning Outcomes
Overview
   Hormone Function
   Endocrine Secretions
   Types of Hormones
      Peptide Hormones
      Lipid Hormones
The Endocrine Glands
   Pituitary Gland
   Pineal Gland
   Adrenal Glands
Thyroid Gland and
   Parathyroid Glands
Pancreas
Thymus Gland
Gonads
Wellness and Illness over the
  Life Span
   Pathology of the Endocrine
System
   Aging of the Endocrine System
CSI Conclusion
Study Guide

Applied Learning Outcomes

- Use the terminology associated with the endocrine system.
- Learn about hormones, glands, and their functions.
- Understand the aging and pathology of the endocrine system.

OVERVIEW

Key Terms: ductless glands, endocrine glands, endocrine secretion, endocrine system, environmental signal, exocrine gland, exocrine gland secretion, hormone, receptor, signal, target cell

What types of jobs does the endocrine system carry out? Many people are aware of the endocrine system's ability to control blood sugar after a meal. In addition, maintaining salt and water balance is an important function of the endocrine system. Endocrine gland secretions control development and growth throughout childhood and adolescence. Blood pressure and heart rate are intimately tied to the endocrine system. The nervous system works closely with the endocrine system in responding to danger and stress. The endocrine system's interactions with activity, diet, and environmental conditions regulate a person's metabolism. The immune system's ability to fight infection is closely tied to the endocrine system. Last, endocrine secretions, which determine gender and fertility, control human reproduction.

As mentioned in Chapter 4, glands are groups of cells that manufacture secretions. The human body produces two types of glandular secretions: **exocrine** and **endocrine**. Exocrine secretions are deposited into body cavities or the surface of the skin through a tunnel of cells, or ducts (Figure 7.1). The digestive enzymes secreted into the stomach are exocrine secretions. Endocrine secretions are typically sent directly into the bloodstream (Figure 7.2). These secretions help coordinate body functions. In effect, they communicate information between cells. Endocrine glands do this by responding to signals from the environment and from other cells. Signals vary greatly and include a diverse array of stimuli. **Environmental signals** originate outside the body and include such things as atmospheric gases, gravity, nutrients, sunlight, and temperature. Cellular signals, or **hormones**, originate inside the body; they are usually chemicals secreted by healthy cells or leaked by damaged cells.

Glandular secretions can be deposited into tissue fluids and body cavities or they can be released into the blood. Almost any organ in the body can produce endocrine secretions. For example, the digestive system is mostly involved in breaking down food using exocrine secretions and muscular mixing actions. However, the digestive system also has endocrine cells that produce endocrine secretions for coordinating digestion and regulating appetite. These pockets of endocrine cells are called diffuse endocrine structures. Various other organs have a small collection of endocrine cells. However, such organs as the pituitary gland are totally dedicated to producing endocrine secretions. The endocrine system is composed of 10 endocrine glands (Figure 7.3). Glands of the endocrine system are predominantly involved in endocrine function.

Exocrine Secretion A secretion deposited into a body cavity through a duct

Endocrine Secretion A secretion sent directly into the bloodstream

Environmental Signals Signals that originate outside the body

Hormones Signals that originate inside the body

CHAPTER 7

without ducts

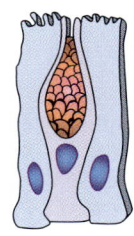
unicellular gland
{ mucous or goblet cell of intestinal mucosa }

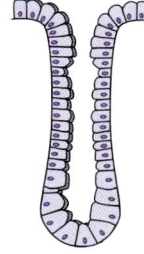
secretory sheet
{ i.e., gastric pit chief and parietal cells }

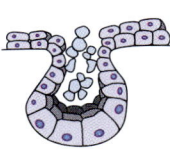
intraepithelial gland
{ i.e., efferent ducts of urethra }

Figure 7.1 Exocrine Glands
Exocrine glands use ducts to carry secretions into body cavity or its surface. Note the different shaped ducts found in various exocrine glands.

with ducts

simple tubular
[intestinal gland of Lieberkuhn]

simple-coiled tubular
{ sweat gland }

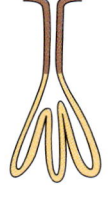
simple-branched tubular
{ fundis glands of stomach }

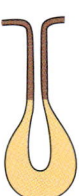
simple alveolar (acinus)
{ mucous glands }

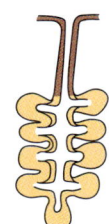
simple branched alveolar
{ sebacous glands }

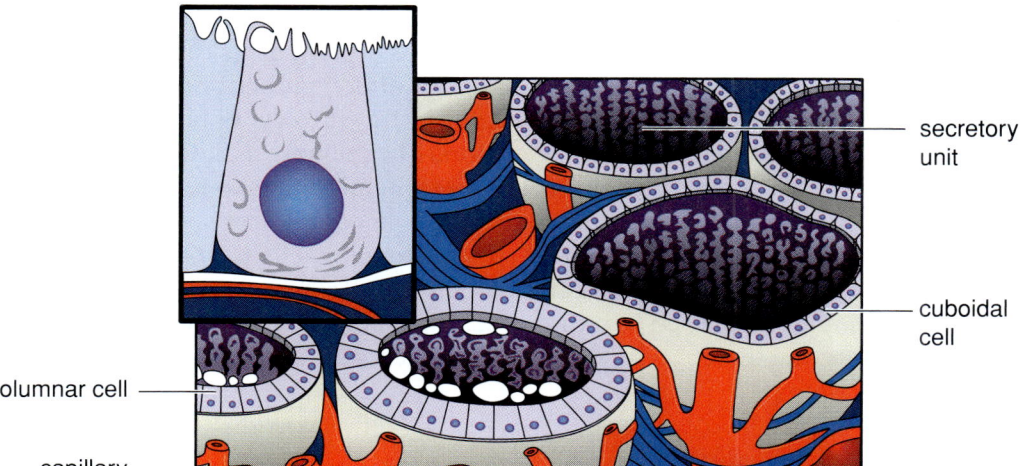

columnar cell

capillary

secretory unit

cuboidal cell

Figure 7.2 Endocrine Glands
Endocrine glands are collections of cuboidal or columnar cells that produce secretions directly into the blood stream without using ducts. Note how the capillaries branch out around the secretory units.

Figure 7.3 Endocrine Systems
The human endocrine system is primarily composed of distinct glands that produce hormones.

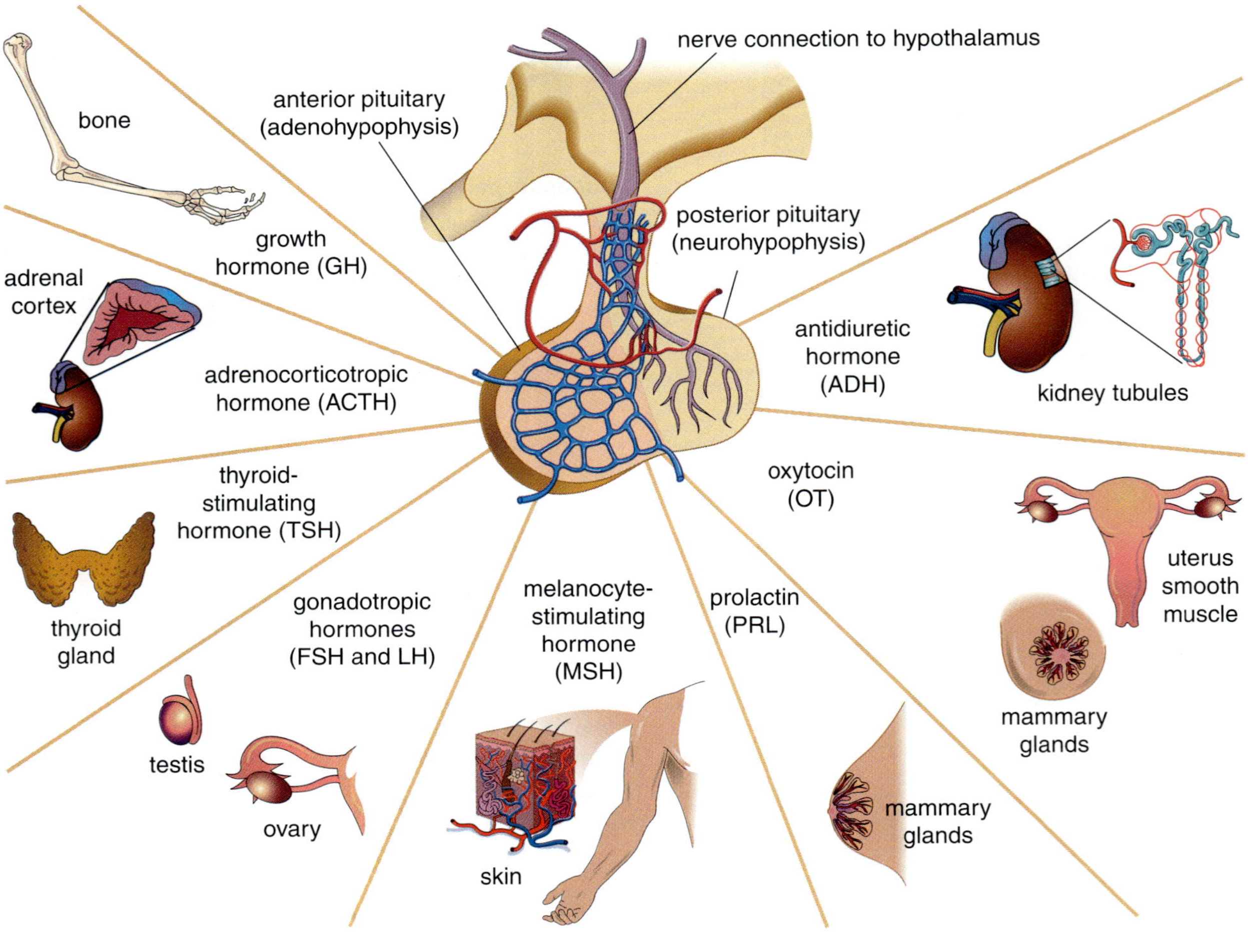

Ductless Glands Glands without ducts to transport secretions

Receptor A protein that permits a cell to detect stimuli

Target Cells Cells with receptors sensitive to endocrine secretions

Endocrine glands are also called **ductless glands** because the secretions do not travel through narrow tubes. The secretions usually enter the blood directly through capillaries. Capillaries do not direct endocrine secretions to any particular part of the body. The secretions travel to every body organ. However, not every cell of the body responds to the information endocrine secretions convey. So, how does a cell know when it needs to react to a particular secretion? Certain cells have **receptors**. Receptors are special proteins that permit the cell to detect various types of stimuli. Cells with receptors sensitive to endocrine secretions are called **target cells** (Figure 7.4). Target cells are genetically programmed to modify their metabolism when they detect a specific endocrine secretion. In effect, only target cells are "listening" to the endocrine system.

✓ Concept Check

1. What are the two types of human glands, and how do they differ?
2. What is the overall function of an endocrine gland?
3. Distinguish between the terms hormone and environmental signal.

CHAPTER 7

Civic Responsibility

CIVIC RESPONSIBILITY: HELPING OTHERS WITH YOUR KNOWLEDGE

It is valuable to use what you have learned about the endocrine system to help others to better understand the world around them. It is very important to check your facts and seek further information about certain topics before discussing health and science issues. Here are some suggestions to foster a better public awareness of the endocrine system:

1. Speak to schoolchildren about the indicators and risks for diabetes.
2. Work with sports clubs to educate players about performance hormones.
3. Help elderly persons to better understand the limitations of antiaging supplements.
4. Volunteer at a school health day to teach children the basics of the endocrine system.

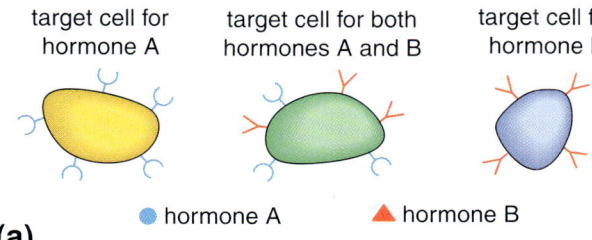

Figure 7.4 Target Cells and Receptors
(a) The endocrine system works by communicating with target cells.
(b) Target cells have receptors that change the cell's metabolism when an endocrine secretion attaches to the receptor.

THE ENDOCRINE GLANDS AND HORMONES

Hormone Function

Key Terms: effector, internal receptor, ligand, surface receptor

Ligand A chemical that attaches to a receptor

Surface Receptor A receptor located on the surface of a cell

Internal Receptor A receptor located within a cell

Effector Another name for a target cell

Ligand is the term for the general group of chemicals that attach to receptors; hormones are a category of ligands. Hormones can attach to receptors on target cells either internally or externally depending on the location of the receptor. Receptors located on the surface of the cell are called **surface receptors**. The receptors located within a cell are **internal receptors**, also called intracellular receptors.

Target cells with surface receptors are stimulated differently than cells with internal receptors. These surface receptors rely on the blood's watery fluid to carry the hormones to the cell membrane. The hormone must then have the correct chemical properties and proper shape to attach to the receptor. Attachment to the receptor initiates a series of complex chemical reactions within the cell (Figure 7.5). The metabolic changes within the target cell are called biological effects; in turn, target cells are also referred to as **effectors**.

Carrier proteins bring hormones that use internal receptors to target cells. One group of carrier proteins is created primarily by the liver and help to transport these hormones in the blood to the cell. Another group of carrier proteins produced in the cell attach the hormone to internal receptors on the genetic material. The transported hormones bind to a special receptor that interacts with a particular region of the DNA. This interaction either switches on or switches off a distinct trait controlled by that region of the DNA (Figure 7.6).

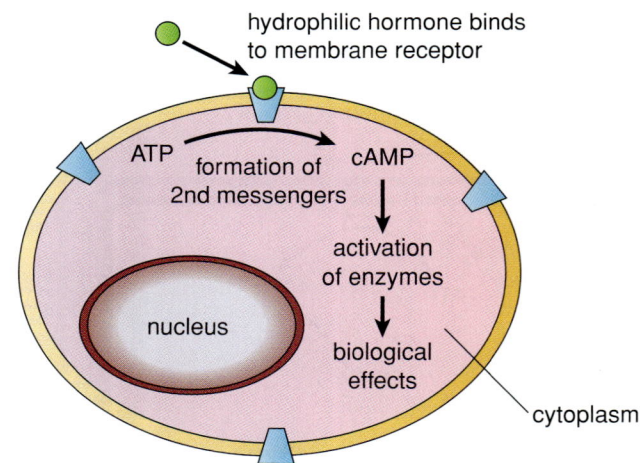

Figure 7.5 Hydrophilic Hormone Binds to Membrane
Note how the hormone binds to the surface receptor. By attaching to the receptor, the hormone turns on one of several types of chemical reactions in a cell. These chemical reactions cause metabolic changes in the target cell.

✓ Concept Check

1. Define the term effector.
2. What is the function of a carrier protein?
3. What are the categories of hormone receptors?

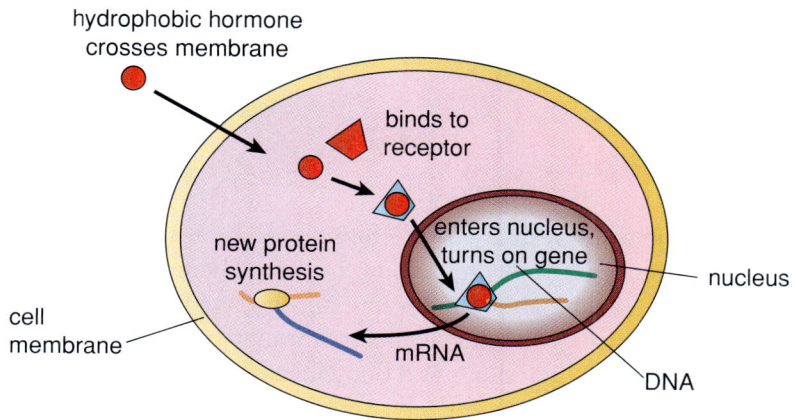

Figure 7.6 Hydrophobic Hormone Crosses Membrane Note how the hormone readily enters the cell membrane. Once in the membrane, the hormone attaches to an internal receptor that directly interacts with DNA.

Endocrine Secretions

Key Terms: autocrine, negative feedback, paracrine, pheromones, thyroxine

Cell secretions involved in body system coordination are named according to their target cells (Figure 7.7). **Autocrine** secretions are self-governing and usually do not travel in the blood. They are secreted into body fluids where they interact with the cell that produced them. Autocrine control allows cells to control their own activities. **Paracrine** secretions also travel short distances via the blood or body fluids to their target cells. The most important function of these secretions is to coordinate the cells within an organ. **Pheromones** are secretions that leave the body and signal the cells of other organisms. They play a role in sexual behavior and the identification of an individual.

Autocrine A type of secretion that is self-governing and does not usually travel far in the blood

Paracrine A type of secretion that travels to target cells via the blood or other body fluids

Pheromones Secretions that leave the body and signal the cells of other organisms

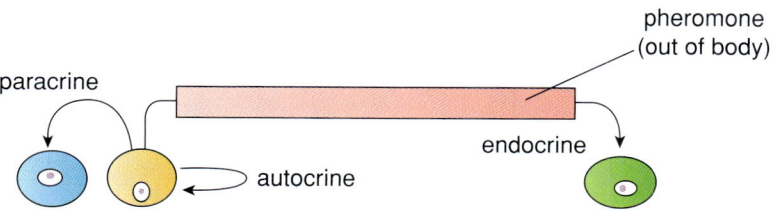

Figure 7.7 Cell Communication Note the ways the different types of cell-communication secretions travel to the target organ.

Each endocrine gland "knows" its job from information that the body provides. The glands typically use a type of communication called **negative feedback**. Negative feedback functions in a way that is similar to temperature control in a building. Turning a dial on a thermostat sets a building's temperature. The thermostat communicates the maximum temperature for the heating system. The heating system is switched on if the temperature drops below the set point of the thermostat. Once the desired temperature is reached, the heating system turns off until the temperature again drops below the setting. The endocrine system works in much the same way. For example, the body secretes **thyroxine** to increase heat production when the internal body temperature drops below a certain setting. Secretions of the hypothalamus stimulate

Negative Feedback A signal that inhibits an endocrine gland, preventing further secretion of a particular hormone

Thyroxine A hormone secreted by the thyroid gland, which controls metabolism

THE ENDOCRINE GLANDS AND HORMONES

Good Choice Bad Choice

A young man is frustrated because he is not getting the desired muscle increase, known as gain, from his long weightlifting workouts and high-protein diet. He knows it is illegal and dangerous to take testosterone injections and other anabolic steroids. He decides to buy prohormones, which a magazine is marketing as "safe and legal steroids." Prohormones are chemicals that are either converted to testosterone directly by the body, or that impersonate testosterone when converted into androgen-like compounds, such as nandrolone. Most professional and amateur sports organizations ban prohormones, but they are readily available to recreational athletes through many nutritional supply stores and bodybuilding magazines. Some prohormones are commonly found in food, so they are considered the "natural" way to gain muscle. What are your views about prohormone use for recreational athletes? In what situations would it be appropriate to take prohormones?

thyroxine production in response to cold internal body temperatures. The role of thyroxine is to increase the metabolic rate, which causes the body to produce heat. Thyroxine production is stimulated as long as the body temperature is low. Thyroxine shuts off its own production as the body temperature rises to a normal level. No thyroxine will be produced as long as the hormone is present and the body temperature stays at the normal setting.

✓ Concept Check

1. What is the function of a receptor?
2. What is the role of a target cell?
3. Give an example of endocrine system negative feedback.

Types of Hormones

Key Terms: agonist, antagonist, lipid hormone, peptide hormone, protein hormone

Agonist A chemical that behaves like a hormone

Antagonist A chemical that blocks the action of a hormone

As mentioned earlier, a hormone is any chemical that signals a cell to alter its metabolism. The signal is effective only if it is able to stimulate a particular target cell into carrying out a specific function. Cells secrete particular types of hormones to communicate specific information to the target cells. A wide array of chemicals can serve as hormones. Chemicals that act like hormones are called **agonists**. For example, certain foods contain hormones called phytoestrogens, which act like the estrogen in the body. Chemicals that block the action of hormones are called **antagonists**. A drug called Finasteride, a testosterone antagonist, is given to people with high testosterone levels because it blocks the effects of too much testosterone in the body. Unfortunately, many chemical pollutants can interfere with hormonal communication because they also mimic hormones. Current research studies show that arsenic, a poison, disrupts the action of the hormone estrogen. Chemicals such as arsenic are called hormone mimics, or endocrine disruptors.

The body typically produces two categories of hormones: **peptide hormones** and **lipid hormones**. Peptides are biological molecules composed of amino-acid chains. The DNA contains a program for building peptide hormones piece by piece from dietary amino acids. Usually, some type of signal from another cell, or from the environment, stimulates endocrine cell DNA to build a particular peptide hormone. Lipid hormones are made from existing lipids in the body or taken in through the diet. They are created by enzymes that convert fatty acids, or cholesterol, into a particular lipid hormone. Some enzymes are capable of converting one lipid hormone into another.

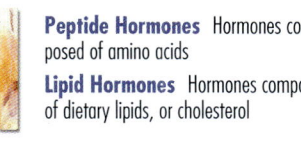

Peptide Hormones Hormones composed of amino acids

Lipid Hormones Hormones composed of dietary lipids, or cholesterol

Peptide Hormones Peptide hormones range in complexity from a single amino acid to a small chain of amino acids called a polypeptide. Many scientists refer to larger polypeptides as proteins. Peptide hormones are usually responsible for rapid changes in the body. Most of them are involved in short-term and immediate changes needed for regulating the body's overall metabolism. However, some peptide hormones have permanent effects on the body. For example, growth hormone causes the rapid height increase associated with puberty.

Peptide hormones usually stay in the blood, binding to receptors on the surface of target cells. The body breaks down the peptides so that they do not accumulate in the blood. Enzymes in the blood help to degrade the peptides. Smaller peptide hormones are taken into the cell where they bind with internal receptors. Enzymes in the cell neutralize these hormones. The hormones would have a long-lasting effect if they were not broken down and removed from the blood. (You can find a list of common peptide hormones in Table 7.1.)

Lipid Hormones While there are fewer types of lipid hormones than peptide hormones, the lipid hormones play valuable roles in body fluid regulation and sexual reproduction. Lipid hormones are made from cholesterol and other fats in the diet. The body produces two types of lipid hormones: hormone-like lipids and steroids. Hormone-like lipids are made of a single chain of fatty acid. It includes a group of hormones called prostaglandins, which are produced by almost any body cell. Prostaglandins carry out a variety of signaling functions, including immune-system control and blood-pressure regulation. Steroids are more chemically complex than hormone-like lipids. They are made from cholesterol molecules, and they carry out specific signaling functions. Lipid hormones do not travel easily in the blood because they cannot dissolve well in the water in the blood serum. Therefore, a carrier molecule secreted into the blood moves them through the body. However, they readily dissolve across the cell membrane once they reach the target cell, where they bind to receptors inside the cell. These receptors interact with the cell's DNA. Many lipid hormones have long-term effects on the body because they directly control DNA function. For example, the results of one's sexual development are retained throughout adulthood because the long-term effects that steroids provide compromise the sex hormones. Like peptide hormones, lipid hormones are also broken down by enzymes within the cell to prevent accumulation. They can cause dangerous metabolic problems, including cancer, if they stay at high levels in the body. (You can find a list of common lipid hormones on this text's Internet Resource Center.)

DISCOVERY SCENE PLEASE ENTER DISCOVERY SCENE PLEASE ENTER

Do you see a problem with the body's hormone feedback? What type of hormone do you think may be causing the problem? Could the girl's condition be due to signal errors or problems with the receptor?

THE ENDOCRINE GLANDS

The human endocrine system is composed of 10 endocrine glands. The major hormones produced by these glands are listed in Table 7.1. Some endocrine glands act individually, while others are under the hormonal control of other endocrine glands. Each endocrine gland needs some type of feedback signal to control its level of hormone production. As mentioned earlier, the signals can be hormones from other endocrine cells, nerve cell activity, or nutrients in the blood or body cavities. Endocrine glands can also carry out functions other than hormone production. The ovaries and testes, collectively called the gonads, produce reproductive cells and various secretions in addition to hormones.

Table 7.1 Some Common Peptide Hormones

Angiotensin	Gonadotropin	Prolactin
Bradykinin	Insulin	Prolactoliberin
Calcitonin	Isotocin	Prolactostatin
Choriogonadotropin	Kallidin	Relaxin
Choriomammotropin	Lipotropin	Secretin
Corticoliberin	Luliberin	Somatoliberin
Corticotropin	Lutropin	Somatomedin
Erythropoietin	Melanoliberin	Somatostatin
Folliberin	Melanostatin	Somatotropin
Follitropin	Melanotropin	Thymopoietin
Gastrin	Mesotocin	Thyroliberin
Gastrin sulphate	Oxytocin	Thyrotropin
Glucagon	Pancreozymin	Urogonadotropin
Glumitocin	Parathyrin	Vasopressin
Gonadoliberin	Proangiotensin	Vasotocin

Pituitary Gland

Key Terms: adrenocorticotropic hormone, anterior pituitary, antidiuretic hormone, estrogen, follicle-stimulating hormone, growth hormone, hypophysis, leuteinizing hormone, melanocyte-stimulating hormone, oxytocin, pituitary gland, posterior pituitary, progesterone, prolactin, releaser hormone, releasers, thyroid-stimulating hormone, testosterone

Pituitary Gland or Hypophysis A gland that controls most of the other endocrine glands in the body

Hypothalamus A region of the brain that controls endocrine activity

The **pituitary gland**, or **hypophysis**, is known as the master endocrine gland because its numerous hormones control most of the other endocrine glands, and it is intimately linked to the overall coordination of the body's organ systems. The pituitary gland is located beneath a region of the brain called the **hypothalamus**. The hypothalamus is a part of the brain that controls endocrine activity as well as other functions, including appetite, body temperature, and sleep. Therefore, it is often identified as one of the endocrine organs. This location near the hypothalamus allows the pituitary gland to serve as an important communication center, synchronizing the brain with the endocrine system.

Figure 7.8 Pituitary Gland
The pituitary gland is divided on the anterior and posterior regions

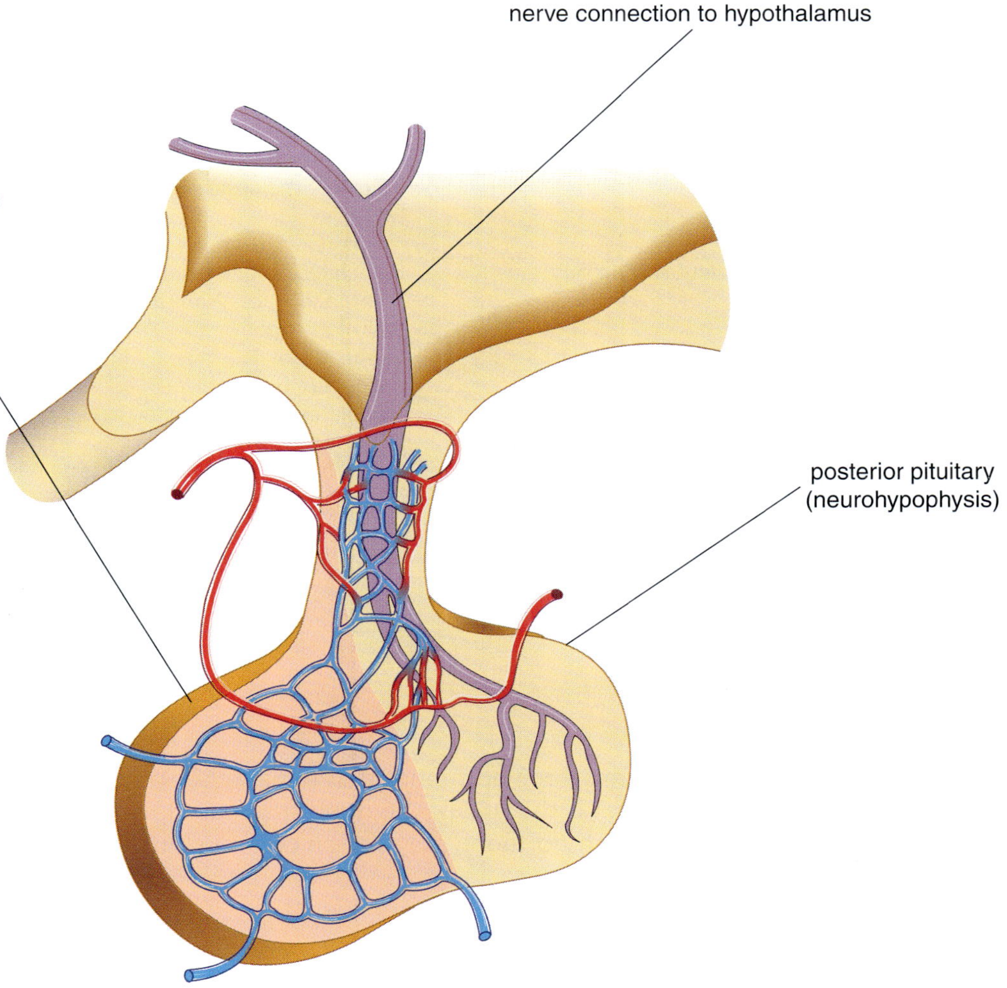

- nerve connection to hypothalamus
- anterior pituitary (adenohypophysis)
- posterior pituitary (neurohypophysis)

The pituitary gland is actually two endocrine glands attached to each other (Figure 7.8). The **anterior** (front) region of the pituitary gland forms from epithelial cells lining the upper throat. The anterior pituitary is controlled by chemicals called **releasers**, or **releasing hormones**. Nerve cells of the hypothalamus produce these releasers in response to signals from various regions of the body, including many types of environmental conditions detected by the skin, and the circulatory, digestive, and excretory systems. Releasers are secreted into capillaries that travel from the hypothalamus to the anterior pituitary.

The **posterior pituitary** (rear region of the pituitary gland) is formed by and derives information from the nerve cells of the hypothalamus. These nerve cells carry information from the brain directly to the posterior pituitary. Table 7.2 shows the major secretions of the pituitary gland.

Anterior Pituitary Gland The forward region of the pituitary gland

Release or Releasing Hormones Hormones released by the hypothalamus that control the anterior pituitary

Posterior Pituitary Gland The rear region of the pituitary gland

THE ENDOCRINE GLANDS AND HORMONES

Table 7.2 Major Secretions of the Pituitary Gland

Anterior Pituitary Gland	
adrenocorticotropic hormone	Stimulates cortisol and other steroid hormone production by the adrenal glands. These hormones help the kidneys control the water and salt content of the body.
follicle-stimulating hormone	Along with growth hormone, assists with egg formation and estrogen production by the ovary in females. Estrogen is the hormone involved in female sexual characteristics. In males, this hormone is responsible for sperm production by the testes.
growth hormone	Needed for skeletal and muscular system growth. It also ensures that all of the organ systems keep up with each other as the body develops. It releases fat needed for organ growth and maintenance from the liver.
leuteinizing hormone	Stimulates progesterone production and helps egg development in females. Progesterone functions in a manner similar to that of estrogen.
melanocyte-stimulating hormone	Stimulates melanocytes (skin cells), which darken the skin with melanin (a brown protein secretion).
prolactin	Stimulates milk production in females and assists luteinizing hormone in males.
thyroid-stimulating hormone	Controls the body's metabolic rate by regulating thyroxine production by the thyroid gland.
Posterior Pituitary Gland	
antidiuretic hormone	Causes the kidneys to retain water.
Oxytocin	Stimulates muscle contractions in the uterus and assists with milk release.

✓ Concept Check

1. Name three hormones that the anterior pituitary gland produces.
2. What is the role of the hypothalamus in the endocrine system?
3. Which hormone only has an effect in females?

Pineal Gland

Key Terms: melatonin, serotonin, pineal gland

Pineal Gland A gland responsible for producing melatonin and serotonin

Melatonin A hormone responsible for regulating the body's daily rhythms

Serotonin A hormone involved with digestion, appetite, moods, and sleep

The **pineal gland** is a small structure located above and behind the hypothalamus. It is responsible for producing the hormones **melatonin** and **serotonin**. A primary role of melatonin is to regulate body rhythms throughout the day. Because sunlight stimulates the production of melatonin, in effect, it is an indicator of exposure to daylight. In humans, low levels of melatonin can contribute to feelings of depression. This explains why people tend to feel unhappy during seasons when daylight is scarce due to cloudy skies or shorter winter days. Melatonin causes other effects, which include controlling the onset of puberty in boys and girls. Serotonin is also involved with appetite, emotions, moods, and sleep.

✓ Concept Check

1. What two hormones does the pineal gland produce?
2. How is daylight related to the production of melatonin?

Adrenal Glands

 **Key Terms:** adrenal cortex, adrenal glands, adrenal medulla, adrenaline, aldosterone, angiotensin II, cortisol, epinephrine, glucocorticosteroids, mineralocorticosteroids, noradrenaline, norepinephrine

Above the kidneys lie the **adrenal glands**, sometimes called the suprarenal glands. Each adrenal gland is actually two glands fused together. An outer shell called the **adrenal cortex** surrounds an inner region called the **adrenal medulla** (Figure 7.9). The adrenal cortex produces two groups of hormones—**glucocorticosteroids** and **mineralocorticosteroids**—known collectively as corticosteroids, and a third group, the androgens.

Adrenal Glands Glands that produce glucocorticosteroids and mineralocorticosteroids

Adrenal Cortex The outer shell of the adrenal glands

Adrenal Medulla The interior region of the adrenal glands

Glucocorticosteroids Steroid hormones produced by the adrenal cortex that affect glucose metabolism in the body

Mineralocorticosteroids Steroid hormones produced by the adrenal cortex that affect salt and water balance in the body

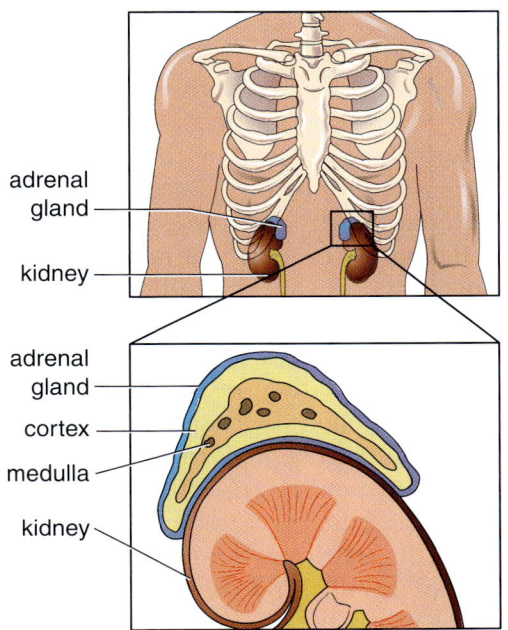

Figure 7.9 Adrenal Glands
The adrenal glands are located above the kidneys. They are composed of the adrenal cortex and the adrenal medulla.

The key functions of glucocorticosteroids are to regulate metabolism and inhibit the release of adrenocorticotropin (adrenocorticotropic hormone) from the anterior pituitary. Low blood sugar and stress are the main factors that call for the release of adrenocorticotrophic hormone.

Adrenocorticotropin stimulates the secretion of **cortisol**, the primary glucocorticosteroid. Cortisol helps metabolize lipids and proteins to produce energy. Diets high in protein and fat require cortisol for these molecules to be used in place of carbohydrates, which the cells normally burn for fuel. Scientists have discovered that the adrenal cortex will diminish without regular adrenocorticotrophic hormone secretions. Consequently, diseases that affect the pituitary could permanently shrink the adrenal cortex.

Mineralocorticosteroids regulate the balance of electrolytes and water in the body. **Aldosterone** is the major mineralocorticosteroid that the adrenal cortex secretes. The minerals potassium and sodium are the two major salts that aldos-

 **Cortisol** A type of glucocorticosteroid associated with stress-fighting and anti-inflammatory responses

Aldosterone A steroid hormone that the adrenal cortex secretes to control sodium and potassium in the blood

Angiotensin II A peptide produced from angiotensin I; it is involved in the maintenance of blood volume and pressure

Adrenaline or Epinephrine A hormone produced by the adrenal glands in response to exercise, fear, or stress

Noradrenaline or Norepinephrine A hormone produced by the adrenal glands that has a stimulatory effect on the nervous system

terone regulates. A healthy balance of potassium and sodium are necessary for normal muscle and nerve function. Increased potassium and decreased sodium in the blood are two factors that signal the release of aldosterone to regulate salts and retain water in the digestive system, kidneys, salivary glands, and sweat glands. Another method of stimulating aldosterone is the secretion of the hormone renin. When excessive bleeding or kidney damage causes blood pressure to drop, the kidneys secrete renin, which stimulates the formation of angiotensin I (a protein) and is immediately converted to **angiotensin II** (a peptide). Angiotensin II signals aldosterone production, which, in turn, raises the blood pressure by increasing the volume of water in the blood. Angiotensin I is the inactive form of the angiotensin II, while angiotensin II is the active form. Scientists believe that the conversion of angiotensin I to angiotensin II protects the body from accidentally stimulating aldosterone production. The adrenal cortex also secretes small amounts of hormones known as androgens, primarily dehydroepiandrosterone (DHEA). Normal levels secreted by the male adrenal cortex usually have minimal effects. However, minor increases in androgen levels released by the female adrenal cortex may lead to prepubertal growth and the development of facial and pubic hair.

Adrenaline is the main hormone produced by the adrenal medulla. Another name for adrenaline is **epinephrine**. This term is commonly used when adrenaline is administered as a drug therapy. The adrenal medulla produces another hormone called **noradrenaline**, or **norepinephrine**, although in smaller amounts than adrenaline. A variety of signals stimulate the adrenal medulla to secrete adrenaline. Heavy physical exertion and stress are the two most common signals that cause adrenaline secretion. Low blood sugar also promotes adrenaline secretion by signaling the hypothalamus. Adrenaline and noradrenaline have many effects on the body. They increase the discharge of glucose and fats into the blood. This provides the body with the immediate energy needed to fuel metabolism. Adrenaline diverts blood away from many of the body organs and directs it to muscle. This is why it is often referred to as the hormone of the reflexive "fight or flight" response. It elevates the metabolic rate of muscle cells, and increases blood pressure and heart function. Noradrenaline primarily affects nerve cell function related to the body's response to activity and stress.

✓ Concept Check

1. Name and explain the functions of the hormones produced by the adrenal cortex.
2. Name and explain the function of the hormone produced by the adrenal medulla.
3. What is the major function of aldosterone?

Thyroid Gland and Parathyroid Glands

Key Terms: calcitonin, parathyroid gland, parathyroid hormone, thyroid gland, thyroid hormone, thyroxine (T4)

The **thyroid** and **parathyroid glands** work so closely together that they will be described as paired endocrine organs. Both are located just below and in front of the voice box in the middle of the neck (Figure 7.10).

Thyroid Gland A gland located in the front of the neck that plays a role in regulating metabolism

Parathyroid Gland One of four small glands located behind the thyroid that increase calcium levels in the blood

Figure 7.10 Thyroid and Parathyroid Glands
The thyroid and parathyroid glands not only work together, but they are also located next to each other.

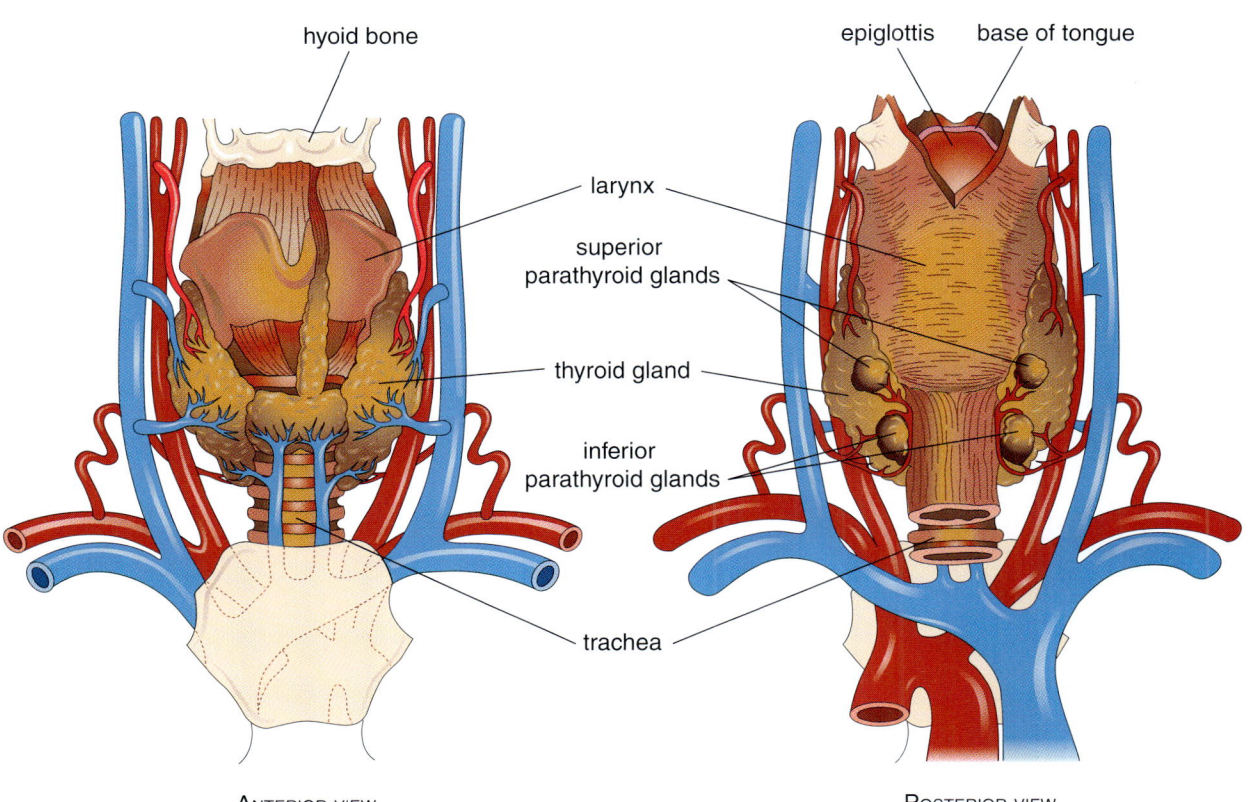

ANTERIOR VIEW   POSTERIOR VIEW

The thyroid gland is under the direct control of thyroid-stimulating hormone (thyrotropin), which is produced by the anterior pituitary. Low body temperature and stress are major factors that prompt the release of thyroid-stimulating hormone, which, in turn, causes the thyroid gland to produce thyroid hormone. Thyroid hormone's main role is to increase the cellular metabolic rate. This speeds up the conversion of food into energy and helps elevate the body temperature (Figure 7.11). Thyroid hormone is really two separate hormones: **thyroxine**, or tetraiodothyronine (T_4), and triiodothyronine (T_3). Thyroid secretion is primarily of thyroxine (T_4), but the latter form, T_3, which results from the loss of an iodine molecule, is actually the active form. Too much thyroxine can cause an accelerated heart rate, fatigue, hair loss, light or absent menstrual periods, muscle weakness, trembling hands, and weight loss. Too little thyroxine can produce cold intolerance, constipation, fatigue, irritability, memory loss, muscle cramps, and weight gain.

THE ENDOCRINE GLANDS AND HORMONES

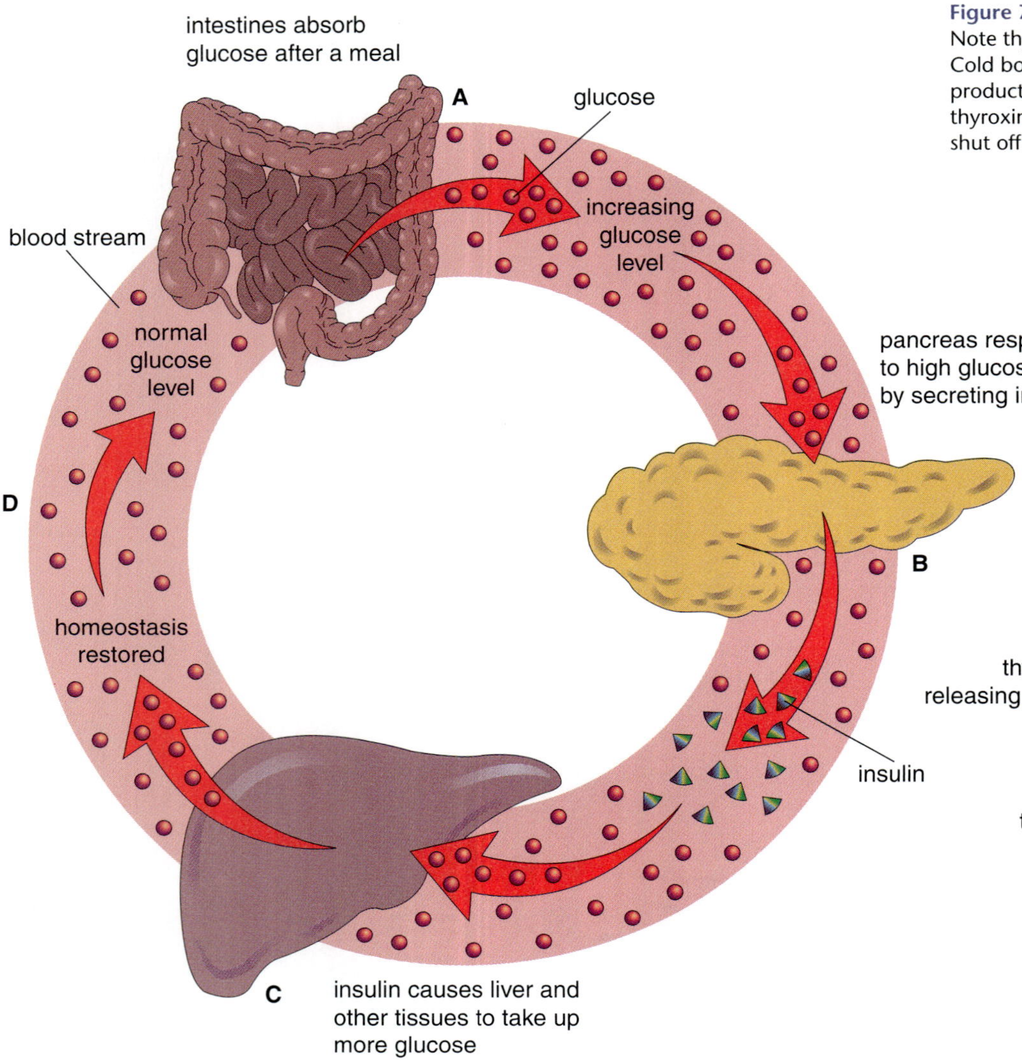

Figure 7.11 Negative Feedback Loop
Note thyroxine controls its own secretion. Cold body temperatures stimulate thyroxine production. However, increased levels of thyroxine without the cold temperatures shut off thyroxine production.

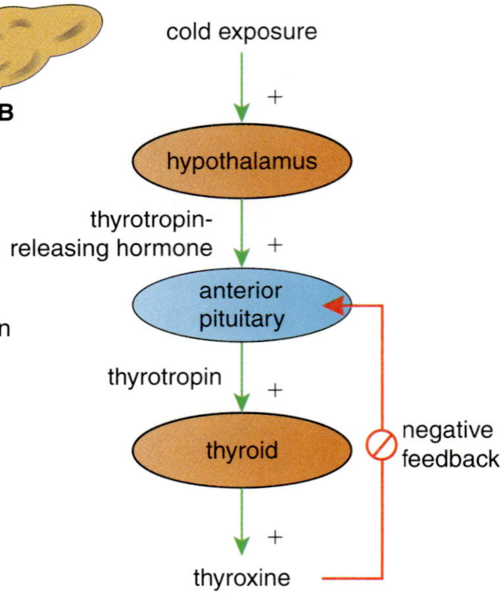

Calcitonin A hormone secreted by the thyroid gland that lowers blood calcium

Parathyroid Hormone (PTH) A hormone secreted by the parathyroid gland that regulates calcium levels in the blood

Calcitonin is produced predominantly by the thyroid gland. It works together with the parathyroid glands to adjust calcium levels in blood and bone. A specific amount of calcium is needed in the blood to ensure proper firing of nerve cells and muscle cell contraction. Calcitonin is produced when the amount of calcium circulating in the blood is very high. This hormone lowers calcium levels by encouraging the retention of calcium in bone. In contrast, the parathyroid glands secrete **parathyroid hormone** (called parathormone in older references), which increases calcium in the blood by removing it from bone tissue. It also stimulates vitamin D production, which helps with calcium absorption. Parathyroid hormone (**PTH**) also encourages the kidneys to retain calcium and helps the digestive system to absorb calcium.

✓ Concept Check

1. Describe the two hormones produced by the thyroid gland.
2. What are the relative roles of calcitonin and parathyroid hormone in the body?

Pancreas

Key Terms: glucagon, insulin, insulin receptor, islet, islets of Langerhans, pancreas

The **pancreas** is a large gland located near the stomach (Figure 7.12). The pancreas primarily produces digestive enzymes for the small intestine. However, clusters of endocrine cells within the pancreas are involved in regulating blood sugar (glucose). This makes the pancreas a dual gland with both exocrine and endocrine function. **Insulin** and **glucagon** are the two major hormones produced by these endocrine cell clusters, which are called **islets** (small islands). The islets of the pancreas that secrete insulin and glucagon are the **islets of Langerhans**. Insulin is produced by beta cells, while glucagon is produced by alpha cells. Glucose itself has hormone-like properties. Working together with insulin, glucose signals the brain that the body has eaten a meal. This helps reduce the feeling of hunger and produces a satiation, or satisfied, response.

Insulin is generated in response to high glucose levels, usually following a meal rich in carbohydrates. The primary job of insulin is to lower glucose levels by stimulating the **insulin receptor**. Insulin receptor activation causes an increase in glucose uptake, or absorption, by all body cells; however, liver and muscle cells have abundant insulin receptors. Insulin secretion speeds the conversion of glucose into glycogen, a storage carbohydrate. Glycogen in the liver helps regulate blood sugar, while glycogen in the muscles is essential as a metabolic fuel following heavy activity or anaerobic exertion. Insulin also prompts the body to store unused glucose in the form of fat cells. Improper insulin function can result in high blood glucose levels and cause blood vessel decay, dehydration, and kidney damage.

Pancreas A gland located near the stomach that produces digestive enzymes and hormones that regulate blood glucose levels

Insulin A peptide hormone produced by the pancreas that lowers blood glucose levels

Glucagon A peptide hormone produced by the pancreas that elevates blood glucose levels

Islets Clumps or islands of cells in the pancreas that secrete hormones

Islets of Langerhans Clumps of cells that secrete Insulin and glucagon

Insulin Receptor A membrane receptor activated by insulin that promotes glucose uptake, or absorption, by cells

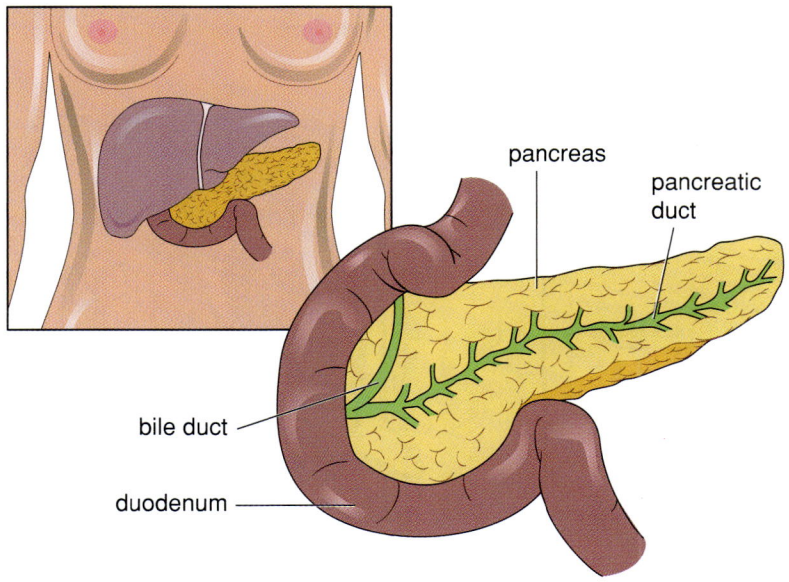

Figure 7.12 The Pancreas The pancreas is a complex collection of exocrine and endocrine glands located next to the liver and stomach.

THE ENDOCRINE GLANDS AND HORMONES

Glucagon has the opposite role of insulin. Produced in response to hypoglycemia (low blood sugar), glucagon encourages the metabolism of fats as a way of preserving existing glucose until blood sugar levels return to normal. Low blood sugar stimulates the metabolism of fats and proteins for energy, which results in the accumulation of ketones and ammonia in the blood. Both ketones and ammonia are waste products that must be metabolized by the liver and removed by the kidneys through urination. Ketones can poison body cells and produce kidney damage if they are allowed to accumulate in the blood.

✓ Concept Check

1. Describe how insulin interacts with glucagon in the blood.
2. What is the role of the insulin receptor?
3. How does glucagon affect the body's metabolism?

Thymus Gland

Key Terms: T cell, thymosin, thymus gland

Thymus Gland A gland located in the chest area that assists the immune response

Thymosin A peptide hormone secreted by the thymus gland that causes the immune system to mature

T Cells A type of blood cell that protects the body from cancer cells, foreign materials, and viruses

Do not confuse the **thymus gland** with the thyroid gland; it is a completely different organ located just above the heart. It is the most specialized of the endocrine glands (Figure 7.13). It secretes **thymosin** stimulating the development of white blood cells called **T cells**. T cells, in turn, produce several hormones that help the body fight disease. The thymus gland is proportionally larger in children than in adults. It becomes smaller as a person ages. Much of the thymus gland's job is to "educate" the immune system. It ensures that the body is able to distinguish disease organisms from body tissues. So, it is busiest when the person is young and encountering a variety of childhood diseases. The thymus of various animals is eaten as a delicacy in many cultures and is thought by some to be a source of youthful vigor.

Figure 7.13 Thymus Gland The thymus is located just above the heart. Its hormones play an important role in fighting disease.

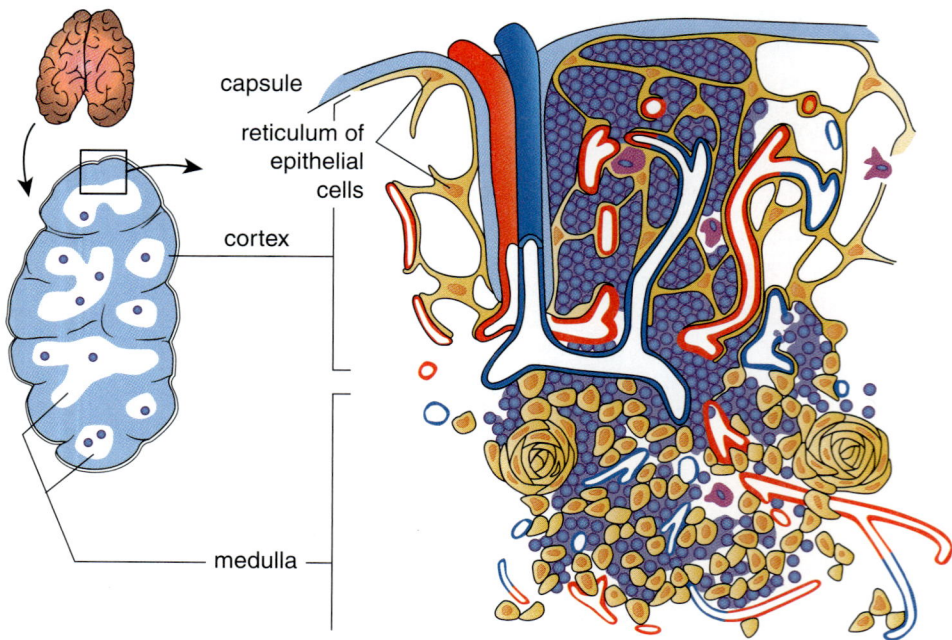

✓ Concept Check

1. What is the function of the thymus gland?
2. What part of the immune system does thymosin stimulate?

Gonads

Key Terms: estrogen, gonads, ovary, progesterone, testes, testosterone

In the young fetus, the **gonads** are no more than modified outgrowths of the kidneys, able to create sexual reproductive cells called gametes. Early in fetal development, the DNA determines whether the structure called an uncommitted gonad will develop into an **ovary** or a **testis** (Figure 7.14). People with two X chromosomes develop ovaries. In those with one X and one Y chromosome, the gonads convert into testes. Ovaries produce gametes called eggs, and the testes produce gametes called sperm.

Ovaries and testes are recipients of hormones, in addition to being involved in hormone production. Follicle-stimulating hormone, produced by the anterior pituitary, promotes the formation of eggs and sperm. Leuteinizing hormone, also produced by the anterior pituitary, enables the maturation of eggs and sperm in preparation for reproduction. Ovaries produce two steroid hormones, **estrogen** and **progesterone**. Estrogen is the major hormone of the ovary. It is primarily responsible for providing the sexual characteristics of a female. It is also essential for bone maintenance. The adrenal gland of females is able to convert steroids into male hormones, which work with estrogen to promote muscle tone. Progesterone functions similarly, and works with estrogen to produce the menstrual cycle and induce changes in the body during pregnancy.

Gonad An organ of the reproductive system that produces eggs or sperm

Ovary A gonad that produces eggs in females

Testis A gonad that produces sperm in males

Estrogen A group of sex hormones produced by the ovary

Progesterone A female steroid sex hormone secreted by the ovary, which prepares the reproductive tract for pregnancy

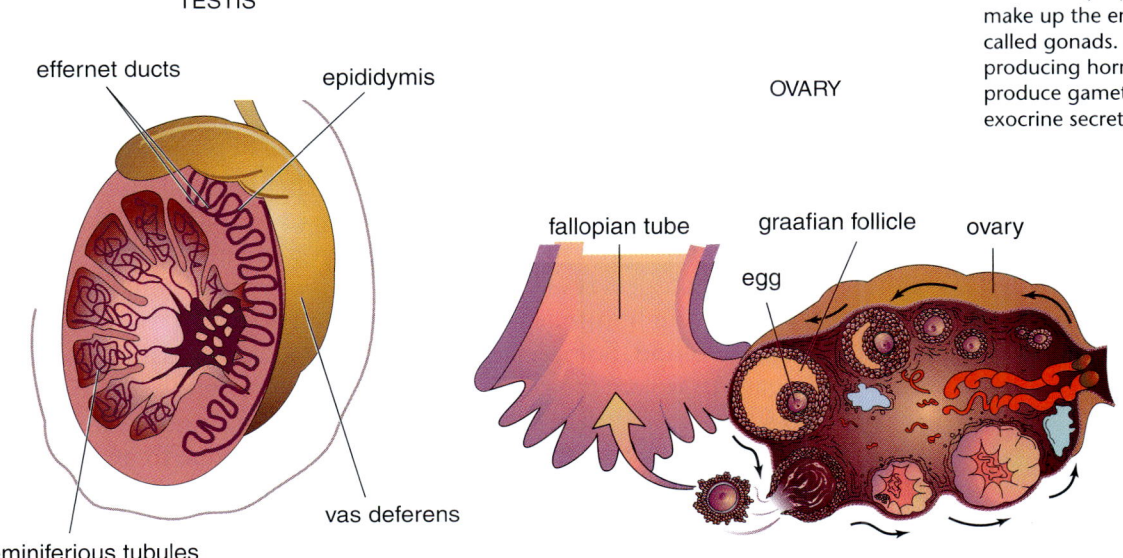

Figure 7.14 **Testes and Ovary** The testes (left) and ovary (right) make up the endocrine glands called gonads. Aside from producing hormones, the gonads produce gametes and a variety of exocrine secretions.

THE ENDOCRINE GLANDS AND HORMONES

Testosterone A hormone that promotes the development and maintenance of male sex characteristics

Testosterone is the primary hormone produced by the testes. It produces male sexual characteristics and ensures sperm maturation. Its most dramatic effect is during puberty when it promotes rapid body growth and significant development of the male reproductive system. Testosterone is responsible for muscle development by helping the muscle cells take up nutrients and build proteins needed for muscle growth. This fact has encouraged some athletes to abuse drugs called anabolic steroids. These steroids are made up of a variety of chemicals resembling testosterone. Unfortunately, high levels of testosterone can induce brain cancers and cause cardiovascular problems. Many males may be affected by certain types of baldness caused by the response of hair cells to testosterone. These cells stop producing hair when testosterone levels rise after puberty. This type of balding starts when a person is about 20 years old and continues until the age of about 40 years.

✓ Concept Check

1. Name three hormones produced by the gonads.
2. What is the role of testosterone in the body?
3. Which pituitary hormone assists the gonads with gamete production?

DISCOVERY SCENE PLEASE ENTER DISCOVERY SCENE PLEASE ENTER

Can you determine which of the endocrine glands may be causing the girl's problems? Is it one gland, or can it be several glands, that are responsible for her disease?

Wellness and Illness over the Life Span

Pathology of the Endocrine System

Key Terms: acromegaly, Addison's disease, Cushing's syndrome, diabetes insipidus, diabetes mellitus, Graves' disease, hyperparathyroidism, hypothyroidism

Diseases of the endocrine system are so numerous that they fill volumes of medical books. However, there are a handful of diseases that physicians come upon regularly and for which they have developed standard effective treatments. Endocrine disorders result from either the overproduction or underproduction of one or more hormones. This is easily detectable by monitoring hormone levels in the blood. Sometimes it is necessary to collect samples of the endocrine organs to isolate the cause of the abnormal hormone production. It is not always simple to figure out why an endocrine gland is not doing its job. For example, a decrease in hormone production might be due to an inability to detect signals, a lack of signals, diminished blood flow to the gland, or diseased endocrine cells. Sometimes tumors can cause the production of too little or too much of a given hormone. Some common endocrine disorders are briefly discussed in Table 7.3.

Table 7.3 Common Endocrine Disorders

	Description	Effects
acromegaly	An increase in growth hormone production that occurs in adulthood.	Produces thickening of the skull and enlargement of the hands, feet, and tongue, and many body organs.
Addison's disease	A decrease in adrenal cortex hormones brought on by malfunctions of the adrenal glands or the pituitary gland.	Results in darkening of the skin due to excess melanin production, dehydration, low blood pressure, low blood sugar, and sodium loss.
Cushing's syndrome	Overproduction of adrenal cortex hormones brought on by malfunctions or tumor of the adrenal glands, or excess adrenocorticotropic hormone (ACTH) secretion of the pituitary gland.	Results in high blood pressure, sodium retention, swelling of body tissues, and water retention.
diabetes insipidus	Inadequate production of antidiuretic hormone, brought on by a malfunction of the posterior pituitary gland.	Produces extreme water loss resulting in frequent urination. The person has to drink large volumes of water to prevent dehydration and to fight the constant feeling of thirst.
diabetes mellitus	This is the commonly known form of diabetes, which is usually caused by decreased insulin production (Type I) or the body's inability to detect insulin signals (Type II). Many things may effect diabetes mellitus, including childhood viral infections, high carbohydrate diets, and obesity.	It results in high blood sugar and a variety of disorders due to abnormal glucose levels. A condition called glycosylation occurs when blood sugar is too high. During glycosylation, the excess sugars stick to various tissues, causing the immune system to destroy the affected tissues. This can ultimately lead to blindness, blood vessel destruction, and kidney failure.
Graves' disease	Inflammation of the thyroid gland due to elevated thyroid hormone production (hyperthyroidism). It is caused by an autoimmune disease.	Results in an elevated metabolic rate and feelings of nervousness or tension. People with this condition usually feel tired throughout the day.
hyperparathyroidism	Overproduction of parathyroid hormone caused by immune system disorders, kidney diseases, parathyroid tumors, pregnancy, or malfunctions of the parathyroid gland. It produces elevated blood calcium levels and, in turn, calcium loss from the bone.	Results in kidney problems and a weakening of the bones. Nervous system and muscle malfunctions are also attributed to too much parathyroid hormone.
hypothyroidism	A condition in which the thyroid gland does not produce enough thyroxine. In children, it is due to a genetic defect. In adults, it is due to a thyroid or pituitary gland malfunction.	In children, it results in mental retardation and short stature. In adults, it causes lethargy, weight gain, dry hair and skin, and sensitivity to cold.

✓ Concept Check

1. What is the difference between diabetes insipidus and diabetes mellitus?
2. Describe two diseases of the thyroid gland.
3. Describe two diseases of the adrenal cortex.

Cutting Edge Research
THYROID HORMONE TO THE RESCUE

Veterinary researchers at the University of Bologna, in Italy, found a new medical use for thyroid hormone. In their 2005 research paper entitled *Thyroid hormone and remyelination in adult central nervous system: a lesson from an inflammatory-demyelinating disease*, they reported how thyroid hormone can undo nerve damage. The researchers were looking for ways to reverse the nerve decay of a disease called multiple sclerosis (MS). In MS, myelin, a nerve cell covering, disappears. Myelin is needed for the nerves to conduct nerve information throughout the body. Eventually, people with MS lose many body functions. Plus, the lack of myelin ultimately causes the nerve cells to deteriorate. By studying nerve development in young animals, the researchers learned that thyroid hormone has a role in nervous system formation. They were then able to use thyroid hormone to stimulate the growth of the oligodendroglia cells that give rise to myelin. Using these findings, they hope to find the correct applications of thyroid hormone needed to rejuvenate myelin in people with MS.

Aging of the Endocrine System

Key Terms: hormone replacement therapy (HRT)

The activity of the endocrine system of children varies greatly with the age of the child. It begins functioning in the fetus to produce the rapid growth needed over the developmental period. Growth hormone, insulin, and thyroxine are particularly important in growth. The thymus starts out very small and grows until a person reaches puberty. It becomes smaller and less active as a person ages.

The sex hormones are formed at about 13 weeks of development to assist the formation of the sexual characteristics. However, the levels of sex hormones reach their highest at puberty and then taper off after age 30 years for males and in the 40s for females. Hormone levels in children usually fluctuate with growth. Any abnormalities in a child's hormones during critical growth periods can cause permanent damage to the body. In particular, children are highly sensitive to small amounts of chemicals that act like sex hormones. Boys exposed to abnormally high levels of estrogen will develop defects in the genitals and may have diminished sperm production. Girls with higher estrogen levels may enter puberty earlier and are thought to be subject to breast cancer. In some children, defective blood vessels may cut off blood flow to a particular body part. This decreases the amount of hormones reaching that body part, leading to a reduction in its size and function.

Much of the endocrine system's aging can be attributed to a decrease in hormone production due to a natural decrease in size of the endocrine glands. Shrinkage of the endocrine glands can be accelerated in people with cardiovascular problems and diabetes.

Human aging is usually accompanied by diminished blood flow through the capillaries. Many of these capillaries provide the endocrine glands with the atmospheric gases, hormones, and nutrients needed for hormone production. Thus, the reduction in capillary action decreases the supply of hormones to the target cells. The uptake of nutrients by the digestive system is another factor leading to endocrine system aging. Elderly people absorb fewer amounts of the lipids and proteins that the endocrine organs need for hormone production. Each endocrine gland ages individually. However, those controlled by the pituitary gland age faster, as the pituitary gland loses some of its functions with age.

Many people seek **hormone replacement therapy** (**HRT**) as a way to counteract aging of the endocrine system. Estrogen is the most common hormone used in HRT. Some individuals use "natural" hormone replacement rather than prescription medications. For example, chemicals similar to estrogen are found in a variety of plants. Soybeans have higher levels of estrogen than any other food. Some women consume them in hopes of reducing the side effects of diminished estrogen production during menopause. Some scientists are not sure about the beneficial effects of HRT. They feel that it is difficult for a persons to properly regulate their hormone levels using oral supplements. Recent experiments show few benefits in many people for various age-related changes. A series of long-term studies called the Rancho Bernardo Studies were conducted on hundreds of middle-age and elderly people. In one group of studies, most of the data showed that there were no significant differences between estrogen supplement users and nonusers for metabolic changes leading to organ system aging and obesity. However, many women had fewer reproductive-tract problems and showed a decrease in osteoporosis with estrogen therapies.

Hormone Replacement Therapy (HRT) Treatment to replace hormones

✓ Concept Check

1. What is the role of blood flow in the aging process of the endocrine system?
2. What happens to the secretion of many hormones from birth until old age?
3. What is hormone replacement therapy (HRT)?

DISCOVERY SCENE PLEASE ENTER DISCOVERY SCENE PLEASE ENTER

What other information have you gathered after reading about diseases of the endocrine system? Does the girl have too much or too little of a particular hormone? Did the information about aging of the endocrine system provide any more clues about the girl's condition?

CSI – Case Study Investigation Conclusion

What can you conclude about the girl's condition after reading the information in this chapter? Is it evident which part of the endocrine system is not functioning properly? What makes the function of her endocrine system unusual for a young female?

Answer:

The girl is suffering from a condition called virilism. She has too much testosterone in her body. This is why she is showing male characteristics. The problem for the physicians is to determine where the testosterone is coming from. Is it entering her body through diet or exposure to chemicals that act like

testosterone? Or, is it due to abnormal production of testosterone by the endocrine system? Upon superficial examination, it could be due to one of two probable conditions: a tumor of the ovary that produces testosterone, or overproduction of testosterone by the adrenal glands. The physicians would have to remove pieces of each endocrine gland to examine in the laboratory. In this case, the girl has tumors on her adrenal glands. These tumors are producing large amounts of testosterone.

This CSI was adapted from the following articles:

1. Powell JL, Dulaney DP, Shiro BC. Androgen-secreting steroid cell tumor of the ovary. *South Med J.* 2000;93(12):1201-1204.
2. McKenna TJ, Cunningham SK, Loughlin T. The adrenal cortex and virilization. *Clin Endcrinol Metab.* 1985;14:997-1020.

Chapter Summary

It helps to envision the endocrine system as a bunch of friends with e-mail. Each friend has personal e-mail, which is very much like the receptors on a cell. One friend is in charge of coordinating a party, so she sends her friend Jane an e-mail telling her to buy cake. Think of the e-mail as a hormone, and Jane as a specific target cell. The instruction to buy a cake is the metabolic change expected of Jane. Jane then goes about getting a cake that she feels all of the other friends would like. Her knowledge of the likes and dislikes of her friends is similar to the overall function of the body. In actuality, the endocrine system is a complex collection of hormones that coordinate many of the body's functions. It coordinates daily activities, such as blood pressure and eating. But, it is also responsible for long-term changes related to aging and puberty.

Overview
- The endocrine system is composed of glands that produce endocrine secretions.
- Endocrine secretions go directly into the blood.
- Exocrine secretions enter body cavities.
- Endocrine secretions are cellular signals.

Hormone Function
- Hormones work by attaching to receptors on target cells.
- Ligands carry chemicals that attach to receptors.
- Carrier proteins transport certain hormones.
- Effectors are cells that change in response to a signal.

Endocrine Secretions
- Endocrine glands are ductless.
- Target cells have receptors that bind to specific endocrine secretions.
- Autocrine secretions target the cells that produce them.
- Paracrine secretions target nearby cells.
- Endocrine secretions usually target distant cells.
- Most endocrine secretions control the body through negative feedback.

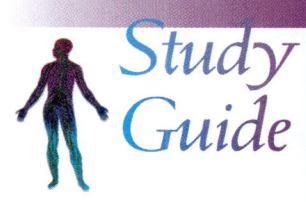

Types of Hormones

- A hormone is any secretion that signals a cell to alter its metabolism.
- Chemicals that carry out the job of a hormone by turning on a cell response are called agonists.
- Chemicals that carry out the job of a hormone by turning off a cell response are called antagonists.
- There are two types of hormones: peptide and lipid.
- Peptide hormones are usually involved in rapid body changes.
- Lipid hormones play a role in body fluid control and sexual reproduction.

The Human Endocrine Glands

- There are 10 distinct endocrine glands.
- The pituitary gland, or hypophysis, is known as the master gland.
- The pituitary gland is composed of the anterior and posterior pituitaries.
- Chemicals called releasers, or releasing hormones, are secretions of the hypothalamus, and they control the anterior pituitary gland.
- Nerve cells from the hypothalamus control the posterior pituitary.
- The pineal gland produces melatonin and serotonin.
- The adrenal gland is divided into an outer region called the adrenal cortex and an inner part called the adrenal medulla.
- The thyroid gland helps control metabolic rate.
- The pancreas produces hormones and digestive enzymes.
- The beta cells of the pancreas produce insulin.
- The alpha cells of the pancreas produce glucagon.
- The thymus gland produces secretions that stimulate the immune system.
- The gonads produce hormones and gametes.
- The female gonads, or ovaries, produce estrogen and progesterone.
- The male gonads, or testes, produce testosterone.

Pathology of the Endocrine System

- Diseases of the endocrine system can cause too much or too little hormone secretion.

Aging of the Endocrine System

- Changes in hormone production contribute to aging.
- Most hormones decrease in amount as adults age.
- Some individuals seek hormone replacement therapy in an attempt to counteract the effects of aging of the endocrine system.

Study Guide

Key Terms

Overview
Ductless gland
Endocrine gland
Endocrine secretions
Endocrine system
Environmental signal
Exocrine gland
Exocrine gland secretion
Hormone
Receptor
Signal
Target cell

Hormone Function
Effector
Internal receptor
Ligand
Surface receptors

Endocrine Secretions
Autocrine
Negative feedback
Paracrine
Pheromones
Thyroxine

Types of Hormones
Agonist
Antagonist
Lipid hormone
Peptide hormone
Protein hormone

Pituitary Gland
Adrenocorticotropic hormone
Anterior pituitary
Antidiuretic hormone
Estrogen
Follicle-stimulating hormone
Growth hormone
Hypophysis
Luteinizing hormone
Melanocyte-stimulating hormone
Oxytocin
Pituitary gland
Posterior pituitary
Progesterone
Prolactin
Releaser hormones, releasers
Thyroid-stimulating hormone
Testosterone

Pineal Gland
Melatonin
Serotonin
Pineal gland

Adrenal Glands
Adrenal cortex
Adrenal gland
Adrenal medulla
Adrenaline
Aldosterone
Angiotensin II
Cortisol
Epinephrine
Glucocorticosteroids
Mineralocorticosteroids
Noradrenaline
Norepinephrine

Thyroid Gland and Parathyroid Gland
Calcitonin
Parathyroid gland
Parathyroid hormone
Thyroid gland
Thyroid hormone
Thyroxine (T4)

Pancreas
Glucagon
Insulin
Insulin receptors
Islets
Islets of Langerhans
Pancreas

Thymus Gland
T cell
Thymosin
Thymus gland

Gonads
Estrogen
Gonads
Progesterone

Ovary
Testis
Testosterone

Pathology of the Endocrine System
Acromegaly
Addison's Disease
Cushing's Syndrome

Diabetes insipidus
Diabetes mellitus
Graves' Disease
Hyperparathyroidism
Hypothyroidism

Aging of the Endocrine System
Hormone Replacement Therapy (HRT)

Check Your Understanding

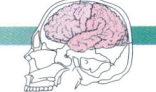

1. Secretions sent directly into the blood are called:
   a. exocrine secretions
   b. endocrine secretions
   c. ducted secretions
   d. environmental signals

2. Which of the following is not considered to be a true endocrine organ, but contains hormone-producing endocrine cells?
   a. pancreas
   b. adrenal glands
   c. pituitary gland
   d. stomach

3. Digestive enzymes are an example of:
   a. exocrine secretions
   b. endocrine secretions
   c. ductless secretions
   d. environmental signals

4. Secretions produced by ductless glands are called:
   a. enzymes
   b. digestive fluids
   c. hormones
   d. excretions

5. External receptors that bind hormones are called:
   a. intrinsic receptors
   b. internal receptors
   c. ligands
   d. surface receptors

6. Carrier proteins transport hormones to these types of receptors:
   a. internal receptors
   b. surface receptors
   c. nicotinic receptors
   d. slow receptors

7. Secretions that travel to nearby target cells are called:
   a. autocrine
   b. paracrine
   c. pheromones
   d. exocrine

THE ENDOCRINE GLANDS AND HORMONES

Study Guide

8. Most hormones use _____ communication to carry out their "jobs" for the body.
   a. physical
   b. negative feedback
   c. positive feedback
   d. intermediary

9. Cells that respond to hormones are called:
   a. target cells
   b. binders
   c. receptor cells
   d. ligands

10. Which statement is true of peptide hormones?
    a. They bind to internal receptors.
    b. They do not dissolve in the blood.
    c. They are secreted into ducts.
    d. They bind to surface receptors.

11. Which statement is true of lipid hormones?
    a. They bind to internal receptors.
    b. They dissolve readily in the blood.
    c. They are secreted into ducts.
    d. They bind to surface receptors.

12. Which of the following hormones is a steroid?
    a. estrogen
    b. glucagon
    c. thyroxine
    d. insulin

13. The hypothalamus has a dual purpose in the body for the following reason:
    a. It sends nerve impulses and also makes hormones.
    b. It is both a nervous and excretory organ.
    c. It belongs to both the nervous and circulatory systems.
    d. It communicates messages between paired endocrine organs.

14. Acromegaly is caused by the following:
    a an increase in thyroid activity
    b. an increase in growth hormone production
    c. a decrease in insulin production
    d. a decrease in thyroxine levels

15. Aging is usually accompanied by the following event:
    a. increased hormone production
    b. decreased blood flow to endocrine glands
    c. diabetes insipidus
    d. hyperthyroidism

A Case Study

Study Guide

ENVIRONMENTAL HORMONES

Imagine what Dr. Louis Guillette was thinking in 1990 when he went to investigate a curious incident at Lake Apopka in Florida. All the resident alligators in the lake were sexually deformed. Most interesting was that something caused the male population to have genitals resembling those of female alligators. Based on Guillette's past research, he surmised that something was interfering with these alligators' endocrine systems. Before coming to Florida, he studied a group of pollutants called endocrine disrupters. These chemicals, collectively called environmental hormones, cause developmental and reproductive problems in a variety of organisms. Surprisingly, almost all of the chemicals he studied interacted in the body in the same way as estrogen. Endocrine disrupters polluting the environment were named environmental estrogens. Chemicals categorized as endocrine disrupters bind to the estrogen receptor. Depending on their chemistry and concentration, they either act like estrogen or inhibit the action of estrogen.

Endocrine disrupters, like any hormone, cause significant changes to the body when consumed in small amounts. They can act as either agonists or antagonists. Barely detectable levels of environmental estrogens can render many of the animals in a lake reproductively sterile. This has been shown to be irrefutably true for fish in Minnesota streams and frogs in California, Louisiana, and Texas. Health officials and scientists are concerned that continuous exposure to endocrine disrupters may cause developmental problems in children and reproductive troubles in adults. There is significant evidence that endocrine-disrupter pollution may be responsible for an increase in the incidence of breast cancer and a decrease in sperm counts in humans. Deviations of human penis development in a significant number of males in Japan and other nations are another indicator of endocrine-disrupter pollution. Some people are intentionally taking in endocrine disrupters though food and dietary supplements. Many plants produce chemicals called phytoestrogens, which are very similar to estrogen.

Use the information in this chapter and in the following Web sites to answer the following questions about environmental hormones:

- What types of chemicals act as endocrine disrupters, and what types of hormones do they interfere with?
- How do endocrine disrupters affect the human body?
- Should the government control and monitor endocrine-disrupter pollution?
- What, if any, actions should the government take to make people aware of endocrine disrupters?
- Would it be advisable for the government to regulate the sale of dietary supplements and foods containing endocrine-disrupting chemicals and hormones?

1. e.Hormone
   http://e.hormone.tulane.edu/

2. Green Facts – Endocrine Disrupters
   http://www.greenfacts.org/endocrine-disruptors/endocrine-disrupters.htm

THE ENDOCRINE GLANDS AND HORMONES

Study Guide

3. National Institutes of Health – Hormones and Health
   http://ehp.niehs.nih.gov/qa/105-5focus/focus.html

4. Institute of Food Science & Technology – Phytoestrogens
   http://www.ifst.org/hottop34.htm

Where Do We Go from Here?

People in health fields can use their knowledge of the endocrine system to solve everyday problems. You may wish to use other resources in addition to you textbook to investigate the answers to each of the following situations:

1. A pollutant was discovered that acts like estrogen in the body. What effects would this have on pregnant women?

   e.Hormone
   http://e.hormone.tulane.edu/

2. You are working with a person who is recovering from a car wreck. His pancreas was severely damaged and is not functioning completely. What dietary restrictions would he expect to have as a result of the accident?

   EMedicine
   http://www.emedicinehealth.com/articles/37539-1.asp

3. You heard a friend talking about a child who is being treated with human growth hormone and human insulin produced by genetic engineering. For what diseases is this child likely being treated? Use the Internet to find out why the physician preferred to use human hormones produced by genetic engineering.

   Kids Health
   http://kidshealth.org/parent/general/body_basics/endocrine.html

4. You were asked if it were possible to take cortisol injections as a treatment to help with weight loss. How would this treatment be effective, and what are its risks?

   WebMD
   (http://www.webmd.com/

5. A runner came across an herbal tea whose manufacturers claim that it increases thyroxine production in the body. What would be the benefits and risks of taking this tea, assuming it does what it claims?

   Medline Plus
   http://www.nlm.nih.gov/medlineplus/endocrinesystemhormones.html

Chapter 7

Skills Activities

1. Microscopic Identification of Normal Endocrine Glands

Materials
- microscope with high-power capability (400X)
- prepared slide of the pancreas
- prepared slide of a thyroid gland
- prepared slide of an adrenal gland

Histologists and pathologists look at microscopic preparations of endocrine glands to help diagnose disease. They compare normal slides with the tissue under examination. From this they are able to determine any visible differences that may cause an abnormality. However, before they can do this, they must practice comparing the normal samples with pictures to help them determine the different components and cell types in the particular gland.

Place a microscopic slide of the adrenal gland under the microscopic. Start with the low magnification, and move to the higher magnification to see details and individual cells. Can you find the hormone-secreting cells after comparing the slide with a photograph taken through the microscope? How do the hormone-secreting cells differ from other cells of the adrenal gland?

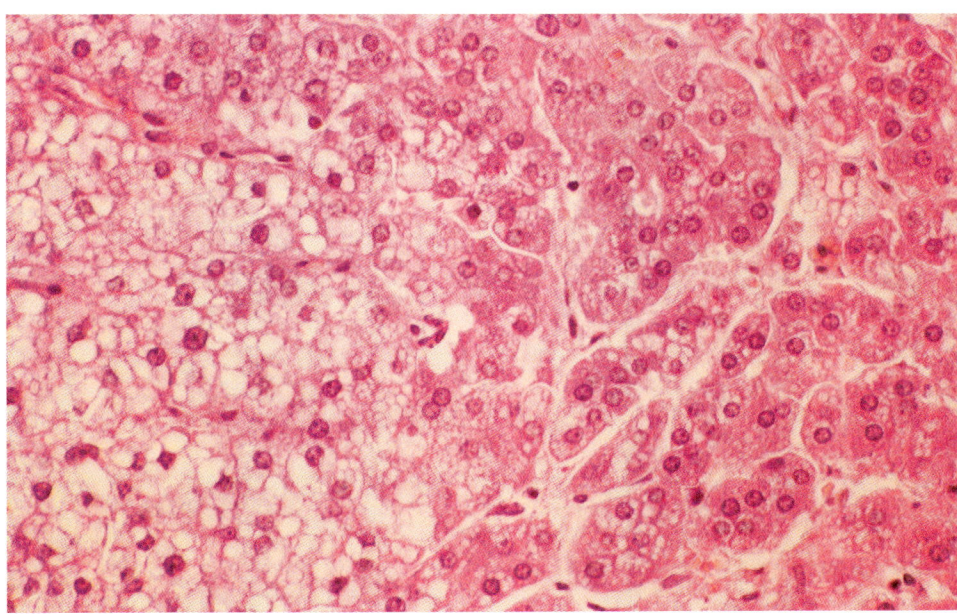

Adrenal Gland

Next, look at a prepared slide of the pancreas under the microscope. Can you identify the endocrine and exocrine cells making up this gland? What major structures are located near the hormone-secreting cells of the gland?

THE ENDOCRINE GLANDS AND HORMONES

Study Guide

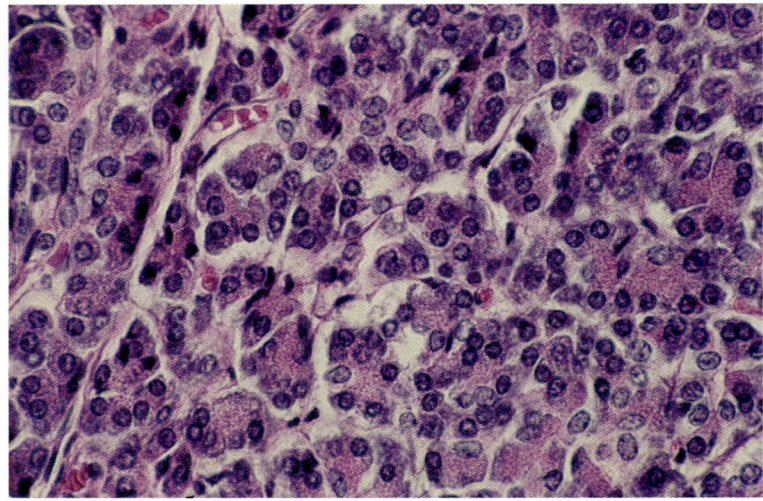

Pancreas

Now, look at a specimen of the thyroid gland under the microscope. How do the hormone-secreting cells of the thyroid differ from those of the pancreas? What would you expect the thyroid to look like if the patient had hyperthyroidism? What would you expect the gland to look like in a patient with hypothyroidism?

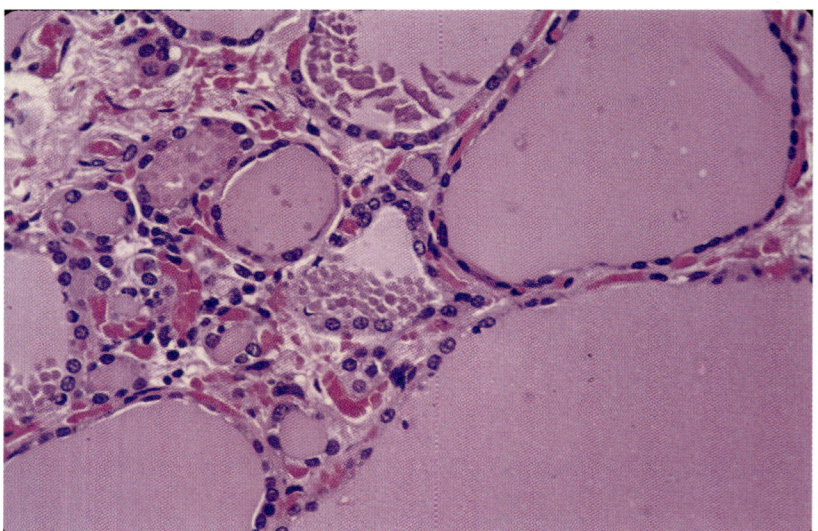

Thyroid Gland

Skills Activities

2 Effects of Adrenaline and Caffeine on Daphnia

Materials
- daphnia kept in a container of clean fresh water
- a large container of clean water for used daphnia
- droppers for collecting and transporting the daphnia
- microscope

- droppers for collecting samples of epinephrine, caffeine, and coffee or tea
- three clean microscope slides per observation
- over-the-counter epinephrine preparation (asthma pills) soaked in 100 mL of a 50% ethyl alchohol/50% water mixture
- 100-mg caffeine pill soaked in 100 mL of a 50% ethyl alchohol/50% water mixture
- half-strength coffee or tea
- surgical gloves (optional)

Adrenaline is a hormone found in almost all animals. It basically has the same effects on human metabolism as it does on that of other animals. This laboratory activity looks at the effect of adrenaline and caffeine on the endocrine control of metabolic rate. The metabolic rate of small animals, called daphnia, will be used as a model of the human endocrine response. Daphnia are minute freshwater organisms related to crabs and shrimp. They have a round body enclosed in a transparent shell. Animal models such as the one being conducted in this activity are very important in obtaining preliminary laboratory data on humans.

Use a dropper to place some daphnia on the microscope slide, and observe their normal behavior. Make sure you keep adding water to the slide to keep them from drying out and dying. Keep track of the speed at which they move around and how fast they move their feet. First, add one drop of the epinephrine solution to the daphnia. Note what happens to their movement. Place the daphnia in the used daphnia container, and place another set of daphnia on a new clean slide. Now, repeat the steps for the caffeine pills. Record your observations. Again, recycle the daphnia, and collect another batch to record the effects, on a drop of coffee or tea on the daphnia. What effects does the epinephrine have on the daphnia? How does the effect of the caffeine pill compare with adrenaline? How does the effect of tea or coffee differ from that of epinephrine and caffeine? What else is present in coffee or tea that might affect the metabolic rate of animals?

Precautions It is recommended that you wear surgical gloves when handing the epinephrine and caffeine solutions. All of the solutions can be stored in labeled bottles in the refrigerator for 2 or 3 weeks. Excess solution should be flushed down a drain.

Note Brine shrimp or other small invertebrates can be used in place of daphnia. Brine shrimp must be kept in the salt solution used to raise them. So, remember not to use fresh water on the slides used to observe brine shrimp.

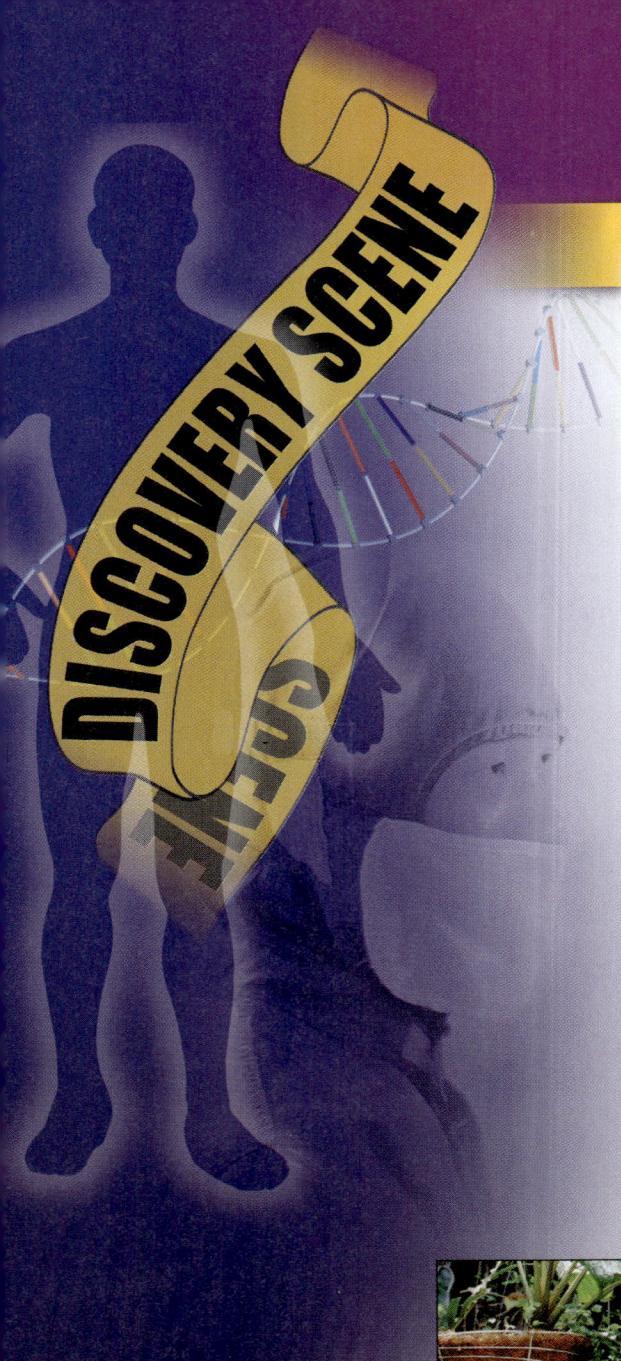

Case Study Investigation

Case Study Investigation #8

Imagine that you are at a friend's backyard party when her uncle starts to stumble, and complain of dizziness and tiredness. It looks like he is having difficulty breathing. He mentions that he was spraying for insects in the front yard and may have breathed in too much spray. He sits and relaxes, but does not feel any better after 15 minutes. His blue complexion provides a good signal that you should take him to the hospital. On the way to the hospital, his muscles go from tense to limp. Now his breathing has become shallow, and his speech is slurred. The emergency-room physician at the hospital finds out that your friend's uncle had a busy day before the party. He was involved in cleaning algae out of the large pond behind the house before spraying the yard. He ate some old beef stew that was sitting in the refrigerator. The stew did not look or smell bad. Your friend is very concerned about her uncle and asks you to explain what the physicians are looking for as the cause of her uncle's illness. You explain that the physicians must use a process of elimination to determine what caused her uncle's problems. She asks you to tell her the possibilities. At the end of the chapter, you will be asked to determine the possible nervous system problems causing the illness.

CHAPTER 8
Function of the Nervous System

Chapter Outline

Case Study Investigation (CSI)
Applied Learning Outcomes
Overview
Types of Nervous System Cells
    Neurons
    Neuroglia and Stem Cells
Neuron Physiology
Types of Neuron Communication
Reflexes
Wellness and Illness over the
  Life Span
    Pathology of the Nervous
      System
    Aging of the Nervous System
CSI Conclusion
Study Guide

Applied Learning Outcomes

- Use the terminology associated with the nervous system.
- Learn about the following:
  - different types of nervous system cells
  - nervous system development
  - nerve cell structure
  - nerve cell function
  - nerve cell physiology
- Understand the biological basis of aging and the pathology of nerve cell function.

Overview

Key Terms: external stimuli, internal stimuli, neural tube

External Stimuli A signal from outside the body that generates nervous system activity

Internal Stimuli A signal from inside the body that generates nervous system activity

Neural Tube A structure in early fetal life that develops into the nervous system

Organ system coordination would not progress at a feasible level without the function of the nervous system. It communicates critical information between the various organ systems and provides information about environmental conditions to all internal organs. It has the capability of translating environmental stimuli into messages understood by the cells. The nervous system is best defined as a system of cells, tissues, and organs that regulates the body's responses to **external** and **internal stimuli**. External stimuli are environmental factors that influence metabolic changes in a cell or physiological changes in tissues and organs. Internal stimuli include a variety of cell secretions used to communicate information about a cell's jobs and needs.

The nervous system starts out as a hollow **neural tube** of ectoderm that runs lengthwise along the back of the developing fetus. This tube produces stem cells that form the two main cell lines of the nervous system: neurons and neuroglia (Figure 8.1a). These two types of cells work together to carry out the jobs of the nervous system (Figure 8.1b). Neurons are the operational component of the nervous system. They are excitable cells that receive, interpret, and transmit external and internal stimuli. Neuroglia are not directly involved in nervous system communication; rather, they are supportive cells. Their major job is to maintain the excitability and health of neurons. Neuroglia are also important for the development and repair of the nervous system. This chapter focuses on the structure and function of nervous system cells. The way in which they are integrated into the anatomy of the nervous system is described in Chapter 9.

Civic Responsibility

CIVIC RESPONSIBILITY: HELPING OTHERS WITH YOUR KNOWLEDGE

It is valuable to use what you have learned about the nervous system to help others to better understand the world around them. It is very important to check your facts and seek further information about certain topics before discussing health and science issues. Here are some suggestions to foster a better public awareness of the nervous system:

1. Speak to schoolchildren about the function of their nervous system.
2. Work with sports clubs to educate the players about factors that affect nervous system health and performance.
3. Help the elderly persons to better understand how maturity affects the nervous system.
4. Assist a child with a science-fair project or a report about the human nervous system.

Figure 8.1 **Types of Nervous System Cells**
(a) Development of nervous system cells
(b) Neuron structure

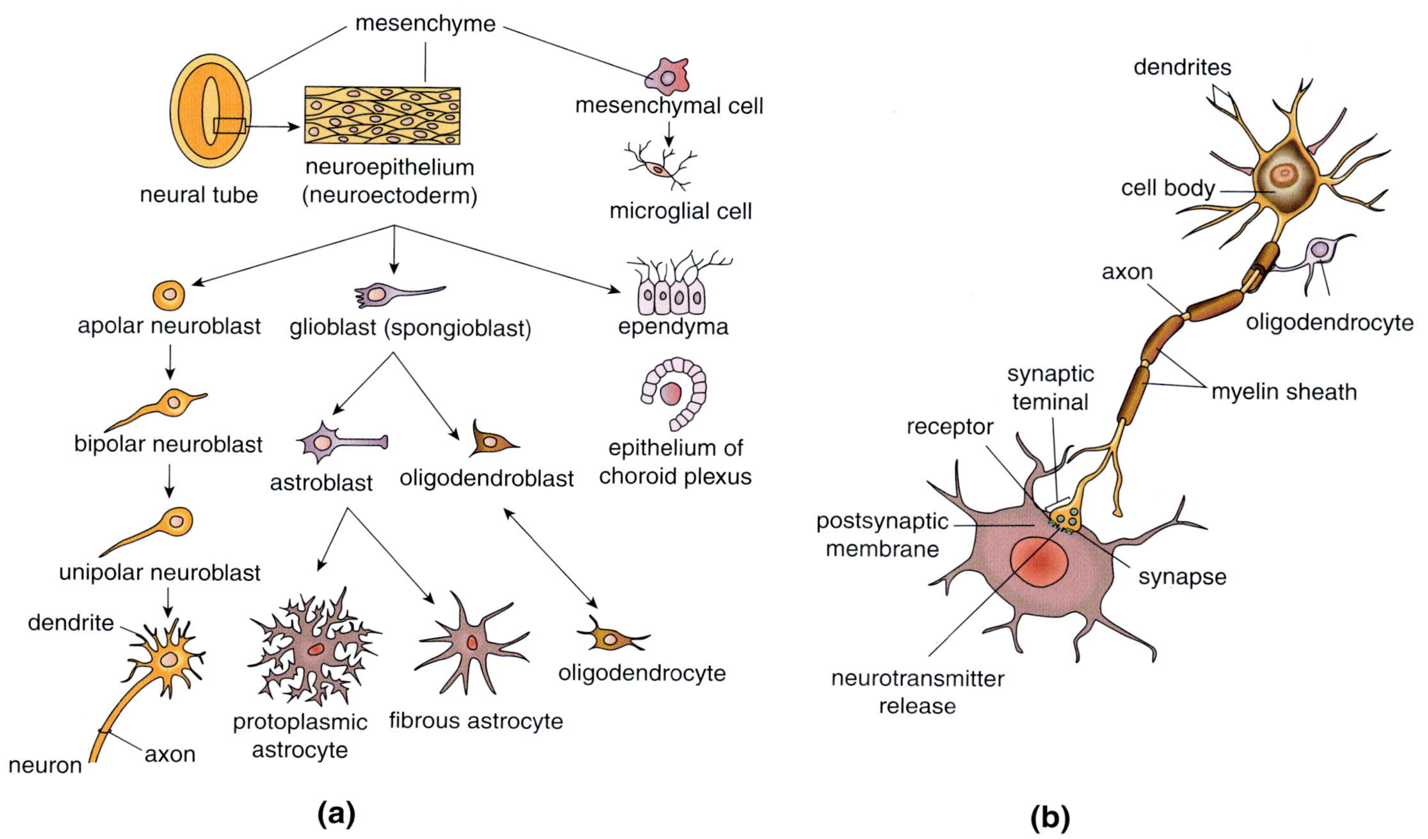

(a)

(b)

✓ Concept Check

1. What are two main functions of the nervous system?
2. Describe the function of neurons and neuroglia.
3. Distinguish between external and internal stimuli.

TYPES OF NERVOUS SYSTEM CELLS

Key Terms: bidirectional communication, neural crest cell

In addition to neurons and neuroglia, there is evidence of a third group of cells called **neural crest cells**. Neural crest cells are derived from the embryonic neural tube. Neurons develop simultaneously in the embryo with the neuroglia. Both start out with the ability to communicate with each other. This **bidirectional communication** between neuroglia and neurons permits the formation and maintenance of the nervous system. Neurons are categorized by their cell anatomy and mode of communication. Neuroglia are classified by how they assist nerve cells. Neural crest cells play a role in the development of the nervous system.

Neural Crest Cells Fetal cells that are derived from the neural tube

Bidirectional Communication Cell signals that pass in either direction between two cells

FUNCTION OF THE NERVOUS SYSTEM

Neurons

Key Terms: axon, axon hillock, bipolar, collaterals, dendrite, interneuron, multipolar, nerve cell body, neurotransmitter receptor, presynaptic, postsynaptic, soma, synaptic cleft, terminus, unipolar

Nerve Cell Body The part of the neuron containing the nucleus

Dendrite Extensions of the nerve cell body that carry information to the nerve cell body

Axon The fiber-like extension of a neuron that sends information to terminus

Terminus The part of the neuron that transfers information to an adjacent neuron or muscle cell

Soma Another name for the nerve cell body

Neuron structure was a mystery to science until Santiago Ramon y Cajal of Spain performed detailed examinations of animal nervous systems in the 1880s. By studying the brain, Cajal observed that the "stringy" substance making up the nervous system was actually a linkage of cells that communicate to each other. German anatomist Heinrich Waldeyer coined the term "neuron" in 1891 while describing Cajal's findings. Today scientists know that there are a variety of neurons that have four common features: the **nerve cell body**, **dendrite**, **axon**, and **terminus**. Neurons are primarily identified by the arrangement of these four components (Figure 8.2). The nerve cell body, or **soma**, houses the nucleus and many of the organelles needed for neuron function. A neuron's

Figure 8.2 **Neurons**

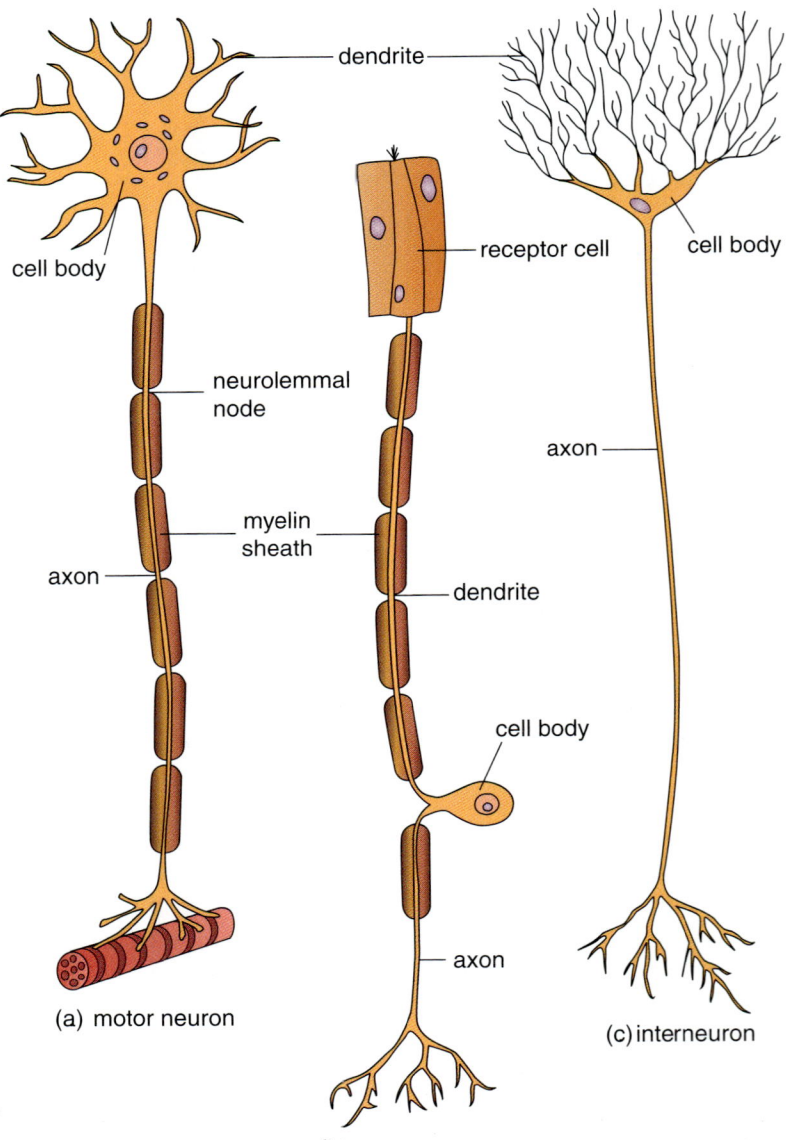

(a) motor neuron

(b) sensory neuron

(c) interneuron

body is filled with endoplasmic reticulum and Golgi bodies that produce the specialized enzymes and secretions needed for nerve cell communication.

Dendrites are the "antennae" of the neuron. They receive a majority of the stimuli that communicate with the neuron. The nerve cell body and axon can also receive stimuli, but the dendrites permit the communication of stimuli from several sources with a single neuron. The axon is a long process that comes off the nerve cell body from a region called the **axon hillock**. It transmits stimuli from the nerve cell body to the terminus, which forms the end of the axon. There is usually only one axon for each neuron. However, some axons have branches, or **collaterals**, that reach out to other neurons. The axon hillock's job is to initiate the electrical signal that will be transmitted from the axon to glands, muscles, and other neurons. The terminus releases chemicals called neurotransmitters, which are so named because they transmit information from neurons to other cells. In order to respond to neurotransmitters, cells must possess membrane proteins called **neurotransmitter receptors**. There are a variety of neurotransmitters that react with specific neurotransmitter receptors.

Neurons do not make direct contact with the cells with which they communicate. The ends of the terminus stop short of touching nearby cells, forming a **synapse**. A gap exists between the cells called the **synaptic cleft**. The neuron whose terminus ends at the synaptic cleft is called the **presynaptic neuron**. It is defined as the neuron that produces the neurotransmitter needed to communicate nervous system information. On the receiving end of the synaptic cleft is the **postsynaptic neuron**. The postsynaptic neuron contains the neurotransmitter receptors. Most presynaptic neurons produce only one particular type of neurotransmitter. The type of neurotransmitter produced depends on the function and location of the neuron in the nervous system. However, postsynaptic neurons can contain a variety of neurotransmitter receptors. This gives certain neurons the ability to respond in different ways based on the neurotransmitters released into the synaptic cleft. The sensitivity of a postsynaptic neuron to a presynaptic neuron for the most part depends on the number of neurotransmitter receptors located around the synaptic cleft. Neurons with fewer receptors are less sensitive to neurotransmitters. Neurons can adjust the number of neurotransmitter receptors they have in response to a variety of conditions. Drugs such as morphine, which mimic natural neurotransmitters, alter the receptors of brain cells. High levels of morphine stimulate "pleasure receptors" in the brain. After long-term morphine use, the brain reduces the number of receptors to desensitize the neurons to the continuous stimulation. The person then develops an addiction to morphine because the body now needs it for the new level of normal receptor function.

Neurons have three primary shapes that can be viewed under the microscope: **bipolar**, **multipolar**, and **unipolar**. Bipolar cells have one dendrite and one axon. The bipolar neurons are sometimes called **interneurons** because they communicate information only from one neuron to another. Multipolar neurons have many dendrites that attach directly to the nerve cell body. They are the most numerous type of neuron in the body, and they usually communicate information from the brain to glands and muscles. Multipolar neurons are also called motor neurons because they are implicated in muscle movement. Unipolar neurons have a long axon that directly connects the dendrites to the terminus. The nerve cell body is positioned to the side of the axon. Unipolar neurons usually carry sensory information. This

Axon Hillock A swelling at the cell body where the axon begins

Collateral A branched axon

Neurotransmitter Receptor A cell membrane protein that binds to a signaling molecule to set off a cell response

Synapse The junction where an impulse is transmitted from one neuron to another

Synaptic Cleft The gap at the junction where a signal is transmitted from one nerve cell to another

Presynaptic Neuron A nerve cell that sends a signal across a synapse

Postsynaptic Neuron A nerve cell that receives a signal across a synapse

Multipolar Refers to a neuron that has one axon with many dendrites

Unipolar Refers to a neuron that has only one process extending from the cell body

Bipolar Refers to a neuron that has two processes, one axon and one dendrite

Interneuron A neuron that communicates only with other neurons

means that they receive and interpret various types of stimuli from the body and the environment.

> ### ✓ Concept Check
>
> 1. What are the three components of a neuron?
> 2. What is the function of the neuron terminus?
> 3. Describe the three primary types of neurons.

Neuroglia and Stem Cells

Key Terms: blood-brain barrier, cerebrospinal fluid, nodes of Ranvier, radial glia, satellite cell, Schwann cell

Neuroglia cells make up the bulk of the cell types in the nervous system, most of which are found in the brain. They carry out crucial roles in regulating the environment of neurons. Many neuroglia closely assist neurons with many of their functions. Neuroglia usually have a very high lipid content. This gives them a shiny, white appearance and also makes them vulnerable to deterioration when the diet contains inadequate amounts of the phospholipid fats needed for their production. There are many types of neuroglia that have a variety of unique cell shapes and specialized functions (Figure 8.3). The major types are summarized below:

- **Astrocytes**, or macroglia, make up the largest class of neuroglia. Their name comes from the fact they have many branches on the cell body giving them a star-like shape. These branches, or "feet," attach to neurons or form covers over small blood vessels. Most astrocytes maintain the chemical environment of neurons. Astrocytes associated with blood vessels control the types of materials that pass from the blood to neurons. This, in effect, protects the neurons from many types of harmful agents and helps form a protective feature called the **blood-brain barrier**. Astrocytes are mostly found in the brain and spinal cord.
- **Ependymal cells** are primarily secretory cells that line the cavities of the brain and spinal column. They produce **cerebrospinal fluid**, which bathes, nourishes, and protects the brain and spinal cord. Cilia on the ependymal cells help to circulate the cerebrospinal fluid.
- **Microglia** make up a highly variable cell type found throughout the nervous system. Many microglia carry out phagocytosis, a process that removes infectious agents and repairs nervous system damage. Other types of microglia produce secretions that maintain the health of neurons and assist neuron healing. Microglia malfunctions are the basis of many nervous system disorders.
- **Oligodendrites** are large cells with numerous branching processes. These branches wrap around the axons of neurons to form an insulating cover called the myelin sheath. The myelin sheath formed by oligodendrites is limited to neurons in the brain and spinal cord. This sheath facilitates the transmission of stimuli along the axon.

Blood-Brain Barrier A barrier between brain blood vessels and brain tissues that restricts the passage of materials from the blood into the brain

Cerebrospinal Fluid A nutrient-rich fluid that circulates around and through the brain and spinal cord

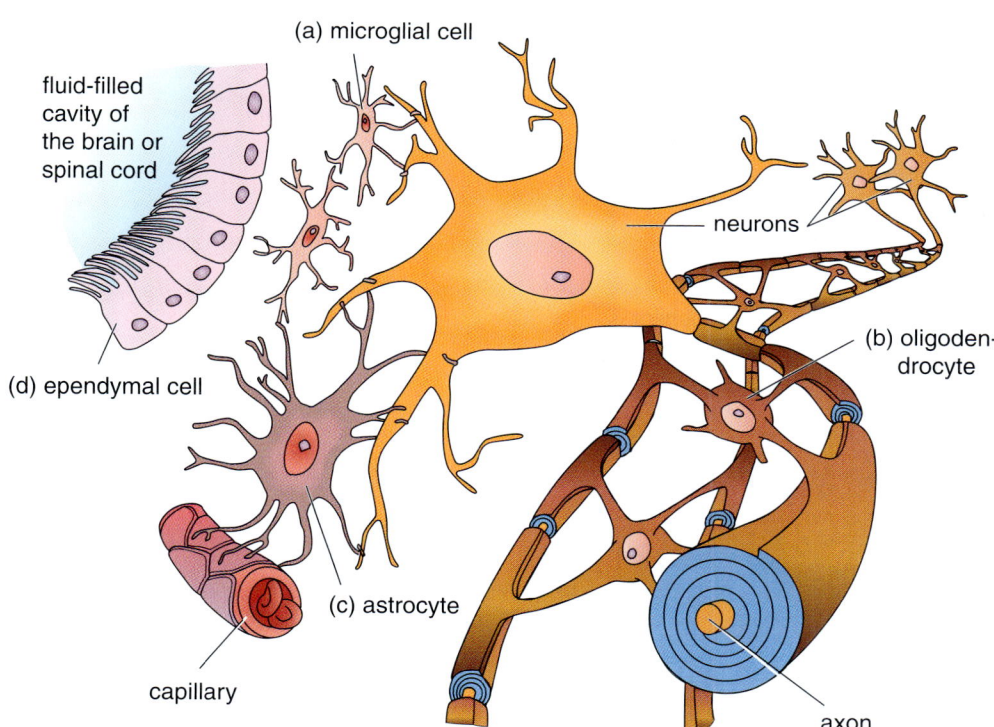

Figure 8.3 Nervous Tissue–Neuroglia Types of glial cells

- **Radial glia** are found in the developing nervous system. They provide a framework that organizes the interconnections of neurons. In the brain and eyes of adults, radial cells assist with nervous system maintenance by carrying out bidirectional communication with neurons. In effect, radial glia communicate the needs of certain neurons.
- **Satellite cells** are very numerous small cells that cover the surface of neurons outside of the brain and spinal cord. They help to maintain the chemical environment of neurons and may also assist with nerve cell repair.
- **Schwann cells** form a myelin sheath around neurons located outside of the brain and spinal cord. Series of these flattened cells almost completely cover the axon. Slight gaps between Schwann cells are called **nodes of Ranvier**. These gaps, which are found at regular intervals along the axon, are responsible for carrying stimuli along the axon's length. Some medical references call these gaps Ranvier's nodes.

Radial Glia Neuroglia that assist with the formation and maintenance of the developing nervous system

Satellite Cells Neuroglia that cover the surface of neurons

Schwann Cells Neuroglia that surround axons

Nodes of Ranvier Gaps found between the Schwann cells

Scientists recently discovered evidence supporting the presence of neural crest cells, a type of stem cell in adult humans. The discovery was confirmed in rats; however, recent studies have provided strong evidence that hair follicles also contain these cells. Rats contain neural crest cells in their digestive systems. It is known that these cells spread out from the brain of the embryo and migrate to various parts of the body. In the embryo, these cells assist

FUNCTION OF THE NERVOUS SYSTEM

Figure 8.4 Stem Cells

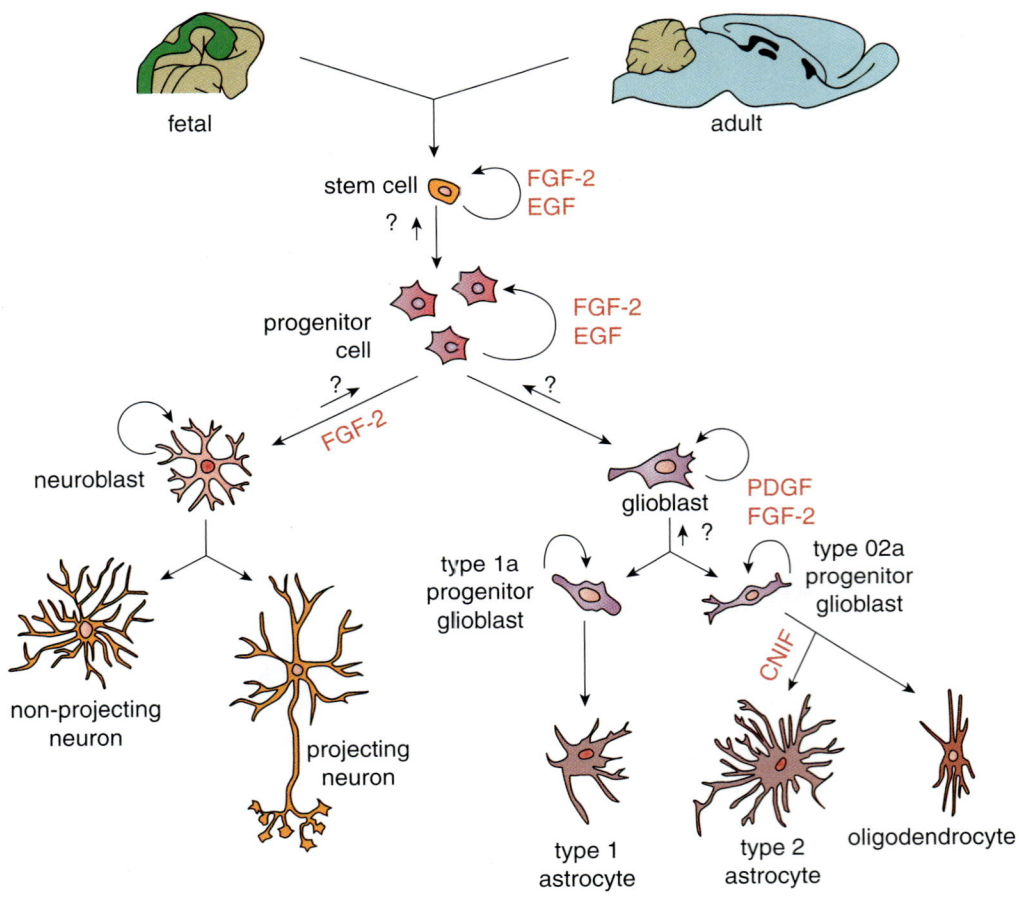

with the formation of the nervous system. Neural crest cells are capable of producing neuroglia, neurons, and more stem cells (Figure 8.4). It is hoped that these cells eventually will be found around components of the nervous system. They can then be used to regenerate neuroglia and neurons destroyed by disease or injury. It is not yet possible to relocate neural crest cells as a means of repairing nervous system damage.

✓ Concept Check

1. Describe the different types of neuroglia.
2. What is the function of myelin?
3. What is the importance of finding neural crest cells in adult humans?

DISCOVERY SCENE PLEASE ENTER DISCOVERY SCENE PLEASE ENTER

So, what have you learned about nerve cell types that helps you to resolve the CSI? Do you envision a particular cell type as the cause of the uncle's condition?

CHAPTER 8

NEURON PHYSIOLOGY

Key Terms: action potential, affector, aspartate, depolarization, dopamine, effector, excitatory, gamma-aminobutyric acid (GABA), gate, glutamate, glycine, histamine, hyperpolarization, inhibitory, propagate, recovery, refractory period, repolarize, resting potential, threshold

The concept of cell excitability was described in Chapter 6 in the discussion on muscle cell contraction. Neuron excitability is not much different. However, the structure of neuron anatomy provides the nervous system with the ability to more efficiently carry out the physiological processes necessary for excitability of the cell. Neurons must be able to transmit the excitable response that started either in the dendrites or the nerve cell body down through the axon. In addition, the cell's cytoplasms must be modified in order to communicate their excited state to other cells. A resting neuron is truly not at rest in the same way a person would be lounging around on a couch. It must maintain an excitable condition called **resting potential**. The resting potential is a chemically unstable condition in which the neuron has a higher concentration of sodium ions outside of the cell than on the inside of the cell. This produces a diffusion potential in which sodium would rush into the cell across the cell membrane if given the chance to do so. In contrast, potassium ions are found at higher concentration inside of the cell. As with muscle, the continuous energy-consuming activity of the sodium/potassium pump maintains this diffusion potential. A neuron at resting potential is not communicating to other cells. It can only communicate when a suitable stimulus disrupts the resting potential.

Scientists measure resting potential using instruments that distinguish the electrical differences created by the unequal concentrations of sodium and potassium ions. The cytoplasm of a resting nerve carries a small negative electrical charge. The charge inside the cell is usually -70 millivolts (mV). A millivolt is one-thousandth of a volt. A neuron becomes unstable if the internal charge of a portion of cytoplasm is elevated to -55 mV by an influx of sodium into the cell. The instability causes nearby cell membrane proteins, called sodium and potassium channels, to open, permitting the free movement of the ions in that region of the cell. A chain reaction follows, which causes the whole cell to lose its resting potential. This chain reaction is called an **action potential**. The action potential usually travels from the dendrites down the length of the axon to the terminus (Figure 8.5).

Neurons are able to transmit information to other cells by exploiting the action-potential effect. An action potential takes place as follows:

1. Dendrites receive stimulus from either the environment or another cell. This causes the sodium channels to open and allows sodium to rapidly move into the cytoplasm. If enough channels open, it brings the cytoplasm's charge to -55 mV. At this charge, the cell reaches its **threshold**. A threshold must be obtained for the action potential to travel, or **propagate**, across the cell membrane.

Resting Potential The difference between the two sides of the neuron's membrane when the cell is not conducting an impulse

Action Potential The electrical signal that rapidly travels along the axon of neurons

Threshold The minimum potential needed to stimulate an action potential

Propagate The transmission of an action potential across a neuron cell membrane

FUNCTION OF THE NERVOUS SYSTEM

Figure 8.5 Neuron Physiology

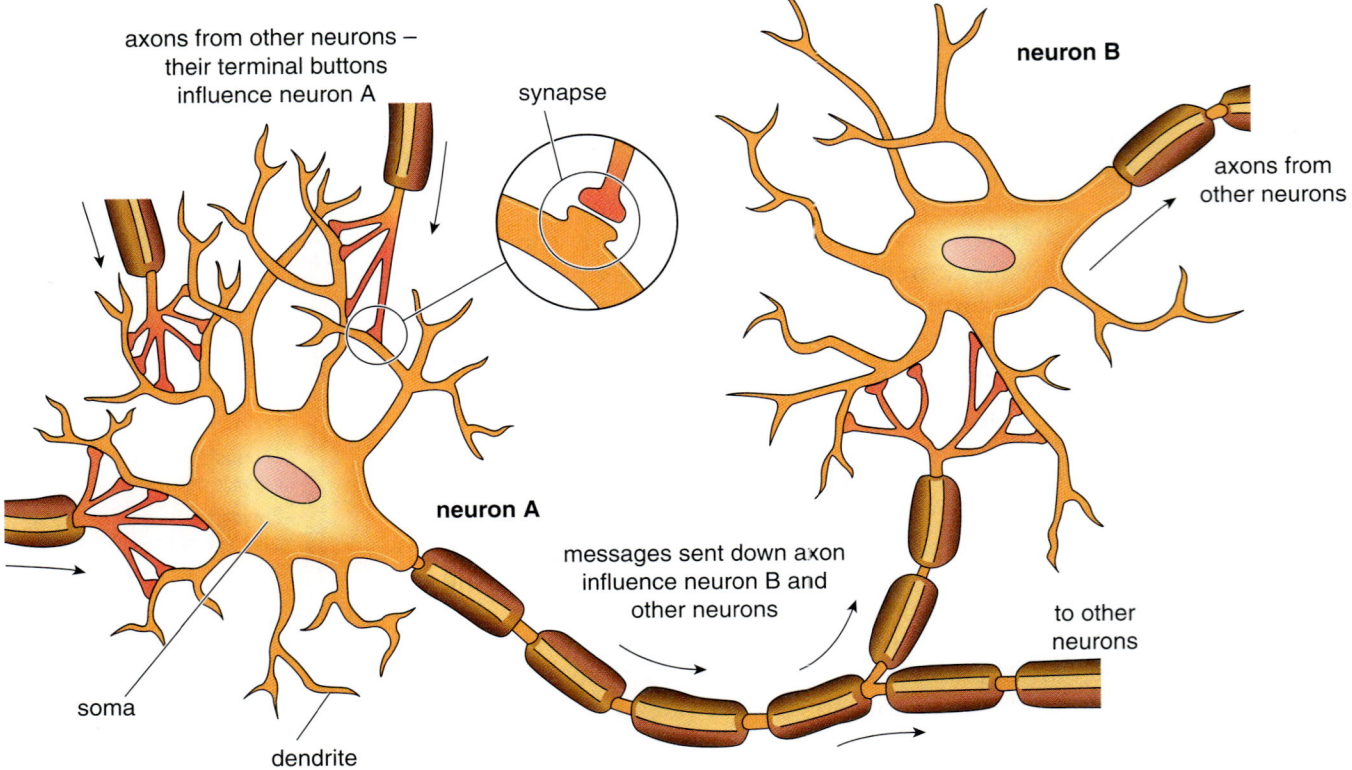

Depolarization The change in a neuron's membrane potential to a more positive potential

Repolarize A change in a neuron's membrane potential that returns the potential to a negative value

Hyperpolarization A condition in which a neuron's membrane potential becomes more negative than the resting potential

Refractory Period The time following an action potential during which normal stimulation will not cause another action potential

Recovery Phase The return of the resting potential following a refractory period

2. At threshold, many more sodium channels start to open starting a second series of sodium channel openings. This increased flow of sodium changes the cytoplasm's charge to +30 mV. The cell is now at a stage called **depolarization**. Depolarization means that the cell has lost its -70-mV resting potential.
3. Depolarization continues across the length of the neuron producing the action-potential propagation down the axon to the terminus.
4. Following depolarization, the sodium channels close, and the potassium channels open, resulting in the outflow of the positively charged potassium ions from the cell's interior. This loss of positively charged ions causes the negative value of the inside charge to increase, and the neuron begins to **repolarize** back to its resting potential.
5. This repolarization stage usually moves so quickly that it goes beyond the resting potential and reaches a -90-mV charge. This event is called **hyperpolarization**. Hyperpolarization serves two purposes: It prevents the neuron from receiving another stimulus during the repolarization period (the **refractory period**), and it prevents the action potential from traveling back in the opposite direction after it reaches the terminus. Any normal stimulation will not cause another action potential when a cell is in the refractory period.
6. The end of hyperpolarization kicks in the sodium/potassium pumps that eventually return the cell to the resting potential by pumping out sodium and pumping in potassium. This is called the **recovery phase**. If this does not occur, the neuron stays in a refractory period and cannot communicate with other cells.

CHAPTER 8

7. The terminus carries out an additional reaction as it undergoes depolarization. Calcium rushes into the terminus, which stimulates the movement of vesicles that contain the neuron's neurotransmitters. The vesicles move to the synaptic membrane, then release the neurotransmitters into the synaptic cleft.
8. The released neurotransmitters travel across the synaptic cleft to the postsynaptic nerve. The binding of the neurotransmitters to appropriate neurotransmitter receptors starts the action potential in the postsynaptic neuron.
9. The cell has completed the full sequence and is ready to pass another potential. Some cells can take on another impulse even before the sequence is completed. The continuous excitation with little or no recovery is called tetany.

The propagation of an action potential down the axon is actually a relatively slow process. Long neurons could not pass messages at the rate needed for the body to successfully respond to many stimuli if it were not for the myelin sheath on the axon. The speed of propagation can be considerably accelerated by the presence of the myelin sheath. The speed of an action potential is 10 to 120 meters per second (m/sec) for a myelinated neuron and 5 to 25 m/sec for an unmyelinated neuron.

The myelin sheath stops all of the channels along the axon from opening and exchanging ions. Therefore, the action potential is restricted to the exposed gaps of cell membrane known as the nodes of Ranvier. This, in effect, decreases the amount of membrane that needs to depolarize for propagation of the action potential. The action potential literally jumps from gap to gap, speeding up the propagation.

BANANAS AND THE NERVOUS SYSTEM

Chiquita™ Brands, L.L.C. is now marketing bananas as the perfect food for nerve cell health. For years, physicians have recommend eating bananas as a way to increase potassium levels in the blood and tissues. Potassium is known to safely reduce the activity of nerve cells linked to high blood pressure and heart disease. The results of a research study of 5000 obese hospital patients showed that a diet high in bananas reduced the subject's high blood pressure. Extensive medical studies in India showed decreased levels of acetylcholine production due to bananas in the diet. Chiquita claims that bananas contain other critical components that assist neurons. They assert that the carbohydrates found in bananas are a good fuel for nerve cell action. Bananas provide a quick boost of glucose and a continuous supply of carbohydrates because of their high polysaccharide content. Fresh bananas are also high in vitamin B_6, which is needed for neuron physiology. An informal study of 200 California high school students showed that eating one banana with each meal improved exam scores. Bananas have also been used to reduce the nerve cell effects that counteract nicotine withdrawal. Nutritionists have long known that bananas are almost a "complete food," supplying most of the nutrients needed for proper cell function. In the future, children may be replacing a well-known proverb with the following adage "A banana a day keeps the doctor away."

Affector A neuron that transmits sensory information

Effector A neuron that transmits motor information

Excitatory A stimulus that encourages an action potential

Inhibitory A stimulus that discourages an action potential

Neurotransmitters are chemical signals that transfer the action potential from an **affector**, or sensory neuron receptor, to a motor neuron of the **effector**, or muscle or gland, that responds to the sensory input. Affectors receive stimuli, while effectors carry out a job for the body. Neurotransmitters do this by binding to the neurotransmitter receptors on the postsynaptic cell (Figure 8.6). A neurotransmitter can have either an **excitatory** or an **inhibitory** effect on the postsynaptic cell. An excitatory neurotransmitter effect helps the postsynaptic neuron reach threshold, thereby making it more likely to produce an action potential. The opposite is true for an inhibitory response. Inhibitory neurotransmitters make it more difficult for the neuron to achieve an action potential. Excitatory neurotransmitters cause the sodium channels to open, creating the influx of sodium that drives the action potential.

Neurons carry out a four-stage sequence of events when using neurotransmitters to communicate with other cells: 1) synthesis and storage of neurotransmitters; 2) neurotransmitter release; 3) neurotransmitter binding to postsynaptic receptors; and 4) inactivation of neurotransmitters. Neurotransmitters are synthesized in the nerve cell body and then transferred in vesicles to the terminus where they are stored. The influx of calcium during the action potential causes neurotransmitter release, and promotes the movement and exocytosis of vesicles containing the neurotransmitters. After release, the neurotransmitters bind to the neurotransmitter receptors. Then, the cell inactivates the neurotransmitters by degrading them in the synapse or taking them back up into the terminus for recycling. Many drugs work at this stage of the neurotransmitter cycle by preventing the breakdown or reuptake of specific neurotransmitters.

Figure 8.6 **Neurotransmitter Action**

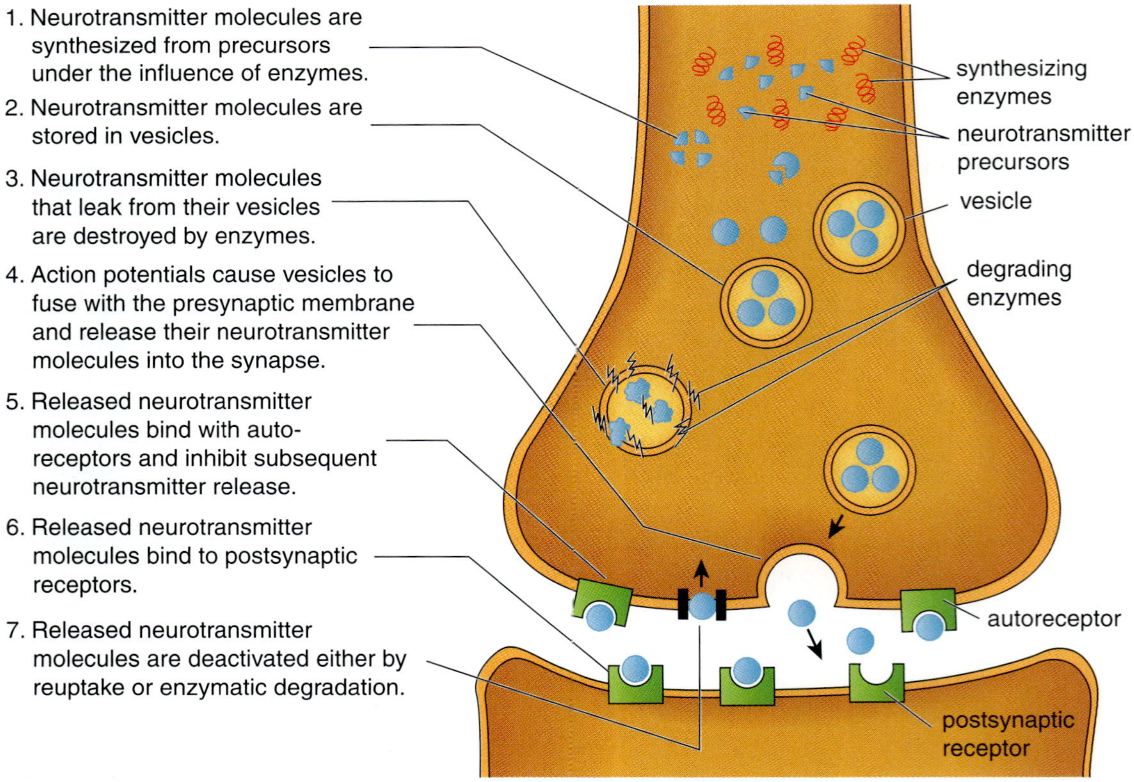

1. Neurotransmitter molecules are synthesized from precursors under the influence of enzymes.
2. Neurotransmitter molecules are stored in vesicles.
3. Neurotransmitter molecules that leak from their vesicles are destroyed by enzymes.
4. Action potentials cause vesicles to fuse with the presynaptic membrane and release their neurotransmitter molecules into the synapse.
5. Released neurotransmitter molecules bind with autoreceptors and inhibit subsequent neurotransmitter release.
6. Released neurotransmitter molecules bind to postsynaptic receptors.
7. Released neurotransmitter molecules are deactivated either by reuptake or enzymatic degradation.

There are many types of neurotransmitters. Their greatest diversity is found in the brain. It is helpful to know some of the more common neurotransmitters because their actions are the basis of many diseases, and many drugs modify their actions. The following neurotransmitters are responsible for most of the nervous system communication in the body:

- **Amino acids:** These are specialized amino acids usually used in the brain and spinal column. The common ones are **aspartate**, **gamma-aminobutyric acid** (**GABA**), **glutamate**, and **glycine**. Glutamate is excitatory, while GABA and glycine are inhibitory.
- **Catecholamines:** These excitatory neurotransmitters are made from the amino acid tyrosine. Examples of catecholamines are epinephrine (adrenaline), norepinephrine (noradrenaline), and **dopamine**. Dopamine can be inhibitory as well as excitatory. High levels of catecholamines in the in blood are associated with stress.
- **Cholinergics:** Acetylcholine is the most common cholinergic neurotransmitter in humans. It is a simple molecule made from dietary fats and compounds produced during cellular metabolism. It is the neurotransmitter used to excite muscle cells.
- **Monoamines:** These neurotransmitters are related to the catecholamines. They include serotonin, which is made from the amino acid tryptophan, and histamine. Serotonin inhibits catecholamine neurotransmitters, while histamine is associated with pain sensations and the stress response.

Aspartate An excitatory amino acid neurotransmitter

Gamma-Aminobutyric Acid (GABA) An inhibitory amino acid neurotransmitter

Glutamate An excitatory amino acid neurotransmitter

Glycine An inhibitory amino acid neurotransmitter

Catecholamines A class of neurotransmitters related to amino acids

Dopamine A type of catecholamine neurotransmitter

Cholinergic Pertaining to the acetylcholine neurotransmitter

Monoamine A neurotransmitter containing only one amino group

✓ Concept Check

1. Distinguish between the resting potential and the action potential.
2. Describe how myelin assists the propagation of the action potential
3. Explain the role of neurotransmitters in neuron communication.

DISCOVERY SCENE PLEASE ENTER DISCOVERY SCENE PLEASE ENTER

So, what information about neuron physiology is most helpful for understanding the uncle's condition in the CSI? Do you see any relationship between his condition and the role of the neurotransmitters in stimulating or inhibiting an action potential to a particular group of neurons?

Types of Neuron Communication

Key Terms: axoaxonic synapse, axodendritic synapse, axosomatic synapse, epilepsy, excitatory postsynaptic potential (EPSP), inhibitory postsynaptic potential (IPSP), innervate, neural pathway, reverberating pathway

The nervous system's ability to coordinate the body is determined by the placement and timing of neurons. Neurons are placed into various types of arrangements that permit them to communicate with one or more structures. For example, some motor nerves **innervate** a gland and muscle simultaneously. The term "innervate" means to supply a body part with nervous stimulation. Other arrangements allow one neuron to receive communications from

Innervate To supply nerves to an organ or other body part

FUNCTION OF THE NERVOUS SYSTEM

Good Choice Bad Choice

A friend discovered that his sister's amphetamine medication for attention-deficit hyperactivity disorder (ADHD) causes weight loss. With his sister's permission, he is thinking of taking some of her medication for a short time to accelerate his weight loss. What are the benefits and risks of his decision? Are the consequences of being overweight worse than the consequences of taking amphetamines?

Neural Pathways An arrangement of neurons connecting one part of the nervous system with another

Axodendritic Synapse A synapse formed with a dendrite

Axosomatic Synapse A synapse formed with a nerve cell body

Axoaxonic Synapse A synapse formed with an axon

Reverberating Pathway A neuron pathway that stimulates itself continually

Epilepsy A neurological disorder that produces recurrent seizures

several other neurons. All of these different types of neuron communication patterns are collectively called **neural pathways**. The simplest pathway pattern describes the location on the cell where a postsynaptic neuron receives a stimulus from another neuron (Figure 8.7). An **axodendritic synapse** describes a terminus that communicates with the dendrite of a postsynaptic neuron. A terminus that links to the nerve cell body of another neuron forms an **axosomatic synapse**.

In complex pathways, it is common to find a terminus communicating with an axon. This is called an **axoaxonic synapse**. This unusual type of neuron arrangement produces a **reverberating pathway** (Figure 8.8). Reverberating pathways are most commonly found in the brain. Neurons in a reverberating pathway are capable of stimulating themselves over and over again until another stimulus comes along to stop the stimulation. They form very important pathways in the parts of brain associated with emotions, learning, and memory. Breakdowns in these pathways can lead to a variety of disorders, including **epilepsy**. Epilepsy is caused by uncontrolled excitatory activity in certain brain pathways.

Another way of identifying neuron arrangement has to do with the type of communication taking place between two neurons. Postsynaptic neurons can respond in an excitatory or inhibitory way to neurotransmitters from a

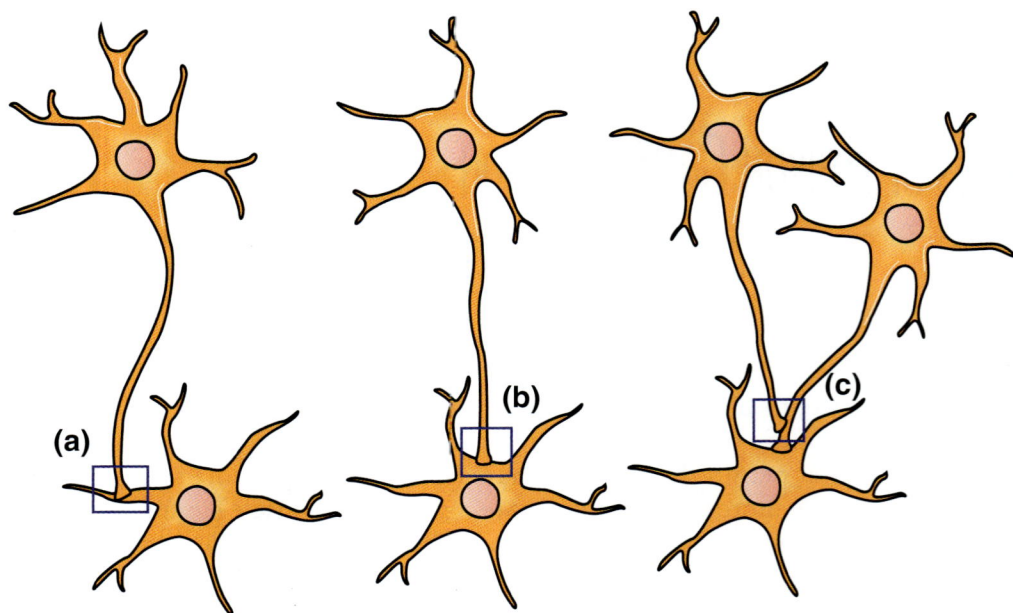

Figure 8.7 **Types of Neuron Communication**
Synaptic arrangements in the central nervous system: **a.** an axodendritic synapse; **b.** an axosomatic synapse; and **c.** an axoaxonic synapse

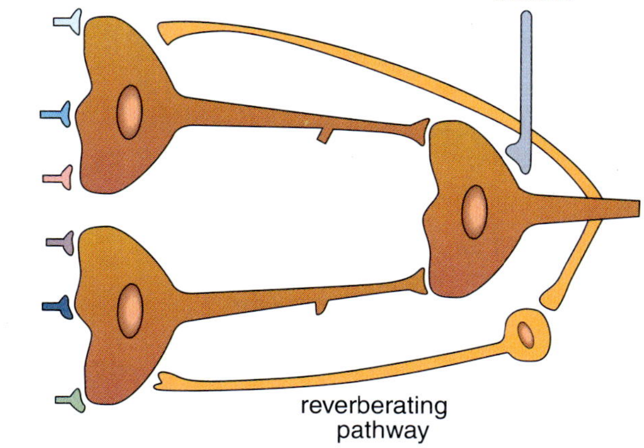

Figure 8.8 Types of Neuron Communication

inhibitory neuron

reverberating pathway

presynaptic neuron. An **excitatory postsynaptic potential** (**EPSP**) causes the postsynaptic nerve to carry out an action potential. For certain nerve pathways, it may take the simultaneous communication of two presynaptic neurons to excite a postsynaptic neurons. An **inhibitory postsynaptic potential** (**IPSP**) reduces the ability of a postsynaptic neuron to respond to further stimulation. Many neurons are organized into pathways that interlink EPSP and IPSP connections (Figure 8.9). With this type of nervous system arrangement, a neuron's response may depend on the input of many other neurons that each have a different degree of influence. Complex EPSP and IPSP pathways are important in decision-making components of the brain. So, certain neurotransmitters can have an excitatory or an inhibitory effect depending on the receptors present on the postsynaptic neuron.

Excitatory Postsynaptic Potential (EPSP) A neural pathway that excites a postsynaptic neuron

Inhibitory Postsynaptic Potential (IPSP) A neural pathway that inhibits a postsynaptic neuron

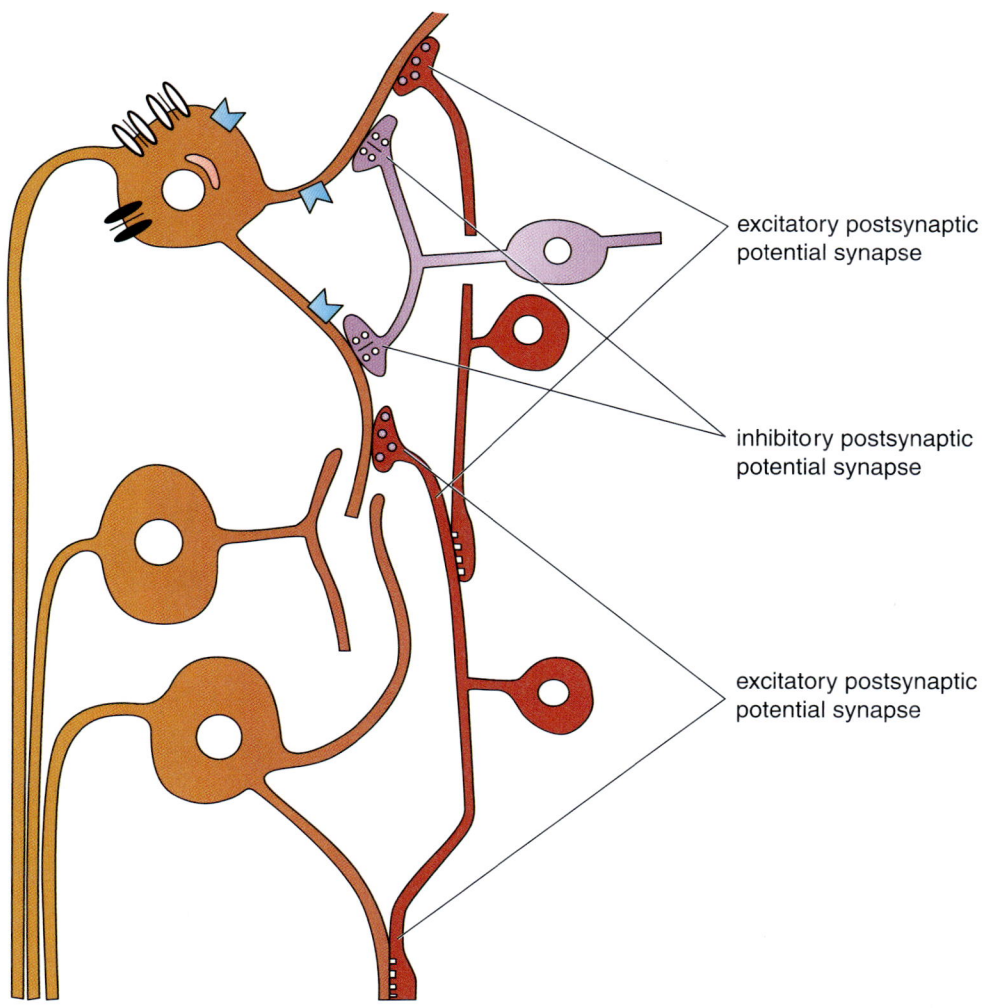

Figure 8.9 Types of Neural Communication
By activating different types of receptors, a neurotransmitter can be excitatory (promoting the firing of action potentials) or inhibitory suppressing the firing of the postsynaptic cell). The effect of the receptor on the cell determines the neurotransmitter's role.

FUNCTION OF THE NERVOUS SYSTEM

✓ Concept Check

1. Describe three types of neural innervation.
2. What is the importance of reverberating pathways?
3. Distinguish between EPSP and IPSP.

REFLEXES

  Key Terms: reflex, reflex arc, transduction

Reflex An involuntary response to a stimulus

Reflex Arc A neural pathway that links a sensory receptor, an interneuron, and an effector

People are not always conscious of many things that the nervous system must respond to. This is evident when people trip and automatically put out their hands in front of them in an attempt to break the fall. The instantaneous response of putting the hand forward is one of many types of nervous system **reflexes**. Reflexes are defined as involuntary responses to a stimulus. Nervous system reflexes constitute a control system that automatically links a stimulus to response without requiring the intervention of the conscious control areas of the brain. This neuron arrangement is called a **reflex arc** (Figure 8.10).

Most reflexes are built into the structure of the nervous system. However, certain types of reflexes can be ignored or learned given intervention by the brain. The simplest reflex arc involves an affector, a sensory neuron, an interneuron, a motor neuron, and an effector. An affector is a specialized neuron or structure that conveys a stimulus to a sensory neuron. Effectors are structures, such as glands or muscles, which carry out a job for the body. They receive neural signals from motor neurons.

Figure 8.10 **Reflexes**

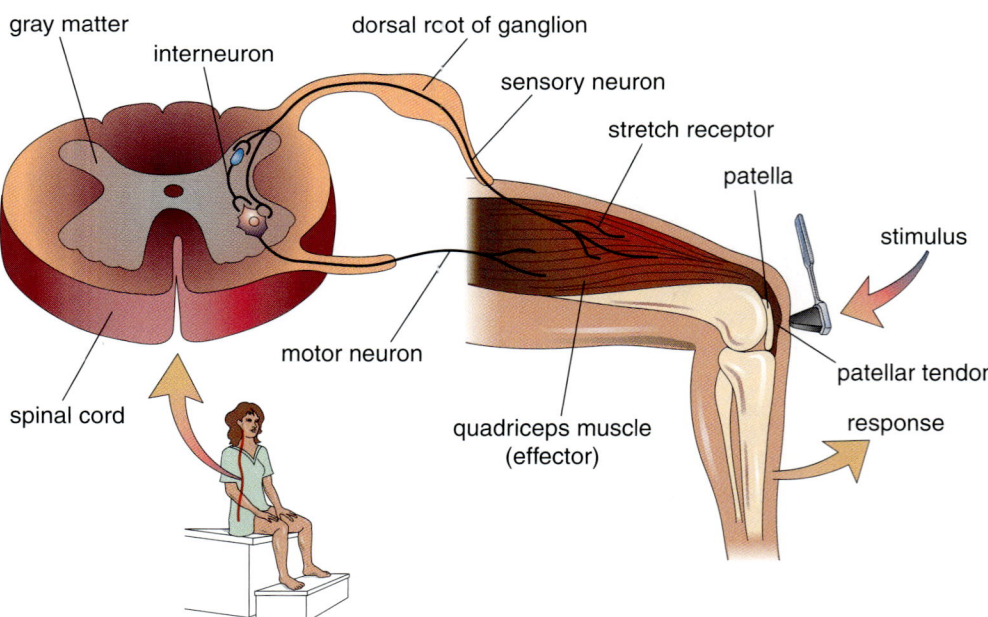

CHAPTER 8

Neuron Trivia

- The average human has more neuroglial cells than neurons.
- The range of speed for a signal transmitted through a neuron is 1.2 to 250 miles per hour.
- The average human brain has 100 billion neurons.
- The average growth rate of neurons forming during the development of a fetus is 250,000 neurons per minute.
- If lined up all together, all the neurons of an average human would extend over 47 miles (76 kilometers).
- The nervous system uses over 25% of the body's energy each day.
- The longest axon of any land creature is found in the neck of a giraffe. It extends over 15 feet. The longest nerve cells in the human body extend down the leg.
- Because of its effects on critical nerve cells, approximately 2 ounces of milk chocolate can be poisonous to a 10-pound puppy.
- Cocaine works differently from other narcotics by stimulating neurons.
- Nine out of 20 common chemicals used in fragrances are linked to nervous system disorders.
- There are 30 times more neurons that feel pain than neurons that detect temperature.

A reflex is initiated when a stimulus excites an affector. Affectors carry out a physiological job called **transduction**. This means they can convert a stimulus, such as touch, into a message that can be relayed to cells. The affector transmits neurotransmitters to convey the information to a sensory neuron. In reflexes, the affector is part of the sensory nerve's dendrites. The sensory neuron now transfers the response to the interneuron. In many reflex arcs, interneurons are linked to several other neurons. The interneuron's primary role is to pass the information along to a motor neuron. However, interneurons can also communicate to the brain and other body regions. This alerts the body to the stimulus, providing, in some cases, the sensation of pain or emotion. Excitation of the motor neuron leads to stimulation of the effector. The effector now carries out the intended task of the reflex. For example, a gland will produce a particular secretion, or a muscle will contract to draw back a limb.

Transduction The conversion of a stimulus from one form to another

Signals sent to the brain and other regions give the body the liberty to control a reflex. This takes place during toilet training when children learn to control urination and bowel movements. They are taught to consciously inhibit the reflexes that cause the urinary bladder to release urine when it starts filling up and stretching. People can also train themselves not to gag when something goes down their throat and not to blink when a rapidly moving object approaches their face. Athletes can improve certain tasks, such as balance and running, by eliminating voluntary and involuntary factors that inhibit certain reflexes.

✓ Concept Check

1. Describe the components of a reflex arc.
2. Define the term transduction.
3. Explain the various roles of interneurons in a reflex arc.

DISCOVERY SCENE PLEASE ENTER DISCOVERY SCENE PLEASE ENTER

Could the uncle in the CSI be having problems with some critical reflexes? If so, what could be causing a breakdown in reflex communication?

WELLNESS AND ILLNESS OVER THE LIFE SPAN

Pathology of Nervous System Function

Key Terms: amyotrophic lateral sclerosis (ALS), botulism, bovine spongiform encephalopathy, congenital, Creutzfeldt-Jakob disease, degenerative, demyelination, encephalitis, endotoxin, galactosylceramide beta-galactosidase, herpes virus, Hirschsprung's disease, Krabbe's disease, Lou Gehrig's disease, mad cow disease, meningitis, multiple sclerosis (MS), neurotrophic, rabies, tetanus, tetrodotoxin, toxicological, traumatic

There are many types of disorders that affect nerve cell function. Nerve cells can be injured directly by infectious agents or by damage to other body parts. They are also subject to disorders caused by DNA defects and poisoning. Neuroglia as well as neuron components, particularly the axon, are subject to specific nervous system pathologies (Figure 8.11). Nerve cell diseases can be categorized into the following major groups:

- **Infectious:** These are diseases caused by microorganisms that are capable of invading or infecting nervous system cells.
- **Degenerative:** This term refers to the progressive deterioration of a cell or tissue over time.
- **Congenital:** These are diseases caused by embryological and maturation errors that affect nerve cell communication, development, and growth.
- **Toxicological:** This type of nervous system damage is caused by poisons that affect cell metabolism or communication.
- **Traumatic:** Injuries resulting from a wound that was caused by an external force or violence.

A variety of infectious agents can harm nerve cells when they invade the body. Bacteria that release toxins into the blood cause the most common type of damage. Certain types of soil bacteria can make their way into foods or wounds. They produce a variety of secretions that can inflame or kill neuroglia and neurons. Some types of secretions can affect the communication between nerve cells. For example, a bacterium that sometimes grows in spoiled foods causes **botulism**, a disease that blocks the action of acetylcholine. Botulism produces flaccid paralysis, or the inability of a muscle to contract. In contrast, the bacterium *Tetanus* produces a secretion that enhances the effects of acetylcholine. This causes the disease known as tetanus, which prevents the muscles from relaxing following contractions (Figure 8.12). Certain types of bacteria release poisons called **endotoxins** as they replicate and die in the body. Endotoxins are composed of lipids, poly-

---

Congenital A condition that appears at birth

Toxicological Pertaining to the toxic effects of substances

Traumatic Pertaining to an injury or wound

Botulism A serious illness caused by a toxin produced by certain bacteria

Endotoxin A toxic substance found in certain bacteri

CHAPTER 8

Figure 8.11 Pathology of the Nervous System

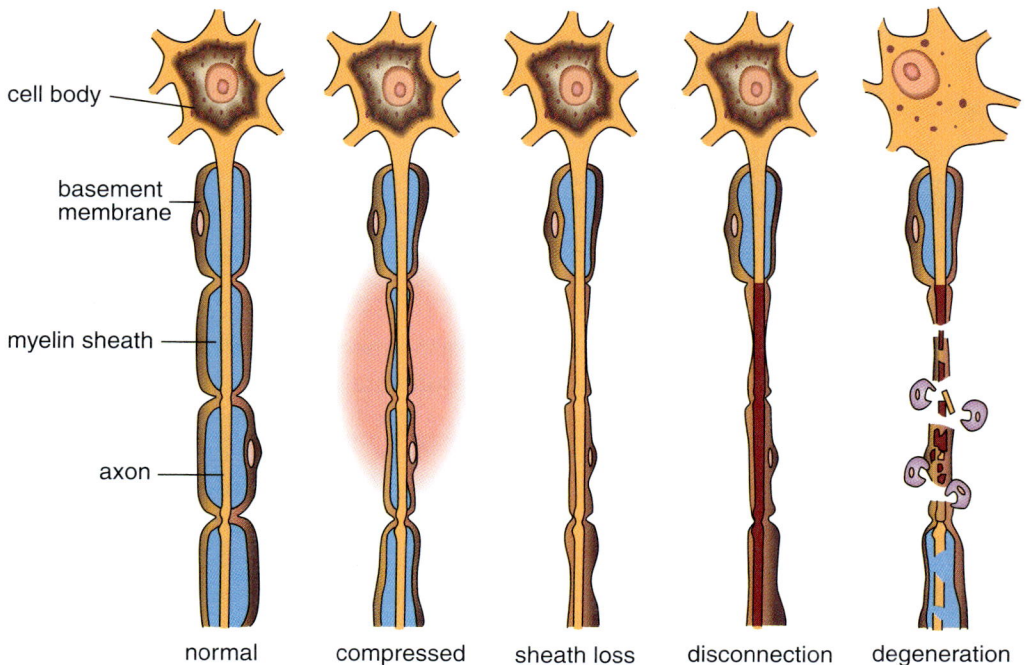

Figure 8.12 Tetanus

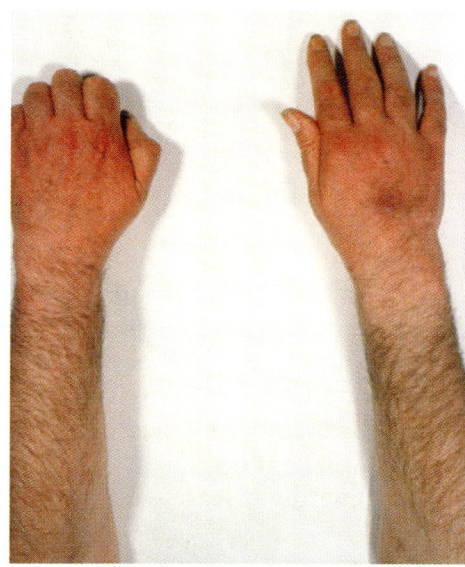

saccharides, and proteins. They can cause immediate death to neuroglia and neurons. Endotoxins commonly cause the diseases **encephalitis** (inflammation of the brain) and **meningitis** (inflammation of the membranes surrounding the brain and spinal cord). Certain fungal toxins produce effects similar to those produced by bacteria.

The term **neurotrophic** refers to microbes that are capable of entering and infecting nervous system cells. These microbes include many types of protista, viruses, and simpler organisms called viroids and prions. They either kill cells outright or produce an inflammation that results in encephalitis or meningitis. Mosquitoes and other biting insects commonly spread encephalitis and meningitis viruses. **Herpesviruses** can live in nerve cells for the life of the person, causing mild to severe nerve cell damage. **Rabies** is a notorious virus that disrupts nerve cell function. **Mad cow disease**, or **bovine spongiform encephalopathy**, is a variation of **Creutzfeldt-Jakob disease**. It is caused by infectious prions that can be contracted through exposure to the blood and meat of infected animals. Humans may be able to spread this disease through blood contact.

A common type of genetic degenerative disorder causes inherited **amyotrophic lateral sclerosis** (**ALS**), or **Lou Gehrig's disease**. It is caused by faulty mitochondria passed down by the egg. It particularly affects motor

Encephalitis Inflammation of the brain

Meningitis Inflammation of the coverings around the brain]

Neurotrophic Refers to an organism that infects nerve tissue

Herpesvirus An inflammatory virus that grows in nerve tissue

Rabies A neurotrophic virus that is spread through the saliva of infected animals

Mad Cow Disease or Bovine Spongiform Encephalopathy A neurotrophic virus that causes degeneration of the nervous system

Creutzfeldt-Jakob Disease A nervous system disease caused by a prion

Amyotrophic Lateral Sclerosis (ALS) or Lou Gehrig's Disease A progressive neurological disease that causes the loss of neuron function

Demyelination The loss of the myelin sheath of a neuron

FUNCTION OF THE NERVOUS SYSTEM

Cutting Edge Research

ACUPUNCTURE—NO MORE POKING FUN AT THIS TREATMENT

Acupuncture is a Chinese therapy that relies on the insertion of thin, sterile needles into the body at specific points. These points are based on a map of "energy pathway" locations. In 5000-year-old acupuncture philosophy, these energy pathways calm or excite the flow of energy, or qi (pronounced "chee"), in the body. Today acupuncture is practiced as a natural healing philosophy by applying electrical stimulation, heat, or motion to the needles. Much of the medical community gives little credibility to the healing powers of acupuncture. They believe its healing powers are nothing more than placebo effects. A placebo effect is an apparently beneficial result of a therapy due to a patient's expectation that the therapy will help. However, current medical and scientific studies are showing that acupuncture may be effective at relieving the pain associated with many diseases.

Studies conducted at the Oregon Health & Science University and at the University of Maryland School of Medicine show that acupuncture has true medicinal value. The results of a series of studies of patients with chronic pain showed that acupuncture provided pain relief comparable with that of traditional pain-killing medications. The studies also showed that acupuncture's pain-killing, or analgesic, properties were not due to the placebo effect. When physicians used traditional acupuncture points on arthritis patients, the pain did not recur when they were taken off their pain-killing medications. These studies only confirm the effectiveness of acupuncture for reducing certain types of pain; they do not support any healing properties of acupuncture. It is believed that the needles somehow block the nerves that normally transmit the feeling of pain to the brain.

Berman BM, Lao L, Langenberg P, Lee WL, Gilpin AM, Hochberg MC. 2004. Effectiveness of acupuncture as adjunctive therapy in osteoarthritis of the knee: a randomized, controlled trial. *Ann Intern Med.* 141(12): 901-910.

Multiple Sclerosis (MS) A potentially debilitating disease caused by demyelination

Galactosylceramide Beta-Galactosidase An enzyme that removes certain waste products from nerve cells

Krabbe's Disease A degenerative disorder that affects the nervous system

Hirschsprung's Disease A congenital disorder in which the intestines do not have the normal network of nerve

nerves, causing a gradual cessation of muscle function. Some type of DNA damage causes many degenerative diseases of neurons. It is also possible for neuroglia to deteriorate over time. **Demyelination** diseases result from the loss of neuroglia around the axons and bodies of neurons. Such diseases can result from metabolic disorders or a loss of blood flow to neuroglia. The characteristic loss of myelin in these diseases is usually followed by slower neural impulses and the eventual degeneration of the axon. **Multiple sclerosis** (**MS**) is a predominant demyelinating disease whose cause is unknown.

Certain congenital disorders can alter the metabolism of nervous system cells. An important enzyme called **galactosylceramide beta-galactosidase** prevents the accumulation of toxic wastes in nerve cells. **Krabbe's disease** usually starts out in the embryo as a lack of this enzyme, causing a buildup of harmful fats in nervous system cells. This fatal condition causes abnormal functioning of neurons and diminishes neuroglia maturation. There are hundreds of enzyme defects that can prevent the normal function of the nervous system.

Hirschsprung's disease, which also develops before a child is born, is specific to the neurons of the large intestine. Nerve cells normally grow in the intestine of the fetus as it develops. In Hirschsprung's disease, the nerve cells stop growing, causing a loss of large-intestine function.

CHAPTER 8

Figure 8.13 Tetrodotoxin
A Norwegian study explained how tetrodotoxin blocks the action of the sodium channels on the nerve cell membrane needed for an action potential. Puffer fish consumption is becoming more common in countries that consume fish as a large part of the diet.

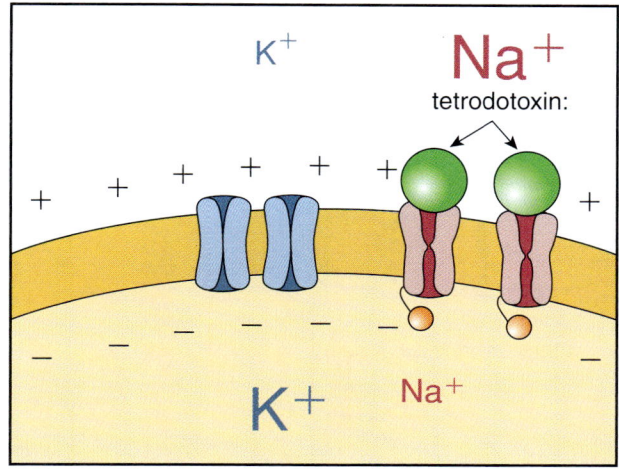
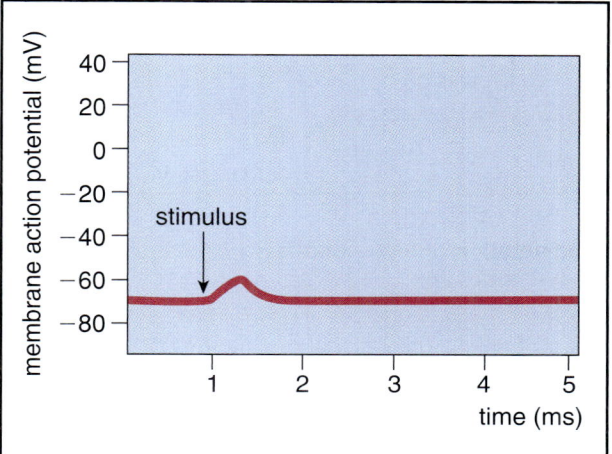

Nervous system poisoning can come from a variety of sources. The plumbing systems of older houses may contain metals, such as lead, which slow neuron function. It is possible for the metals to accumulate to toxic levels, killing neuroglia and neurons. This type of poisoning may be initially misdiagnosed as a degenerative disorder before blood and tissue tests are performed. Arsenic and cyanide are found in many types of pesticides. They can make their way into the body through food, water, or nearby pesticide treatment operations. Both of these poisons block cellular respiration and disable neurons. An exotic form of neuron poison is **tetrodotoxin** (Figure 8.13). It is found in tropical frogs and an edible fish called the puffer. This potent toxin inhibits the flow of sodium into nerve cells, thereby stopping the action potential. Imbalances of calcium, potassium, and sodium can also have toxic effects on neurons. Concentrations of these ions that are too high or too low can alter the excitability to the point that neuron communication is impaired.

Traumatic damage to the nervous system is a common type of pathology. Athletic injuries, automobile accidents, and work-related falls top the list of causes for severe nervous system damage. Neurons cannot be replaced once they die. However, injured neurons can be repaired as long as intact neuroglia are nearby. Luckily, it is possible for most neuroglia to replicate if only a small number are killed. Undamaged neuroglial cells carry out neuron repair (Figure 8.14). The neuroglia rebuild damaged components of the neurons and redirect the axons to their original position. Physicians and researchers are developing an arsenal of strategies to encourage neuron repair following trauma. Some techniques involve the use of growth factors. Surgical methods are also commonly applied. Future developments in stem cell research may one day provide ways of replacing many neurons damaged by traumatic injuries.

Tetrodotoxin A potent nerve toxin derived from certain fish and frogs

FUNCTION OF THE NERVOUS SYSTEM

Figure 8.14 Consequences of Neural Injury

(a) consequences of neural injury

intact nervous system

demyelination
cell death   axon retraction   scar formation

(b) transplantation or endogenous induction

migrational differentiation
glial pathways
neuronal pathways
proliferation

(c) targets for cell replacement

remyelination

prevention of scar/axon guidance

interneuron/relay circuitry

modulation of intact circuitry

replacement of primary neuronal loss

key: oligodendrocyte   neuron   axon   astrocyte   stem cell   progenitor

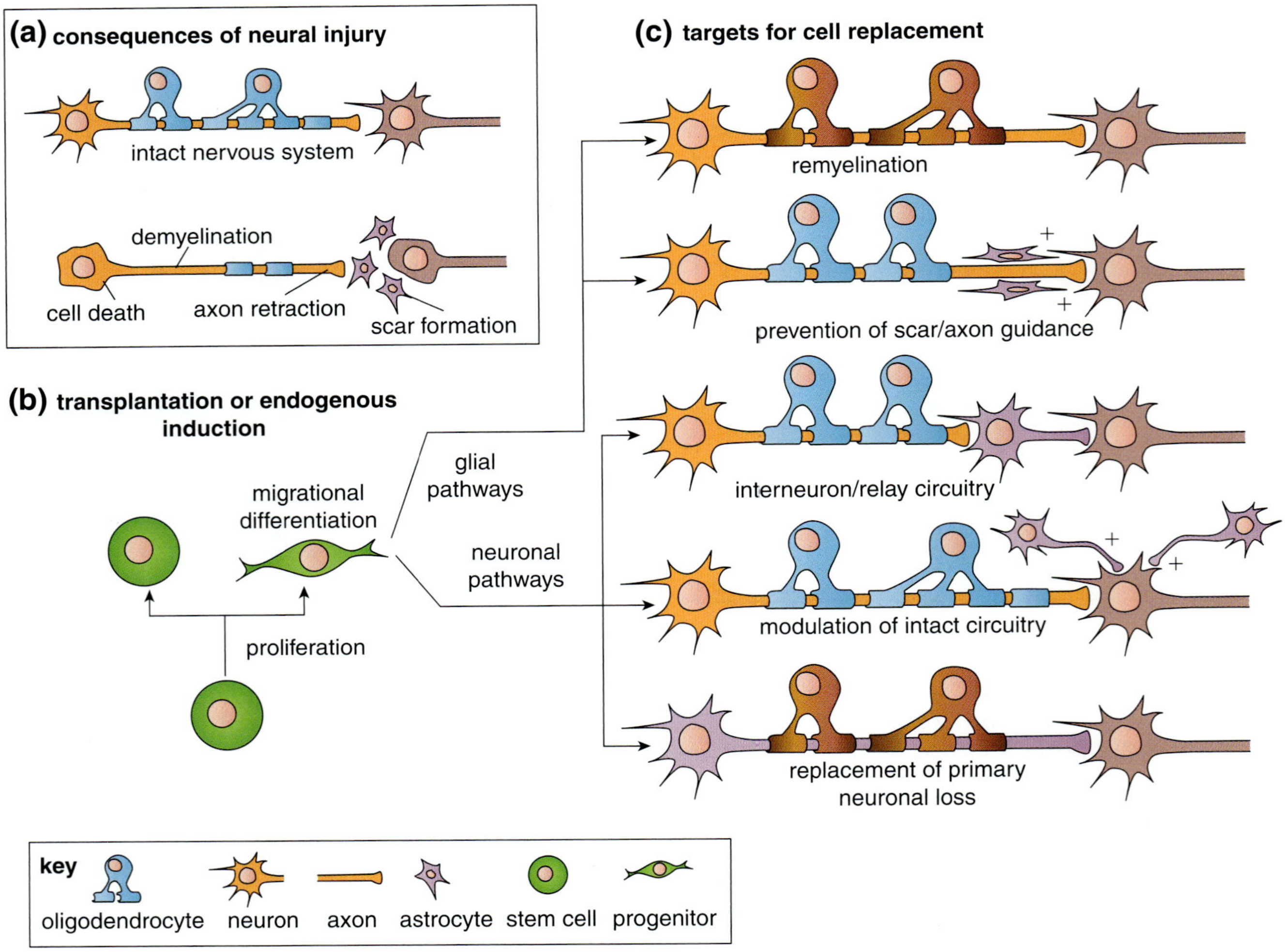

✓ Concept Check

1. Describe five categories of nerve cell pathology.
2. Explain two ways that infectious agents can harm nerve cells.
3. Distinguish between congenital and degenerative nervous system disorders.

Aging of the Nervous System

Key Terms: cytokine, lipofuscin, plaque, tangle, tonic control

The complexity and extreme differentiation of neurons make it almost impossible for them to undergo mitosis. In the fetus, all nerve cells are derived from a core of nerve stem cells that are capable of cell division (Figure 8.15). Neurons cease mitosis as they begin to form mature nervous system structures. This means that a person has the same nerves cells in maturity that are present in childhood. Neuroglia and neurons accumulate a significant amount of injury over a person's lifespan. Damage from cells that

CHAPTER 8

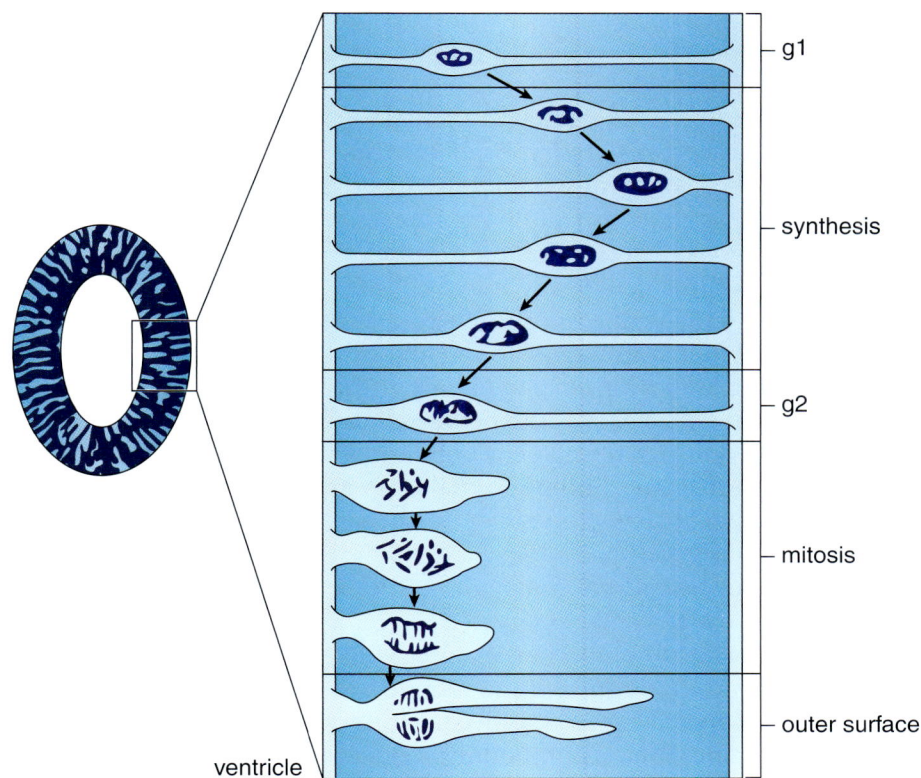

Figure 8.15 Cell Cycle of Embryonic Stem Cell
The figure shows the cell cycle of an embryonic stem cell. The nervous system ages because adult neurons cannot carry out the cell cycle, and so cannot regenerate dead or injured neurons.

are capable of replication is less likely to be long-term because new cells can replace such cells over the lifetime. The high metabolic rate of nervous system cells puts them at higher risk for cellular damage from oxidizing chemicals, which are common byproducts of mitochondrial activity. These oxidizing chemicals can alter the DNA, resulting in a variety of metabolic errors that may be fatal to the cell. Regular use of alcohol and drug abuse can accelerate nervous cell aging. Smoking and air pollution also contribute to cellular damage.

One consequence of maturity for most people is the loss of consistent blood flow to various tissues. This affects many parts of the body; however, the high metabolic needs of nervous system cells make them more susceptible to problems associated with decreased blood flow. Neurons slow down because they cannot obtain adequate nutrients to fuel an action potential. In addition, they may run low on the raw materials needed to build neurotransmitters. Neuron receptors also lose their responsiveness. This affects glands and muscles as well as neurons. A decline in the digestive system's ability to provide the blood with nutrients aggravates blood flow problems. The body's production of chemicals called **cytokines**, which is essential for maintaining the activity and health of cells, also declines with maturity. Cytokines are proteins that stimulate or inhibit the growth and activity of particular cells.

A typical consequence of nervous system cell aging is the reduction of **tonic control** of the body. Tonic control is the result of regular nerve cell communication to certain glands and most muscles. These small, continuous impulses are particularly important in keeping muscles slightly tense, which assists with movement and posture. Without adequate tonic control, a person loses some mobility and has difficulty with balance and posture. Regular neu-

Cytokines Proteins that act as chemical messengers between cells

Tonic Control Regular neural stimulation that maintains the health of glands and muscles

FUNCTION OF THE NERVOUS SYSTEM

Plaque The buildup materials in a nerve cell

Tangle A conditon that causes neurons to lose their shape

Lipofuscin A brown waste material deposited in neurons

ral signals are also needed to maintain sarcomere structure within muscle fibers. Sarcomere degradation resulting from a reduction of tonic control leads to a loss of muscle mass.

The aging of neurons also causes an increase in a nerve cell's latency period, which disables the cell's ability to recover quickly and carry out another action potential. This slows down the impulses to muscles, and delays sensory communication to the brain and body. Waste products from aging nerve cells collect in various neurons causing **plaques** and **tangles**. A plaque is the buildup of materials in a cell. Many nerve cells build up plaques made of amyloid proteins, which can contribute to a decline in nerve cell function. This buildup can block normal metabolic functions of the cell and associated tissues. Tangles are alterations of the nerve cell's cytoplasm, which causes the cell to lose its shape. As a result, the nerve cell cannot adequately conduct impulses to other cells. Maturing cells commonly accumulate a fatty, brown pigment called **lipofuscin**. Lipofuscin builds up in many body cells causing a darkened appearance. It has little effect on the cell, but it can be an important indicator of nerve cell pathology.

Various infectious organisms contribute to nerve cell aging. Herpesviruses and other neural viruses damage nerve cells over a person's lifetime. Chicken pox and cold sores are common herpes infections. Bacterial infections deposit poisons in the blood that diminish the function of neuroglia and neurons. Almost any infection can cause an older person to become severely confused and weak because the disease compounds reduce nervous system function. Diabetes and other metabolic disorders further speed up nervous system aging. Neuroglial cells also decline in function with age.

✓ Concept Check

1. What is the role of cytokines in nerve cell aging?
2. What are some environmental factors that contribute to neuron aging?
3. What are the consequences of losing neuroglial cell function?

DISCOVERY SCENE PLEASE ENTER DISCOVERY SCENE PLEASE ENTER

CSI Break

Have you solved the CSI yet? Did the information about pathology and aging of the nervous system provide any more clues about the uncle's condition? Could his age or any prior disease be related to his illness?

CSI – Case Study Investigation Conclusion

What can you conclude about the uncle's condition? How do you explain the flaccid paralysis? Which event of the day most likely created his symptoms? Does your conclusion explain the blue coloration of the uncle's skin? What do you think the physician found after her analysis?

Answer:

Three things that took place during the day could have explained the uncle's illness. First, he was using pesticides. Many chemicals used to control insects contain a group of poisons called organophosphates. They are neurotoxins that lock onto acetylcholine receptors. Blocking these receptors will eventually cause the muscles to relax, and suppresses breathing and heartbeat. This could produce the uncle's illness. However, there was no evidence of pesticide residues in the blood. It is also possible that the uncle developed botulism from the old stew. Botulism is caused by a bacterium called *Clostridium botulinum*. The bacterium produces a poison that also sits on the acetylcholine receptor. However, the physician found no evidence of this toxin in the body. Normally, this would leave the physician with few other options to investigate immediately. Luckily, the physician was familiar with cases of animals being poisoned after drinking water from stagnant ponds. Certain procaryotic organisms called blue-green algae produce a poison called anatoxin-a. The uncle was exhibiting symptoms similar to those of animals poisoned by anatoxin-a. Anatoxin, a potent alkaloid toxin that blocks the acetylcholine receptors, was the cause of the uncle's condition. He picked up the toxin while cleaning the pond. George Francis, an Australian scientist, first reported this condition in 1878. Anatoxin poisoning is much more common in animals than in humans.

This CSI was adapted from the following articles:

1. Behn D. Coroner cites algae in teen's death. Milwaukee Journal Sentinel [JSOnline]. September 6, 2003. Available at: http://www.jsonline.com/news/state/sep03/167645.asp.
2. Elder GH, Hunter PR, Codd GA. Hazardous freshwater cyanobacteria (blue-green algae). *Lancet*. 1993;341:1519-1520.
3. Mahmood NA, Carmichael WW. The pharmacology of anatoxin-a(s), a neurotoxin produced by the freshwater cyanobacterium *Anabaena flos-aquae* NRC 525-17. *Toxicon*. 1986;24: 425-434.

Chapter Summary

The nervous system communicates external and internal stimuli using electrical impulses called action potentials. It is the major rapid-response regulatory system in the body. It takes stimuli from the environment and converts it into useful information needed for appropriate body responses. Without this function, the body could not quickly adapt to environmental changes and stresses. The nervous system's ability to communicate internal stimuli helps regulate the body's internal conditions. It is responsible for how each cell is able to perform its job in the body. This internal communication is responsible for movement and all mental activities, including thought, learning, and memory.

Types of Nervous System Cells

- The nervous system is an arrangement of cells, tissues, and organs that regulates the body's responses to external and internal stimuli.
- The nervous system cells are neuroglia, neurons, and neural crest stem cells.
- Neuroglia are the most common cells in the nervous system. They assist, protect, and support neurons.
- Neurons are excitable cells that rapidly communicate information about the body and the environment.
- Neurons are composed of dendrites, a nerve cell body, axon hillock, axon, and terminus.
- The synaptic cleft separates neurons from the cells with which they communicate.
- Neural crest cells may be involved in nervous system maintenance and healing.

Neuron Physiology

- Neurons communicate to other cells with neurotransmitters.
- Neurotransmitters can be excitatory (stimulate a neuron) or inhibitory (hinder a neuron).
- A neuron must be excited past its threshold before propagating an action potential.
- The action potential goes through four stages: depolarization, repolarization, hyperpolarization, and recovery.
- The action potential travels in the following order: dendrites → nerve cell body → axon → terminus.

Types of Neuron Communication

- Neurons can be linked in axodendritic, axosomatic, or axoaxonic synapse arrangements.
- An excitatory postsynaptic potential is a neuron arrangement in which one neuron excites another neuron.
- An inhibitory postsynaptic potential is a neuron arrangement in which one neuron inhibits another neuron.
- Reverberating pathways recycle the excitation of a pathway until an inhibitory stimulus shuts it down.

CHAPTER 8

Reflexes
- Reflexes are involuntary responses associated with survival.
- Reflexes start with the stimulation of an affector.
- The flow of information in a reflex is as follows: affector → sensory neuron → interneuron → motor neuron → effector.
- Interneurons can transmit information about the reflex to the brain and other body regions.

Pathology of the Nervous System
- Microorganisms capable of invading or infecting nervous system cells cause infectious pathology.
- Degenerative diseases are due to the progressive deterioration of a cell or a tissue over time.
- Congenital pathologies are caused by embryological and maturation errors that affect nerve cell communication, development, and growth.
- Traumatic injuries result from a wound caused by an external force or violence.
- Toxicological diseases are caused by poisons that affect cell metabolism or communication.

Aging of the Nervous System
- Much of nervous system aging is due to reduced neuroglia and neuron function.
- Neurons cannot be replaced when damaged or lost.
- People carry most of their original neuroglia and neurons from childhood.
- Neurons accumulate damage from oxidization carried out by metabolism.
- Neuron aging can be hastened by alcohol use, drug use, pollution, and smoking.
- Neurons naturally age by accumulating plaques and forming tangles.

 ## Key Terms

Overview
External stimuli
Internal stimuli
Neural tube

Types of Nervous System Cells
Bidirectional communication
Neural crest cell

Neurons
Axon
Axon hillock
Bipolar
Collateral

Dendrite
Interneuron
Multipolar
Nerve cell body
Neurotransmitter receptor
Presynaptic
Postsynaptic
Soma
Synaptic cleft
Terminus
Unipolar

Study Guide

Neuroglia and Stem Cells
Blood-brain barrier
Cerebrospinal fluid
Nodes of Ranvier
Radial glia
Satellite cell
Schwann cell

Neuron Physiology
Action potential
Affector
Aspartate
Depolarization
Dopamine
Effector
Excitatory
Gamma-aminobutyric acid (GABA)
Gate
Glutamate
Glycine
Histamine
Hyperpolarization
Inhibitory
Propagate
Refractory period
Repolarize
Resting potential
Threshold

Types of Neuron Communication
Axoaxonic synapse
Axodendritic synapse
Axosomatic synapse
Epilepsy
Excitatory postsynaptic potential (EPSP)
Inhibitory postsynaptic potential (IPSP)
Innervate
Neural pathway
Reverberating pathway

Reflexes
Reflex
Reflex arc
Transduction

Pathology of the Nervous System
Amyotrophic lateral sclerosis (ALS)
Botulism
Bovine spongiform encephalopathy
Congenital
Creutzfeldt-Jakob disease
Degenerative
Demyelination
Encephalitis
Endotoxin
Galactosylceramide beta-galactosidase
Herpesvirus
Hirschsprung's disease
Krabbe's disease
Lou Gehrig's disease
Mad Cow Disease
Meningitis
Multiple sclerosis (MS)
Neurotrophic
Rabies
Tetanus
Tetrodotoxin
Toxicological
Traumatic

Aging of the Nervous System
Cytokine
Lipofuscin
Plaque
Tangle
Tonic control

Check Your Understanding

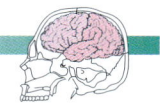
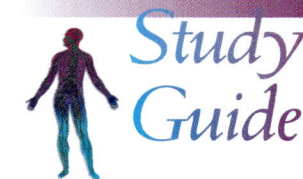

Study Guide

1. What is the major role of the nervous system?
   a. Permit involuntary movement
   b. Detect stimuli from the environment
   c. Maintain the body's physiology
   d. All of the above are major roles.

2. A drug that stops bidirectional communication in the nervous system would have the following effect:
   a. Prevent neurotransmitter release
   b. Stop neuroglia from assisting neurons
   c. Cause neurons to communicate backwards
   d. Cause excitability of neuroglia

3. Which of the following is *not* a feature of neurons?
   a. Excitability
   b. Neurotransmitter production
   c. High energy needs
   d. Capability of mitosis

4. Which of the following is *not* a function of neuroglia?
   a. Control neuron's external environment
   b. Protect neurons from possible toxins in the blood
   c. Stimulate neurons
   d. Insulate the axon

5. Which is the most common type of neuron structure?
   a. Multipolar
   b. Unipolar
   c. Apolar
   d. Bipolar

6. The blood-brain barrier is formed by:
   a. Blood vessels and neuroglia
   b. Neurons and neuroglia
   c. Blood vessels and neurons
   d. Neural crest cells and neurons

7. You are observing a multipolar neuron under the microscope. Which statement is most likely true of this neuron?
   a. It carries out sensory functions.
   b. It is a motor neuron.
   c. It is not involved in reflexes.
   d. It is exhibiting pathology.

8. Which best describes the function of dendrites:
   a. They produce neurotransmitters.
   b. They lack neurotransmitter receptors.
   c. They receive communication from other neurons.
   d. They slow down the action potential.

FUNCTION OF THE NERVOUS SYSTEM

Study Guide

9. You read about research on neuroglial cells that form myelin on adult axons. It is most probable that you are reading about the following type of cell:
   a. Astrocytes
   b. Ependymal cells
   c. Microglia
   d. Schwann cells

10. The death of these cells could directly cause an increase in nervous system infections.
    a. Microglia
    b. Oligodendrites
    c. Astrocytes
    d. Macroglia

11. Which of the following best describes the flow of an action potential?
    a. Dendrite → Axon → Nerve Cell Body → Terminus
    b. Terminus → Axon → Nerve Cell Body → Dendrite
    c. Nerve Cell Body → Dendrite → Axon → Terminus
    d. Dendrite → Nerve Cell Body → Axon → Terminus

12. Which statement is true of a resting nerve?
    a. It has no membrane potential.
    b. Sodium is more abundant outside of the neuron.
    c. Potassium is more abundant outside of the neuron.
    d. The sodium/potassium pump is turned off.

13. The last structure to respond to a reflex is the:
    a. Motor neuron
    b. Effector
    c. Affector
    d. Interneuron

14. Which neuron pathway occurs most often in the part of the brain associated with learning and emotions?
    a. Reverberating
    b. Inhibitory postsynaptic potential (IPSP)
    c. Axodendritic
    d. Neuropathic

15. A virus that damages adult nerve cells is described as this type of pathology:
    a. Congenital
    b. Toxicological
    c. Infectious
    d. Degenerative

A Case Study

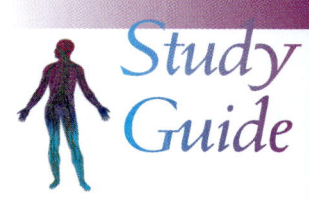

CHEMTRAILS: A REAL PUBLIC HEALTH CONCERN?

Chemtrails is a term used to describe the high-altitude release of chemicals by airplanes. This should not be confused with contrails. Contrails are the exhaust from jet engines that forms above 33,000 feet altitude. Contrails are produced when hot engine exhaust condenses as ice crystals forming "clouds" or vapor trails that quickly fade away. Jet exhaust also contains a certain amount of materials, including metals that form the chemtrails. These chemtrails last longer than contrails and persist in the sky. It is believed that the pollutants eventually settle to the ground or are washed to the earth by rain. The highest concentration of chemtrail contaminants is obviously near airports and military bases with aircraft. Some chemtrails are due to fuel combustion or fuel release. Other chemtrails may be due to the intentional spraying of chemicals associated with atmospheric research and military operations.

People concerned about chemtrail pollution have some evidence that neurological damage can result from this form of pollution. They note that people in high-chemtrail areas have an increase in neural conditions, such as aching joints, anxiety, depression, headaches, loss of bladder control, and nervous spasms. Many organic chemicals and metals are known to effect nerve cell function even in small amounts. A mixed group of petroleum breakdown products called hazardous air pollutants (HAP) are known to cause nervous system disorders. They can also cause developmental defects that affect hearing and vision in children. There is evidence that some types of chemtrails contain aluminum and other metals.

Use the information in this chapter and the following web sites to answer the following questions about the probability of neural diseases due to chemtrails:

- What are the differing views about the health effects of chemtrails?
- How should public health officials respond to possible illnesses due to chemtrails?
- What is the responsibility of the government to monitor the health effects of chemtrails?
- Should the government reduce chemtrail pollution despite conflicting evidence of its effects on neural development in children?
- Who should pay for any illnesses of people living in areas with large amounts of chemtrail pollution?

1. Aluminum and Chemtrails
   http://www.rense.com/general20/alum.htm

2. Chemtrail Central
   http://www.chemtrailcentral.com/

3. Chemtrails Data Page
   http://www.rense.com/politics6/chemdatapage.html

4. Green Futures
   http://www.greenfutures.org/projects/nbairport/nba_airpollute.html

FUNCTION OF THE NERVOUS SYSTEM

Study Guide

5. EPA Mobile Source Emissions
   http://www.epa.gov/otaq/invntory/overview/definitions.htm

6. National Safety Council
   http://www.nsc.org/ehc/mobile/acback.htm

7. Science News On-line (Jet Pollution)
   http://www.sciencenews.org/pages/sn_arch/7_6_96/bob1.htm

Where Do We Go from Here?

People in health fields can use their knowledge of nervous system function to solve everyday problems. You may wish to use other resources, such as the web sites below, in addition to your textbook to investigate the answers to each of the following situations:

1. An elderly neighbor read that foods high in tryptophan would improve his memory. Explain the validity of that claim.
   http://www.emedicinehealth.com/index.asp

2. A female neighbor is being encouraged by friends to seek Botulinum toxin type A (usually prescribed as Botox® by Allergan, Inc.) treatments for facial wrinkles. She asks you to explain how the treatments work.
   http://www.nlm.nih.gov/medlineplus/botox.html

3. You were asked to explain to some young athletes why too much sodium is dangerous during heavy exercise.
   http://www.nlm.nih.gov/medlineplus/ency/article/003927.htm

4. A friend asks you why *Ephedra sinica* or ma huang is no longer available in foods and health supplements.
   http://www.webmd.com/

5. You are volunteering as a personal trainer working with an elderly client. She is recovering from a stroke that killed the nerve cells controlling her arm. She wants you to explain why she was told she would never regain the full function of her arm.
   http://www.strokeassociation.org/

Skills Activities

1 Pupil Reflex

Materials
- small flashlight with a dim bulb
- small ruler
- 12-inch ruler
- meter stick
- ability to darken the room or use of a dark location
- patella reflex hammer (if possible)
- stopwatch or clock with a second hand
- stable table (to sit upon)
- board for recording class results

One way of monitoring the health of particular nerves is by observing reflex responses. The body possesses a diversity of reflexes that respond to various stimuli. Physicians can track specific nerves by investigating reflexes unique to that nerve. The pupil reflex response is used to indicate the health of a particular nerve in the brain. This nerve can be an indirect indicator of brain damage due to disease or injury. The knee-jerk reflex is a general indicator of nervous system function. It involves two sets of long nerve cells that communicate information up and down the length of the upper leg into the lower lumbar region. Observation of the "catch" reflex involves measuring the time it takes a person to communicate information from the eye as it detects a dropped object to the arm and hand in movement to catch it. Reflexes are important indicators of aging and pathology. The speed at which a reflex occurs differs from person to person, and depends upon his or her experiences and current level of attentiveness.

1. Sit the subject in a chair so that you can easily view their eyes.
2. Dim the lights in the room and wait for 3 minutes.
3. Place the ruler near the subject's left eye, and measure and record the diameter of the pupil
4. With the bulb off, place the flashlight 12 inches from the subject's left eye. Aim it straight at the center of the eye.
5. Set the stopwatch to zero, and click on the stopwatch as you turn on the flashlight.
6. Measure and record the time it takes for the pupil to close tightly.
7. Again, measure and record the diameter of the pupil.
8. Repeat steps 1 through 7 for the right eye.
9. This experiment should be repeated three times for each eye to obtain good data.
10. Record the following data for each pupil:
    a. average time it takes to respond to light
    b. average change in diameter
11. Place the data on a chart with the data from other subjects.
    - Are there any differences in the reflex between the right and left eyes of the subjects?
    - Are there any differences between individuals in the average time it takes for the pupils to respond to light?

Study Guide

FUNCTION OF THE NERVOUS SYSTEM

Study Guide

- Are there any differences between individuals in the average change in diameter of the pupils?
- What are some possible reasons for any differences in response between the pupils of one individual?
- What are some possible reasons for any differences in response between the pupils of different individuals?

Skills Activities

2 Knee-Jerk Reflex

1. Have a subject sit on a table with the left leg crossed over the right leg so that the crossed leg swings freely.
2. Set the stopwatch to zero.
3. Click on the stopwatch as you quickly, but gently, hit the crossed leg just below the knee with the side of your hand.
4. Measure and record the time it takes for the subject's leg to kick out and return to the resting position.
5. Repeat the procedure at least three times.
6. Carry out steps 1 through 5 for the right leg.
7. Place the data on a chart with the data from other subjects.
   - Are there any differences in reflexes between the right and left legs of the subjects?
   - Are there any differences in reflexes between the subjects?
   - What are some possible reasons for any differences in response between the left and right knee-jerk reflexes of one individual?

Skills Activities

3 Catch Reflex

1. Have a subject stand upright at least 2 feet away from any chairs or tables.
2. Stand facing the subject.
3. Hold a 12-inch ruler by the upper end, and let it hang down in front of the subject. You should be holding the ruler by the highest number with the lowest number down toward the floor.
4. Have the subject put his or her left hand near the bottom of the ruler.
5. The subject should have his or her hand ready to grab the ruler, and should not be touching the ruler at this point.
7. Tell the subject that you will drop the ruler sometime within the next 5 seconds.
8. Now, tell the subject to watch the ruler because he or she is supposed to catch it when it is dropped.
9. Release the ruler.
10. Record the inch or centimeter marking at which the subject caught the ruler.
11. Use a meter stick if the subject cannot catch the regular ruler.
12. Repeat this test at least three times.

Study Guide

13. Convert the marking into reaction time using the chart below.
14. Record the reaction time.
15. Repeat steps 3 through 11 for the subject's right hand.
16. Place the data on a chart with the data from other subjects.

Distance Marking		Time
Inches (in)	Centimeters (cm)	Seconds (sec)
2	5.1	0.10
4	10.2	0.14
6	15.3	0.17
8	20.4	0.20
10	25.5	0.23
12	30.0	0.25
17	42.5	0.30
24	60.0	0.35
31	77.5	0.40

- Are there any differences in reflexes between the left and right arm catch reflex of the subjects?
- Are there any differences in reflexes between the subjects?
- What are some possible reasons for any differences in response between the left and right arm catch reflex of one individual?
- What are some possible reasons for any differences in response between the left and right arm catch reflexes of different individuals?

FUNCTION OF THE NERVOUS SYSTEM

Case Study Investigation

Case Study Investigation #9

You just heard that a former professional football player living in your neighborhood was taken to the police station after a disturbance at a local grocery store. You find out that he was wandering around the store aimlessly and making aggressive comments as if he were drunk. A friend then calls to tell you that the football player was admitted to the hospital because he started having seizures. Later in the week, some neighbors tell you that the football player was showing signs of weakness and often acted confused during conversation. There were also some concerns mentioned about his excessive alcohol consumption and weight gain. Your friend asks you what might be wrong with the 50-year-old former athlete. At the end of the chapter, you will be asked to determine the possible nervous system problems causing the illness.

CHAPTER 9

STRUCTURE OF THE NERVOUS SYSTEM

Chapter Outline

Case Study Investigation (CSI)
Applied Learning Outcomes
Overview
Nerve Structure
Nervous System Components
   Central Nervous System
      Brain
      Spinal Cord
   Peripheral Nervous System
      Cranial and Spinal Nerves
      Autonomic Nervous System
Human Senses
   Taste
   Smell
   Vision
   Hearing and Balance
Wellness and Illness over the Life Span
   Pathology of the Nervous System Structure
   Aging of the Nervous System Structure
CSI Conclusion
Study Guide

Applied Learning Outcomes

- Use the terminology associated with the nervous system.
- Learn about the following:
  - nerve structure
  - types of nerve pathways
  - nervous system components
  - central nervous system structure and function
  - peripheral nervous system structure and function
- Understand the aging and pathology of the nervous system.

Overview

Key Terms: central nervous system (CNS), peripheral nervous system (PNS)

Central Nervous System (CNS) A major division of the nervous system composed of the brain and spinal cord

Peripheral Nervous System (PNS) The part of the nervous system made up of neurons and neuroglia outside of the brain and spinal cord

The nervous system is a complex arrangement of neuroglia and neurons bundled into two major components, the central and peripheral nervous systems. Equally important in the operations of the nervous system are blood vessels and connective-tissue structures. Within the skull and spinal column lies the **central nervous system (CNS)**. It is composed of the brain and spinal cord, which work as a controlling network for the entire body. Portions of the nervous system that extend beyond the brain and spinal cord form the **peripheral nervous system (PNS)**. The PNS provides motor and sensory communication between the CNS and the body. Sensory information usually travels from the PNS to the CNS. In contrast, motor information travels from controlling centers in the CNS to the PNS.

Good Choice Bad Choice

It is valuable to use what you learned about the nervous system to help others to better understand the world around them. It is very important to check your facts and seek further information about certain topics before discussing health and science issues. Here are some suggestions to foster a better public awareness of the nervous system:

1. Speak to schoolchildren about nervous system function.
2. Volunteer at an assisted-living center with neurologically impaired people.
3. Help elderly persons to better understand dietary supplements claimed to improve memory.
4. Volunteer at a school health day to teach children how to prevent nervous system damage during play and sports.

In the fetus, the CNS is the first component of the nervous system to form. It starts out as a hollow tube called the neural tube, which lies along the middorsal region of the developing fetus. Neuroglia in the neural tube spread to various parts of the body to form the PNS. Growth factors secreted by the other body components determine the pattern of the PNS. These growth factors direct the patterns that produce basic nervous system communication, including reflexes. The organs of the nervous system are structures called nerves. Nerves are the neuron pathways bundled together to form the various parts of the CNS and PNS.

✓ Concept Check

1. What is the CNS?
2. Define the roles of the PNS.
3. Describe the basic formation of the nervous system.

Nerve Structure

Key Terms: afferent nerve, efferent nerve, endoneurium, epineurium, neurofascicle, perineurium

A nerve is defined as an enclosed bundle of neurons and associated neuroglia running to various structures throughout the body (Figure 9.1). Nerves primarily constitute the PNS. The brain and spinal cord have **nerve tracts**, or neurons, bunched together in pathways that are not indistinct bundles. Nerves can be categorized as **afferent** or **efferent**. Afferent nerves carry sensory information from the body to the brain (Figure 9.2). Efferent nerves transmit signals from the brain primarily to the glands and muscles.

Nerve Tract A bundle of neurons following a path through the body or brain

Afferent Nerve The part of the PNS that carries information from the body to the central nervous system

Efferent Nerve The part of the peripheral nervous system that carries information from the central nervous system to muscles and glands

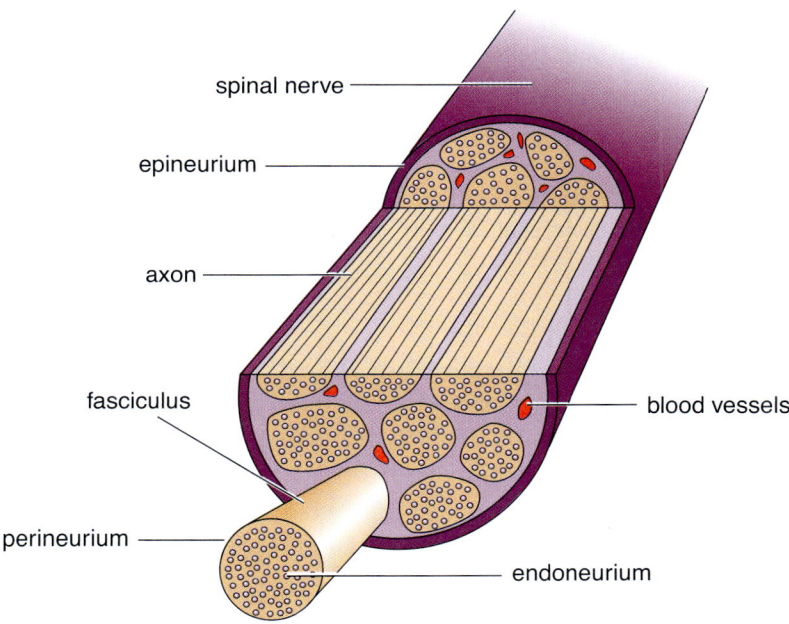

Figure 9.1 Nerve Structure

A typical nerve is covered by a continuous sheet of areolar connective tissue called the **epineurium**. It forms a protective sheath around the nerve and is an entryway for blood vessels that assist the nerve. Fat cells within the epineurium cushion the underlying tissue from damage. The epineurium is a dynamic layer involved in healing and is subject to diseases than can cause it to become inflamed. Within the epineurium are two anatomical regions of the nerve. The predominant region is composed of numerous **neurofascicles**. Each neurofascicle is a tight bundle of axons held in place by the epineurium. Neurofascicles generally contain a mixture of myelinated and unmyelinated axons. Myelinated axons contain Schwann cells that expedite the speed of neural impulses. Unmyelinated axons do not contain Schwann cells. However, the axons of unmyelinated axons can share the benefits of nearby myelinated axons that they make contact with. Contact with the surface of another axon's Schwann cell will speed up impulses. Surrounding the neuro-

Epineurium The outermost layer of connective tissue covering a nerve

Neurofascicle A bundle of neurons and associated neuroglia that form a part of a nerve

Figure 9.2 Afferent Nerve Paths

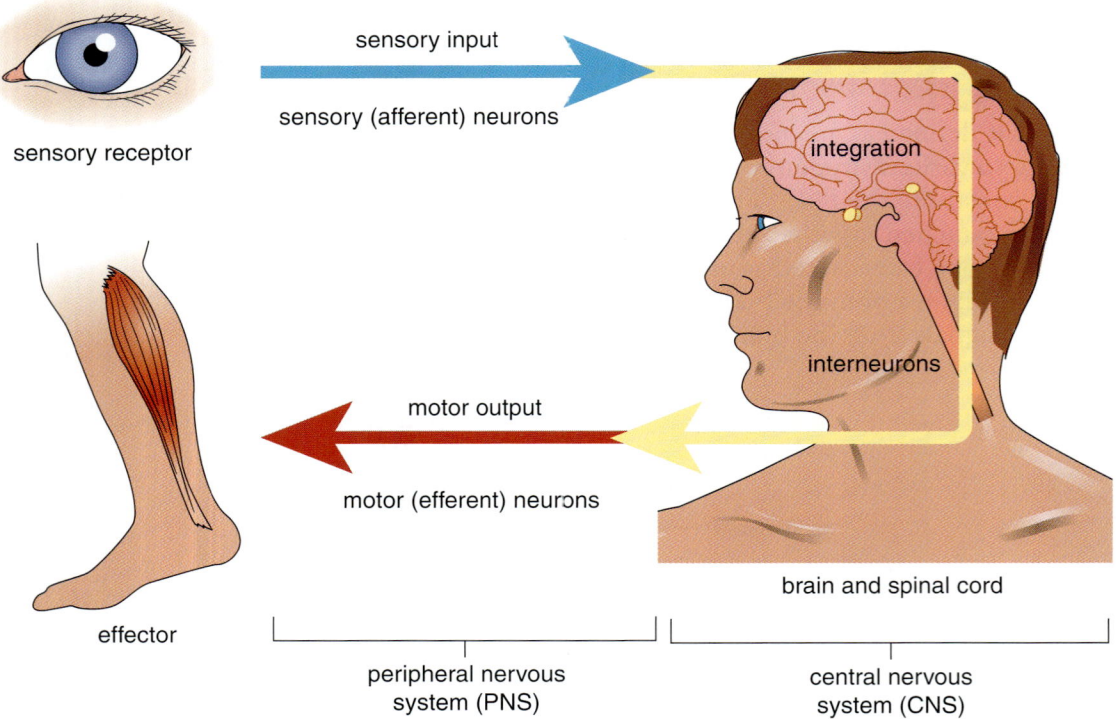

Perineurium The layer of connective tissue covering a nerve neurofascicle

fascicle is a covering called the **perineurium**. The perineurium is composed of a dense layer of metabolically active cells that assist with the diffusion of ions needed for the action potential. It also provides the underlying cells protection from damage that can tear the axons.

CAFFEINE: ITS EFFECTS ON THE NERVOUS SYSTEM

Is caffeine truly effective at stimulating brain activity, or is it merely a myth? The chemistry of caffeine provides clues that it acts like a neurotransmitter in the brain. Caffeine, or 3,7-dihydro-1,3,7-trimethyl-1H-purine-2,6,-dione, belongs to a group of chemicals called xanthines. Other types of xanthines are found in chocolate, cola nuts, and tea. Caffeine has been shown to interfere with the function of the neurotransmitter adenosine in the brain. Adenosine, a CNS neuromodulator, inhibits excitatory neural impulses. Caffeine's observed mental stimulation effects come from the fact that it blocks the inhibitory effects of adenosine. However, adenosine inhibition is not linked to mental function. It works on parts of the brain and nervous system pathways affecting motor function. The typical effects of caffeine are the following:

- diminished fine motor coordination
- dizziness
- increased mental alertness
- increased sensory alertness
- insomnia
- restlessness

Caffeine is absorbed rapidly through the stomach and small intestine into the blood where it is carried throughout the body. It will affect the body within about 15 minutes after being consumed and may continue its influence for hours. Caffeine can be fatal if the average adult takes in more than 10 grams (g) within 1 hour. One cup of coffee can contain up to 0.1 g of caffeine. This is the equivalent of the dosages found in "wake-up pills" and energy-boosting drinks. At one time, it was thought that caffeine was not truly addictive. Current research shows that there is some degree of dependence if 0.4 g of caffeine is consumed daily. The traditional "caffeine-withdrawal headache" is a sign of its mildly addictive effects.

Neurofascicles work as independently functioning groups of neurons. They can branch away from the main nerve to innervate particular structures. Smaller nerves are no more than an epineurium surrounding one neurofascicle. Each neuron and its associated neuroglia within a neurofascicle are surrounded by an **endoneurium**. The endoneurium is a thin connective tissue that surrounds each axon with its myelin sheath. It functionally separates the nerve cells so that the action potential from one does not interfere with the functioning of other axons. In addition, the endoneurium plays a very important role in maintaining the physiological environment of individual axons. It controls the levels of electrolytes needed for carrying out the action potential, and helps to supply nutrients and remove wastes.

Endoneurium The connective-tissue sheath that surrounds each neuron and associated neuroglia, such as myelin

Ganglia A collection of nerve cell bodies

Some parts of the epineurium form coverings around collections of nerve cell bodies called **ganglia**. Ganglia are defined as accumulations of nerve cell bodies located outside the CNS. Concentrations of nerve cell bodies called nuclei are found in the CNS as well; however, they are not covered by sheaths. The ganglia of the PNS are composed of unipolar neurons and are usually associated with sensory function. The epineurium surrounding a ganglion forms a capsule made of the outer portion of connective tissue and the inner portion of satellite cells. These satellite cells provide insulation and maintain the physiological environment of the neurons.

✓ Concept Check

1. Describe the structure of a nerve.
2. What are the roles of afferent and efferent nerves?
3. Define the structure and function of a ganglion.

DISCOVERY SCENE PLEASE ENTER DISCOVERY SCENE PLEASE ENTER

What information about nerve structure could be helpful in better understanding the former football player's condition? Could it be due to a problem with one nerve or many nerves? Could a particular region of a nerve, such as a ganglion, be responsible for the symptoms?

STRUCTURE OF THE NERVOUS SYSTEM

Nervous System Components

The networks of nerves making up the human nervous system ensure that each region of the body is in regular communication with the CNS (Figure 9.3). Afferent peripheral nerves act as trunks that feed sensory information to the brain through their entry into the spinal cord. The spinal cord uses this sensory

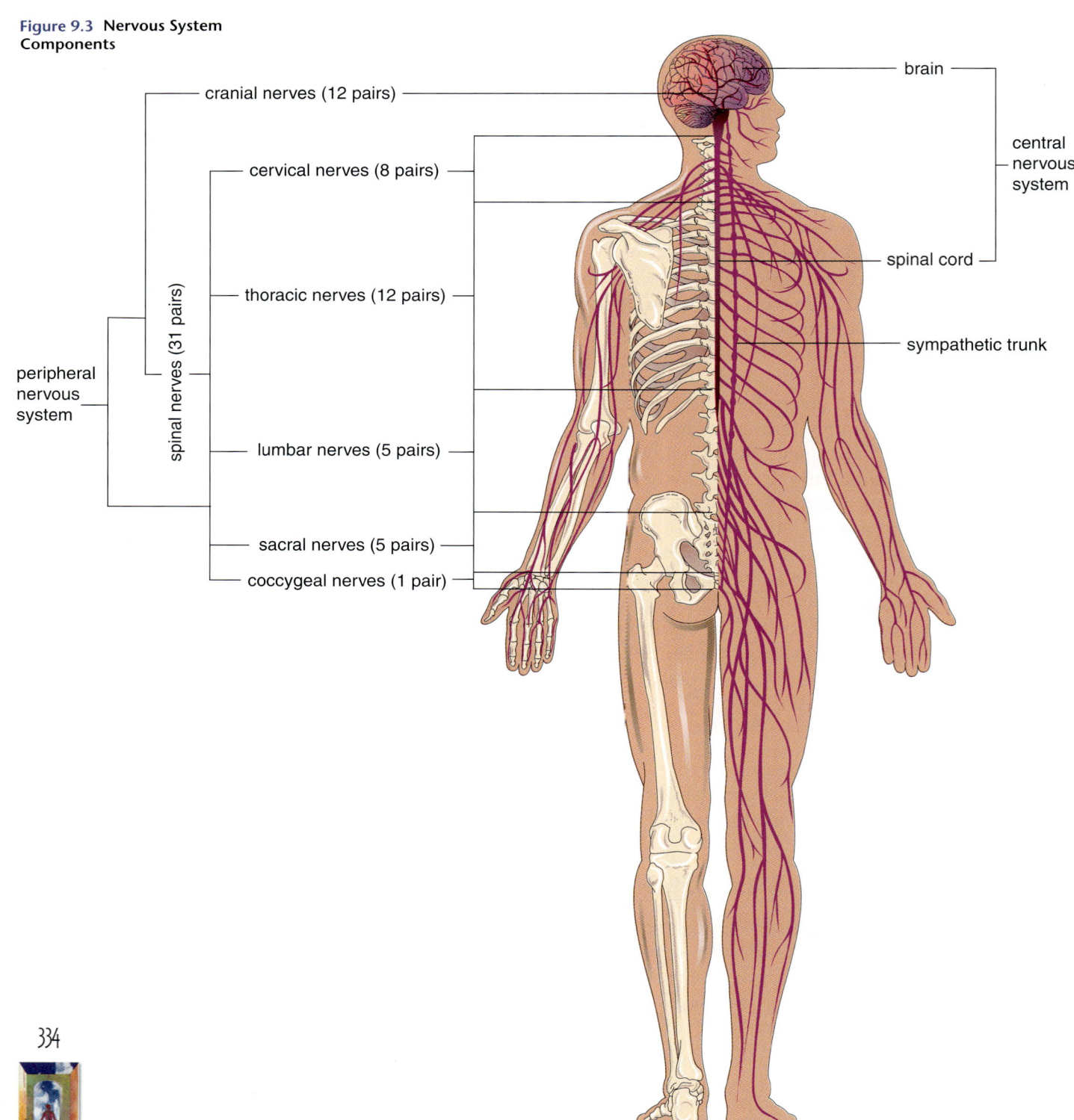

Figure 9.3 Nervous System Components

information to conduct a reflex or convey crucial information to the brain. Peripheral nerves provide information from the skin and all of the major internal organs. The brain has a pivotal role in human survival by coordinating and responding to the large amounts of information it is constantly being fed. Messages from the brain travel down efferent nerves that branch out from the brain and neuron bundles running throughout the spinal cord. Most of the brain's information stimulates motor nerves that regulate the functions of glands and muscles.

Central Nervous System

> **Key Terms:** ascending tracts, arachnoid mater, basal nuclei, brainstem, central canal, central sulcus, cerebellum, cerebral hemispheres, cerebrum, choroid plexus, convolutions, corpus callosum, cortex, descending tracts, dura mater, extrapyramidal tracts, forebrain, frontal lobe, gray matter, gyri, hindbrain, hydrocephalus, insula, intermediate gray matter, lateral sulcus, limbic system, longitudinal cerebral fissure, medulla oblongata, meninges, midbrain, motor area, occipital lobe, parietal lobe, pia mater, pons, sensory area, subarachnoid space, temporal lobe, ventricles, vertebral lumen, white matter

In spite of being the regulatory hub of the body, the CNS is isolated from the rest of body by layers of membranes encased in a bony covering. The brain lies within the cranial cavity. It connects directly to the spinal cord through the foramen magnum on the inferior portion of the occipital bone. The spinal cord runs through a flexible tunnel called the **vertebral canal**. It is formed by the vertebral foramen of the various vertebrae stacked into the vertebral column. Major blood vessels enter and exit the skull and vertebral column to ensure an adequate exchange of nutrients and wastes needed for normal nervous system function. Three layers of tissue called the **meninges** separate the brain and spinal cord from their bony covering. The three layers are the **dura mater**, **arachnoid mater**, and **pia mater**.

The outermost layer of the meninges is the dura mater, which is pressed tightly against the bony surface on the interior of the cranium and the vertebral column. It is a tough, dense connective tissue that envelopes the underlying layers of the meninges. The dura mater acts as a barrier against trauma, which can wound the CNS. It prevents the CNS from rubbing against the skull and vertebral column. Below the dura mater is the thin, delicate arachnoid mater. Arachnoid means "like a spider's web." The arachnoid mater is composed of a loosely organized, irregular connective tissue. A major function of the arachnoid mater is its ability to cushion the CNS from rapid movements and blunt hits to the skull and vertebral column. Underneath the arachnoid mater is the **subarachnoid space**, a cavity filled with cerebrospinal fluid. Many blood vessels run throughout the subarachnoid space.

Directly making contact with the brain and spinal cord is the delicate pia mater. The connective tissue of the pia mater adheres to the surface of the brain and spinal cord. It carries out several jobs, including carrying the blood vessels into the CNS, forming sheaths over nerves passing through the outer meninges layers, and assisting with the production of cerebrospinal fluid.

Vertebral Canal A tube-like cavity that runs the length of the vertebral column

Meninges Membranes surrounding the brain and spinal cord

Dura Mater The tough outermost layer of the meninges protecting the brain and spinal cord

Arachnoid Mater The middle layer of the meninges, which nourishes and protects the brain and spinal cord

Pia Mater The inner layer of the meninges, which nourishes and protects the brain and spinal cord

Subarachnoid Space A cavity that lies between the arachnoid and pia mater of the meninges

Associated with the pia mater is the **choroid plexus**. It is a mass of blood vessels and associated glial cells that secretes cerebrospinal fluid into specific regions of the brain. Ependymal cells are the neuroglia that produces the cerebrospinal fluid.

The CNS is noted for its patterns of gray and white color when viewed without a microscope. These patterns represent the **gray** and **white matter**. Gray matter is composed of concentrated areas of neuron cell bodies. It is a gray color because the neurons accumulate the dark-pigmented fat called lipofuscin. White matter refers to areas composed primarily of axons and neuroglia. The white color comes from the light-color fats making up myelin cell membranes. Gray matter in the brain is distributed predominantly on the surface, or **cortex**. The cortex of the brain is noticeably ridged in patterns called **convolutions** or **gyri**. It is believed that the convolutions increase the surface area of the gray matter, making the brain more compact. Small pockets of gray matter called nuclei are found in specialized interior portions of the brain. They control specific brain functions. In contrast to the brain, the spinal cord has its gray matter concentrated on the inside. Its surface is a thick covering of white matter.

> **✓ Concept Check**

> 1. Describe the three layers of the meninges.
> 2. What is the function of the choroid plexus?
> 3. Describe the differences between the gray and white matter of the CNS.

Brain The brain is an incredibly complex accumulation of neurons organized into the body's neural command center (Figure 9.4). It contains billions of neurons designed to send out hundreds of commands each second throughout the body. Even something as simple as maintaining one's posture requires an orchestration of neural signals to dozens of muscles that help a person stand upright. Most of the neurons making up the brain are categorized as multipolar neurons, which have a single axon and many dendrites. This permits these neurons to receive and process a great amount of information from other neurons. The multipolar neurons of the brain vary greatly in their number of dendrites. Tracts of axons are also important components of brain structure. They transfer the information to other regions of the brain and throughout the body.

The brain is developmentally divided into three parts: the **forebrain**, **midbrain**, and **hindbrain**. Each component of the embryonic brain develops a specific function as it matures into the adult brain. However, these parts work closely together as a single, coordinated center of control. The forebrain forms the **cerebrum** and **diencephalon**, which is the largest and most-anterior portion of the brain. This part of the brain is responsible for emotions, memory, motor movement, and thought. It contains the **basal nuclei** and the cerebral cortex. The basal nuclei are four pockets of gray matter buried deep in the anterior portion of the cerebrum. Neurons of the basal nuclei relay motor information from the cerebrum to the spinal cord. They assist with motor function, permitting coordinated, steady body movements. The cerebral cortex is the convoluted gray matter covering of the brain. If stretched out, the cortex would cover an area of 18 square feet.

The cerebral cortex is nourished by blood vessels in the arachnoid layer that form the blood-brain barrier (Figure 9.5). The blood-brain barrier pre-

---

Choroid Plexus A collection of blood vessels in regions of pia mater in the brain

Gray Matter The parts of the brain and spinal cord composed primarily of nerve cell bodies

White Matter The parts of the brain and spinal cord composed primarily of axons and myelin

Cortex The gray matter covering the surface of the brain

Convolutions or Gyri Folded ridges in the cortex of the brain

Forebrain The largest division of the embryonic brain

Midbrain An arrangement of neurons that connects with the forebrain and organizes sensory information

Hindbrain The lowermost portion of the embryonic brain, just above the spinal cord

Cerebrum The part of the brain responsible for emotions, memory, motor movement, and thought

Diencephalon The part of the forebrain that contains the thalamus and hypothalamus

Basal Nuclei Four regions of gray matter found deep in the forebrain

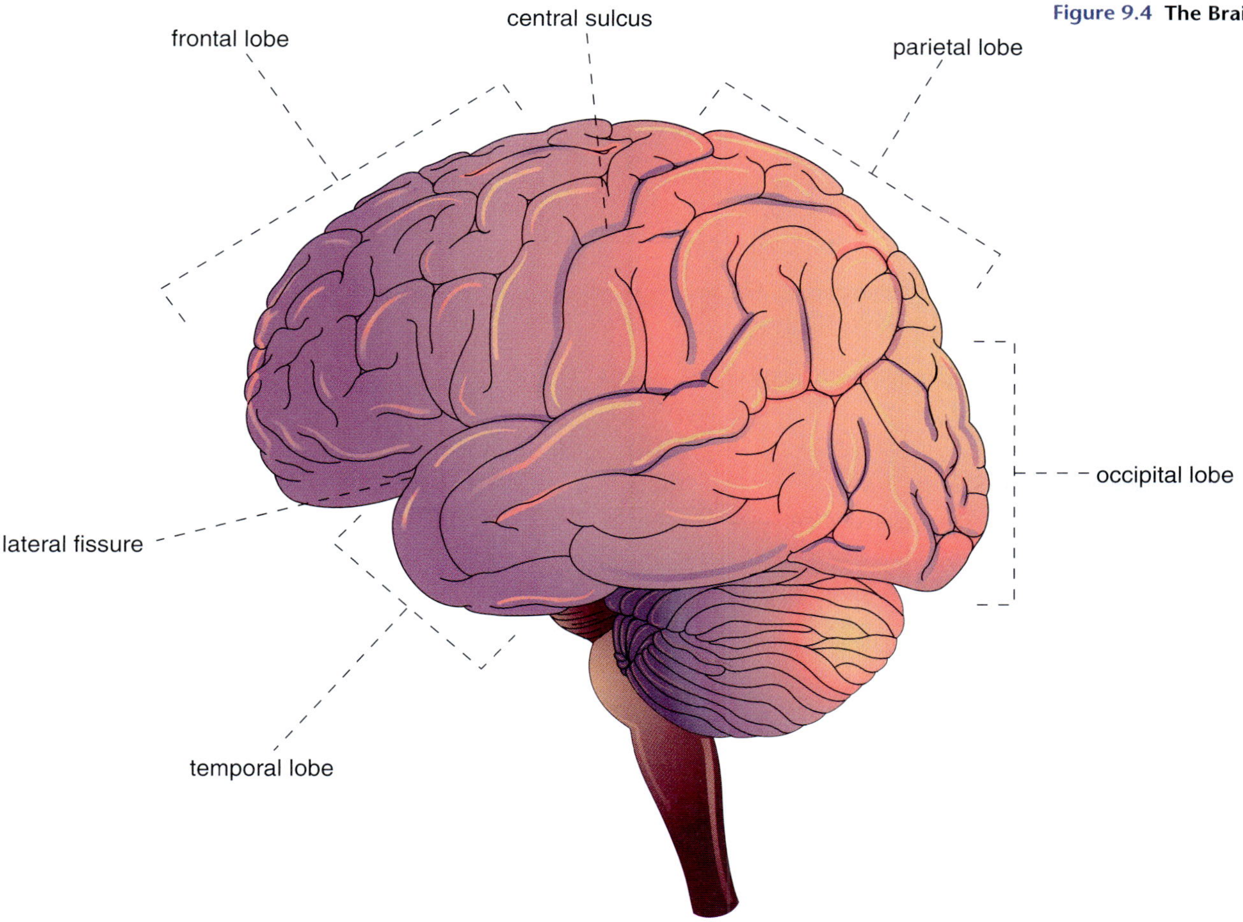

Figure 9.4 The Brain

- frontal lobe
- central sulcus
- parietal lobe
- occipital lobe
- temporal lobe
- lateral fissure

vents many potentially harmful substances from entering the brain. In effect, the blood-brain barrier is a selective filter that keeps certain substances from entering the nervous system. Unfortunately, the blood-brain barrier also blocks many medications, so drug designers and physicians have to find creative ways of bypassing it. Sometimes this includes injecting medications directly into the cerebrospinal fluid.

The cerebrum is actually composed of left and right halves called the **cerebral hemispheres**. They are separated by a midsagittal crease called the **longitudinal cerebral fissure**. Each hemisphere carries out slightly different motor and sensory functions. For example, the regions of the left hemisphere are specialized for language and speech. All people use both hemispheres. However, certain functions of a hemisphere dominate a person's behavior. A person with a dominant left hemisphere would be more adept at language skills and the analytical processing skills required for scientific and mathematical reasoning. The right hemisphere is considered to be the site of more "creative" perception. Artists and musicians are, therefore, considered to have right-brain dominance. Between the hemispheres is a band of white matter called the **corpus callosum**. It connects the hemispheres deep within the base of the cerebrum. The corpus callosum ensures communication between the hemispheres. It also allows one hemisphere to partly compensate for loss of function if the other hemisphere is damaged. The hemispheres can be further divided into four distinct anatomical regions called lobes (Figure 9.4). Making up the anterior portion of the cerebrum is the **frontal lobe**. It primarily processes

Cerebral Hemispheres The left and right halves of the cerebrum

Longitudinal Cerebral Fissure A crevice that separates the left and right hemispheres of the cerebrum

Corpus Callosum A band of white matter connecting the left and right hemispheres of the cerebrum

Frontal Lobe The region of the brain situated directly behind the forehead

STRUCTURE OF THE NERVOUS SYSTEM

Figure 9.5 **Blood-Brain Barrier**

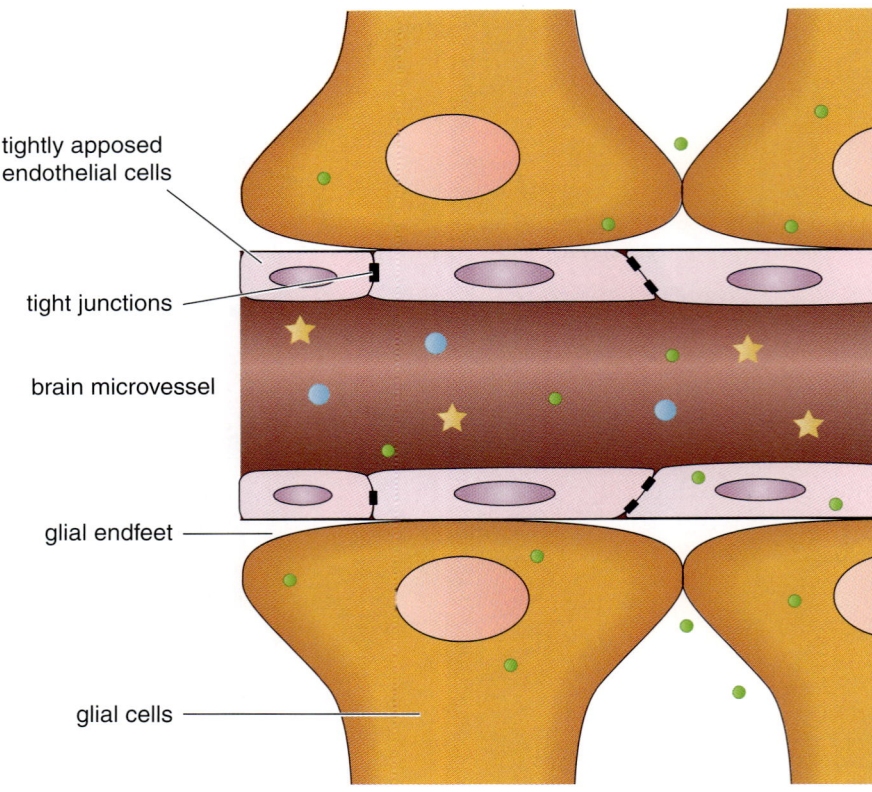

Motor Cortex A strip of cortex involved in voluntary skeletal muscle control

intellectual information that helps with the organization of thoughts. A strip of cortex making up the posterior region of the frontal lobe is the **motor cortex**. The motor cortex has voluntary control over the skeletal muscles.

CREATIVE TREATMENTS FOR PARKINSON'S DISEASE

Parkinson's disease is a degenerative brain disorder caused by the loss of cells that produce the neurotransmitter dopamine. It was named after Dr. James Parkinson of London who first described the disease in 1817. The most common form of Parkinson's disease results from some type of destruction to dopamine-producing cells in the substantial nigra (or black substance). This brain region is located in the midbrain and is involved in controlling posture and voluntary movement. It provides us with the ability to carry out smooth, coordinated actions. The disease causes breathing difficulties, limb and joint stiffness, speech problems, and tremors. A drug called levodopa is administered to replace the lost dopamine. However, the drug is not equally effective in all people, and is very difficult to match the dose to the levels the body regularly needs.

Physicians are trying new ways to treat Parkinson's disease without the need for levodopa treatments. One approach is to transplant parts of the adrenal glands, which produce dopamine, into the brain near the substantial nigra. The adrenal glands are located just above the kidneys. This transplantation surgery is not always successful. Plus, it is difficult to get the adrenal glands to produce the correct amounts of dopamine that the brain needs. Another strategy being tested involves the use of fetal brain cells. It is hoped that the fetal cells will fill in the lost dopamine-producing cells of the substantia nigra, in effect, curing the disease.

A groove called the **central sulcus** separates the motor area of the frontal lobe from the **parietal lobe**. The parietal lobe is involved mostly in emotions and sensory interpretation. Running along the central sulcus is a strip of cortex called the **sensory cortex**. The sensory cortex is specialized to interpret sensory information from the lips, skin, and tongue. Inferior and lateral to the frontal and parietal lobes is the **temporal lobe**. It is separated from the other lobes by a distinct groove called the lateral sulcus. The temporal lobe organizes and stores the memories of sounds and vision. Forming the posterior region of the cerebrum is the **occipital lobe**. The occipital lobe interprets vision and assists with eye function. Look at a diagram of the skull, and note how the lobes correspond to the names of the skull bones under which they predominantly lie. A small region called the **insula** is found buried midway under the lateral sulcus. It plays a very important role in processing memories.

The deep, medial portion of the forebrain also contains the hypothalamus, thalamus, and **ventricles**. An area called the **limbic system** comprises the combined functioning of the basal nuclei, hypothalamus, and thalamus. The ventricles are four connected cavities within the forebrain that contain cerebrospinal fluid. They are associated with the choroid plexus and continue into the spinal cord. The ventricles protect the brain from trauma by acting as an internal cushion when the head is hit or violently moved. The ventricles provide a pathway for the cerebrospinal fluid. Swelling of the ventricles due to the abnormal production of cerebrospinal fluid can cause severe head pain and brain damage. In children, this causes a disease called **hydrocephalus**, which is characterized by an enlarged cranium.

The midbrain is a small strip of neurons that connects the cerebrum to the hindbrain. It also possesses auditory and visual reflex areas. The midbrain controls the ability of the eye to adjust to changes in light intensity or sound. Two small nuclei in the midbrain are involved in coordinating motor function. These nuclei produce the neurotransmitter dopamine. A decrease in dopamine from this part of the brain is responsible for Parkinson's disease. Directly attached to the midbrain is the **pons**, the first component of the hindbrain. Nuclei in the pons organize and transmit sensory information received from the body. Just below the pons is the **medulla oblongata**. The medulla oblongata regulates involuntary body functions, such as blood pressure, breathing, heart rate, and swallowing. Posterior to the pons is the **cerebellum**, the third component of the hindbrain. The term means "little cerebrum," which is indicative of its appearance, not its function. The cerebellum plays a very important role in the part of the brain concerned with balance, posture, and the coordination of body movement. Many people use the term **brain stem** to describe the midbrain and hindbrain collectively.

Central Sulcus A groove separating the frontal lobe from the parietal lobe of the cerebrum

Parietal Lobe A region of the cerebrum posterior to the frontal lobe

Sensory Cortex A strip of cortex involved in sensory interpretation

Temporal Lobe A region of the cerebrum located inferior and lateral to the frontal and parietal lobes

Occipital Lobe The posterior region of the cerebrum

Insula A region of the cerebrum associated with memories

Ventricles A collection of cavities within the forebrain containing cerebral spinal fluid

Limbic System A collection of nuclei at the base of the cerebrum associated with emotions

Hydrocephalus Excess fluid in the ventricles

Pons The topmost portion of the hindbrain

Medulla Oblongata A section of the hindbrain connecting the pons to the spinal cord

Cerebellum A large portion of the base of the brain that coordinates voluntary movements

Brain Stem The part of the brain that is connected to the spinal cord

✓ Concept Check

1. Describe the components of the forebrain.
2. Explain the role of the midbrain.
3. Describe the components of the hindbrain.

Spinal Cord Arising from the base of the medulla oblongata is the spinal cord. The spinal cord is a cylinder of nervous tissue enclosed in the spinal canal of the vertebral column (Figure 9.6). Meninges cover its outer surface and it contains a hollow core called the **central canal**. The central canal con-

STRUCTURE OF THE NERVOUS SYSTEM

Figure 9.6 **Cross-Section of the Spinal Cord**

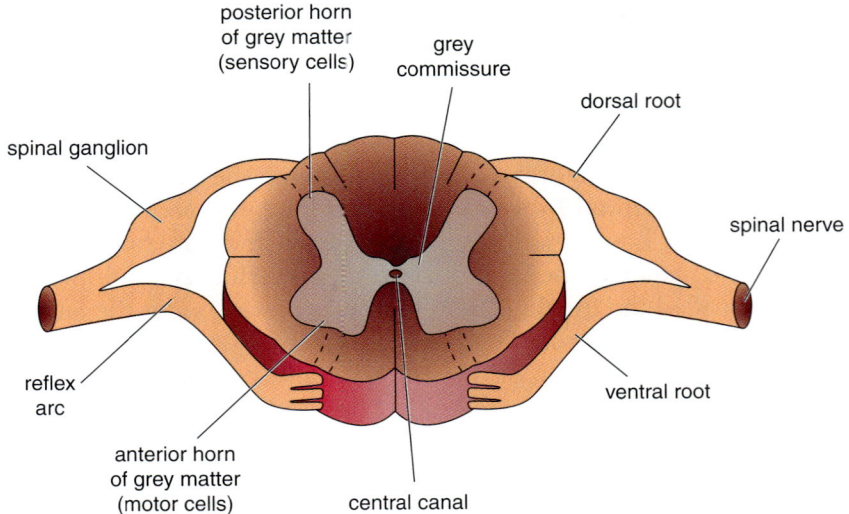

tains cerebral spinal fluid and is continuous with the ventricles of the brain. A structure similar to the blood-brain barrier also protects the spinal column from potentially harmful materials in the blood. The spinal cord does not run the full length of the vertebral column. It narrows down to several loose strands of nerve tracts that exit the sacral and coccogeal vertebrae.

The surface of the spinal cord is covered with white matter. This white matter is composed of 15 pairs of longitudinal strips of axons, or nerve tracts, which convey specific types of information to and from the brain. The tracts found along the dorsal portion of the spinal cord carry sensory information. They are called **ascending tracts** because they carry information up the spinal cord to the brain. **Descending tracts** run along the ventral and lateral portions of the spinal cord. They carry motor information down from the brain to various effectors. The axon tracts transmit information directly to the PNS and interneurons involved in reflex arcs. Some tracts are highly specialized, such as the **extrapyramidal tracts**, which carry information from the medulla oblongata to assist with posture and muscle tone. Ascending tracts in a specific ventral band of the spinal cord are important for communicating sensations of pain to the brain.

Spinal cord gray matter is organized much differently from white matter. It is predominantly made of nerve cell bodies that carry information for reflexes and transfer information to the white-matter nerve tracts. Basically, the gray matter establishes the information flow for the spinal cord. The gray matter looks like a butterfly when viewed from a transverse section. Each wing of the butterfly is divided into a dorsal horn and a ventral horn. The dorsal horn consists primarily of interneurons that convey sensory information. Some of these interneurons send pain signals to the brain. The ventral horn contains the nerve cell bodies of motor neurons. They take information from the white matter and transfer it to muscles. The **intermediate gray matter** lies between the butterfly wings. It conveys sensory and motor information for glands and muscles. In the center of the gray matter between the butterfly's wings is the central canal.

Central Canal A centrally located cavity that runs the length of the spinal cord

Ascending Tract A longitudinal band of white matter in the spinal cord that carries sensory information to the brain

Descending Tract A longitudinal band of white matter in the spinal cord that carries motor information to the body

Extrapyramidal Tract A band of white matter on the ventral portion of the spinal cord that carries motor information

Intermediate Gray Matter A band of gray matter between the dorsal and ventral horns of the spinal cord

Good Choice Bad Choice

An elderly neighbor who is taking memory classes was told to take a "natural compound" called procaine hydrochloride (HCl). She was told that the classes would not be completely effective if the procaine HCl was not taken daily. What are the risks and benefits of taking procain HCl? In which situations would it be appropriate for people to take compounds that purportedly enhance brain function?

The spinal column gives rise to 31 pairs of laterally projecting spinal nerves that form part of the PNS. Each pair of nerves corresponds to a segment of the spinal cord that relays information for specific sections of the body. For example, the spinal nerves of the cervical vertebrae carry motor and sensory information for the upper appendages. The nerve pairs also conduct reflex arcs for their respective body regions.

✓ Concept Check

1. What types of information do ascending tracts carry?
2. What types of information do descending tracts carry?
3. Describe the functions of the spinal nerves.

Peripheral Nervous System

Key Terms: abducens, accessory, autonomic, cranial nerve, dorsal root, dorsal root ganglion, facial, glossopharyngeal, hypoglossal, iris, oculomotor, olfactory, optic, parasympathetic, postganglionic, preganglionic, somatic, sympathetic, sympathetic ganglion chain, trigeminal, trochlear, vagus, ventral root, vestibulocochlear

The PNS is composed of nerves that branch out from the brain and spinal column (Figure 9.7). They are not covered with a meninges and do not have a cavity containing cerebrospinal fluid. The PNS is generally divided into the **somatic** and **autonomic** nervous branches. Somatic nerves enable the voluntary control of body movements through their communication with skeletal muscles. They are composed of afferent neurons, which collect sensory information, and efferent neurons, which relay commands for muscle action. The autonomic nervous system controls involuntary body functions. Its primary purpose is to maintain a stable internal environment for the body.

Somatic The part of the peripheral nervous system associated with the voluntary control of body movements

Autonomic The part of the peripheral nervous system responsible for such involuntary body functions as heartbeat, blood pressure, and digestion

STRUCTURE OF THE NERVOUS SYSTEM

Figure 9.7 **The Peripheral Nervous System**

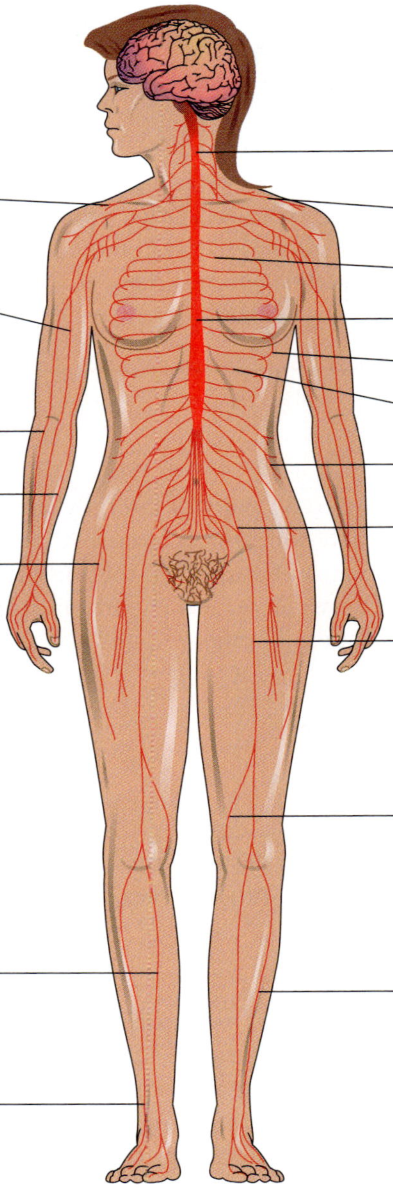

Labels (clockwise from upper right): spinal accessory nerve, vagus nerve, phrenic nerve, spinal cord, intercostal nerve, iliohypogastric nerve, ilioinguinal nerve, sacral plexus, sciatic nerve, common peroneal nerve, tibial nerve, deep peroneal nerve, superficial peroneal nerve, femoral nerve, median nerve, ulnar nerve, radial nerve, brachial plexus

Cranial Nerves Nerves of the PNS that arise from the brain

Spinal Nerves Nerves that originate from the spinal cord and pass out of the vertebral column

Cranial and Spinal Nerves The **cranial** and **spinal nerves** carry various messages for both branches of the PNS. Twelve cranial nerves arise from various regions of the brain and pass out of the skull through numerous fissures and foramina (Figure 9.8). They carry out a variety of specialized tasks associated with the somatic and autonomic nervous system. Cranial nerves are identified by both a Roman numeral and an anatomical name. Their functions are summarized in Table 9.1. Cranial nerves I, II, and VIII contain only special sensory neurons. The remaining cranial nerves have a mixture

CHAPTER 9

Figure 9.8 Cranial Nerve Attachment

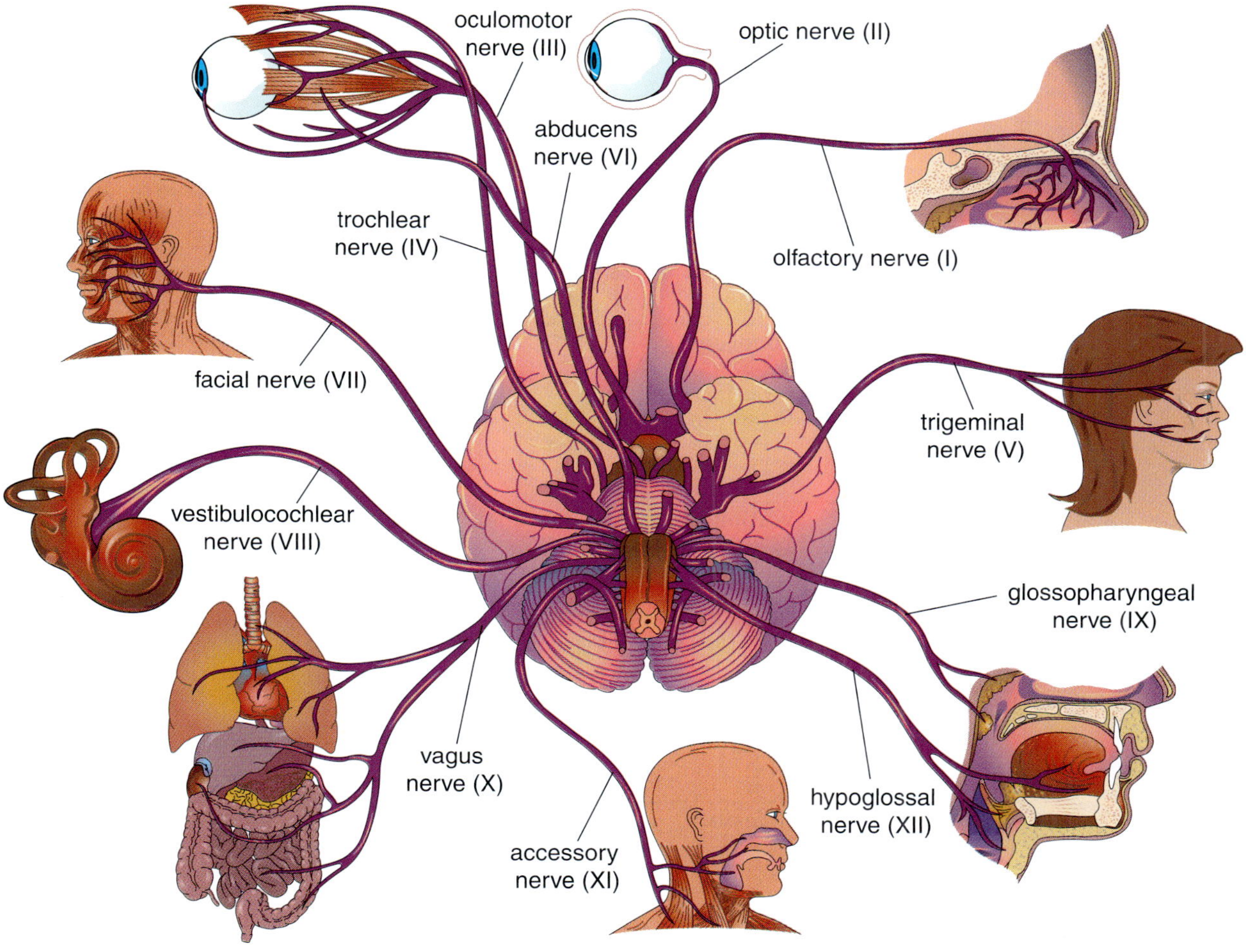

of motor and sensory neurons. Several of the cranial nerves contain branches that carry out a particular function. For example, the trigeminal nerve has three branches. One branch serves the eyes, while the other two branches split up to innervate the mandible and maxilla. Head injuries and nervous system viruses can hinder the action of cranial nerves. Physicians can locate the cranial nerve damage by evaluating the particular body functions that are affected.

STRUCTURE OF THE NERVOUS SYSTEM

Table 9.1 Cranial Nerves

Designation	Nerve	Sensory Function	Motor Function
I	olfactory	transmits smells to the brain	
II	optic	transmits vision to the brain	
III	oculomotor		controls eye and eyelid movement
IV	trochlear		controls downward and lateral eye movement
V	trigeminal	transmits sensory information from the face and mouth to the brain	
VI	abducens		transmits signals for lateral eye movement
VII	facial		controls facial expressions, and mouth and eye secretions
VIII	vestibulocochlear	transmits sensations of balance and hearing to the brain	
IX	glossopharyngeal	transmits sensory information from skin and tongue	controls swallowing and movement of food through the digestive system
X	vagus	sensory: transmits cardiovascular reflexes	controls heart rate and digestion
XI	accessory		controls swallowing movements, movements of the neck and shoulder
XII	hypoglossal		controls tongue movements

The 31 spinal nerves branch off of the various regions of the spinal cord. Eight nerves pass through the cervical vertebrae, 12 exit the thoracic vertebrae, five exit the lumbar vertebrae, another five are associated with the sacrum, and the coccyx has only one spinal nerve. Spinal nerves carry motor and sensory information for reflex control and two-way communication with the brain. Each spinal nerve exits the spinal cord from two short, lateral branches. A sensory branch called the **dorsal root** goes into the dorsal portion of the spinal column. Each dorsal root has a collection of nerve cell bodies located along the branch, forming a bulge called the **dorsal root ganglion**. Motor nerves come off of the **ventral root**. Nerve cell bodies of the ventral root are located in the ventral horn of the gray matter.

Autonomic Nervous System The autonomic nervous system uses cranial and spinal nerve pathways to carry out its wide array of tasks (Figure 9.9). It performs these tasks by transmitting regulatory information from the brain. This information controls the glands, cardiac muscle, and smooth muscles neces-

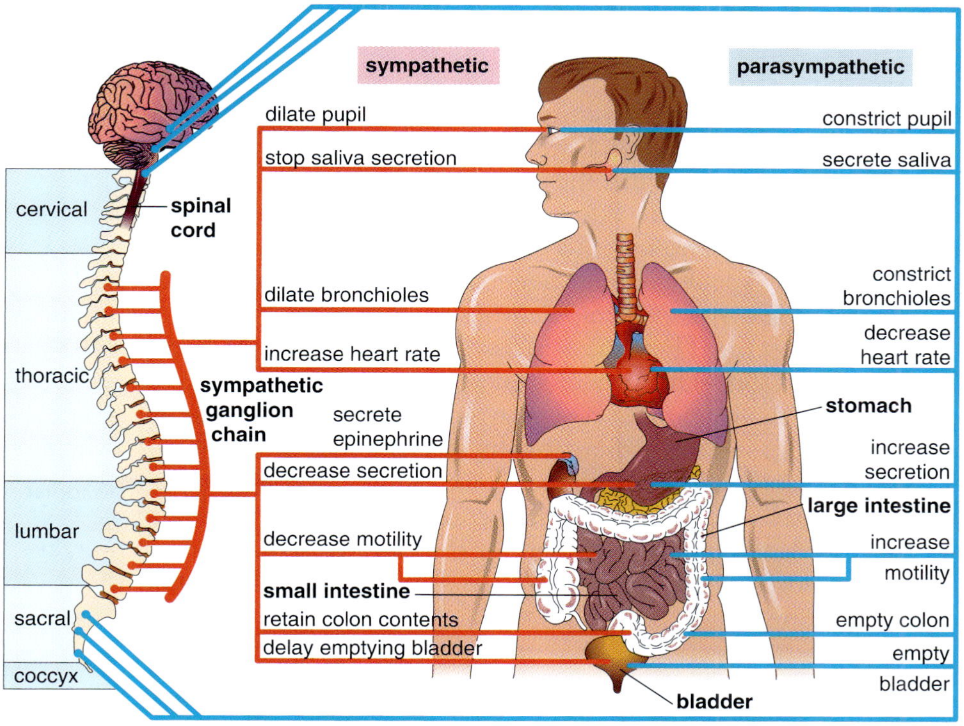

Figure 9.9 Autonomic Nervous System

sary for maintaining homeostasis. Its activities do not require conscious thought. The autonomic nervous system is integrated with the endocrine system to assist the body with digestion, sexual functions, and stress responses. The autonomic nervous system has two anatomically and functionally distinct regions. The **parasympathetic nerve** division emerges from the cranial nerves and the sacral spinal nerves. **Sympathetic** nerves arise from the thoracic and lumbar spinal cord. These two components typically counteract each other's actions to fine tune the body's responses to internal changes and environmental stimuli. The autonomic nervous system is sometimes called the "fight or flight, and rest to digest" system. This name comes from the fact that signals from the sympathetic system prepare the body to react to stress, while the parasympathetic system promotes relaxation and digestion.

Somatic nerves are a bundle of nerve cells whose full length connects the spinal cord to a body component. For example, a typical somatic motor neuron has its body located in the gray matter and its terminus contacting an effector. This is not true for nerves of the autonomic nervous system. The neural pathway is composed of two neurons working in tandem. A **preganglionic neuron** originates in the brain or spinal cord. This presynaptic neuron's terminus forms a synapse with the **postganglionic neuron**. The preganglionic neurons of both the sympathetic and parasympathetic systems have myelinated axons. In contrast, the postganglionic axons are unmyelinated. Parasympathetic preganglionic neurons are very long and terminate close to the effector. The corresponding postganglionic neuron is much shorter and runs from the synapse to the effector. In sympathetic nerves, the preganglionic neurons are usually relatively short and often terminate in a region called the **sympathetic ganglion chain**.

Parasympathetic Nerves The cranial and sacral divisions of the autonomic nervous system

Sympathetic Nerves The thoracic and lumbar divisions of the autonomic nervous system

Preganglionic Neuron A neuron situated before a synapse between two neurons

Postganglionic Neuron A neuron situated after a synapse between two neurons

Sympathetic Ganglion Chain A chain composed of the nerve cell bodies of sympathetic postganglionic neurons

STRUCTURE OF THE NERVOUS SYSTEM

Parasympathetic and sympathetic postganglionic neurons often innervate the same effectors; yet, they have opposite effects on the body structure they have innervated. Some of these effects are shown in Table 9.2. The differences are due to the neurotransmitter secreted by the postganglionic neurons. Preganglionic neurons of both systems secrete acetylcholine into the synapse. Acetylcholine excites the postganglionic neurons, causing them to produce an action potential. Postganglionic neurons of the parasympathetic neurons release acetylcholine into the synapse it forms with the effector. This produces the specific effects needed during parasympathetic stimulation. Sympathetic postganglionic neurons use norepinephrine to communicate with the effector. There are various types of norepinephrine receptors on different effectors. This provides an assortment of effects based on the number of a particular type of receptor on the effector. For example, some blood vessels respond more dramatically to norepinephrine than do others because of the composition of the receptor. Some norepinephrine receptors of the heart and blood vessels strongly counteract the parasympathetic system's effects. Many medications used for the cardiovascular system interact with specific norepinephrine receptors to produce desired results without many side effects.

Table 9.2 Autonomic Nervous System Action

Effector	Sympathetic Action	Parasympathetic Action
iris	opens the pupil	closes the pupil
salivary glands	reduces saliva production	increases saliva production
oral/nasal mucosa	reduces mucus production	increases mucus production
heart	increases heart activity	decreases heart activity
lung	relaxes respiratory muscles	constricts respiratory-tree muscles
stomach	reduces digestive action	increases digestive action
small intestine	reduces digestive action	increases digestive action
large intestine	reduces digestive action	increases digestive action
kidney	decreases urine production	increases urine production
bladder	relaxes the bladder	contracts the bladder

✓ Concept Check

1. Describe the different types of cranial nerves.
2. Distinguish between the functions of the cranial and spinal nerves.
3. Distinguish between the autonomic and somatic nervous systems.

DISCOVERY SCENE PLEASE ENTER DISCOVERY SCENE PLEASE ENTER

What components of the CNS could play a role in the former football player's condition? Could his condition be due to problems in a particular region of the brain? Could a particular component of the spinal cord be responsible? Has information about the PNS contributed to your understanding of the CSI? What aspects of the PNS could be involved in the former football player's ailments?

Human Senses

 Key Terms: audition, chemoreceptor, equilibrium, external auditory meatus, gustation, olfaction, photoreceptor

The human head contains a concentration of sensory structures that are collectively called the special senses. Each of these sensory structures uses special receptors that detect environmental stimuli specific to that sense. The senses of smell (**olfaction**) and taste (**gustation**) rely on **chemoreceptors** to detect chemicals dissolved in the air or water. Vision makes use of **photoreceptors**, which are capable of converting light into a neural signal. Hearing (**audition**) uses a series of receptors that are sensitive to different types of sounds or vibrations. The sense called balance, or **equilibrium**, uses a combination of receptors to detect the body's position as it is moving and standing still.

> **Olfaction** The sense of smell
> **Gustation** The sense of taste
> **Chemoreceptor** A sensory receptor that detects chemical stimuli
> **Photoreceptor** A sensory receptor that detects light stimuli
> **Audition** The ability to hear
> **Equilibrium** A sensory system that detects the orientation of the body and head

Taste

 Key Terms: bitter, papillae, saliva, salt, sour, sweet, taste bud, umani

Taste is not merely a luxury that permits a person to enjoy a variety of foods. It is very important in stimulating appetite and helping the body determine whether something is safe to eat. Taste is detected by chemoreceptors located in structures called **taste buds**, which are distributed throughout the upper surface of the tongue. Elevations of the tongue surface, called **papillae**, contain the taste buds. An individual taste bud contains 50 to 100 neurons that represent all five taste sensations: **bitter**, **salt**, **sour**, **sweet**, and **umani** (Figure 9.10). Tastes can be detected only when the chemicals making up flavor are dissolved in water. One role of **saliva** is to ensure a wet environment for gustation.

Certain parts of the tongue have a higher concentration of one type of taste receptor. The taste of sweet is most concentrated in the front of the tongue. It stimulates appetite and is used to determine the palatability of food. Salt receptors are found on the tips and sides of the tongue. Sour receptors are generally found on the proximal edges of the tongue. Bitter receptors are located on enlarged papillae along the back of the tongue. These receptors detect spoilage, or potential toxins, in food and can cause a person to gag when the receptor is overstimulated. Umani receptors are distributed throughout the tongue; they are thought to stimulate appetite.

> **Taste Bud** A small structure on the upper surface of the tongue that contains chemoreceptors
> **Papillae** Elevated areas of the tongue that contain taste buds
> **Bitter** The taste that occurs when bases, such as soap, dissolve in the mouth
> **Sour** The taste that occurs when acids dissolve in the mouth
> **Salt** The taste that occurs when salt dissolves in the mouth
> **Sweet** The taste that occurs when sugar dissolves in the mouth
> **Umani** The taste that occurs when the food ingredient monosodium glutamate (MSG) dissolves in the mouth
> **MSG** The abbreviation for "monosodium glutamate," which is a seasoning often found in Chinese food
> **Saliva** A watery secretion from glands in the mouth

347

STRUCTURE OF THE NERVOUS SYSTEM

Figure 9.10 Taste

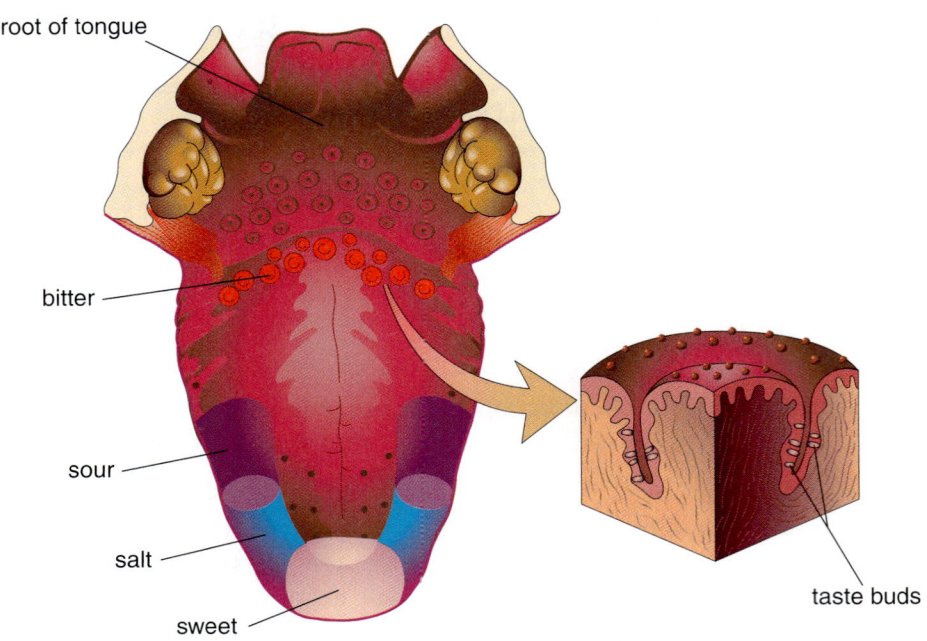

Smell

Key Term: olfactory bulb, vomeronasal gland

Olfactory Bulb An enlargement of the olfactory nerve that senses smell

Vomeronasal Gland A structure containing chemoreceptors, which are believed to sense chemicals associated with sexual behavior

Olfaction serves two main purposes: Its primary role is to detect potentially harmful or valuable chemicals in the air; and it is an important supplement to taste. Much of the flavor of food is due to its smells as it is served and eaten. Chemoreceptors for smell are located within the **olfactory bulb** in the upper region of the nasal cavity (Figure 9.11). The olfactory bulb is an extension of the olfactory nerve, which is one of the cranial nerves. It is difficult to categorize smell into distinct odors in the same way taste sensations can be classified. Scientists have recently identified an amazingly large number of genes that program for smell receptors. These genes may produce from 500 to 1000 different types of smell receptors, some of which detect chemicals that alter human behavior, particularly behavior related to sexual responses. Odors are strong signals that bring out memories associated with a particular smell. They are also important in protecting the body from harm because certain odors are remembered as being associated with flammable materials, such as gasoline, or rotting foods. There is still much to learn about the way the brain interprets smells.

Scientists have recently identified what they believe is the **vomeronasal gland** at the base of the nasal cavity. It is thought to sense body chemicals associated with sexual behavior It is debated whether humans have a fully functional vomeronasal gland. Many scientists believe the lack of nerve structures associated with the gland would prevent any affect on behavior. However, several studies show that people who are past puberty very likely use the vomeronasal gland to determine differences in body odor between men and women.

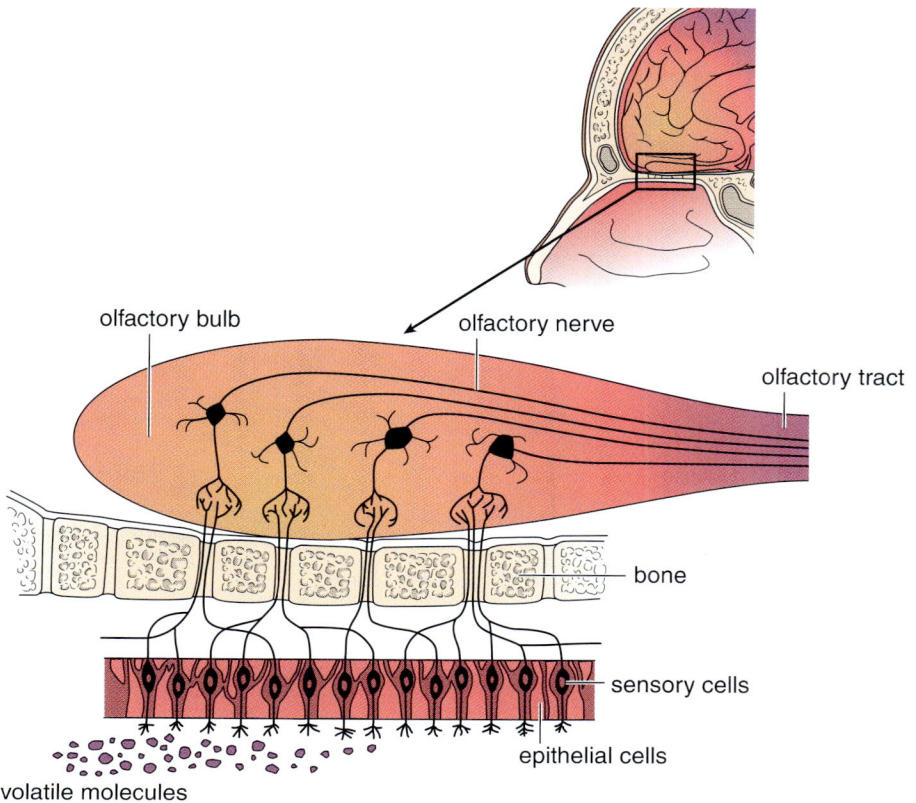

Figure 9.11 **Olfaction**

Vision

Key Terms: aqueous humor, choroids, ciliary body, cone, conjunctiva, cornea, extrinsic muscles, fovea, lacrimal duct, lacrimal gland, lens, optic cup, palpebrae, retina, rhodopsin, rod, sclera, vitreous humor

Eyes are highly specialized organs that give people the ability to see objects and perceive movement. The eyes sit protected in the bony orbits of the skull and send their information directly to the brain through the optic nerve. Each eye is protected on its exposed anterior surface by the eyelids, or **palpebrae** (Figure 9.12). They are thin flaps of skin controlled by orbicular muscles. Eyelids have a reflex arc that forces them to close when objects are placed near the eye surface. Eyelashes on the edges of the eyelids are believed to protect the eye surface from dirt. The **conjunctiva**, a thin, transparent epithelium, tightly covers the outer anterior surface of the eye. The conjunctiva contains many minute blood vessels that are nearly invisible until the eye becomes irritated. Superior and lateral to each eye is a **lacrimal gland** (Figure 9.13). Lacrimal glands produce tears that lubricate the conjunctiva and protect the eye from infection by certain bacteria. Tears usually drain off the eye surface into the **lacrimal duct** of each eye. The lacrimal ducts lead into the nasal cavity. This produces the "runny-nose effect" associated with crying.

Eyes develop from two outgrowths of the brain that form the optic nerves and the **optic cup**. The optic cup is the posterior lining of the eye, which contains the photoreceptors. In the mature eye, the optic cup is called the

Palpebrae Eyelids

Conjunctiva A thin, transparent epithelium that covers the eye

Lacrimal Gland A gland that produces tears

Lacrimal duct A tube connecting the orbit with the nasal cavity

Optic Cup A cuplike depression that develops into the sensory layer of the eyes

Figure 9.12 External Features of the Eye

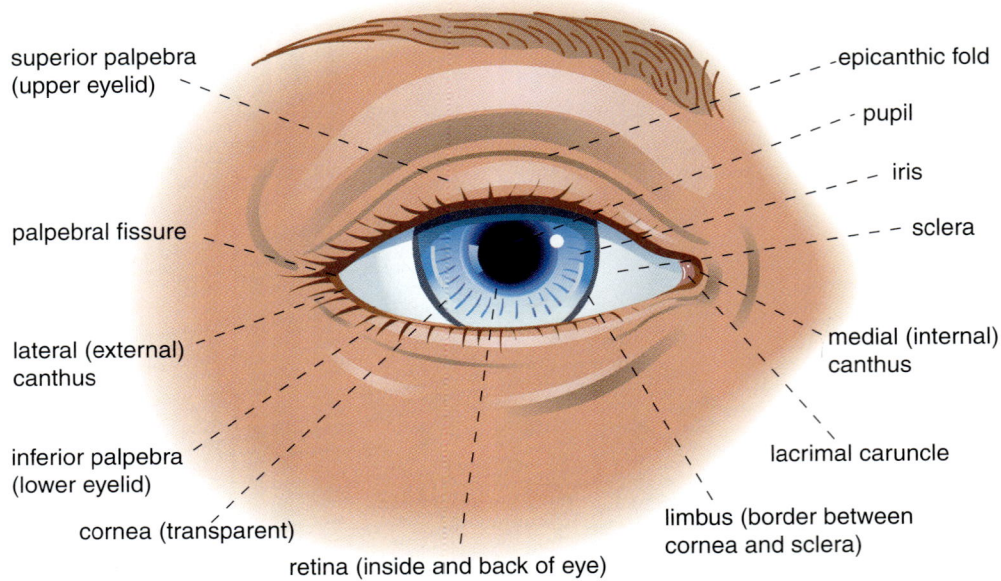

Figure 9.13 Flow of Tears

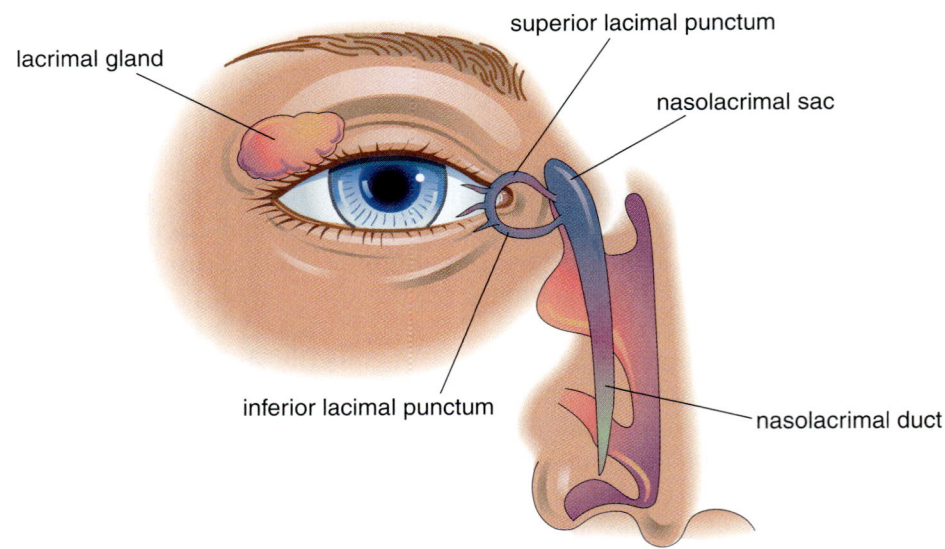

Retina The inside layer located at the back of the eye that contains photoreceptors

Sclera The tough outermost layer of the eye

Cornea The clear covering at the front surface of the eye that permits light to enter

Extrinsic Muscles The six muscles that move the eye

Choroid A layer of blood vessels that lines the inner surface of the sclera

retina. Forming the outer surface of the eye is a white-colored, tough, fibrous connective-tissue covering called the **sclera**. The sclera covers approximately the posterior five-sixths of the surface of the eye. It connects with the **cornea** on its anterior surface and with the optic nerve on its posterior surface. The cornea is a clear connective-tissue covering that permits light to enter the eye. It is positioned so that the light falls on the surface of the retina. Attached to the anterior outer surface of the sclera are six eye muscles, called the **extrinsic muscles**, which are anchored to the bones of the orbit. These muscles are organized in antagonistic arrangements, which permit the eyes to move freely in various directions. Lining the inside of the sclera is the **choroid** (Figure 9.14).

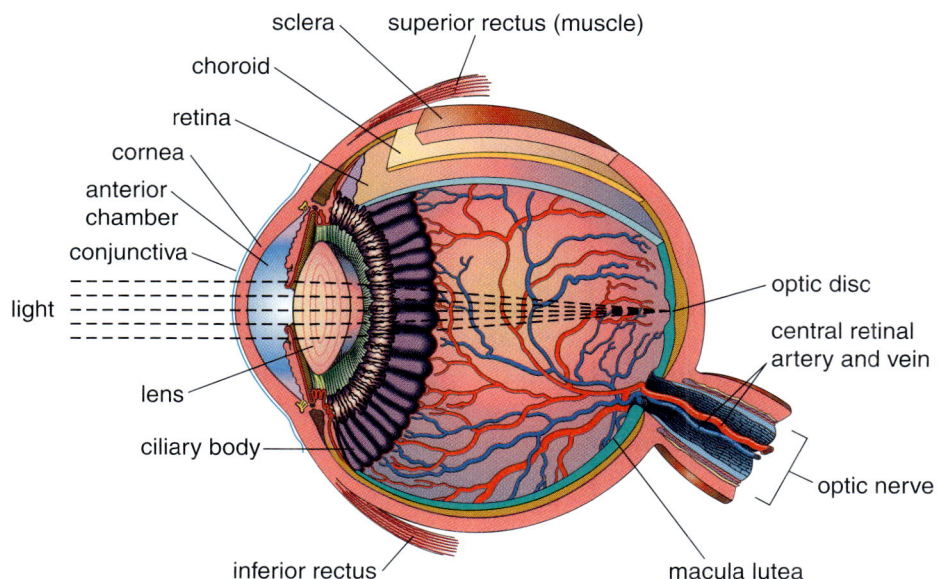

Figure 9.14 Internal Features of the Eye

The choroid is a darkly pigmented layer of blood vessels that supplies blood to the photoreceptors. It also absorbs excess light in the eye, preventing washed-out vision.

A ring of muscles and connective tissue called the **ciliary body** attaches the choroid layer on its anterior portion. Attached to the ciliary body is the **lens**. The lens is composed of flexible, crystal-like epithelial cells. Actions of the ciliary body change the shape of the lens, which allows the eye to focus incoming light on the photoreceptors. The lens is flatter when the ciliary muscle is in a relaxed state as it is when viewing distant objects. To properly direct light rays from an object close to the eyes, the ciliary muscle constricts and the attached lens thickens. The loss of elasticity of the lens, which occurs with aging, diminishes this ability of the lens. The common need for "reading glasses" around the age of 40 years is attributed to this change. The colored part of the eye behind the cornea and in front of the lens is called the iris. Melanin in the iris provides the various eye colors. An adjustable, central opening in the iris is called the pupil. It controls the amount of light entering the eye. The ciliary body and iris separate the eye into two fluid-filled chambers. A thin fluid located in the anterior-most chamber is called **aqueous humor**. The aqueous humor helps maintain the shape of the cornea for properly focused vision. It is secreted continuously by the ciliary body. An abnormal increase in the volume of this fluid is indicative of the well-known disorder glaucoma, which can greatly damage vision. **Vitreous humor** fills the posterior cavity. The vitreous humor maintains the spherical shape of the eye. Surrounding the eyes are fat cells that support the eye's position in the orbit.

Making up the innermost posterior layer of the eye is the retina. The retina contains the photoreceptors that convert light into a neuron impulse, which the optic nerve transmits to the brain. Two types of photoreceptors are found in the human eye: **cones** and **rods**. Cones respond to bright light and perceive color vision. They are mostly located in the central area of the retina and permit day vision. A central point at the back of the eye called the **fovea** contains only the densest concentration of cones. It is believed that there are about 7 million cones; they can be divided into red cones (64%), green cones (32%),

Ciliary Body Part of the eye that contains a focusing muscle and connective tissue

Lens The transparent structure inside the eye that focuses light rays for clear vision

Aqueous Humor A clear, watery fluid in the front of the eyeball

Vitreous Humor A gel-like fluid that fills the cavity behind the eye lens

Cone A photoreceptor sensitive to bright light and color

Rod A photoreceptor sensitive to dim light

Fovea A depression in the retina that contains only cones

STRUCTURE OF THE NERVOUS SYSTEM

and blue cones (2%). The cones blend these three colors into the wide array of colors that the brain perceives. People with one or two types of cones that do not function properly are called color blind. Rods are primarily located along the lateral edges of the retina. They are responsible for night vision, motion detection, and peripheral vision. Rods contain a light-sensitive chemical called **rhodopsin**, which the body manufactures from vitamin A.

Rhodopsin A light-sensitive chemical contained in the rods of the retina

Hearing and Balance

Key Terms: auditory canal, auricle, cochlea, eardrum, eustachian tube, external auditory meatus, external ear, incus, inner ear, malleus, middle ear, organ of Corti, ossicle, oval window, pinna, round window, semicircular canals, stapes, tympanic membrane, vestibule

The human ear converts sounds and body orientation into neuron signals for the brain. Located lateral to and within the temporal bone, the ear is composed of three regions: the **external ear**, **middle ear**, and **inner ear** (Figure 9.15). The external ear is composed of two features. The **auricle**, or **pinna**, which is the most obvious, is the flexible component. The auricle is composed of skin covering a funnel-shaped strip of cartilage and fat cells. It has an opening that leads to the second structure of the outer ear, a bony canal called the **external auditory meatus**, or **auditory canal**. The auricle collects sound waves and funnels them into the auditory canal. Fine hairs and ceruminous glands that produce cerumen, or earwax, line the auditory canal.

External Ear The part of the ear that is visible

Middle Ear The part of the ear inside the skull that contains the ear bones

Inner Ear The part of the ear that contains both the organs for hearing and balance

Auricle or Pinna The external portion of the ear

External Auditory Meatus or Auditory Canal A bony structure that leads from the outer ear to the middle ear

Figure 9.15 **External, Middle, and Inner Ear**

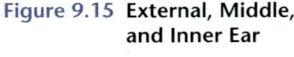

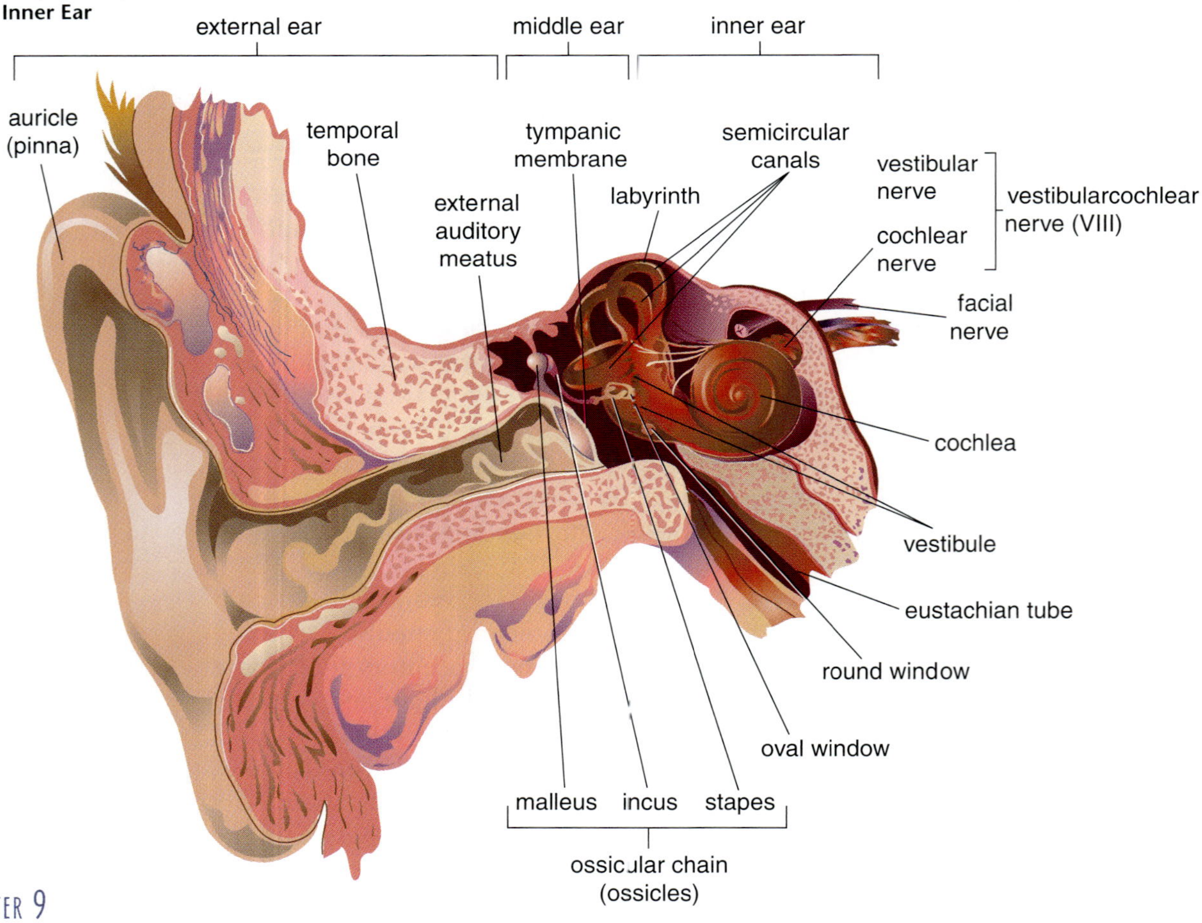

At the entrance of the middle ear is the **tympanic membrane**, or **eardrum**. The eardrum vibrates in response to sounds transmitted by the auditory canal. It vibrates against three ear bones, or **ossicles**, found in the middle ear. Directly attached to the eardrum is the **malleus**. Its name means "like a hammer." The eardrum vibrates the malleus and, in turn, the malleus vibrates the **incus**, which is attached to the malleus by connective tissue. The incus, or anvil, then vibrates against the **stapes**. Vibrations from the stapes, or stirrup, are then transmitted to the inner ear. Small skeletal muscles attached to the ossicles adjust the tension of the bones to the intensity of sound. These muscles adjust the volume of sound going to the brain and prevent damage of the ossicles. A canal called the **eustachian tube** connects the middle ear to the throat. It prevents damage to the middle ear by adapting to changes in air pressure.

The inner ear is housed in a bony cavity of the temporal bone (see Figure 9.16). It is composed of the **cochlea** and the **semicircular canals**. Sound is transmitted to the inner ear when the stapes vibrates against a portion of the cochlea called the **oval window**. When vibrated, the oval window produces pressure waves in the fluid-filled cochlea. This triggers neurons that send signals to the brain via the vestibulocochlear nerve. The neurons are located in a structure called the **organ of Corti**, which gives a person the ability to distinguish different pitches or tones. Sound waves then vibrate against the **round window** of the cochlea. The round window prevents vibrations in the fluids of the cochlea from reverberating (Figure 9.17).

Tympanic Membrane or Eardrum A thin epithelial layer at the end of the auditory canal

Ossicles The three tiny bones in the middle ear

Malleus The first bone in the series of three ossicles in the middle ear

Incus The second bone in the series of three ossicles in the middle ear

Stapes The third bone in the series of three ossicles in the middle ear

Eustachian Tube The tube connecting the middle ear to the throat

Cochlea A coiled organ in the inner ear that converts vibrations into neuron impulses that are sent to the brain

Semicircular Canals Structures of the inner ear that detect movement of the body in space

Oval Window A membrane separating the middle ear from the inner ear

Organ of Corti The part of the inner ear that responds to different sounds

Round Window A membrane in the cochlea that permits pressure from sound waves to be released

Figure 9.16 Inner Ear

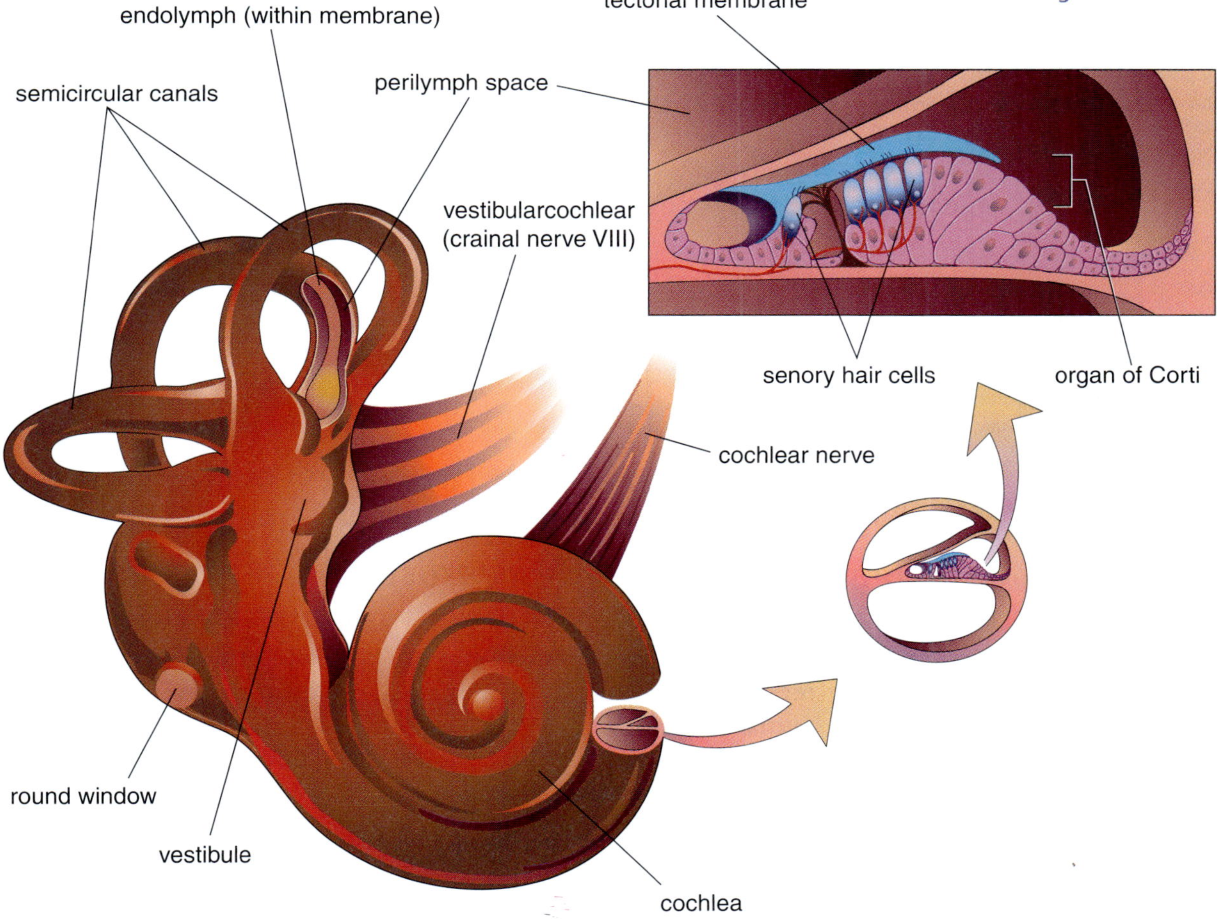

STRUCTURE OF THE NERVOUS SYSTEM

Figure 9.17 **Movement of Sound Waves**

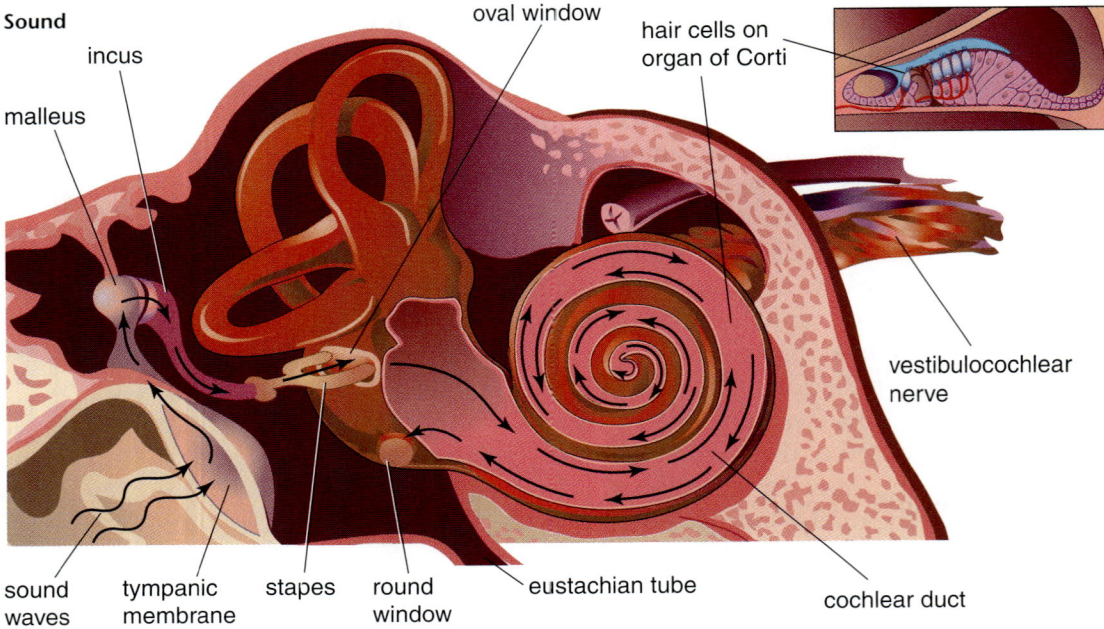

The semicircular canals are fluid-filled chambers that respond to body position. Fluids in the canals flow in response to body movement. Changes in the direction of flow stimulate neurons, which the brain interprets as movement. The brain is able to perceive the direction of movement by assessing the relative flow of fluids in the three semicircular canals. Each canal is offset from the others so that the brain can detect a wide array of movements. The superior canal detects forward and backward motion. The lateral canal senses forward and side-to-side motions. Side-to-side motions are also detected by the posterior canal. The **vestibule** in the base of the semicircular canals detects body position. Special neurons in the vestibule respond to the settling of calcium salt grains floating in a thick fluid. The grains settle on a certain group of neurons depending on the body's position.

Vestibule A portion of the inner ear between cochlear and semicircular canals that detects body position

✓ Concept Check

1. Compare and contrast gustation and olfaction.
2. Describe the structure and function of the eyes.
3. Describe the structure and function of the ears.

Wellness and Illness over the Life Span

Pathology of Nervous System Structure

Key terms: aneurysm, arteriovenous malformation, athetosis, cerebrovascular, chorea, glioma, ischemic attack, lymphoma, meningioma, metastatic, ministroke, neuroblastoma, neurodegenerative, neuroma, neurovascular, palsy, peripheral neuropathy, shaken-baby syndrome, stroke, tremor, whiplash

Structural pathology of the nervous system is categorized by the type of factor causing nerve tract loss. Each category impairs nerve function in ways that can be difficult to diagnosis. It usually takes a battery of tests to confirm the

particular cause of nervous system pathology. The common categories are as follows:
- trauma
- **cerebrovascular** and **neurovascular** diseases
- nervous system tumors
- developmental disorders
- metabolic and toxic diseases
- CNS infection
- **neurodegenerative** disease

Cerebrovascular Disease Disorders of blood vessels in the brain

Neurovascular Disease Disorders of blood vessels in the nerves

Neurodegenerative Disease A condition characterized by the deterioration of nervous tissue

Peripheral Neuropathy A condition of the nervous system that causes numbness, pain, tingling, or weakness in the peripheral nervous system

Nervous system trauma can result from a variety of causes ranging from severe blows to the body to gunshot wounds. The outcome of trauma in its simplest form can be a "pinched" nerve that causes myelin or axon damage. Severe trauma can completely break an axon or destroy the nerve cell bodies in a ganglion. Trauma can be categorized by its location in the nervous system. Traumatic **peripheral neuropathy**, as the name implies, is found in the PNS. The effects on the body vary greatly depending on the nature of the nerve tract that is damaged. Some types of peripheral neuropathy can cause numbness, muscle weakness, pricking sensations, sensation loss, tingling, and sensitivity to touch. Severe damage can include burning pain, muscle wasting, organ and organ system failure, and paralysis. Death can result from the long-term effects of severe peripheral neuropathy.

HOLES IN THE BRAIN

Nervous system disorders are fairly common, yet few people understand the fear of knowing that they carry the gene(s) for a rare and severely debilitating neurological disorder. The Shmaefskys (the author's immediate family) knew they had neuromuscular disorders in the family; however, the condition was misdiagnosed first as polio, a neurological viral disease. Later, the disorder was identified as cerebral palsy. A study of the author's niece, Nicole Shmaefsky,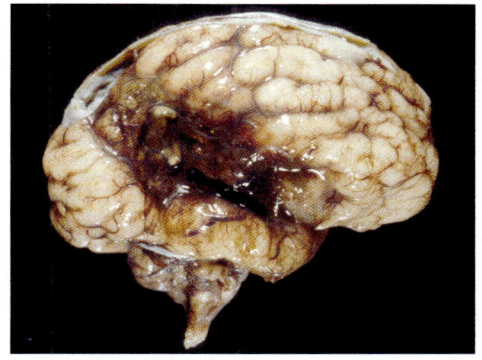
finally revealed the true nature of the disorder that affected various family members in different degrees of severity. It was determined to be a rare genetic condition called autosomal-dominant porencephaly. The term porencephaly means "holes in the brain." Autosomal dominant refers to the fact that the disease is usually expressed in people who carry the gene.

Autosomal-dominant porencephaly is characterized by cysts or cavities forming in one or both cerebral hemispheres. These cysts or cavities are usually the result of abnormal blood vessel development. The lack of blood flow kills off the young neuroglia and neurons, leaving empty spaces behind. It usually starts immediately before or just after birth. Most infants show signs of the disorder within days after birth. The disease causes a variety of conditions, including epilepsy, incomplete paralysis, low muscle tone, poor speech development, postural problems, seizures, and spastic muscle contractions. Nicole's severe condition was the factor that alerted the physicians that the family was carrying the disorder. This is a concern to the author and his children because the condition could be carried in the family for many generations.

Whiplash Nerve damage in the neck caused by an abrupt, forced movement of the head

Shaken-Baby Syndrome The severe injuries that result when a young child is violently shaken

Aneurysm An abnormal swelling of a blood vessel wall

Arteriovenous Malformation An abnormal tangling of blood vessels in the brain that disrupts blood flow

Ischemic Attack A condition caused by insufficient blood flow to a body part

Stroke A cerebrovascular disorder that occurs when the blood supply to part of the brain is suddenly interrupted

Transient Ischemic Attack (TIA) A "ministroke" caused by temporary loss of blood flow

Ministroke A mild and temporary cut-off of the blood supply to the CNS

Traumatic neuropathies of the CNS affect the brain and spinal cord. Brain damage can be caused by violent movement of the head as well as other forms of traumatic damage. **Whiplash** is a common cause of brain trauma. It results from a sharp back-and-forth movement of the head and neck. Whiplash can cause trauma to the cranial nerves and upper peripheral nerves. Whiplash is related to **shaken-baby syndrome**, in which a child is violently shaken without the head being supported. Nerve tracks in the brain can also be damaged if the brain strikes sharply upon the skull during the movement. Abnormal sensations, dizziness, headache, and muscle stiffness can result from minor whiplash. Serious whiplash can sever nerves in the neck resulting in paralysis and loss of feeling. Damage to the brain can cause the loss of the ability to learn, remember, and think. Blows to the back occurring during athletic injuries, automobile crashes, and occupational accidents commonly cause traumatic damage to the spinal cord. Damage to specific tracts running up and down the spinal cord produces a variety of outcomes. For example, damage to nerve tracts in the dorsal lumbar region of the spinal cord could cause a loss of feeling in the abdominal region.

Cerebrovascular and neurovascular diseases are blood vessel disorders that impair nervous system function. Disruption of blood flow to the brain produces cerebrovascular diseases. Blood flow irregularities of the CNS and PNS cause neurovascular diseases. There are three major types of vascular diseases affecting the CNS and PNS: **aneurysms**, **arteriovenous malformations**, and **ischemic attacks**. Aneurysms are bulges in the vascular system caused by the stretching and thinning of blood vessels. Their occurrence is prevalent in larger blood vessels commonly found in the CNS. They may cause a slowing of blood flow to the brain and nerves. Some aneurysms can rupture, completely cutting off the blood supply to large parts of the nervous system. A ruptured aneurysm can cause bleeding in the brain and spinal column, which usually leads to severe impairment or death, and is a common cause of **strokes**. A stroke is brain damage resulting from a cutoff of the blood supply to the neuroglia and neurons. Many aneurysms can be diagnosed and repaired before they create significant problems.

Arteriovenous malformations (AVMs) are abnormal connections between the blood vessels within a particular body part. They occur just after birth, but may not produce problems until a person reaches adulthood. It is not unusual to find them in the CNS and major peripheral nerves. Most people with AVMs experience few noteworthy symptoms; however, AVMs in the brain can cause constant headaches or seizures. Spinal cord and PNS AVMs can lead to progressive neurological problems that eventually debilitate a person. Less common types of fatal strokes can result from AVMs in children and adults.

Transient ischemic attacks (**TIAs**) are abnormalities that prevent adequate blood flow to various parts of the body. The nervous system is not exempt from this condition. TIAs can damage groups of nerves or individual nerve tracts. Many physicians consider them to be a warning sign that a person is susceptible to strokes. Approximately 200,000 to 500,000 TIA cases occur each year in North America. Many go unnoticed and, therefore, are not reported to physicians. A common cause of TIAs is the narrowing of blood vessels that result from the accumulation of fatty deposits on the inner walls of the vessels. Any factors that further narrow the blood vessels can increase blood-flow blockage. The first evidence of TIAs is what physicians call a **ministroke**. It causes mild and temporary neural problems that usually last only a few minutes. Ministrokes can worsen and cause a full stroke if not treated.

Nervous system tumors are abnormal growths that develop from neuroglia (**glioma**), cells of the meninges (**meningioma**), immune cells in the nervous system (**lymphoma**), and neurons (**neuroblastoma** and **neuroma**). Neuroblastomas develop from immature nerve cells and are found throughout the nervous system. They most commonly form in children. Neuromas can form in the mature nerve cells of adults and can appear anywhere in the body. These nervous system tumors do their damage by crushing nearby nerves and blood vessels that serve other nerves. Many nervous system tumors are benign and cause damage without spreading throughout the body. Cancerous, or **metastatic**, tumors can spread and multiply throughout the body.

Many brain tumors come from cancer cells that spread from other body regions. Tumors that travel to the brain can affect many parts of the brain depending on the blood flow that transported the cancerous cells. A tumor that forms in the brain affects a particular function associated with its location. Approximately 44% of all brain tumors that form within the brain are benign. The specific location of the tumor within the brain determines the severity of the damage it causes. Although benign and cancerous tumors both cause damage by destroying neurons, cancerous tumors pose more problems because they can spread to other regions of the body.

Developmental disorders are the result of some factor that interferes with the DNA's ability to form or carry out the normal functions of a body component. This can be caused by damage to the DNA or by chemicals that obstruct the actions of the DNA. It is estimated that 2.5% to 5% of children are born with developmental disorders. Up to 50% of these children have nervous system defects resulting from the disorder. There are many disorders caused by an abnormal number of chromosomes passed along by the eggs or sperm. These conditions are usually accompanied by mental retardation and the loss of many body functions. Some developmental disorders occur when blood vessels fail to form in an area of the fetus, which prevents the normal formation of nerves and sensory structures.

Any of the nervous system disorders mentioned above can produce four characteristic problems with nerve function: **athetosis**, **chorea**, **palsy**, and **tremor**. Athetosis is a repetitive, involuntary, twisting movement usually isolated to the arms and hands (Figure 9.18), but sometimes it

Figure 9.18 Athetosis

affects posture. It is usually caused by damage to particular nuclei of the brain. Chorea is characterized by short, irregular, nonrepetitive muscle contractions (Figure 9.19). Drugs, metabolic disorders, and vascular problems that affect dopamine production in the brain can also cause this condition. Palsy is a term used to describe muscle paralysis caused by nerve loss. It can result from damage to the brain or certain nerve tracts. Tremors are rhythmic, involuntary muscle

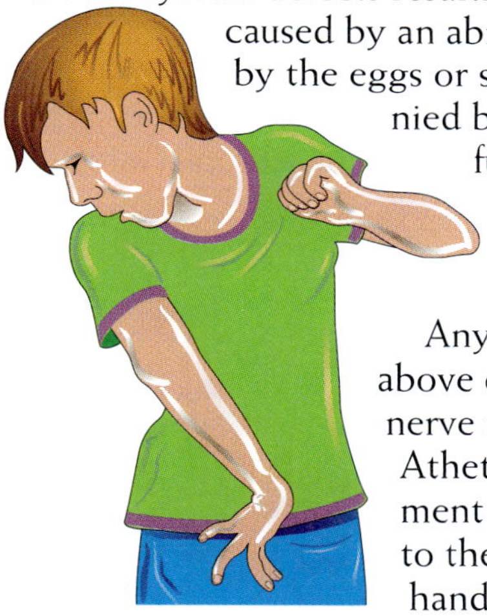

Figure 9.19 Chorea

Glioma A tumor that develops from neuroglia in the brain

Meningioma A tumor that develops from the meninges

Lymphoma A tumor that develops from cells of the immune system

Neuroblastoma A tumor that develops from immature nervous system cells

Neuroma A tumor that develops from nervous system cells

Metastatic Pertaining to the ability of a cell or a group of cancerous cells to move throughout the body

Athetosis A nervous system disorder that causes slow, involuntary movements of the hands and feet

Chorea A nervous system disorder that causes muscular twitching of arms, legs, and face

Palsy A nervous system disorder that causes paralysis of a muscle or group of muscles

Tremor A nervous system disorder that causes uncontrollable, rhythmic, shaking movements

STRUCTURE OF THE NERVOUS SYSTEM

contractions characterized by back-and-forth movements of a body part. These movements are not as exaggerated as those caused by athetosis. Physicians categorize five types of tremors by the types of movement they impede (Figure 9.20). Many types of brain and spinal cord disorders can cause tremors.

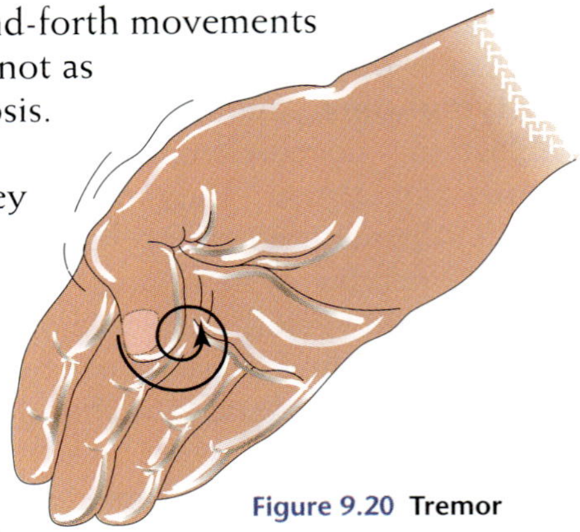

Metabolic and toxic nervous system diseases are caused by poisons that impede the functions of neuroglia or neurons. Large amounts of calcium, potassium, and sodium affect the action potential of the neurons that make up all nerve tracts. Too much calcium and sodium sensitizes neurons so that they are inappropriately stimulated. In large amounts, potassium will inhibit action potential. Many medications can block or mimic the action of neurotransmitters, causing a loss of control of nervous system function. Many types of metals interfere with the metabolism of nervous system cells. Arsenic, cadmium, mercury, and lead can build up in the body and cause long-term neurological disorders. Many CNS infections have a similar effect on the nervous system. Bacterial, fungal, and protistan infection of the nervous system damage cells because the toxins they produce invade the body. Infectious organisms will also cause inflammation of the connective tissues of nerves and meninges. Viruses, such as herpes, directly disrupt cell metabolism. This is due to the fact that the cell's metabolism is redirected to replicate the virus (Figure 9.21). Herpesvirus travels along nerve tracts to other parts of body, including the skin, causing inflammation.

The physiological effects of neurodegenerative diseases discussed in chapter 8 can lead to deterioration of nervous system structure. The neuroglia affected

Figure 9.20 Tremor

Figure 9.21 Herpes in Nerve Tract

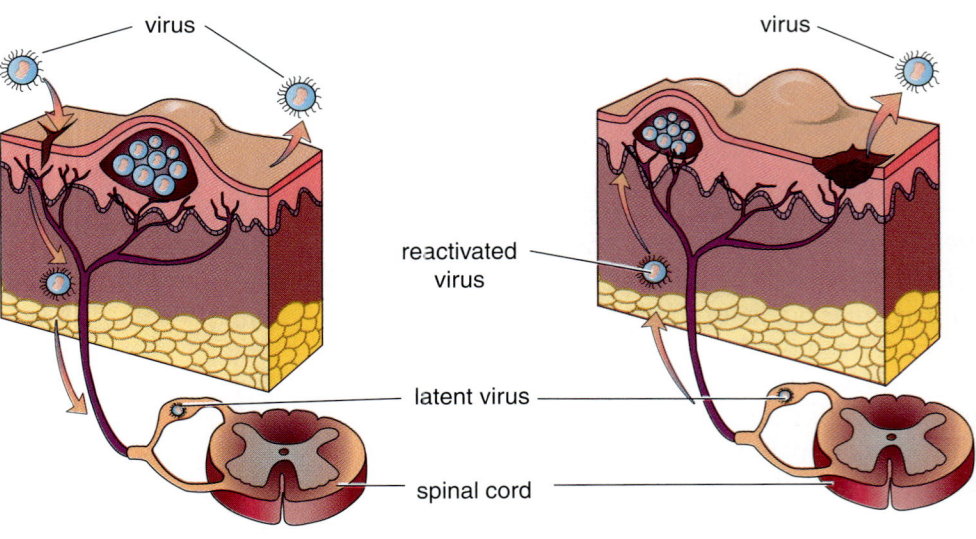

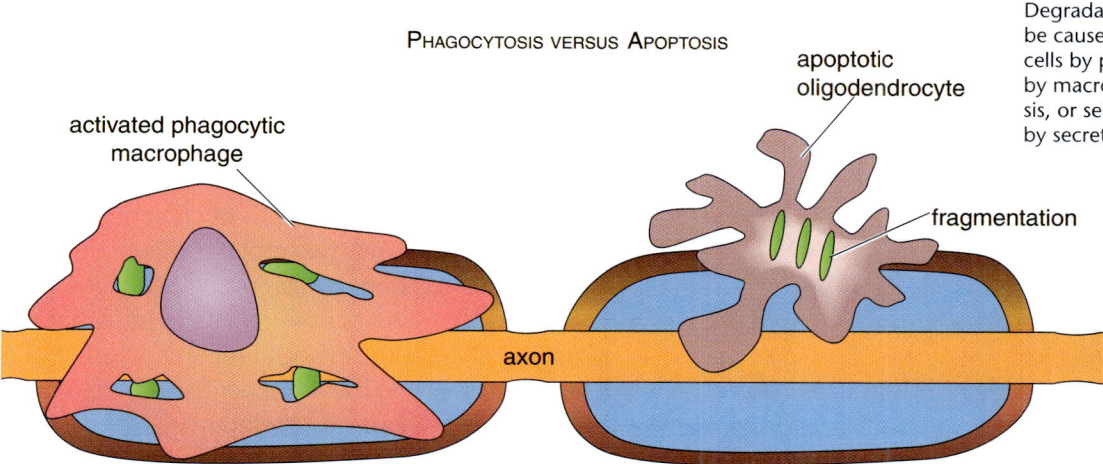

Figure 9.22 **Degradation of Neuroglia**
Degradation of the neuroglia can be caused by destruction of the cells by phagocytosis carried out by macrophages, and by apoptosis, or self-destruction, stimulated by secretions from oligodendrites.

by Alzheimer's disease and other degenerative disorders of the brain eventually kills the neurons. These neurons are not replaced, leaving small holes in the brain. Neurovascular disorders can also have the same effect leaving gaps in the CNS and PNS. Neurodegenerative changes to neuroglia and neurons can also be caused by toxic substances and by unintentional attack of the immune system. Certain infectious organisms stimulate the immune system to destroy nervous system components leading to neurodegenerative changes. The destruction of myelin by the immune system as occurs in multiple sclerosis is thought to be caused by such viral invasion (Figure 9.22). The signs and symptoms of neurodegenerative loss of nervous system structure vary greatly with the particular component that is destroyed.

✓ Concept Check

1. What are the major categories of nervous system structural disorders?
2. Describe the possible causes of strokes.
3. Compare and contrast metabolic and toxic nervous system diseases with infectious nervous system diseases.

Aging of the Nervous System Structure

Key terms: plasticity, redundancy

The nerve cell mass of the brain and spinal cord steadily decreases after a person reaches about age 40. Recent studies show that neuron loss due to aging is not as significant as once thought. It was once believed that people lost 85,000 brain cells a day after age 40. This means they would have lost billions of cells by age 80. The misunderstanding of neuron loss was due to the inclusion of people with Alzheimer's disease in the research data. Now, it is known that the massive brain cell loss is more a factor of degenerative brain diseases than the normal aging process. The neuron loss that does take place varies between different regions of the CNS. Brain weight, however, does decrease significantly

Cutting Edge Research
THE GENETICS OF DEMENTIA

A popular area of medical research investigates the roles of more than a dozen genes that are now thought to be involved in the aging process. One gene called apolipoprotein E (*APOE*) [changed per AMA style manual, p. 381] creates certain fat-metabolism pathways in astrocytes. Aberrations of the gene can create abnormal pathways that result in the buildup of potentially lethal waste products in the astrocytes. Early in 2005, researchers at the University of New Mexico discovered that people with a gene *APO4* (a variety of the *APOE* gene) have an increased risk for developing Alzheimer's disease. However, even if they do not develop Alzheimer's disease, these people are more likely to become forgetful as they age. The study included 32 people between the ages of 60 and 87 years who showed no signs of dementia. The subjects were divided into two groups based on the presence of the *APO4* gene. Each group was given a memory test that evaluated their ability to remember everyday tasks. The performance on the memory tests of the subjects carrying the *APO4* gene was much worse than that of those with the normal *APOE* gene. There is also evidence that *APO4* is associated with problems remembering past events. The gene very likely alters the functions of the astrocytes necessary to maintain healthy nerve pathways in the brain.

Driscoll I, McDaniel IA, Guynn MJ. Apolipoprotein E and prospective memory in normally aging adults. *Neuropsychology*. 2005;Vol 19(1):28-34.

with the age. The average person loses 5% to 10% of his or her brain's weight between the ages of 20 and 90 years. It is hypothesized that much of this is due to a loss of myelination in the CNS, which, of course, still contributes to an inhibition of the rate at which information can be processed. Decreased blood flow to the brain may also account for some brain shrinkage and weight loss.

The slight neuron loss associated with aging does not necessarily translate into any observable decrease in brain function. It is surprising to learn that people aged 90 to 98 years have as many brain cells in the memory regions of their brain as those aged 57 to 60 years. However, people in the older age groups have diminished and slower memory capabilities. Some of this memory loss is most likely due to a decrease in neuroglial function. It is believed that the reduced blood flow in aging brains robs the neuroglia of nutrients and may cause a buildup of potentially toxic waste products. Not all functions of the brain diminish as a person matures. Verbal intelligence does not decrease in most people. In addition, the metabolic rate of neurons stays the same through the life span of a healthy person.

Any neuron loss associated with aging is compensated for by physiological processes called **plasticity** and **redundancy**. Plasticity is the ability of neurons to alter their function as a result of experience and usage. Connections between neurons can change to compensate for pathways that were lost. For example, the decay of one pathway associated with one aspect of speech can be replaced by a neighboring pathway. This does not appear to cause a loss of the body functions carried out by the pathways that are replacing another. Redundancy occurs when two neural pathways carry out equivalent functions. The loss of

Plasticity The ability of neurons to alter their function as a result of experience and usage

Redundancy Two neural pathways that carry out equivalent functions

one pathway is compensated for by use of the other pathway. Another factor in the aging process is the reduction of brain complexity. It is normal for each neuron to have fewer dendrites as a person ages. Excessive dendrite loss can reduce the brain's plasticity and redundancy capacities. Dendrite loss makes it more difficult for the brain to carry out complex tasks and may slow memory retrieval. It is believed that regular mental and physical activity throughout the lifetime slows dendrite loss.

A significant degree of peripheral nerve degeneration is evident in elderly people. This generally causes a reduction or loss of reflexes. These changes produce problems with balance and mobility. In contrast, certain reflexes that were present at birth, such as sucking and grasping, reappear as a person matures. It is believed that inhibitory pathways that suppress these reflexes are diminished with age. A loss of body-function control, confusion, dementia, and severe memory loss are common changes associated with aging. These changes are most likely associated with degenerative brain disorders. In people who have many illnesses or injuries throughout their lifetime, changes in brain and spinal cord function accelerate or mimic other conditions with age.

Sensory loss leading to problems with gustation, hearing, olfaction, and vision is most likely due to degradation of the sensory structures. The regions of the brain associated with sensory perception show no decrease in function in older people. Hearing loss affects 83% of men and 55% of women aged 70 to 79 years. This increases to 97% of men and 86% of women whose age is greater than 80 years. Vision loss occurs equally in each gender. There can be a 21% loss of vision after age 70 years and 42% in persons in their 80s. The threshold of olfactory receptors, taste buds, and sensory structures of the skin increases with age. This creates a loss of sensitivity that reduces sensory input to the brain and to reflexes.

✓ Concept Check

1. Describe two changes to the brain related to normal aging.
2. Define the terms plasticity and redundancy.
3. What factor most likely leads to the age-related loss of the senses?

DISCOVERY SCENE PLEASE ENTER DISCOVERY SCENE PLEASE ENTER

Have you solved the CSI yet? Did the information about aging and the pathology of the nervous system's structure provide any more clues about the football player's condition?

CSI — Case Study Investigation Conclusion

What did you determine about the football player's condition? What was the most important factor in determining the nature of his memory problems? Were they related to the other conditions he was exhibiting? Does he have one medical condition or a combination of two or more diseases?

Answer:

The football player has a neurovascular condition called vascular dementia. The condition is highly variable, affecting different people in dissimilar ways. It is caused by a loss of blood flow to general regions of the brain and other parts of the CNS. The progression of this condition varies from person to person. It is commonly mistaken for other categories of dementia, including prion-related disorders and Alzheimer's disease. People with vascular dementia may experience confusion, dementia, depression, paralysis, general muscle weakness, and speech problems. Many factors contribute to this disease; however, much of it is due to a person's genetic makeup. Vascular dementia is more common in people with a family history of heart disease, high blood pressure, and stroke. The condition is aggravated by alcohol abuse, high-fat diets, inactivity, and obesity. At this point of his life, the condition can be slowed, but the brain damage cannot be reversed. Neurovascular dementia can be treated with drugs and surgical procedures that improve blood flood to the CNS. People likely to develop neurovascular dementia can slow the progression of the disease by adopting life styles that reduce the risk factors.

This CSI was adapted from the following articles:

1. Dominguez J, Morris JC. Differentiating vascular dementia and Alzheimer's disease: making the differential diagnosis. *Alzheimer's Disease Management Today*. 1999;2:3-10.
2. Inouye S, Robison J, Froehlich T, et al. 1999. The time and change test: a simple screening test for dementia. *J Gerontol*. 1999;53A:M281-M286.
3. Strub R. Vascular dementia. *S Med J*. 2003;96(4):363-366.

Chapter Summary

The human nervous system is formed of two components that work together to coordinate body functions. Information from the environment is transmitted to the CNS by nerves of the PNS. Within the CNS, this information is integrated into a reflex or cognitively processed by the brain. Specialized regions of the brain interpret sensory information. The brain then uses the sensory information to formulate a response. Responses of the brain are channeled to the body through either the autonomic or the somatic nervous system. Special sensory structures in the head add to variety of environmental stimuli that the body receives. Nervous system structure is subject to damage resulting from a variety of diseases. These diseases usually cause the decrease or loss of one or more body functions. Aging induces body changes that diminish nervous system function, or causes the death of neurons and neuroglia.

Nerve Structure
- Nerves are bundles of neurons.
- Afferent nerves carry sensory information.
- Efferent nerves carry motor information.

Nervous System Components
- The CNS is composed of the brain and spinal cord.
- The spinal cord coordinates reflexes and signals to and from the brain.
- The PNS is composed of somatic nerves, autonomic nerves, and ganglia.
- The autonomic nervous system is divided into the parasympathetic and sympathetic nervous systems.

Human Senses
- The special senses are concentrated in the head.
- Chemoreceptors on the tongue sense taste (gustation).
- Chemoreceptors in the nose sense smell (olfaction).
- Photoreceptors in the retina of the eye sense light (vision).
- Cones are color photoreceptors, and rods detect dim light.
- Neurons in the cochlea sense sound vibrations (hearing).
- Neurons in the semicircular canals and vestibule sense position (equilibrium).

Wellness and Illness over the Life Span
- Structural diseases of the nervous system are categorized as follows:
  - trauma
  - cerebrovascular and neurovascular diseases
  - CNS tumors
  - developmental disorders
  - metabolic and toxic diseases
  - nervous system infection
  - neurodegenerative disease
- Aging is accompanied by some neuron loss in the brain.
- The CNS loses weight as a person matures.
- Most brain aging is due to loss of myelinization and decreased blood flow.
- Most people show a decrease in complex brain functions as they age.
- Plasticity and redundancy compensate for neuron loss in the CNS.
- Sensory structure thresholds increase with age.

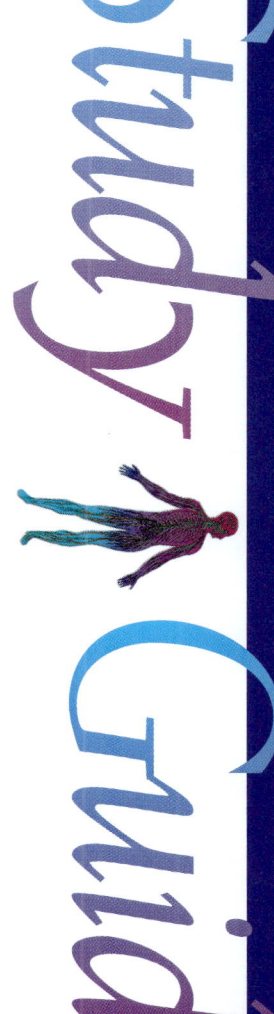

Study Guide

Key Terms

Overview
Central nervous system (CNS)
Peripheral nervous system (PNS)

Nerve Structure
Afferent nerve
Efferent nerve
Endoneurium
Epineurium
Neurofascicle
Perineurium

Central Nervous System
Ascending tracts
Arachnoid
Basal nuclei
Brain stem
Central canal
Central sulcus
Cerebellum
Cerebral hemispheres
Cerebrum
Choroid plexus
Convolutions
Corpus callosum
Cortex
Descending tracts
Dura mater
Extrapyramidal tracts
Forebrain
Frontal lobe
Gray matter
Gyri
Hindbrain
Hydrocephalus
Insula
Intermediate gray matter
Lateral sulcus
Limbic system
Longitudinal cerebral fissure
Medulla oblongata
Meninges
Midbrain
Motor area
Occipital lobe
Parietal lobe
Pia mater
Pons
Sensory area
Subarachnoid space
Temporal lobe
Ventricles
Vertebral lumen
White matter

Peripheral Nervous System
Abducens
Accessory
Autonomic
Cranial nerve
Dorsal root
Dorsal root ganglion
Facial
Glossopharyngeal
Hypoglossal
Iris
Oculomotor
Olfactory
Optic
Parasympathetic
Postganglionic
Preganglionic
Somatic
Sympathetic
Sympathetic ganglion chain
Trigeminal
Trochlear
Vagus
Ventral root
Vestibulocochlea

Human Senses
Audition
Chemoreceptor
Equilibrium
External auditory meatus
Gustation
Olfaction
Photoreceptors

Taste
Bitter
Papillae
Saliva

Study Guide

Salt
Sour
Sweet
Taste bud
Umani

Smell
Olfactory bulb
Vomeronasal gland

Vision
Aqueous humor
Choroids
Ciliary body
Cone
Conjunctiva
Cornea
Extrinsic muscles
Fovea
Lacrimal duct
Lacrimal gland
Lens
Optic cup
Palpebrae
Retina
Rhodopsin
Rod
Sclera
Vitreous humor

Hearing and Balance
Auditory canal
Auricle
Cochlea
Eardrum
Eustachian tube
External auditory meatus
External ear
Incus
Inner ear
Malleus

Middle ear
Organ of Corti
Ossicle
Oval window
Pinna
Round window
Semicircular canals
Stapes
Tympanic membrane
Vestibule

Pathology of Nervous System Structure
Aneurysm
Arteriovenous malformation
Athetosis
Cerebrovascular
Chorea
Glioma
Ischemic attack
Lymphoma
Meningioma
Metastatic
Ministroke
Neuroblastoma
Neurodegenerative
Neuroma
Neurovascular
Palsy
Peripheral neuropathy
Shaken-baby syndrome
Stroke
Tremor
Whiplash

Aging of the Nervous System Structure
Plasticity
Redundancy

STRUCTURE OF THE NERVOUS SYSTEM

Study Guide

Check Your Understanding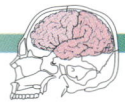

1. The outermost covering of a nerve is called the:
    a. endoneurium
    b. perineurium
    c. epineurium
    d. ectoneurium

2. The sense of pain travels through this type of nerve:
    a. motor
    b. efferent
    c. reverberating
    d. afferent

3. Which is *not* part of the CNS:
    a. cranial nerves
    b. forebrain
    c. hypothalamus
    d. medulla oblongata

4. Humans have _____ pairs of cranial nerves and _____ pairs of spinal nerves.
    a. 5 and 25
    b. 12 and 31
    c. 20 and 20
    d. 31 and 50

5. Trauma to the surface of the brain directly damages:
    a. nerve cell bodies
    b. myelin sheathes
    c. deep ganglia
    d. axons

6. Nerve cell bodies are most likely found in the:
    a. white matter of the brain
    b. white matter of the spinal cord
    c. subarachnoid space
    d. gray matter of the spinal cord

7. Defects of the hindbrain affect the development of the:
    a. cerebrum
    b. basal bodies
    c. medulla oblongata
    d. thalamus

8. Damage to this part of the brain would directly affect respiration:
    a. cerebrum
    b. medulla oblongata
    c. cerebellum
    d. hypothalamus

CHAPTER 9

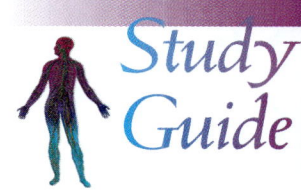

9. The innermost layer of the meninges is the:
   a. dura mater
   b. arachoid
   c. gray matter
   d. pia mater

10. Cerebrospinal fluid is found in the:
    a. dura mater
    b. ventricles
    c. white matter
    d. spinal nerve sheathes

11. Which of the following is *not* a cranial nerve?
    a. abducens
    b. thoracic
    c. vestibulocochlear
    d. facial

12. Postganglionic neurons of the sympathetic nervous system use as a neurotransmitter:
    a. acetylcholine
    b. dopamine
    c. gamma-aminobutyric acid
    d. norephinephrine

13. Olfaction takes place in the:
    a. eye
    b. nose
    c. mouth
    d. ears

14. Color vision is detected by the:
    a. cones
    b. rods
    c. choroid
    d. lens

15. Palsy is a type of:
    a. ganglioma
    b. paralysis
    c. spasm
    d. muscle atrophy

Study Guide

A Case Study

Aluminum

Many conveniences of modern life would not be possible were it not for lightweight materials, such as aluminum and plastic. Almost every consumer item would be much heavier and larger if these substances were not used in their construction. Think about lugging around a CD player made of solid steel. Unfortunately, each new innovation introduced to society brings with it a new health risk. Aluminum is known to inhibit cellular respiration needed for nerve cell function. Many studies are showing the prevalence of aluminum in many food containers, household items, and personal hygiene products, which may lead to serious health risks. Aluminum is a common pollutant found in air and water. It is also a common constituent of personal hygiene products, and it is found in some foods. Aluminum cans are also a potential source of aluminum contact with the body.

Various public interest groups are petitioning the US government to reduce aluminum use, particularly among children. They are protesting the method of disposal of aluminum products to ensure that it these materials do not pollute the water. Some groups are pushing for legislation to reduce aluminum air pollutants produced in manufacturing processes. In addition, some people want to see the elimination of aluminum cans and food packaging, fearing that children can be harmed by small amounts of contamination. It is possible for pregnant women to accidentally harm their developing fetus by taking in aluminum.

Use the information in this chapter and the following websites to answer the following questions about aluminum contamination:

- Which ways can aluminum get into the body?
- What evidence is available to show that aluminum is a serious public health issue?
- How much aluminum must be taken into the body to harm an adult? How much for a child?
- What are the major sources of aluminum contamination in the environment?
- Should the government be responsible for regulating the amount of aluminum that a person can be exposed to in a lifetime?
- What, if any, actions should be taken to control potential aluminum contamination of an individual?
- Who should be responsible for any problems related to aluminum packaging used for beverages and foods?

1. Alzheimer's Society Fact Sheet
   http://www.alzheimers.org.uk/Facts_about_dementia/Risk_factors/info_aluminium.htm

2. Leading Edge Research Group
   http://www.cco.net/~trufax/general/aluminum.html

3. Health Canada: Aluminum
   http://www.hc-sc.gc.ca/ewh-semt/water-eau/drink-potab/aluminum-aluminium_e.html

4. Occupational and Environmental Medicine Journal
   http://oem.bmjjournals.com/cgi/content/full/58/7/453

Chapter 9

Where Do We Go from Here?

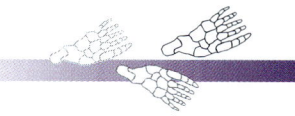

People in health fields can use their knowledge of the nervous system's structure to solve everyday problems. You may wish to use other resources in addition to your textbook to investigate the answers to each of the following situations:

1. A friend heard that eating Chinese food could cause migraine headaches. What would you tell him about the possible truth or inaccuracy of the claim?
   http://www.emedicinehealth.com/articles/9385-1.asp

2. A neighbor's wife is concerned about her husband's emotional lows and excessive alcohol consumption. She is afraid to have him quit drinking because she thinks he would become more depressed. What advice would you give her about her decision to let his drinking continue?
   http://www.fhsu.edu/kellycenter/dawn/bal.shtml

3. A friend is pregnant. She is being pestered by her mother about staying healthy. Her mother claims that having a fever during pregnancy can harm the baby. What would you explain to your friend about her mother's worries?
   http://www.fetal-exposure.org/HYPERTH.html

4. A friend's father suffered from a virus that attacks the nerves of the brain. Your friend overheard the physician saying that the virus damaged the facial and vestibulocochlear nerves. He asks you what impact it has on his father.
   http://www.neuroanatomy.wisc.edu/cn/cn.htm

5. A salesperson is trying to sell a friend an electrical nerve device for reducing muscle pain after too much exercise. Explain to your friend the effectiveness of this type of treatment.
   http://www.emedicine.com/pmr/topic206.htm

Skills Activities

1 Stroop Effect and Brain Function

Materials
- team of two students
- lists 1 and 2 from the book or written as is on a board
- stopwatch or clock with a second hand
- paper for recording individual student results
- board or projector for showing class data

The brain is responsible for linking information from all the senses into organized thoughts that lead to appropriate responses to incoming stimuli. In this activity you will be testing a brain function phenomenon called the Stroop effect. It is a form of neural interaction called interference. Interference occurs when information from one nerve tract or nervous system function obstructs the action or impulses of another neural pathway. Dr. John Ridley Stroop of the United States wrote the first medical study on brain function interference in 1935. He discovered that the ability to make an appropriate response when given two conflicting signals interferes with the function of an organizing

STRUCTURE OF THE NERVOUS SYSTEM

Study Guide

region of the brain called the anterior cingulate. It is located in a region of the brain between the right and left halves of the frontal lobe. The anterior cingulate helps organize emotional responses and learned thought processes. The activity described below is one way to measure the Stroop effect.

This activity will require two people to work as partners. The partners will each take turns being timed as they recall the colors of three printed lists of words. Person A will be the recorder of person B in the first set of trials. Then the partners will swap roles as person B records the responses of person A. The colors must be read aloud as fast as possible with as few errors as possible. The following steps should be carried out for consistent results:

Step 1: Person A should have the stopwatch or clock ready for timing Trial 1.
Step 2: Person A says "go" as person B reads the color of the words in List 1 out loud.
Step 3: Person A records how long it took person B to read the list.
Step 4: Repeat steps 1 through 3 four times and record the average.
Step 5: Person A now records the average time it took person B to read List 2 four times.
Step 6: Person A then records the average time it took person B to read List 3 four times.
Step 7: Person B now becomes the reader and repeats steps 1 through 6 on person A.
Step 8: Record these data on a chart where they can be compared with the results from the rest of the class.

List 1	List 2	List 3
RED	RED	Word
YELLOW	YELLOW	Word
GREEN	GREEN	Word
BLUE	BLUE	Word
RED	RED	Word
BLUE	BLUE	Word
YELLOW	YELLOW	Word
GREEN	GREEN	Word
BLUE	BLUE	Word
RED	RED	Word

Data Analysis Is there evidence that a person had trouble naming the color of the word when the word didn't match its color? Were there any differences in the time it took to say the colors for each list? What factors interfered with the ability to quickly say the name of the color? Were there any individual differences in the abilities of people to name the colors for List 2? What are some possible explanations for the differences?

Skills Activities

2 Brain Pathology Interpretation

Materials
- microscope with high-power capability (400X)
- prepared slide of the human brain

Many diseases cause visible changes to the brain's structure, which are easily identifiable under the microscope.

Place a microscopic slide of the human brain under the microscopic. Start with the low magnification, and move to the higher magnification to see details and individual cells. Can you distinguish between the nerve cell bodies, axons, and neuroglia? How many nerve cell bodies can be seen in the viewing field? Does the number vary from one part of the brain to another?

Now, look at the microscopic photographs below and answer the questions for each image.

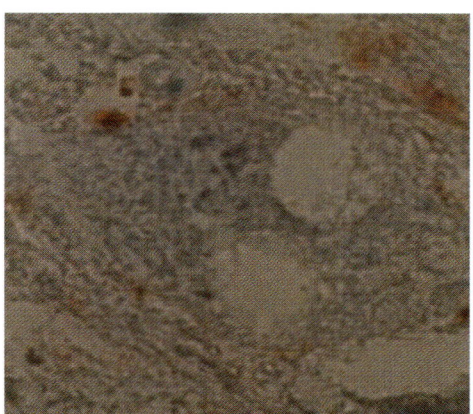

Image 1

How does this brain specimen differ from normal brain tissue? Can you explain the large, round, dark cells? Are the irregular small holes found in normal brain tissue?

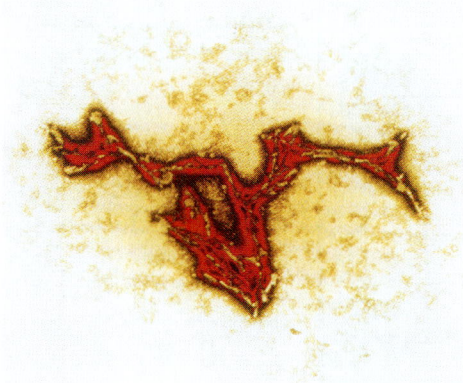

Image 2

How does this brain specimen differ from normal brain tissue? How does it differ from Image 1? Explain the presence of the brown "stringy" cells present in this brain specimen.

Images 1 and 2 are sections of a sheep brain with a disease called scrapie, or transmissible spongiform encephalopathy (TSE). Scrapie is caused by prions and is related to mad cow disease and Creutzfeldt-Jakob disease. Humans very likely do not contract scrapie. However, it is possible to get related diseases from cattle, horses, and pigs. More information, including images of Creutzfeldt-Jakob disease can be found at:
http://www.ninds.nih.gov/disorders/cjd/cjd.htm.

STRUCTURE OF THE NERVOUS SYSTEM

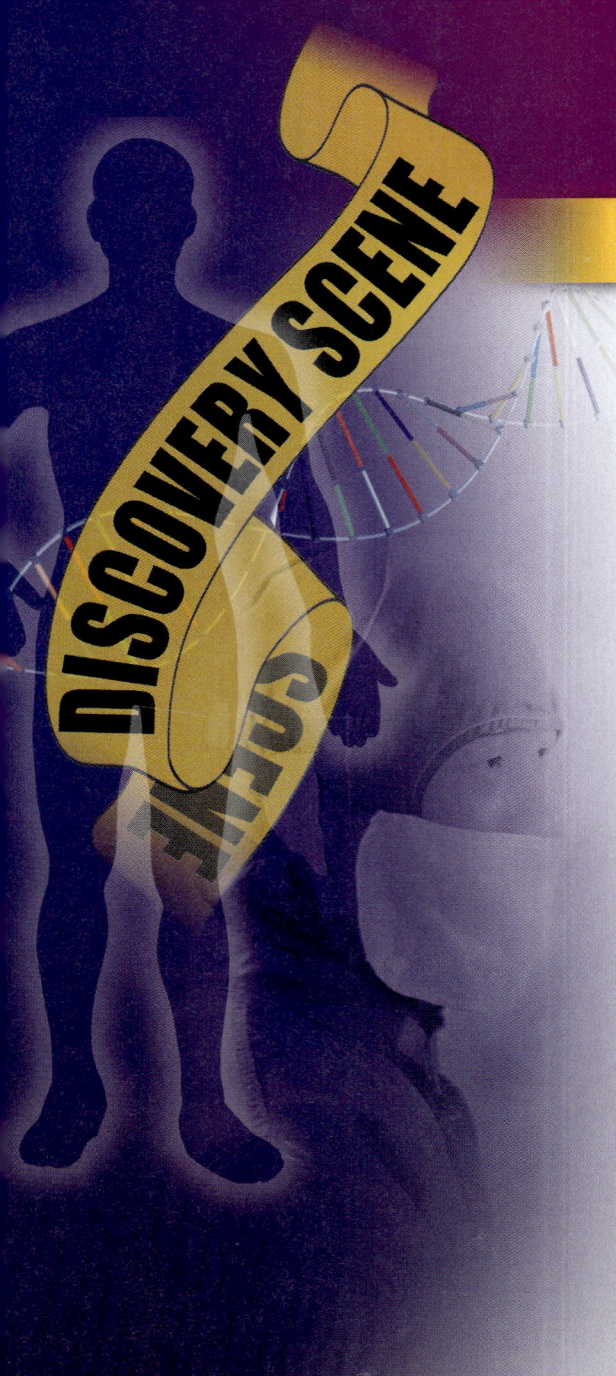

Case Study Investigation

Case Study Investigation #10

After a routine workout at your health club, you decide to soak in the hot tub. A young man in the tub tells you how he just finished a heavy weightlifting routine that lasted 2 hours. He says that he soaks in the hot tub for 20 minutes to ward off muscle cramping. The man does this routine 7 days a week. The next day at the health club, you hear that he was hospitalized that morning with the following symptoms: chills, coughing, fatigue, fever, shortness of breath, and tightness in his chest. He appeared healthy until that incident. A friend of his who also works out at the club said he had no history of heart disease or respiratory ailments. However, lately he was losing weight because of the heavy workout. People at the health club are speculating what could have caused the illness. What information can you provide to the conversation?

CHAPTER 10

The Respiratory System

Chapter Outline

- Case Study Investigation (CSI)
- Applied Learning Outcomes
- Overview
- Components of the Respiratory System
  - Nose
  - Nasal Cavity
  - Paranasal Sinuses
  - Pharynx
  - Larynx
  - Trachea
  - Bronchial Tree
  - Lungs
- Breathing
  - Mechanics of Breathing
  - Gas Exchange
- Wellness and Illness over the Life Span
  - Pathology of the Respiratory System
    - Developmental Diseases
    - Infectious Diseases
  - Aging of the Respiratory System
- CSI Conclusion
- Study Guide

Applied Learning Outcomes

- Use the terminology associated with the respiratory system.
- Learn about the following:
  - respiratory system components
  - development and histology of the respiratory system
  - respiratory system function
  - breathing process
- Understand the aging and pathology of the respiratory system.

Overview

Key Terms: breathing, lung, ventilation

Lung One of two large organs in which gas is exchanged between the blood and the environment

Ventilation or Breathing The process of transporting air to the surface of the lungs

The body's ability to carry out cellular activities depends on exchanging materials with the environment. Cells are in constant need of nutrients and oxygen to sustain metabolic pathways. These pathways provide cell energy, build cell components, and produce secretions. Metabolic wastes must also be removed. A buildup of carbon dioxide, the waste of cellular respiration, will shut down the metabolic functions of cells. The respiratory system remedies this need to exchange certain materials by facilitating the passive diffusion of carbon dioxide and oxygen. It accomplishes this task by bringing air to the **lungs** where diffusion can take place.

There is the belief that the skin can "breathe," or carry out gas exchange. Human skin is not a good surface for diffusion because it is covered with an impermeable barrier. It is designed to reduce water loss and prevent the uncontrolled entry of molecules from the environment. The skin is also so thick that diffusion could not take place at the rate needed to meet the demands of metabolism. In addition, the skin is too dry to permit the wet surface needed for the diffusion of gases. The internal location of the human respiratory system allows it to stay moist without dehydrating the body. The delicate surfaces that are used for the diffusion of gases are protected because they are inside the body. The only drawback to having an internal diffusion surface is that air must be moved to come in contact with the lung surface. **Ventilation**, or **breathing**, is the method of bringing air from outside the body to the lungs. However, ventilation does not ensure that diffusion will take place. Other factors described in this chapter and in Chapter 12 will discuss the conditions that promote diffusion once ventilation has taken place.

✓ Concept Check

1. What is the main function of the respiratory system?
2. What conditions are needed for the diffusion of materials into and out of the body?
3. Describe ventilation.

Civic Responsibility

CIVIC RESPONSIBILITY: HELPING OTHERS WITH YOUR KNOWLEDGE

It is valuable to use what you learned about the respiratory system to help others to better understand the world around them. It is very important to check your facts and seek further information about certain topics before discussing health and science issues. Here are some suggestions to foster a better public awareness of the respiratory system:

1. Speak to schoolchildren about the effects of smoking on the respiratory system.
2. Volunteer with the American Lung Association to promote a healthy respiratory system.
3. Help elderly persons to better understand how air quality affects their health.
4. Work with local or national environmental groups to reduce air pollution.

CHAPTER 10

COMPONENTS OF THE RESPIRATORY SYSTEM

Key Terms: bronchial tree, larynx, lower respiratory system, nose, paranasal sinuses, pharynx, trachea, upper respiratory system

The respiratory system (Figure 10.1) is composed of the **nose**, nasal cavity, **paranasal sinuses**, **pharynx**, **larynx**, **trachea**, **bronchial tree**, and **lungs**. It is important to remember that each of these structures is an organ composed of a particular arrangement of tissues, which provides that organ with its function. Physicians usually divide the respiratory system into the **upper respiratory system** and the **lower respiratory system**. The upper respiratory system is composed of the nose, nasal cavity, paranasal sinuses, and larynx. Also included are the mouth and the eustachian tubes of the middle ear. The lower respiratory system is composed of the trachea, bronchial tree, and lungs. It is debated whether the larynx is part of the upper or lower respiratory system. Each system is plagued by a different array of microorganisms that cause disease. Many of the microorganisms that thrive in one part of the respiratory system do not necessarily survive in the other component. Upper respiratory infections are usually less serious than lower respiratory infections. However, it is very likely that certain upper respiratory infections can become lower respiratory infections if the bacteria spread into the lungs. This is true for the bacterium *Streptococcus pneumoniae*, which irritates the throat and can inflame the lungs.

Nose The entrance to the respiratory tract

Paranasal Sinuses Air cavities within the facial bones

Pharynx The throat, or cavity, behind the mouth

Larynx The area of the throat that houses the vocal cords

Trachea The windpipe; a passage for the admission of air to the lungs

Bronchial Tree A network of passages that supplies the lungs with air

Upper Respiratory System The part of the respiratory system composed of the nose, nasal cavity, paranasal sinuses, eustachian tubes, and larynx

Lower Respiratory System The part of the respiratory system composed of the trachea, bronchial tree, and lungs

Nose

Key Terms: nares, nostril, quadrangular cartilage

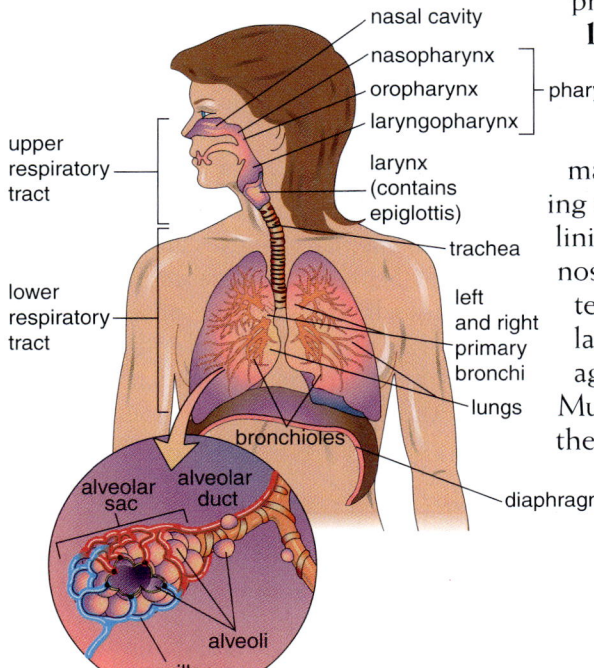

Figure 10.1 Components of Respiratory System

The nose (Figure 10.2) is a projection of skin shaped by a flexible cartilage projection called the **quadrangular cartilage**. The bones of the nasal cavity support the quadrangular cartilage. The **nostrils**, or **nares**, are the major point of entry for air entering the respiratory system. Hairs lining the interior surface of the nostrils protect the respiratory system by restricting the passage of large particles, which could damage the lower respiratory system. Mucus covering the hairs reduces the number of microorganisms that can cause disease. The microorganisms stick to the mucus and flake off the hairs when the mucus dries.

Quadrangular Cartilage A vertical partition of cartilage that supports the nose

Nostrils or Nares The openings in the nose through which air enters the nasal cavity

THE RESPIRATORY SYSTEM

Figure 10.2 The Nose

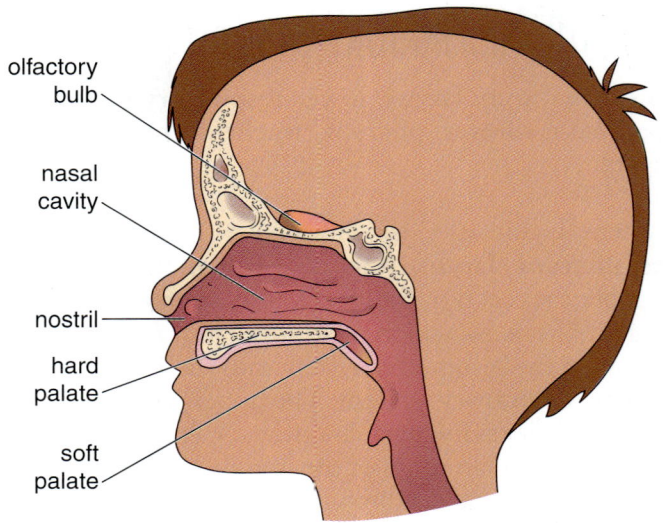

Nasal Cavity

Key Terms: turbinates

The nasal cavity is a large, air-filled space formed by the ethmoid, maxilla, and nasal bones. It is lined with a mucous membrane with a rich blood supply. The mucus assists in cleaning and moistening the air, while the blood supply warms the air that enters the nose. Cilia on the mucous membrane move the mucus into the digestive system. Microorganisms and materials trapped in the mucus are then degraded in the digestive system.

A vertical bony structure called the nasal septum separates the nasal cavity into left and right airways. The nasal septum is composed of the ethmoid bone, quadrangular cartilage, and vomer bone. Each nostril serves one airway. On the lateral walls of the nasal cavity are three structures called conchae, or **turbinates**. It is believed that they direct air to the olfactory gland to facilitate smell. They also play a role in circulating air to be cleaned, warmed, and moistened.

Turbinates Bony structures in the nasal cavity that clean and moisten air

Paranasal Sinuses

The paranasal sinuses are connected to the nasal cavity through small openings in the bones. They are composed of four pairs of sinuses lined with mucuous membranes and located in the facial bones. The maxillary sinuses are located under the eyes in the maxillary bone. The frontal sinuses are located just above the eyes in the frontal bone. The ethmoid sinuses lie between the nose and the eyes. The sphenoid sinuses are located at the base of the skull, behind the nasal cavity. The role of the paranasal sinuses is not fully known; however, it is believed that they help to warm and moisten the air, and provide resonance for speech.

Laryngopharynx The lower part of the pharynx
Nasopharynx The upper part of the throat behind the nose
Oropharynx The area of the throat at the back of the mouth

Pharynx

Key Terms: adenoids, laryngopharynx, nasopharynx, oropharynx, tonsils

The pharynx is a large cavity formed by the facial bones. It is divided into the **nasopharynx**, **oropharynx**, and **laryngopharynx** (Figure 10.3). The

Good Choice Bad Choice

An asthmatic friend wants to go on a hot air balloon ride in Taos, New Mexico. She tells you that she lied on the form asking about health limitations. Your friend indicated that she did not have any respiratory conditions. What would you say to your friend about the risks associated with going up in a balloon for a person with asthma?

Figure 10.3 **Pharynx**

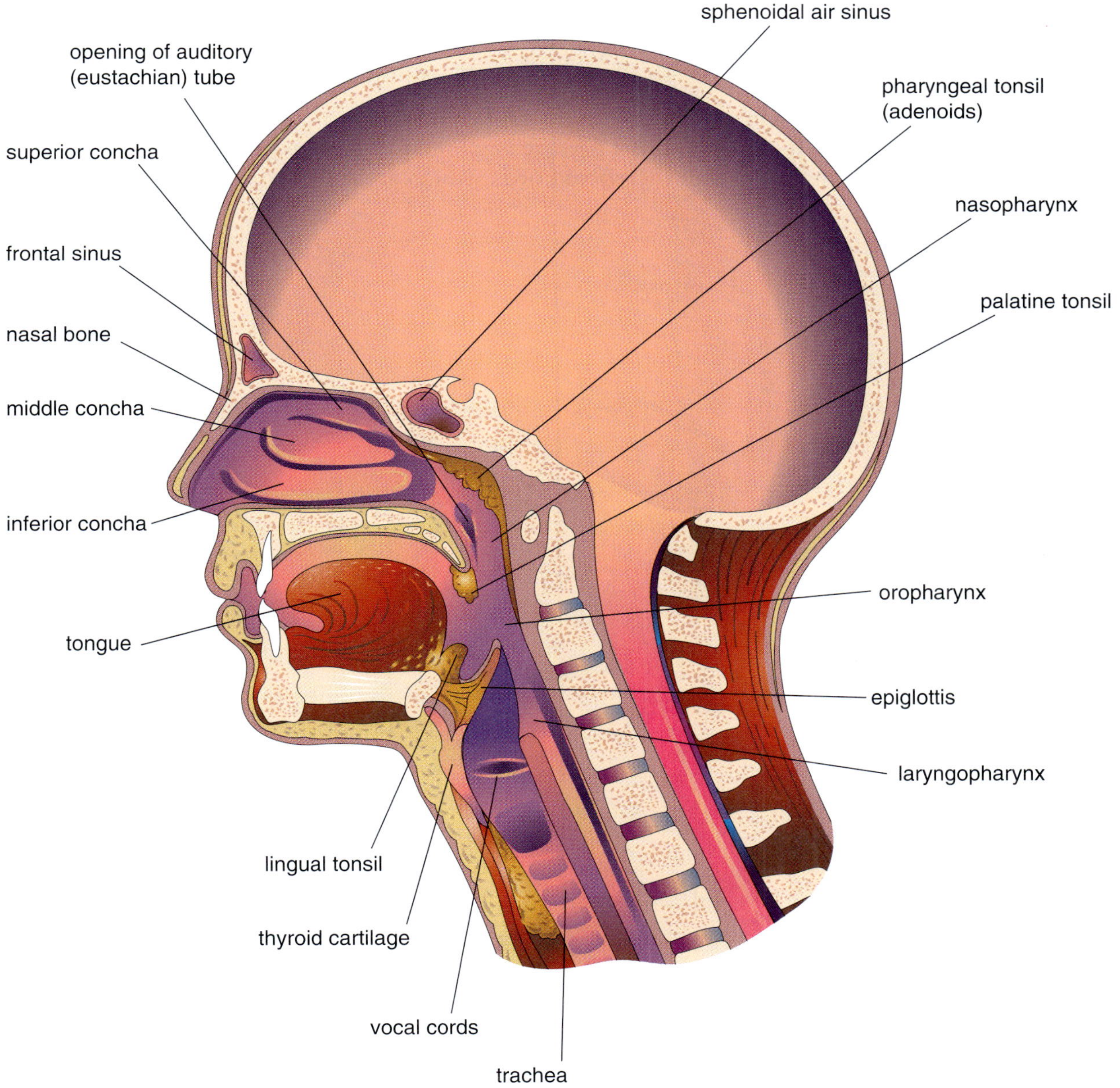

THE RESPIRATORY SYSTEM

nasopharynx lies posterior to the nasal cavity and is lined with stratified squamous epithelia that secrete mucus. The mucus helps to clean the air entering the lower respiratory system. Two openings in the nasopharynx lead to the eustachian tubes of the middle ears. They equalize air pressure in the ear and drain fluids from the ear into the throat. Two large glandular structures called **adenoids** are also found in the posterior region of the nasopharynx. Adenoids are components of the immune system that prevent organisms that cause infection from entering the lower respiratory system. As the nasopharynx extends down into the throat it meets the oropharynx. The anterior part of the oropharynx is a component of the digestive system. It serves as a common passageway for air and food. It is also the location of the **tonsils** which are related to the adenoids of the nasopharynx. The distal portion of the pharynx is called the laryngopharynx.

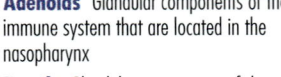
Adenoids Glandular components of the immune system that are located in the nasopharynx

Tonsils Glandular components of the immune system located in the oropharynx

Larynx

Key Terms: arytenoid cartilage, cricoid cartilage, epiglottis, glottis, Heimlich maneuver, laryngeal prominence, thyroid cartilage, vocal cords

Many people refer to the larynx as the voice box (Figure 10.4). The larynx lies below the pharynx; it is a passageway for air to enter into and out of the trachea. It also houses the **vocal cords**, which are responsible for producing

Vocal Cords Two small bands of muscle within the larynx that produce vocal sounds

Figure 10.4 **Larynx**

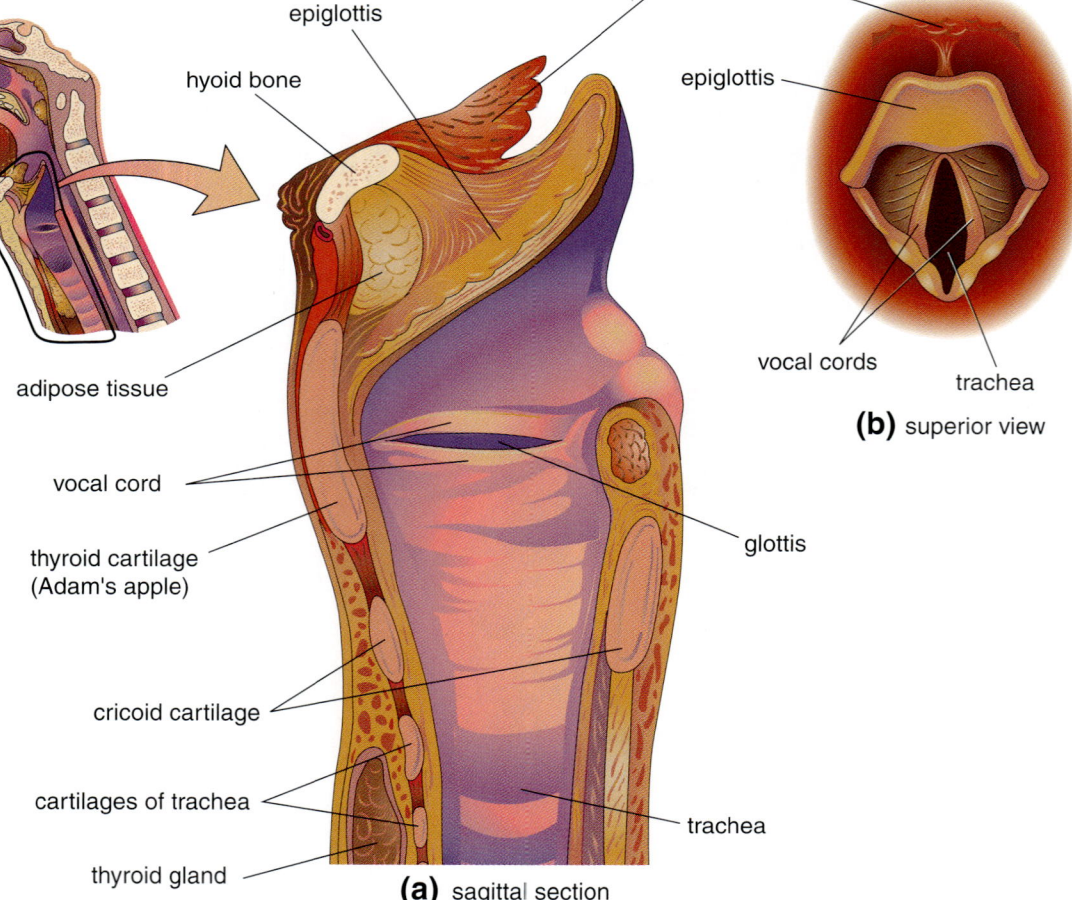

(a) sagittal section

(b) superior view

voice sounds. The larynx is most noted for the vocal cords. The vocal cords are located in a region of the larynx called the **glottis**. Within the glottis are two folds that contain the vocal cords. The vocal cords are two muscles supported anteriorly by the thyroid cartilage. A posterior attachment is formed with the arytenoid and cricoid cartilages. Sound is produced when air rushing up from the lungs vibrates the vocal-cord. Contraction and relaxation of the laryngeal muscles alter tension on the vocal cords and produce the different pitches of sounds. Males have thicker vocal cords, which accounts for the lower pitch of the voice compared with that of females. The formation of sounds that are converted into words is accomplished by modifying the air flowing through the pharynx, paranasal sinuses, mouth, and nose. The larynx is also needed for effective cough response. In addition, the larynx acts like a valve that prevents food and other objects from entering the lungs. The larynx is composed of the hyoid bone, muscles, and nine pieces of irregularly shaped cartilage.

Attached to the thyroid cartilage is a flaplike structure called the **epiglottis**. The epiglottis is a plate of cartilage covered by mucous membrane. It projects over the larynx, covering the opening of the trachea and assisting with swallowing. The epiglottis is attached to the cartilage and muscles making up the larynx. During swallowing, the epiglottis is pulled forward as the tongue presses the epiglottis over the trachea. This lets food slide over the smooth mucous-membrane covering into the digestive system. The epiglottis relaxes and flaps up during breathing. During a cough, the epiglottis remains closed while the chest is pushing out air. Rapid opening of the epiglottis produces a rush of air out of the respiratory system, forcing foreign objects out into the mouth. The **Heimlich maneuver**, a technique used to help a person who is choking, acts like a strong cough by forcing air out of the lungs to dislodge an obstruction. It is carried out using an abrupt and strong lifting action on the upper abdomen.

Three major cartilages form the larynx: **thyroid**, **cricoid**, and **arytenoid**. The thyroid cartilage forms the upper anterior and lateral portions of the larynx. An anterior ridge of the thyroid cartilage forms the **laryngeal prominence**, or Adam's apple. A difference in the shape of the thyroid cartilage accounts for the more prominent Adam's apple in most males. Male hormones early in puberty create this difference. The ring-shaped cricoid cartilage lies below the thyroid cartilage. Its larger posterior portion forms the back of the larynx. It is a strong, complete ring of cartilage that protects the larynx from damage. The two arytenoid cartilages are the attachment for the vocal cords.

Trachea

Key Terms: primary bronchi, tracheal cartilage

The trachea, or windpipe, is an airway that connects the larynx to the two **primary bronchi** (Figure 10.5). It is a single tube that branches into two at its inferior end. The right bronchus is shorter and more vertical than the left bronchus. This condition makes it more likely that objects accidentally entering the lungs will end up in the right lung. The trachea is typically 9 to 12 cm long and 2 to 2.5 cm in diameter. The diameter of the trachea is usually wider in males than in females. The trachea is composed of 15 to 20 regularly spaced rings of cartilage called **tracheal cartilages**. The cartilage does not form a complete ring. There is a slight gap on the posterior surface that allows some expansion of the trachea's diameter. The cartilage ring prevents the collapse

Glottis The opening at the superior end of the larynx

Epiglottis A flap of cartilage that covers the trachea while swallowing

Heimlich Maneuver First aid given to a choking victim

Thyroid Cartilage The largest cartilage; it makes up the ventral and lateral part of the larynx

Cricoid Cartilage A ring of cartilage that lies below the thyroid cartilage

Arytenoid Cartilage Two small cartilages at the back of the larynx

Laryngeal Prominence The anterior region of the thyroid cartilage commonly called the Adam's apple

Primary Bronchi The first division of the respiratory tree following the trachea

Tracheal Cartilage A ring of cartilage that supports the trachea

Figure 10.5 **Trachea**

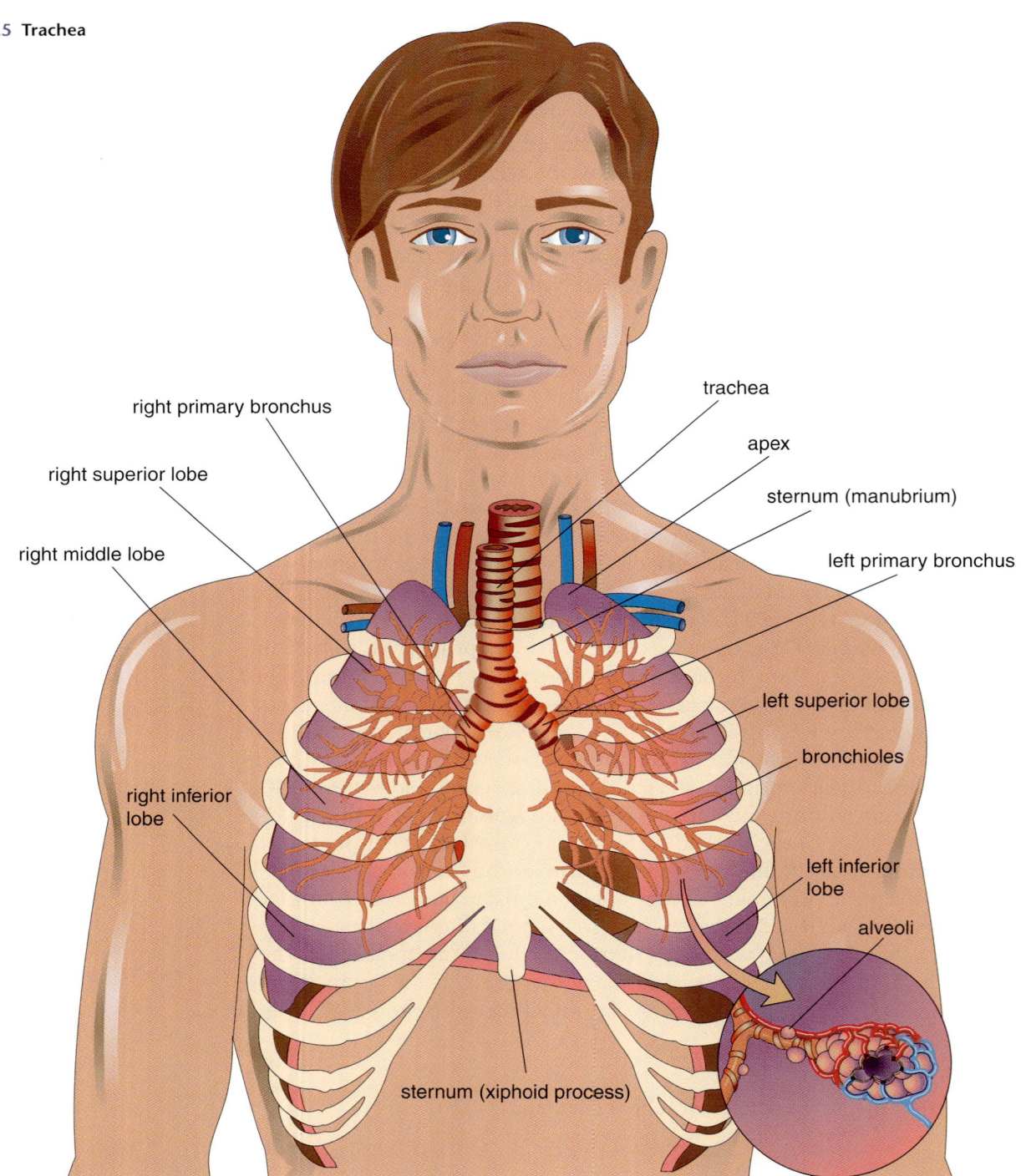

and overexpansion of the trachea as a person breathes. A soft tube would close during breathing and block the passage of air (Figure 10.6). The inner surface of the trachea is lined with ciliated columnar epithelia that cover a cylinder of areolar connective tissue with networks of blood vessels and nerves. The cilia move mucus containing small particles upward and out of the trachea. Irritation to the lining of the trachea stimulates the cough response. Research studies show that smoking alters the cytoskeleton component of these cells, which slows the function of the cilia and makes the lungs more susceptible to disease.

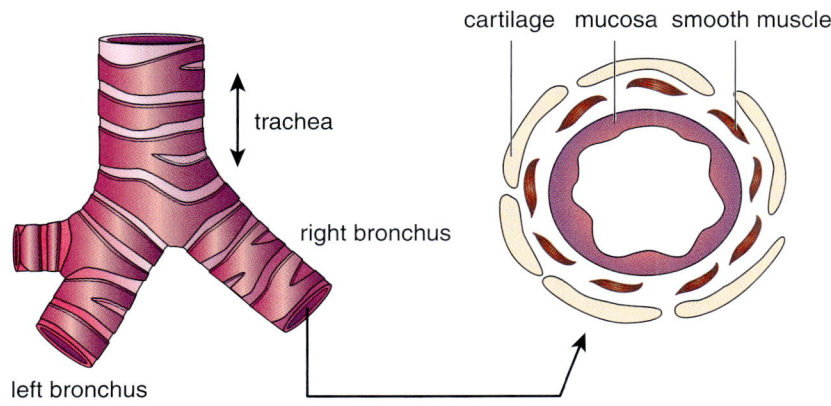

Figure 10.6 The Cartilage Ring of the Trachea

Bronchial Tree

Key Terms: bronchiole, bronchoconstriction, bronchodilation, bronchospasm, respiratory bronchiole, secondary bronchi, terminal bronchioles, tertiary bronchi

Each bronchus is a conduit for air into and out of an individual lung. The right bronchus branches into three **secondary bronchi** and 10 **tertiary bronchi**, which branch out further into the right lung. The left bronchus leads into two secondary bronchi and eight tertiary bronchi. Tertiary bronchi branch out into a network of very small tubes called **bronchioles**. Bronchioles have no glands or cartilage, and the epithelial cells become more cuboidal in shape. The bronchioles spread throughout the lung tissue and keep dividing into **terminal bronchioles**. Each terminal bronchiole subdivides into two or more **respiratory bronchioles**. The walls of the respiratory bronchioles have almost no smooth muscle but contain numerous sac-like alveoli where gas exchange occurs. Smooth muscles lining the terminal bronchioles control the flow of air through the lung by constricting and dilating. The constriction of the terminal bronchioles is called **bronchoconstriction**, while the expanson of terminal bronchioles is called **bronchodilation**. The parasympathetic nervous system stimulates bronchoconstriction, whereas the sympathetic nervous system produces bronchodilation. Certain chemicals and medical conditions produce **bronchospasms**. Bronchospasms occur when the smooth muscles of the terminal bronchioles constrict, severely narrowing their diameter. This condition is usually treated with oxygen therapy and drugs that dilate the terminal bronchioles.

Figure 10.7 Bronchial Tree

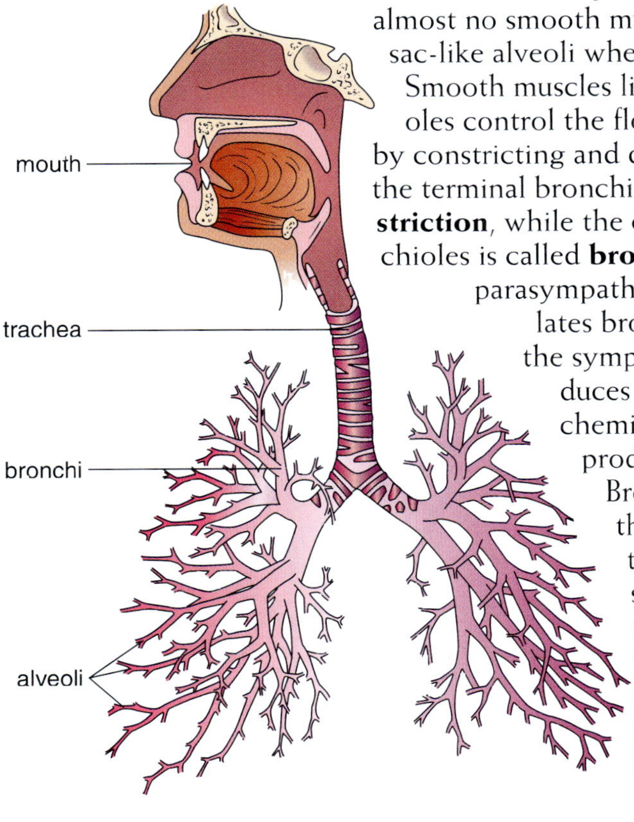

Secondary Bronchi A branch of the bronchi

Tertiary Bronchi A branch of the secondary bronchi

Bronchiole The smaller subdivision of the bronchi

Terminal Bronchiole The smallest branches of the bronchioles that lead into the alveoli

Respiratory Bronchiole The airways in the lung that branch off from the larger bronchi

Bronchoconstriction The constriction of bands of smooth muscle in the terminal bronchioles

Bronchodilation The expansion of smooth muscle in the terminal bronchioles

Bronchospasm The tightening of bands of smooth muscle in the terminal bronchioles

THE RESPIRATORY SYSTEM

Lungs

Key Terms: alveolus, cardiac notch, lobe, lobule, parietal pleura, pleura, pleuritis, surfactant, visceral pleura

The lungs are paired organs (Figure 10.8) that exchange atmospheric gases with the blood. They exchange air with the environment using the bronchial tree. The lungs lie within the thoracic cavity, which is surrounded by two layers of serous membranes called the **pleura**. The outer layer, which lines the chest cavity, is called the **parietal pleura**. The inner layer, or **visceral pleura**, is composed of two membrane sacs that cover each lung. These membranes secrete fluid that reduces abrasion as the lungs rub against the thoracic cavity during the breathing process. Infections and thoracic injury can inflame the pleura, causing a painful condition called **pleuritis** or pleurisy.

Each lung is divided into units called **lobes**. The right lung is composed of three lobes, which are named according to their anatomical location: the superior, middle, and inferior lobes. The left lung has two lobes: the superior and inferior lobes. In the lower region of the left lung's inferior lobe is a region called the **cardiac notch**. The heart is located within this region. Each lobe is associated with a distinct connection to a secondary bronchus. The lobes are further subdivided into smaller and smaller segments, ending in a structure called a **lobule**. Lobes and lobules provide functional independence to different parts of the lung; one part can be exchanging atmospheric gases while other areas are not. This controls the efficiency of gas exchange and reduces the effect of lung damage. Thus, damage to one part of the lung does not impair the function of the whole lung. Surgical procedures involving removal of a lobe are conducted for certain diseases, including lung cancer.

Pleura The membranes lining the lungs and thoracic cavity

Parietal Pleura The outer membrane of the pleura

Visceral Pleura Two inner membranes of the pleura

Pleuritis Inflammation of the pleura, also called pleurisy

Lobe A subdivision of the lung structure

Cardiac Notch A region of the left lung where the heart sits

Lobule The smallest subdivision of the lung lobes

Figure 10.8 Lungs

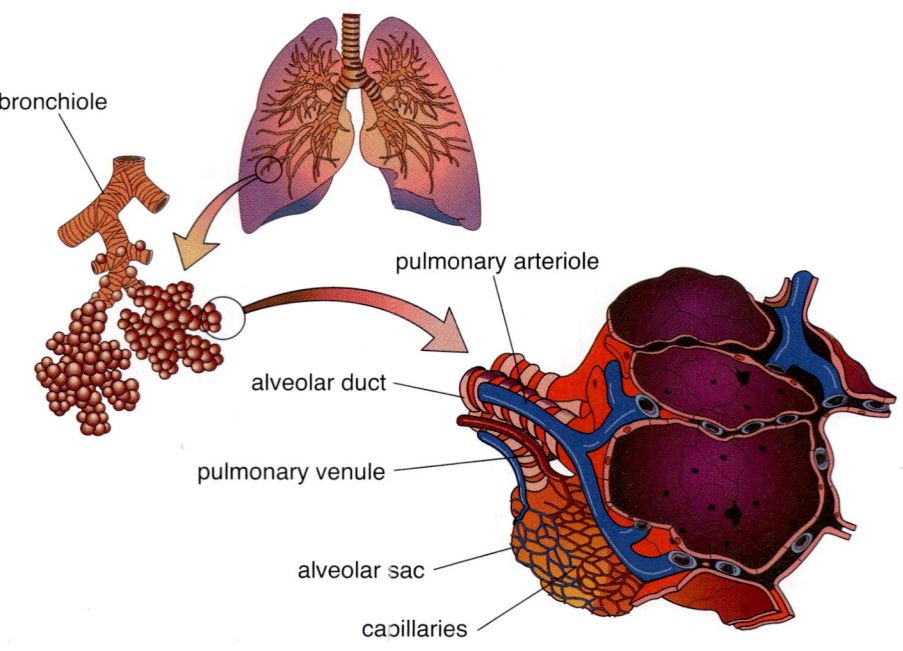

Cutting Edge Research
An Emergent Respiratory Disease

The news abounds with stories of bird flu outbreaks throughout the world. This illness is caused by a variety of avian influenza viruses that occur naturally in birds and other animals generally used as livestock. Birds in the wild normally carry these viruses in their intestines and usually do not become ill. However, in domesticated chickens, ducks, and turkeys, they can cause fatal illnesses. To date, the virus has passed only from animals to humans. And, although many strains of avian flu will not pass along to humans, some variations of the virus can invade the human respiratory system. The names given to these viruses designate their genetic tendency to cause human disease, such as H1N1, H1N2, and H3N2. Humans who come in contact with body fluids and wastes of infected birds, such as people who raise or slaughter chickens, ducks, turkeys, and wild birds, can contract the avian flu. Symptoms are similar to those of traditional human influenza infections: cough, fever, muscle aches, and sore throat. The disease can be fatal in those with preexisting diseases or weakened immune systems. Although no cases have been reported of humans transmitting the virus to other humans, it is believed that it would most likely be spread by coughing and sneezing. Currently available drugs reduce the virus's ability to spread throughout the body; however, the utility of these drugs may be short-lived. Researchers discovered a human form of the virus called H5N1, which resists drug treatment. In response to this finding, scientists are developing vaccines to prevent the spread of the virus. Unfortunately, vaccines can lose their effectiveness when viruses undergo genetic changes when transmitted from one person to another. Avian flu should not be confused with severe acute respiratory syndrome (SARS), which is caused by several dangerous forms of cold viruses.

Le QM, Kiso M, Someya K, Sakai YT, et al. Avian flu: isolation of drug-resistant H5N1 virus. *Nature*. 2005;437(7062):1108.

The lobules are divided into two functional and structural components: The conducting portion and the respiratory portion. The conducting portion is composed of the terminal bronchioles previously described. Special cells of the terminal bronchioles called Clara cells remove toxins from the lung surface. They contain two types of cells: Type I aveolar cells are simple squamous epithelial cells responsible for gas exchange. Type II alveolar cells are simple cuboidal epithelial cells responsible for the secretion of surfactant. **Surfactant** reduces the evaporation of water from the lung's wet surface. In addition, it prevents the respiratory portion of the lung from collapsing during ventilation. Gas exchange with the blood takes place in the respiratory portions. At the end of each respiratory bronchiole is a small sac of epithelial cells called an **alveolus**. The alveoli make up a bulk of the respiratory portion. Each alveolus is surrounded by a network of capillaries in which gases pass back and forth between the lungs and blood.

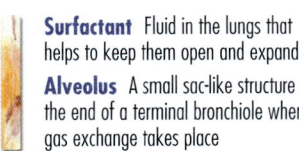

Surfactant Fluid in the lungs that helps to keep them open and expanded

Alveolus A small sac-like structure at the end of a terminal bronchiole where gas exchange takes place

✓ Concept Check

1. Describe the airways of the respiratory system.
2. What is the function of the larynx?
3. Describe the structure and function of the lung lobules.

DISCOVERY SCENE PLEASE ENTER DISCOVERY SCENE PLEASE ENTER

What part of the respiratory system is likely causing the young man's complications? How could he have developed the condition? What about his daily routine could contribute to respiratory system problems?

Diaphragm The muscle between the thorax and the abdomen that is used for breathing

Inspiration or inhalation The action of breathing in

Expiration or Exhalation The action of breathing out

Respiration The process of taking in oxygen and releasing carbon dioxide for cell metabolism

External Respiration The process of inhalation and exhalation

BREATHING

Key Terms: diaphragm, exhalation, expiration, external respiration, inhalation, inspiration, internal respiration, respiration

Breathing depends on the action of various structures that are anatomically part of the thorax. The **diaphragm**, intercostal muscles, and ribs work together with the respiratory system to carry out breathing. Air enters the lungs using an action called **inspiration**, or **inhalation**. **Expiration**, or **exhalation**, is the expulsion of air from the lungs. Breathing is sometimes called **respiration**. However, the term respiration should be reserved for the process of taking in oxygen and releasing carbon dioxide. Breathing is more precisely defined as the process of inhalation and exhalation. Many references call breathing **external respiration** (Figure 10.9). The exchange of gases between the blood and

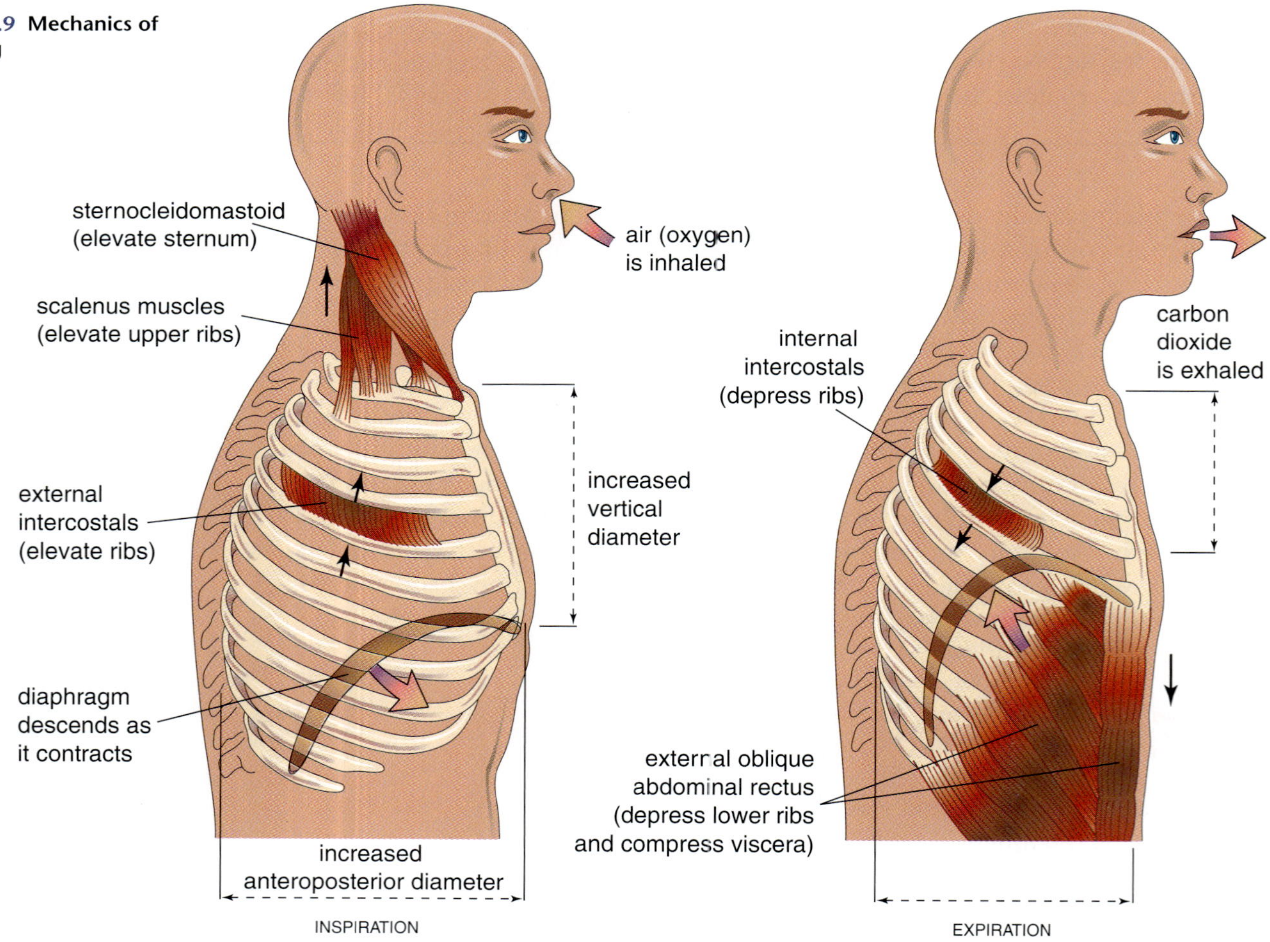

Figure 10.9 Mechanics of Breathing

Figure 10.10 Internal Respiration

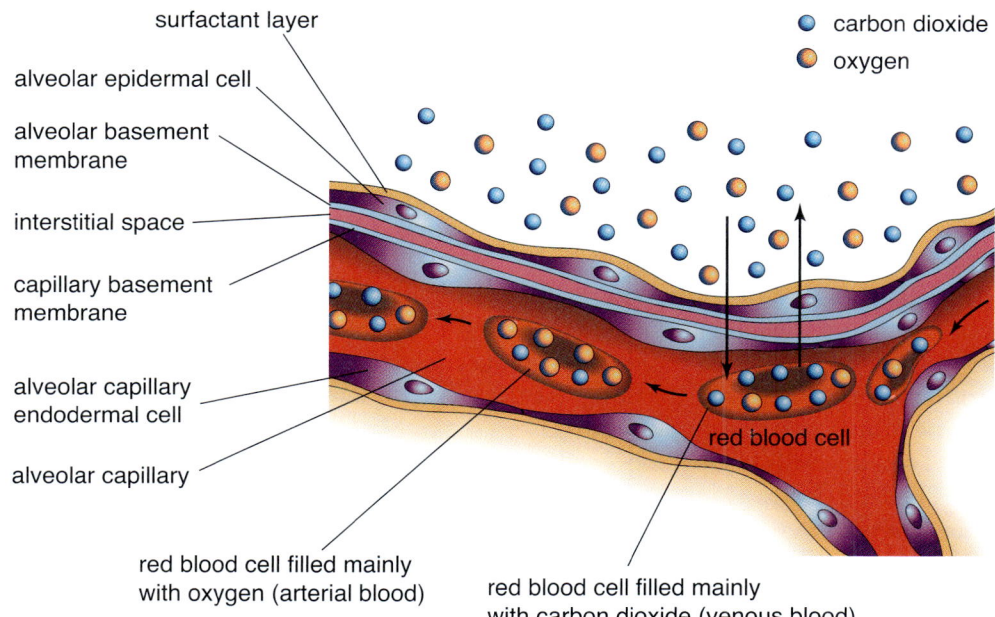

body cells is referred to as **internal respiration** (Figure 10.10). The oxygen and carbon dioxide exchange does not always take place during breathing. Gas exchange, or respiration, depends on the conditions that permit the diffusion of oxygen into the body and carbon dioxide out of the body. Breathing simply brings air to the lung surface where the diffusion of gases with the blood takes place.

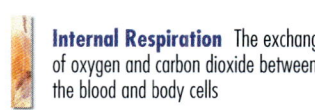

Internal Respiration The exchange of oxygen and carbon dioxide between the blood and body cells

Mechanics of Breathing

 Key Terms: intrapleural pressure, phrenic nerve

The breathing process is divided into inspiration and expiration. Inspiration is primarily carried out by the diaphragm and intercostal muscles. Neural signals from the respiratory center of the brain stem have involuntary control over breathing. These signals work in coordination with **phrenic nerve** reflex arcs associated with the diaphragm and intercostal muscles. However, voluntary control is possible using signals from the brain. During inspiration, the diaphragm contracts and moves downward. At the same time, the external intercostal muscles contract. This action expands and lifts up the rib cage increasing the volume of the thoracic cavity. Each lung stretches out and increases in volume accordingly. Air pressure within the lungs drops as they increase in volume. Air from the environment then flows through the nose or mouth into the lungs because gas flows from a region of high pressure to a region of low pressure.

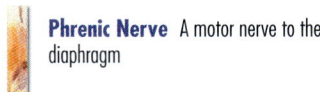

Phrenic Nerve A motor nerve to the diaphragm

Safe Smoke?

Each year in the United States, over 400,000 people die from diseases associated with cigarette smoking. Smoking-related respiratory diseases are the most preventable causes of premature death. Tobacco companies are required to place warnings on tobacco products about the dangers of smoking. They are also limited by adverting restrictions, and they often lose lawsuits to people who have suffered health complications from smoking. Despite the negative health effects attributed to smoking, tobacco companies are not likely to discontinue making their products. However, they are looking into producing safer cigarettes. Researchers at various tobacco companies are analyzing the chemistry of cigarette smoke to determine the harmful compounds that produce disease. With this information, they intend to grow tobacco plants with lower concentrations of the dangerous chemicals. The companies can also process the tobacco to remove harmful substances. Recent attempts to make safer cigarettes include low-tar products and those with special filters that remove many of the smoke particles that would directly enter the mouth. Currently, it is unlikely to see a tobacco product labeled "safe." It is difficult to remove all the hazardous substances without altering the flavor and quality of the tobacco. The United States Federal Trade Commission ensures that no tobacco product will be labeled "safe" unless the claim is backed by rigorous scientific evidence provided by federal and university laboratories. A history of the search for safer cigarettes can be found at the Public Broadcasting Service Web site at: http://www.pbs.org/wgbh/nova/cigarette.

Intrapleural Pressure The air pressure within the pleural membranes

Inspiration is effective only if the lungs and alveoli are expanded. The pleural membranes and the shape of the thoracic cavity ensure that the lungs remain pressed tightly against the walls of the visceral pleura. Alveoli are kept inflated by the expansion of the lungs. A layer of surfactant lining the surface of the alveoli keeps them from completely deflating. The surfactant, in effect, makes the walls of the alveoli rigid, which supports their shape. Intrapleural pressure is a term that describes a low-pressure environment that is maintained between the pleural membranes. It is this **intrapleural pressure** that keeps the lungs inflated, and capable of being expanded and contracted without collapsing.

Expiration forces air out of the lungs. It is initiated by relaxation of the diaphragm and external intercostal muscles. Upon relaxation, the diaphragm returns to its resting curved position. Relaxation of the intercostal muscles causes the expanded rib cage to recoil, returning to its original smaller volume. Both of these actions cause the volume of the lungs to decrease, forcing air out of the lungs and into the airways. Intercostal muscle attached to internal surfaces of the ribs assist with forced exhalation associated with heavy breathing. The air then exits the airways and is expelled into the environment.

Intrapleural pressure and surfactants prevent the collapse of the lungs and alveoli during expiration. Damage to the pleural membranes can cause one or both lungs to collapse after expiration. Puncture wounds to the chest can produce this condition. Heavy breathing involves the efforts of muscle in the upper part of the rib cage that further expand the thoracic cavity. Abdominal muscles may also be involved in inspiration during heavy breathing. Many people who are unaccustomed to rigorous exercise will feel pains in the neck and sides of the abdomen after strenuous activity.

Gas Exchange

Key Terms: millimeters of mercury (mmHg), partial pressure

Gas exchange, or internal respiration, requires gases to be transported by diffusion across the cell membranes of the alveoli cells. Diffusion can only occur if the concentration of a gas in the atmosphere and the body differ. For example, the concentration of oxygen must be higher into the alveoli for the oxygen to pass into the blood. Similarly, carbon dioxide must be more concentrated in the blood for it to diffuse into the alveoli for expulsion from the body during expiration (Figure 10.11). The proportion of a particular gas in a mixture of gases, such as air, is called **partial pressure**. Partial pressure for a particular gas is represented by the symbol "P" placed before the chemical designation for the gas. For example, the partial pressure of oxygen is written "PO_2," while carbon dioxide is expressed as "PCO_2." The level of partial pressure for a gas determines the rate at which it will diffuse into or out of the blood and cells. Partial pressure is measured in millimeters of mercury (**mmHg**), the abbreviation for millimeters of mercury. The pressure a column of mercury exerts at 0°C at sea level is equal to 1 mmHg. The same unit of measure is used for blood pressure and atmospheric pressure. Table 10.1 shows the partial pressure of air in the atmosphere and in the respiratory system.

> **Partial Pressure** The individual pressure exerted by a particular component of a gas mixture
>
> **mmHg** The abbreviation for millimeters of mercury used to measure pressure

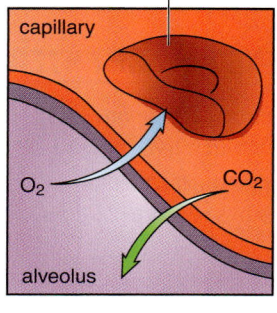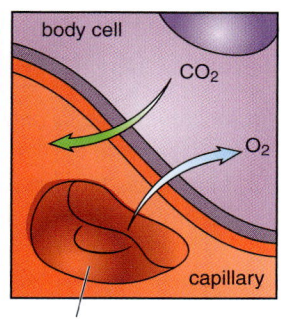

Figure 10.11 Gas Exchange

The partial pressure of oxygen needed in the blood to keep cells functioning properly is 70 to 100 mmHg. Air entering the lungs must have a PO_2 greater than 100 mmHg for oxygen to enter the blood. The partial pressure of oxygen is very low at high altitudes, which reduces the diffusion rate of oxygen entering the blood. This is why people tire more easily at elevations greater than 5000 feet: They cannot get oxygen fast enough to keep up with many activities. Most people cannot adequately exchange oxygen above 20,000 feet because the partial pressure of the oxygen is less than that in the blood. In effect, oxygen will leave the blood and enter the alveoli upon inspiration. Oxygen is then lost from the body during expiration. The partial pressure of carbon dioxide in the blood circulating in the body is 35 to 45 mmHg. It must remain higher than atmospheric carbon dioxide for the gas to diffuse out of the blood into the alveoli. Fires produce higher levels of carbon dioxide compared with those of the normal atmosphere. High levels of this gas created by a fire

Table 10.1 Partial Pressure and Percent Composition of Air

Gas	Dry Air		Humid Air		Air in Alveoli		Expired Air	
	mmHg	%	mmHg	%	mmHg	%	mmHg	%
nitrogen	600.2	78.98	563.4	74.09	569.0	74.9	566.0	74.5
oxygen	159.5	20.98	149.3	19.67	104.0	13.6	120.0	15.7
carbon dioxide	0.3	0.04	0.3	0.04	40.0	5.3	27.0	3.6
water vapor	0.0	0.0	47.0	6.20	47.0v	6.2	47.0	6.2

contained within a building can accumulate to produce a situation in which the carbon dioxide entering the alveoli is higher than that in the blood. As a result, the carbon dioxide enters the blood instead of leaving the blood. Body cells usually have 40 mmHg of oxygen and 45 mmHg of carbon dioxide.

Blood gases are measured by taking a sample of blood that circulates from the lungs to the body. A small needle attached to a syringe is used to collect the blood. The sample is then placed in a medical instrument called a blood gas detector that measures the partial pressure of oxygen and carbon dioxide. A newer method called transcutaneous monitoring measures oxygen and carbon dioxide in the skin. It does not require a syringe to collect the blood sample. Instead, the instrument heats a small area of the skin and measures the partial pressure of oxygen and carbon dioxide as they evaporate from the skin. It is becoming the preferred way to monitor blood gases in infants and children.

✓ Concept Check

1. Distinguish between the terms ventilation and respiration.
2. Describe the steps involved in inspiration and expiration.
3. What is the role of partial pressure in determining the exchange of gases between the alveoli and the blood?

DISCOVERY SCENE PLEASE ENTER DISCOVERY SCENE PLEASE ENTER

Did the information about breathing provide any more clues about the CSI? Could something have happened to the man's ability to carry out proper inspiration or expiration? Is his problem due to a neural condition that affects breathing, or could his condition be due to some type of damage to the lung tissue?

WELLNESS AND ILLNESS OVER THE LIFE SPAN

Pathology of the Respiratory System

 Key Terms: bronchitis, emphysema

A large number of respiratory system diseases can be characterized as either developmental or infectious. Developmental diseases result from a condition that produces pathology of one or more respiratory system components. Cardiovascular system diseases commonly cause developmental respiratory system disorders. Infectious diseases of the respiratory system are produced by microorganisms. The upper and lower respiratory systems usually acquire different pathogenic microorganisms. Microorganisms of the lower respiratory system survive best in the moist environment of the bronchial tree and lungs. Some respiratory system diseases, such as **bronchitis**, are general conditions whose origins are developmental or infectious. Bronchitis is commonly due to smoking or living in areas with large amounts of air pollution (Figure 10.12). Bronchitis can lead to another general condition called **emphysema** (Figure 10.13). Alveoli will deteriorate over time if they do not receive an adequate air supply. Ultimately, they will collapse and the capillaries will cut off the blood supply to the alveoli.

 **Bronchitis** Inflammation of the bronchi
Emphysema Enlargement and damage of the alveoli

Figure 10.12 Chronic Bronchitis and Emphysema

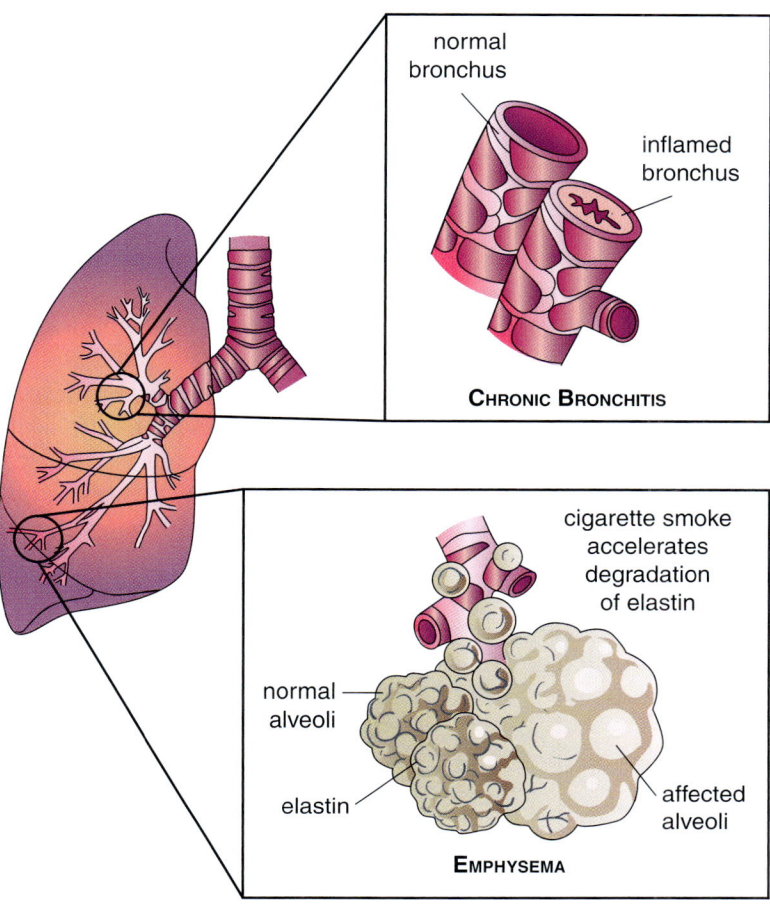

Developmental Diseases

Key Terms: acute respiratory distress syndrome (ARDS), atelectasis, bronchiectasis, chronic obstructive pulmonary disease (COPD), lung cancer, pneumothorax, restrictive lung disease, sleep apnea

Acute Respiratory Distress Syndrome (ARDS) is the rapid development of respiratory system failure. Widespread inflammation of the lungs and capillaries of the alveoli typically causes ARDS. A variety of factors can cause lung inflammation, such as breathing irritating fumes. Respiratory failure usually occurs within 24 hours to 3 days after the start of the conditions causing the inflammation. Even with medical intervention, ARDS kills approximately 30% to 40% of those afflicted. Death usually results from the failure of other organ systems, which results from the lung failure. Anti-inflammatory drugs and artificial ventilation are usually used to treat ARDS.

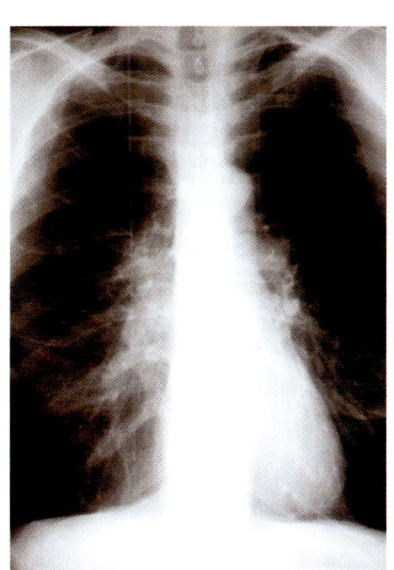

Figure 10.13 Emphysema

Acute Respiratory Distress Syndrome (ARDS) The rapid onset of respiratory failure

Atelectasis The complete or partial collapse of a lung

Pneumothorax The presence of air in the pleural membranes

Bronchiectasis The abnormal stretching and dilation of the bronchi or bronchioles

Atelectasis is the complete or partial collapse of a lung. This condition results from the absence of gas in one or more parts of the lungs. It may be due to excessive pressure on lung tissue or obstruction of the airways. Air in the pleural cavity, fluid buildup in the pleura, a foreign object in the airways, obesity, and bronchial tumors can cause atelectasis. The condition causes breathing difficulties, chest pain, and coughing. Treatment involves expansion of the affected lung, which is usually done only if a large amount of lung tissue is affected. **Pneumothorax** is a condition in which air enters the pleural membranes, the space between the lungs and chest wall. It is one cause of atelectasis, and it can be due to lung damage produced by severe lung infection or thorax injury.

Bronchiectasis is the abnormal stretching and dilation of the bronchi or bronchioles. Long-term mucous blockage of the airways weakens the walls of these structures, leading to excessive dilation. The airways can become scarred and weakened, reducing the effectiveness of air flow into the alveoli. A buildup of mucus increases as the disease progresses. This further aggravates the condition. The disease is characterized by abnormal breathing sounds, coughing, fatigue, shortness of breath, weakness, and weight loss. Bronchiectasis is usually treated with bronchodilators and medications specific to the cause of the mucous buildup. Drugs called antispasmodics and corticosteroids are prescribed in these situations.

ASTHMA FROM OZONE?

Asthma was thought to be a genetic disorder affecting the smooth muscles of the bronchioles. Symptoms of the disease were brought about by allergies, cold temperatures, and stress. Most physicians and scientists shunned the idea that asthma could be acquired from environmental factors; however, a 10-year study completed in 2002 by the California Environmental Protection Agency's Air Resources Board (ARB) and the University of Southern California (USC) provides evidence that ozone can cause asthma to develop in children. Ozone is a molecule composed of three atoms of oxygen. It is an extremely reactive form of oxygen produced in the presence of sunlight by reacting pollutants emitted from burning fuels. It is mostly found in a type of air pollution called smog. Previous evidence had shown that asthma was common in areas with ozone pollution; however, the ozone was thought to aggravate mild cases of asthma that were difficult to detect. The ARB/USC study compared new asthma cases in 3535 children who were tracked for 5 years in 12 Southern California communities. The researchers were initially investigating the potential health effects of growing up in polluted air. Six of the communities had higher-than-average ozone concentrations, while the other six had lower-than-average concentrations. Children in the high-ozone areas were more likely to develop asthma. The researchers concluded that ozone and other pollutants could damage children's lungs, causing conditions identical to those of the inherited form of asthma. The study also showed that children with asthma who are exposed to large amounts of air pollution are more likely to develop bronchitis and other respiratory disease.

Chronic Obstructive Pulmonary Disease (COPD) is caused by long-term lung disorders due to genetic conditions or certain lifestyles. Chronic obstructive pulmonary disease is usually due to long-term chronic bronchitis or emphysema. It leads to permanent narrowing of the small airways, which can cause high blood pressure in the lungs. Shortness of breath characterizes COPD. Treatment depends on the symptoms and the cause of the disease. Various types of bronchodilators and antispasmodics are prescribed to reduce the effects of COPD. Patients are also advised to avoid smoking and any irritating chemicals in the air. COPD should not be confused with **restrictive lung disease**, which is caused by a decrease in the amount of air the lungs can hold. This is due to stiffness of the tissue, which a variety of lung diseases can cause. Inhalation is difficult for persons with this disease because the lungs resist expansion.

Sleep apnea means "without breath during sleep." People with this condition stop breathing repeatedly while sleeping. This can occur hundreds of times during the night, and the cessation of breathing can last a minute or longer. The most common type of sleep apnea is due to blockage of the bronchial tree. This makes it difficult to inspire and expire during sleep, and causes breathing to pause. Another type of sleep apnea results from problems with the neural pathways from the brain that control breathing. A periodic cessation of the motor signals from the brain stops inspiration and expiration. A combination of obstruction and neural malfunction can also cause sleep apnea. It is estimated that over 12 million people in North America have some form of sleep apnea. Untreated sleep apnea can cause loss of sleep, headaches, high blood pressure, and cardiovascular diseases.

Lung cancer most often occurs in people aged 55 to 60 years. It is prevalent in people who smoke and/or those continually exposed to irritating chemicals in the air. Most lung cancers originate in the lining of the bronchi and can spread throughout the body by entering nearby blood vessels. Cancers from other parts of the body can lodge in the lungs, causing lung cancer. The progression of the disease varies greatly with the particular type of lung cancer. Treatment depends on the type of cancer and the degree to which it has spread in the body. It is estimated that nearly 200,000 cases of lung cancer are detected each year in North America.

Infectious Diseases

Key Terms: bronchopneumonia, flu, hantavirus pulmonary syndrome, hydatid lung disease, influenza, lobar pneumonia, pneumonia, tuberculosis (TB)

Flu, or **influenza**, is a contagious respiratory disease caused by a wide variety of influenza viruses. Each year in the United States, 5% to 20% of the population contracts the flu; over 200,000 people are hospitalized and approximately 36,000 die. The influenza virus is spread though contact with respiratory system fluids. Symptoms include a dry cough, diarrhea, headache, high fever, muscle aches, nausea, sore throat, runny nose, fatigue, and vomiting. It is possible for flu to spread from various animals to humans. There is no cure for influenza, however, vaccination against specific types of flu reduces the risk for developing serious complications. The Centers for Disease Control and Prevention (CDC) has to decide which types of influenza may be prevalent during a particular flu season. Vaccinations are recommended for

---

Chronic Obstructive Pulmonary Disease (COPD) Any disorder that persistently obstructs bronchial airflow

Restrictive Lung Disease A disease that restricts the expansion of the lungs

Sleep Apnea Cessation of breathing during sleep

Lung Cancer The uncontrolled growth of abnormal cells in lung tissue

Flu or Influenza A contagious disease that is caused by the influenza virus

Hantavirus Pulmonary Syndrome A viral disease characterized by a sudden onset of fever, pain, and vomiting

Pneumonia Inflammation of the lungs caused by an infection

Lobar Pneumonia A type of pneumonia limited to one lobe of the lung

Bronchoneumonia A type of pneumonia scattered throughout the lung

Tuberculosis A bacterial infection that usually affects the lungs

Hydatid Lung Disease A lung infection caused by the inactive stage of a worm

children aged 6 to 23 months, adults aged over 65 years, and individuals with heart and respiratory diseases. **Hantavirus pulmonary syndrome** is a little known viral disease that was once confused with influenza. People contract the virus through contact with rodent feces, saliva, and urine. In recent cases, it also was contracted from prairie dogs and exotic pet rodents. The disease has a high fatality rate, and there is no well-established treatment or vaccination strategy. Other viruses cause common colds, headaches, and sinus infections.

Pneumonia is an inflammation of the lungs caused by bacteria, fungi, protista, and viruses. The disease usually develops in people with weakened immune defenses or in individuals who have contracted other respiratory diseases. Pneumonia is categorized by either its distribution in the lung or the type of microorganism causing it. **Lobar pneumonia** occurs in one lobe of the lung, whereas **bronchopneumonia** is scattered throughout the lungs. Viruses cause approximately 50% of the pneumonias in North America. These tend to be mild forms of the disease. Bacterial pneumonias are more severe, but they can be treated with antibiotics. Pneumonia is characterized by breathing difficulties, chills, chest pains, and coughing.

Tuberculosis (TB) is one of the longest-known infectious respiratory diseases. It is caused by a bacterium called *Mycobacterium tuberculosis*. Improved health and sanitation, in addition to the use of vaccinations and antibiotics, has limited the spread of TB in the United States, although it is still common in underserved US populations and in many nations throughout the world. The bacterium is readily spread by coughing, sneezing, and spitting. Most people who contract TB are able to fight off the infection; however, the bacteria can lie dormant in the body, causing no symptoms for long periods of time. The bacteria can become active later as the person's immune system weakens with age or other illnesses. A long-lasting cough, chest pain, fever, and night sweats characterize TB. In advanced cases, it is typical to cough up blood. The disease is fatal if it is not treated with antibiotics. *Mycobacterium tuberculosis* irritates the lung, causing damage to the alveoli. Other microorganisms can also enter the lungs and remain inactive until a later time.

Hydatid lung disease is caused by a worm that forms large inactive cyst structures in the lung. The cyst is the inactive stage of the worm, and it remains in the lung until the person dies. It does not usually cause problems, but severe cases produce cough, fever, and chest pain. Chest x-rays readily reveal the cysts and resultant inflammation of the lungs (Figure 10.14). Fungal diseases of the lungs are very serious and more common in people with weakened immune systems. Some of the fungi are commonly found in soil and may be spread by bird droppings.

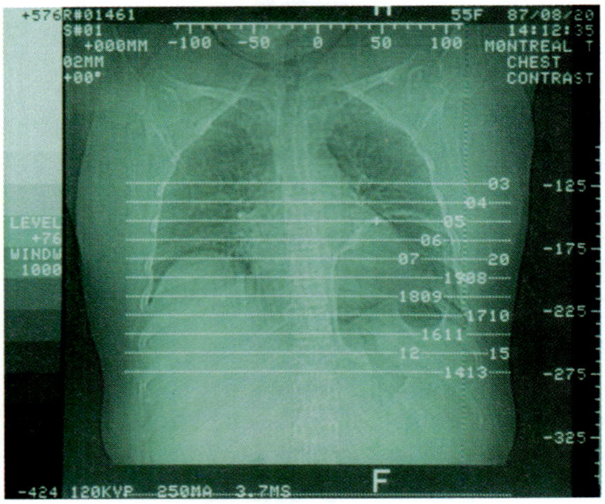

Figure 10.14 Hydatid Lung Disease

✓ Concept Check

1. Describe the two major categories of respiratory diseases.
2. What is the difference between obstructive and restrictive respiratory diseases?
3. Describe the causes of respiratory system infections.

Aging of the Respiratory System

Key Terms: expiratory reserve volume, functional residual capacity, inspiratory reserve volume, minute respiratory volume, residual volume, tidal volume, total lung capacity, vital capacity

The respiratory system is not active until after birth. Its role in the fetus is carried out by a special blood supply that exchanges gases with the mother's blood (Figure 10.15). Just before birth, surfactant secretion increases, allowing full expansion of the lungs. Today, surfactant is used as a medical treatment in infants born with underdeveloped lungs. Many babies died from the inability to produce surfactant until the 1980s, when surfactant treatments became regularly available. The stress of childbirth causes the release of adrenalin in the infant that further matures the lungs. Most aging of the lungs that takes place from birth until middle-age is due to lifestyle. Stress and high-fat diets encourage cardiovascular system diseases that accelerate aging of the respiratory system. Smoking and exposure to polluted air degrade the walls of the alveoli, further worsening any age-related changes. Research shows that aging reduces the capacity of all respiratory functions. The typical signs of respiratory system aging are as follows:

- a decrease in elasticity of the lung tissue and rib cage, which makes inspiration more difficult
- a loss of skeletal muscle mass, which weakens the inspiration muscles
- a decrease in smooth-muscle control, which makes it more difficult to regulate air flow to the alveoli
- a decrease in alveolar gas exchange surface, which reduces the rate of diffusion
- a decrease in nervous system control, which makes it more difficult to coordinate the rate of breathing with body activity
- a decrease in laryngeal function, which diminishes the cough and gagging responses
- a decrease in mucus production, which makes the lungs susceptible to infection
- a decrease in blood flow through the alveoli, which reduces gas exchange
- a decrease in the diameter of the bronchi, which reduces the flow of air into and out of the lungs

Figure 10.15 Fetal Circulation

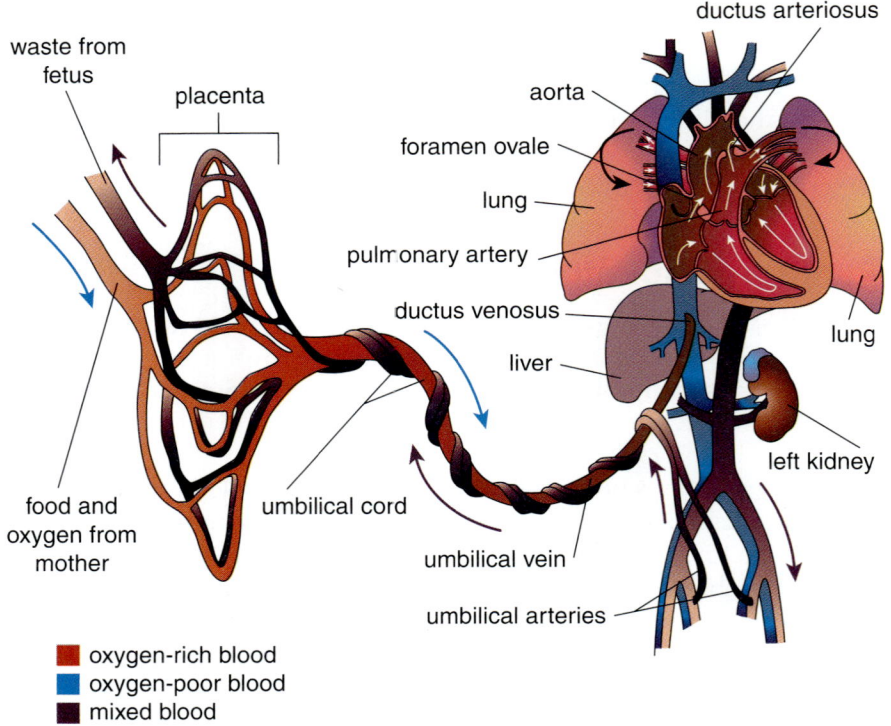

Tidal Volume The amount of air moved into and out of the lungs during normal breathing

Vital Capacity The maximum amount of air that can be moved into and out of the lungs

Minute Respiratory Volume The amount of air moved into and out of the lungs in 1 minute

Inspiratory Reserve Volume The amount of air forcefully inspired once normal tidal inspiration is reached

Expiratory Reserve Volume The amount of air that is forcefully expired after a normal tidal expiration

Functional Residual Capacity The amount of air remaining in the lungs after normal expiration

Residual Volume The amount of air remaining in the lungs after a forced expiration

Total Lung Capacity The maximum amount of air the lungs can hold

Respiratory system aging contributes to the decline of other organ systems. The cardiovascular system has to work harder to compensate for deficiencies in respiratory system function. Skeletal muscle loses endurance and strength if gas exchange is reduced. Cells throughout the body slow down and are less likely to carry out their functions adequately. Tissues are less likely to maintain themselves and heal when gas exchange is insufficient.

Physicians can readily measure and monitor respiratory system health and aging by calculating a person's efficiency at air exchange, or breathing capacity values. **Tidal volume** measures the quantity of air moved into and out of the lungs during a normal breath. **Vital capacity** is the maximum quantity of air that can be moved into and out of the lungs. **Minute respiratory volume** is the quantity of air moved into and out of the lungs in 1 minute. **Inspiratory reserve volume** is the quantity of air forcefully inspired once normal tidal inspiration is reached. The **expiratory reserve volume** measures the quantity of air that is forcefully expired after a normal tidal expiration. **Functional residual capacity** is the quantity of air remaining in the lungs after normal expiration. **Residual volume** is the quantity of air remaining in the lungs after a forced expiration. **Total lung capacity** is the maximum quantity of air the lungs can hold after forced breathing. The normal values for breathing capacity are provided in Table 10.2. One or more of these values normally declines with age.

Table 10.2 Breathing Capacity Values

Measurement	Normal Measure
tital volume (TV)	500 mL
minute respiratory volume (MRV)	6000 mL/min
inspiratory reserve volume (IRV)	3000 mL
expiratory reserve volume (ERV)	1200 mL
vital capacity (VC)	4500-5500 mL
functional residual capacity (FRC)	2400 mL
residual volume (RV)	1200 mL
total lung capacity (TLC)	4400-6400 mL

✓ Concept Check

1. What are two factors that contribute to respiratory system aging?
2. Describe the age-related changes in the respiratory system.
3. Explain how aging of the respiratory system affects other organ systems.

DISCOVERY SCENE PLEASE ENTER DISCOVERY SCENE PLEASE ENTER

CSI Break

Have you solved the CSI yet? Did the information about aging of the respiratory system provide any more clues about the man's condition?

CSI – Case Study Investigation Conclusion

What can you conclude about the man's condition? What factors led to his illness? Were they related to anything he did at the health club?

Answer:

The young man developed a condition called "hot-tub lung." It is an unusual infection contracted from inhaling the bacterium *Mycobacterium avium* from hot-tub water. *Mycobacterium avium* is related to the bacteria that cause tuberculosis. These bacteria find hot tubs to be an ideal environment for growth and a means of entry into the lungs. The bacteria tend to grow in hot tubs that are not cleaned according to public health recommendations. In addition, the chlorine added to hot tubs loses much of its disinfectant properties at hot-tub temperatures. The bubbling water of the hot tubs sprays the bacteria into the air as the bubbles burst. This permits the bacteria to enter the lungs when a person is sitting in a hot tub. The warm chlorine in the air is also believed to irritate the respiratory tree and lung lining, making it vulnerable to infection. Symptoms usually develop 4 to 12 hours after exposure to the bacteria.

No one else developed hot-tub lung in the club because the person's immune system fought off the disease. The extreme exercise routine ran down the young man's immune system, making him vulnerable to the disease. Physicians treat the disease with antibiotics and anti-inflammatory drugs. Allergy treatments are sometimes given because some people are allergic to chemicals produced by the bacteria. The incidence of hot-tub lung can be reduced by regularly cleaning the hot-tub water. It is important to add 2 to 5 parts per million (ppm) of chlorine to the water and keep it at a pH level of 7.2 to 7.8. Many physicians recommend that children, elderly individuals, and people with respiratory diseases should avoid exposure to hot tubs. Hot-tub lung often goes undiagnosed because it is mistaken for asthma and bronchitis.

This CSI was adapted from the following articles:

1. Cappelluti E, Fraire AE, Schaefer OP. A case of "hot tub lung" due to *Mycobacterium avium* complex in an immunocompetent host. *Arch Intern Med*. 2003;163:845-848.
2. Rickman OB, Ryu JH, Fidler ME, et al. Hypersensitivity pneumonitis associated with *Mycobacterium avium* complex and hot tub use. *Mayo Clin Proc*. 2002;77:1233-1237.
3. Rose C, Martyny J, Huitt G. Hot tub associated granulomatous lung disease from mycobacterial bioaerosols. *Am J Resp Crit Care Med*. 2000;161:A730.

Chapter Summary

The respiratory system is responsible for exchanging atmospheric gases with those in the body's internal environment. Oxygen is taken up by the blood to fuel metabolism carried out by all of the body's cells. Carbon dioxide, a metabolic waste, is removed from the cells to keep metabolism from seizing up. Ventilation is carried out to ensure that atmospheric air comes in contact with the alveoli of the lungs. The alveoli are the site of gas diffusion. However, diffusion can only take place if the partial pressure of each gas is at the appropriate ratio to diffuse oxygen into the blood from the lungs and carbon dioxide out of the blood into the lungs. Lung diseases either affect the flow of air through the airways to the lungs, or they are caused by deterioration of the alveoli. Some lung disorders block or degrade the alveolar lining, preventing the diffusion of gases. Certain life style habits increase the chances of lung disease. Aging mostly affects lung elasticity and contributes to diseases that impair the airways. Microbial infections and pollution contribute to lung aging. Diseases of the cardiovascular system also affect the respiratory system.

Overview
- The respiratory system provides cells with the gas exchange needed for metabolism.
- Breathing, or ventilation, brings air into the respiratory system.
- Gases are exchanged between the respiratory system and the blood by diffusion.

Components of the Human Respiratory System
- The respiratory system is composed of the nose, nasal cavity, paranasal sinuses, pharynx, larynx, trachea, bronchial tree, and lungs.
- Air is cleaned and moistened in the nose, nasal cavity, paranasal sinuses, and pharynx.
- The nose, nasal cavity, paranasal sinuses, and pharynx make up the upper respiratory system.
- The larynx, trachea, bronchial tree, and lungs make up the lower respiratory system.
- The larynx regulates coughing and separates the air and food pathways.
- The trachea and bronchial tree direct air to the lungs.
- Bronchioles regulate the flow of air to lobules in the lungs.
- Lobules are pockets of alveoli and blood vessels.
- Diffusion of gases takes place on the walls of the alveoli.

Breathing
- Breathing is due to the action of the muscles and bones of the thorax.
- The autonomic and somatic nervous systems control breathing.
- Inspiration is due to contraction of the diaphragm and expansion of the rib cage.
- Expiration is due to relaxation of the diaphragm and external contraction of intercostal muscles.
- Alveoli expand and fill with air upon inspiration.
- The partial pressure of gases in the air determines the direction of diffusion during breathing.

Study Guide

Pathology of the Respiratory System
- Diseases of the respiratory system are developmental or infectious.
- Developmental diseases are due to genetic conditions or life style factors.
- Developmental diseases affect the airways or the alveoli.
- Microorganisms cause infectious diseases.
- Most microorganisms cause disease specific to only the upper respiratory system or only the lower respiratory system.

Aging of the Respiratory System
- Some aging is due to wear and tear on the body.
- Lifestyle plays a significant role in respiratory system aging.
- Aging reduces the ability to carry out inspiration and expiration.
- Aging reduces the diffusion of gases across the alveoli.
- Aging of the respiratory system can be measured as different aspects of breathing capacity.

 ## Key Terms

Overview
Breathing
Lung
Ventilation

Components of the Human Respiratory System
Bronchial tree
Larynx
Lower respiratory system
Nose
Paranasal sinuses
Pharynx
Trachea
Upper respiratory system

Nose
Nares
Nostril
Quadrangular cartilage

Nasal Cavity
Turbinate

Pharynx
Adenoids
Laryngopharynx
Nasopharynx
Oropharynx
Tonsils

Larynx
Arytenoid cartilage
Cricoid cartilage
Epiglottis
Glottis
Heimlich maneuver
Laryngeal prominence
Thyroid cartilage
Vocal cords

Trachea
Primary bronchi
Tracheal cartilage

Bronchial Tree
Bronchiole
Bronchoconstriction
Bronchodilation
Bronchospasm
Respiratory bronchiole
Secondary bronchi
Terminal bronchioles
Tertiary bronchi

Lungs
Alveolus
Cardiac notch
Lobe
Lobule
Parietal pleura

Chapter 10

Pleura
Pleuritis
Surfactant
Visceral pleura

Breathing
Diaphragm
Exhalation
Expiration
External respiration
Inhalation
Inspiration
Internal respiration
Respiration

Mechanics of Breathing
Intrapleural pressure
Phrenic nerve

Gas Exchange
Millimeters of mercury (mmHg)
Partial pressure

Pathology of the Respiratory System
Bronchitis
Emphysema

Developmental Diseases
Acute respiratory distress syndrome (ARDS)

Atelectasis
Bronchiectasis
Chronic obstructive pulmonary disease (COPD)
Lung cancer
Pneumothorax
Restrictive lung disease
Sleep apnea

Infectious Diseases
Bronchopneumonia
Flu or influenza
Hantavirus pulmonary syndrome
Hydatid lung disease
Lobar pneumonia
Pneumonia
Tuberculosis (TB)

Aging of the Respiratory System
Expiratory reserve volume
Functional residual capacity
Inspiratory reserve volume
Minute respiratory volume
Residual volume
Tidal volume
Total lung capacity
Vital capacity

Check Your Understanding

1. Which is *not* a function of the nose?
    a. warms air
    b. removes particles from air
    c. moistens air
    d. regulates air flow rate

2. Much of the respiratory system is lined with _____:
    a. mucous membrane
    b. fibrous connective tissue
    c. hyaline cartilage
    d. flagella

3. The respiratory system develops from the embryonic _____ system.
    a. endocrine
    b. cardiovascular
    c. skeletal
    d. digestive

Study Guide

4. Damage to the larynx would affect the _____.
   a. voice
   b. warming of air
   c. secretion of nasal mucus
   d. humidification of air

5. Blocking the trachea would prevent the passage of _____ into the body.
   a. air only
   b. food only
   c. air and food
   d. neither air nor food would be affected

6. The _____ prevent(s) the passage of food into the lungs:
   a. bronchi
   b. nares
   c. epiglottis
   d. nasopharynx

7. The shape of the bronchial tree is supported by _____:
   a. adipose tissue
   b. cartilage rings
   c. bands of skeletal muscle
   d. bony projections of the vertebrae

8. In the respiratory tree, the _____ has the smallest diameter:
   a. trachea
   b. bronchus
   c. oropharynx
   d. bronchiole

9. A total obstruction of air into the lungs is possible if the _____ is blocked:
   a. trachea
   b. left bronchus
   c. bronchus
   d. bronchiole

10. In the respiratory system, gas exchange takes place in the _____:
    a. larynx
    b. bronchiole
    c. tertiary bronchus
    d. alveolus

11. _____ assists with the inspiration of air:
    a. recoiling of the ribs
    b. lowering of the diaphragm
    c. closing of the bronchioles
    d. closing of the epiglottis

12. In the respiratory system, _____ is most affected by rib damage:
    a. bronchodilation
    b. movement of mucus from the lungs
    c. humidification of the air
    d. inspiration and expiration

13. The lungs are surrounded by the _____:
    a. pleural membranes
    b. a cartilaginous capsule
    c. mucosa
    d. ciliated columnar cells

14. Foreign objects that enter the trachea are most likely to end up in the _____:
    a. left lung
    b. right lung
    c. thyroid cartilage
    d. nasopharynx

15. A hard punch to the stomach region is likely to produce _____ in the respiratory system:
    a. expiration
    b. collapse of the bronchi
    c. pneumothorax
    d. alveolar collapse

A Case Study

THE WHITE LUNG CONTROVERSY

The human respiratory system is not equipped to remove large, heavy particles that make their way into the bronchioles and alveoli. These molecules remain in the lungs and accumulate, causing inflammation, scarring, and cell damage. Asbestos, a fiber that was in the past commonly used as a component of many building materials, is one of the particles that will accumulate in the lungs. The accumulation of asbestos fibers in the lung is called white lung disease. Asbestos was banned from use in buildings and consumer products in 1989 by the US Environmental Protection Agency (EPA). The ban was based on evidence that white lung disease led to lung cancer. However, a recent study by the Office of Solid Waste and Emergency Response found conflicting evidence about the dangers of asbestos inhalation. They held the Asbestos Mechanisms of Toxicity Workshop in 2003 in Chicago, IL. The workshop determined that the size and shape of asbestos fibers do not fully explain how asbestos causes cancer. They believed that lifestyle factors, such as smoking, produce the cancer. Asbestos weakens the lungs and makes it more prone to developing cancer. As a result of these findings, many companies are hoping to reverse the ban on asbestos. It was at one time commonly used to make fireproof and heatproof building materials. Groups such as the Asbestos Disease Awareness Organization (ADAO) want to uphold the ban. They feel that asbestos is too dangerous to use because it makes people susceptible to other respiratory diseases.

Use the information in this chapter and the following Web sites to answer the following questions about white lung disease:

- What would be some justifications for lifting the ban on asbestos?
- Should the government ban a product if it is harmful only to people who have lifestyles that increase the risk for a specific disease?

Study Guide

- How does a government weigh the benefits and risks of permitting the use of certain materials, such as asbestos?
- What are the rights of people who do not want to be exposed to asbestos products?
- What are the possible consequences of maintaining the ban on asbestos?
- Who should make decisions about banning the use of materials that have conflicting safety information?

1. Asbestos Disease Awareness Organization
   http://www.asbestosdiseaseawareness.org/

2. Asbestos Victims Organization
   http://www.avoglobal.org/

3. National Institutes of Health – Asbestos & Cancer
   http://ehp.niehs.nih.gov/members/2003/6704/6704.html

4. International Ban Asbestos Secretariat
   http://www.ibas.btinternet.co.uk/

Where Do We Go from Here?

People in health fields can use their knowledge of the respiratory system to solve everyday problems. You may wish to use other resources in addition to your textbook to investigate the answers to each of the following situations:

1. A friend was told that she has a nasal septum defect. She wants to know if you feel it is a condition that is serious enough to justify having it surgically repaired.
   http://www.ama-assn.org/ama/pub/category/7166.html

2. A neighbor tells you that he saw an advertisement of a spirometer that can be used at home. The company selling the meter claims that everybody needs to monitor their spirometer measurements regularly. What would you tell your neighbor about the claim?
   http://training.seer.cancer.gov/module_anatomy/unit9_1_resp_intro.html

3. A relative asks you why his child, who has asthma, cannot play outside when the air quality is bad. What would you explain to him?
   http://www.kidshealth.org/teen/asthma_basics/lungs/lungs.html

4. An attorney approaches you and states that you should file a lawsuit against a former employer because you were required to work around a lot of road dust. What is the medical basis of her intent to pursue a lawsuit?
   http://www.nlm.nih.gov/medlineplus/respiratorydiseases.html

5. A neighbor is hesitant about taking a business trip to Asia because he is afraid of developing a dangerous respiratory disease. What information could you provide to help him with his decision?
   http://www.webmd.com/

Chapter 10

Skills Activities

1. Histology of Lung Pathology

Materials

- microscope with high-power capability (400X)
- prepared slide of a human lung
  http://www.med.uiuc.edu/PathAtlasf/framer2/path3.html

Histologists and pathologists look at microscopic preparations of lungs to better understand respiratory diseases. They compare normal slides and prepared slides of diseased lungs to the tissue under examination. Before a histologist can make conclusions about lung disease, he or she must practice comparing the normal samples with samples showing diseases to gain experience determining the different components and cell types in the respiratory system.

Activity

Place a microscopic slide of the normal human lung under the microscopic. Start with low magnification and move to higher magnification to see details and individual cells. Can you find the terminal bronchioles and the alveoli? Now, compare the slide findings with the diseased lung specimens shown in the Urbana Atlas of Pathology Web site at: http://www.med.uiuc.edu/PathAtlasf/framer2/path3.html. What are the similarities and differences between the normal lung and each diseased lung? How does each disease affect the histological appearance of the lung?

Skills Activities

2. Lung Function Models: Part 1. Lung Capacity Model

Materials

- an empty 1-gallon milk jug with the cap
- a metric measuring cup or beaker
- masking tape
- a black, fine-point permanent marker
- access to water
- a 1-m or 3-ft length of aquarium tubing
- a large, flat-bottomed, aluminum baking pan that can hold 1 gal of water

Physicians and scientists regularly use models to better understand body functions. Biological models are representations of body functions. In this activity, students will build a model for measuring the amount of air displaced by the lung during expiration. It provides information about human lung capacity.

Study Guide

Activity

1. Obtain a milk jug, masking tape, and marker.
2. Place a strip of masking tape down the side of the milk jug from the top to the bottom.
3. Fill the jug with water using the measuring cup or beaker to keep track of the amount of each liter of water it takes to fill the jug. Mark each liter on the tape.
4. Place the cap on the jug.
5. Fill the baking pan about half full with water.
6. Slowly place the jug upside down in the water and remove the cap while the mouth of the jug is submerged. Do not allow air bubbles to enter the milk jug.
7. Lift the jug carefully to place one end of the aquarium tubing inside the mouth of the jug, making sure that the opening stays submerged. Make sure there are at least 3 in of tubing in the jug.
8. Take a normal breath, and exhale through the tubing. The volume of air entering the jug should displace the same volume of water. Mark the water level on the tape. Record the data.
9. Refill the jug with water and return it to the baking pan.
10. Breathe in deeply and make an effort to exhale all of the air in your lungs through the tubing. Mark the water level on the tape, and record the data.
11. Repeat steps 8 and 9 three times, and take an average of the readings.

Replace the tubing with fresh tubing for each person if the setup is being shared. Compare your data with those of other students in the class.

Skills Activities

3 Lung Function Models: Part 2. Inspiration and Expiration Model

Materials
- one soda straw cut into 2-in lengths
- one scissors
- two small, round party balloons
- one large, rubber party balloon
- fast-drying glue
- one large, clear plastic cup
- two small rubber bands
- one rubber band, large enough to just fit over the lip of the cup
- cellophane tape

This biological model represents the action of the diaphragm in human respiration. It can be adapted to model the way certain respiratory diseases affect inspiration and expiration.

Chapter 10

Activity

1. Obtain all of the materials needed for the setup.
2. Take the 2-in piece of straw and cut a small triangle in the center without cutting through to the opposite side
3. Place one small balloon over each end of the straw and secure it with a small rubber band. Test the setup to make sure that air will go into each balloon when air is blown into the straw through the triangular opening.
4. Bend the straw in the middle of the hole so that the opening is facing up.
5. Take a second piece of straw and cut a V-shape on the end. Fit the slanted points of the straw into the opening of the hole on the bent straw.
6. Use the glue to hold the two pieces of the straw together. Allow it to dry. This is called the lung model. The straight straw is the "trachea," while the bent straw serves as the two bronchi leading to each lung (balloon).
7. Cut a hole in the bottom of the clear plastic cup using the diameter of the straw as a guide to the size.
8. Push the "trachea" of the lung model into the hole of the plastic cup from the inside. Cement the "trachea" into hole.
9. Cut the neck of the balloon off. Carefully stretch the cut balloon over the opening of the cup. Try not to crack the cup. Secure the edges of the balloon with the large rubber band.
10. Now, pull the bottom balloon gently, and observe what happens to the model.
11. Place one finger just above the trachea and repeat step 10. What happened to the flow of air through the trachea?
12. Slowly release the balloon and again observe what happens to the lung model.
13. Again, place one finger just about the trachea, and repeat step 12. What happened to the flow of air through the trachea?
14. How does this model represent the role of the diaphragm in inspiration and expiration?
15. Rebuild the setup with cellophane tape placed on one balloon. Observe what happens when you model the action of the diaphragm. What disease does this model?
16. Rebuild the setup with a hole poked into one balloon. Observe what happens when you model the action of the diaphragm. What disease does this model?

THE RESPIRATORY SYSTEM

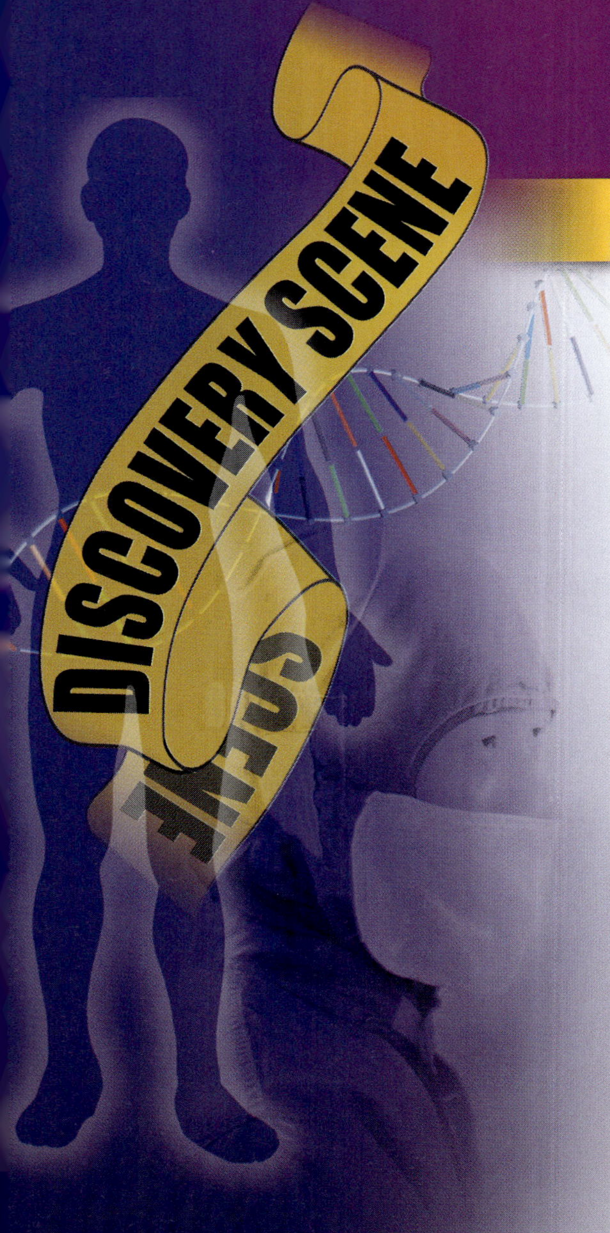

Case Study Investigation

Case Study Investigation #11

The local YMCA is having a 5-killometer (K) Run, Walk, or Crawl fundraiser. You decide to volunteer at the first-aid station to take care of blisters and other minor injuries associated with these events. It is a cool, overcast morning, making it a perfect day for a race, so you do not expect any problems due to hot weather. A call comes in on the two-way radio that a young, adult, male runner just passed out at the 2-K point of the run. You alert the emergency medical technician (EMT) on duty, and you both head out to see the person. When you get there, the EMT starts questioning the young man, and you record the conversation for the EMT's report. The 25-year-old runner commented that he had tightness in his chest followed by chest pain before feeling dizzy and fainting. He also said that he had been short of breath after reaching the 1-K mark. This was the first race he had ever run; however, he said that he takes short walks each night. You learn that he has no history of heart attacks or high blood pressure. He claims that his physical examinations always show low levels of cholesterol and fats in his blood. He also stated that he had a reasonable breakfast and was well-hydrated that morning. The EMT decides to test your knowledge of anatomy and physiology by asking you what may be wrong with the runner. At the end of the chapter, you will be asked to determine the possible cardiovascular system problems causing the illness.

CHAPTER 11

THE CARDIOVASCULAR SYSTEM

Chapter Outline

Case Study Investigation (CSI)
Applied Learning Outcomes
Overview
Circulatory System Vessels
    Arteries and Veins
    Small Vessels and Capillaries
Structure of the Human Heart
    The Adult Heart
    The Fetal Heart
Human Heart Function

Electrocardiography Basics
Wellness and Illness over the
   Life Span
    Pathology of the Cardiovascular
      System
    Aging of the Cardiovascular
      System
CSI Conclusion
Study Guide

Applied Learning Outcomes

- Use the terminology associated with the cardiovascular system.
- Learn about the following:
  - blood vessel function and structure
  - circulatory system pathways
  - heart function and structure
  - electrocardiography principles
- Understand the aging and pathology of the cardiovascular system.

Overview

Key Terms: angiogenesis factors, blood pressure, blood vessel, cardiovascular system, circulatory system, heart, pulse, vascularization

Cardiovascular System Refers to the heart and blood vessels

Heart The hollow muscular organ that pumps blood throughout the body

Blood Vessels A part of the cardiovascular system that carries blood throughout the body

Circulatory System Pertains to blood circulation, blood vessels, and the heart

Blood Pressure The force of blood pushing against blood vessel walls

Pulse A throbbing of the blood vessels produced by the heart beat

Vascularization The formation of blood vessels

Angiogensis Factors Chemicals that stimulate the growth of new blood vessels

The **cardiovascular system** gets it name from its two main components. Cardio comes from the Greek word "Kardia," or **heart**. The term "vasculum" is Latin for a vessel, or tube, referring to the cylindrical nature of **blood vessels**. Another name for the cardiovascular system is the **circulatory system**.

Ancient Greeks understood that the heart had an important role in the body, although they did not fully understand its role or the various blood vessels. Greek philosophers and physicians argued back and forth whether it was the liver or the heart that housed emotions and the mind. Little did they know that the brain actually carried out this function. It was not until the 1400s that experiments conducted on live animals showed the true nature of the cardiovascular system. These early studies identified such features as **blood pressure** and the different characteristics of the blood flowing in various vessels. Modern studies of the cardiovascular system still require anatomical studies much like those conducted long ago. However, modern medical devices that use sound waves, such as ultrasound, can be used to study blood flow patterns and even blood chemistry in more detail (Figure 11.1). Ancient people were even aware of **pulse** fluctuations and equated them to illness. Today, health professionals still measure pulse at various parts of the body to determine cardiovascular fitness (Figure 11.2).

Figure 11.1 Blood Vessel Ultrasound

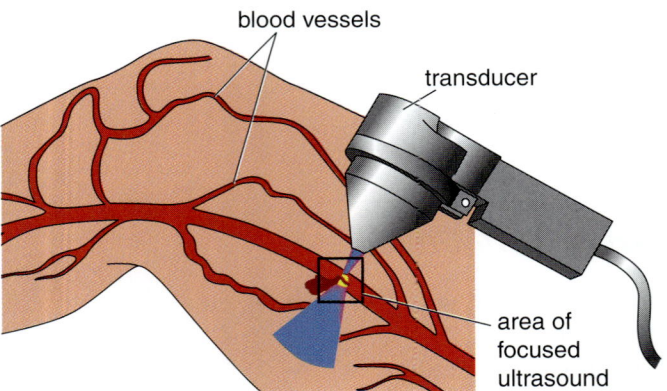

Evidence of cardiovascular system formation is present at the beginning of the third week of embryological development. It is one of the first organ systems to develop. A thick blood vessel that will ultimately develop into the heart starts beating at three weeks. It is derived from the same type of mesoderm tissue that forms bone and muscle. At this time, the heart is not connected to blood vessels. Its job is to mix the fluids filling the young embryo. Several blood vessels form soon after the heart begins beating. At first, these blood vessels transport materials from stored food reserves in the embryo to the rapidly growing cells. After 24 days of development, the cardiovascular system is composed of a complex network of vessels that transport materials throughout the embryo.

The early cardiovascular system provides a way of exchanging nutrients and wastes between the embryo and the mother's body. The embryonic heart does not take on the characteristics of the adult until the baby is born. **Vascularization** is a term used to describe the development and growth of blood vessels. Chemicals called **angiogenesis factors** direct the growth of

Figure 11.2 Circulatory System

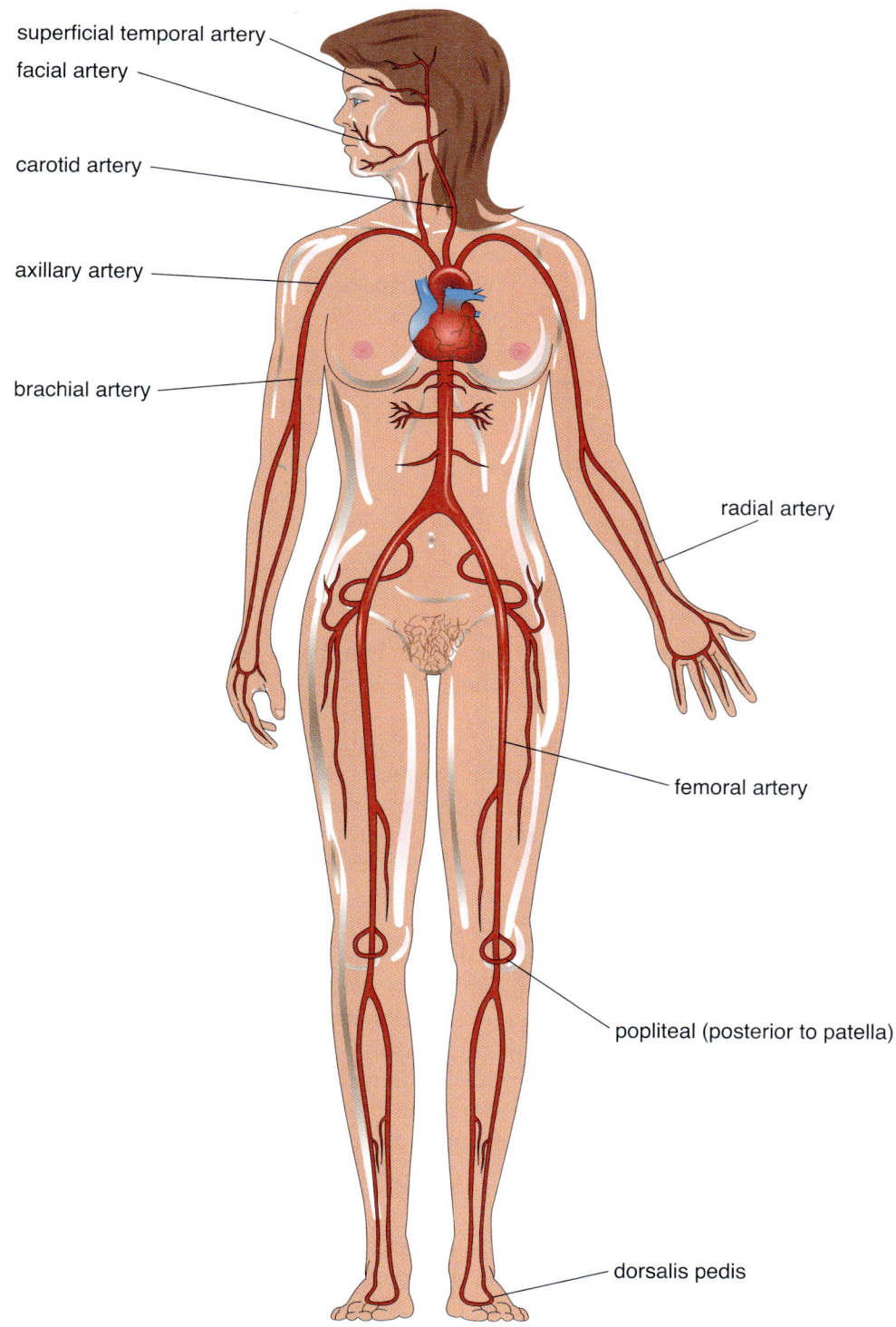

Civic Responsibility

CIVIC RESPONSIBILITY: HELPING OTHERS WITH YOUR KNOWLEDGE

It is valuable to use what you have learned about the cardiovascular system to help others to better understand the world around them. It is very important to check your facts and seek further information about certain topics before discussing health and science issues. Here are some suggestions to foster a better public awareness of the cardiovascular system:

1. Speak to young adults about the effects of smoking on the heart.
2. Work with sports clubs to educate the players about cardiovascular health.
3. Speak to a civic group about the risk factors for cardiovascular diseases.
4. Volunteer at a hospital to work with cardiac patients.

blood vessels. Growing or healing tissues secrete angiogenesis factors. Cancer cells also produce angiogenesis factors.

CIRCULATORY SYSTEM VESSELS

Key Terms: artery, hydrostatic pressure, lymphatic vessel, microcirculation, vein

Artery A blood vessel that carries blood from the heart to the body

Vein A blood vessel that carries blood from the body to the heart

Hydrostatic Pressure The pressure of the water that is circulated in the blood and tissues

Lymphatic Vessel A thin vessel that carries lymph fluid

Microcirculation The flow of minute amounts of blood and body fluids through blood vessels

Tunica Adventitia The outer layer of a blood vessel

The circulatory system is composed of three major types of blood vessels: **arteries**, **veins**, and capillaries. Arteries are muscular blood vessels that carry blood from the heart to the body. They are usually not visible through the skin. Veins are flexible vessels that carry blood from the body to the heart. These vessels are often easily seen just under the surface of the skin. Capillaries are small vessels that connect the arteries to the veins. They form networks that exchange materials between the blood and the cells (Figure 11.3). Capillaries carry out certain functions in association with **lymphatic vessels** that are part of the immune system. The major role of capillary and lymphatic vessel interaction is to control fluid levels, or **hydrostatic pressure**, of the body tissues (Figure 11.4). The flow of minute amounts of blood and body fluids is called **microcirculation**. Microcirculation carries out the exchange of atmospheric gases and nutrients between the blood and body cells. A certain hydrostatic pressure is required to ensure an adequate exchange rate.

Arteries and Veins

Key Terms: constriction, dilation, lumen, tunica adventitia, tunica intima, tunica media, vasoconstriction, vasodilation

Arteries and veins are the major conduits for moving blood around the body. Both are composed of three distinct tissue layers (Figure 11.5). An outer layer called the **tunica adventitia** is composed of a tough, fibrous connective tissue. It contains collagen fibers for strength, and elastin fibers for flexibility. Fibroblasts in the tunica adventitia assist with healing and maintenance of this

Figure 11.3 Lymph Capillaries

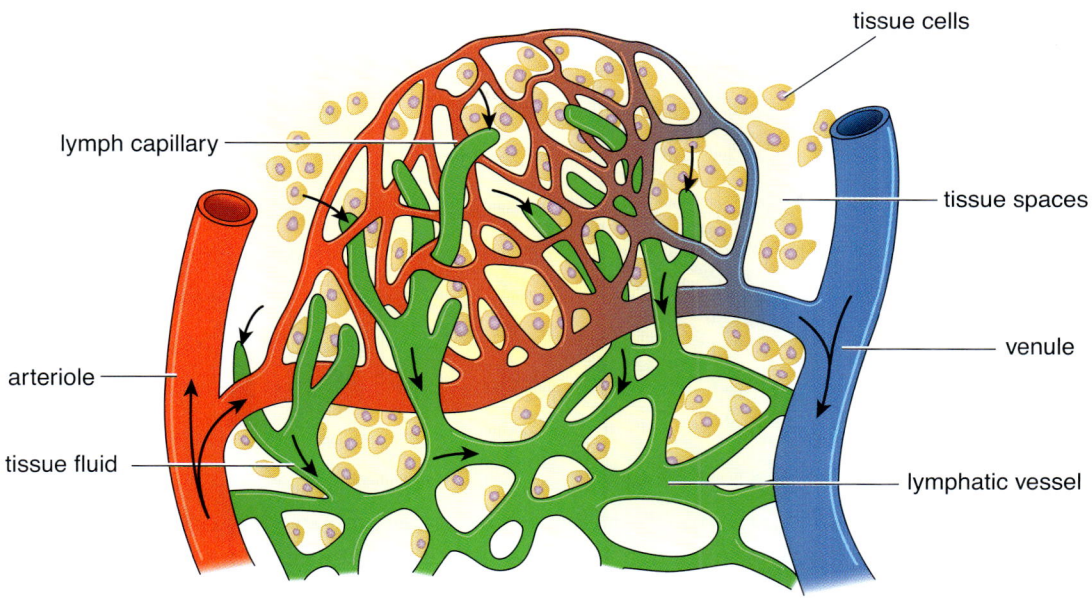

Figure 11.4 Hydrostatic Pressure

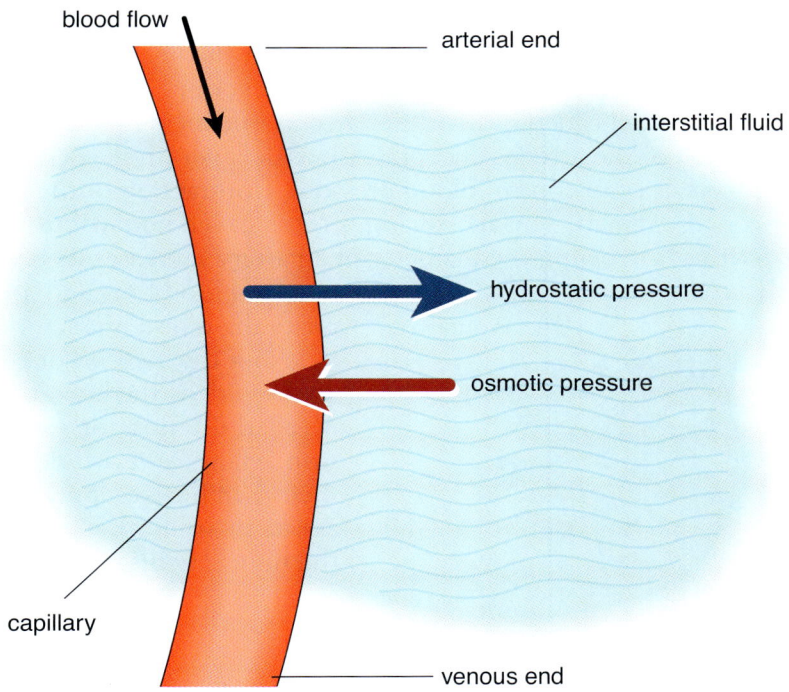

THE CARDIOVASCULAR SYSTEM

Figure 11.5 Vein and Artery

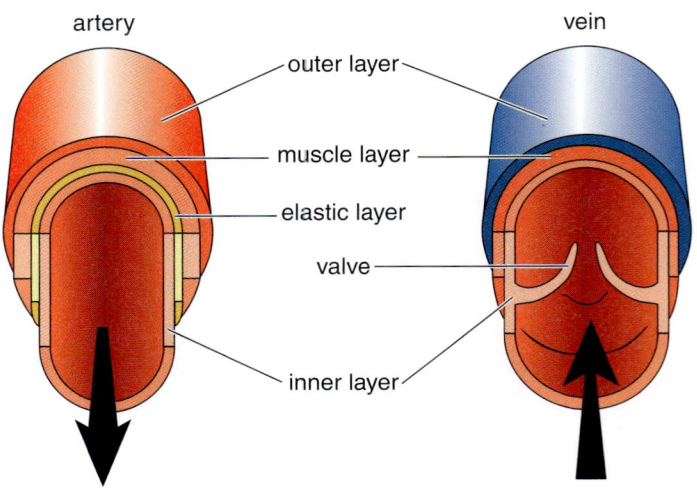

Tunica Media The middle layer of a blood vessel

Tunica Intima The inner layer of a blood vessel

Lumen The space within the interior of a tubular body structure

Constriction or Vasoconstriction The narrowing of the diameter of a blood vessel

Dilation or Vasodilation The widening of the diameter of a blood vessel

layer. Underlying the tunica adventitia is the **tunica media**, or middle layer. The tunica media is primarily composed of smooth muscle interspersed with collagen and elastin fibers. Parasympathetic and sympathetic nerves control muscle contraction of certain vessels. The smooth muscles also respond to hydrostatic pressure and certain chemicals circulating in the blood. Making up the innermost layer of arteries and veins is the **tunica intima**. The tunica intima is composed of a layer of simple, squamous, endothelial cells attached to a thin layer of loose connective tissue. It provides a smooth lining called the endothelium for the **lumen**, or central cavity, of the blood vessel.

Arteries distribute blood throughout the body (Figure 11.6). They differ from veins in the composition of the three blood vessel layers. Arteries are proportionately stronger and thicker than veins (Figure 11.7). They need these properties to sustain the force of the blood that the heart pumps into the lumen. The tunica adventitia is only slightly thicker in arteries than it is in veins of a similar size. Much of the difference in thickness is due to the larger tunica media of the arteries. This thick muscle layer provides strength and permits the arteries to control blood pressure. Contraction of the arterial muscles narrows the lumen. This narrowing is often called **constriction**, or **vasoconstriction**. Narrowing the diameter of an artery increases its resistance to blood flow and, in turn, raises the blood pressure. Relaxing the muscles of an artery widens the lumen. This is called **dilation**, or **vasodilation**. Many large arteries have their own blood supply, which assists the smooth muscles of the tunica media.

Figure 11.6 Arteries

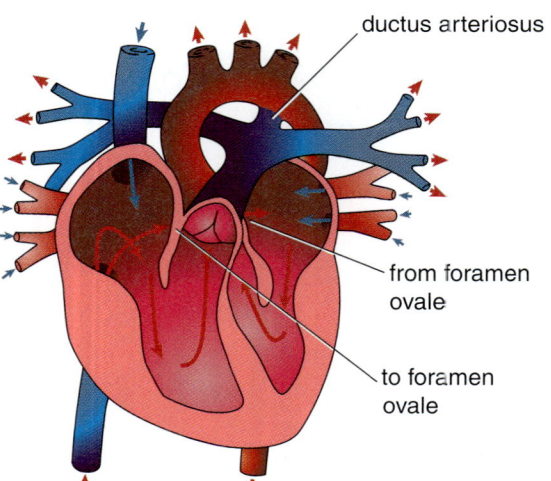

412

CHAPTER 11

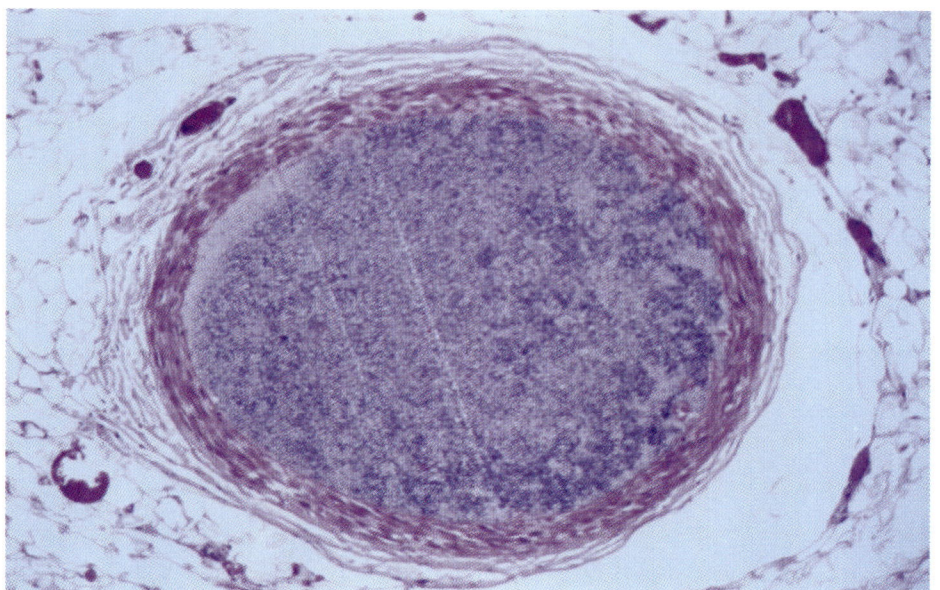

Figure 11.7 An Artery

> **VIRTUAL PHYSICIAN**
>
> Much of a physician's education and knowledge comes from "on-the-job training." Physicians learn much of what they need to know about disease from their experiences with diseased patients. A medical student may not observe certain rare medical conditions until they encounter them in clinical practice. This adds an element of difficulty when trying to make a correct diagnosis. In response to this potential problem, many medical schools have designed "virtual patients" to provide medical students with the experiences that they may not get in medical school or in the first years of their practice.
>
> One interesting virtual medical education application was developed at Blaufuss Multimedia Laboratories of Palo Alto, California. They created a way for physicians to study abnormal heart sounds and electrocardiograms (ECGs). Go to their Web site at *http://www.blaufuss.org/*, and see if you can "play doctor" by interpreting the normal and abnormal heart sounds and ECGs on virtual patients.

Veins return blood from the arteries back to the heart (Figure 11.8). Compared with arteries, veins generally have less elastin in their tunica adventitia. They also have a thinner muscle layer, which functions primarily to maintain the rigidity of the vessel (Figure 11.9). Veins transport blood under low pressure toward the heart. This pressure is partly due to the force of blood passing through the capillaries. Respiratory activity and the contraction of skeletal muscles also contribute to blood flow in the veins. Muscles in the extremities squeeze blood through the veins as they alternate between relaxation and contraction. Because veins contain blood under low pressure, they require a mechanism that prevents the downward pull of gravity on the blood. The pull of gravity can reverse blood flow in the veins, causing fluids to build up in the body tissues. Many veins contain special one-way valves that prevent the backflow of blood. In addition to mov-

Figure 11.8 Veins

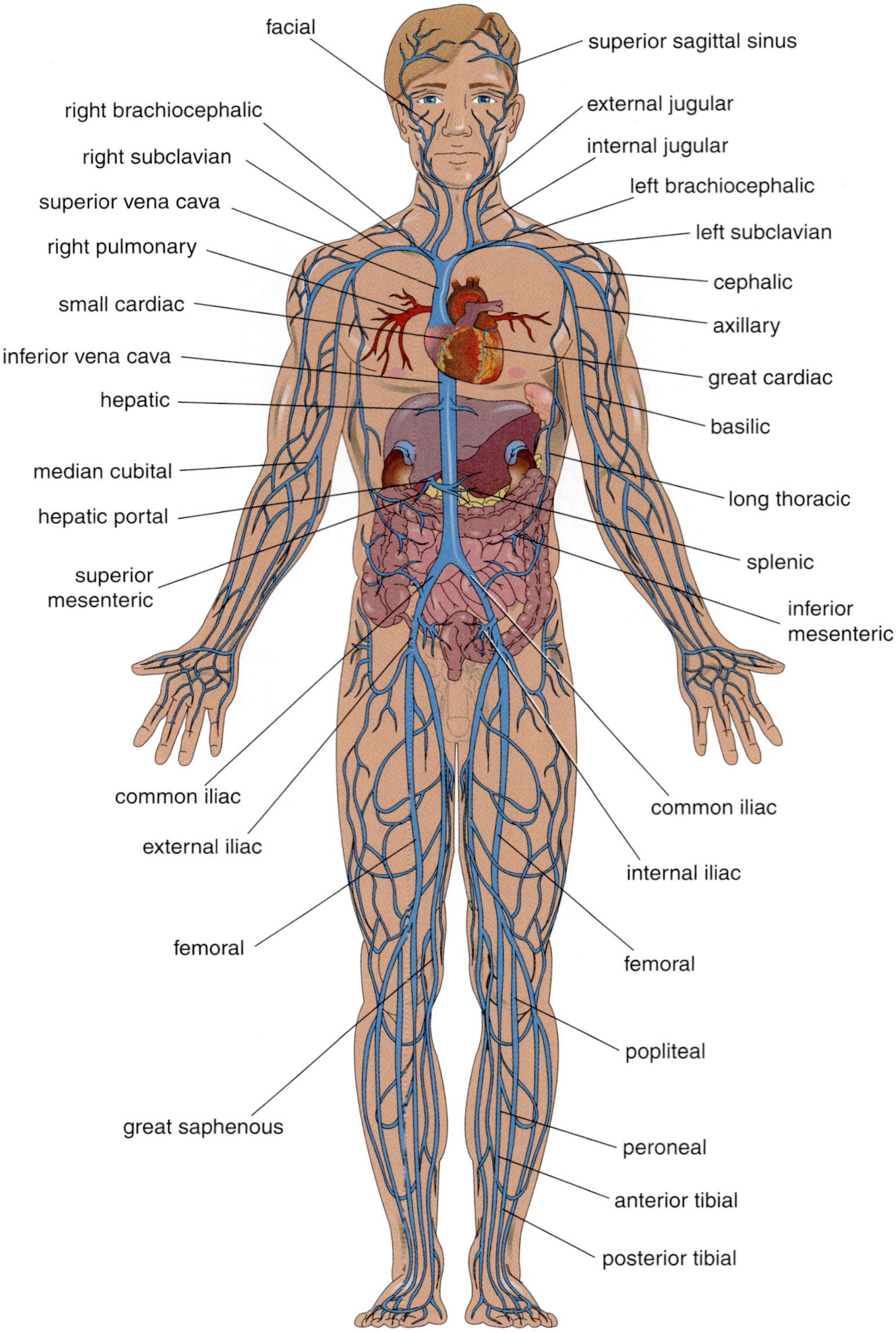

ing blood to the heart, veins are able to stretch outward, permitting them to function as a reservoir for blood. The liver contains special thin-walled veins called sinuses. These sinuses are important reservoirs of blood. They allow the body to store over 3 liters or about 6 pints of blood to make up for excessive blood loss from an injury.

CHAPTER 11

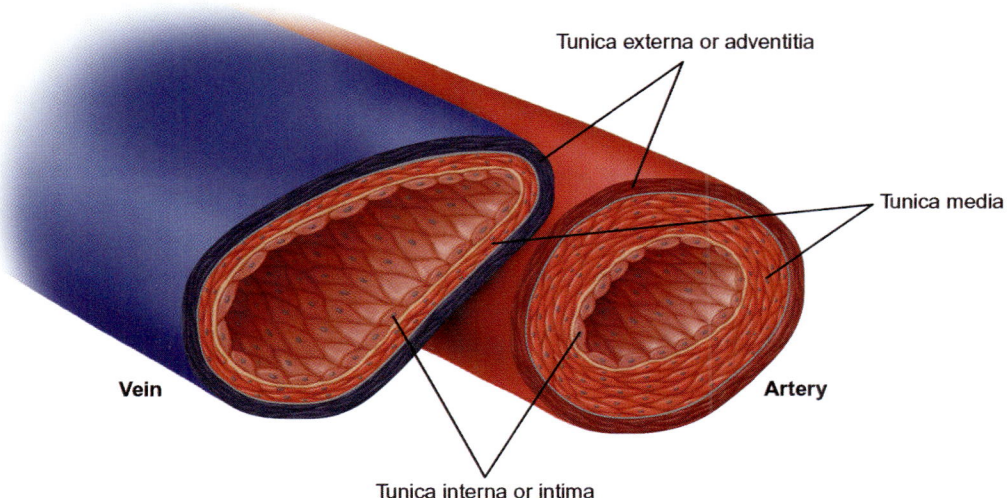

Figure 11.9 Vein

Small Vessels and Capillaries

Key Terms: arteriole, continuous capillary, fenestrated capillary, venule

Arteries and veins branch off and taper down into smaller blood vessels called **arterioles**, **venules**, and capillaries. Arterioles are distinguished from arteries by their very small size and simple structure. The tunica adventitia of arterioles is a thin sheath of connective tissue. In addition, the tunica media usually consists of two layers of smooth muscle. Arterioles are important for controlling the flow of blood to particular parts of an organ or tissue. They constrict and dilate in response to various chemicals that the body produces. Venules are typically larger in diameter than arterioles. They have a thin tunica adventitia and an inconspicuous smooth-muscle layer. Capillaries are very small vessels that are just large enough to allow the passage of blood cells. They are composed of a cylinder of endothelial cells held together by a thin membrane composed of fibrous proteins. A small capillary is made of a single endothelial cell, while larger capillaries are composed of two or three cells. Capillaries have no tunica adventitia or smooth muscles. Two types of capillaries can be distinguished: **Continuous** and **fenestrated**. Continuous capillaries are most common and are composed of endothelial cells tightly connected to each other (Figure 11.10). The tight connections between the cells form a barrier that limits the types of materials that pass into and out of the blood. Continuous capillaries are most often found in the central nervous system (CNS), lungs, muscles, and skin. Fenestrated capillaries have fenestrations, or openings, and materials are readily exchanged through them (Figure 11.11). They are commonly found in the digestive, endocrine, and urinary systems. Capillaries and venules are the major vessels through which materials are exchanged between the blood and tissues.

Arteriole A small branch of an artery

Venule A small branch of a vein

Continuous Capillary A capillary that forms a continuous covering around the lumen

Fenestrated Capillary A capillary that has small openings around the endothelial covering

Figure 11.10 **Continuous Capillaries**

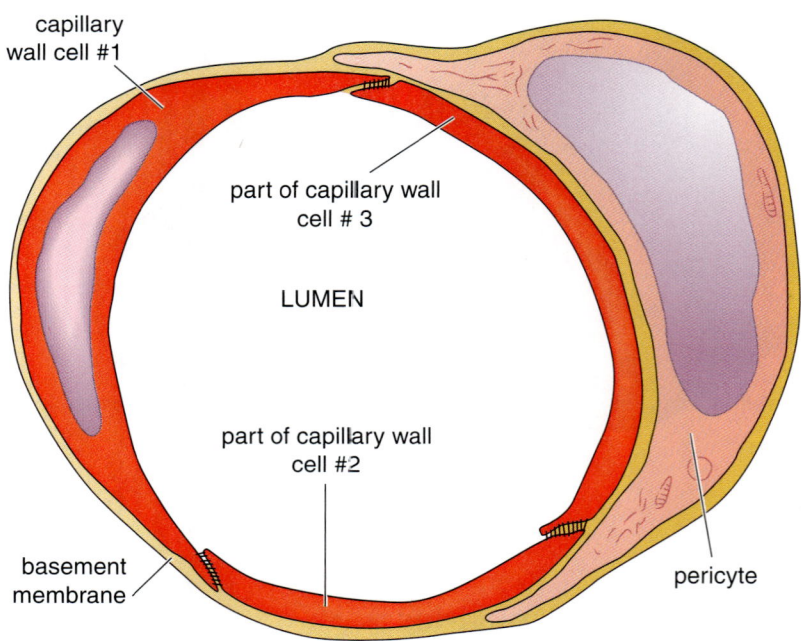

Figure 11.11 **Fenestrated Capillaries**

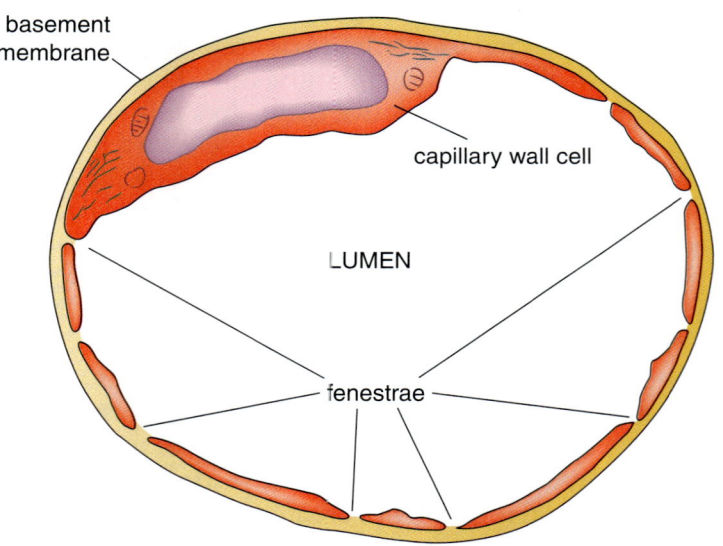

✓ Concept Check

1. What are the structural differences between arteries and veins?
2. Describe the different functions of arteries and veins.
3. Distinguish between arterioles, venules, and capillaries.

DISCOVERY SCENE PLEASE ENTER DISCOVERY SCENE PLEASE ENTER

CSI Break

Does the information about blood vessels provide any insight into the CSI? Could a particular type of blood vessel be responsible for the young runner's problems? Is there a relationship between the man's breathing difficulty and the function of his arteries, veins, or capillaries?

STRUCTURE OF THE HUMAN HEART

Key Terms: cardiac infarction, cardiac ischemia, coronary artery, coronary vein, endocardium, epicardium, fibrous pericardium, myocardium, pericardium, pulmonary circulation, serous pericardium, systemic circulation, visceral layer

The heart is a muscular, two-part pump that forces blood throughout the body (Figure 11.12). One part of the pump, the left side of the heart, sends blood at high pressure to all parts of the body, except the lungs. This is called **systemic circulation**. The right side of the heart pumps blood through the lungs at a lower pressure than does the left side. This is called **pulmonary circulation**. Delicate capillaries in the lungs would burst if blood were pumped into them with too high a pressure. The heart is nestled in the mediastinum, which is located between the lungs. It is separated from the lungs by the cavity membrane called the **pericardium** (Figure 11.13).

Systemic Circulation Circulation that supplies blood to all parts of the body, except the lungs

Pulmonary Circulation Circulation that supplies blood to the lungs

Pericardium A membranous sac that encloses the heart

Figure 11.12 **Human Heart**

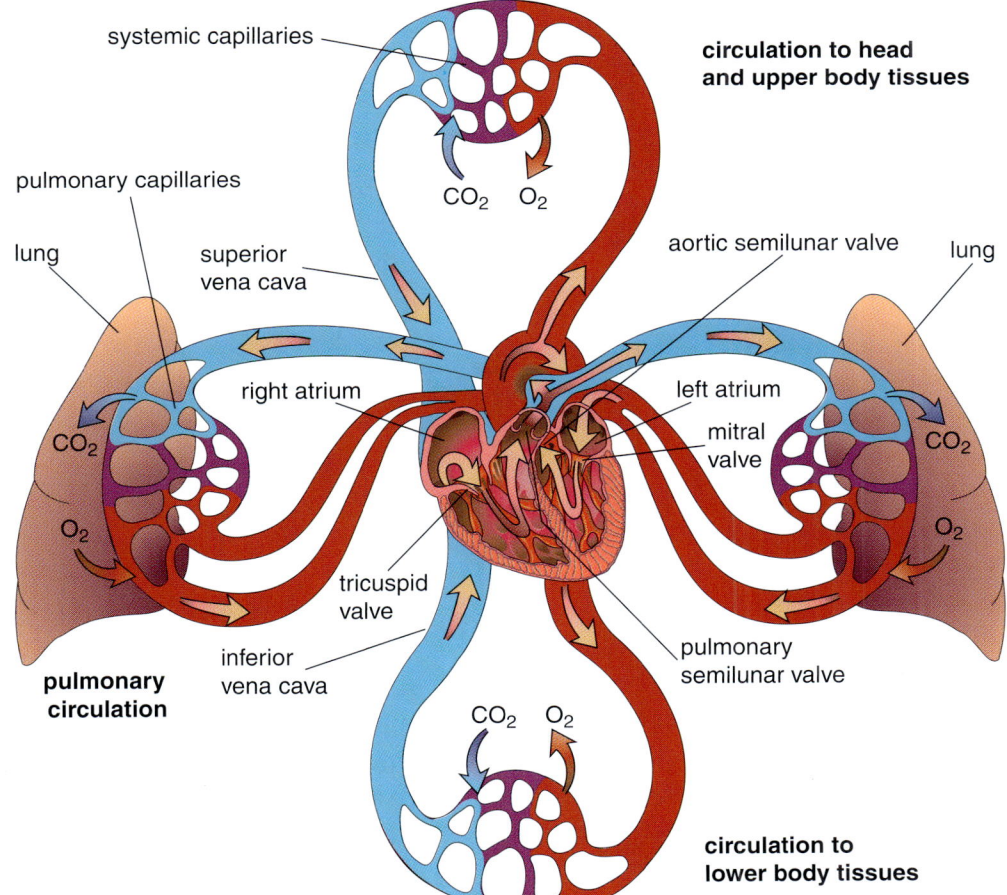

THE CARDIOVASCULAR SYSTEM

Figure 11.13 **Structure of the Heart**

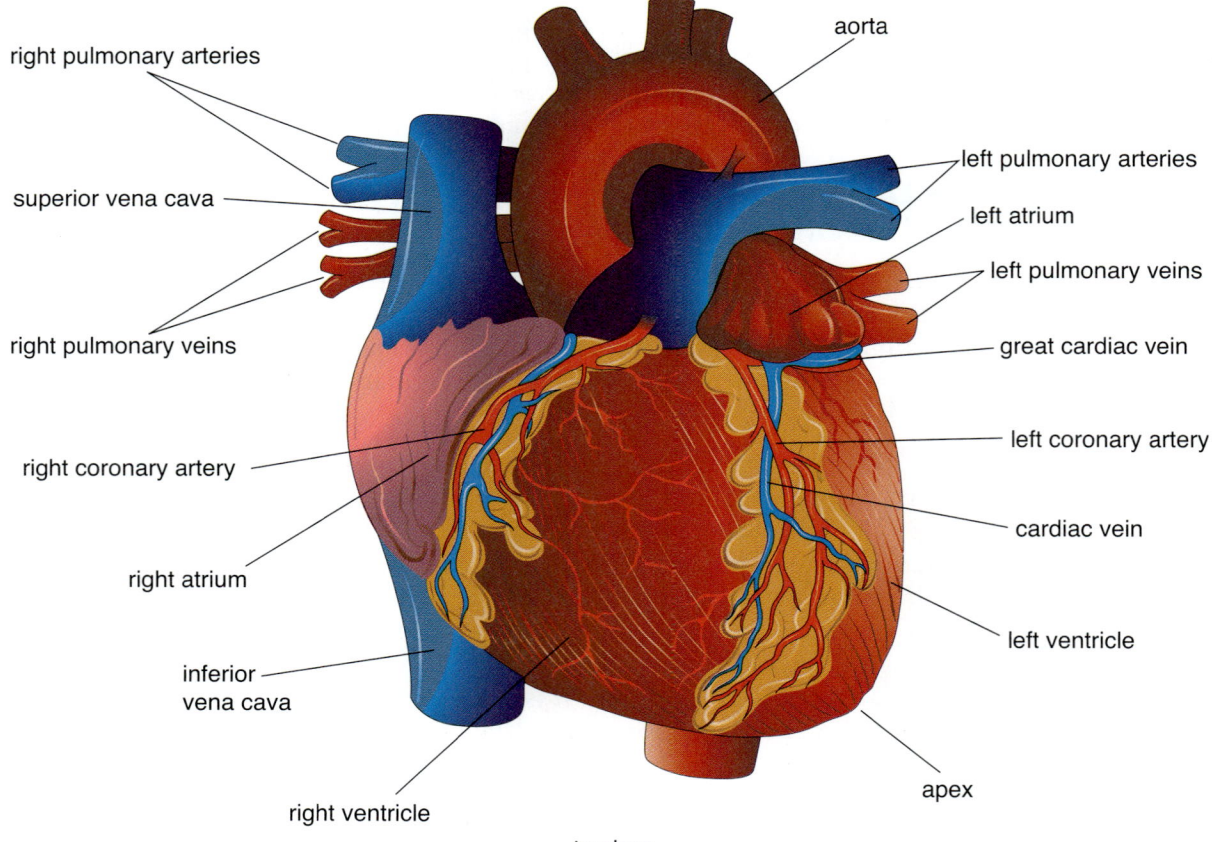

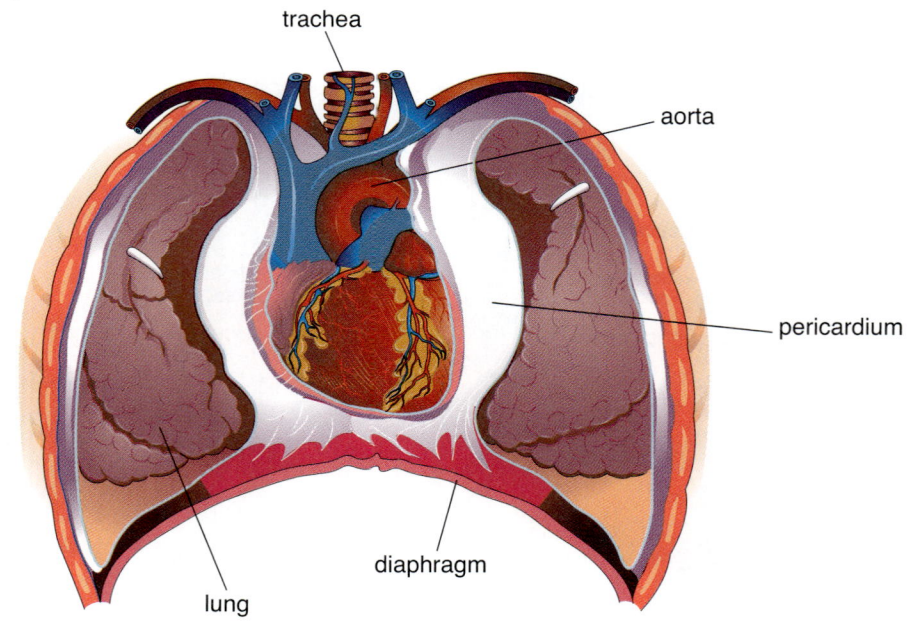

Fibrous Pericardium The outermost layer of the pericardium
Serous Pericardium The innermost layer of the pericardium

The pericardium is a membranous sac filled with serous fluid that encloses the heart. It lubricates and protects the heart as it beats in the thoracic cavity. There are three layers of the pericardium: the **fibrous pericardium**, **serous pericardium**, and pericardial cavity. The fibrous pericardium is the outermost layer. It is a connective-tissue layer that works to protect the heart and anchor it to the surrounding structures. The serous pericardium is underneath the fibrous pericardium. It is composed of two epithelial layers that lubricate the heart to prevent friction during heart activity. The layer next to the fibrous pericardium is the parietal layer of the serous pericardium. Next to the heart is the visceral layer of the serous pericardium or **epicardium**. Between these two layers is the pericardial cavity, which is filled with serous fluid.

The heart is a hollow structure composed primarily of thick sheets of cardiac muscle called the **myocardium** (Figure 11.14) The myocardium is tightly

> **Epicardium** The outer layer of the heart formed by the visceral layer of the pericardium
>
> **Myocardium** The muscle of the heart wall that contracts to pump blood

Figure 11.14 Adult Heart

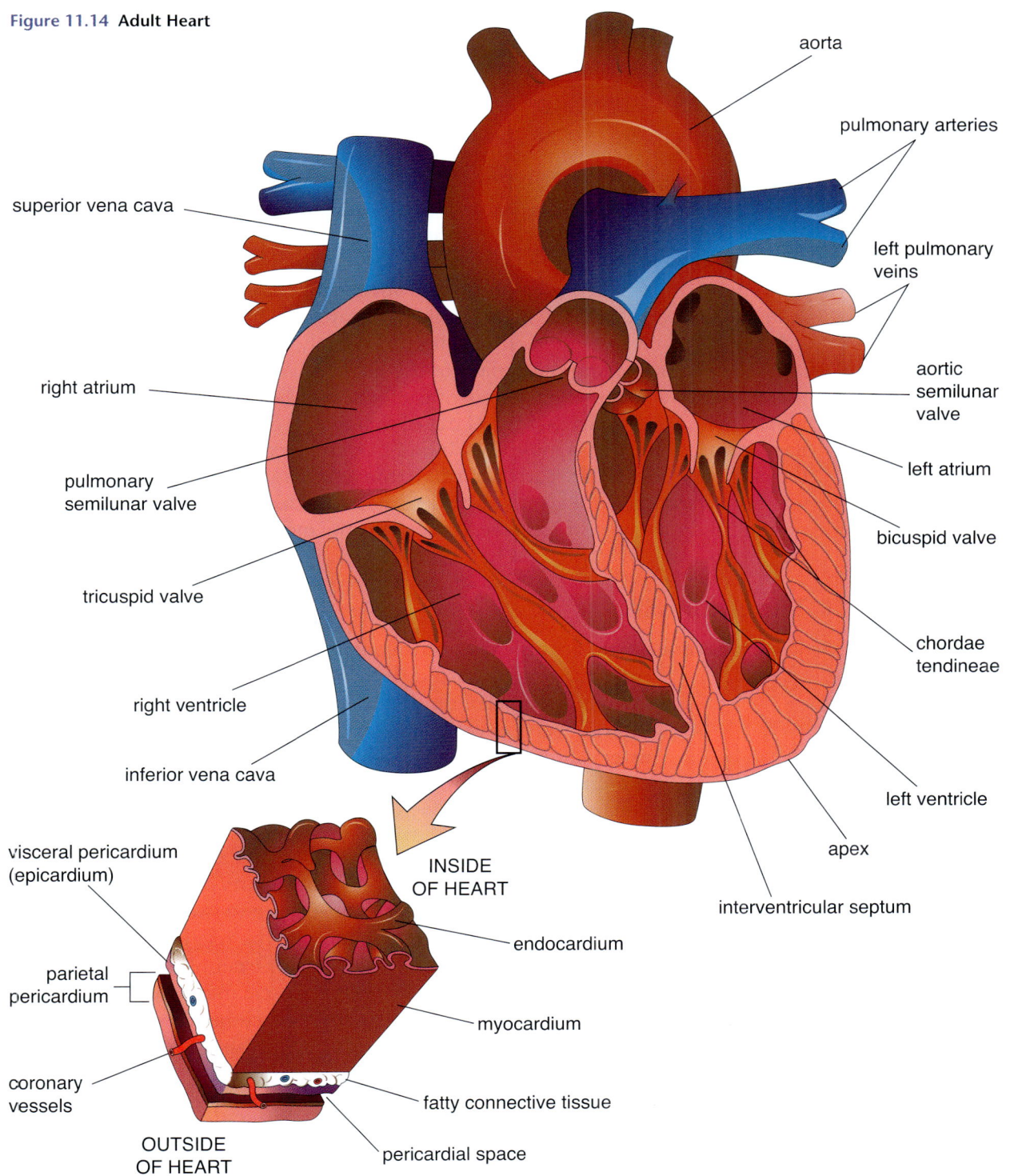

THE CARDIOVASCULAR SYSTEM

Good Choice Bad Choice

A friend wants to give up smoking; however, he wants to replace his smoking habit with chewing tobacco. What are the differences and similarities between smoking and chewing tobacco products? Is one safer than the other? How would you advise your friend about his decision?

Endocardium The inner lining of the heart

Coronary Artery Vessels that supply oxygenated blood to the heart muscle

Coronary Veins Vessels that collect blood from the heart muscle

Cardiac Ischemia A lack of sufficient oxygen for normal function of heart muscle

Cardiac Infarction Death of heart muscle due to lack of oxygen from the blood

attached to the epicardium on the outer surface of the heart. Fatty connective tissue can be found in this area, too. The inner surface of the myocardium is covered with a very smooth lining of epithelium called the **endocardium**. To keep a person alive, the heart must continuously contract and relax. This constant activity means that the heart must have a constant supply of oxygen and nutrients so it has a rich supply of blood vessels. The **coronary arteries** are a system of vessels that carry blood to the myocardium. These vessels branch into progressively smaller vessels, which, ultimately, lead to capillaries. The larger vessels pass along the surface of the heart. Smaller branches penetrate the myocardium, providing abundant capillaries for heart activity. The capillaries feed into small veins that travel out of the myocardium to form the **coronary veins**. Coronary veins collect blood from the heart muscle. Blockage of the coronary vessels can lead to **cardiac infarction** and **cardiac ischemia**. Cardiac infarction is the death of cardiac muscles after prolonged or repeated periods of diminished blood flow. Ischemia is the malfunction of cardiac muscle due to a lack of oxygen. Ischemia can lead to infarction.

The Adult Heart

Key Terms: aorta, aortic valve, atrioventricular (AV), atrioventricular node, atrium, bicuspid, bundle of His (pronounced hiss), chamber, chordae tendineae, electrical conduction system, inferior vena cava, mitral, papillary muscles, pulmonary artery, pulmonary valve, pulmonary vein, Purkinje system, semilunar valve, septum, sinoatrial (SA) node, superior vena cava, tricuspid, vasculature, venae cavae, ventricle

The anatomy of the adult heart develops from genetic and environmental factors. Genetic factors determine the general functional parts of the heart. Environmental factors are the result of lifetime exposure to various conditions that affect heart function and health. They determine the relative proportions of the components making up the functional parts. For example, the lower portion of the left side of the heart becomes larger in muscle mass because it has to worker harder pumping blood throughout a person's lifetime. People who partake in strenuous activities develop larger hearts in general for the same reason. The adult heart can be divided into five functional parts:

Chambers Four sections of the heart through which blood is pumped

Septum A muscular wall that divides the heart chambers

Vasculature A network of blood vessels in an organ or body part

- The four heart muscle **chambers**, **septum**, and **vasculature**.
- the heart valves
- the vessels that circulate blood to and from the heart

- the electrical conduction system
- the autonomic nervous system innervations

As mentioned earlier, the heart is composed of left and right pumping systems that are separated by a thick muscular wall called the septum. Each half of the heart contains two chambers: The **atrium**, a thin chamber on the superior surface of the heart; and the **ventricle**, a thick muscular chamber located below the atrium. Atria transfer blood into the ventricles, and the two ventricles then pump blood out of the heart. In adults, the myocardium of the left ventricle is much thicker than that of the right ventricle. This is due to the fact that the left ventricle must work harder to pump blood through the great network of blood vessels all over the body, while the right ventricle needs to pump it only to the lungs, which are in the nearby vicinity of the heart. (Figure 11.15)

Four heart valves facilitate the heart's pumping action: The **mitral, or bicuspid**, valve, the **tricuspid valve**, the **pulmonary valve**, and the **aortic valve**. The mitral, or bicuspid, valve and the tricuspid valve are the two **atrioventricular (AV)** valves. The pulmonary valve and the aortic valve are the two **semilunar valves**.

The mitral, or bicuspid, valve is composed of two rigid flaps that separate the left atrium from the left ventricle. It opens to allow blood collected in the left atrium to flow into the left ventricle. The mitral valve closes as the left ventricle contracts. This action keeps the blood flowing in one direction by preventing blood from returning to the left atrium. The tricuspid valve is composed of three rigid flaps that separate the right atrium from the right ventricle. It opens to permit blood collected in the right atrium to flow into the right ventricle. The tricuspid valve closes as the right ventricle contracts, which prevents blood from returning to the right atrium. **Papillary muscles** attached to the ventricle walls control the AV valves. **Chordae tendineae** attach the papillary muscles to the valves. The chordae tendineae are threadlike bands of fibrous tissue that attach edges of the tricuspid and mitral valves to the papillary muscles. When the ventricle contracts, the papillary muscles also contract and pull on the chrodae tendinae attached to the valves. This acts to prevent the valves from going up into the atria because of the pressure produced during ventricular contraction.

Each ventricle pumps blood out of the heart through a semilunar valve. These valves prevent blood from backing up into the heart. The blood flow produced by the heart completely controls the semilunar valves.

Five major blood vessels direct blood flowing to and from the heart chambers: the **superior vena cava**, the **inferior vena cava**, the **aorta**, the **pulmonary artery**, and the **pulmonary vein**. The **venae cavae** are composed of two large veins that bring blood to the right atrium. Blood from the head and upper body feeds into the heart through the superior vena cava. The inferior vena cava transports blood from the lower body to the right atrium. Blood exits the left ventricle through a large artery called the aorta. It is the largest blood vessel in the body. The semilunar valve associated with the aorta is sometimes referred to as the **aortic valve** because it separates the left ventricle from the aorta. The aorta and venae cavae are part of the systemic circulation. The other vessels of the heart are the pulmonary artery and pulmonary vein. Blood from the right ventricle is passed along to the lungs through the pulmonary artery. The semilunar valve associated with the aorta is sometimes referred to as the pulmonary valve because it separates the right ventricle from the pulmonary artery. Blood from the lungs enters the left atrium through the pulmonary veins.

Atrium One of the two thin upper chambers of the heart

Ventricle One of the muscular lower chambers of the heart

Mitral or Bicuspid Valve One of the two atrioventricular valves; it lies between the left atrium and left ventricle

Tricuspid Valve One of the two atrioventricular valves; it lies between the right atrium and right ventricle

Pulmonary Valve The right semilunar valve

Aortic Valve The left semilunar valve

Atrioventricular (AV) Valves Relating to the atria and ventricles of the heart; the mitral and tricuspid valves

Semilunar Valve A heart valve that prevents blood from flowing back into the heart

Papillary Muscles Muscles in the wall of the ventricles that control the atrioventricular valves

Chordae Tendineae Fibrous threads that attach papillary muscles to the atrioventricular valves

Superior Vena Cava The large vein that carries blood to the right side of the heart from the head and upper body

Inferior Vena Cava The large vein that carries blood from the lower body and organs to the right side of the heart

Aorta The large artery that carries blood from the heart to the body

Pulmonary Artery The blood vessel that takes blood from the heart to the lungs

Pulmonary Veins The blood vessels that take blood from the lungs to the heart

Venae Cavae Large veins that bring blood into the right side of the heart

THE CARDIOVASCULAR SYSTEM

Figure 11.15 Atria and Ventricles

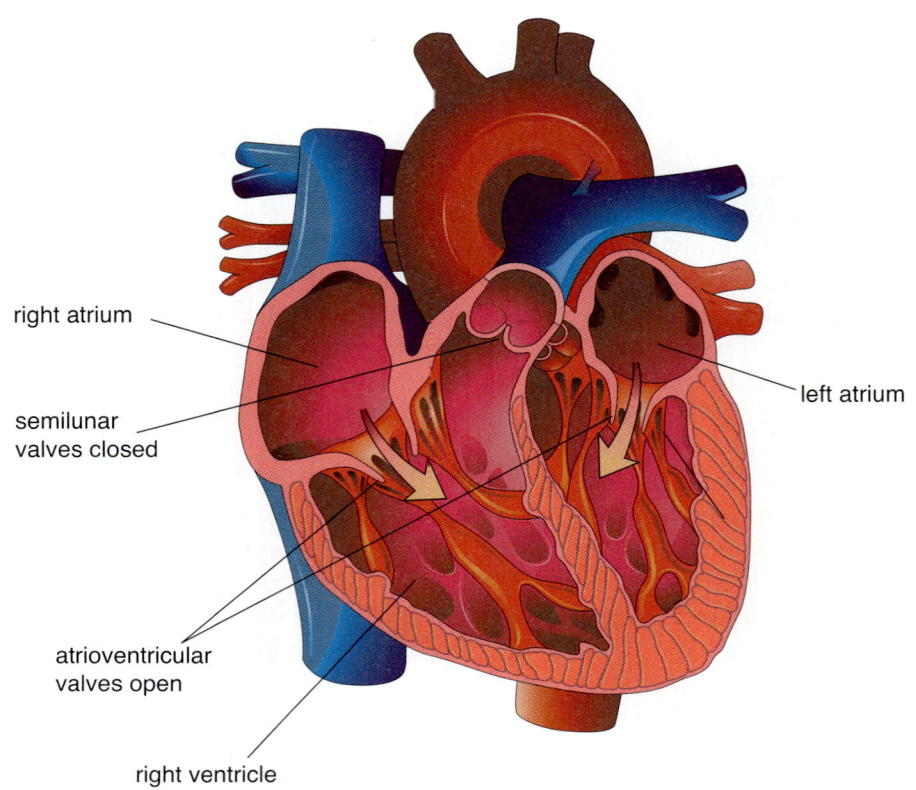

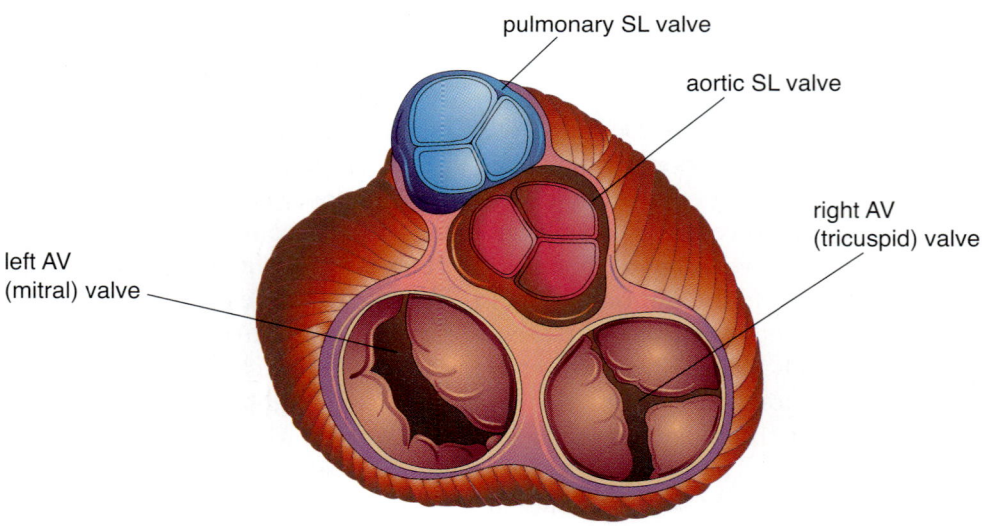

The **electrical conduction system** of the heart is composed of specialized cardiac muscle cells that act like a miniature nervous system. These cells produce electrical signals that stimulate the contraction of particular regions of the heart. It is common for the system to produce 60 to 100 signals per minute, which means that the heart beats this many times per minute. The electrical conduction system is composed of the **sinoatrial (SA) node**, **atrioventricular (AV) node**, the **bundle of His** (pronounced hiss), and the **Purkinje system**. The SA node, which is located in the right atrium, controls heart rate. It is called the "pacemaker" of the heart because it determines heart rate. The AV node is responsible for the contraction of ventricles. It receives signals from the SA node and then stimulates the bundle of His and Purkinje system to carry out contraction of the ventricles. The autonomic nervous system regulates heart rate using the vagus nerve. Sympathetic nerve endings innervate the muscles of the atria, ventricles, AV node, and SA node. Parasympathetic nerve endings mainly innervate the atrial muscle and the AV node.

Electrical Conduction System Specialized cardiac muscle cells that stimulate heart contraction

Sinoatrial (SA) Node Specialized cells of the heart that initiate the heart beat

Atrioventricular (AV) Node A relay station between the atria and ventricles that coordinates atrial and ventricular connections

Bundle of His Specialized cells at the superior interventricular septum that receive nerve impulses from the atrioventricular node

Purkinje System Specialized muscle cells that carry the electric impulses through the ventricles

The Fetal Heart

Key Terms: ductus arteriosus, foramen ovale

The fetal heart starts out as a large blood vessel that folds in upon itself as it rapidly grows during the first 2 weeks of development. This folding initiates the formation of the two halves of the heart. Atria and ventricles develop and send out blood vessel branches that become the major blood vessels of the heart. The septum also begins to form, dividing the two atria and the two ventricles into left and right. By the end of week 8, the chambers are completed and the fully functional fetal heart is formed (Figure 11.16). The fetal heart is not identical to the adult heart. It possesses two different structures that adapt it to the conditions inside the mother's body: the **ductus arteriosus** and **foramen ovale**. The ductus arteriosus usually diverts blood from the pulmonary artery to the aorta. This keeps large amounts of blood from entering the capillaries of the lungs. After all, the fetus does not use the lungs. The ductus arteriosus closes usually a day after birth and is completely gone by 8 weeks. Respiratory diseases in infants can reopen the ductus arteriosis or prevent it from fully closing. This condition often corrects itself within several months of

Ductus Arteriosus A blood vessel that connects the pulmonary artery to the aorta in the fetus

Foramen Ovale An opening between the right and left atria

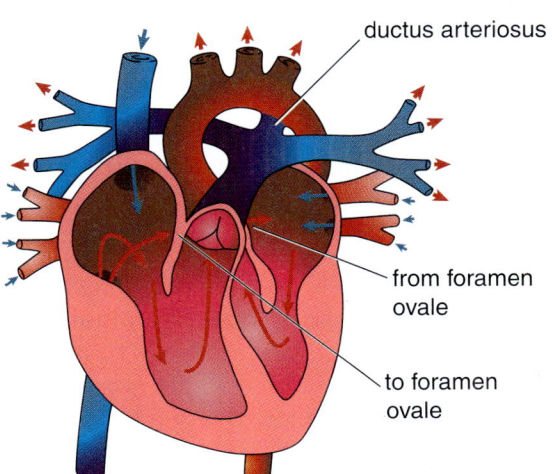

Figure 11.16 **Fetal Heart**

birth. If it does not, surgical closure is performed. An incompletely closed ductus arteriosus would not allow adequate oxygen to be delivered to the body.

The foramen ovale is a flap-like opening within the septum between the atria. It directs blood flow from the right atrium to the left atrium. Like the ductus arteriosus it reduces blood flow to the lungs. Pressure in the left atrium assists with the closure of the flap after birth. For no known reason, the foramen ovale may fail to close in approximately one of five people. This condition is not severe and can be corrected surgically, if necessary. Usually, a remnant flap of the foreman ovale persists for a year after birth. Blood flow in takes a different path in the fetal heart than in the adult heart. Approximately 30% of the blood from the right atrium flows to the left atrium through the foramen ovale. The blood then passes into the left ventricle and to the aorta. This reduces the blood going to the pulmonary artery. Much of the blood entering the pulmonary artery passes into the ductus arteriosis where it is diverted to the aorta and into the systemic circulation. The blood then returns to the heart by way of the venae cavae.

Physicians can detect potential heart formation defects in the fetus using ultrasound, which allows them to view the shape of the heart, the chambers, and the condition of the valves. Other imaging technologies permit the physician to determine blood flow patterns. Fetal heart defects, if found early, can be corrected before birth. Fetal surgery is regularly used to improve the chance for survival after delivery in babies with serious heart defects.

✓ Concept Check

1. Distinguish between pulmonary and systemic circulation.
2. Describe the five components of the adult heart.
3. What are the main differences between the adult heart and the fetal heart?

DISCOVERY SCENE PLEASE ENTER DISCOVERY SCENE PLEASE ENTER

CSI Break

What information about heart structure is important in understanding the CSI? Could the young man's condition be due to some malfunction of a component of the heart? If so, what part of the heart would produce illness only during the race?

CARDIOVASCULAR TRIVIA

The cardiovascular system is an amazingly complex organ system that provides the resources for the other organ systems to operate. Here is some interesting, and maybe somewhat useful, trivia about the human cardiovascular system:

- The average human heart beats 36.5 million times a year.
- An average baby's heart will beat about 60 million times before it is born.
- An average adult human heart weighs about 10 ounces (283 grams) and is the size of a fist. The heart of a blue whale is as big as a car!
- Yawning may be due to inadequate oxygen being carried by the cardiovascular system.

- Carotid arteries got their name from the Greek word karotides, which means "to stupefy." The ancient Greeks learned that pressing these arteries closed causes a person to become unconsciousness or stupefied.
- Ancient Greek physicians believed that the heart was the center of emotion, intelligence, and memory.
- According to German researchers, the risk for heart attack is higher on Monday than on any other day of the week.
- In the overall population, the heart beats faster in women than in men.
- Well-trained endurance athletes can have a resting heart rate of 35 to 40 beats per minute.
- The first known heart medicine was discovered in 1799 by physician John Ferriar. The drug was digitalis, made from the dried leaves of the foxglove plant.
- Each square inch of human skin contains 20 feet (6.1 meters) of blood vessels
- The human body has approximately 70,000 miles of blood vessels.
- Approximately 7% of a person's body weight is blood.

Heart Function

Key Terms: cardiac cycle, cardiac output, diastole, heart rate, stroke volume, systole

The term **cardiac cycle** means a single cycle of cardiac activity. It can be divided into two basic stages that coordinate blood flow through the heart and body (Figure 11.17). The first stage is called **diastole**. During diastole, the ventricles fill with blood delivered by contractions of the atria. At this point, the ventricles are relaxed. **Systole** is the second part of the cardiac cycle. It involves the contraction and discharge of blood from the ventricles.

The heart's electrical conduction system controls the coordination of the cardiac cycle. It synchronizes contraction of the chambers, and the subsequent opening and closing of the valves in each region of the heart. The electrical conduction phases of the cardiac cycle are described below (Figure 11.18):

- An electrical signal from the SA node causes contraction of the atria and stimulation of the AV node. As this occurs, the AV valves begin to open, letting blood from the atria enter the ventricles. The right atrium contracts slightly before the left atrium.
- The AV valves then open fully, causing the ventricles to fill with blood.
- A short delay of the electrical signal at the AV node allows the ventricles to completely fill with blood.
- The AV node then sends an electrical signal to the bundle of His and Purkinje system, which starts the systole phase. This causes the ventricles to contract from the bottom of the heart toward the semilunar valves.
- Contraction of the ventricles forces the blood upward and pushes the flaps of the AV valves back up to their closed position. The length of the chordae determines the proper position of valve closure. The atria then begin to relax and refill with blood.

Cardiac Cycle One complete contraction and relaxation of the heart

Diastole A part of the cardiac cycle during which the heart muscle relaxes and fills with blood

Systole A part of the cardiac cycle during which the heart muscle contracts and forces blood out

THE CARDIOVASCULAR SYSTEM

Figure 11.17 Blood Flow

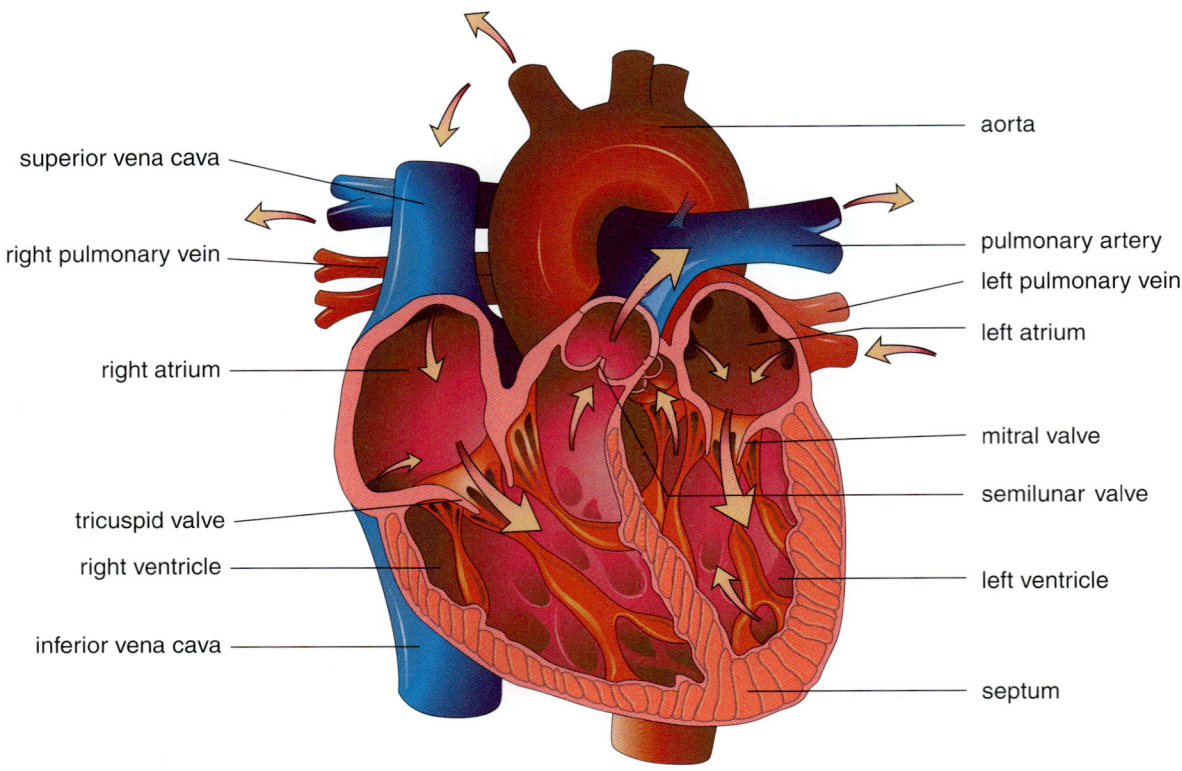

Figure 11.18 Electrical Conduction System

- The pulmonary and aortic valves open as the pressure in the ventricles exceeds the pressure in the pulmonary artery and aorta. Blood then enters into the pulmonary and systemic circulations.
- The ventricles begin to relax, and the pulmonary and aortic valves close. This leads to diastole and a repeat of the cycle.

The SA node has a built-in excitation pattern that automatically stimulates the cardiac cycle. Stimulation from the sympathetic nervous system can increase the speed of this rhythm. Signals from the parasympathetic nervous system slow the rate of the cycle. The autonomic nervous system, which is responsible for cardiac control, receives signals from the brain stem and hypothalamus. Cardiac muscle also responds to signals from baroreceptors and chemoreceptors in major blood vessels and the heart. The efficiency of a cardiac cycle is determined using various measures. **Heart rate** refers to the number of ventricular contractions per minute. **Stroke volume** is another measure of the cardiac cycle. It refers to the amount of blood pumped by the ventricle of the heart during one cycle. The volume of blood that the heart pumps in 1 minute is called **cardiac output**. It is calculated by multiplying the heart rate by the stroke volume. The following formula expresses cardiac output:

CO = HR X SV

Heart Rate The number of times the heart beats in 1 minute

Stroke Volume The amount of blood pumped by the ventricle of the heart with each beat

Cardiac Output The amount of blood the heart pumps each minute

The abbreviation CO stands for cardiac output, SV represents stroke volume, and HR refers to heart rate. Cardiac output can be measured as milliliters per minute (mL/min) or liters per minute (L/min). For the average person, the heart beats 75 times per minute. Stroke volume averages 70 mL of blood for each heartbeat. Therefore, the cardiac output is 5.25 L/min.

✓ Concept Check

1. Define and describe the cardiac cycle.
2. Describe the sequence of events carried out by the electrical conduction system.
3. Describe cardiac output.

DISCOVERY SCENE PLEASE ENTER DISCOVERY SCENE PLEASE ENTER

Could cardiac function play a role in the young man's condition? What could be the role of the cardiac cycle in producing illness? How could the man's participation in the race impact cardiac function?

ELECTROCARDIOGRAPHY BASICS

Key Terms: electrocardiogram (ECG), electrocardiography, P-R interval, P wave, Q wave, QRS complex, Q-T interval, R wave, S wave, S-T interval, T wave

Electrocardiography A procedure that measures the electrical activity of the heart

Minute electrical impulses are discharged every time the heart goes through a cardiac cycle. The electrical conduction system produces the impulses, which can be measured using a procedure called **electrocardiography**.

Figure 11.19 ECG Basics

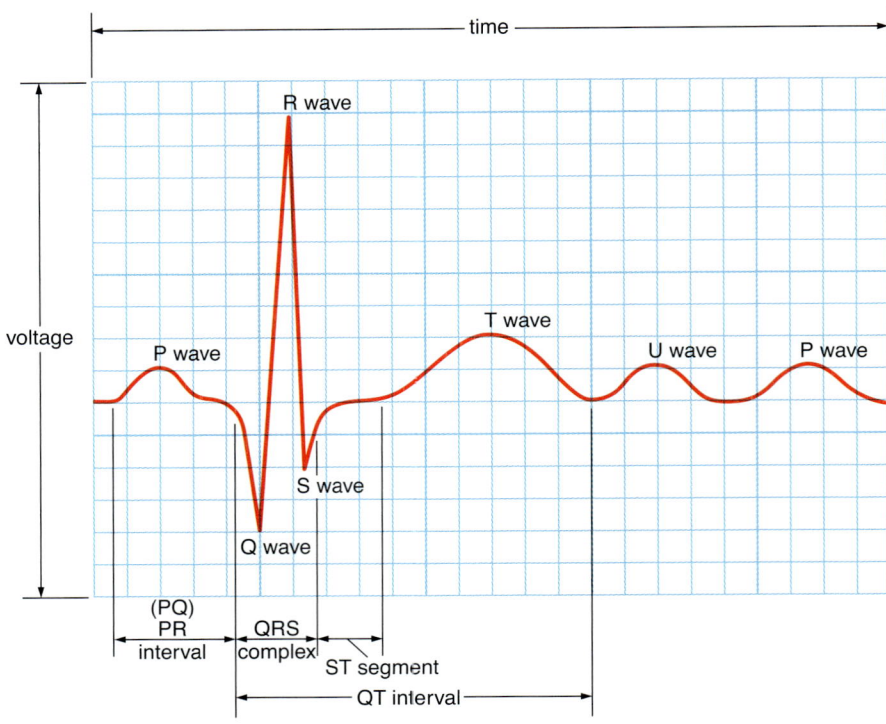

Electrocardiogram (ECG) A record of the heart's electrical activity

P Wave Electrical activity of the SA node and atria

Q Wave The beginning of ventricular depolarization

R Wave The electrical activity of ventricular contraction

S Wave The end of ventricular contraction

T Wave The beginning of ventricular repolarization

QRS Complex A combined reading of the Q, R, and S waves of an electrocardiogram

P-R (PQ) Interval The time interval between atrial depolarization and ventricular depolarization

Electrocardiography is recorded in an **ECG** (Figure 11.19). The abbreviation **ECG** is now used to indicate **electrocardiogram** (EKG is the abbreviation that previously was more commonly used). Electrical discharges of the electrical conduction system follow a pattern that can be used to determine whether the heart is healthy or diseased.

The electrical activity of the heart can be detected through the skin using small metal discs called electrodes. Electrocardiography requires the use of several wires attached to electrodes that are placed on particular regions of the body. They are usually placed on the chest, arms, and legs. The wires conduct the electrical charges into an instrument that measures and records the impulses as an ECG. Initially, ECGs are conducted when the patient is at rest. However, when heart disease is suspected, electrocardiography can be performed while the patient is under the stress of strenuous activity. Usually, the patient walks on a treadmill while heart readings are measured.

A typical ECG pattern can be correlated with other heart activities and cardiac function measurements (Figure 11.20). The ECG represents a particular pattern of the electrical conduction system based on the placement of the electrodes on the skin. Different patterns, or waves, on the ECG shows the sequence of depolarization and repolarization of the atria and ventricles. The electrocardiography pattern is divided into waves that are known as the **P, Q, R, S,** and **T waves**. The consistency, rate, and shape of the waves provide specific information about any heart damage that might be present.

The P wave represents the impulse that travels from the SA node throughout the atria. It is usually 0.08 to 0.1 seconds in duration. The Q, R, and S waves are lumped together in a reading called the **QRS complex**, which is a measure of ventricular activity. The **P-R (PQ) interval** represents the timing between

Figure 11.20 Systole and Diastole

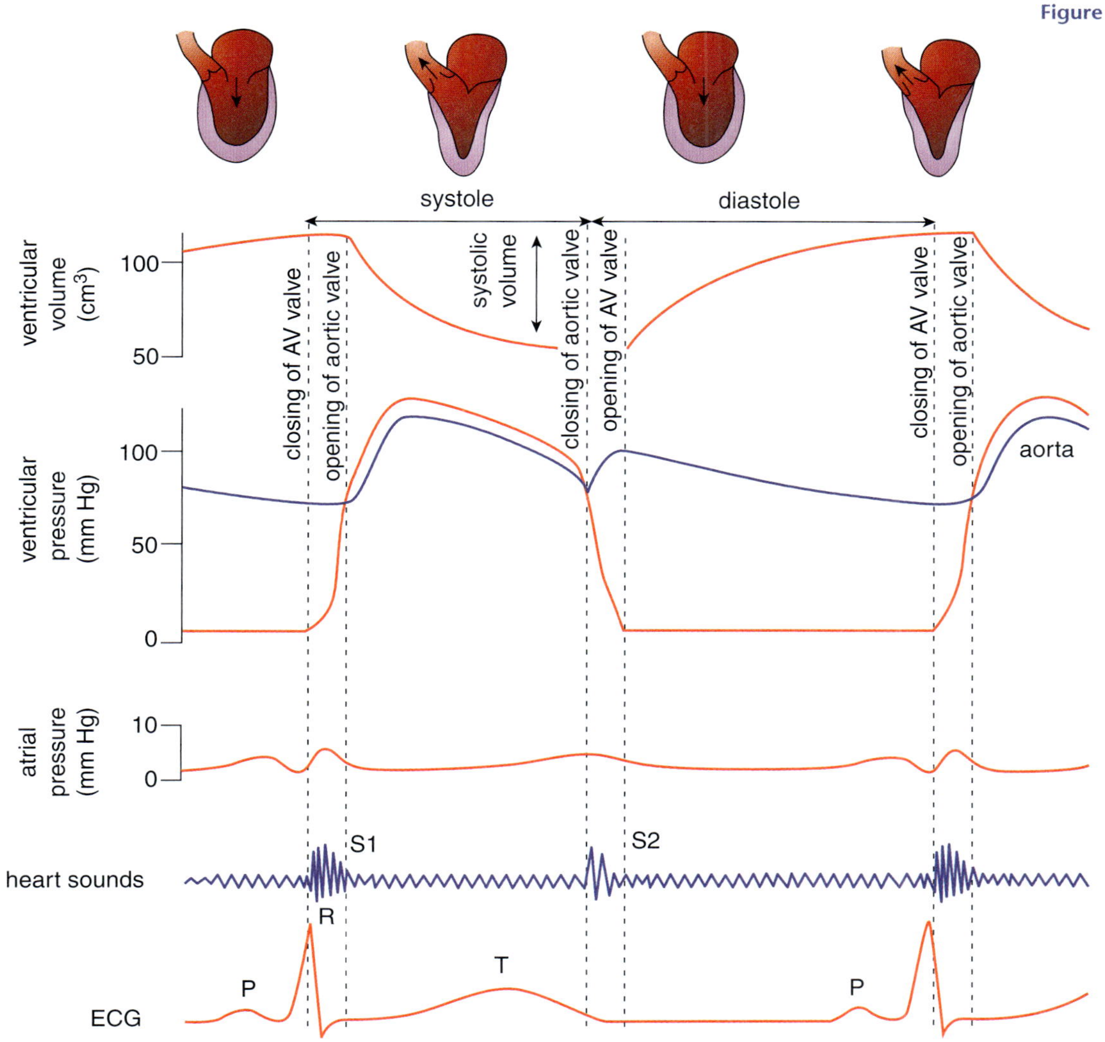

atrial depolarization and ventricular depolarization. The interval normally lasts 0.12 to 0.20 seconds. The **S-T interval** corresponds to the entire ventricular action potential. It is important for diagnosing ventricular ischemia. The T wave represents ventricular repolarization. It is combined into a measure called the **Q-T interval**, which indicates the time between ventricular depolarization and repolarization. This interval ranges from 0.2 to 0.4 seconds depending upon the heart rate. The wave for atrial repolarization is not evident because it is hidden by the electrical impulse of ventricular depolarization.

Other useful information is available from the ECG. Heart rate is determined from the number of waves per minute. The distance between the consecutive waves is the heart rhythm. The shapes of the waves on the readout reveal the health of the cardiac muscle, the size of the heart, and the degree of coordination between the components of the heart. The consistency or inconsistency of the waves provides information about any heart damage that might be present.

S-T Interval The time interval between atrial depolarization and ventricular depolarization

Q-T Interval The time interval between ventricular depolarization and repolarization

429

THE CARDIOVASCULAR SYSTEM

✓ Concept Check

1. Define electrocardiography.
2. Explain the components of an ECG.
3. What information does the ECG provide?

DISCOVERY SCENE PLEASE ENTER DISCOVERY SCENE PLEASE ENTER

What information could electrocardiography provide for identifying the condition that the young runner exhibited?

Wellness and Illness over the Life Span

Pathology of the Cardiovascular System

Key Terms: aneurysm, angina pectoris, arrhythmia, atherosclerosis, arteriosclerosis, cardiac disorders, clot, congenital heart disease, endocarditis, enlarged heart, fibrillation, hypertension, murmur, pericarditis, plaque, prolapse, regurgitation, rheumatic heart disease, sudden cardiac death, tamponade, thrombosis, vascular disorders

There are many types of disorders that affect various components of the cardiovascular system. The greatest variety of diseases affects the heart because of its incredible complexity. In addition, the heart is one of the few muscular structures that must work continuously without any rest. Cardiovascular diseases are divided into two categories: **vascular disorders** and **cardiac disorders**. Vascular disorders are diseases of the arteries, capillaries, and veins. Cardiac disorders affect the heart conduction system, muscle, pericardium, valves, and vasculature. Below is a summary of the major cardiovascular conditions common in North America and Western Europe.

An **aneurysm** is a bulging of the wall of a blood vessel. Aneurysms usually form in large arteries. The bulge forms around a local weakening of the blood vessel wall. Congenital defects, infectious diseases, and injuries can cause blood vessel weakening, which can lead to an aneurysm. Aneurysms are currently treated with surgical procedures that involve clipping the bulge or packing the bulge with flexible wire coils. Clipping the bulge involves surgical removal of the weakened area of the vessel. The vessel is sewn together to produce the normal cylindrical shape. Packing is used when it becomes too dangerous to remove the bulge. An aneurysm is packed by using a needle to insert a flexible wire into the bulge. The wire is used as a packing that eventually fills the bulge. Filling the bulge prevents blood from entering and consequently expanding the bulge.

Many diseases of the heart create pain in the chest area. This pain is called **angina pectoris**. Angina pectoris is usually felt when the heart needs more blood. The severity of the pain depends on the person's tolerance for pain and the degree of heart disease. There is no treatment for angina. It is an indicator of disease rather than a disease in itself. Physicians investigate the causes of angina pectoris with a medical examination that includes a compre-

Vascular Disorder A disease of the blood vessels

Cardiac Disorder A disease of the heart

Aneurysm A bulging of a blood vessel wall

Angina Pectoris Chest pain due to coronary heart disease

Cutting Edge Research
THE "TYPE D" PERSON

Is it possible that a person's character and personality are hazardous to his or her health? Johan Denollet, a psychologist at Tilburg University in the Netherlands, believes this may be true when it comes to heart disease. He developed a personality test that correlated certain behaviors with an increased risk for heart disease. People who have high levels of anxiety, hostility, and a sense of hopelessness are categorized Type D personalities. In the 1960s, other psychologists developed instruments to assess and classify personalities. Competitive perfectionists were classified as Type A personalities, and noncompetitive, relaxed persons were classified as Type B. It was believed that Type A people were prone to heart disease despite the lack of scientific evidence supporting this view. The Type C category was then added in the 1980s. Type C people appear calm, but suppress their anger and feelings. Again, it was erroneously thought that this behavior was harmful to the health of the heart. Cardiologists are now finding that Type D people are less likely to recover fully from cardiovascular treatments. Dr. Denollet found that 27% of cardiac rehabilitation patients classified as Type D had died after intervention. Only 7% of people not classified as Type D died after given similar rehabilitation treatment. Most of the people in both groups died from heart diseases or strokes. Scientists currently have no solid physiological explanation for the effects of Type D personality on the body. Many physicians believe that increased anxiety makes these people more susceptible to disease in general. Physicians have known for years that a patient's negative attitude can counteract even the best medical interventions. Psychologists are currently developing strategies to reduce the high-risk emotions of people with a Type D personality.

Denollet, J, Vaes, J, Brutsaert, DL. 2000. Inadequate Response to Treatment in Coronary Heart Disease Adverse: Effects of Type D Personality and Younger Age on 5-Year Prognosis and Quality of Life Circulation. 102(6):630-5.

---

hensive physical exam, an investigation of the patient's medical history, an ECG, and a chest X-ray.

An **arrhythmia** is any deviation in the normal heartbeat rhythm. It is usually caused by disturbances in the heart's electrical impulse system. This condition is usually not serious unless the heartbeat becomes too fast or too slow to maintain normal body function. Vascular disorders could damage the parts of the electrical impulse system that are producing the irregular heartbeat. There is evidence that alcohol, caffeine, diet pills, high-protein diets, stress, tobacco, and certain medications can produce temporary arrhythmias. The most common symptom is a trembling feeling in the chest or the feeling that the heart is skipping beats. Many arrhythmias require no medical treatment. Severe conditions are treated with medications that regulate cardiac muscle contraction.

Atherosclerosis develops when **plaque** (cellular waste products, calcium, cholesterol, and fats) builds up in the inner lining of an artery (Figure 11.21). Many resources use the term **arteriosclerosis** to explain conditions in which calcium deposits form in the vessels. They reserve the term atherosclerosis for fat deposits in the arterial walls. Immune system cells within the vessel lining accumulate these materials, which results in a narrowing of the lumen of the artery. Narrowing of the arteries restricts blood flow, resulting in tissue damage or death. The heart in particular can be harmed by atherosclerosis. There is strong evidence that certain genetic conditions produce the ability of the

Arrhythmia Irregular rhythmic beating of the heart

Atherosclerosis A progressive narrowing and hardening of arteries due to plaque formation

Plaque Hardened deposits that form on the inner walls of blood vessels]

Arteriosclerosis A gradual stiffening of arterial walls due to aging

Figure 11.21 **Normal and Blocked Arteries**

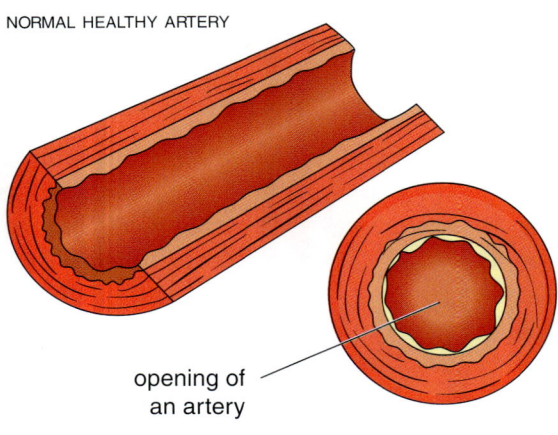

opening of an artery

narrowed opening of an artery

immune cells to take up these materials. Diets high in cholesterol and fats can aggravate the condition by accelerating the formation of plaque and the subsequent narrowing of the artery. Early treatment includes changing one's lifestyle to reduce plaque formation. Surgery is often performed to reduce or remove the plaque. One procedure involves the insertion of a stent (a supporting mesh), which keeps the lumen of the vessel open (Figure 11.22). The stent is a hollow wire mesh tube that props open an artery. It is inserted in a collapsed form and then expanded using a balloon-type device. The stent is expanded to fit tightly against the inner wall of the blood vessel.

Bacterial **endocarditis** is caused by a bacterial infection that inflames the lining of the heart. Bacteria entering the blood can lodge in the heart and produce secretions that irritate the heart lining. It is not unusual for bacteria to enter the blood during dental work and injuries that produce bleeding. The condition can damage heart valves and produce irregular blood flow. Many cases of endocarditis can be prevented with antibiotics. Administering antibiotics is now a common practice during dental procedures. Unfortunately, there is little that can be done to remedy the effects of this disease once it has damaged the heart.

Congenital heart disease is a defect that occurs when the heart, or blood vessels near the heart, does not develop normally before birth. It can be caused by a variety of genetic conditions. There are at least 35 types of congenital heart conditions. Most congenital defects restrict blood flow to the heart

Endocarditis Inflammation of the inner lining of the heart
Congenital Heart Disease A heart problem present before birth

Figure 11.22 **Stent**

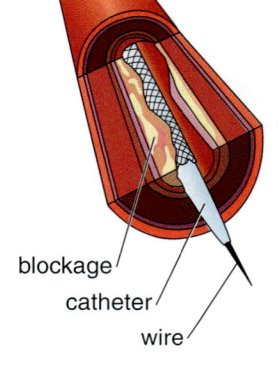 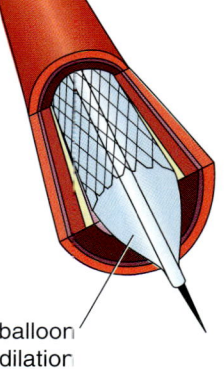 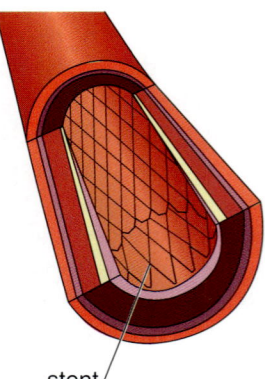

blockage
catheter
wire

balloon dilation

stent

Figure 11.23 Pathology of the Cardiovascular System

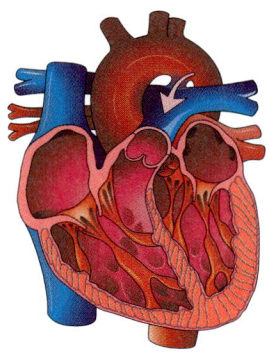

(a) patent ductus arteriosus (PDA)

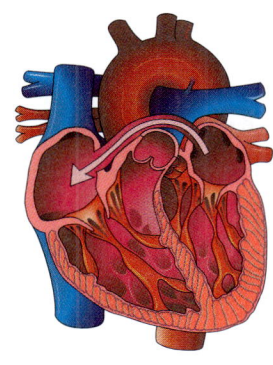

(b) atrial septal defect (ASD)

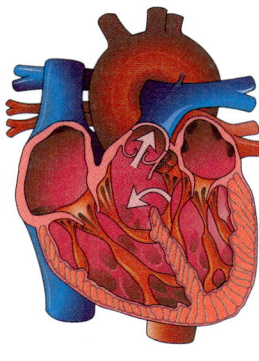

(c) ventricular septal defect (VSD)

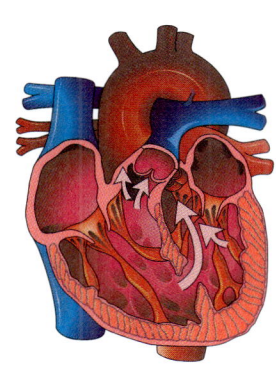

(d) tetralogy of Fallot

(Figure 11.23). This can result in abnormal heart tissue development or irregular heartbeat patterns. Currently, over 4000 people die each year in North America from complications of congenital heart defects. Treatments vary greatly with the type of condition.

Congestive heart failure (CHF) is a general term used to describe the heart's loss of pumping ability. Blood enters the heart faster than it can be pumped out during CHF. It can be caused by a variety of conditions, including diabetes, high blood pressure, and lung diseases. Physicians have to carefully study the underlying measures to prevent death from CHF. Usually, people with certain lifestyles or underlying conditions that promote CHF are presented strategies to prevent sudden death from the disease. For example, smokers are asked to reduce or stop smoking to lessen the chances of heart failure. Similarly, people with high pressure or stress are advised to avoid salty foods because salt elevates blood pressure, which makes the heart work harder.

An **enlarged heart** is caused by a thickening, or hypertrophy, of the heart muscles. In some cases, the heart can enlarge due to dilation of the ventricles. An enlarged heart can result from vascular disorders that overwork the heart. In addition, obesity and excessive exercise are two other ways of overworking the heart muscle. An enlarged heart is more susceptible to heart disease. Currently, there is no treatment.

Fibrillation is the rapid contraction of either the atria or ventricles of the heart. Atrial fibrillation has a variety of causes and is most likely to occur as a person ages. The rapid beating prevents proper emptying of the atria. This

Congestive Heart Failure (CHF) A condition in which the heart cannot pump out all of the blood entering the chambers

Enlarged Heart A condition in which the heart is larger than normal

Fibrillation Rapid contractions of the heart muscles

causes a pooling of blood in the atria, leading to the formation of blood clots. These clots can then lodge in the small vessels of the brain or lungs. Ventricular fibrillation is a more serious condition that can to lead to rapid heart failure. Rapid contractions of the ventricles reduce blood flow to the body because the ventricles have trouble filling with adequate amounts of blood. Treatment varies with the cause and the type of fibrillation. Currently, defibrillator devices used to reverse ventricular fibrillation are becoming a common component of the first-aid practices of emergency responders. Home defibrillators are also available. However, defibrillators save lives only if used by people trained to use the machines. First-aid training is also important when carrying out cardiac resuscitation.

Heart **murmurs** are usually the result of defective heart valves. The murmur is a sound the heart produces when a valve is not properly closing. Valve defects are caused by a variety of conditions, including fevers and pregnancy. In many cases, they are due to congenital heart conditions. Untreated murmurs can lead to heart failure. Treatments vary with the cause and severity of the condition causing the murmur.

Hypertension is most commonly known as high blood pressure. It is commonly indicated by blood pressure readings of 120 to 139 for the systolic pressure and 80 to 89 for the diastolic pressure on multiple readings There is evidence that one in three adults in North America have hypertension. It can be caused by a variety of congenital cardiovascular conditions, and by diseases of the kidney and lungs. However, lifestyle has an important role in causing hypertension. Improper diets, obesity, and smoking are linked to high blood pressure. Untreated high blood pressure can lead to other conditions including heart failure. Preventive measures and treatments vary greatly with the cause of the hypertension.

Pericarditis is an inflammation of the pericardium. Many factors can cause pericarditis. In most cases, the cause is not known, although it can occur after a bacterial infection, a heart attack, or heart surgery. Pericarditis can last for weeks or months. It rarely persists for more than a year. The condition produces chest pain and is usually accompanied by fever. Severe pericarditis can put pressure on the heart, restricting the ability of the chambers to pump blood. This squeezing effect on the heart is called **tamponade**. Pericarditis is usually treated with drugs that reduce the inflammation. It is most common in 20- to 50-year-old males.

Rheumatic heart disease is the result of a bacterial infection called rheumatic fever. The condition usually starts out as an infection in the throat and is commonly referred to as strep throat. If left untreated, the bacteria can enter the blood stream and damage various organs, including the heart. In many cases, the bacteria primarily damage the heart valves. The condition is becoming less common in North America, however, it is still prevalent in populations with little access to healthcare. It can be prevented with antibiotics; however, treatment varies with the severity of the heart damage.

Sudden cardiac death is caused by an abrupt loss of heart function. It is sometimes referred to as cardiac arrest or heart attack. Sudden cardiac death has many causes and can occur in people with no diagnosed heart disease. Symptoms appear only minutes before death, making it a difficult condition to prevent. It is most likely the result of long-term cardiovascular disorders. Atherosclerosis is believed to be the most common factor. Recently, sudden cardiac death has occurred in athletes immediately following physical activities. There is much concern that athletes today are overtraining and becoming

---

Murmur An abnormal sound of the heart

Hypertension Blood pressure above the normal range

Pericarditis Inflammation of the membrane that surrounds the heart

Tamponade Pressure on the heart that obstructs filling of the chambers

Rheumatic Heart Disease Damage to heart valves caused by a bacterial infection

Sudden Cardiac Death An abrupt and unexpected death due to a loss of heart function

susceptible to sudden cardiac death. Health officials are concerned that sudden cardiac death is more likely to happen in athletes who take performance-enhancing supplements, which were not readily available years ago.

A **thrombosis** is a **clot** that forms in blood vessels or the heart. Clots are plugs of proteins and blood cells that form at the wound site to stop bleeding. Thrombosis can be a dangerous condition if the clot breaks loose, and alters the flow of blood through blood vessels and the heart. Deep-vein thrombosis is a blood clot that develops in deep veins of the body (Figure 11.24). These clots usually form in the leg. A deep-vein thrombosis can cause pain in the affected body part. Sometimes these clots can break free and travel in the blood stream to the lungs. Deep-vein thrombosis is most common in people aged over 40 years who have sedentary lifestyles.

Prolapse refers to the stretching of the heart valves. It occurs most often in the mitral valve and is usually detected when a physician hears the murmur the condition produces. **Prolapse** mostly results in reduced blood-pumping capacity by the heart. Long-term valve prolapse can produce thickening of the affected ventricle. This occurs because of an incomplete closure of the valve, which results in **regurgitation**, or a back-flow of blood into the atrium. Mitral valve prolapse is a common heart problem in North America that affects almost 20% of the population. This condition is most likely to occur in females and may be linked to hormonal differences. Surgical procedures are used to repair the heart valve.

Thrombosis A blood clot that forms in a blood vessel or in the heart
Clot A plug that forms in the blood to stop bleeding
Prolapse Stretching of the heart valves
Regurgitation Backflow of blood through a heart valve

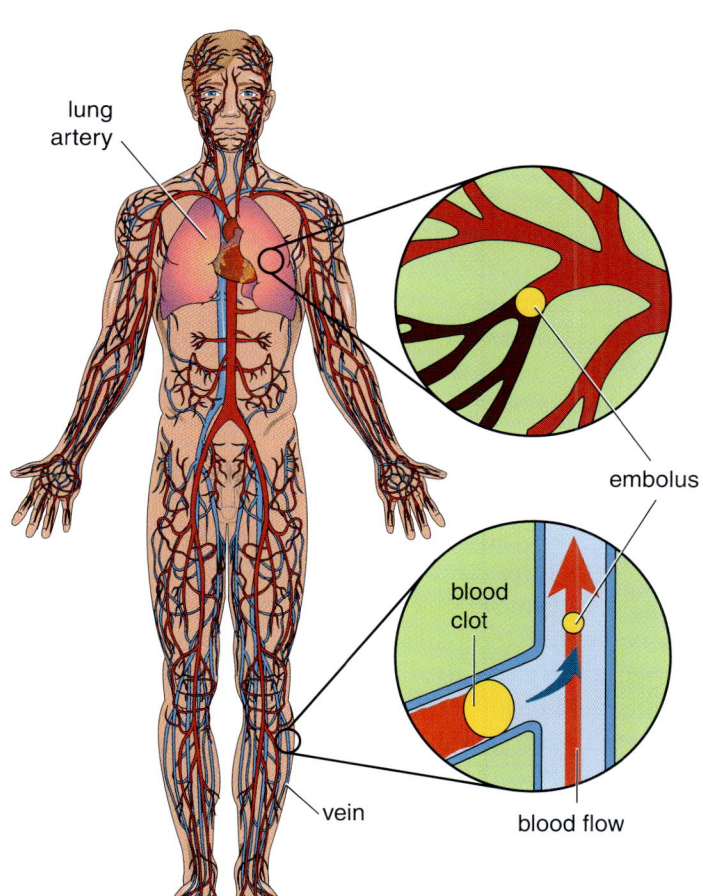

Figure 11.24 Deep Vein Thrombosis

Monitoring cardiovascular diseases is now becoming a high-tech science. Physicians today can place instruments that monitor a variety of cardiovascular functions on or within patients. The IBM corporation developed a heart monitor that can send a warning message to a preprogrammed cell phone number when it detects cardiac distress. The device transmits data in the form of a text message to the cell phone number of a physician or emergency medical care service. Monitors can be designed to detect blood pressure, blood flow patterns, blood gases, heart sounds, pulse, and enzymes associated with a heart attack.

✓ Concept Check

1. Describe the major types of cardiovascular disorders.
2. How do vascular diseases affect the function of the heart?
3. How do congenital cardiovascular disorders differ from those caused by lifestyle?

Aging of the Cardiovascular System

Key Terms: arterial stiffness, cardiovagal baroreflex, dyspnea, maximal heart rate, varicose

There is an abundance of research documenting the differences in cardiovascular function and structure between older and younger people. Most of these changes that occur throughout the life span make the cardiovascular system more susceptible to failure and malfunction. The research shows that much of the cardiovascular aging seen in many people is not due to the blood vessels and heart wearing out. Rather, most cardiovascular aging is caused by interactions between age, disease, and lifestyle. It is now believed that the greater occurrences of vascular diseases and heart failure associated with aging are the result of longer exposure to the risk factors for these diseases than to degeneration of body system structures.

Arterial stiffness is one of the major progressive changes that take place during aging. The arteries of young people are very flexible and stretch slightly as the ventricles pump blood through them. This stretching reduces the amount of work the heart needs to do to move blood through the arteries. Arteries undergo a loss of elastin in their connective-tissue composition as a person ages. This makes the vessels stiffer, which produces resistance to blood flow and increases blood pressure. In turn, the heart has to work harder as a person ages, making it more prone to cardiac disorders. In addition, it makes the blood vessels more susceptible to atherosclerosis and arteriosclerosis. Veins become **varicose**, or stretched out, when their vessel walls weaken. This reduces their ability to carry blood back to the heart (Figure 11.25). Elastin also decreases in the skin and connective tissue of other organ systems. It appears that age-related changes in the genetic material cause this decrease.

Maximal heart rate also naturally decreases with age. The highest heart rate reached during exercise is called the maximal heart rate. Young children can achieve a maximal heart rate of 220 beats per minute. This rate decreases to 160 beats per minute by age 60 years. The reduction in maximal heart rate is believed to be caused by changes to the cardiac electrical conduction sys-

Arterial Stiffness A progressive loss of flexibility of the arterial walls

Varicose Refers to a condition of being dilated or swollen

Maximal Heart Rate The fastest heart rate possible under normal, maximal exercise conditions

tem. Research also shows that an inability to regulate cardiac muscle receptors decreases the heart's sensitivity to neurotransmitter stimulation. The inability to raise the heart rate makes it difficult for the body to coordinate heart rate with body activity, thereby reducing blood flow to muscle and other organs. These changes cannot be fully explained by a person's lifestyle. It is assumed that the changes are due to alterations in DNA structure over time. There is ample research showing that the DNA of highly active cells, such as cardiac muscle, is damaged by metabolic waste products. Active cells damage the DNA faster than do other cells because they produce high concentrations of metabolic wastes.

Ventricles also thicken with time. This natural event is caused by hypertrophy of cardiac muscle, which is a response to years of pumping blood. Ventricular hypertrophy reduces the amount of blood being pumped by the heart. The size of the atria also increases because they swell with blood that is not able to readily enter the ventricles. Enlarged atria are more subject to fibrillation. These changes, in turn, can produce breathing difficulties known as **dyspnea**. Thickening of the ventricular walls also increases blood pressure and aggravates the problems caused by arterial stiffness. Researchers also note that the type of myosin making up cardiac muscle also changes as DNA is damaged. This reduces the pumping action of the ventricles, further decreasing blood flow. Along with this is a decrease in the **cardiovagal baroreflex**. This reflex permits the heart to regulate its pumping strength to the demands of the body. A decrease in the cardiovagal baroreflex makes the heart more susceptible to heart failure.

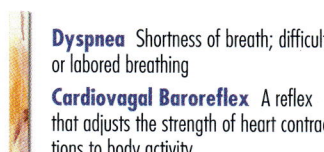

Dyspnea Shortness of breath; difficult or labored breathing

Cardiovagal Baroreflex A reflex that adjusts the strength of heart contractions to body activity

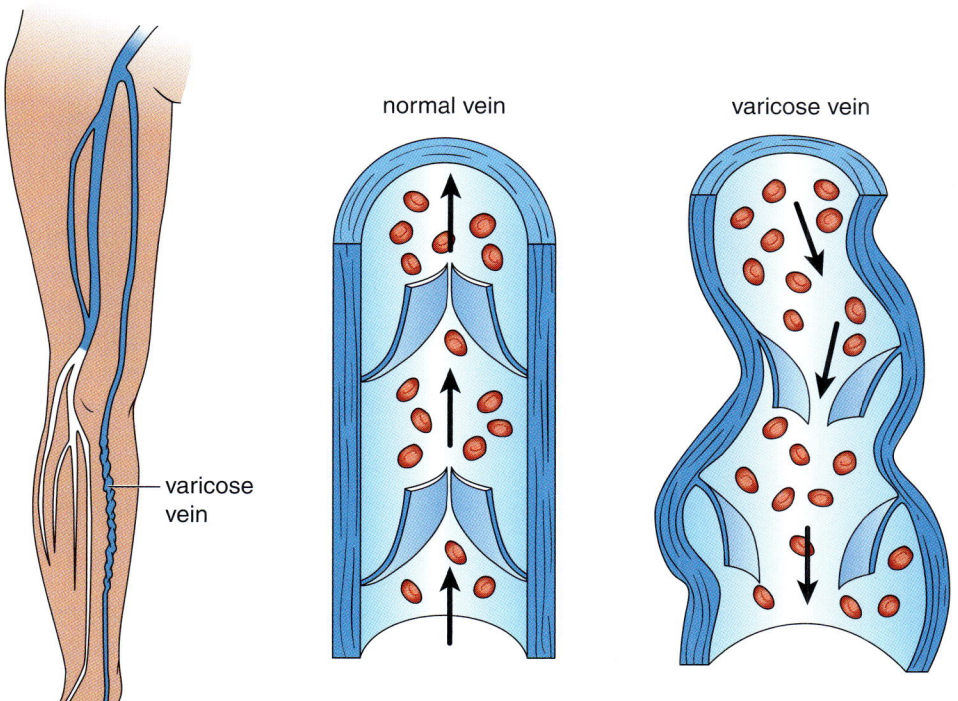

Figure 11.25 Varicose Vein

Scientists now know that the progressive changes that decrease cardiovascular function can be altered by a change in lifestyle. Imbalanced diets, a lack of exercise, obesity, and smoking accelerate degenerative change to the blood vessels and the heart. In particular, diets high in fats and proteins contribute to atherosclerosis and arteriosclerosis. Physical conditioning slows cardiovascular aging by increasing cardiac output and improving vascularization of the heart. Studies now show that older people in good physical condition can match the cardiovascular capacity of younger people who do not exercise regularly. Moderate exercise is the most important factor in maintaining cardiovascular health.

✓ Concept Check

1. What are the primary causes of cardiovascular system aging?
2. What role does genetics play in aging of the cardiovascular system?
3. What is the role of lifestyle in cardiovascular aging?

DISCOVERY SCENE PLEASE ENTER DISCOVERY SCENE PLEASE ENTER

Have you resolved the CSI yet? Did the information about the pathology and aging of the cardiovascular system provide any more clues about the young runner's condition?

CSI — Case Study Investigation Conclusion

What can you conclude about the reason the runner passed out during the race? Was it a vascular problem or a disease of the heart? Why didn't the runner know that he had such a potentially harmful condition?

Answer:

The runner was unknowingly suffering from mitral stenosis. His mitral valve opening was narrowed and restricted blood flow from the left atrium into the left ventricle, which reduced blood flow to the brain and body. The condition is most prominent during heavy exertion, such as running in a race. He could have developed the condition from a bacterial infection that inflames or distorts the mitral valve. Mitral stenosis is almost always the result of rheumatic fever caused by bacteria. It could also have been a genetic condition that slowly developed and did not cause problems until this incident. Symptoms of mitral valve stenosis usually develop between the ages of 20 and 50 years.

This CSI was adapted from the following article:

1. Carabello BA, Crawford FA. Valvular heart disease. *N Engl J Med.* 1997;337(1):32-41.
2. Gilon D, Buonanno FS, Joffe MM, Leavitt M, Marshall JE, Kistler JP, et al.. Lack of evidence of an association between mitral-valve prolapse and stroke in young patients. *N Engl J Med.* 1999;341(1):8-13.
3. Rahimtoola SH, l. Mitral stenosis section of mitral valve disease. In: Fuster, V, O'Rourke, RA, Roberts, R. King, SB, Prystowsky, EN. *Hurst's The Heart.* 10th ed. New York, NY: McGraw-Hill. 2001;2:1697–1707.

Chapter Summary

The cardiovascular system is responsible for distributing such resources as nutrients and oxygen to the other organ systems. Its ability to do this depends on the proper functioning of blood vessels and the heart. The blood vessels ensure the distribution of resources based on the needs of the particular tissue or organ. Various chemical signals alter the diameter of the blood vessels causing them to constrict or dilate. This then diverts the blood flow. Constricting a blood vessel diverts blood away from the structure being served by that vessel. Dilating an artery diverts blood flow to the structure it serves. Veins are essential for returning "used blood" to the heart to replenish nutrients and remove metabolic wastes. The heart relies on nervous system impulses and coordinated signals from the heart's conduction system. Cardiovascular diseases are the most common cause of death in North America. Some cardiovascular degeneration is due to genetic changes that occur with age. However, lifestyle is the major contributing factor to cardiovascular system aging.

Overview

- The cardiovascular system is composed of the heart and blood vessels.
- Blood vessels form the circulatory system.
- Pulse is an indicator of heartbeat.
- Blood pressure is produced by the heartbeat.

Circulatory System Vessels

- Arteries and arterioles carry blood from the heart to the rest of the body.
- Veins and venules carry blood from the body to the heart.
- Capillaries are small blood vessels that exchange materials with tissues.
- Hydrostatic pressure permits the exchange of materials between the blood and tissues.
- Vasoconstriction is the narrowing of a vessel; vasodilation is the widening of a vessel.

Structure of the Human Heart

- The heart is made of myocardium covered by a protective pericardium.
- The heart is composed of four chambers that are separated by a septum into two halves.
- The left half of the heart controls systemic circulation.
- The right half of the heart controls pulmonary circulation.
- Four valves facilitate the pumping of blood by the heart chambers.
- The four major blood vessels are the aorta, pulmonary artery, pulmonary veins, and venae cavae or "vena cava."
- The electrical conduction system of the heart stimulates and coordinates the heartbeat.
- The fetal heart is designed to keep blood away from the lungs.

Human Heart Function

- One pumping action of the heart is a called the cardiac cycle.
- Diastole is the filling of the atria and ventricles; systole is the emptying of the ventricles.
- Heart rate is the number of cycles per minute.
- Stroke volume is the amount of blood pumped per cycle.
- Cardiac output is the volume of blood pumped per minute.

Study Guide

Electrocardiography Basics
- Electrocardiography measures the electrical activity of the heart.
- An ECG is a recording of electrocardiography.
- ECGs are recorded as waves.
- The P, Q, R, S, and T waves make up the wave of a cardiac cycle
- The QRS complex, P-R interval, Q-T interval, and S-T intervals are indicators of depolarization and repolarization.

Pathology of the Cardiovascular System
- Diseases of the cardiovascular system affect either blood vessels or the heart.
- Cardiovascular diseases are either congenital, or produced by lifestyle factors and microorganisms.
- Common vascular diseases disrupt blood flow.
- Common heart diseases prevent the chambers and/or valves from working properly.

Aging of the Cardiovascular System
- Lifestyle plays a significant role in cardiovascular aging.
- Arterial stiffness is a common event associated with cardiovascular system aging.
- The heart becomes more susceptible to damage as a person ages.

 ## Key Terms

Overview
Angiogenesis factors
Blood pressure
Blood vessel
Cardiovascular
Circulatory system
Heart
Pulse
Vascularization

Circulatory System Vessels
Artery
Hydrostatic pressure
Lymphatic vessel
Microcirculation
Vein

Arteries and Veins
Constriction
Dilation
Lumen
Tunica adventitia
Tunica intima
Tunica media

Vasoconstriction
Vasodilation

Small Vessels and Capillaries
Arteriole
Capillary
Continuous capillary
Fenestrated capillary
Venule

Structure of the Human Heart
Cardiac infarction
Cardiac ischemia
Coronary artery
Coronary vein
Endocardium
Epicardium
Fibrous pericardium
Myocardium
Pericardium
Pulmonary circulation
Serous pericardium
Systemic circulation
Visceral layer

Study Guide

Adult Heart
Aorta
Aortic valve
Atrioventricular (AV)
Atrioventricular node
Atrium
Biscuspid
Bundle of His
Chamber
Chordae tendineae
Electrical conduction system
Inferior vena cava
Mitral
Papillary muscles
Pulmonary artery
Pulmonary valve
Pulmonary vein
Purkinje system
Semilunar valve
Septum
Sinoatrial (SA) node
Superior vena cava
Tricuspid
Vasculature
Vena cava or Venae cavae
Ventricle

Fetal Heart
Ductus arteriosis
Foramen ovale

Human Heart Function
Cardiac cycle
Cardiac output
Diastole
Heart rate
Stroke volume
Systole

Electrocardiography Basics
Electrocardiogram (ECG)
Electrocardiography
P-R interval
P wave
Q wave
QRS complex
Q-T interval
R wave
S wave
S-T interval
T wave

Pathology of the Cardiovascular System
Aneurysm
Angina pectoris
Arrhythmia
Atherosclerosis
Arteriosclerosis
Cardiac disorders
Clot
Congenital heart disease
Endocarditis
Enlarged heart
Fibrillation
Hypertension
Murmur
Pericarditis
Plaque
Prolapse
Regurgitation
Rheumatic heart disease
Sudden cardiac death
Tamponade
Thrombosis
Vascular disorders

Aging of the Cardiovascular System
Arterial stiffness
Cardiovagal baroreflex
Dyspnea
Maximal heart rate
Varicose

Study Guide

Check Your Understanding

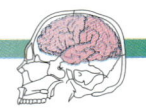

1. Most of the exchange between the blood and tissues takes place across these vessels:
   a. capillaries
   b. arteries
   c. veins
   d. arterioles

2. Blood exits the heart through:
   a. veins
   b. arteries
   c. venules
   d. arterioles

3. A blockage of this vessel would prevent a tissue from receiving blood:
   a. venule
   b. lymph vessel
   c. fenestra
   d. arteriole

4. Valves prevent a backup of blood flow in the following vessels:
   a. arteries
   b. veins
   c. arterioles
   d. capillaries

5. Blood flow to a tissue decreases when an artery supplying that tissue undergoes the following:
   a. countercurrent flow
   b. contraction
   c. vasodilation
   d. vasoconstriction

6. Which blood vessels have the thickest muscle lining?
   a. veins
   b. venules
   c. arteries
   d. arterioles

7. Which of the following blood vessels do not have three tissue layers?
   a. veins
   b. venules
   c. capillaries
   d. arteries

8. Blood enters the heart through the:
   a. atria
   b. septum
   c. ventricles
   d. myocardium

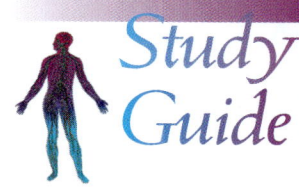

9. The protective covering of the heart is called the:
   a. epicardium
   b. pericardium
   c. myocardium
   d. endocardium

10. Atria pump blood across these valves:
    a. semilunar
    b. aortic
    c. pulmonary
    d. mitral and tricuspid

11. Blood from the left ventricle enters the:
    a. aorta
    b. pulmonary artery
    c. pulmonary vein
    d. inferior vena cava

12. Which component of the electrical conduction system initiates a heartbeat?
    a. the atrioventricular (AV) node
    b. the sinoatrial (SA) node
    c. the bundle of His
    d. the Purkinje system

13. This ECG feature indicates abnormalities of ventricular depolarization:
    a. T wave
    b. P wave
    c. QRS complex
    d. P-Q interval

14. The volume of blood leaving the heart per minute is called:
    a. stroke volume
    b. cardiac output
    c. heart rate
    d. cardiac ischemia

15. Parasympathetic stimulation would have the following effect on the heart:
    a. increase stroke volume
    b. decrease heart rate
    c. decrease coronary flow
    d. increase blood pressure

Study Guide

A Case Study

SMOKING AND HEART DISEASE

The Centers for Disease Control and Prevention in Atlanta, Georgia, estimates that in 2002 almost 22% of the American population smoked a tobacco product. This number is down from the 1965 data, which showed that 41.9% of people smoked. From 1995 to 2002, the decrease in smoking was significantly greater among males than among females. The American Heart Association claims that cigarette smoking is the most preventable cause of premature death in the United States. The data show that cigarette smoking is responsible for upwards of 440,000 of the more than 2.4 million annual deaths in America. Many public health authorities cite the results of research studies that implicate cigarette smoking as a major cause of heart diseases that lead to heart attack. Tobacco companies accept the fact that cigarette smoking poses health risks; however, they believe that research about these hazards does not reflect the newer "safe cigarettes" being sold. The tobacco companies are currently researching new types of cigarettes that further reduce the hazards of smoking.

Use the information in this chapter and the following Web sites to answer the following questions about the effects of smoking on the heart:

- How does cigarette smoking damage the heart?
- How strong is the evidence linking smoking to fatal heart attacks?
- Who is responsible for paying the medical bills of people who have been harmed by smoking?
- Should the government ban smoking in public areas because of the health risks of exposure to second-hand smoke?
- What are the rights of nonsmokers who fear the risks of exposure to second-hand smoke?

Do people have the right to smoke even if they know it could harm their health, and cost the government and insurance companies money for their medical care?

1. NOVA: "Safer" Cigarettes: A History
   http://www.pbs.org/wgbh/nova/cigarette/history.html

2. BBC News
   http://news.bbc.co.uk/1/hi/health/background_briefings/smoking/289211.stm

3. National Institutes of Health; Smoking and Heart Disease Document
   http://ehp.niehs.nih.gov/members/1999/suppl-6/841-846thun/thun-full.html

4. Texas Heart Institute
   http://www.texasheartinstitute.org/riskfact.html

CHAPTER 11

Where Do We Go from Here?

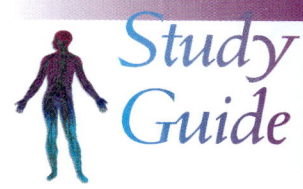

People in health fields can use their knowledge of the cardiovascular system to solve everyday problems. You may wish to use other resources, such as the Web sites listed below, in addition to your textbook to investigate the answers to each of the following situations:

1. A relative asked your advice about diets that reduce the risk for heart disease. What information could you give him about the benefits and risks of various diets designed to control heart disease?
   Medline Plus: Diet and Heart Disease
   http://www.nlm.nih.gov/medlineplus/ency/article/002436.htm

2. A female friend has a history of cardiovascular disease in her family. She asks you if it safe for her to use birth control pills, or whether she should use an alternative method of contraception. Her fear is that birth control pills will aggravate any potential cardiovascular problems.
   American Hearth Association (Search "birth control pill")
   http://www.americanheart.org/

3. A young boy living next door to you tells you he was diagnosed with a heart murmur. He asks you if this puts an end to his dreams of playing high-school football. What advice would you give him?
   KidsHealth: Heart Murmurs
   http://kidshealth.org/parent/medical/heart/murmurs.html

4. A neighbor heard on the news that drinking red wine reduces the chances of developing cardiovascular disease. She asks your opinion about the claim. How would you respond?
   Virtual Hospital (Search "red wine heart")
   http://www.vh.org/index.html

5. A classmate told you that heart disease would not be a problem in the future because of a new therapy called angiogenesis. How would you respond to his assertion?
   The Angiogenesis Foundation
   http://www.angio.org/

Skills Activities

1 Using a Stethoscope to Identify Heart Sounds

Physicians listen to heart sounds because they serve as an indicator of cardiovascular disorders. A trained physician can listen to the sounds of blood flowing through the heart to determine whether there are slight abnormalities in the atria, ventricles, and/or valves. Physicians can correlate heart sounds with the waves of an ECG to get a full picture of a person's cardiac health. This activity provides practice listening to both abnormal and normal heart sounds.

Study Guide

Materials
- alcohol swabs
- stethoscope
- stopwatch or clock with second hand
- access to BioScience.org—Heat Sounds and ECG
  http://www.bioscience.org/atlases/heart/

Activity

1. Sit with a laboratory partner at a workstation.
2. Obtain a stethoscope, stopwatch, or timing device, and alcohol swabs.
3. Wipe the ear pieces of the stethoscope with an alcohol swab. Let the alcohol dry and then place the ear pieces in the ears.
4. Have your laboratory partner place the bell of the stethoscope on his or her chest just over the heart.
5. Listen for the heart sounds, and ask your laboratory partner to move the bell around until you hear the sounds clearly. Notice the two-beat rhythm of the heartbeat.
6. You will hear a pattern of "lubb-dupp" sounds that are recognized as the heartbeat cycle. These sounds are produced when the heart valves close.
7. Identify the first sound. It is the sound of atrial contraction being completed as the valves close. Ventricular contraction begins at this point.
8. Next, listen for the second sound. It marks the end of ventricular contraction as the valves leading to the aorta and pulmonary arteries closes.
9. Use the stopwatch or timer to measure the time interval between the first and second sounds.
10. Time the interval between the second sound and the next first sound. After five of these intervals, estimate the average time the heart is at rest in 1 minute, and the average time the heart is in contraction in 1 minute.
11. Repeat steps 3 through 10 for the other lab partner.
12. Go to the BioScience.org Web site. Scroll down to the "HEART SOUND" section, and click on the "Normal" link.
13. Determine whether the sounds you heard and measured from your lab partner match the normal heart sounds from the recording.
14. Next, try to identify how the sounds of the abnormal heart differ from those of the normal heart.

Skills Activities

2 Identification of Venous Valves

Venous valves ensure that the blood returning to the heart does not flow back to the various parts of the body. Certain venous disorders prevent the valves from doing their job. This can cause a variety of problems ranging from impaired blood flow to a buildup of fluids in various organs and limbs. This activity will guide you through a simple test to investigate the integrity of the venous system. Proper functioning of the venous system depends on the ability of the valves to prevent reverse blood flow.

Activity

1. If wearing a long-sleeve shirt, roll up the sleeve of one arm to expose the veins on the forearm.
2. Now, hold your arm with the hand down, and clench your fist several times until the veins appear larger.
3. Place the forefinger of your other hand on one of the larger veins, and push your thumb along the vein toward your shoulder. Observe whether part of the vein "disappears" as the blood is milked out of the vein.
4. Remove your thumb, leaving your forefinger in place on the vein. Observe whether the blood flows back into the vein. Blood should not flow back if the valve is working properly.
5. Now, remove your forefinger and observe what happens.
6. Place your finger on one of the veins (preferably the same vein), and push your thumb along the vein toward your hand. Leave your finger in place, and observe whether the blood flows back into the vein. Now, remove your finger, and observe what happens.
7. Based on your observations, predict where the valves are in the vein.

Case Study Investigation

Case Study Investigation #12

A friend's father has been spending his retirement restoring a 1942 Ford pickup truck. He started working on the truck 4 years ago and spent the first 2 years stripping off the old paint and cleaning the engine components using a variety of chemicals. He works on the truck in the enclosed garage, sometimes with the engine running for short periods of time. He feels stressed because he is behind schedule in getting the truck ready for an upcoming car show in the area. Recently, your friend's father visited his physician because of persistent headaches and sore throats. He also complained of fatigue and was catching colds more frequently than usual. The physician admitted the father to the hospital because of an unusual red blood cell (RBC) count and white blood cell (WBC) abnormalities. Your friend is very concerned about her father's health and she is seeking your advice about his illness. What can you tell her about the probable cause of her father's ailments?

CHAPTER 12

The Lymphatic System and the Blood

Chapter Outline

Case Study Investigation (CSI)
Applied Learning Outcomes
Overview
Blood Cells
    Red Blood Cells
    White Blood Cells
    Platelets
Blood Cell Function
    Red Blood Cells
    White Blood Cells
    Platelets
Blood Cell Formation
Lymphatic System
    Structures of the Lymphatic System

Immune Response
    Innate Immunity
    Acquired Immunity
        Primary and Secondary Response
        Inflammation
Immunization and Vaccination
Wellness and Illness over the Life Span
    Pathology of the Blood and Lymphatic System
    Aging of the Blood and Lymphatic System
CSI Conclusion
Study Guide

Applied Learning Outcomes

- Use the terminology associated with the blood and lymphatic system.
- Learn about the following:
  - blood components
  - lymphatic system components
  - immune system function
  - mechanisms of immunization and vaccination
- Understand the aging and pathology of the lymphatic system.

Overview

Key Terms: centrifuge, erythrocyte, hematocrit, leukocyte, packed cell volume (PCV), plasma, platelet, red blood cell (RBC), thrombocyte

Plasma The fluid portion of the blood

Erythrocytes or Red Blood Cells (RBCs) Blood cells responsible for transporting oxygen throughout the body

Leukocytes or White Blood Cells (WBCs) Colorless blood cells that fight disease and maintain immune function

Thrombocytes or Platelets Blood cell fragments that control bleeding

Hematocrit or packed cell volume (PCV) The percentage of packed RBCs in a unit volume of whole blood

Centrifuge A machine that rapidly spins liquid samples and separates the materials by density

Blood is sometimes called the "fluid of life" because of the many vital functions it carries out for the body. An embryo lacks blood until week 5 of development. Pockets of special cells start to produce blood 2 weeks after the blood vessels start to form from the mesoderm. Blood is classified as a connective tissue, as it connects and supports the functions of other tissues. Connective tissues are also defined by their structure, which is composed of a matrix surrounding a dispersed arrangement of cells (Figure 12.1). Unlike other connective tissues, which have a solid or semisolid matrix material, the matrix of blood is a watery material called **plasma**. Plasma is the clear, yellowish fluid portion of the blood through which cells and other materials are transported. It is composed of approximately 90% water, 7% protein, 1% minerals, and 2% other materials, including atmospheric gases, chemical signals, and nutrients. The primary blood gases are oxygen, carbon dioxide, and nitrogen. Many of the blood proteins carry out functions that assist with healing and immune system activities. Approximately 55% of blood volume is plasma, and the remaining 45% is composed of formed elements. There are three categories of blood cells: **erythrocytes** (**red blood cells or RBCs**), **leukocytes** (**white blood cells or WBCs**), and **thrombocytes** (**platelets**). A measurement called a **hematocrit**, or **packed cell volume** (**PCV**), calculates the volume of RBCs making up blood. Hematocrit is determined by placing the blood in a tube and spinning it rapidly in a machine called a **centrifuge**. This packs the cells into a solid pellet, making it easy to measure the total volume. Blood cells are sometimes called formed elements. The packed volume of WBCs and platelets can be similarly determined.

The lymphatic system is made up of tissues and organs that produce, store, and carry lymphocytes throughout the body. This system is responsible for fighting infection and controlling the fluid levels in tissues. The system includes the bone marrow, specialized organs, and a network of thin tubes. The tubes branch out to all the tissues of the body in a similar manner to blood vessels. The lymphatic system begins to develop from mesoderm around week 5 of embryological development. Its full growth and function are not complete until after birth.

✓ Concept Check

1. Describe the composition of blood.
2. Define hematocrit.
3. Explain the basic structure and function of the lymphatic system.

Figure 12.1 Components of Blood

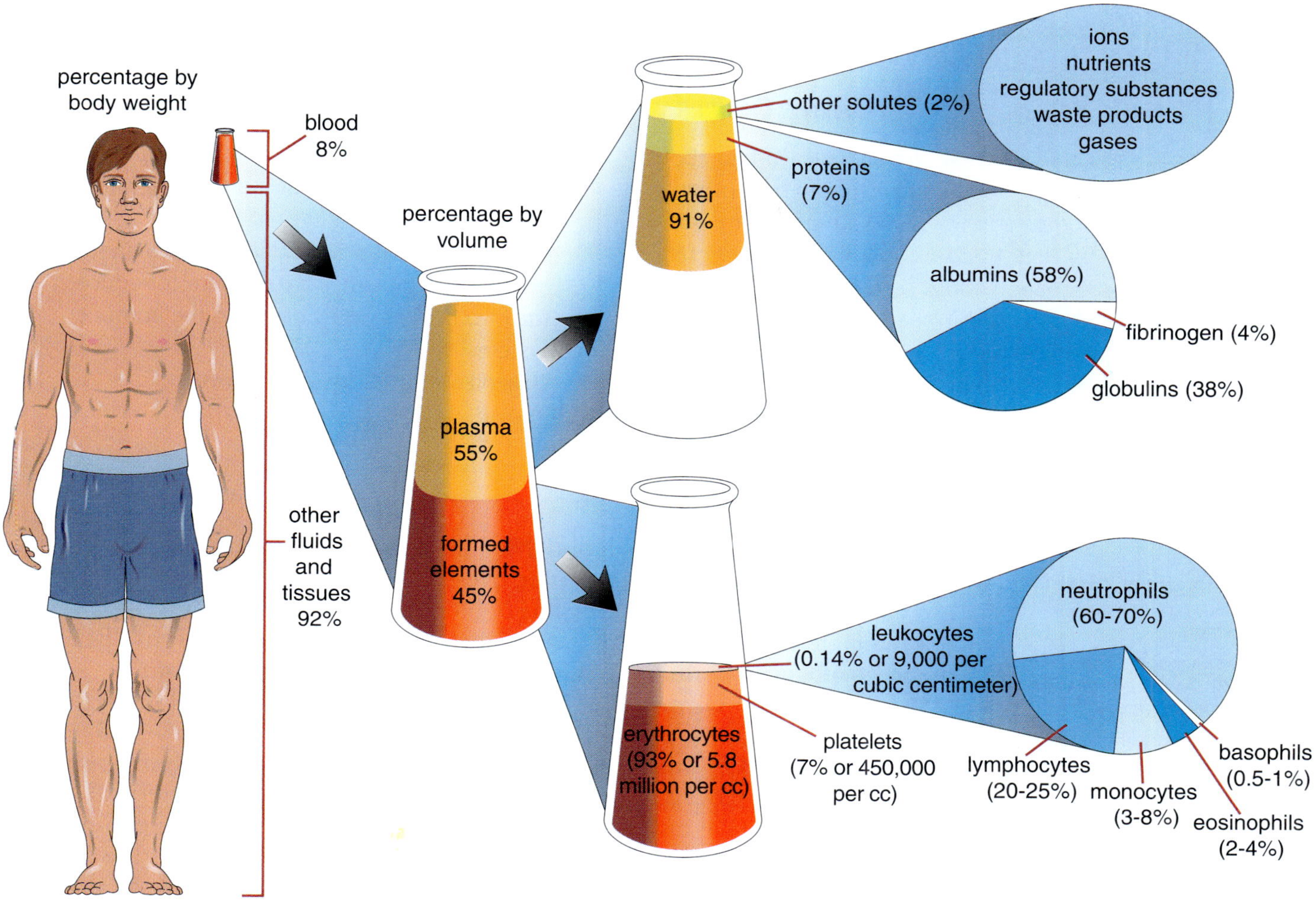

Blood Cells

Red Blood Cells (RBCs)

Key Terms: ABO blood group system, blood type, complete blood count (CBC), erythroblastosis fetalis, hemoglobin, full blood count (FBC), hemogram, mean corpuscular hemoglobin (MCH), mean corpuscular hemoglobin concentration (MCHC), mean corpuscular volume (MCV), reticulocyte, Rh factor, transfusion

Red blood cells are atypical in that the mature cells lack a nucleus. This means they have no DNA, and they carry out their tasks using the enzymes produced when they were developing RBCs called **reticulocytes**. Their lack of a nucleus causes them to live only a maximum of 120 days. They have no way to repair and replace damaged or expended cellular components. The name "red blood cell" comes from the fact that they appear red when viewed through the microscope. These disc-shaped small cells contain a protein called **hemoglobin**, which is bright red (Figure 12.2). Hemoglobin contains iron, which facilitates the transport of oxygen and carbon dioxide. Women average about 4.8 million

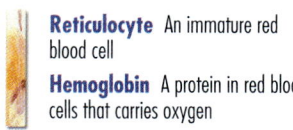

Reticulocyte An immature red blood cell

Hemoglobin A protein in red blood cells that carries oxygen

THE LYMPHATIC SYSTEM AND THE BLOOD

Civic Responsibility

CIVIC RESPONSIBILITY: HELPING OTHERS WITH YOUR KNOWLEDGE

It is valuable to use what you have learned about blood and the lymphatic system to help others to better understand the world around them. It is very important to check your facts and seek further information about certain topics before discussing health and science issues. Here are some suggestions to foster a better public awareness of blood and the lymphatic system:

1. Volunteer to do education programs on such topics as HIV at an area clinic or hospital.
2. Speak to children in an after-school group about the role proper nutrition plays in keeping the immune system healthy.
3. Provide information to elderly persons about nutritional supplements that claim to reduce the effects of aging by boosting the immune system.
4. Organize a Red Cross blood drive.

Mean Corpuscular Volume (MCV) A measure of the average volume of red blood cells

Mean Corpuscular Hemoglobin (MCH) The mass of hemoglobin molecules in each red blood cell

Mean Corpuscular Hemoglobin Concentration (MCHC) The average concentration of hemoglobin in red blood cells

Complete Blood Cell Count (CBC) or Full Blood Cell Count (FCB) A series of tests that examine components of the blood

Hemogram Another name for complete blood cell count

RBCs per cubic millimeter of blood. Men have about 5.4 million RBCs per cubic millimeter. This means there are over 23 trillion RBCs circulating in the average person. The hematocrit value of blood is slightly below 45%.

Physicians use three types of measurements to monitor the health of RBCs: **mean corpuscular volume** (**MCV**), **mean corpuscular hemoglobin** (**MCH**), and **mean corpuscular hemoglobin concentration** (**MCHC**). The three measurements are part of the **complete blood count** (**CBC**), or **full blood count** (**FBC**), a test that provides information about the cells in a patient's blood. A CBC is also known as a **hemogram**. These characteristics are measured by a clinical instrument called a blood cell analyzer that determines the values along with other blood measurements. The MCV indicates the cytoplasmic volume of the RBCs. A normal range is typically 80 to 96 femtoliters (fL) per RBC; 1 trillion fL equal 1 milliliter (mL). It would take the cytoplasm of 18 trillion RBCs to fill a teaspoon. The MCH value is the mass of hemoglobin molecules in an average RBC. The MCHC measures the concentration of hemoglobin in an average RBC. This measurement is determined by the intensity of the red color. These values help in the diagnosis of different types of RBC disorders.

Figure 12.2 Red Blood Cells

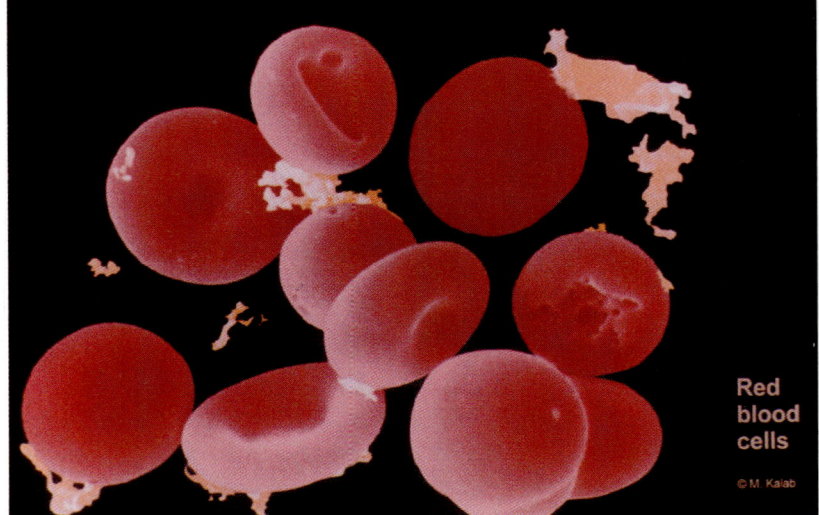

Blood type is another feature of RBCs. Blood type is a way of categorizing RBCs according to variations in proteins on the cell membrane surface. These proteins are called agglutinogens and can be classified as types A, B, and D. The A and B proteins make up what is called the **ABO blood group system**. Each person receives one ABO blood group protein gene from each parent. People with type O blood are missing the genes for the types A and B proteins. People who are type A have one or two genes for the type A protein. Similarly, people with type B blood have one or two genes for the type B protein. The D protein is called the **Rh factor**. People with Rh-negative blood lack the Rh factor genes. Blood cells that are Rh-positive have one or two of the D protein genes. The immune system is capable of attacking blood that does not match the ABO blood group type and the Rh type of the person. A person with O- blood will have an immune response against every other blood type. However, a person with AB+ will not produce an immune response against any other blood type. The possibility of immune attack against blood poses difficulties when people require blood for surgical procedures or **transfusions**. A serious condition called **erythroblastosis fetalis** occurs when a pregnant woman's body produces an immune response against Rh proteins on the RBCs of the developing fetus. This can be fatal to a newly born baby and, in severe cases, requires a blood transfusion. It can be prevented if the mother takes a drug called Rhogam during and after pregnanacy.

> **Blood Type** Classification of a red blood cell according to the composition of proteins on its membrane
>
> **ABO Blood Group System** A classification system for the proteins on human red blood cells
>
> **Rh Factor** The type of D protein on the surface of red blood cells in some blood types
>
> **Transfusion** The transfer of blood from one person to another
>
> **Erythroblastosis Fetalis** An immune response against a fetus' red blood cells

White Blood Cells (WBCs)

> **Key Terms:** agranulocyte, B-lymphocyte, basophil, differential WBC count, eosinophil, granulocyte, lymphocyte, monocyte, mononuclear WBC, neutrophil, polymorphonuclear WBC, T-lymphocyte

The number of WBCs is measured as cells per cubic millimeter of blood. The total WBC count normally ranges between 4000 and 11,000 cells per cubic millimeter. Therefore, there is only one WBC for every 500 to 1000 RBCs. The WBCs use the blood as a way to move from the bone marrow to the regions of the body where they carry out their jobs. Most WBC functions occur when they leave the blood and enter other tissues. There are five types of WBCs: **neutrophils**, **lymphocytes**, **eosinophils**, **monocytes**, and **basophils**. The relative proportions of WBCs are as follows: neutrophils, 40% to 75 %; lymphocytes, 20% to 50 %; eosinophils, 5 %; monocytes, 1% to 5 %; and basophils, 0.5 %. A measurement called the **differential WBC count** evaluates the percentage of each WBC type. This measure is a very important indicator of particular causes of infectious diseases.

Neutrophils, eosinophils, and basophils are called **granulocytes**, or **polymorphonuclear WBCs** (Figure 12.3). They have noticeable granules in the cytoplasm that produce specialized secretions for fighting infections. The nucleus of these cells is polymorphic, or has an unusual shape. The shape of the nucleus may be due to the specialized functions

> **Neutrophil** The most common type of white blood cell
>
> **Lymphocyte** A white blood cell present in the blood and lymphatic system
>
> **Eosinophil** A blood cell associated with parasitic infections and allergies
>
> **Monocyte** A large white blood cell involved in immune defense
>
> **Basophil** A white blood cell associated with allergies
>
> **Differential White Blood Cell Count** A measure of the percentage of each white blood cell type
>
> **Granulocyte or Polymorphonuclear White Blood Cell** A WBC with conspicuous granules in the cytoplasm

Figure 12.3 Granulocytes

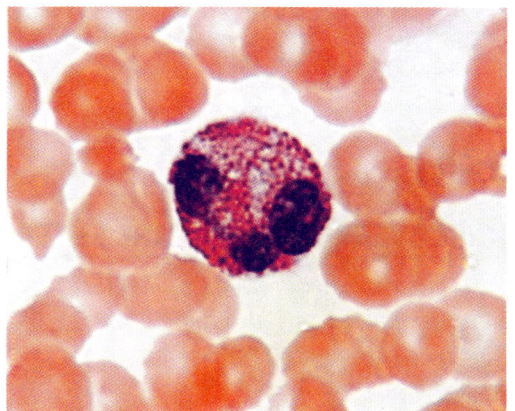

Figure 12.4 Lymphocytes

Figure 12.5 Monocytes

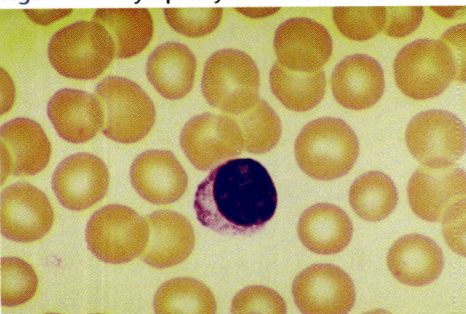
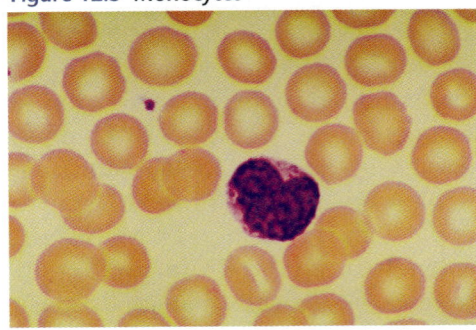

Agranulocyte or Mononuclear White Blood Cell A WBC that lacks granules

B-Lymphocyte A white blood cell that assists with the immune response

T-Lymphocyte A white blood cell responsible for stimulating the immune system

of these cells. **Agranulocytes**, or **mononuclear WBCs**, include the lymphocytes (Figure 12.4) and monocytes (Figure 12.5). Agranulocytes lack visible granules in their cytoplasms. Monocytes are the largest of the WBCs and have a U-shaped nucleus. Lymphocytes are divided into two functional types: **B-lymphocytes** and **T-lymphocytes**. Both of these cells play fundamental roles in the immune system.

Platelets

Key Term: megakaryocytes

Figure 12.6 Platelets

Megakaryocyte A bone marrow cell responsible for the production of blood platelets

Platelets are not true cells but instead are cell fragments derived from larger cells called **megakaryocytes** (Figure 12.6). One cubic millimeter of blood contains 250,000 to 500,000 platelets. The cell membranes of megakaryocytes are covered with many proteins that allow the platelets to stick to a variety of materials. This stickiness helps to reduce blood flow to a damaged area during injuries that break blood vessels. Platelets are sometimes called the "Band-Aids® of the blood" because their sticky properties reduce blood loss. Platelet formation is sensitive to many types of hazardous chemicals and pollutants. In addition, an abnormal number of platelets may indicate a bone marrow disorder, including certain types of cancers.

✓ Concept Check

1. Describe the structure of RBCs.
2. Explain the different types of WBCs.
3. Define the term thrombocyte.

DISCOVERY SCENE PLEASE ENTER DISCOVERY SCENE PLEASE ENTER

What information about the cells of the immune system is helpful for understanding the CSI? What types of blood cells are being affected by the condition? Is the ailment due to a blood cell disorder, or could the blood cell disorder be the result of another factor causing the condition?

Blood Cell Function

Red Blood Cells (RBCs)

Key Terms: acute, carbon dioxide intoxication, carbonic anhydrase, percent saturation, subacute

The main function of RBCs is to carry atmospheric gases throughout the body. It is intended that they carry oxygen from the lungs to the body cells and carbon dioxide from the body cells to the lungs. Oxygen enters the RBCs as they squeeze through the capillaries of the alveoli (Figure 12.7). The oxygen can only enter the cells if the concentration or partial pressure of oxygen is greater in the alveoli than in the cells. Once in the cytoplasm, the oxygen molecules attach to the hemoglobin. Four oxygen molecules bind to one hemoglobin molecule. The oxygen-carrying efficiency of hemoglobin is measured as **percent saturation** (Figure 12.8). Muscle myoglobin binds oxygen in a similar manner. However, myoglobin fills with oxygen faster than does hemoglobin. This gives hemoglobin the advantage of being able to collect oxygen even at low partial pressures. As the blood passes through the body's tissues, the hemoglobin releases the oxygen to the cells. Again, the partial pressure must be appropriate for the oxygen to diffuse into the cells. In addition, certain cellular metabolic waste products stimulate the release of the oxygen from the hemoglobin. This permits the body to provide more oxygen to the cells that have high metabolic needs.

Carbon dioxide is carried three ways in the blood. Approximately 10% of the carbon dioxide never enters the RBCs and is transported as a gas in the blood plasma. The empty hemoglobin molecules in the body tissue bond with the carbon dioxide and transport it away from the alveoli. Carbon dioxide and oxygen each attach to hemoglobin molecules in a different region. Hemoglobin transports about 22% of the carbon dioxide released by tissues.

Percent Saturation The amount of oxygen that is dissolved in a solution of hemoglobin molecules

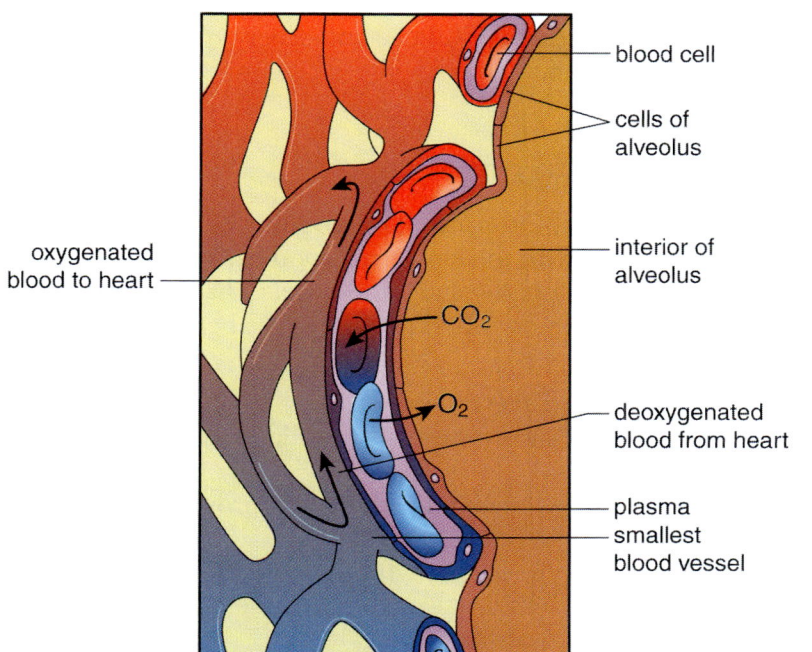

Figure 12.7 Red Blood Cell Function

Figure 12.8 Percent Saturation

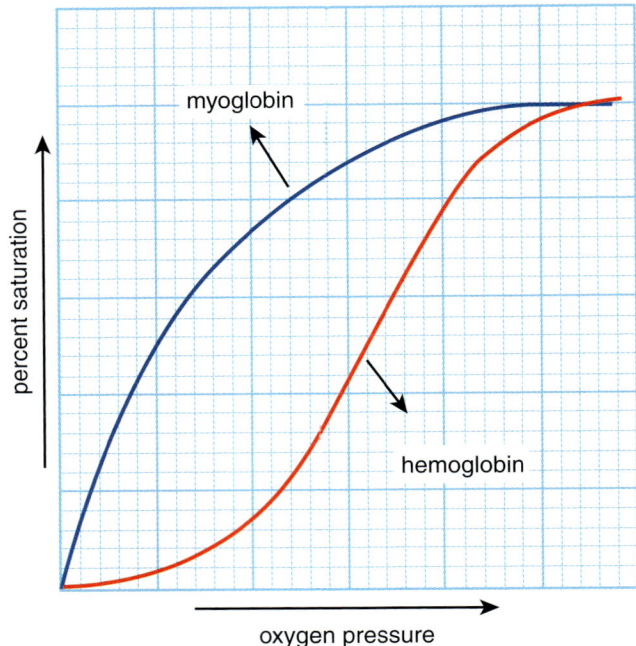

Most of the carbon dioxide is transported as a bicarbonate ion, while a small amount is carried in the blood as a gas. In water, carbon dioxide is converted into bicarbonate ions, which readily dissolve in the blood plasma and cytoplasm of RBCs. Red blood cells stimulate the formation of bicarbonate ions using an enzyme called **carbonic anhydrase**. Bicarbonate ions in the blood are very important in maintaining the pH of the blood serum.

Carbon dioxide diffusion into the blood requires that the partial pressure of the carbon dioxide is higher in the tissues than in the blood. **Carbon dioxide intoxication** occurs when carbon dioxide is extremely high in the environment or in the blood. It may be **acute** or **subacute**. Acute carbon dioxide intoxication is caused mostly by high levels of carbon dioxide in the air. Subacute toxicity is likely caused by the body's failure to eliminate carbon dioxide. Chronic obstructive pulmonary diseases (COPDs) and other types of respiratory disorders produce this condition.

Carbonic Anhydrase An enzyme that converts carbon dioxide and water into bicarbonate ions

Carbon Dioxide Intoxication A state resulting from excessive carbon dioxide in the environment or the blood

Acute Rapid onset with a short but severe course

Subacute Describes a condition with a moderate duration and/or severity

BLOOD DONATION TRIVIA

Many people are aware of the concept of blood types. However, it is likely that they are unfamiliar with the concerns blood banks have when considering blood donations. Here is some "blood trivia" that you can share with family, friends, and coworkers when discussing the benefits of donating blood:

- Someone will use blood every 3 seconds!
- There currently is no substitute for the cellular components in human blood.
- Scientists have not developed a substitute that provides all the functions of blood serum.
- Percentages of blood type for each 100 blood donors in North America is as follows:
  - 0.5%   AB negative
  - 1.5%   B negative

- 3.5%   AB positive
- 6%   O negative
- 7%   A negative
- 8.5%   B positive
- 34%   A positive
- 39%   O positive

- Blood comprises 7% of a person's body weight.
- People with type O negative blood can receive only their own specific type, even in an emergency, although they are the universal donors.
- Type AB plasma has been considered the universal blood plasma recipient because it does not contain immune proteins that damage any blood type. Therefore, AB plasma is given to patients with any blood type.
- Cows have over 800 blood types.
- Platelets cannot be frozen and have only a 5-day shelf life.
- The average adult body has 10 to 12 pints of blood.
- 50 to 100 units of blood components can be used during organ transplantation.
- A person can donate whole blood every 56 days.
- A person can donate platelets every 2 weeks.
- A healthy person can donate plasma as often as twice in a 7-day period.
- Frozen plasma can be stored for up to 1 year.

Information courtesy of the University of California–Irvine Medical Center and the Blood Cord Registry

White Blood Cells (WBCs)

Key Terms: antibiotic, circulating monocyte, Kupffer cell, macrophage, major basic protein, mast cell, tissue monocyte

The different WBCs carry out a variety of jobs. Most of their functions are involved in repairing the body and fighting infectious diseases. Granulocytes contain granules of toxic chemicals in the cytoplasm that are released to kill microorganisms and regulate reactions to foreign materials in the body. Neutrophils pass through capillary walls into infected tissues where they kill bacteria. Damaged tissues and immune system cells secrete factors that attract neutrophils to the area. The neutrophils adhere to the injured tissues and use phagocytosis to engulf the remains of bacteria and damaged cells. Many of the chemicals secreted by neutrophils are potent **antibiotics**. Other chemicals stimulate inflammation. Inflammation encourages blood flow to the injured area, attracting more WBCs that help with healing.

Eosinophils produce secretions that defend against large parasitic organisms in the body. Their secretions selectively kill protista and worms. A blood test is used to detect a great increase in eosinophil numbers, which indicates parasitic infection. The granules of eosinophils contain a toxin called **major basic protein**. Eosinophils also have surface proteins that allow them to secrete chemicals associated with allergies. They will normally produce theses secretions as long as the chemicals are present in the blood. Basophils are the least common of the WBCs. They secrete histamine, which is involved in stimulating the immune response. The overproduction of histamine is associated with the symptoms of allergies, such as a runny nose, sneezing, and watery eyes.

Antibiotic A chemical that harms or kills bacteria

Major Basic Protein A chemical that harms or kills parasites

Good Choice Bad Choice

A neighbor told you that he wants his young children exposed to various diseases as a way to naturally boost their immune systems. He is adamantly opposed to using medications or vaccines. What are the benefits and dangers of following this plan?

Mast Cell A type of basophil that causes inflammation

Circulating Monocyte A large phagocytic cell found in the blood

Tissue Monocyte Large phagocytic cells found in connective tissue

Macrophage A large phagocytic cell found in connective tissue

Kupffer Cell A large phagocytic cell found in the liver

A special type of basophil called a **mast cell** releases chemicals that initiate the immune response. Mast cells are responsible for the inflammation of tissues. Some of these secretions attract neutrophils to an injured tissue. Mast cells are distributed throughout the body, but they are usually found along the walls of small blood vessels

Agranulocytes visibly differ from granulocytes in that they lack granules in their cytoplasm. Monocytes contain numerous clear granules that give their cytoplasm a uniform grayish appearance. Once monocytes leave the bone marrow, they develop into either **circulating monocytes** or **tissue monocytes** called **macrophages**. Circulating monocytes carry out a variety of roles. Their major role is to detect infectious agents traveling in the blood. However, some circulating monocytes are involved in bone growth and maintenance. Tissue macrophages remove dead cells and attack microorganisms that are difficult to kill, such as fungi. The body contains specialized types of tissue monocytes in different organ systems. For example, the liver contains a monocyte called the **Kupffer cell**. The roles of lymphocytes will be described later in the discussion about immune response. Lymphocytes carry out most of the duties of the immune system.

Platelets

Key Terms: clotting cascade, clotting factor, fibrin, fibrinogen, plasmin, plasminogen, prostacyclin, prothrombin, thrombin, tissue plasminogen activator (TPA)

Prostacyclin A lipid that prevents platelet activation

Clotting Cascade A series of chemical reactions that cause clotting

Clotting Factor A chemical that initiates clotting

Platelets play a central role in the blood clotting process. They carry out a two-step process: the platelets adhere to an injured area, which is followed by the activation of clot formation. It is important that platelets are alerted to form clots only when there is an injury requiring a cessation of blood loss. Intact blood vessels secrete a lipid called **prostacyclin**, which prevents platelet activation. Damage to a blood vessel or body tissue produces a reaction called the **clotting cascade**. The cascade involves the following sequence of events:

1. Blood vessel or tissue damage releases a variety of tissue components into the blood.
2. Collagen and chemicals called **clotting factors** stimulate other factors that communicate the presence of damaged tissues.
3. These other factors associated with clotting encourage platelets to adhere to the damaged tissue and to each other.

Figure 12.9 Clotting Cascade

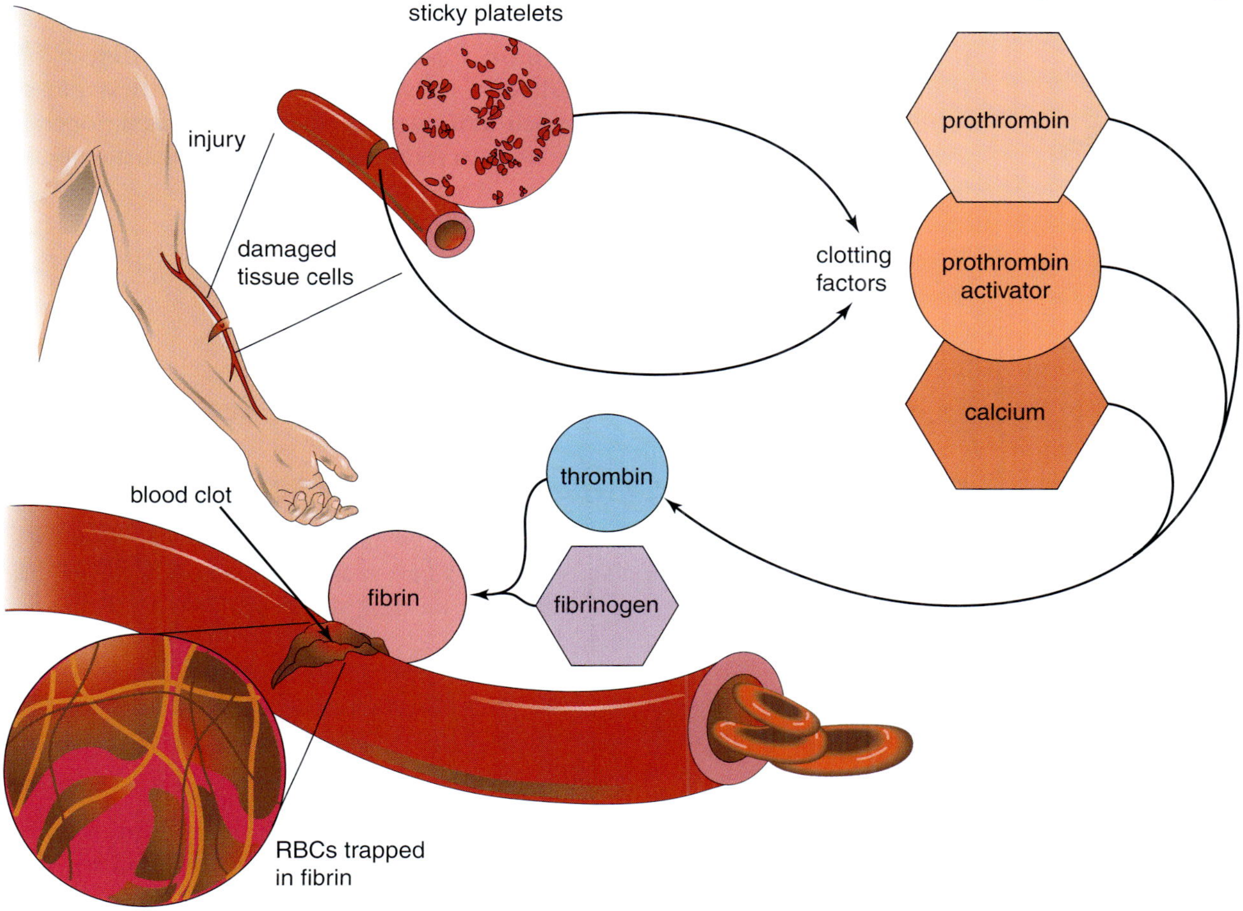

4. The factors also stimulate the conversion of a protein called **prothrombin** into **thrombin** (Figure 12.9).
5. Prothrombin is secreted by platelets that were activated by adhesion to damaged tissues and other platelets.
6. Platelets also release prothrombin when they fall apart after exposure to air.
7. The cascade causes the conversion of prothrombin into thrombin.
8. Thrombin converts a protein called **fibrinogen** into **fibrin**.
9. Fibrin then forms a sticky meshwork that adheres to thrombocytes and other blood components.
10. This sticky meshwork, or clot, forms a temporary barrier, which prevents blood loss. It also forms a plug that impedes the passage of microorganisms into damaged body tissues.
11. The scab that forms after cutting the skin is an example of a clot. Calcium and vitamin K are very important in clot formation. The calcium assists with the stickiness of fibrin. Vitamin K is needed for the synthesis of the clotting factors.

This complex cascade of chemical reactions needed to activate clotting prevents the blood from clotting unintentionally. Certain people develop blood clots in their capillaries and other small blood vessels when the clotting cascade is unintentionally activated.

Prothrombin An inactive form of an enzyme that stimulates blood clotting

Thrombin An enzyme involved in blood clotting

Fibrinogen An inactive form of a protein necessary for blood clotting

Fibrin A protein necessary for blood clotting

THE LYMPHATIC SYSTEM AND THE BLOOD

Plasmin An enzyme that dissolves the fibrin of blood clots

Plasminogen The inactive form of an enzyme that dissolves the fibrin of blood clots

Tissue Plasminogen Activator (TPA) An enzyme that dissolves the fibrin in blood clots

Clots are not permanent structures. Rather, they temporarily patch the damage until the injured blood vessels or tissues are repaired. A protein called **plasminogen** is converted into **plasmin**. Plasmin is an enzyme that digests fibrin and thereby dissolves a clot. When healthy cells near the clot, they release a chemical called **tissue plasminogen activator** (**TPA**). Tissue plasminogen activator and related chemicals are used to reduce the damage of cardiovascular diseases caused by blood clots formed in arteries and veins.

✓ Concept Check

1. How are oxygen and carbon dioxide carried in the blood?
2. What are the roles of the different WBCs?
3. Describe how platelets protect the body from excessive blood loss.

DISCOVERY SCENE PLEASE ENTER DISCOVERY SCENE PLEASE ENTER

Do you see a relationship between the function of a particular blood cell type and the condition experienced by your friend's father? What are some reasons for the abnormal WBC number? What conditions are the unusual RBC counts causing?

Blood Cell Formation

 Key Terms: band, bilirubin, erythroblast, erythropoiesis, erythropoietin, hematopoietic stem cell, lymphoid progenitor, multipotent stem cell, myeloid progenitor, stab

All blood cells in adults are produced in the bone marrow (Figure 12.10). In the embryo, the liver is responsible for blood cell formation. Approximately 11 million new blood cells are produced each second in an adult human. The ratio of WBC production to RBC production is one WBC for every 700 RBCs produced. They are derived from a stem cell type called the **hematopoietic**, or **multipotent**, **stem cell**. The hematopoietic stem cell can develop into a **lymphoid** or a **myeloid progenitor**. Lymphoid progenitor cells give rise to WBCs of the lymphatic system. Myeloid progenitor cells give rise to platelets, RBCs, and WBCs that mainly circulate in the blood. Chemicals called growth factors stimulate the production of a particular blood cell lineage. A variety of organs secrete these growth factors, which regulate the formation of blood cells.

A growth factor called **erythropoietin** stimulates RBC formation. Erythropoietin is commonly abbreviated **EPO**. It is secreted by the kidneys to control the production of RBCs. The process by which RBCs are produced is called **erythropoiesis**. Moving to elevated altitudes above 5000 feet for at least 3 weeks stimulates enhanced erythropoietin production. This, in turn, increases RBC formation, which improves the efficiency of oxygen transport. Red blood cells mature in approximately 7 days and live about 120 days. They start out as cells called reticulocytes, or **ethrythroblasts**, which are immature cells that contain a large nucleus (Figure 12.11). Aging RBCs swell and are destroyed in the liver and spleen. The digestive system then removes the hemoglobin as a yellowish chemical called **bilirubin**. A fetus produces

Hematopoietic or Multipotent Stem Cell A stem cell from which all RBCs and WBCs develop

Lymphoid A stem cell that gives rise to WBCs

Myeloid Progenitor A stem cell that gives rise to platelets

Erythropoietin (EPO) A protein secreted by the kidneys that stimulates RBC production

Erythropoiesis The formation of RBCs

Erythroblast An immature red blood cell

Bilirubin A chemical breakdown product of hemoglobin

Chapter 12

Figure 12.10 Human Blood Cell Formation

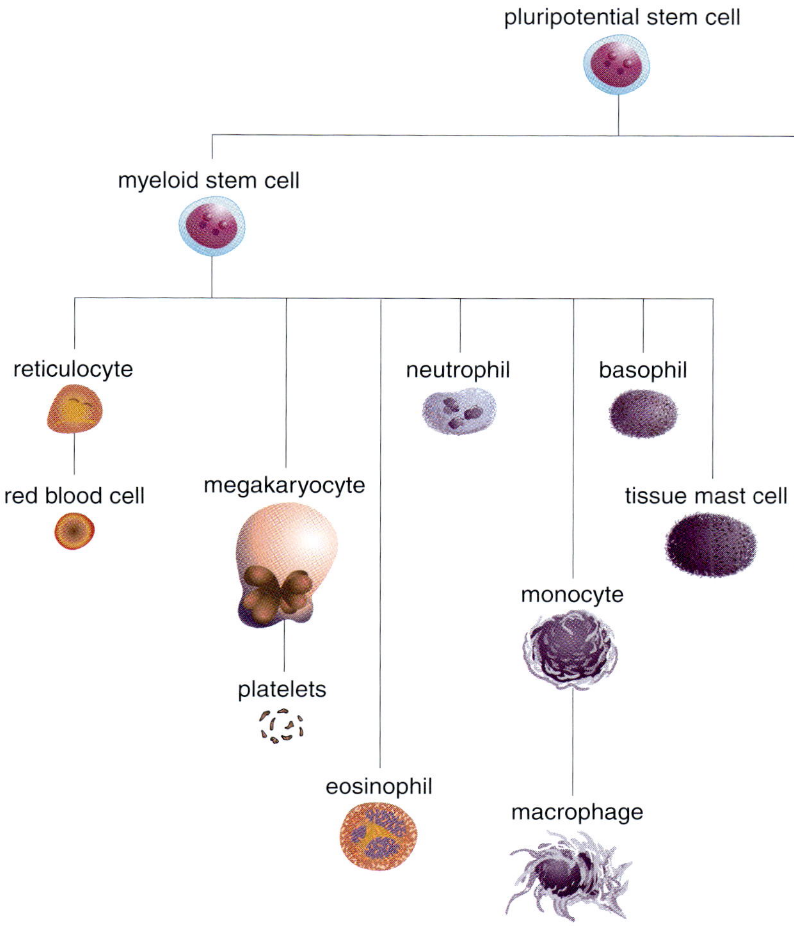

pluripotential stem cell

myeloid stem cell

lymphoid stem cell

reticulocyte

red blood cell

megakaryocyte

platelets

eosinophil

neutrophil

monocyte

macrophage

basophil

tissue mast cell

pre-B cell

B lymphocyte

plasma cell

pre-T cell

T lymphocyte

Figure 12.11 Erythroblast

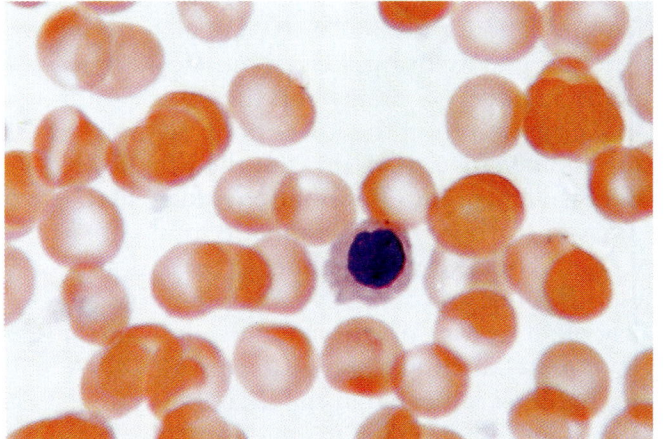

Figure 12.12 Bone and Blood

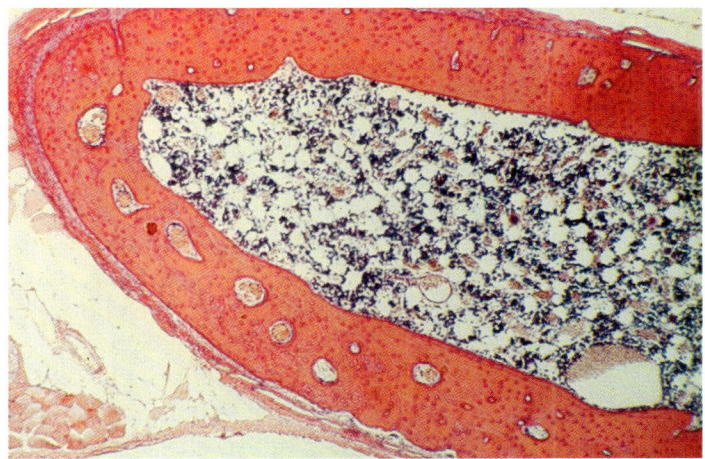

THE LYMPHATIC SYSTEM AND THE BLOOD

several types of hemoglobin throughout its development. These different types of hemoglobin adapt the fetus to its varying metabolic needs for oxygen. The lifetime of WBCs ranges from 13 to 20 days. After this time, they are destroyed in the lymphatic system. When immature WBCs are released from the bone marrow they are called **bands**, or **stabs**. Normal bone marrow shows an abundance of newly formed RBCs and the megakaryocytes that produce platelets (Figure 12.12). The nature of blood cell disorders can be determined by collecting bone marrow samples and viewing the types of stem cells under a microscope.

Band or Stab An immature white blood cell

Did the information on blood cell formation provide any more hints about the causes of the abnormal blood cell numbers? What are some factors that could affect blood cell formation or cause a loss of blood cells?

Lymphatic System

Structures of the Lymphatic System

Key Terms: hilum, lymph, lymph gland, lymph node, lymphatic sinuses, lymphatic trunk, Peyer's patches, red pulp, spleen, trabeculae, white pulp

Lymph Node or Lymph Gland A small round gland that is part of the immune system

Spleen A large lymphatic organ in the abdominal cavity

Lymph A colorless fluid that bathes cells and moves through lymphatic vessels

Lymphatic Trunk A network of lymph vessels that drains regions of the body

Peyer's Patches Lymphoid tissues in the lining of the digestive system

Lymphatic Sinuses A fluid-filled sack located in a lymph node

Hilum A region of the lymph node where blood vessels enter and exit

The structural components of the lymphatic system include the lymphatic vessels, **lymph nodes** (**lymph glands**), **spleen**, thymus, tonsils, and WBCs (Figure 12.13). Lymph vessels are thin ducts that carry a clear fluid called **lymph**. Lymph carries out a variety of jobs, including the removal of foreign bacteria and foreign material from cells. It also transports fat from the digestive system and moves mature lymphocytes to the blood. A very important function of the lymph vessels is that they drain excess fluids that accumulate in tissues, thus preventing a type of swelling called edema. The collected fluid is deposited into veins and carried throughout the circulatory system.

Lymph flow is produced either by gravity or by skeletal muscle action. Lymph from upper parts of the body passively drains into the vessels using gravity, while muscle action moves lymph from the lower body. Lymph vessels contain valves, similar to those in veins, which prevent the backflow of lymph. The lymph vessel system is divided into **lymphatic trunks**, which drain lymph from the larger areas of the body. The lymphatic trunks are named for the areas that they drain. For example, the lumbar trunk drains lymph from the abdominal region, pelvic area, and legs. Each lymphatic trunk has collections of small swollen regions of lymphatic tissue. A tonsil is one of a pair of clusters of lymphatic tissue in the throat (Figure 12.14). A related structure called **Peyer's patches** is located in the digestive system. Lymph nodes are complex collections of lymphatic tissue covered by connective-tissue capsules (Figure 12.15). Their funcion is to eliminate antigens from the lymph. They are generally found in groups along the larger lymphatic vessels. Lymph nodes have four components: Blood vessels, **lymphatic sinuses**, lymphocytes, and filler tissue (stroma). The blood vessels enter a region of the lymph node called the **hilum**.

Figure 12.13 Lymphatic System

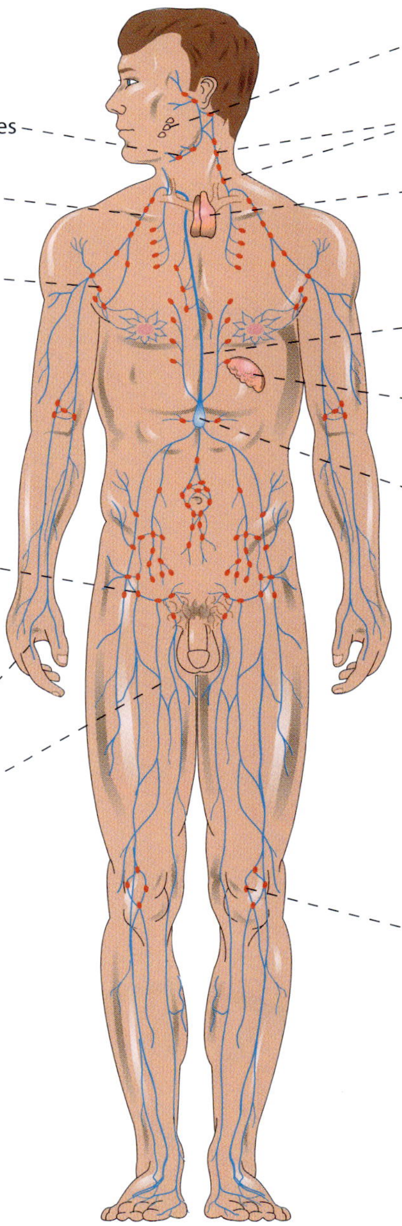

- tonsils
- submandibular lymph nodes
- cervical lymph nodes
- right lymphatic duct
- thymus
- axillary lymph nodes
- thoracic duct
- spleen
- cisterna chyli
- inguinal lymph nodes
- lymph vessels
- popliteal (behind the knees) lymph nodes

Lymphatic vessels also exit the lymph node from the hilum. They enter the lymph node through various openings in the capsule. The stroma is divided into regions by walls in the capsule called **trabeculae**. These small regions contain B-lymphocytes and T-lymphocytes that originated in the bone marrow. B-lymphocytes mature in the lymph nodes. Macrophages are found in the lymphatic sinuses.

The spleen is located in the upper left region of the abdomen near the stomach. It is divided into two functional regions: **red pulp** and **white pulp**. Red pulp is a storage area for RBCs. Red pulp is also responsible for removing old or damaged RBCs from the circulation. White pulp is lymphatic tissue and contains B-lymphocytes and T-lymphocytes. B-lympho-

Trabeculae Partitions that divide the lymph node into regions

Red Pulp An interior area of the spleen that contains RBCs

White Pulp A region of the spleen composed of lymphatic tissue

THE LYMPHATIC SYSTEM AND THE BLOOD

Figure 12.14 **Tonsils**

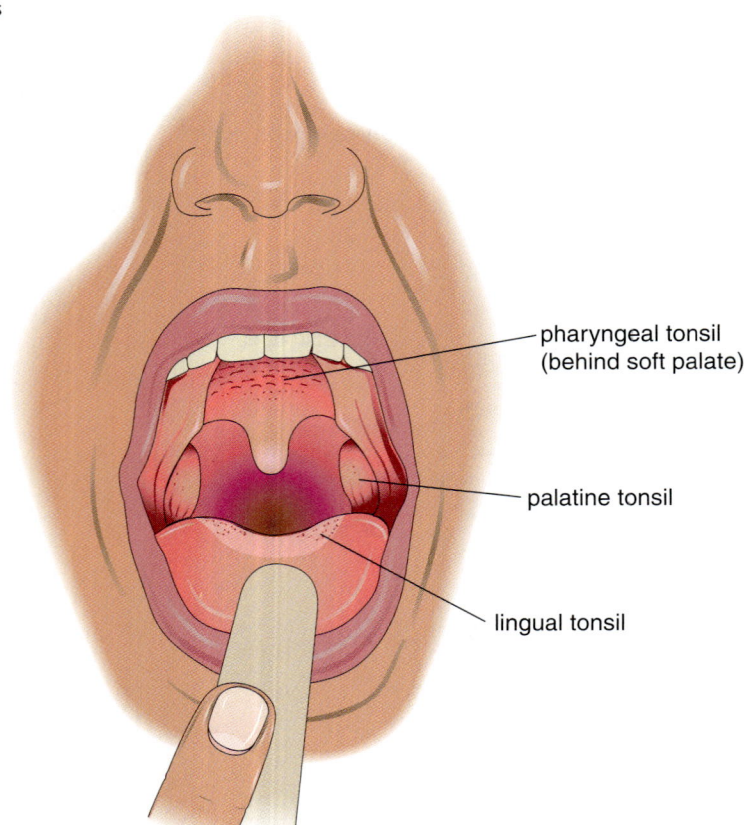

cytes and T-lymphocytes assist the body with infections that require a large immune response. The thymus gland is a mass of lymphatic tissue that gradually enlarges during childhood until puberty. After puberty, it undergoes a reduction of mass and gradually loses some of its functions. The thymus produces secretions that mature T-lymphocytes. As part of this maturation process, the thymus "educates" T-lymphocytes so that they do not accidentally attack the body. This property of T-lymphocytes is called self-tolerance. Self-intolerance is induced by secretions of the thyroid that modify T-lymphocyte activity.

Figure 12.15 **Lymph Node**

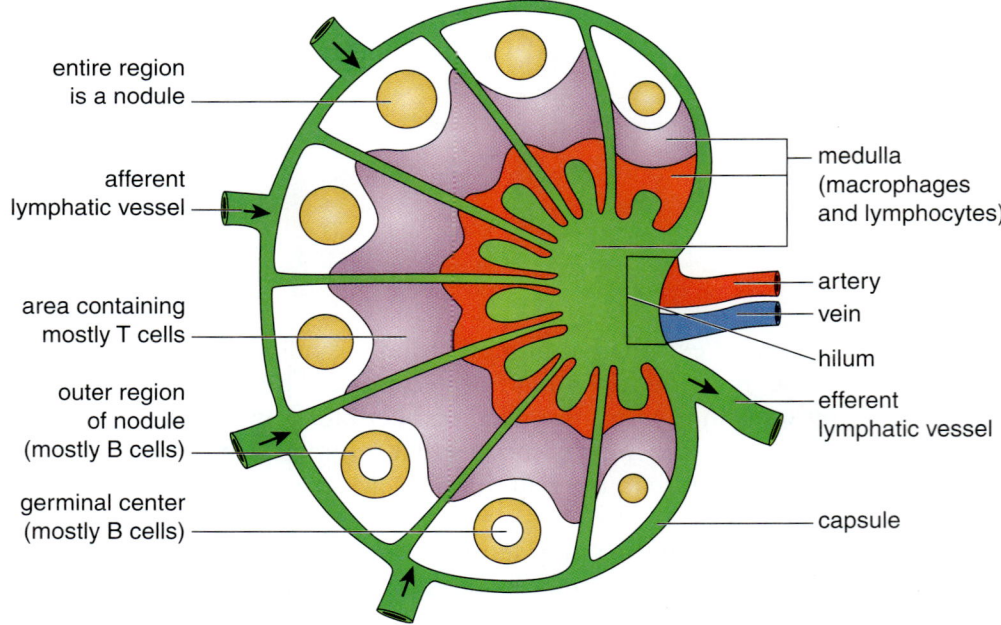

Immune Response

Key Terms: acquired immunity, antibody, antigen, cell-mediated immunity, complement, hapten, helper T-lymphocyte, humoral immunity, IgA, IgD, IgE, IgG, IgM, immunoglobulin, inflammatory response, innate immunity, interferon, memory cell, natural killer (NK), NK cells, nonspecific immunity, plasma cell, primary response, secondary response, suppressor T-lymphocyte

The immune response is a complex array of anatomical structures and cellular events that protect the body from foreign chemicals called **antigens**. An antigen is any chemical that can produce an immune response. The most common types of antigens are composed of proteins. **Haptens** are chemicals that bind to blood proteins to form antigens. Haptens alone are too simple in their chemistry to be identified as antigens by the immune system. The immune system uses two mechanisms to respond to disease: **Innate immunity** and **acquired immunity**.

Antigen A substance that can induce an immune response

Hapten A molecule that can cause an immune response when attached to blood proteins

Innate Immunity or Nonspecific Immunity A general way in which the body responds to disease

Acquired Immunity A specific set of events that the body uses to fight a particular infection

Innate Immunity Innate immunity is sometimes called **nonspecific immunity** because it uses barriers that block a variety of infections to protect the body (Figure 12.16). Anatomical barriers are features that block microorganisms from entering the body. Skin provides a physical barrier by blocking microorganisms from entering underlying tissues. The regular shedding of the epidermis removes microorganisms from the skin. Skin is renewed every 15 to 30 days. Harmless microbes on the skin inhibit the growth of pathogenic microbes. Skin secretions form a chemical barrier that disrupts the growth of microorganisms. Skin has an acidic pH due to lactic acid and fatty acids, secretions that are unfavorable to many bacteria. Mucous membranes of the digestive, respiratory, and urogenital systems use mucus to block and remove microorganisms. Saliva, stomach acids, and tears are other chemical barriers that reduce infection.

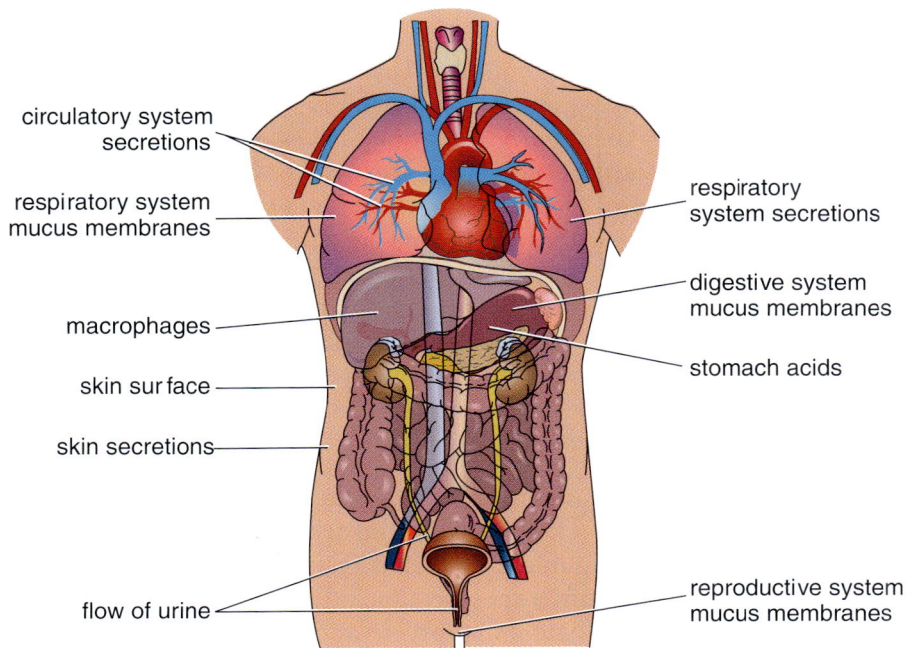

Figure 12.16 Innate Immunity

THE LYMPHATIC SYSTEM AND THE BLOOD

Cutting Edge Research

BLOOD SUBSTITUTES

Safe blood needed for blood products, surgery, and transfusions is not as abundant as it should be for emergency situations. Military operations in particular unexpectedly need large volumes of blood at a moment's notice. This shortage has created the need for scientists to develop blood substitutes. Currently, blood substitutes have been developed by six pharmaceutical companies. However, these substitutes do not fully replace all the roles that blood plays in the body. Blood substitutes are oxygen-carrying chemicals known as perfluorocarbons. They act mainly as plasma expanders. This means that they maintain blood volume.

Perfluorocarbons do the job of RBCs by carrying oxygen; however, they do not facilitate the transport of carbon dioxide and other blood components needed for homeostasis. Most of the research emphasis today is on artificial hemoglobin-containing solutions. Hemoglobin is more effective than perfluorocarbons at regulating the levels of atmospheric gases. New experiments are under way to make blood substitutes using artificial RBCs made through biotechnology. Researchers at the Quebec Heart and Lung Institute at Laval University in Quebec, Canada, are investigating body factors that influence the effectiveness of blood substitutes. They have used rats to test the effects of blood substitutes used as hemorrhage treatments. Initial studies have shown that the substitutes reduce the stress of excess blood loss. In addition, the oxygen-carrying ability of the substitutes reduces tissue damage from ischemia which is caused by inadequate oxygen to the tissues. It is believed that hemoglobin-based blood substitutes will further reduce body stress from artificial blood therapies.

For more information, please follow this Web link: *http://www.med.unipi.it/patchir/bloodl/bmr/tools/tools10.htm*.

1. Winslow RM. Blood substitutes. A review of the literature 1997-1998. *Drugs*. 1999;2(4):340-354.
2. Daull P, Blouin A, Cayer J, et al. Profiling biochemical and hemodynamic markers using chronically instrumented, conscious and unrestrained rats undergoing severe, acute controlled hemorrhagic hypovolemic shock as an integrated *in vivo* model system to assess new blood substitutes. *Vasc Pharmacol*. In press.

Interferons A group of proteins that interfere with viruses

Natural Killer (NK) Cells T-lymphocytes that kill tumor cells and microorganisms

Fever is another defense against microbial disease. Elevated body temperatures inhibit the growth of most microorganisms. Fevers are also accompanied by physiological reactions that deprive microorganisms of resources in the blood. For example, long-term fevers reduce the iron in the blood needed for microbial growth. An array of secretions produced by various body organs and WBCs obstruct microbial growth. **Interferons** are a group of proteins that cells produce following a viral infection. The interferons then bind to other cells and inhibit viral replication. **Complements** are a group of plasma proteins that can be activated to destroy microorganisms. Monocytes are also part of the innate immunity. They travel the body destroying any material that is not recognized as part of the body. Special T-lymphocytes called **natural killer (NK) cells** secrete proteins that kill tumor cells and microorganisms.

Acquired Immunity Acquired immunity occurs when the body adapts to specific infections (Figure 12.17). The body uses the acquired immune response to learn the nature of the invading microorganisms (Figure 12.18). Using this information, the body can mount a specific attack that fights the infection more effectively than would innate immunity. In addition, specific immunity

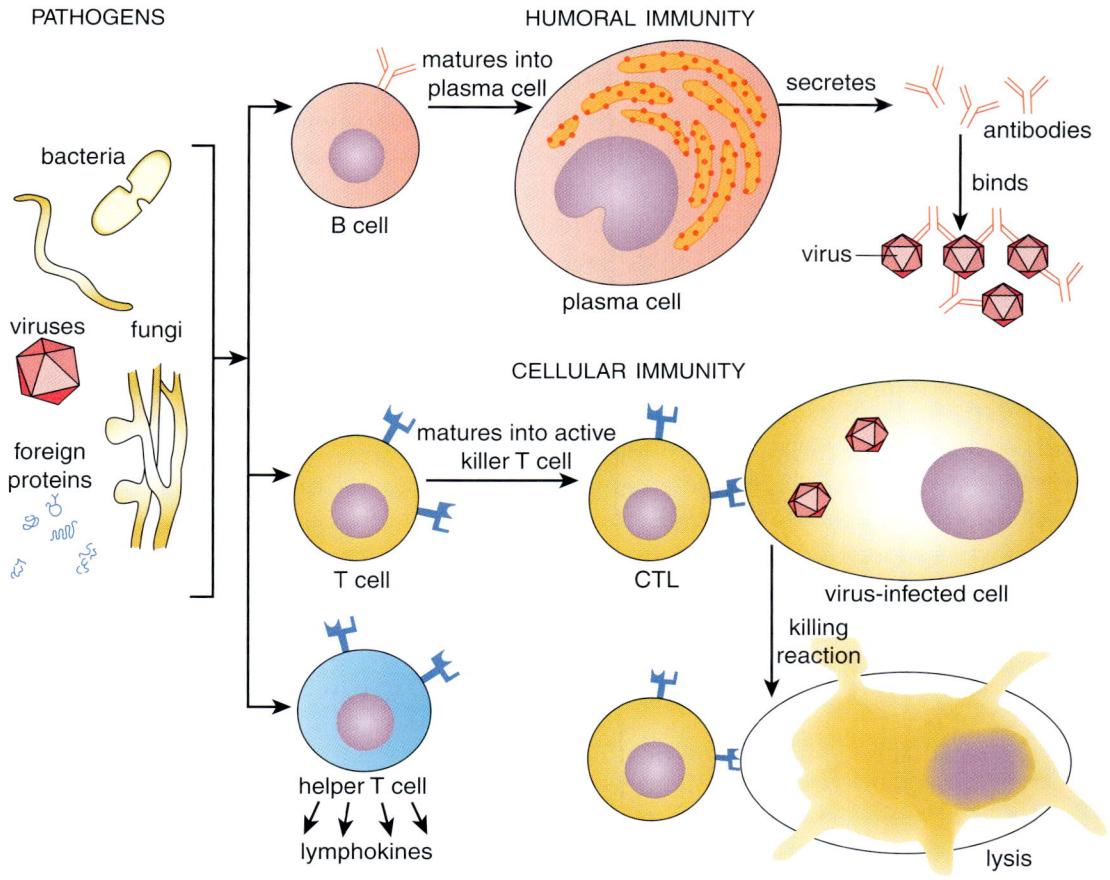

Figure 12.17 Acquired Immunity

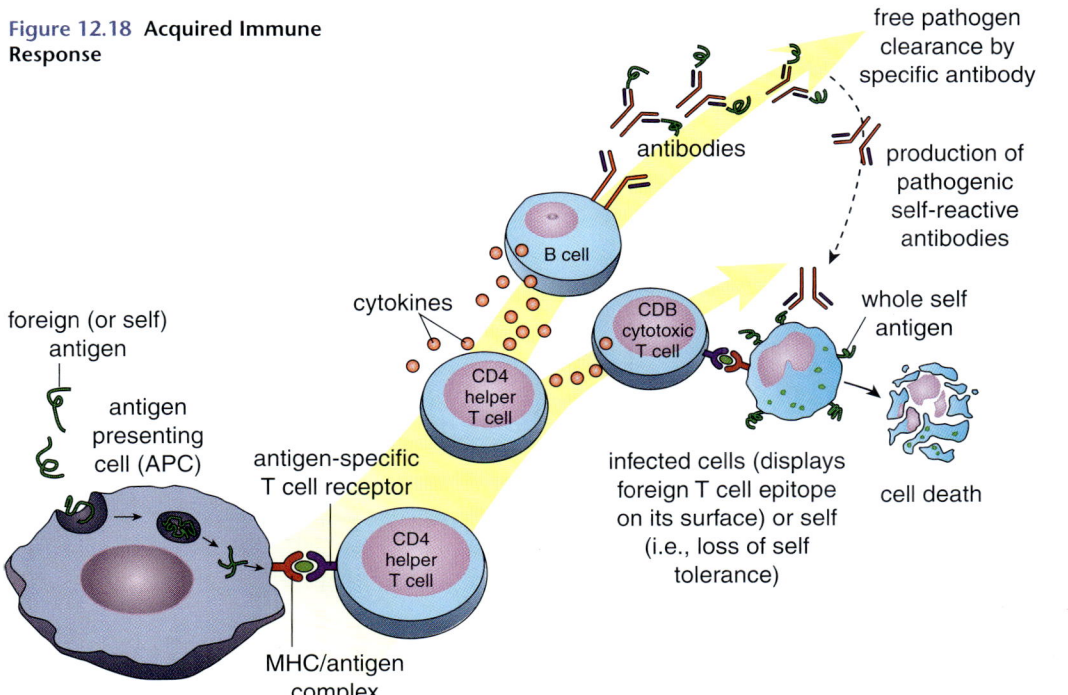

Figure 12.18 Acquired Immune Response

Primary Response A response that occurs the first time an antigen enters the body

Secondary Response A response that takes place when a subsequent encounter with an antigen occurs

Helper T-lymphocyte A type of T-lymphocyte that cooperates with B-lymphocytes

Plasma Cell A B-lymphocyte that produces immune proteins

Memory Cell A B-lymphocyte that stores the information to produce immune proteins

Antibody or Immunoglobulin A protein produced by the immune system that binds to specific foreign antigens

provides a way for the body to prevent future infections of the same microorganism from causing serious disease. Acquired immunity is divided into a **primary response** and a **secondary response**.

Primary and Secondary Responses

The primary immune response begins when macrophages detect the presence of new antigens in the body (Figure 12.19). These antigens are usually cell components of invading organisms. However, dust particles, food, haptens, pollen, and altered body molecules also stimulate the primary response. Macrophages then take up the foreign antigen by phagocytosis and produce a protein on their own surfaces that resembles the antigen. In most cases, the macrophages that encountered the antigen will travel to the lymph nodes where they encounter the B-lymphocytes and T-lymphocytes. Several types of T-lymphocytes are found in lymphatic tissue. One type, called the **helper T-lymphocyte**, assists with the primary response. The macrophages then attach to both the B-lymphocytes and helper T-lymphocytes, activating them to respond to the particular antigen. B-lymphocytes respond by multiplying and subsequently dividing into **plasma cells** and **memory cells**. Plasma cells start producing immune proteins called **antibodies**, or **immunoglobulins**. Memory cells circulate around the body and are involved in the secondary response.

B-lymphocytes produce a variety of antibodies. All antibodies have a component that permits them to bind to antigens. An individual antibody has two antigen-binding regions. Some antibodies work as a single unit, while others are attached into a complex structure that can bind to many antigens at a time. Antibodies are given immunoglobulin designations, abbreviated Ig, followed by a letter that indicates a particular category of antibody. The common types

Figure 12.19 Primary Response

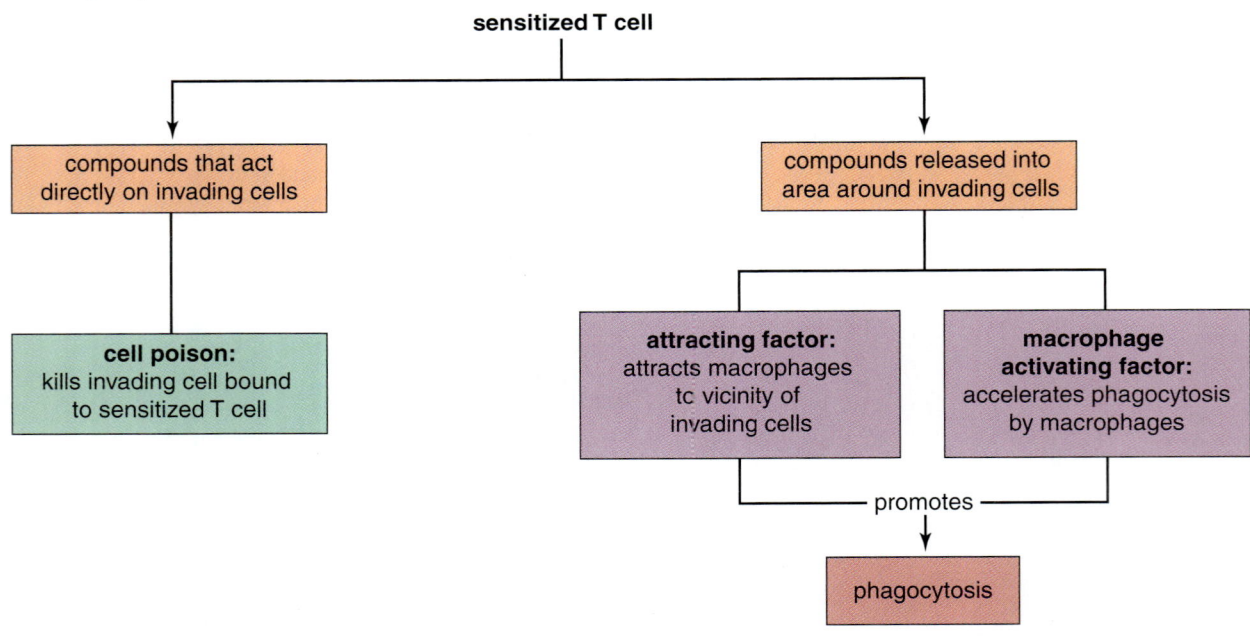

of antibody classes in humans are **IgG**, **IgE**, **IgA**, **IgM**, and **IgD**. IgGs are the most common type of antibody produced for fighting general infections. They are the only major immunoglobulin able to pass from the mother to the child during pregnancy. These antibodies carry out a variety of jobs that activate the immune response throughout the body. IgEs specifically stimulate basophils and mast cells in connective tissue. They defend against parasites, such as fungi and worms. IgA antibodies make up about 15% to 20% of immunoglobulins in the blood. They are mostly found in mucous membranes of the digestive system. IgAs are also found in milk, tears, and saliva. IgMs are found in the plasma without any evidence of prior contact with an antigen. They may serve as natural defenses against general bacterial infections. IgD antibodies are found mostly on the membranes of B-lymphocytes. They are probably involved in the maturation of B-lymphocytes into plasma and memory cells.

IgG The predominant type of immunoglobulin existing in the blood

IgE The type of antibody most instrumental in allergic reactions

IgA An immunoglobulin found in blood, tears, saliva, and mucous membranes

IgM An immunoglobulin that gathers in clusters and is involved in killing bacteria

IgD An antibody present on the B-lymphocyte membrane

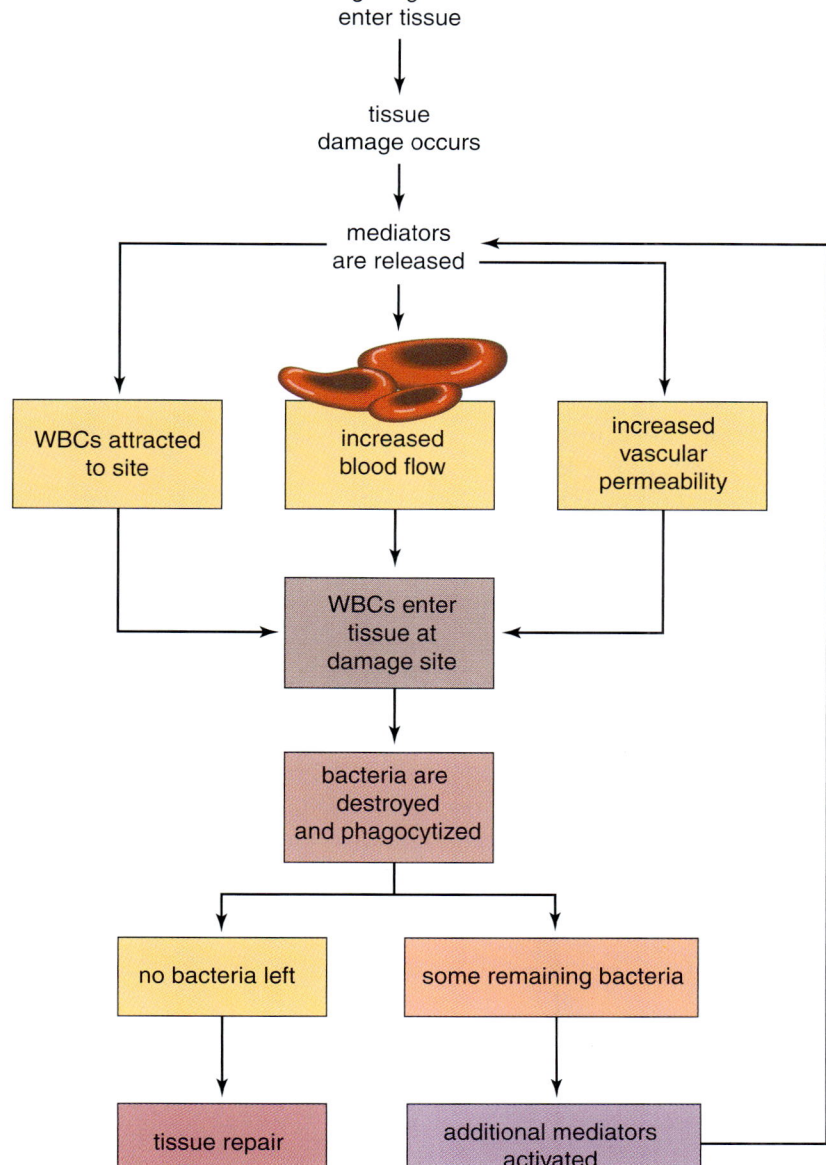

Figure 12.20 Inflammation

THE LYMPHATIC SYSTEM AND THE BLOOD

Humoral Immunity The immunity that results from antibody production by B-lymphocytes

Cell-Mediated Immunity The immune response coordinated by T-lymphocytes

Suppressor T-lymphocyte A T-lymphocyte that inhibits the immune response

Inflammatory Response The swelling, redness, warmth, or pain in the area of an immune response

Antibody binding to antigens stimulates many body reactions that help with the removal of the foreign antigen or microorganism. This response is called **humoral immunity**. The term refers to the antibodies circulating through body fluids, or "humors." T-lymphocytes also assist with the removal of the foreign antigen. This response is called **cell-mediated immunity**.

Most of the T-lymphocytes react by stimulating other components of the immune system. However, there is a T-lymphocyte that inhibits the immune response, the **suppressor T-lymphocyte**. These cells are needed to prevent the immune system from overreacting to foreign antigens. Too much immune system activity can damage healthy body cells. Hormones, such as adrenaline and cortisone, also reduce the immune response.

The body responds more quickly to antigens during the secondary immune response. Memory cells respond to previously encountered antigens by immediately producing antibodies. This eliminates the need for time-consuming macrophage activity. Memory cells eventually die and can leave the body vulnerable to infectious diseases that are not frequently encountered. Continuous exposure to foreign antigens ensures a regular supply of memory cells. A person loses much of the secondary immune response after 5 years of not being exposed to the particular antigen or microorganism.

Inflammation

Inflammation (Figure 12.20) is a common outcome of an acquired immunity response. The **inflammatory response** is an intentional action that increases blood circulation to an affected area. Blood vessels around the site of inflammation dilate, thereby allowing a localized increase in blood flow. Immune system cells are then transported to the injured area and pass into the surrounding tissues. This enhances the immune presence in that particular location. All of the various cell types then carry out the immune response at the site of inflammation. Inflammation is first noticeable when the tissues in the area are red and warm. This is a result of the large amount of blood entering the area. The increase in temperature may have an antimicrobial effect. The tissues swell, and pain caused by pressure on nerve cells follows. Immune system chemicals can also stimulate pain receptors.

The inflammatory process continues until the infection that caused it has been reduced. Phagocytic WBCs continue to consume and destroy bacteria after the inflammatory response slows down. They also are involved in repair of any damaged tissues. Any pus that is produced during inflammation is what remains of the phagocytic activity. The same factors that inhibit the progression of the immune response also regulate the degree of the inflammatory response. During the inflammatory response, the tissue is not capable of carrying out many of its functions. Normal functions return when the tissue is fully recovered.

✓ Concept Check

1. Distinguish between innate and acquired immunity.
2. What is the difference between primary and secondary immunity?
3. Describe what happens during an inflammatory response.

Immunization and Vaccination

> **Key Terms:** active immunity, artificial immunization, globulin, immunization, natural immunization, passive immunity, smallpox, vaccine, vaccination

Immunization is a medical strategy for building up the body's acquired immune response. It is categorized as **natural** or **artificial immunization**. Natural immunization occurs when a person is exposed to foreign antigens as a part of daily life. Before the development of modern medicine it was common for children to build up their immune systems by being exposed to particular diseases, such as chicken pox and measles. This exposure caused childhood illnesses; however, it protected children from developing the diseases over and over again as they aged. It is not feasible today to rely on natural immunization. British surgeon Edward Jenner (1749-1823) developed a way to immunize people against the deadly **smallpox** disease in 1796. He did this by introducing a related virus from cattle into the blood stream of healthy people. The cattle disease did not cause illness in humans; however, its similarity to smallpox imparted smallpox immunity. His highly criticized strategy was the first medical attempt at artificial immunity.

Immunity against disease can be acquired through **active** or **passive immunity**. Jenner applied a strategy that used active immunity to build up the immune system. Active immunity stimulates the primary immune response by introducing foreign antigens into the body. **Vaccines** are drugs that provide antigens for developing active immunity.

Vaccination is the process of artificially building up the immune system using antigens. Vaccines can contain killed or weakened (attenuated) microorganisms. Many vaccines today contain only certain antigens from the microorganism. These are considered safer to use, although they may not be as effective as a vaccine made from whole organisms. Whole organism vaccines provide many more antigens that the body recognizes and to which it responds.

Passive immunity occurs naturally during embryological development when antibodies from the mother's blood stream are passed to the fetus. Breast milk also passes along antibodies to the baby. Some people debate the value of cow's milk and formula because these foods do not give children the passive immunity needed to combat disease. Certain types of immunization involve injecting large amounts of general antibodies into the body to fight disease. **Globulin** injections remove certain microorganisms from the body.

Immunization The process of inducing immunity

Natural Immunization Natural exposure to foreign antigens

Artificial Immunization Therapeutic exposure to foreign antigens

Smallpox A highly contagious, infectious, and often deadly viral disease

Active Immunity Immunity gained by exposure to foreign antigens

Passive Immunity Immunity gained by the introduction of antibodies

Vaccine A drug preparation that contains an antigen

Vaccination The therapeutic use of antigens to build up the immune system

Globulin A blood protein that acts like an antibody

✓ Concept Check

1. Define the term immunization.
2. Distinguish between natural and artificial immunization.
3. Distinguish between active and passive immunity.

DISCOVERY SCENE PLEASE ENTER DISCOVERY SCENE PLEASE ENTER

What is the possible role of the immune system in the illness your friend's father has developed? What features of the immune response would you look for to help determine the nature of the ailment?

FEAR OF BLOOD

Periodic shortages of blood supplies forces agencies, such as the American Red Cross, to aggressively seek a regular stream of blood donors. Unfortunately, most people would probably rather visit the dentist than donate blood. Only about 5% of eligible people in North America donate blood. In addition, there is a growing concern about the potential hazards of blood products. Medical and psychological researchers are currently studying attitudes about donating and receiving blood. Studies conducted in Canada, England, and the United States are revealing several key beliefs about donating and receiving blood or blood substitutes:

- Many people have a great fear of receiving donated blood, even in life-threatening situations.
- Blood substitutes are perceived to be significantly less safe than donated blood.
- Stricter blood testing is believed to reduce the chances of contracting diseases from donated blood; however, more people are skeptical of the safety testing performed on blood substitutes.
- The dislike for donating blood is due primarily to concerns about potential adverse effects.
- Most people understand that they cannot contract a disease through blood donation.
- Altruism, defined as unselfish consideration for other people, dominated as the reason most people donate blood.

Psychologists used these data to conclude that blood-donation education programs should be aimed at helping people to overcome their fear of giving blood.

1. Ferguson E, Leaviss J, Townsend E, Fleming P, Lowe KC. Perceived safety of donor blood and blood substitutes for transfusion: the role of informational frame, patient groups and stress appraisals. *Transfus Med.* 2005;15(5):401-412.
2. Hupfer ME, Taylor DW, Letwin JA. Understanding Canadian student motivations and beliefs about giving blood. *Transfusion.* 2005;45(2):149-161.

WELLNESS AND ILLNESS OVER THE LIFE SPAN
Pathology of the Blood and Lymphatic System

Key Terms: acquired immunodeficiency syndrome (AIDS), allergy, anemia, elephantiasis, hemophilia, Hodgkin's lymphoma, human immunodeficiency virus (HIV), hypersensitivity, immunodeficiency, leukemia, lymphedema, lymphomas, macrocytic anemia, microcytic anemia, non-Hodgkin's lymphoma, pernicious anemia, sickle cell disease, thalassemia

Anemia A deficiency of normal red blood cells in the bloodstream

Pernicious Anemia A type of anemia caused by vitamin B12 deficiency

Anemia is a common ailment of the blood that is actually a group of various RBC disorders. The condition is best described as a deficiency of normal RBCs in the blood stream. Most cases of anemia involve a decrease in the number of RBCs. This can be caused by a variety of factors, including bleeding, malnutrition, iron deficiency, insufficient erythropoietin levels, and kidney damage. A decrease in RBCs due specifically to vitamin B12 deficiency is called **pernicious anemia**.

CHAPTER 12

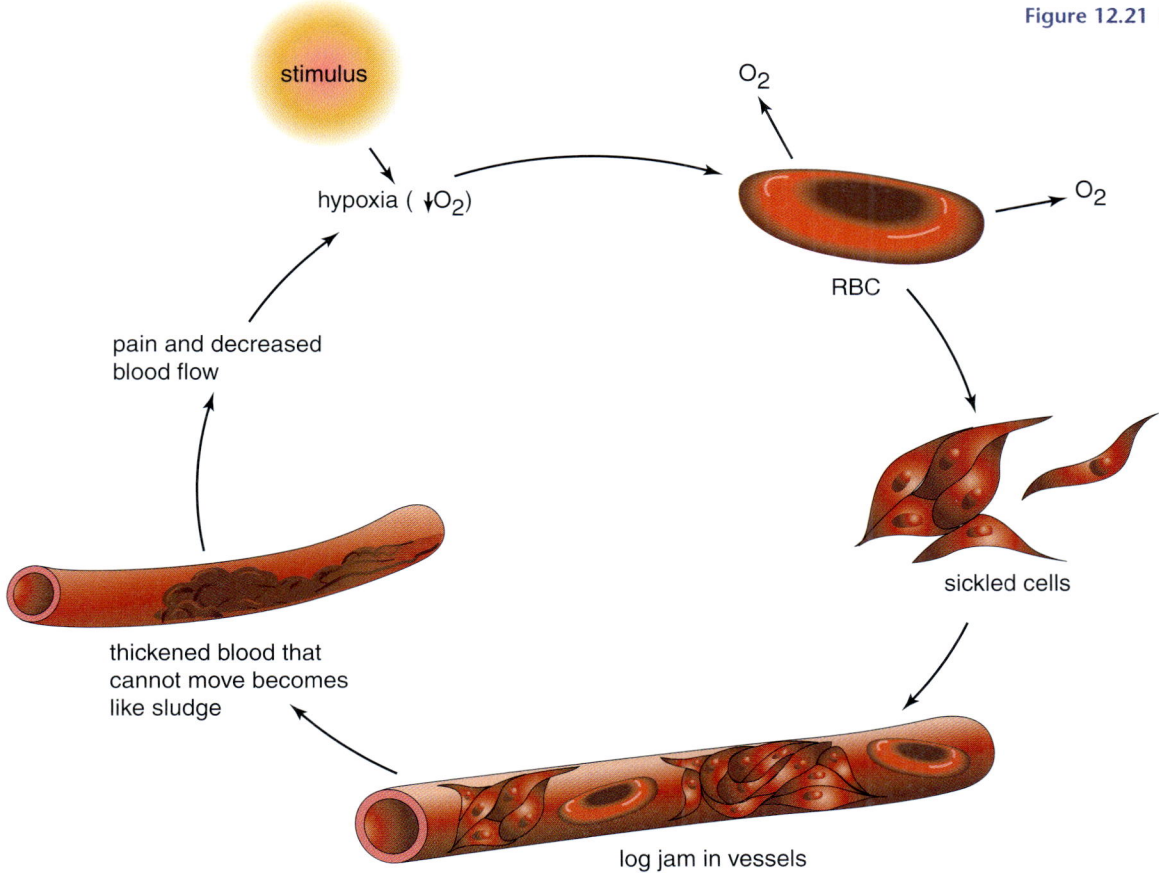

Figure 12.21 Sickle Cell Disease

Genetic disorders can lead to the production of abnormal blood cells. A group of diseases called **thalassemias** alter the hemoglobin molecule. This reduces the ability of RBCs to carry oxygen. **Sickle cell disease** is a form of thalassemia that alters the shapes of the RBCs after they pick up carbon dioxide from the tissues (Figure 12.21). The RBCs take on a crescent, or sickle, shape that causes them to clog capillaries. These abnormal cells are removed by the spleen, reducing the body's ability to carry oxygen. Thalassemias are common in people of Mediterranean and North African descent. Certain conditions result in the production of RBCs that are larger than normal, which is called **macrocytic anemia**. It is caused by a variety of factors that affect bone-marrow erythropoiesis. Lead poisoning and malnutrition can produce a condition called **microcytic anemia**. Microcytic RBCs are not effective at picking up or releasing oxygen in the capillaries.

Blood clotting is also subject to a variety of disorders. **Hemophilia** is a rare, inherited bleeding disorder. People with hemophilia will bleed for a long time following an injury or accident. They also may bleed internally, especially during normal motion of the joints. There are many types of hemophilia. Most types are caused by deficiencies of the blood clotting factors. Others are due to platelet abnormalities.

The WBCs are prone to deficiencies and cancers. Most deficiencies in WBC number are due to malnutrition and exposure to hazardous chemicals. Bone marrow cancers that affect WBCs are caused by a variety of factors. **Leukemia** is a group of four cancer types that produce an increased number of abnormally functioning WBCs. Tens of thousands of cases occur each year in North

Thalassemia A genetic disorder of hemoglobin in red blood cells

Sickle Cell Disease A genetic disorder of hemoglobin that affects the shape of the RBCs

Macrocytic Anemia Anemia characterized by RBCs that are larger than normal

Microcytic Anemia Anemia characterized by RBCs that are smaller than normal

Hemophilia A genetic disease that prevents normal blood clotting

Leukemia A cancer characterized by an increased number of abnormal leukocytes

Hodgkin's Lymphoma Cancers of the lymphatic system cells

Non-Hodgkin's Lymphoma Cancers of the lymphocytes in the lymph nodes

Lymphedema Swelling of the appendages due to blockage of the lymph vessels

Elephantiasis Blockage of the lymphatic system by infectious worms

America. This treatable disease weakens the immune system and can spread cancerous cells throughout the body. **Lymphomas** are cancers that originate in cells of the lymphatic system. They start as cancers of the lymph nodes and spleen that can spread to other parts of the body. Lymphomas are divided into **Hodgkin's** and **non-Hodgkin's**. Hodgkin's lymphomas are formed by cancerous lymphatic system cells other than lymphocytes. Non-Hodgkin's lymphomas are the most common type of lymphomas and are caused by lymphocytes that form into cancerous cells. Myelomas are general bone marrow cancers that affect the development of a variety of WBCs.

Disorders of the lymphatic system affect its ability to drain body fluids and carry out immune functions. **Lymphedema** is a condition in which excess fluid collects in body tissues due to a loss of flow in the lymphatic vessels. The condition causes abnormal swelling of the affected body parts. It usually occurs in the arms and legs, but it can also develop in the groin (Figure 12.22). **Elephantiasis** is a type of lymphedema caused by small worms that block the lymph nodes and vessels. Mosquitoes spread the tropical worm between animals and people. Swelling from elephantiasis is more severe than other lymphedema (Figure 12.23).

Figure 12.22 Lymphedema

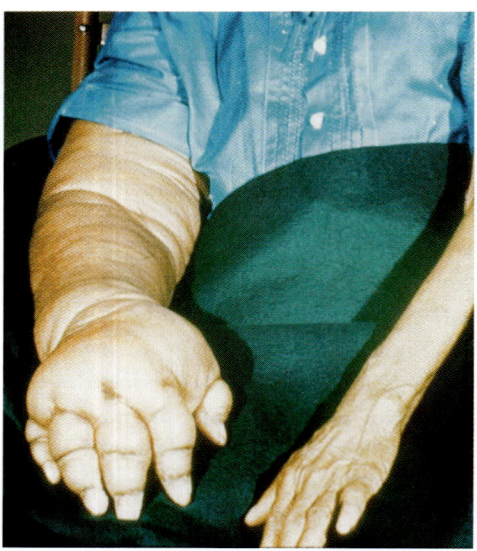

Figure 12.23 Elephantiasis

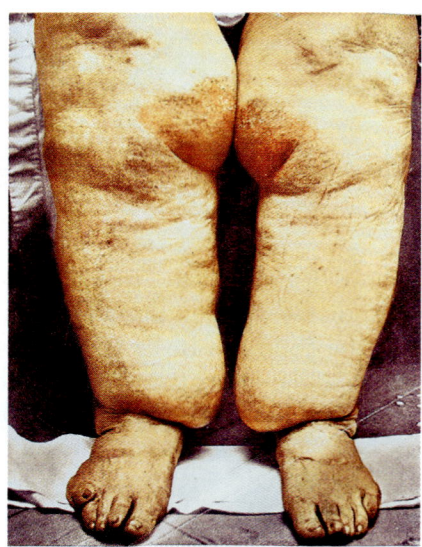

Immunodeficiency A decreased ability of the body to fight infection and disease

Acquired Immunodeficiency Syndrome (AIDS) An infectious disease caused by the human immunodeficiency virus (HIV)

Human Immunodeficiency Virus (HIV) An infectious virus that causes the immunodeficiency disease AIDS

A variety of disorders can diminish immune system function or increase its sensitivity. **Immunodeficiency** disorders cause the body to lose its ability to fight disease. Many conditions can produce immunodeficiency; however, malnutrition and stress are the most common causes. Some types of immunodeficiency disorders are caused by a variety of genetic conditions that impair WBC function. There is global concern about infectious agents that weaken the immune system's defenses. Most notable is **acquired immunodeficiency syndrome** (**AIDS**), which is caused by the **human immunodeficiency virus** (**HIV**). All viruses have the ability to weaken the immune system. However, the AIDS virus specifically attacks the helper T-lymphocytes that facilitate primary immunity. This reduces the body's ability to fight disease and encourages the activity of suppressor T-lymphocytes, which inhibit the immune response. People with AIDS waste away and die from a variety of cancers and infectious diseases (Figure 12.24).

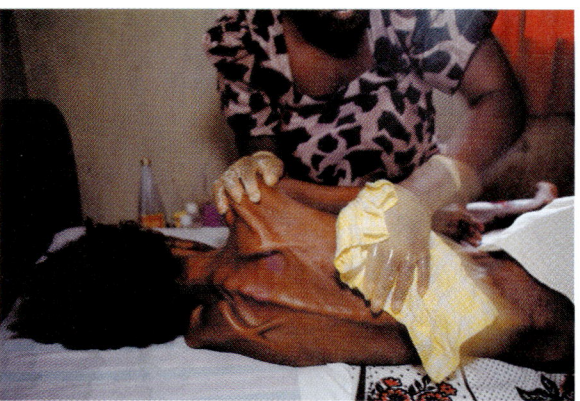

Hypersensitivities are disorders in which the immune system overreacts to an antigen. **Allergies** are the most common form of hypersensitivity. People with allergies are hypersensitive to the inflammatory effects of IgEs. The IgEs cause mast cells to release a variety of immune system chemicals that stimulate secretions, including histamine. An overproduction of histamine causes excessive mucus production, swelling of certain tissues, and, possibly, rashes on certain parts of the body. Other secretions can produce a fever and other ill effects. Allergies are usually treated by avoiding the specific antigen causing the response and by administering anti-inflammatory drugs. Certain treatments can desensitize the immune system to the antigen so that the body reduces its exaggerated response.

> **Hypersensitivity** An exaggerated response by the immune system to an antigen
>
> **Allergy** An overreaction of the body's immune system

Figure 12.24 AIDS

✓ Concept Check

1. Describe the different types of anemia.
2. Explain the different ailments of WBCs.
3. Distinguish between immune deficiencies and hypersensitivity.

Aging of the Blood and Lymphatic System

Key Terms: autoantibody, autoimmunity

Much of the aging of the blood and lymphatic system is due to changes in the digestive and endocrine systems. A decrease in nutrient absorption diminishes the ability for bone marrow and immune system cells to replicate. In addition, some of the hormones needed for body maintenance decline and impair the regulation of blood cell formation. Research shows that the percentage of hematopoietic tissue in the bone marrow declines progressively from birth until about age 30, then levels off until about 70 years of age. After age 70, the production of blood cells declines steadily. This mostly occurs in the long bones and usually does not affect the flat bones. Microscopic studies of the bone marrow indicate a decline in stem cells. An age-related decrease in erythropoietin further reduces RBC production. Because of this, elderly people become more susceptible to anemia and the complications of immunodeficiency.

White blood cell function is likely to become abnormal with age. A lifetime of exposure to chemicals and other factors that damage DNA increases the risks for bone marrow cancers. In addition, the WBCs are more likely to mistake normal cells for foreign antigens as the body ages. This causes the body to attack itself, producing a condition called **autoimmunity**. B-lymphocytes are stimulated to produce antibodies that attack certain proteins on the body's tissues. These antibodies are called **autoantibodies**. It is also likely for the body to produce autoantibodies against abnormal proteins on the surfaces of body cells. The synthesis of abnormal proteins increases with age.

> **Autoimmunity** A condition in which the body produces an immune response against its own organs or tissues
>
> **Autoantibody** An antibody that reacts against the person's own tissues

Figure 12.25 Aging of the Lymphatic System

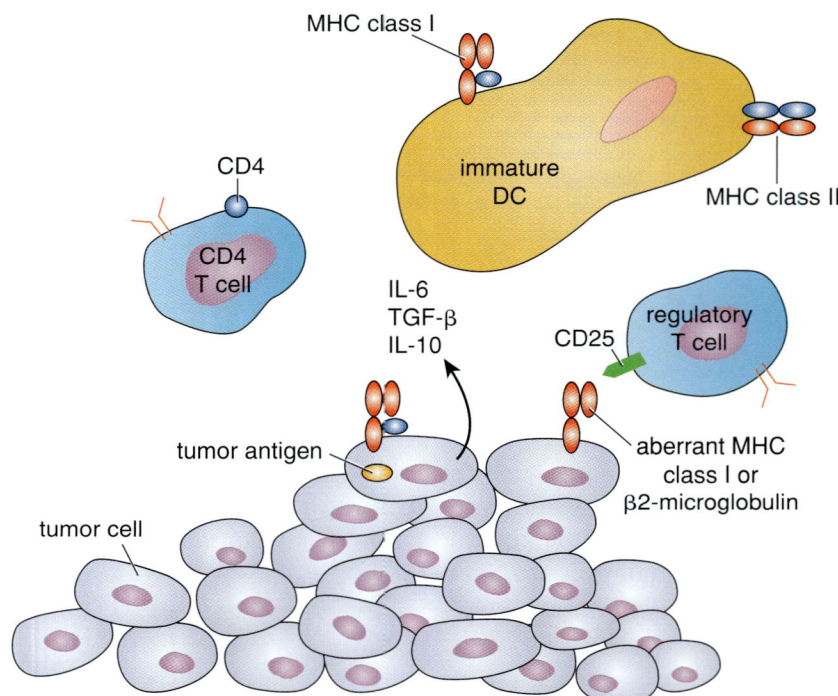

The risk for an autoimmune response increases when a person's body is fighting off tumors, an event that may increase in frequency as the body ages (Figure 12.25).

Immobility and a weakening of the skeletal muscles is another feature of aging. This makes the body more prone to lymphedema. A decline in venous flow also makes it more difficult for the lymphatic system to remove tissue fluids. Sometimes this condition is corrected by placing tight wrappings around the arms and legs. Elevating the feet and hands also assists the lymphatic system in removing tissue fluid.

✓ Concept Check

1. Describe the effects of aging on the bone marrow that produces blood cells.
2. Explain how age can encourage autoimmune disorders.
3. What are two likely causes of lymphedema in aging people?

DISCOVERY SCENE PLEASE ENTER DISCOVERY SCENE PLEASE ENTER

Did you come to a conclusion about the father's situation in the CSI? Does it resemble any type of pathology discussed in the chapter? Could it also have been associated with aging of the blood or immune system?

CSI — Case Study Investigation Conclusion

What can you conclude about your friend's father? Is his condition due to the truck restoration? Can it be aggravating an existing condition, or is his hobby making him ill?

Answer:

Your friend's father is feeling the effects of long-term benzene exposure. Despite its poisonous nature, benzene is a widely used solvent in North America. It is one of the top 20 chemicals in industrial use. Benzene is sometimes used to make automotive cleaning agents and lubricants. Exposure to benzene disrupts cell metabolism. It interferes with cells' ability to produce energy and carry out mitosis. As a consequence, benzene can reduce blood cell production by bone marrow and lead to anemia. It can also inhibit the immune system's ability to produce antibodies by decreasing lymphocyte function. The degree of damage benzene causes depends on the amount, duration, and route of exposure. Benzene exposure can also aggravate any preexisting diseases. There is no treatment for benzene poisoning. People are encouraged to eliminate exposure, and, hopefully, allow the body to recuperate from much of the damage.

This CSI was adapted from the following articles:

1. Brandao MM, Rego MA, Pugliese L, et al. Phenotype analysis of lymphocytes of workers with chronic benzene poisoning. *Immunol Lett.* 2005;101(1):65-70.
2. Goldstein, BD. Benzene toxicity. *Occup Med: State Art Rev.* 1988;3(3):541-554.
3. Kuang S, Liang W. Clinical analysis of 43 cases of chronic benzene poisoning. *Chem Biol Interact.* 2005;30:(153-154):129-135.

Chapter Summary

Blood is a vital fluid that provides many resources for satisfying the metabolic needs of cells throughout the body. The plasma component of blood brings many nutrients to cells and carries away wastes. It also stabilizes the chemical environment of the cell by providing ions that regulate pH and cell transport. The RBCs make up the bulk of the formed elements of blood. They carry oxygen to cells and remove carbon dioxide, so it can be expelled by the lungs. Diseases of the RBCs have far-reaching effects on the body because RBCs transport the gases associated with cell metabolism. The WBCs form part of the immune system that protects the body from disease and assists with repair after an injury. The inflammatory response is one normal feature of tissue repair the WBCs and platelets carry out.

The lymphatic system uses its own components, and cells derived from the blood, to prevent and fight off infections. However, it is just one component of the immune system. Other organ systems that act as barriers to infection fight off many microorganisms. These barriers either inhibit microbial growth or stop microbes from entering the body. The acquired immune response permits the body to tailor its mechanisms for removing specific infectious agents. One general immune response is not adequate for controlling infection by all the different microorganisms. Bacteria, fungi, and viruses must be killed using different strategies. Physicians take advantage of secondary immunity by administering immunizations that permit the body to effectively fight diseases without having to go through a primary immune response. Unfortunately, the blood and lymphatic system are subject to many diseases that reduce their capability to fight disease. Aging also brings about changes to the immune system that make people more susceptible to infections and less likely to heal injuries at a rapid rate.

Overview

- Blood is composed of plasma and formed elements.
- Blood transports materials needed for body homeostasis.
- There are three blood cell types: RBCs, WBCs, and platelets.
- The lymphatic system regulates body fluids and the immune system.

Blood Cells

- All adult blood cells are produced in the bone marrow from stem cells.
- The RBCs contain the protein hemoglobin.
- The WBCs are categorized as granulocytic and agranulocytic.
- Granulocytic WBCs include basophils, eosinophils, and neutrophils.
- Agranulocytic WBCs include lymphocytes and monocytes.
- Platelets, or thrombocytes, are cell fragments that are involved in blood clotting.

Blood Cell Function

- The RBCs transport oxygen and carbon dioxide.
- Granulocytes produce secretions that kill microorganisms.
- Monocytes are phagocytic agranulocytes.
- Lymphocytes are agranulocytes that produce the immune response.
- Platelets assist with blood clotting.

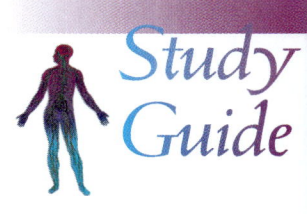

Study Guide

Blood Cell Formation
- Blood cells are produced in the bone marrow by hematopoiesis.
- All blood cells are derived from a multipotent stem cell.
- Erythropoietin stimulates RBC production.

Lymphatic System
- The lymphatic system is composed of lymphatic glands, lymph nodes, and lymph vessels.
- The WBCs provide the immune system function.
- Innate immunity provides barriers against infections.
- Acquired immunity permits the body to recognize specific infections.
- The primary immune response is the first reaction to an antigen.
- The secondary immune response protects against subsequent infections.

Immunization and Vaccination
- Immunization is a strategy for enhancing the immune response.
- Natural immunity is the unintentional stimulation of the immune system.
- Artificial immunity is the use of a therapy to stimulate the immune system.
- Active immunity is caused by exposure to antigens.
- Passive immunity is produced by providing the body with antibodies.
- Vaccines are immunizations that produce active immunity.

Wellness and Illness over the Life Span
- Anemia is a decrease in normal RBCs.
- Cancer is a common WBC disorder.
- Hemophilia is a disorder that prevents clotting.
- Hypersensitivity diseases and allergies cause the body to overreact to antigens.
- The digestive and endocrine systems affect blood and immune system aging.
- Damage to DNA contributes to WBC abnormalities.
- Autoimmunity is more likely to occur with aging.
- Lymphedema is more likely to occur with aging.

Key Terms

Overview
Centrifuge
Erythrocyte
Hematocrit
Leukocyte
Packed cell volume (PCV)
Plasma
Platelet
Red blood cell (RBC)
Thrombocyte

Red Blood Cells
ABO blood group system
Blood type
Complete blood count (CBC)
Erythroblastosis fetalis
Full blood count (FBC)
Hemoglobin
Hemogram

THE LYMPHATIC SYSTEM AND THE BLOOD

Study Guide

Mean corpuscular hemoglobin (MCH)
Mean corpuscular hemoglobin concentration (MCHC)
Mean corpuscular volume (MCV)
Reticulocyte
Rh factor
Transfusion

White Blood Cells (WBCs)
Agranulocyte
B-lymphocyte
Basophil
Differential white blood cell (WBC) count
Eosinophil
Granulocyte
Lymphocyte
Monocyte
Mononuclear white blood cell (WBC)
Neutrophil
Polymorphonuclear white blood cell (WBC)
T-lymphocyte

Platelets
Megakaryocytes

Red Blood Cell Function
Acute
Carbon dioxide intoxication
Carbonic anhydrase
Percent saturation
Subacute

White Blood Cell Function
Antibiotic
Circulating monocyte
Kupffer cell
Macrophage
Major basic protein
Mast cell
Tissue monocyte

Platelet Function
Clotting cascade
Clotting factor
Fibrin
Fibrinogen
Plasmin
Plasminogen
Prostacyclin
Prothrombin
Thrombin
Tissue plasminogen activator (TPA)

Blood Cell Formation
Band
Bilirubin
Erythroblast
Erythropoiesis
Erythropoietin
Hematopoietic stem cell
Lymphoid progenitor
Multipotent stem cell
Myeloid progenitor
Reticulocyte
Stab

Structures of the Lymphatic System
Hilum
Lymph
Lymph gland
Lymph node
Lymphatic sinus
Lymphatic trunk
Lymphatic vessel
Peyer's patches
Red pulp
Spleen
Trabeculae
White pulp

Immune Response
Acquired immunity
Antibody
Antigen
Cell-mediated immunity
Complement
Hapten
Helper T-lymphocyte
Humoral immunity
IgA
IgD
IgE
IgG
IgM
Immunoglobulin
Inflammatory response
Innate immunity or nonspecific immunity
Interferon
Memory cell
Natural killer (NK) cell

Study Guide

Plasma cell
Primary response
Secondary response
Suppressor T-lymphocyte

Immunization and Vaccination
Active immunity
Artificial immunization
Globulin
Immunization
Natural immunization
Passive immunity
Smallpox
Vaccine
Vaccination

Pathology of the Blood and Lymphatic System
Acquired immunodeficiency syndrome (AIDS)
Allergy
Anemia
Elephantiasis
Hemophilia
Hodgkin's lymphoma
Human immunodeficiency virus (HIV)
Hypersensitivity
Immunodeficiency
Leukemia
Lymphedema
Lymphomas
Macrocytic anemia
Microcytic anemia
Non-Hodgkin's lymphoma
Pernicious anemia
Sickle cell disease
Thalassemia

Aging of the Blood and Lymphatic System
Autoantibody
Autoimmunity

Check Your Understanding

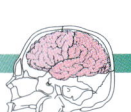

1. Which of the following is *not* a function of blood:
    a. transports gases to cells
    b. carries metabolic wastes from cells
    c. breaks down food particles
    d. is involved in the immune response

2. At least 55% of normal blood is composed of:
    a. WBCs
    b. plasma
    c. metabolic wastes
    d. thrombocytes

3. RBCs do *not* have this characteristic:
    a. a large nucleus
    b. hemoglobin in the cytoplasm
    c. the ability to carry carbon dioxide
    d. a disc-like shape

4. Which statement is true of hemoglobin?
    a. gives WBC granules their color
    b. carries oxygen and carbon dioxide
    c. is commonly found in the blood plasma
    d. is removed from the body as urea

THE LYMPHATIC SYSTEM AND THE BLOOD

Study Guide

5. All adult blood cells are made in the:
   a. liver
   b. spleen
   c. bone marrow
   d. thymus

6. Which of the following is *not* a leukocyte?
   a. lymphocyte
   b. eosinophil
   c. basophil
   d. thrombocyte

7. What is the major protein that forms into a blood clot?
   a. prothrombin
   b. immunoglobulin
   c. albumin
   d. fibrin

8. Iron has the following role in blood function:
   a. helps hemoglobin bind to oxygen
   b. is needed to form platelet stickiness
   c. removes carbon dioxide from tissues
   d. forms the granules in white blood cells

9. A lack of this chemical reduces RBC formation:
   a. leukotriene
   b. vitamin K
   c. erythropoietin
   d. prothrombin

10. Which of the following is not a granulocyte:
    a. monocyte
    b. basophil
    c. neutrophil
    d. eosinophil

11. Which structure is not part of the immune system?
    a. thymus
    b. thyroid
    c. lymph node
    d. spleen

12. Which type of blood cell matures in the thymus?
    a. B-lymphocyte
    b. T-lymphocyte
    c. erythrocyte
    d. macrophage

13. Antibodies are secreted by:
    a. damaged tissue
    b. B-lymphocytes
    c. monocytes
    d. thrombocytes

14. The secondary immune response is primarily initiated by these cells:
    a. platelets
    b. memory cells
    c. macrophages
    d. basophils

15. These cells are killed by AIDS:
    a. T-lymphocytes
    b. B-lymphocytes
    c. basophils
    d. reticulocytes

A Case Study

ENVIRONMENT IMMUNIZAITON

An illusive, controversial immunological condition called multiple chemical sensitivity (MCS) syndrome has been diagnosed as various diseases since the 1940s. No known, measurable clinical signs are associated with MCS syndrome. It is identified by a patient's conviction that his or her symptoms are caused by exposure to a variety of environmental chemicals and materials. People with MCS syndrome claim to have allergic reactions to many types of household chemicals and personal hygiene items. This syndrome has created a great controversy among physicians, public health officials, and medical researchers. Many medical professionals reject MCS syndrome as an established organic disease. The American Academy of Allergy and Immunization, American Medical Association, American College of Physicians, and International Society of Regulatory Toxicology and Pharmacology characterize MCS syndrome as a psychological condition.

The condition was first identified in 1940 by Theron G. Randolph, a physician who specialized in allergies. He claimed that patients with MCS syndrome became ill when exposed to very low levels of chemicals, levels that do not normally cause problems. Randolph claimed that people with the condition suffered behavioral problems, confusion, depression, fatigue, and irritability. He believed the condition resulted from the inability of certain people to adapt to modern synthetic chemicals. The condition has been attributed to a variety of situations ranging from pesticide applications to Gulf War syndrome noted in troops fighting in Kuwait. There is no known immunological explanation for why low levels of chemicals should cause the characteristic symptoms. There are no consistent clinical tests or diagnostic strategies for identifying MCS syndrome; however, that has not stopped many attorneys from taking on lawsuits from people claiming that their homes or work environments are handicapping them because of MCS syndrome.

Use the information in this chapter and the following Web sites to answer the following questions about MCS syndrome:

- What research is needed to confirm the existence of MCS syndrome?
- How should physicians treat an ailment that is not a confirmed medical condition?
- Do physicians have the right to recommend psychiatric visits for people with MCS syndrome?

Study Guide

- Should health insurance companies be required to cover the medical care for people claiming MCS syndrome?
- Should the government put any restrictions on treatments claiming to alleviate the symptoms of MCS syndrome?
- How does medical science resolve the controversy of a purported disease that cannot be detected using current medical testing?

1. MCS Survivors
   http://www.mcsurvivors.com/

2. Multiple Chemical Sensitivity Organization
   http://www.mcsrr.org/

3. National Institute of Environmental Health
   http://www.niehs.nih.gov/external/faq/allergy.htm

4. Quackwatch
   http://www.quackwatch.org/01QuackeryRelatedTopics/nejac.html

Where Do We Go from Here?

People in health fields can use their knowledge of the blood and lymphatic system to solve everyday problems. You may wish to use other resources in addition to your book to investigate the answers to each of the following situations.

1. A friend read on a Web site that the risk for cancer can be reduced by building up the immune system. What would you tell him about the benefits and risks of following this advice?
   http://www.cancer.gov/

2. A female friend had breast cancer surgery that involved the removal of lymph nodes in her armpits. She asks you what specific effects this would have on her health. How would you answer her query?
   http://www.niaid.nih.gov/final/immun/immun.htm

3. You remember hearing as a child that stress causes you to become sick more often. Explain to your friends if there is any truth behind this statement.
   http://www.apa.org/ (Search "stress immune system")

4. A female relative who is a vegetarian heard from one of her friends that she is susceptible to developing anemia because she does not eat meat. Explain to her the basis of the warning that she was given by her friend.
   http://www.emedicinehealth.com

5. One of your friends refuses to be vaccinated against the flu. He claims that the vaccine can cause more harm than the disease. What would you explain to your friend?
   http://www.acsh.org/

Skills Activities

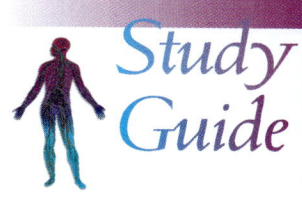

1 Pathology of the Blood and Lymphatic System

Materials
- microscope with high-power capability (400X)
- prepared slide of normal red bone marrow
- prepared slide of normal human blood
- access to this Web site:
  http://www-medlib.med.utah.edu/WebPath/HEMEHTML/HEMEIDX.html

A biopsy and microscopic analysis of blood, lymph nodes, and red bone marrow is sometimes necessary to identify cancers and blood-related diseases. A biopsy is a collection of live tissues from a patient. Pathologists compare normal images of the tissue under examination to the samples from the patient. From this, they are able to determine any visible differences that may cause an abnormality. However, before they can do this, they must practice comparing the normal samples to pictures to help them determine the different components and cell types making up the blood and lymphatic system.

Place a microscopic slide of the red bone marrow under the microscope. Start with low magnification and move to higher magnification to see details and individual cells. Can you find the RBCs (erythroid precursors) and WBCs granulocytic precursors) after comparing the slide to a photograph below from the Hematopathology Index Web site?

- normal bone marrow, medium-power microscopic image
- normal bone marrow, high-power microscopic image
- normal bone marrow smear, high-power microscopic image No. 3
- normal bone marrow smear, high-power microscopic image No. 4
- normal bone marrow smear, high-power microscopic image No. 5

Repeat these steps for the blood slide by comparing its findings with the following images:

- normal RBCs on the smear, microscopic image
- WBC identification, microscopic image

Can you identify the RBCs and the different types of WBCs?

Now, view the photographs of the following diseased specimens found on the Hematopathology Index Web site:

- hypochromic microcytic anemia on smear, microscopic
- CBC with iron deficiency anemia, diagram
- hypersegmented neutrophil with megaloblastic anemia on smear, microscopic image
- CBC with megaloblastic anemia, diagram
- atypical lymphocytes on smear, microscopic image
- sickle cell disease on smear, microscopic image
- acute myeloblastic leukemia on smear, microscopic image
- acute myeloblastic leukemia in bone marrow, low power microscopic image

Can you recognize the differences in the cell types that characterize each disease?

THE LYMPHATIC SYSTEM AND THE BLOOD

Study Guide

Skills Activities

2 Assessing Potential Allergens

Materials
- clear cellophane tape
- small centimeter ruler
- three nutrient agar Petri plates per test
- incubator set at 37°C
- microscope with high-power capability (400X)
- five clean microscope slides per sampling
- access to *http://images.google.com/*

The Environmental Protection Agency (EPA) recognizes that indoor air quality (IAQ) is a major causal factor of many illnesses at home and at work. It is estimated that most Americans spend up to 90% of their time indoors. Much of this time is spent at school or at work. The EPA estimates that poor IAQ costs billions of dollars a year in medical care and missed work. Many of the pollutants that reduce the quality of air in buildings increase a person's risk for illnesses due to allergies, asthma, and respiratory diseases. Microorganisms, such as bacteria and molds, contribute to allergies and respiratory diseases. Cockroach parts, dust mites, fur, and pollen are indoor contaminants that aggravate allergies and asthma. Shed skin cells carry viruses that can cause respiratory diseases. This activity investigates a simple way to detect the presence of potential allergens and other substances that reduce IAQ.

Microorganism Collection

1. Collect three nutrient agar Petri plates and leave them uncovered in three different areas of the room. Try to select areas that differ in air flow, height, human traffic flow, or proximity to doors.
2. Let the plates remain open for 30 minutes.
3. After 30 minutes, put the lid on each Petri plate and place them in a warm area of the room or in an incubator set at 37°C for 2 days.
4. After the 2-day period, remove the Petri plates from the incubator and carefully view the types of microorganisms growing on the agar surface of each Petri plate. Do not remove the lid.
5. Identify your specimens using a Google Image Internet search to find pictures of bacteria and fungi growing on Petri plates (*http://images.google.com/*). Type in the terms "bacteria culture" and "mold culture" in two separate searches.
6. See if you can tell whether bacteria and mold are growing on your Petri plates.
   Do the Petri plates differ in the types of growth for each location?
7. Compare your Petri plates with those of other students.

Dust Collection

1. Collect five 2-cm pieces of tape
2. Place the "sticky" side on five different surfaces, including tabletops, clothing, shoe bottoms, and air vents.
3. Stick each piece of the tape to the center of a separate microscope slides.
4. Place one microscopic slide on the microscope. Start with low magnification and move to higher magnification to see details of the dust particles.
5. Look up photographs of the various dust particles using a Google Image Internet search (<http://images.google.com/>). Type in terms such as "cockroach," "cat hair," "dog hair," "dust mites," and "dust particles" to find appropriate images. Also investigate the forensic fibers Web sites for more photographs of materials found in dirt and dust.
6. Repeat this for each slide.
7. See what types of potential allergens you can find on the different materials you sampled.
8. Compare what you found with other students' samples.

Case Study Investigation

Case Study Investigation #13

A classmate tells you that he missed three classes due to illness. He tells you that he had no health problems until about 2 months ago when he started developing recurring bouts of diarrhea, four times in the past 2 months. Each time, the diarrhea lasted 3 days and then went away. Since you are in an anatomy and physiology class, he had no reservations describing that he had semisolid bowel movements with some blood after 2 days of diarrhea. During the bouts of diarrhea, he experienced headaches, mild cramps, and abdominal pain on the right side, but no fever, nausea, or vomiting. You notice that your classmate smokes regularly. He also has significant stress from school and work. Your classmate mentions that his mother had an intestinal disorder, but he cannot remember the name of the disease. What could be causing your classmate's ailments?

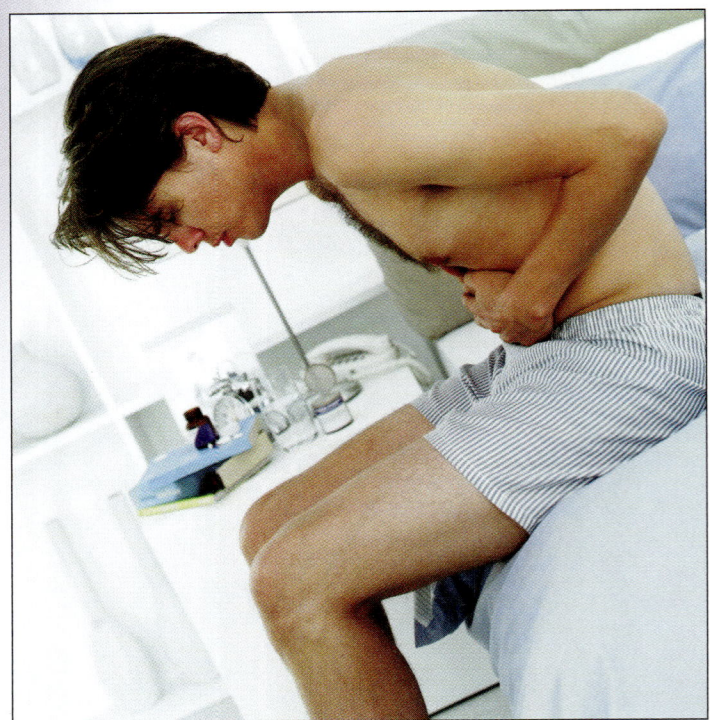

CHAPTER 13

THE DIGESTIVE SYSTEM

Chapter Outline

Case Study Investigation (CSI)
Applied Learning Outcomes
Overview
Components of the Digestive Tract
   Mouth and Pharynx
   Esophagus and Stomach
   Small Intestine
   Large Intestine and Rectum
Glandular Structures of Digestive System
Pancreas
Liver and Gallbladder
The Digestive Process
Wellness and Illness over the Life Span
   Pathology of the Digestive System
   Aging of the Digestive System
CSI Conclusion
Study Guide

Applied Learning Outcomes

- Use the terminology associated with the digestive system.
- Learn about the following:
  - organization of the digestive system
  - structure of the digestive system organs
  - function of the digestive system organs
  - the digestive process
  - waste production
- Understand the aging and pathology of the digestive system.

OVERVIEW

Key Terms: accessory digestive organ, alimentary canal, digestive tract, esophagus, gallbladder, gastrulation, large intestine, meconium, rectum, salivary gland, small intestine, stomach

If all the organ systems had to vote on each system's importance to the body, they would most likely rate the digestive system high on the list. The digestive system allows the body to break down complex molecules into simple molecules. Many of these molecules are needed for body energy. Other simple molecules, built into larger molecules, are used in cell and tissue development. The digestive system is composed of two components: the digestive **tract**, or **alimentary canal**, and the accessory digestive organs (Figure 13.1). The digestive tract comprises the mouth, **esophagus**, **stomach**, **small intestine**, **large intestine**, and **rectum**. Accessory digestive organs produce secretions that

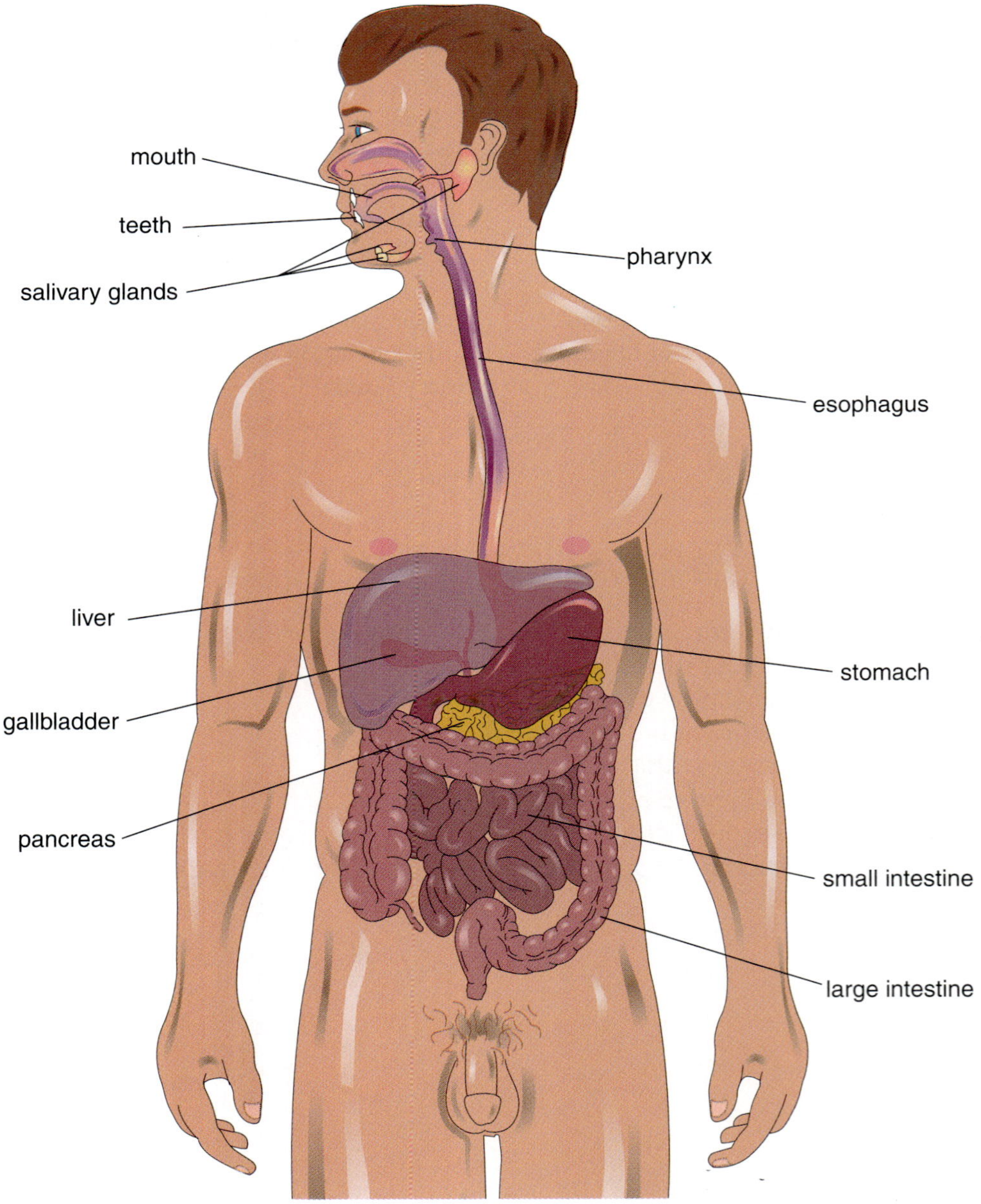

Digestive Tract or Alimentary Canal The tubular organs through which food passes

Accessory Digestive Organs The secretory organs associated with the digestive tract

Esophagus A muscular tube through which food passes from the mouth to the stomach

Stomach An enlarged, muscular sac-like organ of the digestive tract

Small Intestine The part of the digestive tract located between the stomach and the large intestine

Large Intestine A tube that extends from the small intestine to the rectum

Rectum The last section of large intestine

Figure 13.1 Components of the Digestive System

490

CHAPTER 13

help the digestive tract with digestion and food absorption. The **salivary glands**, pancreas, liver, and **gallbladder** make up the accessory digestive organs.

The digestive system is one of the first organ systems to develop in the human embryo. All of the digestive system develops from a pocket of endoderm located inside the embryo. This occurs after week 4 of development during a process called **gastrulation**. After gastrulation, a simple digestive tract called the digestive tube develops. At this point, there are no accessory digestive organs. A simple liver and pancreas start to form after week 5 of development. The small intestine is connected to blood vessels of the umbilical cord. It is difficult to distinguish the small intestine from the large intestine until after week 6 of development. By week 8, the digestive tract grows rapidly and starts to look like a miniature of the adult system. Coiling of the small intestine takes place at this time, as well as the shaping of the large intestine. The accessory digestive organs also grow rapidly and are now capable of producing their particular secretions. Much of the liver is formed by week 40. It begins to produce blood cells for the fetus at this time. At birth, the digestive system is fully formed and already contains a waste product called **meconium**. It is a greenish-black, sticky substance that the baby passes at or after birth. Meconium consists of bile, mucus, and epithelial cells from the digestive tract.

Salivary Gland A gland in the mouth that secretes saliva

Gallbladder A muscular sac attached to the liver

Gastrulation A stage in embryonic development that forms the tissues of the digestive system

Meconium The intestinal contents of the embryo

✓ Concept Check

1. Describe the digestive tract.
2. What are the accessory organs of the digestive system?
3. Describe the development of the digestive system.

DISCOVERY SCENE PLEASE ENTER DISCOVERY SCENE PLEASE ENTER

What major parts of the digestive system are involved in the disorder described in the CSI? What additional information is needed to better understand the cause of the ailment?

COMPONENTS OF THE DIGESTIVE TRACT

Mouth and Pharynx

Key Terms: bicuspid, buccal cavity, canine, cuspid, extrinsic muscle, hard palate, incisor, intrinsic muscle, lingual membrane, lingual tonsil, labia, molar, palate, parotid, premolar, soft palate, sublingual, submandibular, uvula, wisdom teeth

The mouth makes up the first part of the digestive tract (Figure 13.2). It is capable of both chemical and mechanical digestion. The anterior portion of the mouth is covered by the lips, or **labia**. Lips contain sensory receptors that detect the temperature and texture of food. They get their color from the large quantity of blood vessels just underneath the skin. The internal border of the lips, just inside the mouth, is where the facial skin meets the mouth's mucous membranes. The cheeks, which form the lateral walls of the

Labia Another name for the lips

THE DIGESTIVE SYSTEM

Buccal or Oval Cavity The portion of the oropharynx associated with the lips and cheeks

Palate The roof of the buccal cavity

Hard Palate The anterior part of the palate that covers the maxillary and palatine bones

Soft Palate The muscular, posterior part of the palate

Uvula A cone-shape projection posterior to the soft palate

mouth, form the **buccal**, or **oval cavity**. The buccal cavity is the area located inside the teeth and is where food is moistened in preparation for the first steps of digestion. The **palate** forms the roof of the buccal cavity. The anterior portion of the palate covers the maxillary and palatine bones, and forms a hard plate called the **hard palate**. Posterior to the hard palate is the **soft palate**. It is a muscular arch with a cone-shaped projection on its posterior portion, which is called the **uvula**. The posterior part of the buccal cavity opens into the oropharynx (Figure 13.3). At the base of the oropharynx is the epiglottis, which separates the respiratory system from the pharynx.

Figure 13.2 The Mouth

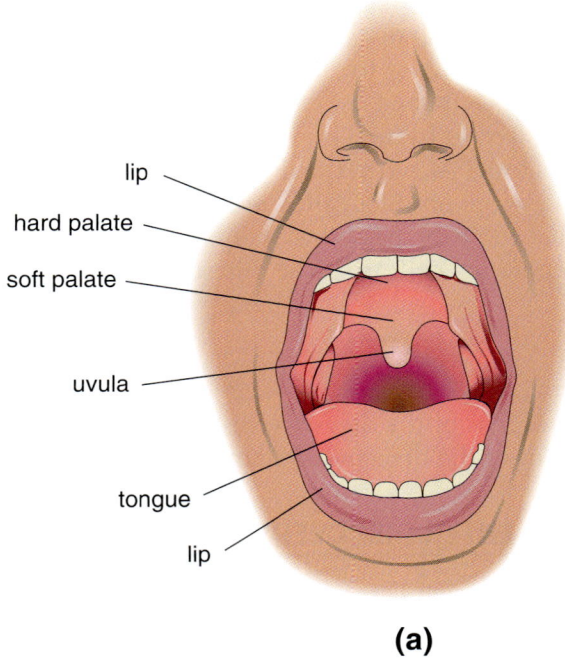

(a)

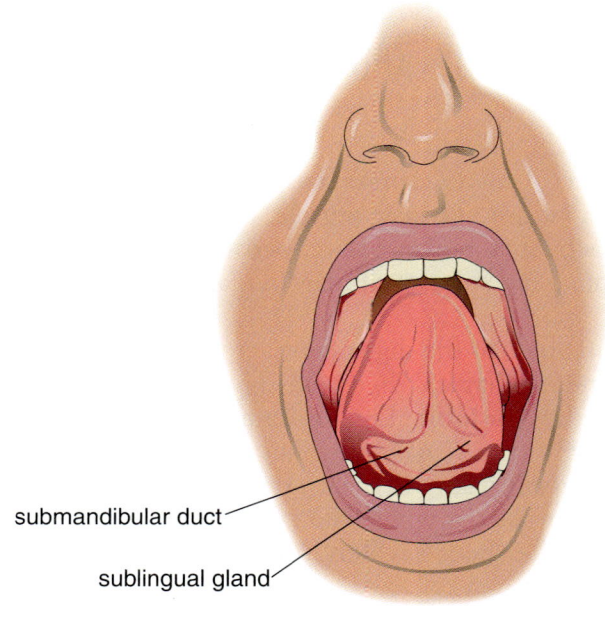

(b)

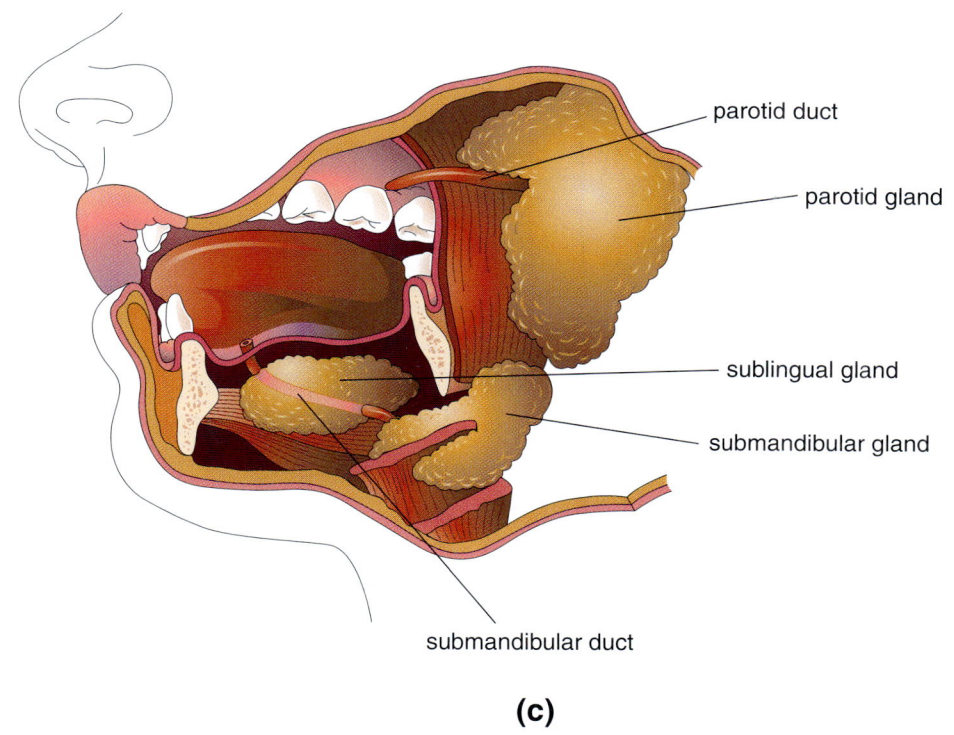

(c)

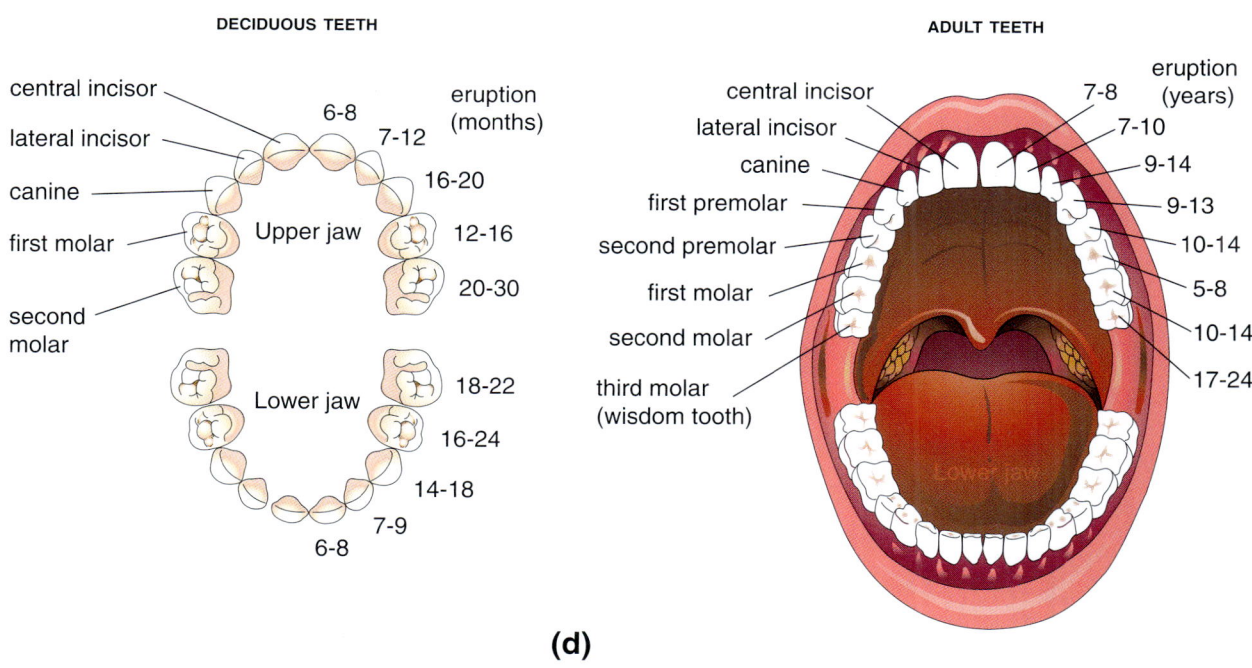

(d)

Figure 13.3 Pharynx and Surrounding Structures

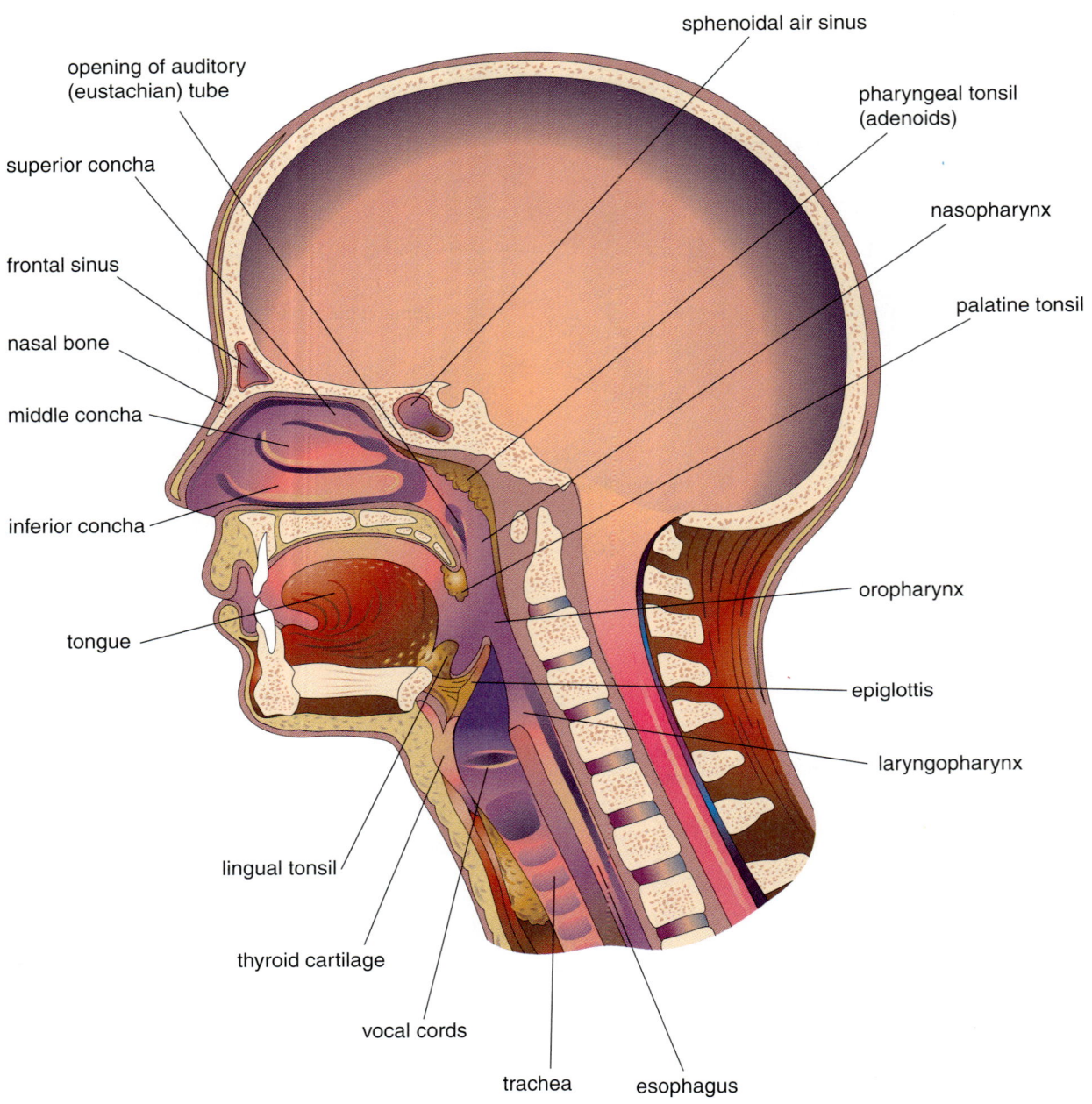

LACTOSE INTOLERANCE: THE DISEASE THAT IS NOT A DISEASE

Various conditions that are identified as a disease are actually a normal genetic trait that produces illness under certain conditions. A prime example is the inability of many adults to produce the enzyme lactase. Lactase breaks down lactose sugar, which is found in milk products, into glucose and galactose. Most mammals lose the ability to digest lactose when they are old enough to feed on regular foods. In some humans, the lactase gene shuts down between ages 5 and 10 years. After this point, it is expected that people are not consuming breast milk as a major nutrient. However, the advent of agriculture introduced milk products into the staple diets of many cultures. This created a problem for people who lack the ability to break down lactose. The result was a disease called lactose intolerance, which developed when these people ate milk products. Lactose intolerance is characterized by bloating, cramping, diarrhea, gas, loose stools, and nausea. These signs are due to the build up of lactose in the intestine and its breakdown by microorganisms. The microorganisms produce gas and irritating waste products as they digest the lactose.

Lactose intolerance is most common in people of African, Asian, and Mediterranean descent (see the table below).

Table 13.1 Lactose Intolerance in Various Populations

Population	Percent of the Population That Is Intolerant
Swedish	2%
European in Australia	4%
Swiss	10%
American Caucasian	12%
Finnish	18%
African Tutsi	20%
African Fulani	23%
African American	75%
Australian Aborigine	85%
African Bantu	89%
Chinese	93%
Thai	98%
American Indian	100%

Kretchmer N. 1972. Lactose and lactase. *Sci Am.* 227(4): 70-78

Various over-the-counter medications can correct for lactose tolerance. They usually contain an enzyme called galactosidase, which breaks down lactose into glucose and galactose.

Lingual Membrane The covering of the tongue

Intrinsic Muscles The tongue muscles that attach to other tongue muscles and mucus membranes

The tongue, which forms the bottom of the buccal cavity, is covered by the **lingual membrane**. The taste buds are on the upper surface of the lingual membrane. Underneath the tongue, the lingual membrane forms a mucous membrane that covers the four **intrinsic muscles** of the tongue: superior longitudinal, inferior longitudinal, verticalis, and transverse. Intrinsic tongue muscles form the bulk of the tongue and attach to mucus membranes and other tongue muscles. They permit the tongue to take on a variety of shapes that assist with speech and swallowing. The superior longitudinal muscle extends along the uppermost surface of the tongue and allows the tongue to curl backward. The inferior longitudinal muscle lines the sides of the tongue. The verticalis muscle attaches to the superior and inferior longitudinal muscles in the middle of the tongue. The transverse muscle runs along the sides of the tongue and divides it at the middle.

Civic Responsibility

CIVIC RESPONSIBILITY: HELPING OTHERS WITH YOUR KNOWLEDGE

It is valuable to use what you have learned about the digestive system to help others to better understand the world around them. It is very important to check your facts and seek further information about certain topics before discussing health and science issues. Here are some suggestions to foster a better public awareness of the digestive system:

1. Volunteer at an eating-disorder clinic.
2. Organize a forum for a civic group to debate traditional versus alternative medical care for treating digestive system disorders.
3. Assist at a resident-care facility helping elderly persons with information about digestive system care.
4. Volunteer at a school health day to teach children about the digestive system.

Directly under the intrinsic muscles of the tongue is the genioglossus muscle. It forms much of the tongue's muscle mass, and acts to protract and depress the tongue. It is attached to the interior portion of the mandible. Below the genioglossus muscle lays the hyoglossus muscle that acts to flatten and pull down the sides of tongue. Located in the back of the tongue are the styloglossus and palatoglossus muscles. The styloglossus retracts the tongue while the palatoglossus raises the tongue to the palate during swallowing. These muscles are called **extrinsic muscles** of the tongue because they attach to bony structures.

The mouth also contains glandular structures. **Lingual tonsils** are a group of lymphatic tissues located at the base of the tongue. They are just anterior to the epiglottis and help fight throat infections. Inflammation of the tonsils makes it difficult to swallow. Salivary glands lie under the facial skin and buccal cavity. They produce a digestive secretion called saliva. Saliva moistens food and starts chemical digestion. The three major salivary glands are the **parotid**, **sublingual**, and **submandibular glands**. Parotid glands, located in the cheeks in front of the ears, are the largest salivary glands, and they produce a watery secretion. Sublingual glands lie underneath the tongue, and they produce a mucous secretion. The submandibular glands, located under the mandible, produce a mixture of a water and mucous secretion.

Extrinic Muscles Muscles that originate outside of a structure and act as a unit

Lingual Tonsils A mass of lymphatic tissue that covers the posterior region of the tongue

Lingual Tonsils A mass of lymphatic tissue that covers the posterior region of the tongue

Parotid Glands The largest salivary glands, located in front of the ears

Sublingual Glands A small salivary gland located under the tongue

Submandibular Glands The salivary gland near the lower edge of the mandible

CHAPTER 13

The two jaw bones and the hard palate form the bony structures of the mouth. They are responsible for the mechanical breakdown of food and assist with speech. The adult, or permanent, teeth consist of 32 teeth. Each jaw has 16 teeth located in the alveolar sockets (Figure 13.4). The central four teeth in each jaw bone are called **incisors**, and they are used to cut up food. The four **canine**, or **cuspid**, teeth are used to hold and tear food; they are found adjacent to the lateral incisors on the upper and lower jawbones. The eight **bicuspids**, or **premolars**, assist with breaking food into fine particles. **Molars** have slightly ridged surfaces and are specialized for grinding food into a fine mash. In most people, each jaw has six molars located at the back of mouth. **Wisdom teeth**, or third molars, are found farthest back in the mouth. They usually appear between ages 18 and 20 years. Wisdom teeth are removed if they crowd the other teeth after they develop.

Incisors The front (cutting) teeth
Canine or Cuspid Teeth Sharp teeth located on each side of the mouth
Bicuspid or Premolar Teeth Double-pointed teeth that are lateral to the canines
Molars The large teeth at the back of the mouth
Wisdom Teeth The last set of molars at the back of the mouth

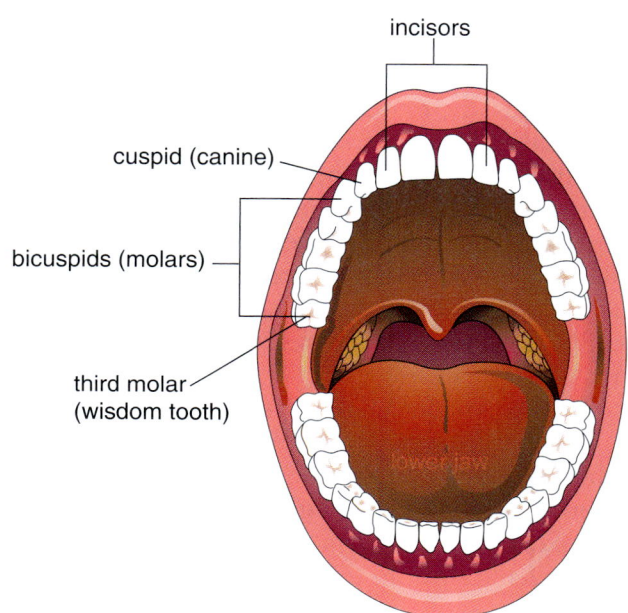

Figure 13.4 The Adult Jaw

✓ Concept Check

1. What are the major structures of the mouth and pharynx?
2. Describe the glandular structures of the mouth.
3. Explain the different types of human teeth.

Esophagus and Stomach

Key Terms: adventitia, cardiac gland, cardiac region, cardiac sphincter, chief cell, fundic gland, fundic region, gastric stem cell, lamina propria, mucosa, mucous neck cell, muscularis layer, parietal cell, pyloric glands, pyloric region, pyloric sphincter, reflux, submucosa, serosa

At the base of the pharynx, just behind the epiglottis, is the opening to the esophagus (Figure 13.5). The esophagus is a muscular tube that carries food and liquids from the pharynx to the stomach. It passes through an opening in the diaphragm just above the stomach. Four distinct tissue layers make up the esophagus and the rest of the digestive tract: **mucosa**, **submucosa**, **muscularis layer**, and **serosa** or **adventitia**. A layer of epithelium that secretes mucus makes up the innermost layer, or mucosa. The epithelial cells are attached to an underlying cylinder of connective tissue called the **lamina propria**. Underneath the lamina propria at the boundary between mucosa and submucosa is a thin layer of smooth muscle and connective tissue called the muscularis mucosae layer. This layer is thicker in the esophagus than in the rest of the digestive tract. In the esophagus, the muscularis mucosae assist with the passage of food. The submucosa is a layer of connective tissue that supports the shape of the esophagus. Underneath the submucosa is a thick cylinder of smooth muscles called the muscularis layer. Contractions of the muscularis layer move food and liquid through the digestive tract. The muscularis layer should not be confused with the muscularis mucosae layer. A thick outer covering of connective tissue forms the adventitia, which becomes serosa once the esophagus passes through the diaphragm. Nerves travel through the serosa to innervate the muscularis.

Two circular groups of muscles called sphincters close off the esophagus from the pharynx and the stomach. The upper sphincter is composed of a bundle of smooth muscle closely associated with the larynx. It is used during swallowing. A lower sphincter called the **cardiac sphincter** surrounds the esophagus at the entrance of the stomach. Both sphincters are normally closed, except during swallowing. This prevents air from the pharynx from regularly entering

Mucosa The mucus membrane making up the inner layer of the digestive tract

Submucosa A connective tissue layer beneath the mucous membrane

Muscularis Layer The smooth muscle layer of the digestive tract

Serosa or Adventitia The outer layer of the digestive tract

Lamina Propria A layer of connective tissue underneath the epithelium of mucosa

Cardiac Sphincter A muscular valve between the esophagus and the stomach

Figure 13.5 Esophagus

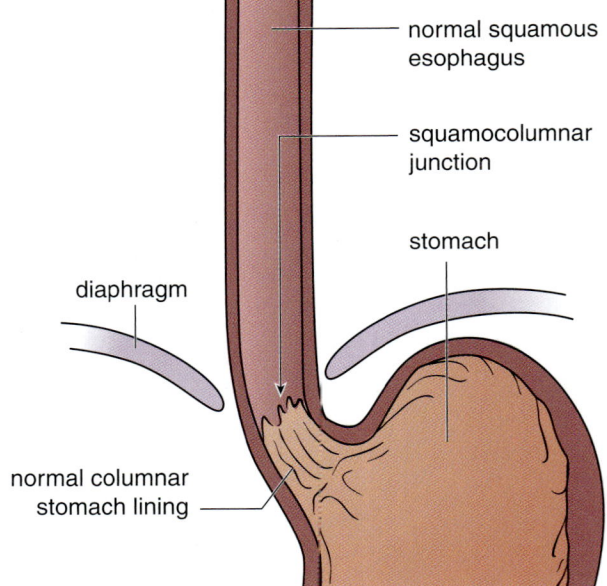

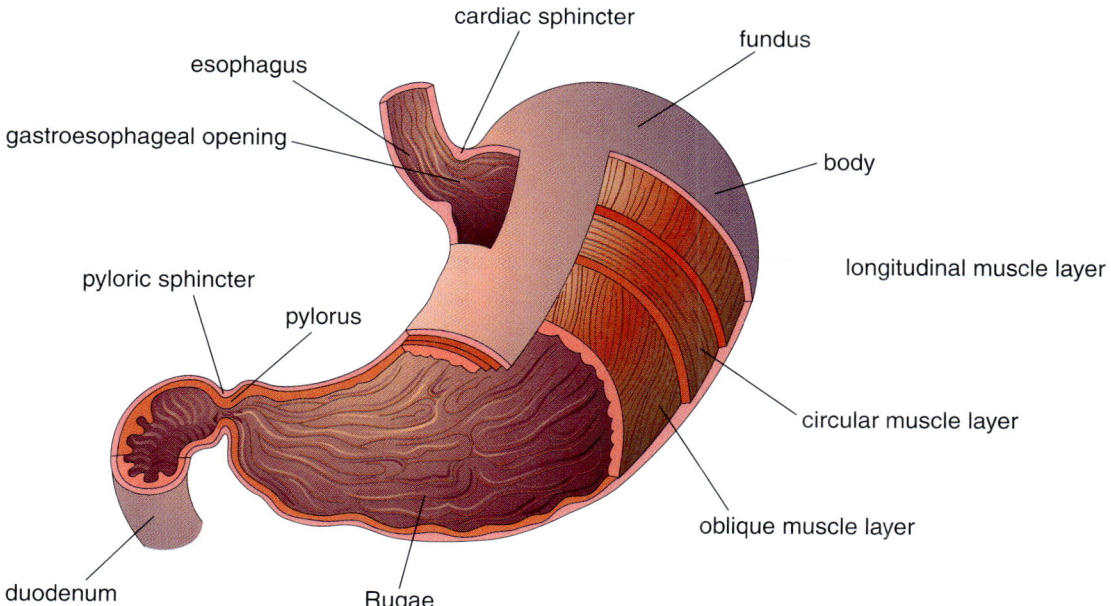

Figure 13.6 Stomach

the stomach. In addition, it prevents a backflow, or **reflux**, of stomach contents that can damage the esophagus and pharynx. The stomach is a large, saclike organ at the distal end of the esophagus (Figure 13.6). Its four layers are formed to withstand corrosive chemical digestion and strong muscular contractions. The stomach's large size permits it to store 3 pints of food for digestion. Specialized cells in the stomach's mucosa secrete acid and protein digestive enzymes. The mucosal surface is a simple columnar epithelium with mucus-secreting cells that protect the stomach from digesting itself. Its lamina propria and muscularis layers are very thin and not easily distinguished.

The stomach is divided into upper (**cardiac**), middle (**fundic**), and lower (**pyloric**) regions. Each region has unique sets of glands: **cardiac**, **fundic**, and **pyloric**. Cardiac glands continuously secrete mucus. Four types of cells are found in fundic glands of the stomach's mucosa: **parietal cells** produce hydrochloric acid by secreting chloride and hydrogen ions into the water of the stomach cavity; **chief cells** secrete digestive enzymes; **mucous neck cells** secrete mucus only when stimulated by the vagus nerve; and **gastric stem cells**, which are responsible for replacing the other fundic-gland cells. Glands in the pyloric region produce mucus.

The stomach's muscularis layer consists of three thick layers of smooth muscle: oblique, circular, and longitudinal. Directly underneath the submucosa is the oblique layer. These thin bands of muscle permit stomach squeezing. The circular layer is composed of thick bands of muscles making up the middle layer. Contractions of these muscles produce a mixing effect. The outermost layer is the longitudinal layer. These muscles assist with mixing and moving digested food out of the stomach. Closing off the upper portion of the stomach is the **cardiac sphincter**, which prevents the reflux of stomach contents into the esophagus. Heartburn results when acids from the stomach enter the esophagus through a faulty cardiac sphincter. The **pyloric sphincter** closes off the bottom of the stomach. Covering the stomach is a thick serosa, which is tightly attached to the peritoneum of the abdominal cavity.

Reflux A backward flow of body fluids
Cardiac Region Upper region of the stomach
Fundic Region Middle region of the stomach
Pyloric Region Lower region of the stomach
Cardiac Glands Pits of secretory cells in the upper region of the stomach
Fundic Glands Pits of secretory cells in lower region of the stomach
Pyloric Glands Mucus-secreting glands in the lower portion of the stomach
Parietal Cells Acid-secreting cells in the fundic stomach lining
Chief Cells Cells of the stomach that secrete digestive enzymes
Mucous Neck Cells Cells of the fundic stomach that secrete mucus
Gastric Stem Cells Cells that replace other glandular cells of the fundic stomach
Pyloric Sphincter The lower sphincter of the stomach

THE DIGESTIVE SYSTEM

✓ Concept Check

1. Describe the three layers of the digestive tract.
2. Compare the composition of the esophagus and stomach.
3. Explain the types of glandular cells found in the stomach.

Small Intestine

Key Terms: crypt, duodenum, enteroendocrine cells, enterocyte, ileocecal valve, ileum, jejunum, lacteal, mesentery, microvilli, paneth cell, villi

Duodenum The first section of the small intestine
Jejunum The middle section of the small intestine
Ileum The last section of the small intestine

The small intestine is a long, narrow tube running from the pyloric region of the stomach to the large intestine (Figure 13.7). It is tightly looped back and forth within the abdominal cavity. The small intestine is divided into three distinct sections: **duodenum**, **jejunum**, and **ileum**. Each section of the small intestine has a distinct histology and carries out a particular function. The duodenum receives partially digested food from the stomach. It is in the duodenum that most of the stomach's digestion takes place. It then passes the food into the jejunum, where most of the nutrients are absorbed into the blood. The ileum is where the remaining nutrients are absorbed before entering the large

Figure 13.7 Lower Gastrointestinal Tract

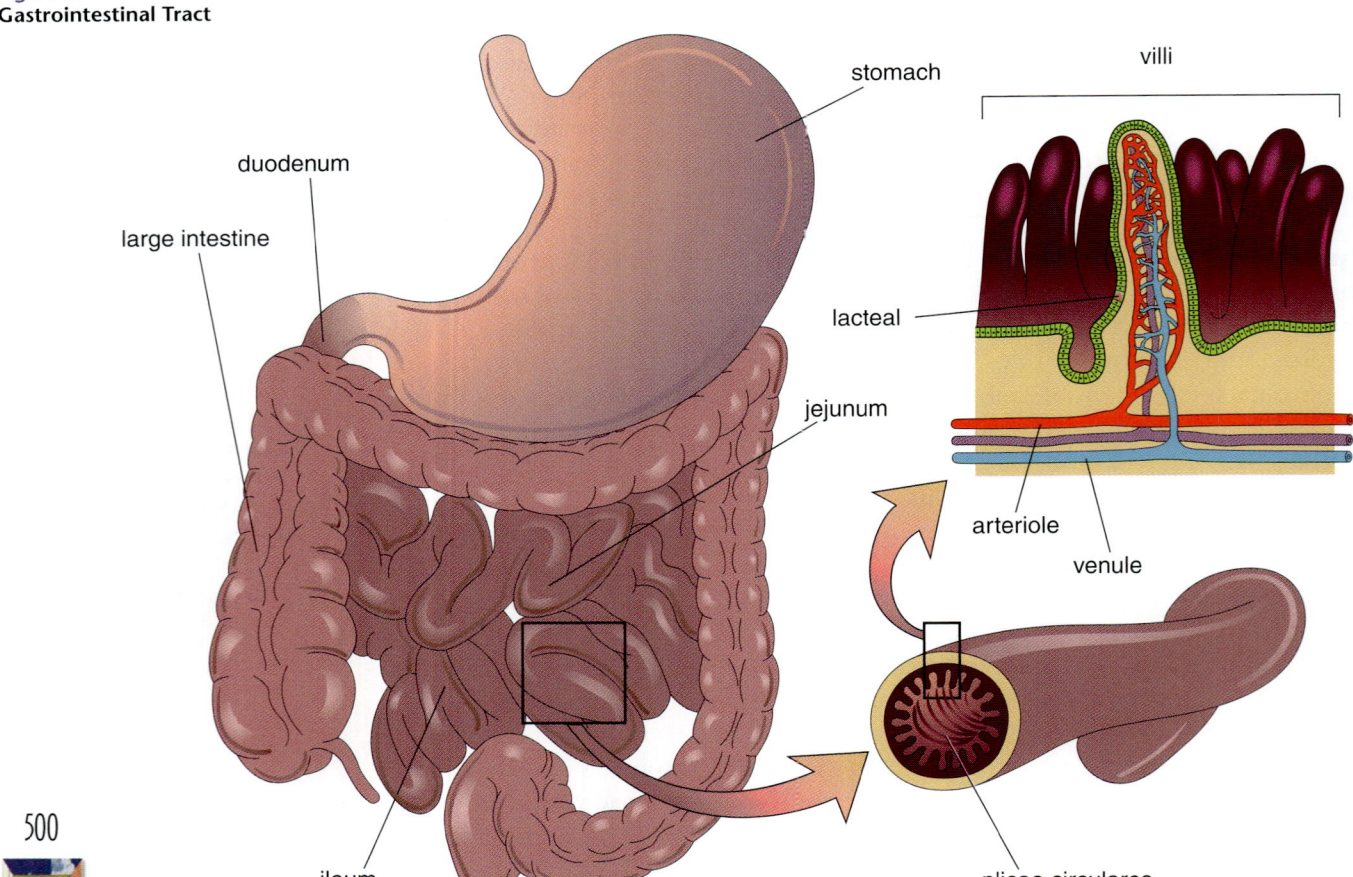

Good Choice Bad Choice

A neighbor brings you an advertisement describing colonic cleansing as a means of "safe" weight loss. The advertisement also claims that the technique washes out poisons that accumulate in the colon. What are the benefits and risks of using colonic cleansing to lose weight?

intestine. Fingerlike projections called **villi** are characteristic of the mucosa of the small intestine. These out-pockets of the mucosa increase the surface area for food digestion and absorption. **Lacteals** are collections of lymphatic vessels inside the villi. They carry absorbed foods, especially fats, to the liver. The columnar epithelial cells covering the villi have fine ridges on the surface facing the lumen called **microvilli**. Microvilli are projections of the cell membrane that greatly enhance the surface area for absorption of digested nutrients.

The mucosa of the small intestine contains a diversity of specialized cells, which vary in composition for each section of the small intestine. The most common cells, found primarily in the jejunum, are absorptive cells called **enterocytes**. **Enteroendocrine cells** produce hormones that regulate digestion. A very important group of cells called **paneth cells** is found in regions of the mucosa called **crypts**. Paneth cells produce antibacterial enzymes and nutrients that favor the growth of beneficial microorganisms in the small intestine. These cells also destroy pathogenic microorganisms by phagocytosis. Peyer's patches are a collection of large patches of lymphatic tissue located in the mucosa of the duodenum. They help fight infections of the small intestine and are associated with the immune response to food allergies.

Unlike the stomach, the muscularis layer of the small intestine is composed of two layers: an inner circular layer and an outer longitudinal layer. The circular layer is involved in mixing food, while the longitudinal muscles transport food through the length of the small intestine. Extensive networks of nerves that control muscle contraction run throughout the muscularis layers. Most of the small intestine is covered with a very thin serosa that is closely attached to the **mesenteries** of the peritoneum. Some of the duodenum is located outside of the peritoneal cavity and is covered with an adventitia. Mesenteries hold the small intestine in place and are attachment sites for blood vessels traveling to and from the serosa. The small intestines enter the large intestine at the **ileocecal valve**. The ileocecal valve separates the small intestine from the large intestine. This forms a barrier that prevents bacteria from entering the small intestine. It also controls the emptying of the small intestine into the large intestine.

Villi Fingerlike projections on the small intestine mucosa

Lacteals A collection of lymphatic vessels inside the villi

Microvilli Small projections on the surface of epithelial cells of the villi in the small intestine

Enterocytes Absorptive cells of the small intestine

Enteroendocrine Cells Hormone-producing cells of the small intestine

Paneth Cells Cells of the small intestine that maintain beneficial microorganisms

Crypts Pits of glandular cells found in the small intestine

Mesentery A membranous tissue that connects the small intestine to the peritoneum

Ileocecal Valve A muscular valve separating the small intestine from the large intestine

✓ Concept Check

1. Describe the three regions of the small intestine.
2. Explain the different glands found in the small intestine.
3. What is the function of the ileocecal valve?

THE DIGESTIVE SYSTEM

Large Intestine and Rectum

Key Terms: anal canal, anal sphincter, anus, appendicitis, appendix, ascending colon, cecum, colon, descending colon, hepatic flexure, sigmoid colon, splenic flexure, transverse colon

Colon Another name for the large intestine

Cecum The first portion of the large intestine

Ascending Colon The upward section of the large intestine on the right side of the body

Transverse Colon A portion of the large intestine running horizontally below the diaphragm

Descending Colon A portion of the large intestine that travels downward

Sigmoid Colon The last portion of the large intestine

Appendix A small pouch protruding from the cecum

Appendicitis Inflammation of the appendix

Hepatic Flexure The last portion of the ascending colon

Splenic Flexure The last portion of the transverse colon

The large intestine, or **colon**, is larger in diameter and much shorter than the small intestine (Figure 13.8). It is composed of five anatomical regions: the **cecum**, **ascending colon**, **transverse colon**, **descending colon**, and **sigmoid colon**. The cecum is a small, swollen region of the intestine that contains the ileocecal valve and the **appendix**. The function of the appendix in humans is unknown, and it is a common site of infection from bacteria and viruses. Sometimes the infection becomes severe, causing an inflammatory disease called **appendicitis**. The ascending colon starts at the ileocecal valve. It travels vertically along the right abdominal wall to a region called the **hepatic flexure**. The hepatic flexure lies below the liver. The next segment, the transverse colon, runs parallel to the diaphragm. It connects to the descending colon at the **splenic flexure**. As the name implies, it is located beneath the spleen. The descending colon passes vertically downward along the left abdominal wall to the pelvis. An s-shaped curve at the end of the descending colon forms the sigmoid colon.

The large intestine's main digestive function is the absorption of electrolytes, vitamins, and water. Its primary job is to remove undigested materials from the digestive tract. The mucosa of the large intestine is smooth and has no villi. It

Figure 13.8 Large Intestine

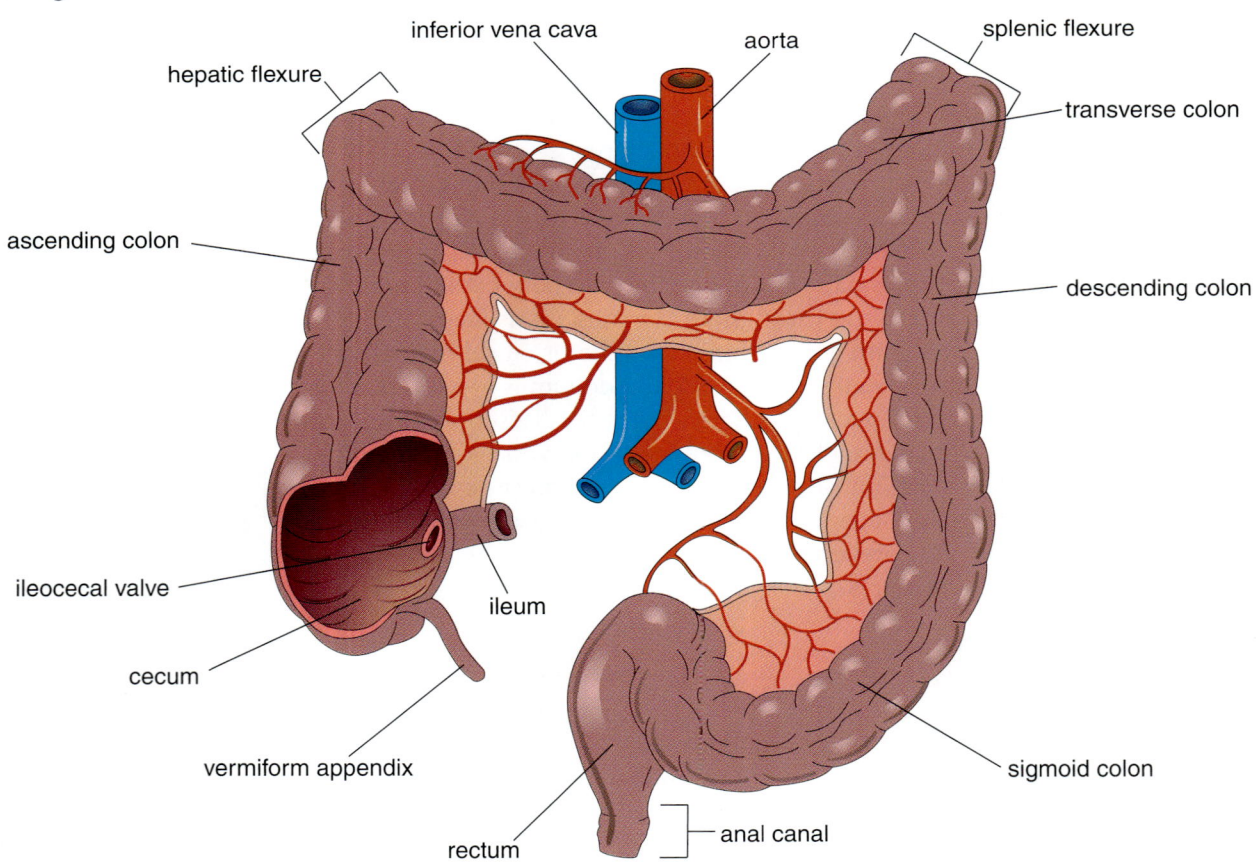

contains numerous absorptive cells that use active transport to move electrolytes into its many capillaries. The large intestine's mucosa is inhabited by many types of bacteria and one type of yeast. Most of these microorganisms break down wastes; they provide little benefit or harm to humans. Some of these microorganisms are a source of vitamins and other simple nutrients, such as certain amino acids and monosaccharides. The bacteria normally inhabiting the large intestine displace any potentially harmful bacteria that may enter the digestive system.

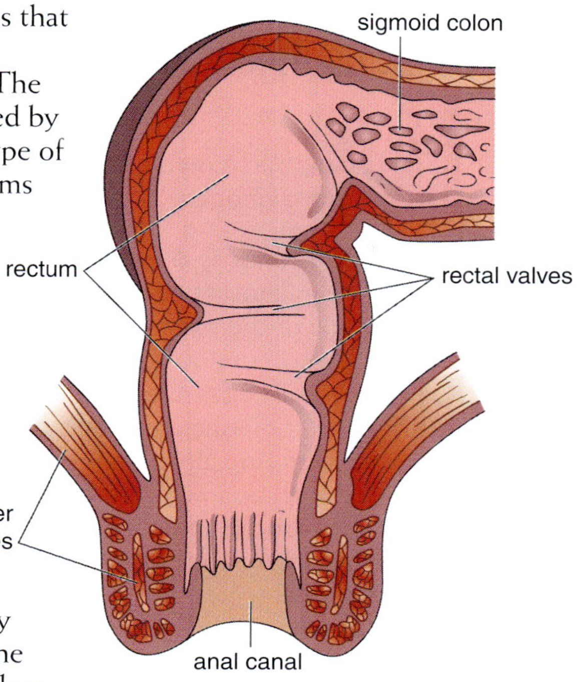

Figure 13.9 Rectum

Underneath the mucosa is a highly cellular lamina propria. Many of the cells are found in large patches of lymphoid tissue that protect the submucosa from invasion by microorganisms. The muscularis layer consists of an inner circular layer and an outer longitudinal layer. However, regions of the longitudinal muscles form three distinct bands called *taenia coli* which are located throughout most of the length of the colon.

The rectum, a muscular storage area for undigested wastes, makes up the final portion of the digestive tract (Figure 13.9). The layers of the rectum have the same structure as the large intestine. The **anal canal** is a transitional region between the digestive tract and the region of skin around the **anus**. The anal canal terminates in skin that is rich in apocrine and eccrine sweat glands. A sphincter at the end of the anal canal helps to form the anus. This **anal sphincter** has a complex anatomy. It is divided into internal and external bands of muscle. The internal component is composed of smooth muscle, while the external portion is made up of the skeletal muscles of the pelvis. Nerves passing out of the sacrum provide autonomic and voluntary control over the anus.

Anal Canal A small portion of the rectum ending in the anus

Anus The opening of the rectum to the outside of the body

Anal Sphincter The sphincter muscle of the anus

✓ Concept Check

1. What are the five portions of the large intestine?
2. Explain the unique nature of the large intestine's mucosa.
3. Describe the structure of the anus.

CSI – Case Study Investigation Conclusion

Can you locate the illness in the case study to one particular region of the digestive tract? Is the condition limited to a problem with one specific layer of the digestive tract?

THE DIGESTIVE SYSTEM

Glandular Structures of the Digestive System

Pancreas

Key Terms: acini, bile, common bile duct, glandular lobule, pancreatic duct, zymogen

The pancreas lies posterior to the stomach, attached by a sheet of mesentery (Figure 13.10). A thin layer of connective tissue forms a capsule around the pancreas. The inside of the pancreas is divided into **glandular lobules** surrounded by connective tissue. Each lobule has blood vessels, nerves, and ducts. The pancreas has endocrine and exocrine functions related to digestion. The endocrine component is made up of islets of Langerhans, which contain the hormone-secreting alpha and beta cells. As mentioned in Chapter 7, alpha cells secrete glucagon, which is actually a group of several related hormones that elevate blood glucose. Insulin from the beta cells lowers glucose in the blood. Approximately 2% of the pancreas is made up of the islets of Langerhans. Hormonal secretions from the pancreas enter directly into capillaries. The exocrine features of the pancreas consist of saclike clusters of serous glands called **acini**.

Glandular Lobule Glandular clusters in an organ
Acini Saclike clusters of exocrine cells

Digestive System Fun Facts

- The average North American male will eat over 50 tons of food during his lifetime.
- The mouth produces approximately 1 to 3 pints of saliva per day.
- An adult esophagus ranges from 10 to 14 inches in length and averages 1 inch in diameter.
- A belch is caused when swallowed gas moves up the esophagus and is released from the mouth.
- An average adult stomach can hold approximately 1.5 liters of food.
- A meal high in fat and protein can stay in the stomach for 2 to 3 hours.
- The average person's stomach produces 2 liters of hydrochloric acid per day.
- The average person secretes about 400 milliliters of gastric juice per meal.
- The typical small intestine is over 20 feet long and 1½ to 2 inches in diameter.
- The large intestine is about 5 feet long and 3 to 4 inches in diameter.
- A typical person's large intestine is home to more than 400 types of bacteria.
- The liver is the largest internal organ in the body.
- The liver performs more than 500 functions.

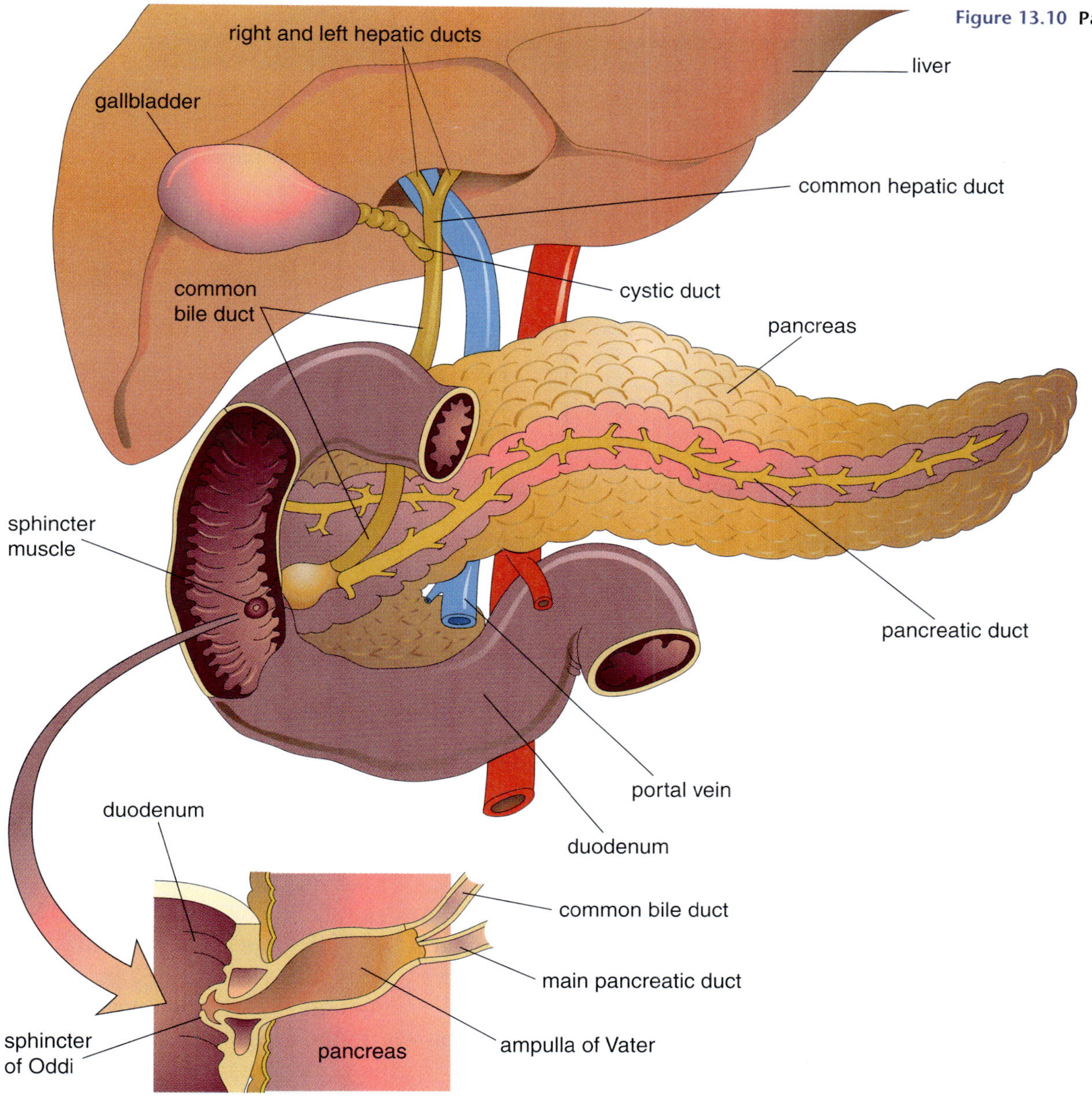

Figure 13.10 Pancreas and Liver

The cells that make up the acini have granules that contain a variety of active and inactive digestive enzymes. Inactive enzymes are called **zymogens** (their name ends in "-ogen"). Zymogens are converted into active enzymes once they enter the digestive tract (Table 13.1). Keeping the enzymes inactive prevents them from degrading the acini cells. The ductal system in the pancreas originates in the acini and feeds into a large duct called the **pancreatic duct**. This duct leads to the **common bile duct**, which empties into the duodenum. The yellow-green **bile** contains acids, cholesterol, glyceride fats, and salts. The major job of the acids, glyceride fats, and salts in bile is to help with fat digestion. Cholesterol is secreted from the liver through the bile and plays no role in fat digestion. Bile also contains bilirubin, pigments, and breakdown products of destroyed RBCs.

Zymogen An inactive form of an enzyme

Pancreatic Duct A tube that collects pancreatic exocrine secretions

Common Bile Duct A tube through which secretions of the liver and pancreas pass into the duodenum

Bile A yellow-green fluid produced by the liver

505

THE DIGESTIVE SYSTEM

Table 13.1 Pancreatic Zymogens

Enzyme	Function of Active Enzyme
chymotrypsinogen	digests protein
deoxyribonuclease	breaks down DNA
elastase	degrades connective tissue proteins
pancreatic amylase	digests starch
phospholipase A2	digests lipids
procarboxypeptidase	digests protein
ribonuclease	breaks down RNA
triacylglycerol lipase	digests lipids
trypsinogen	digests protein

✓ Concept Check

1. What are the two functions of the pancreas?
2. Explain how secretions exit the pancreas.
3. Distinguish between enzymes and zymogens.

Liver and Gallbladder

Key Terms: caudate lobe, falicform ligament, gallstone, heparin, hepatic artery, hepatic portal vein, hepatic vein, hepatocyte, hilum, left lobe, quadrate lobe, right lobe, serum globulin, sinusoid

Left Lobe A large subdivision of the liver

Right Lobe The largest subdivision of the liver

Falicform Ligament A ligament that divides the liver into left and right lobes

Quadrate Lobe A small subdivision of the liver

Caudate Lobe A small subdivision of the liver

Hilum A depression where blood vessels enter the liver

Hepatic Artery The artery that provides resources for the liver

Hepatic Portal Vein A vein that carries blood from the small intestine to the liver

Hepatic Vein A vein that carries wastes from the liver

Hepatocyte A liver cell

Sinusoid A small cavity in the liver filled with blood

The liver carries out the most complex functions of the digestive system. It is divided into four lobes and is surrounded by a capsule of connective tissue. The capsule is covered by a visceral peritoneum. Two of the lobes, the **left** and **right**, make up most of the liver's mass. These two lobes are separated by the **falicform ligament**, which divides the liver longitudinally into two unequal parts. The two other small lobes are buried within the medial, inferior portion of the liver. The **quadrate lobe** is situated in the ventral region. The **caudate lobe** is located in the dorsal area. Two blood vessels enter the liver at a region called the **hilum**. The **hepatic artery** provides blood for the liver cells. A large **hepatic portal vein** carries food from the small intestine to the liver. The common bile duct exits the hilum. A **hepatic vein** carries wastes from the liver lobes to the inferior vena cava.

Liver tissue is composed of **hepatocytes** embedded in a fibrous connective tissue. Hepatocytes are unusual cells because they have two or more nuclei. The typical hepatocyte lives about 5 months and is very slow at carrying out mitosis. It is possible during a liver disease that the cells die off faster then they are replaced. However, current research shows that healthy liver cells are capable of considerable regeneration when the liver is damaged. Physicians are able to grow a complete lobe from small pieces of liver transplanted into a patient. Arteries going to the hepatocytes empty into pockets called **sinusoids** (Figure 13.11). Hepatocytes exchange atmospheric gases, nutrients, and wastes with the blood in the sinusoids. Kupffer cells are commonly found in the sinusoids where they remove invading microorganisms and recognize foreign antigens.

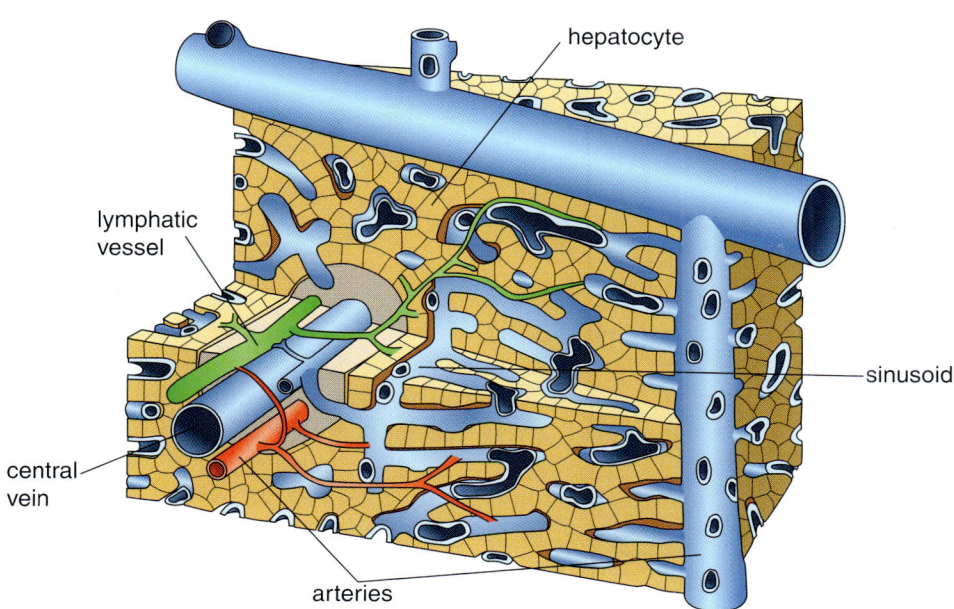

Figure 13.11 **Liver Organization**
Hepatocytes and sinusoids

Lymphatic vessels and large vessels of the hepatic portal system are found throughout the liver tissue.

The liver carries out a wide variety of tasks, including endocrine and exocrine functions. The liver's many jobs include the following:

- glycogen storage for regulating blood sugar
- formation of urea
- formation of blood proteins and clotting factors
- synthesis of **heparin**, a blood anticoagulant
- metabolism of cholesterol and fatty acids
- formation of **serum globulins**
- metabolism of vitamin D
- removal of microorganisms from blood
- breakdown of drugs and many poisons
- breakdown of amino acids
- destruction of bilirubin, a hemoglobin breakdown product of regulation of blood volume in the sinusoids
- storage of iron

The gallbladder is a small pear-shaped sac that stores and concentrates bile. It lies underneath the right lobe of the liver and receives bile through the bile ductwork. Smooth muscles of the gallbladder contract to release bile through the cystic duct, which joins the common hepatic duct to form the common bile duct that enters the small intestine (Figure 13.9). **Gallstones** form when the bile contains too much bilirubin, cholesterol, or bile salts, and these substances accumulate into a mass. They can block the flow of bile and cause considerable pain when they pass through the bile ducts.

Heparin A protein that helps prevent blood clotting

Serum Globulin A general antibody found in the blood serum

Gallstone A solid material that forms in the gallbladder

✓ Concept Check

1. Describe the gross structures of the liver.
2. Explain the fine anatomy of liver tissue.
3. What are the various functions of the liver?

THE DIGESTIVE SYSTEM

DISCOVERY SCENE PLEASE ENTER DISCOVERY SCENE PLEASE ENTER

 *Does the information about the glandular components of the digestive system help you to better understand the CSI? How can the function of the liver or pancreas contribute to the condition?*

The Digestive Process

 **Key Terms:** cholecystokinin (CCK), chyme, enterokinase, feces, flatulence, gastrin, hunger center, ingestion, parenteral, peristalsis, pregastric factor, protease, reverse peristalsis, salivary amylase, satiety center, secretin

Pregastric Factors Conditions that can dramatically affect food intake

Hunger Center A region of the central nervous system that signals hunger

Satiety Center A region of the central nervous system that signals that a person has eaten

Ingestion Taking something into the body by the mouth

Parenteral Nutrition taken into the body by bypassing the digestive tract

Mastication Chewing

Salivary Amylase An enzyme in the saliva that digests starch

Peristalsis Wave-like muscle contractions that push food and liquid

Reverse Peristalsis A peristalsis that forces food and liquid in the opposite direction of flow

Human digestion begins with the intake of food as a result of hunger and thirst. Hunger and thirst are called **pregastric factors**. Pregastric factors include any condition that determines the degree of a person's hunger or thirst. Many physicians and scientists believe that behavior plays a strong role in food intake. Evidence shows that attitudes about eating and one's upbringing help determine a person's food intake. The central nervous system has a **hunger center** in the lateral hypothalamus. A **satiety center** is located in the ventral hypothalamus. Both centers respond to many factors, including blood sugar and the amount of fats transported in the blood. The appearance and taste of food can also affect these centers. Abnormalities of these regions of the hypothalamus help explain some types of eating disorders.

The act of taking food into the body through the mouth is called **ingestion**. In certain medical situations, it is sometimes essential to bypass the digestive system to provide a person with nutrients. This is called **parenteral** nutrition, which means "beyond the digestive tract." Parenteral nutrients are injected into muscles or veins. This strategy gets crucial materials into the body quickly. **Mastication**, or chewing, is the first step of mechanical digestion in the mouth. The action of mastication uses the teeth to break down food into fine particles. Digestion is most effective when more of the food surface is exposed to chemical degradation by enzymes. Chemical digestion begins during mastication as saliva is mixed with the food. Salivation wets the food to facilitate swallowing and to assist with chemical degradation of polymers. The **salivary amylase** enzyme starts the chemical breakdown of starch into glucose.

During swallowing, muscles of the mouth and tongue bring the soft palate and the uvula upward. This prevents food from entering the nasal cavity by closing the opening between the nasal cavity and the pharynx. The upper sphincter of the esophagus then pulls the larynx forward, providing a more direct route to the esophagus. This action occurs with the closing of the epiglottis to prevent food from entering the larynx. Food and liquids travel down the esophagus by rhythmic waves of smooth-muscle contractions called **peristalsis**. Peristalsis permits persons to swallow food even when they are upside down. This gives astronauts the ability to swallow food in outer space. Bitter receptors of the tongue can inhibit the swallowing sequence and even cause vomiting or **reverse peristalsis** of the esophagus.

The cardiac sphincter relaxes to let food pass into the cardiac region of the stomach. Neural signals from the eyes, nose, and tongue prepare the stomach for digestion. Pregastric factors induce signals to release stomach secretions,

CHAPTER 13

which may initiate peristalsis of the stomach's muscularis layer. Signals from the parasympathetic nervous system also stimulate the stomach to carry out digestion. **Protease** zymogens are secreted into the stomach along with hydrochloric acid produced by the stomach lining. Proteases used to digest protein require the acidic environment that hydrochloric acid provides. The pH of the stomach can vary from 1.0 to 3.0 during digestion. This acidic environment helps to kill microorganisms that have been ingested. A concurrent increase in production of bicarbonate and mucus also occurs to protect the mucosa from digestion.

Hormones produced by the stomach and small intestine further stimulate the digestive system. These hormones are released into the blood of the digestive tract and travel through arteries that return to the digestive system. The three major hormones that control digestion are **cholecystokinin (CCK)**, **gastrin**, and **secretin**. Cholecystokinin stimulates the pancreas to produce the enzyme component of pancreatic juice and causes the gallbladder to empty. Gastrin causes the stomach to produce hydrochloric acid that assists with protein digestion. Secretin stimulates the pancreas to secrete digestive juices containing bicarbonate. It also signals the stomach to produce proteases and stimulates the liver to produce bile. Peristalsis of the stomach muscles mixes the food to facilitate its breakdown.

Partially digested food in the stomach, called **chyme**, stimulates the pyloric sphincter to relax. This permits food to enter the duodenum as strong contractions of the stomach muscle force the food downward. Only a few molecules, including alcohol and many types of drugs, are absorbed in the stomach. In the duodenum, the food is mixed with bile and pancreatic secretions. Bile acts to physically break down fat into smaller droplets called micelles. The pancreatic secretions provide for the chemical breakdown of carbohydrates, lipids, nucleic acids, and proteins. The duodenal mucosa releases enzymes called **enterokinases** to activate any zymogens. Peristalsis then carries the mixture into the jejunum. The numerous villi in the jejunum facilitate the absorption of nutrients into the blood and lacteals. Various passive transport methods transfer amino acids, carbohydrates, lipids, and nucleic acids to the blood and lacteals. Amino acids, carbohydrates, and nucleic acids that are absorbed into the blood go to the liver through the hepatic portal system for further digestion. The hepatic portal system carries blood from capillaries of the mesenteries, small and large intestine, spleen, stomach and pancreas to the liver. Enzymes in the liver convert amino acids into ammonia and other molecules and simple lipids by the process called deamination. Alcohol, which is absorbed primarily in the stomach and small intestine, and other potentially toxic chemicals, taken in the small and large intestines, are also broken down in the liver.

Any substances that have not been absorbed in the small intestine enter the large intestine after passing through the ileocecal valve. Bacteria in the large intestine digest and absorb many of the undigested materials. As a result of their metabolic activities, bacteria produce a variety of intestinal gases. Excessive gas production is called **flatulence**. The large intestine slowly moves food from one section to the other by peristalsis. Most fluids and electrolytes are transported into the blood stream by the ascending and transverse large intestine, or colon. Cholesterol and other fats are also absorbed in these regions of the large intestine. The remaining sections of the large intestine compact the wastes into **feces**. Feces are mostly composed of water. The solid matter consists of bacteria, carbohydrate polymers, dried digestive

> **Protease** An enzyme that digests proteins
>
> **Cholecystokinin (CCK)** A hormone the causes the pancreas to produce enzymes
>
> **Gastrin** A hormone that stimulates acid production
>
> **Secretin** A hormone that causes the pancreas to release digestive juices
>
> **Chyme** Partially digested food
>
> **Enterokinase** An enzyme that activates intestinal zymogens
>
> **Flatulence** Excessive gas production in the intestine
>
> **Feces** Waste eliminated from the large intestine

secretions, fat, intestinal cells, and protein. Indigestible carbohydrate polymers form most of the roughage in the diet. Roughage assists the large intestine with peristalsis by providing something solid for the muscles to push upon when squeezing the lumen. The descending and sigmoid sections of the colon also secrete mucus that binds the feces. Mucus lubricates the feces as they pass through the large intestine. The feces then enter the rectum where they are stored until expulsion by muscular contractions of rectal muscles and sphincters. Emptying of the rectum is under voluntary and involuntary control. Autonomic signals stimulate muscle contractions in the rectum as it fills up. However, these contractions can be inhibited by voluntary motor signals that prevent feces from passing at inappropriate times.

✓ Concept Check

1. Describe the passage of food through the digestive tract.
2. Explain three factors that stimulate the digestive processes.
3. Distinguish between chyme and feces.

DISCOVERY SCENE PLEASE ENTER DISCOVERY SCENE PLEASE ENTER

CSI Break

What is the relationship between the digestive process and your classmate's condition? Could something that he eats be affecting his health?

Wellness and Illness over the Life Span

Pathology of the Digestive System

Key Terms: acid reflux, acute diarrhea, adenomatous polyps, amoebic dysentery, celiac disease, chronic diarrhea, colon cancer, colon polyp, diarrhea, dysphagia, food intolerance, gastric reflux, gastroesophageal reflux disease (GERD), hepatitis, hernia, hiatal hernia, inflammatory bowel disease (IBD), inguinal hernia, pancreatitis, *Salmonella*, ulcer

Humans are subject to dozens of digestive system disorders that can affect specific organs or involve the whole digestive system. Diseases of the digestive system have various origins, including psychological disorders, allergies, infections, genetic syndromes, and degenerative changes from toxins or trauma. Nervous system disorders also contribute to many diseases of the digestive system. A common group of disorders results from a lack of certain digestive enzymes. This produces a condition called **food intolerance**. Food intolerances can restrict a person's diet because the undigested or unabsorbed food can produce intestinal irritation and painful gas buildup. Food intolerances are usually the result of a genetic disorder. **Celiac disease** is one such disorder. This disease is caused by the inability to digest a certain protein found in wheat.

Food Intolerance The inability to digest or absorb a particular food

Celiac Disease An inability to digest and absorb a certain protein found in wheat

General abdominal pain is the most common digestive system ailment. It can be very simple to diagnose, or it can result from a complicated sequence of events that are difficult to identify and treat. Dehydration, flatulence, food poisoning, infection, overeating, overexertion, stress, and trauma can produce generalized abdominal pain. **Gastric reflux**, or **acid reflux**, is a common painful condition that is the subject of many television advertisements and e-mail spam that promote various remedies (Figure 13.12). It is caused by the backflow of stomach contents into the lower esophagus, which is due to a weakening or incomplete closure of the cardiac sphincter. Muscle contractions in the stomach can force stomach contents up through the sphincter. The mucous membrane of the esophagus lacks the conditions needed to handle acidic fluids. As a result, the stomach acids erode the esophageal mucosa, causing the burning sensation known as heartburn. **Dysphagia** is a condition that causes difficulty with swallowing. This condition has many causes and is sometimes associated with gastric reflux. A condition called gastroesophageal reflux disease (GERD) is due to a combination of esophageal and gastric reflux.

Diarrhea is another common disorder that has many causes. It can result from allergies, food poisoning, infection, overeating, overexertion, and stress. Bacterial and viral food poisoning is likely the most common cause of diarrhea worldwide. Children and adults weakened by other diseases can die from the dehydration and electrolyte loss that results from diarrhea. Flatulence and anal bleeding can accompany diarrhea. **Acute diarrhea** refers to a rapidly occurring condition that lasts no more than 2 weeks. This condition can be caused by antibiotics, bacterial infection, dietary changes, drugs containing magnesium, and excessive caffeine consumption. Acute bloody diarrhea will accompany severe bacterial and protistan infections spread by contaminated food or water. **Salmonella** is a bacterium that causes severe diarrhea and other symptoms. The bacterium is mostly found in poultry products and foods made with

Gastric Reflux or Acid Reflux The backward flow of stomach contents into the esophagus

Gastric Dysphagia A swallowing disorder

Diarrhea Frequent and watery bowel movements

Acute Diarrhea A short-term, rapid-onset diarrhea

Salmonella A type of bacteria that causes food poisoning

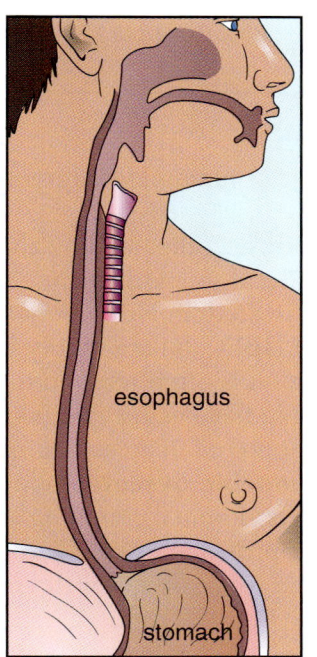
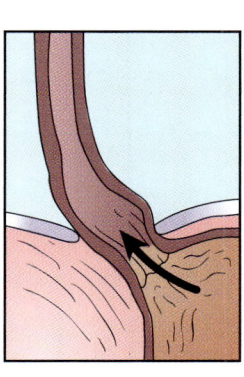

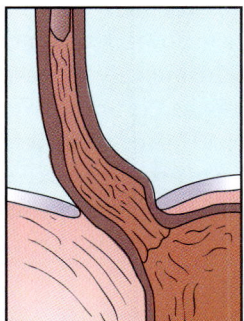

Figure 13.12 **Acid Reflux**

Cutting Edge Research

CAN KISSING CAUSE ULCERS?

Is it possible to get ulcers from kissing? Some scientists believe the answer is yes if one of the people doing the kissing has a *Helicobacter pylori* (*H. pylori*) infection. In the 1980s, Australian scientists Drs. Barry Marshall and Robin Warren discovered an association between a bacterium in the stomach lining and chronic gastritis and peptic ulcers. They identified the bacterium as *H. pylori*. Bacteria are not supposed to survive in the stomach. However, these bacteria survive by becoming embedded in the alkaline mucus of the stomach. The findings of Drs. Marshall and Warren were at first not accepted by the scientific community. Most physicians and scientists believed that the bacterium invaded the ulcer. There was no previous evidence that a bacterium could cause digestive system ulcers. The scientific community stuck with the traditional belief that ulcers were caused by stress or excessive stomach acidity.

Marshall was frustrated by the lack of acceptance of his findings. He decided to test his findings by drinking a culture of *H. pylori*. After 5 days, Marshall was vomiting from severe inflammation of the stomach lining. The condition lasted 2 weeks and then led to stomach ulcers. He had no ulcers before conducting this risky experiment. Throughout medical history, scientists have carried out experiments seeking to discover evidence about disease transmission. The most famous was Dr. James Carroll, who, in 1876, purposely contracted a fatal infection from the virus that caused yellow fever. His lethal heroic action showed that mosquitoes spread the deadly viral disease. Most physicians and scientists now believe that *H. pylori* is the cause of most stomach ulcers. Many researchers believe the bacterium is transmitted through fecal matter in contaminated food or water. In addition, evidence now shows that *H. pylori* is carried from the stomach to the mouth through belching, reflux, and vomiting. There is also evidence that the bacterium can be spread from person to person through kissing.

These new findings have changed the way physicians diagnosis and treat stomach ulcers. They now perform tests that can confirm the presence of *H. pylori*. The urease test can detect an enzyme secreted by this bacterium. This information is used in conjunction with cultures of samples taken from the stomach lining. Ulcer treatments now include antibiotics and other therapeutics that hinder the growth of *H. pylori*.

Kikuchi S, Dore MP. Epidemiology of *Helicobacter pylori* infection. *Helicobacter*. 2005;10(Suppl 1):1-4.

Chronic Diarrhea A long-term, usually painful, diarrhea

Inflammatory Bowel Disease (IBD) Disease that causes irritation to the intestines

Amoebic Dysentery An inflammation of the intestines caused by a protistan called *Endamoeba histolytica*

ground beef. Seafood collected from sewage-contaminated water can also contain *Salmonella*.

Chronic diarrhea is a long-term condition accompanied by abdominal pain. Persistent diarrhea is a major concern to physicians because it can cause severe dehydration and malnutrition if not treated. In North America, it is most likely due to **inflammatory bowel disease** (**IBD**). Inflammatory bowel disease is a peristalsis disorder with no known cause. It is often referred to as irritable bowel syndrome. Syndromes are diseases that have many effects on the body. Some infectious organisms commonly encountered in areas lacking sewage treatment can produce chronic diarrhea. **Amoebic dysentery** is a painful, long-term condition that produces extreme abdominal cramping and regular bouts of diarrhea. Many people confuse the protistan disease with bacterial conditions. Amoebic dysentery is contracted from food or water contaminated with human feces.

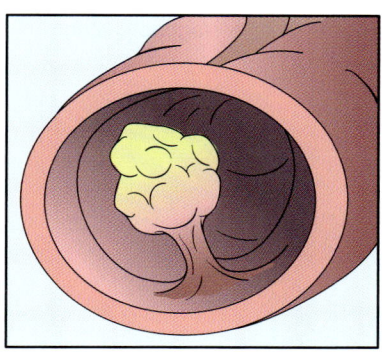

Figure 13.13 Colon Polyp

Colon polyps occur in approximately 20% of the adult population in North America (Figure 13.13). These growths result when cells of the large intestine mucosa project into the lumen. People who smoke, eat high-fat diets, eat very little dietary fiber, or are overweight are at risk for developing colon polyps. Colon polyps can cause rectal bleeding and changes in the frequency of bowel movements. Various surgical procedures are used to remove colon polyps. Most colon polyps cause few serious problems; however, some colon polyps, called **adenomatous polyps**, can lead to **colon cancer**. Over 50,000 people in North America die each year from cancers of the large intestine and rectum, yet advanced cancers of the colon and rectum are highly preventable and easily detectable. Physicians recommend colon cancer screening for people aged 50 years or older.

An **ulcer** is a condition caused by erosion of the inner linings of the digestive tract. Ulcers can form anywhere in the digestive tract. However, they are most common in the lower stomach and duodenum. Ulcers have a variety of origins and can cause mild to severe damage. Certain ulcers of the stomach and duodenum are the result of bacterial infection. **Hernias** are not fully a digestive tract disorder. They are defined as the protrusion of an organ into surrounding tissues. Many hernias involve components of the digestive tract that pass into weak spots of the diaphragm, and abdominal and pelvic muscles. Hernias can develop into a digestive system disorder if the affected digestive tract organ is constricted or damaged. A **hiatal hernia** is a protrusion of the stomach through the diaphragm. It occurs close to where the esophagus passes through the diaphragm. **Inguinal hernias** occur lower in the body when a section of the small intestine passes into areas of muscle weakness or openings in the pelvic cavity.

Liver and pancreatic diseases also plague the human digestive system. Both organs can develop inflammatory diseases due to infection or autoimmune disorders. **Hepatitis** is a liver disease caused by a variety of viruses that attack hepatocytes. There are various types of hepatitis viruses. The different types are designated with letters, such as hepatitis A, B, C, D, and E. Hepatitis B is spread through fecal contamination. The others strains are spread through blood contact or sexual activity. Hepatitis should not be confused with another liver disease called **cirrhosis**. Cirrhosis is a condition in which the liver becomes scarred and filled with fat. It is usually the result of long-term liver damage caused by chronic drug and/or alcohol use, hepatitis, and various toxins. Inflammation of the pancreas, or **pancreatitis**, also has a wide variety of causes. Some diseases of the digestive system organs are caused by parasites (Figure 13.14). These diseases are more common in areas with improper food handling and poor sewage treatment. All organ diseases of the digestive system reduce the body's ability to digest and absorb nutrients, which leads to malnutrition.

Colon Polyp A small, fleshy outgrowth in the colon
Adenomatous Polyp A polyp that can develop into cancer
Colon Cancer A cancer that forms in the large intestine
Ulcer A condition caused by erosion of the digestive tract mucosa
Hernia The protrusion of an organ into surrounding tissues
Hiatal Hernia A protrusion of the upper part of the stomach into the thorax
Inguinal Hernia A protrusion of the small intestine into pelvic muscles
Cirrhosis A disease that causes the liver to become scarred and filled with fat
Hepatitis Inflammation of the liver
Pancreatitis Inflammation of the pancreas

Figure 13.14 Lifecycle of a Parasitic Worm

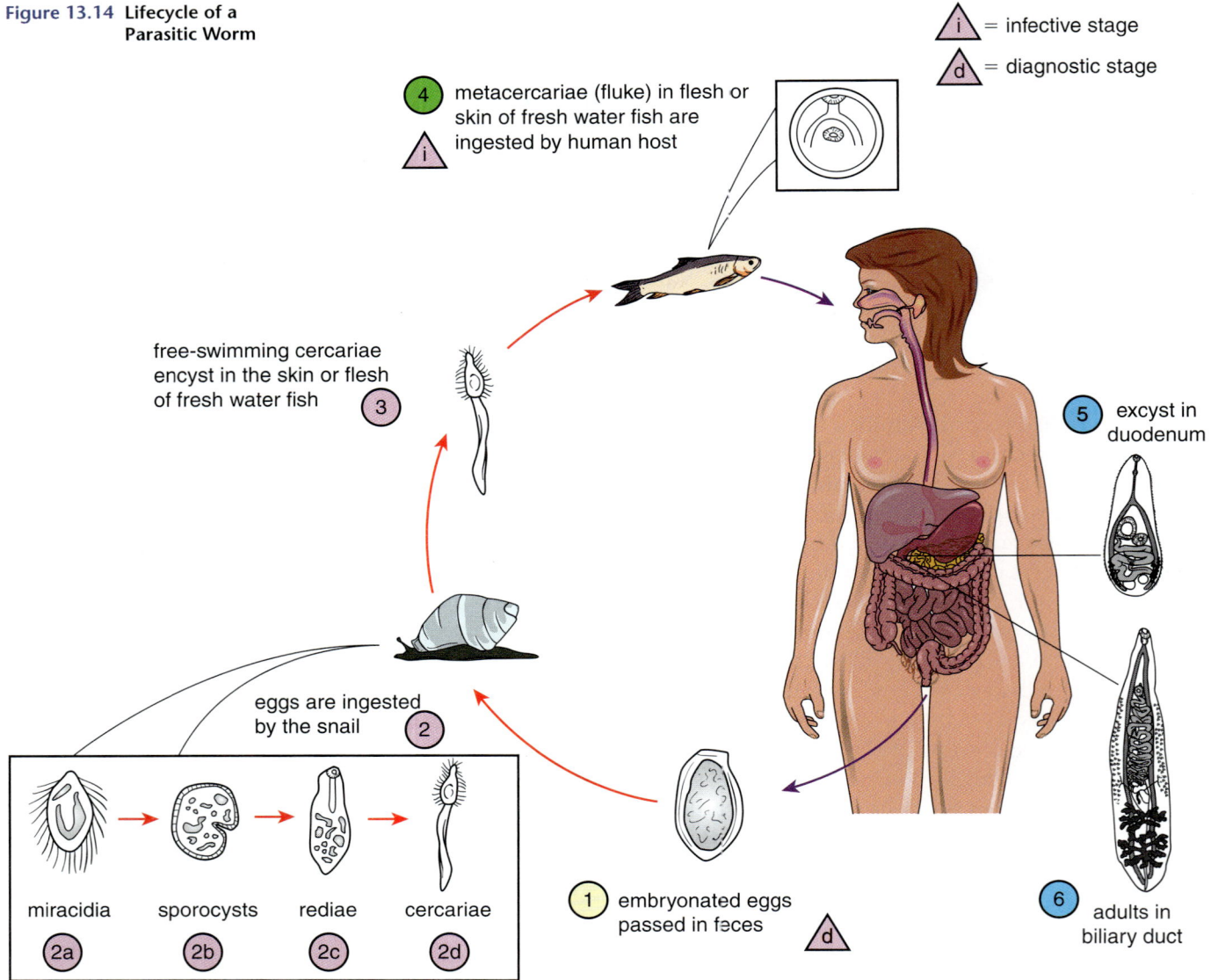

✓ Concept Check

1. Describe the possible causes of abdominal pain.
2. What is the difference between an ulcer and a polyp?
3. Distinguish between digestive tract and digestive system organ disorders.

Aging of the Digestive System

Much of the aging of the digestive system is due to a decrease in the neural activity needed for smooth-muscle function and glandular secretion. Contractions of the esophageal sphincter decrease with age. Peristalsis also becomes weaker. However, these changes have little effect on the movement of food to the stomach. Some people develop other neural conditions as they age that aggravate the reduced peristalsis, making it difficult to swallow food. The lining of the stomach loses its ability to resist damage as a person ages. Therefore, the stomach becomes more susceptible to peptic ulcers with age. Peptic ulcers erode the stomach lining and may reduce the production of enzymes needed for protein digestion. Aging usually has little effect on the secretion of stomach juices, such as acid and pepsin. Aspirin and other anti-inflammatory drugs worsen the erosion of the stomach lining.

Aging has little impact on the structure of the small intestine. The movement of food through the small intestine and the absorption of many nutrients are not significantly affected. However, food intolerances increase because enzymes needed to break down nutrients decrease. Even a slight buildup of undigested nutrients in the small intestine could make the person more susceptible to dehydration. The large intestine and rectum do not undergo much change with age (Figure 13.15). However, the large intestine can lose muscular action and the rectum enlarge somewhat. Certain bacteria become more prevalent with age. This can lead to weight loss as the bacteria compete for food in the large intestine. The weight loss occurs because the bacteria can take in certain nutrients before they are absorbed into the intestine. Consequently, calcium, folic acid, and iron uptake are decreased, affecting overall health and blood cell formation. Loss of smooth-muscle contraction leads to constipation. Older people are more likely to develop **diverticulosis** and other large-intestine disorders. Diverticulosis is characterized by the formation of pouchlike pockets in the large intestine.

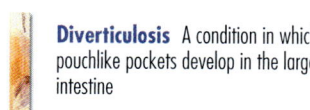

Diverticulosis A condition in which pouchlike pockets develop in the large intestine

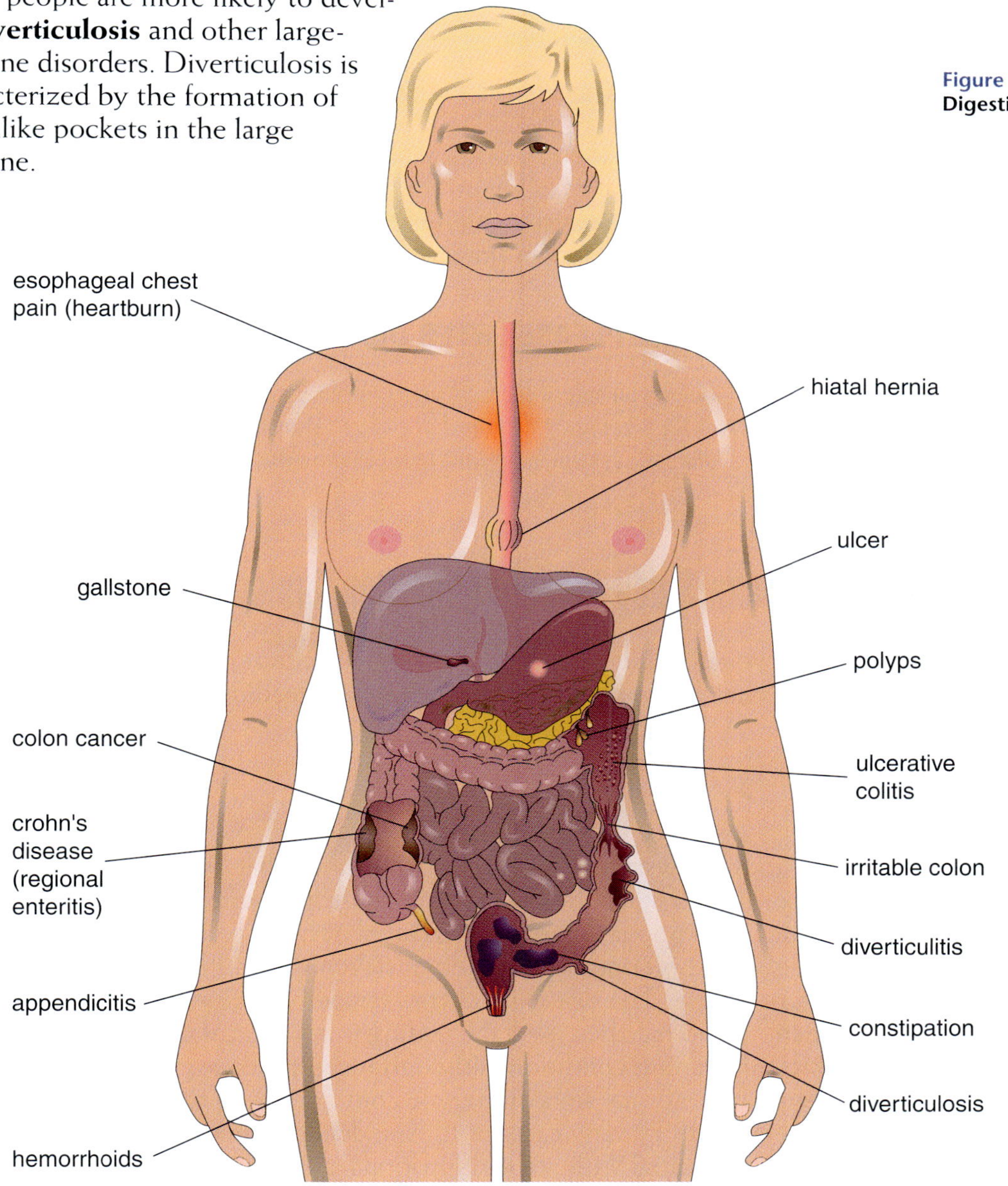

Figure 13.15 **Pathology of the Digestive System**

The glandular organs of the digestive system degenerate with age. The pancreas decreases in mass, and some of its glandular tissues become scarred. However, these changes do not significantly decrease the ability of the pancreas to produce digestive enzymes and sodium bicarbonate, but they may affect its ability to regulate blood sugar. The liver and gallbladder decrease in function with age. This reduces the liver's ability to break down potentially hazardous chemicals in food. A decline in liver and gallbladder function impairs the body's ability to digest and absorb fats.

Because the digestive system runs very efficiently, it is not severely impacted by aging of the body. However, a small decrease in digestion system function can significantly affect aging of the other organ systems. An age-related decrease in digestive system function significantly reduces nutrients and electrolytes, which are needed for muscle and nerve function. This results in muscle atrophy and a slowing of the central nervous system. It also increases the chances for overall malnutrition and undernutrition in the elderly persons. Elderly people may eat less as a result of the loss of taste and reduced hunger signals from the central nervous, system, which may aggravate malnutrition and under-nutrition.

✓ Concept Check

1. What is the impact of aging on the digestive system?
2. How does digestive system aging affect the other organ systems?
3. How does digestive system aging compare with that of other organ systems?

DISCOVERY SCENE PLEASE ENTER DISCOVERY SCENE PLEASE ENTER

Have you solved the CSI yet? Did your classmate's situation resemble any of the previously described disorders? Was the information about aging of digestive system important in understanding your classmate's problem?

CSI — Case Study Investigation Conclusion

What can you conclude about your classmate's condition? Where you able to determine the cause of the diarrhea? Were the bleeding and pain related to the diarrhea?

Answer:

In a visit to the physician, your classmate learned that he had irritable bowel syndrome (IBS). Irritable bowel syndrome affects approximately 30% the North American population. It is also called nervous colon, mucous colitis, and spastic bowel. This condition is often mistakenly identified as ulcerative colitis or Crohn's disease. The disease is characterized by abdominal pain, bloating, constipation, and diarrhea. Rectal bleeding is not typical of the condition. It is not a life-threatening condition, although it can produce considerable discomfort. Periodic, abnormal smooth-muscle contractions of the small and large intestines cause IBS. Many cases of IBS are due to unknown causes; however, it is typically associated with stress. Nutritional and psychological counseling are

used to reduce the recurrence of IBS, and medications are sometimes used to reduce the pain.

This CSI was adapted from the following articles:

1. Lembo A, Ameen VZ, Drossman DA. Irritable bowel syndrome: toward an understanding of severity. *Clin Gastroenterol Hepatol.* 2005;3(8):717-725.
2. Saito YA, Petersen GM, Locke GR, Talley NJ. The genetics of irritable bowel syndrome. *Clin Gastroenterol Hepatol.* 2005;3(11):1057-1065
3. Toner BB. Cognitive-behavioral treatment of irritable bowel syndrome. *CNS Specter.* 200510(11):883-890.

Chapter Summary

All of the other organ systems rely completely on the function of the digestive system to meet their nutritional needs. The digestive system makes it possible for the body to break down the complex molecules of food into the simple molecules needed for energy and raw materials to build cellular components. The blood then transports these nutrients throughout the body by absorbing the digested foods and passing them into the blood stream. Absorption is only possible if the food is mechanically and chemically digested. Organs of the digestive tract work in unison with accessory digestive system organs to carry out these tasks. Most mechanical digestion takes place in the mouth and stomach. Chemical digestion using enzymes at a particular pH level occurs in the stomach and proximal regions of the small intestine. Absorption commonly takes place in the middle and distal small intestine and the large intestine. Much of the water and electrolytes from food is absorbed in the large intestine. Beneficial bacteria residing in the large intestine protect the body from harmful bacteria and assist with further breakdown of certain foods. Diseases and aging of the immune system decrease the digestive or absorptive capabilities of the digestive system which, in turn, harms other organ systems.

Overview

- The digestive system is composed of the digestive tract and accessory digestive organs.
- The digestive tract is composed of the mouth, pharynx, esophagus, stomach, salivary glands, small intestine, large intestine, and rectum.
- The pancreas and liver are glandular organs.
- The digestive system carries out chemical and mechanical digestion.

Components of the Digestive Tract

- The mouth contains the salivary glands, teeth, tonsils, and tongue.
- The pharynx connects the mouth to the esophagus.
- The tubular components of the digestive tract are composed of four layers: mucosa, submucosa, muscularis, and serosa.
- The esophagus is a muscular tube that leads to the stomach.
- The stomach is a muscular sac that stores and digests food.
- The small intestine is a long, narrow tube that digests and absorbs food.
- The large intestine is a short, wide tube that absorbs water and electrolytes.
- The rectum, at the end of the digestive tract, removes undigested materials.

Study Guide

- The anus is composed of muscles that close off the rectum.

Glandular Structures of the Digestive System
- Exocrine functions of the pancreas involve the production of digestive enzymes.
- The endocrine function of the pancreas involves insulin and glucagon, which regulate blood sugar.
- Bile production is a major function of the liver.
- The gallbladder stores and releases bile.

The Digestive Process
- The hypothalamus regulates pregastric factors that, in turn, regulate hunger and thirst.
- Chemical and mechanical digestion begins in the mouth.
- Peristalsis moves food through the digestive tract.
- The stomach uses enzymes and acids to digest proteins.
- Hormones produced by the stomach and small intestine regulate digestion.
- The parasympathetic nervous system stimulates digestion.
- The small intestine is divided into the duodenum, jejunum, and ileum.
- Digestion of most food takes place in the proximal portions of the small intestine.
- Absorption of digested foods takes place in the distal portions of the small intestine.
- The large intestine absorbs water and electrolytes in its proximal components.
- Feces are formed in the distal portions of the large intestine.
- The rectum forces feces out of the body through the anal sphincter.

Wellness and Illness over the Life Span
- Food intolerances are caused by the inability to absorb or digest food.
- General abdominal pain is a common symptom of digestive tract disorders.
- Gastric reflux disease is a common disorder of the esophagus.
- Diarrhea is a common disorder with a variety of causes.
- Polyps are outgrowths of the mucosa that can develop into cancer.
- Ulcers are damage to the digestive tract mucosa.
- Digestive system gland disorders include cirrhosis, hepatitis, and pancreatitis.
- Aging of the digestive system affects the nutrient supply to other organ systems.
- Most digestive tract aging affects peristalsis.
- Aging of the digestive system glands reduces their ability to carry out their functions.

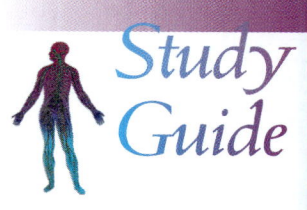

Key Terms

Overview
Accessory digestive organ
Alimentary canal
Digestive tract
Esophagus
Gastrulation
Large intestine
Meconium
Rectum
Salivary gland
Small intestine
Stomach

Mouth and Pharynx
Bicuspid
Buccal or oral cavity
Canine
Cuspid
Extrinsic muscle
Hard palate
Incisor
Intrinsic muscle
Lingual membrane
Lingual tonsil
Labia
Molar
Palate
Parotid
Premolar
Soft palate
Sublingual
Submandibular
Uvula
Wisdom teeth

Esophagus and Stomach
Adventitia
Cardiac gland
Cardiac region
Cardiac sphincter
Chief cell
Fundic gland
Fundic region
Gastric stem cell
Lamina propria
Mucosa
Mucous neck cell
Muscularis layer
Parietal cell
Pyloric glands
Pyloric region
Pyloric sphincter
Reflux
Submucosa
Serosa

Small Intestine
Crypt
Duodenum
Enteroendocrine cell
Enterocyte
Ileocecal valve
Ileum
Jejunum
Lacteal
Mesentery
Microvilli
Paneth cell
Villi

Large Intestine and Rectum
Anal canal
Anal sphincter
Anus
Appendicitis
Appendix
Ascending colon
Cecum
Colon
Descending colon
Hepatic flexure
Sigmoid colon
Splenic flexure
Transverse colon

Pancreas
Acini
Bile
Common bile duct
Glandular lobule
Pancreatic duct
Zymogen

Liver and Gallbladder
Caudate lobe
Falciform ligament
Gallbladder

THE DIGESTIVE SYSTEM

Study Guide

Gallstone
Heparin
Hepatic artery
Hepatic portal vein
Hepatic vein
Hepatocyte
Hilum
Left lobe
Quadrate lobe
Right lobe
Serum globulin
Sinusoid

The Digestive Process
Cholecystokinin (CCK)
Chyme
Enterokinase
Feces
Flatulence
Gastrin
Hunger center
Ingestion
Parenteral
Peristalsis
Pregastric factor
Protease
Reverse peristalsis

Salivary amylase
Satiety center
Secretin

Pathology of the Digestive System
Acid reflux
Acute diarrhea
Adenomatous polyps
Amoebic dysentery
Celiac disease
Chronic diarrhea
Colon cancer
Colon polyp
Diarrhea
Dysphagia
Food intolerance
Gastric reflux
Gastroesophageal reflux disease (GERD)
Hepatitis
Hernia
Hiatal hernia
Inflammatory bowel disease (IBD)
Inguinal hernia
Pancreatitis
Salmonella
Ulcer

Check Your Understanding

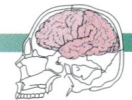

1. Which is not a function of the digestive system?
   a. removal of undigested food
   b. production of urea
   c. enzymatic breakdown of food
   d. absorption of food

2. The innermost layer of the digestive tract is the:
   a. serosa
   b. mucosa
   c. lamina propia
   d. tunica

3. Which of the following is an accessory organ of the digestive system?
   a. liver
   b. stomach
   c. duodenum
   d. rectum

Study Guide

4. A microscopic section of the digestive system shows blood vessels, nerves, and small glands. You are most likely viewing this portion of the digestive tract:
   a. submucosa
   b. serosa
   c. Tunica
   d. muscular layer

5. Peristalsis of the digestive tract is carried out by:
   a. osmosis
   b. fluid dynamics
   c. skeletal muscle contractions
   d. smooth-muscle contractions

6. Damage to the mucosa by microorganisms would affect this digestive system function:
   a. bile secretion
   b. peristalsis
   c. absorption of nutrients
   d. blood flow through the serosa

7. The loss of teeth reduces this digestive system capability:
   a. mastication
   b. saliva production
   c. gastric juice secretion
   d. bicarbonate production

8. Swallowed food does not enter the lungs because of the function of this structure:
   a. pyloric sphincter
   b. epiglottis
   c. muscularis layer of esophagus
   d. duodenum

9. Proteins are primarily broken down in the:
   a. mouth
   b. esophagus
   c. stomach
   d. large intestine

10. The pancreas carries out the following digestive system function:
    a. physical breakdown of food
    b. secretion of digestive enzymes
    c. bile production
    d. absorption of amino acids

11. Removal of the gallbladder directly affects:
    a. bile storage
    b. bile production
    c. pepsin secretion
    d. lipase secretion

THE DIGESTIVE SYSTEM

Study Guide

12. Damage to the hepatic portal vein would directly affect this digestive system function:
    a. transport of the fats to the liver
    b. absorption of nutrients into the mucosa
    c. digestive tract support by the serosa
    d. release of digestive enzymes by the stomach

13. Starch is digested in these two structures:
    a. stomach and small intestine
    b. mouth and large intestine
    c. stomach and mouth
    d. mouth and small intestine

14. Which is not a function of the large intestine?
    a. absorption of water
    b. uptake of electrolytes
    c. conversion of wastes to urea
    d. passage of undigested food

15. Digestive system ulcers are most commonly found in the:
    a. esophagus
    b. lower portion of stomach
    c. large intestine
    d. rectum

A Case Study

IS EXERCISE BAD FOR THE DIGESTIVE SYSTEM?

Most fitness experts and personal trainers assert that exercise is good for the digestive system. Many nutritionists and physicians support this view. It has been shown, however, that many people develop occasional trouble with their digestive systems after moderate to strenuous exercise. Scientifically conducted surveys recently revealed that approximately 30% of all runners suffer from abdominal cramping during practice or after a race. The immediate need to defecate occurs in 30% of these people. Another 25% develop diarrhea during or just after a race. It is not surprising that strenuous exercise could cause intestinal problems at the time of the activity. After all, the parasympathetic nervous system that controls intestinal activity relaxes during exercise. Some physicians believe that this cuts off blood to the digestive system and may produce discomfort if undigested food is in the stomach or intestines.

   Scientists now know that it is not the exercise alone that produces many of the intestinal problems. The surveys show that runners who suffer intestinal cramping very likely have digestive system disorders, such as irritable bowel syndrome or lactose intolerance. Those without any underlying digestive system disorders reported having diets that are probably too high in fiber. Almost half of the runners who defecate after a race have a digestive system function that is modified by exercise. Scientists now believe that the condition humorously called "runners' trots" is caused by a decrease in intestinal transit time. Intestinal transit time is the amount of time it takes for food to pass through the digestive system. So far, this explanation is not fully supported by research.

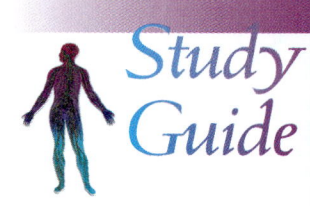

Study Guide

Studies show that people who start regular strenuous exercise routines have a decreased average intestinal transit time. People who normally took 35 hours to pass a meal from mouth to rectum reduced their intestinal transit time to 24 hours. The researchers did not take into account any significant changes in diet. Other scientists take the traditional argument that blood flow to the intestines is reduced by approximately 80% during exercise. This loss of blood may cause digestive system abnormalities, such as cramping and diarrhea. Then, there are scientists who believe that dehydration associated with exercise could be producing the digestive system disorders.

Use the information in this chapter and the following Web sites to answer the following questions about exercise and digestive system health:

- How does average intestinal transit time affect a person's health?
- Is there enough evidence supporting the negative effects of exercise on digestive system function?
- Should physicians caution people about exercise given the recent evidence supporting its negative effects on the digestive system?
- What warnings about exercise and eating should a personal trainer or physician give to people who are considering starting an exercise program?
- What, if any, actions should a person take to avoid intestinal problems during exercise?
- Should people with preexisting digestive system disorders be discouraged from participating in strenuous exercise?
- Should the government require people or agencies that promote athletic events or regular exercise to provide warnings about digestive system disorders associated with activity?

1. e-cobr information
   http://www.cobr.co.uk/e-cobr_information/t_and_r_section/sections/nutrition/Effects%20of%20exercise%20on%20the%20digestive%20tract.shtml

2. Baptist Memorial Healthcare
   http://www.baptistonline.org/health/library/digestive.asp

3. Mayo Clinic (search digestive system)
   http://www.mayoclinic.com/

4. National Digestive Diseases Information Clearinghouse
   http://digestive.niddk.nih.gov/ddiseases/pubs/ibs/

Where Do We Go from Here?

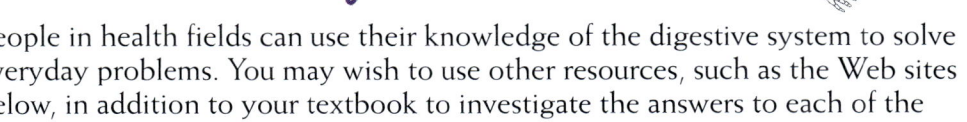

People in health fields can use their knowledge of the digestive system to solve everyday problems. You may wish to use other resources, such as the Web sites below, in addition to your textbook to investigate the answers to each of the following situations:

1. An elderly neighbor wants to significantly increase the fiber content of her diet. What would you tell her about the benefits and risks of increasing dietary fiber intake?
   Web MD (Search digestive system into the search window):
   http://www.Web.md.com/

Study Guide

2. A female friend heard that taking laxatives is a safe way to lose weight rapidly. What effects would this have on her health?
   Food and Drug Administration (Search laxatives into the search window):
   http://www.fda.gov/

3. A neighbor's young daughter has been discovered to be bulimic. Explain to him the possible effects it could have on the girl's body.
   e-Medicine - bulimia: http://www.emedicinehealth.com/Articles/12710-1.asp

4. An older relative tells you that he drinks four glasses of milk per day to relieve his ulcers. What are the benefits or risks of this practice?
   Web MD - Understanding Ulcers:
   http://www.Web.md.com/content/article/90/100627.htm

5. A friend confides that he has been passing blood in his stool. What could you tell him before he pays a visit to his physician?
   Medline Plus –Digestive System:
   http://www.nlm.nih.gov/medlineplus/digestivediseases.html

Skills Activities

1 Microscopic Identification of Normal Digestive Organs

Materials
- access to the Internet
- link to http://www.bu.edu/histology/m/i_main00.htm
- white drawing paper
- yellow highlighter marker
- orange highlighter marker
- green highlighter marker

Looking at microscopic preparations of digestive system structures helps a person to better understand how the various organs function. From this, a person can determine any differences that cause each organ to carry out its particular role. Many physicians and scientists investigate organ function by reviewing photographs of tissues kept in databases of graphics. They can access the databases to view specimens not readily available to them. Plus, they can share the images with colleagues throughout the world to help with a variety of medical investigations.

Go to a computer that has Internet access and is hooked up to a printer. Use a browser to link to http://www.bu.edu/histology/m/i_main00.htm. Click on the Digestive System: Alimentary Canal link. Follow the instructions for each activity.

Esophagus

Click on the link for the esophagus. Study the image and click to view the labeled diagram. Note the relative proportions of mucosa, muscle layers, and any visible glandular tissues. Print out the labeled image. Use the highlighters to color the following structures on the printed image: mucosa, yellow; muscle, orange; and glandular tissues, green.

Stomach

Now, look at the stomach specimens by clicking on the fundic stomach (H and E) and pyloric stomach I (H and E) links. H and E (hematoxylin and eosin) is the type of staining technique used to color molecules in the cell. Note the relative proportions of mucosa, muscle layers, and any visible glandular tissues between the two stomach sections. Print out the labeled image and mark it. Use the highlighters to color the following structures: mucosa, yellow; muscle, orange; and glandular tissues, green. Compare each stomach section to the diagram of the esophagus. Note any differences and similarities. Which stomach section most resembles the esophagus?

Small Intestine

Click on the links leading to the images of the jejunum I (eosin and toluidine blue), ileum I, villi (H and E), and ileum I, Peyer's patches (H and E). How do the sections of intestine vary from each other? Print out the labeled image and mark it. Use the highlighters to color the following structures: mucosa, yellow; muscle, orange; and glandular tissues, green. Compare each section to the diagram of the stomach. Note any differences and similarities.

Large Intestine and Rectum

Look at the images of the appendix (H and E), colon (H and E), and anal canal (H and E). Compare the sections noting any differences or similarities. Print out the labeled image and mark it. Use the highlighters to color the following structures: mucosa, yellow; muscle, orange; and glandular tissues, green. Compare these sections to the images of the small intestine. Note any differences and similarities.

A good way to review the images is to have a classmate print out the unlabeled images, and see if you can identify the organ and its particular features.

Skills Activities

2 Effects of Antacids on Protein Digestion

Materials
- ½-inch-square cubes of egg white from a soft-boiled egg
- 1% pepsin solution
- 0.8% hydrochloric acid solution
- 0.5% baking soda (sodium bicarbonate) solution
- two types of antacid pills, each ground into separate containers holding 100 mL of distilled water
- distilled water
- six calibrated droppers for transferring the different solutions
- droppers for collecting samples of epinephrine, caffeine, and coffee or tea.
- seven clean, large test tubes
- test-tube rack
- strips of universal pH paper
- forceps
- water bath or incubator set at 37° C
- one sheet of lined notebook paper

Study Guide

Antacids are a common over-the-counter remedy used by many people to treat reflux and upset stomach. They reduce the acidity of the stomach contents, thereby reducing the corrosive effects of the acid on the esophagus or an irritated stomach. Its overall effectiveness is debatable; some physicians believe there are risks to taking antacids. The inappropriate use of antacids is believed to cause digestive system problems if taken with a high-protein meal. Information about the chemistry of antacids can by obtained by searching the term "antacids" on the National Institutes of Health Web site at: *http://www.nlm.nih.gov/medlineplus/*. This activity investigates the effects of antacids on the digestion of egg-white protein.

This activity involves setting up a series of conditions that model the chemical environment of the stomach. Scientists use experiments similar to this one to better understand the chemical reactions taking place during digestion. This experiment models the effects of antacids on protein digestion. Carry out the following steps to conduct this experiment:

1. Label the test tubes from No. 1 to 7, and place them in the rack in numerical order.
2. Place a square of egg white into each tube.
3. Add the various chemicals following the grid provided below. Use separate droppers for the water, baking soda, antacid, hydrochloric acid, and pepsin solutions:

Test-Tube Setup Grid

Tube Number	Water (mL)	Baking Soda (mL)	Antacid (mL)	Hydrochloric Acid (mL)	Pepsin Solution (mL)
1	15				
2	12				3
3	12			3	
4	9			3	3
5	6	3		3	3
6	6		3	3	3
7	6		3	3	3

4. Place the rack of test tubes into a water bath or incubator set at 37° C.
5. Gently shake the tubes after every 5 minutes for 25 minutes. Note the consistency of the egg white after shaking the tubes.
6. Write out a table of the setup for recording data.
7. Observe the egg white after 30 minutes.
8. Test the pH of each tube by placing the tip of the pH paper into each one. Use a separate strip for each tube. Record the pH of each solution on the data sheet.
9. Record the appearance of the egg white in each tube on the data sheet. Compare the egg white in tubes No. 2 through 7 with that in test tube No. 1. Answer the following questions:

   a. What effect does pepsin alone (test tube No. 2) have on the digestion of proteins?

CHAPTER 13

Study Guide

b. What effect does hydrochloric acid alone (test tube No. 3) have on the digestion of proteins?
c. How effective was the digestion of egg white in test tube No. 4, which modeled the normal human stomach?
d. What effects did baking soda have on the digestion of protein in the model stomach?
e. What effects did the different antacids have on the digestion of protein in the model stomach?
f. Describe the relationship between pH and the digestion of protein.
g. How would you use this experiment to explain the possible negative health effects of antacids?

Case Study Investigation

Case Study Investigation #14

A 79-year-old female neighbor asks you to drive her to the doctor for a weekly examination. She lives alone, and her children are out of town. She claims that she is afraid to drive herself to the doctor because of "medical reasons." At the doctor's office you discover that she is still recovering from a fall that broke her hip several weeks ago. You think this is the reason that she is hesitant to drive. However, by overhearing the physician, you learn she cannot control her bladder function. Certain movements cause her to urinate without control, and she often has to go several times through the night. On the drive

home she admits her "situation" to you, saying that she never had "that problem" until after the surgery for her broken hip. She does not understand the problem and is hoping you can explain her condition. You find out her medical exams come back normal except for the frequent urination. What is the cause of her urination problem? Did it happen after the fall or after the surgery? At the end of the chapter, you will be asked to determine the possible problems causing her condition.

CHAPTER 14
THE URINARY SYSTEM

Chapter Outline

Case Study Investigation (CSI)
Applied Learning Outcomes
Overview
Gross Anatomy of the Urinary System
    The Kidney – External Anatomy
    The Kidney – Internal Anatomy
    The Ureters
    The Bladder
    The Urethra
Urine Voiding
The Nephrons
Urine Formation
    Filtration in the Glomerulus
    Reabsorption in the Proximal Convoluted Tubule
    Reabsorption in the Loop of Henle
    Reabsorption in the Distal Convoluted Tubule and Collecting Duct
    Tubular Secretion
Hormonal Regulation of Urine Formation
Wellness and Illness over the Life Span
    Pathology of the Urinary System
        Congenital Disorders
        Infection and Inflammation
        Immune Disorders
        Hormonal Disorders
        Degenerative Disorders
        Tumors
    Aging of the Urinary System
CSI Conclusion
Study Guide

Applied Learning Outcomes

- Use the terminology associated with the urinary system.
- Learn about the following:
  - organs and structures of the urinary system
  - physiology of the urinary system
- Understand the biological basis of the pathology and aging of the urinary system.

Overview

As you have learned in previous chapters, various systems aid the body in the excretion of waste products. Carbon dioxide leaves the body through the respiratory system. Perspiration produced by the integumentary system contains certain metabolic wastes that are dissolved in the water that cools the body. The digestive system excretes substances that have been consumed, but are unusable to the body. The urinary system, too, plays a major role in the removal of wastes to maintain homeostasis of the body. It does this by acting as a filtering system of the blood in a series of processes that culminates in the formation of urine. As you know, blood collects many waste products of cellular metabolism in its journey throughout the body. These waste products become interspersed in the blood plasma with other substances, such as water, proteins, ions, nutrients, and gases. The presence of these substances in specific proportions is extremely vital to life. In a healthy individual, the urinary system is able to retain the proper amounts of these desirable substances while ridding the body of the undesirable ones. In doing so, it not only removes wastes, it maintains the pH level, electrolyte composition, and water content of the blood, which, consequently, controls the balance of these components in tissue fluids.

✓ Concept Check

1. List four important functions of the urinary system.
2. What other body systems function in waste removal and how do they accomplish it?
3. Does the urinary system provide homeostasis to the blood only? Explain.

Gross Anatomy of the Urinary System

Key Terms: kidneys, umbilical cord, ureters, urethra, urinary bladder

Kidneys The organs that form urine

Ureters Tubes that send urine from the kidneys to the bladder

Urinary bladder A sac-like organ that serves as a reservoir for urine storage

Urethra The tube that transfers urine from the bladder to the body's exterior

Umbilical Cord The cord that transports blood, oxygen, nutrients, and wastes between the fetus and the mother during pregnancy

The **kidneys**, **ureters**, **urinary bladder**, and **urethra** are the organs and structures that make up the urinary system (Figure 14.1). The pair of kidneys are the true workhorses of the urinary system, as they are the organs that actually perform the physiological processes required to form urine and maintain pH, electrolytes, and water homeostasis. In doing so, they have a great deal of influence on certain functions of other body systems. In addition, you may remember from Chapter 11 that they produce erythropoietin to stimulate RBC formation. The remaining components of the urinary system can be thought of as the "plumbing system"; they form a network of pathways, and a reservoir for the transportation and temporary storage of urine as it exits the body. They play no physiological role in altering the composition of urine.

Kidneys form early in week 5 of human embryological development. They start out as dorsal strips of mesoderm tissue that run from the neck down to the pelvic regions. This early kidney drains fluids from the developing body cavities into a sac that feeds into the **umbilical cord**. Kidneys are not functional until week 9 of development. The urinary bladder and associated structures start to form after week 10 and are not fully functional until after birth. Components of the reproductive tract integrate into the urinary system around week 14 of development.

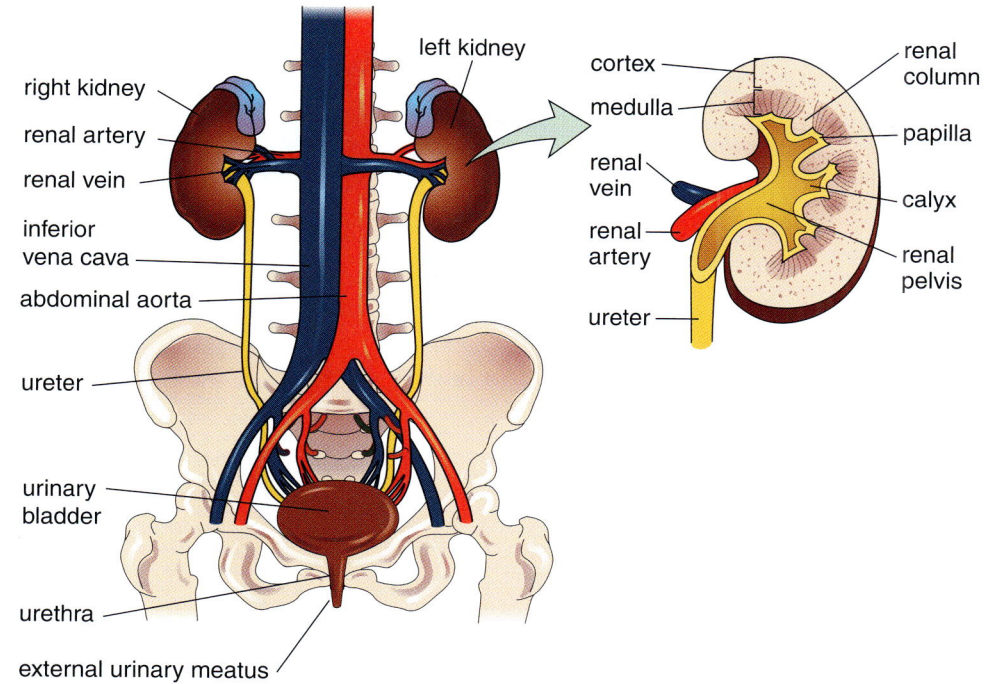

Figure 14.1 Components of the Urinary System

The Kidneys – External Anatomy

Key Terms: adipose capsule, hilus, renal artery, renal fascia, renal vein, retroperitoneal

The bean-shaped kidneys are located bilaterally near the midline of the abdomen with the left kidney positioned a little superior to the right one. They are not, however, within the abdominal cavity, but instead lie **retroperitoneal**, or behind the peritoneum. A cushioning of fat, the **adipose capsule**, encases each of these organs. The kidneys are secured to the abdominal wall by a superficial connective-tissue covering called the **renal fascia**. The kidneys' position high in the abdominal cavity, just beneath the diaphragm, protects them from the inferior ribs. Each kidney is capped by an adrenal gland (as discussed in Chapter 8) and anatomically positioned so that the concave indentation, the **hilus**, is medial to the body. This inward curve is both the entry point for the **renal artery**, which branches off from the abdominal aorta (to carry blood to be filtered), and the exit point for the **renal vein** (Figure 14.2). The veins run parallel to the arteries as they return filtered blood to the inferior vena cava to enter general circulation.

Retroperitoneal Behind the peritoneum

Adipose Capsule A cushioning of fat that encases each kidney

Renal Fascia The connective tissue that secures each kidney to the posterior abdominal wall

Hilus The medial concave curve of each kidney

Renal Artery A branch of the abdominal aorta that carries blood to the kidney

Renal Vein The vein that carries blood from the kidney to the inferior vena cava

Figure 14.2 **Kidney Structure**

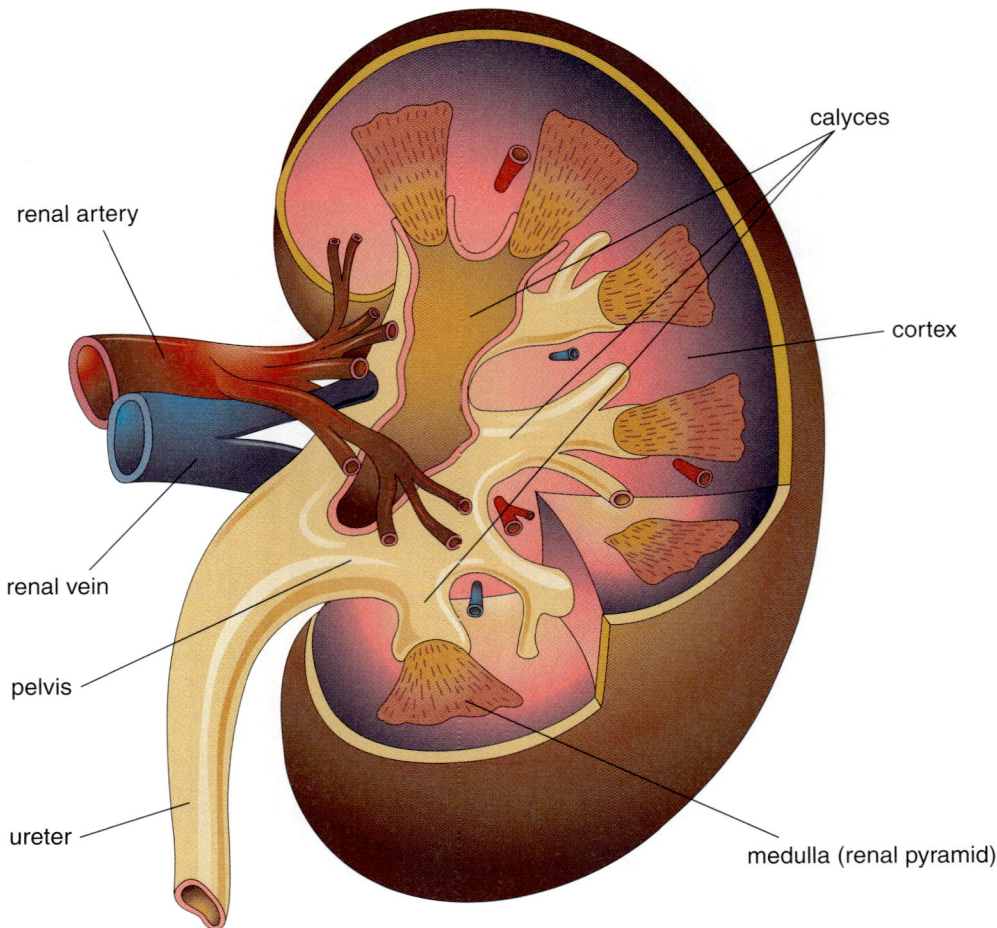

Civic Responsibility

CIVIC RESPONSIBILITY: HELPING OTHERS WITH YOUR KNOWLEDGE

It is valuable to use what you have learned about the urinary system to help others to better understand the world around them. It is very important to check your facts and seek further information about certain topics before discussing health and science issues. Here are some suggestions to foster a better public awareness of the urinary system:

1. Speak to toddlers at a daycare center about toilet training.
2. Work with sports clubs to educate the players about the importance of proper hydration during exercise.
3. Help elderly persons to better understand the condition of incontinence, its physiological basis, and the role that exercise and diet can play in its treatment.
4. Volunteer at a public health awareness day at your college to educate participants against the use of diuretics as a weight loss technique, but in favor of their use in controlling hypertension.

CHAPTER 14

The Kidney – Internal Anatomy

Key Terms: calyces, renal columns, renal cortex, renal medulla, renal pelvis, renal pyramids, ureter

There are three easily distinguishable areas of the gross internal anatomy of the kidney, as revealed in a coronal section (Figure 14.2). The superficial layer is known as the **renal cortex**. It can be distinguished from the inner **renal medulla** by its lighter color. Within the deeper medulla area are several triangular collections of tissue called **renal pyramids**. They are separated by inward extensions of the cortex, or **renal columns**. The renal pyramids are the location of many tubules where urine collection actually takes place. Their parallel arrangement gives this tissue a striped appearance. These tubules empty the collected urine into the third area of the kidney called the **renal pelvis**. Urine collected in the tubules of the pyramid region begins to leave the kidneys by passing through ducts located at the apex (point) of each pyramid. These ducts drain urine into branching extensions of the renal pelvis known as **calyces**. Urine transferred through the calyces enters the renal pelvis cavity, which is connected to a ureter at each kidney's hilus (Figure 14.3).

Renal Cortex The outer portion of the kidneys

Renal Medulla The soft, marrow-like structure in the center of each kidney

Renal Pyramid A triangular collection of tissue in the renal medulla

Renal Pelvis A cavity of the kidney where formed urine is collected before it enters the ureter

Renal Column Renal cortex tissue that extends inward between the renal pyramids

Calyces (singular is calyx) Branching extensions of the renal pelvis that transfer urine from the renal pyramids

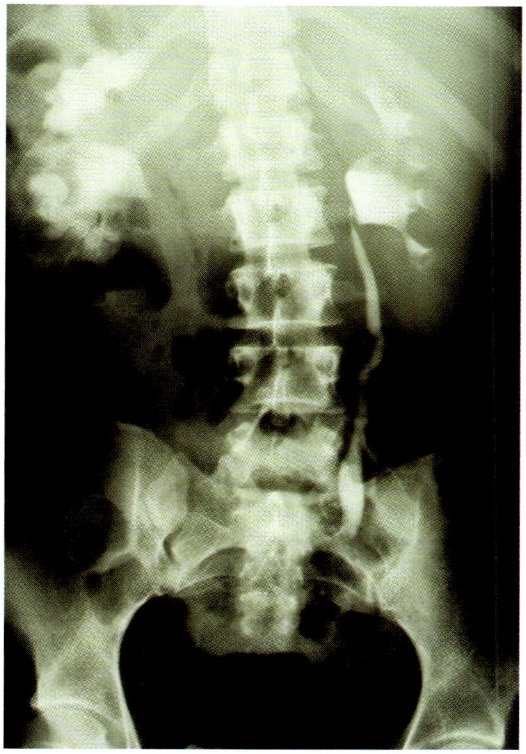

Figure 14.3 The Ureters

The Ureters

The long, thin ureters are muscular tubes that are also in a retroperitoneal position. They extend inferiorly from the hilus and enter the urinary bladder posteriorly at separate locations, one on each side of the bladder floor. They function only in urine transport, helping gravity to move urine through by peristaltic contractions. Receptors in their tissues detect a stretch stimulus as urine enters and elicit this smooth-muscle action. The ureters do not have valves that close to prevent urine from passing into the bladder; however, the bladder exerts upward pressure on them as it fills, which pinches the tube ends and closes them off at their entry points.

THE URINARY SYSTEM

The Urinary Bladder

Key Terms: detrusor muscle, internal urinary sphincter, rugae, transitional epithelium, trigone

The urinary bladder, which is located inferiorly in the pelvic cavity, accumulates and temporarily stores urine before it exits the body. In females, this sac-like structure is located anterior and slightly inferior to the uterus. This explains the need for frequent urination when the uterus is in an enlarged state of pregnancy. In males, it is superior to the prostate gland. This can pose problems with urination when this gland enlarges with age as it so commonly does. The lining of the bladder is composed of **transitional epithelium**. The cells of this tissue can stretch and elongate along with the smooth muscle of the organ wall as it relaxes. This enables the bladder to distend to hold increasing volumes of urine (Figure 14.4). The wall of the bladder is composed of a thick crisscross arrangement of smooth muscle called the **detrusor muscle**. It folds inward to form **rugae**, much like the stomach, when it is empty. The floor of the bladder forms a triangular area known as the **trigone**. This area has three openings: Each ureter opens at the corners of the base of the triangle, and at the apex, or point, is a single opening to the urethra. The **internal urinary sphincter**, an involuntary circular muscle, keeps the ureter closed. In its fully distended state, the bladder can hold up to one liter of urine. However, holding the maximum capacity causes extremely uncomfortable pressure. The need to void is usually detected by stretch receptors when only 20% of the maximum volume is present.

Transitional Epithelium Epithelial tissue that can change shape with expansion and contraction of the underlying tissue

Detrusor Muscle The smooth muscle of the urinary bladder

Rugae Inward folds of an organ

Trigone The smooth triangular area of the urinary bladder floor

Internal Urinary Sphincter An involuntary muscular ring at the exit of the urinary bladder

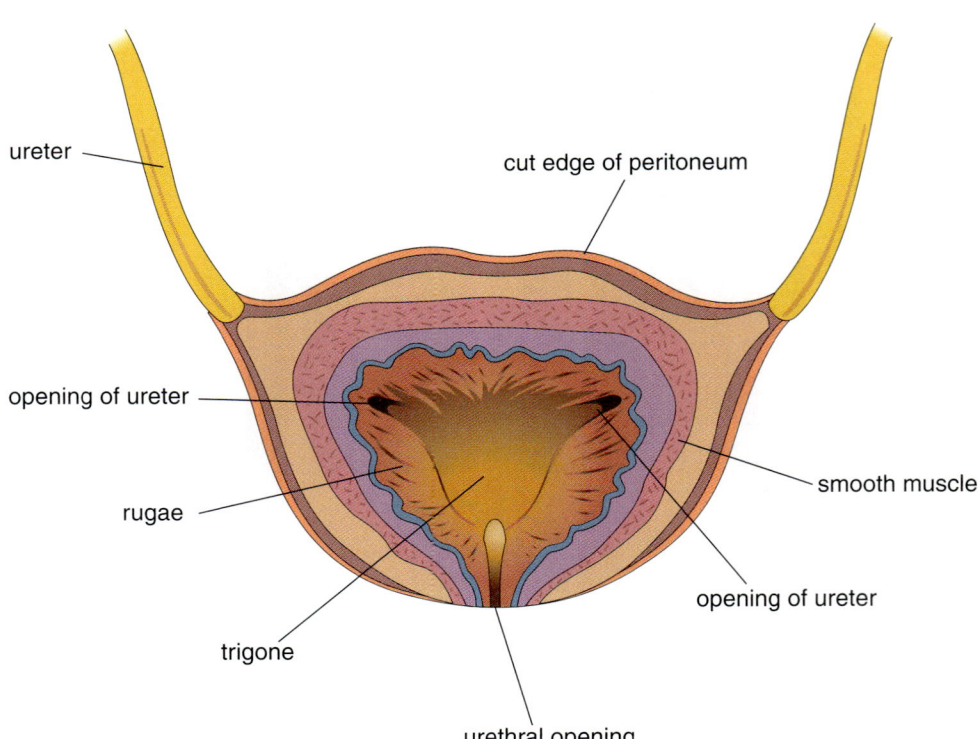

Figure 14.4 The Bladder
The urinary bladder is a muscular sac that stores and controls the flow of urine

The Urethra

Key Terms: external urethral sphincter, urethral orifice

The urethra forms the passageway from the bladder to the body's exterior. It is a single muscular tube whose closure is controlled by a voluntary muscle called the **external urethral sphincter**. This circular arrangement of skeletal muscle is located near the midpoint of its length in females and just beneath the prostate gland in males. The male urethra is considerably longer than the female urethra. It is divided into three distinct portions as it descends through the prostate and the full length of the penis. The female urethra has the sole function of carrying urine to the outside of the body through the **urethral orifice**, which is located anterior to the vaginal opening. In the male, the urethra has the same role, but it also provides a passageway for semen during ejaculation (see Chapter 15). The difference in urethral length between males and females is of clinical interest because the location of the female's bladder is closer to the body's exterior. This makes it much more susceptible to the entry of bacteria because of its proximity to the anal area and, consequently, possible exposure to fecal contamination. It should be no surprise that females are much more prone to urinary bladder infections.

External Urethral Sphincter A voluntary muscular ring located in the urethra

Urethral Orifice The external opening through which urine leaves the body

✓ Concept Check

1. What functional difference exists between the kidneys and the other components of the urinary system?
2. Name and describe the three distinguishable areas of internal kidney anatomy.
3. Describe the pathway that urine follows beginning with its final point of formation in the renal tubules and ending at the body's exterior.

DISCOVERY SCENE PLEASE ENTER DISCOVERY SCENE PLEASE ENTER

What information about the gross structure of the urinary system is helpful for understanding the CSI? What anatomical parts are possibly affected by the woman's condition? Is the ailment due to an abnormality of only one component of the urinary system?

URINE VOIDING

Key Terms: anuria, catheter, incontinence, oliguria, polyuria, micturition, nephron, urinary retention

Emptying the bladder is commonly known as urination, but the medical term is **micturition**. In the immature nervous system of an infant, the external urethral sphincter is not yet under voluntary control, so urination occurs reflexively as stretch receptors in the bladder wall detect the accumulated urine. Effectors of this reflex arc include the detrusor muscles of the bladder. These muscles respond to pressure in the bladder by contracting and forcing urine into the opening of the internal urethral sphincter. The sphincter is innervated by a

Micturition Elimination of urine from the bladder

Incontinence The inability to hold urine

Anuria The inability to produce urine

Urinary Retention The inability to expel urine from the bladder

Catheter A tube inserted into the urethra to expel urine

Oliguria Decreased urine production

Polyuria The production of an increased amount of urine

Nephron Tubular structures that filter the urine in the kidneys

reflex arc, which, conversely, causes it to relax. Thus, urine is allowed to exit the bladder and flow through the urethra to the body's exterior without voluntary control. As the nervous system matures, voluntary control of the external urethral sphincter is gained to control micturition. Loss of this voluntary control, or **incontinence**, can occur from factors that decrease the competence of the urinary sphincter muscles.

The amount of urine voided can be an important indicator of certain health disorders. **Anuria**, or lack of urine production, should be differentiated from **urinary retention**, in which urine is produced, but cannot be expelled from the bladder. Anuria indicates renal failure and is an extremely serious condition; death will occur if enough wastes accumulate to poison the body. Cases of extreme urinary retention can be relieved by a **catheter**, or drainage tube, inserted into the urethra to expel the urine. Decreased urine production is called **oliguria**. If it occurs following adequate fluid intake, it can indicate such disorders as kidney damage or ureter obstruction. **Polyuria** is the production of excessive volumes of urine. It can be an indicator of diabetes mellitus. The root of any disorder that causes an abnormal volume of urine to be produced involves the physiological mechanisms of the **nephrons** (the anatomical structures within the kidneys). The nephrons are actually responsible for the many physiological processes involved in urine formation.

✓ Concept Check

1. What is the medical term for voiding urine?
2. Describe the reflex arc that controls urination.
3. List four medical terms used to describe abnormalities in urine voiding, and explain what each means.

DISCOVERY SCENE PLEASE ENTER DISCOVERY SCENE PLEASE ENTER

Does the information about urine voiding provide more of an understanding of the woman's condition? Could the bladder muscles be the problem? Is it possible that the condition is due to neurological problems?

THE NEPHRONS

Key Terms: afferent arteriole, Bowman's capsule, collecting tubule, distal convoluted tubule, efferent arteriole, epithelial tubule, glomerulus, loop of Henle, peritubular capillary system, proximal convoluted tubule, renal corpuscle, renal tubules

Renal Tubules Small tubes within the kidney

Glomerulus A capillary loop within the nephron

Afferent Arteriole The blood vessel that narrows to become the glomerulus

Thousands of complex microscopic nephrons are present in each kidney (Figure 14.5). The nephrons carry out several of the kidney's many jobs (Figure 14.6). Each nephron is composed of an arrangement of **renal tubules** with an intricate vascular network. The location of the tubules allows the blood to be filtered and the filtered substances to form urine. The nephron consists of a uniquely folded capillary network called the **glomerulus**. The glomerulus originates from the **afferent arteriole**, a blood vessel formed from continuous branching of the renal artery through the renal column of the cortex

CHAPTER 14

Figure 14.5 The Nephron

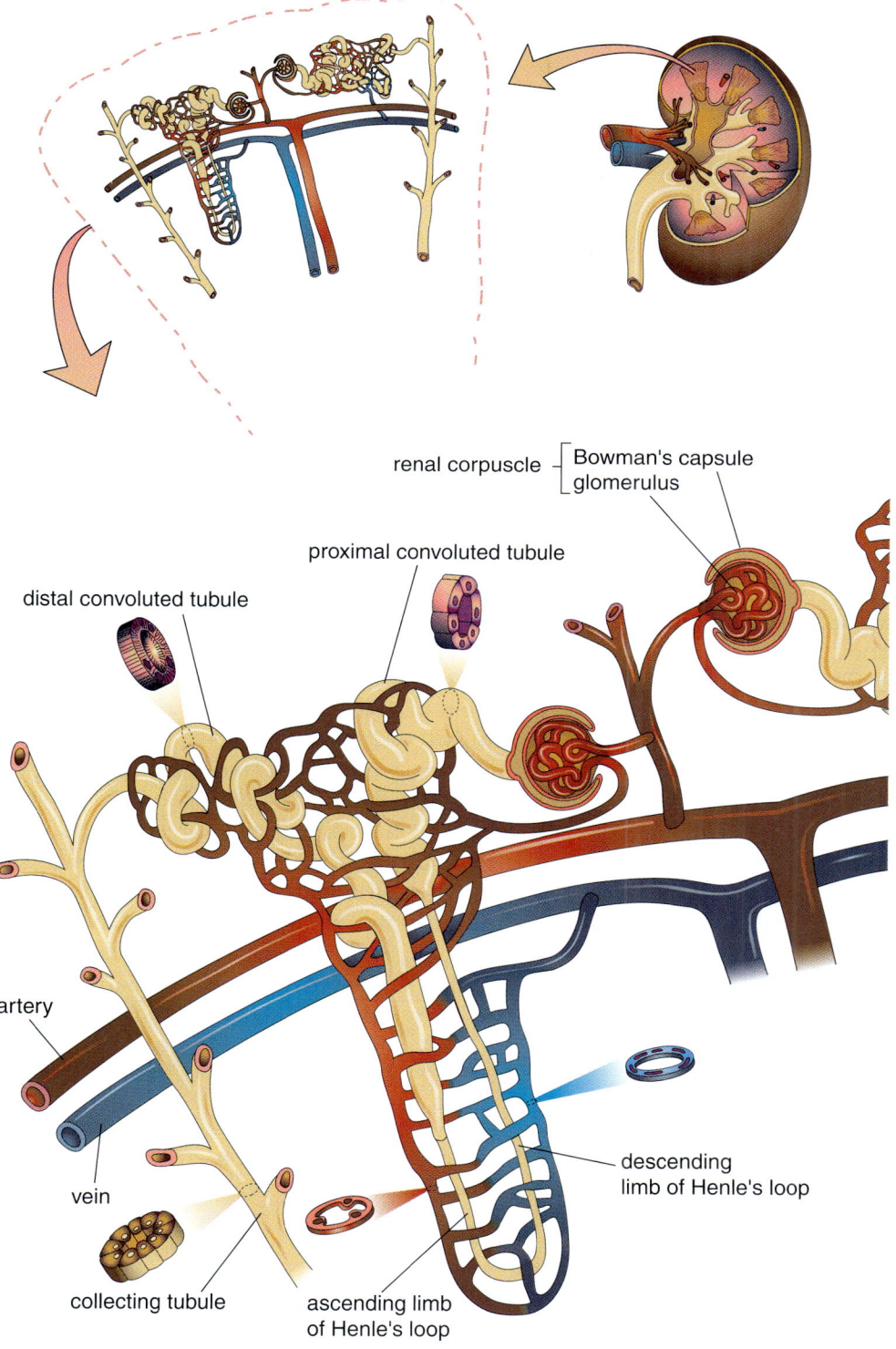

THE URINARY SYSTEM

Figure 14.6 Kidney Function
The kidney is involved in a variety of functions that maintain the body's homeostasis.

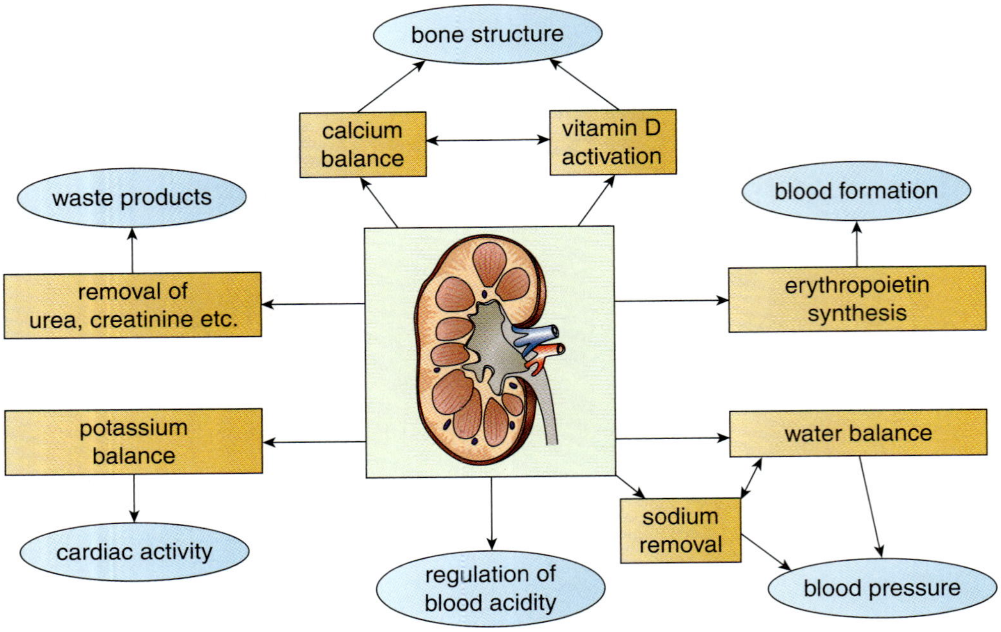

Bowman's Capsule An expanded portion of the renal tubule that contains the glomerulus

Renal Corpuscle The renal structure formed by the glomerulus inside the Bowman's capsule

Efferent Arteriole The blood vessel formed by the glomerulus where it exits the renal corpuscle

Peritubular Capillary System The network of renal tubules and capillaries distal to the glomerulus and Bowman's capsule of the renal corpuscle

Proximal Convoluted Tubule The first segment of the nephron's peritubular arrangement

Loop of Henle The second segment of the nephron's peritubular arrangement

Distal Convoluted Tubule The third segment of the nephron's peritubular arrangement

Collecting Tubule A straight tubule to which the distal tubules of several nephrons lead

tissue. The glomerulus is surrounded by **Bowman's capsule**, an expanded portion of the renal tubule. Bowman's capsule and the the glomerulus tucked within it form a structure called the **renal corpuscle**. The distal end of the glomerulus exits the renal corpuscle and becomes the **efferent arteriole**. This vessel branches into a secondary capillary system. These capillaries wrap around a network of renal tubules that are formed from an extension of Bowman's capsule at its distal end. As you will see in the discussion of nephron physiology, blood filtration occurs in the renal corpuscle, and reabsorption and secretion occur in the **peritubular capillary system**. The nephron's tubular arrangement has four major segments: the **proximal convoluted tubule**; the **loop of Henle**; the **distal convoluted tubule**; and the **collecting tubule**. For most nephrons, the renal corpuscle, and both the proximal and distal convoluted tubules are located solely in the kidney's outer cortex tissue. The loop of Henle and the collecting tubules extend into the inner medullary tissue.

✓ Concept Check

1. Describe the blood vessel components of the nephron.
2. Describe the renal tubule arrangement of the nephron.
3. Differentiate between the glomerulus, Bowman's capsule, and the renal corpuscle.

URINE FORMATION

Key Terms: glomerular filtration, tubular reabsorption, tubular secretion

Glomerular Filtration The process by which plasma and many dissolved substances are moved from the blood into Bowman's capsule

Tubular Reabsorption A process in the peritubular capillary system in which water, nutrients, and electrolytes travel back into the blood

Tubular Secretion The process by which certain waste products and ions are removed from the blood into the tubular fluid

Normal urine formation occurs at a rate of approximately 1 mL/min in the nephrons. It is a three step process: 1) **glomerular filtration**; 2) **tubular reabsorption**; and 3) **tubular secretion**. Each nephron section is anatomically suited to carry out the physiological process for which it is responsible (Figure 14.7).

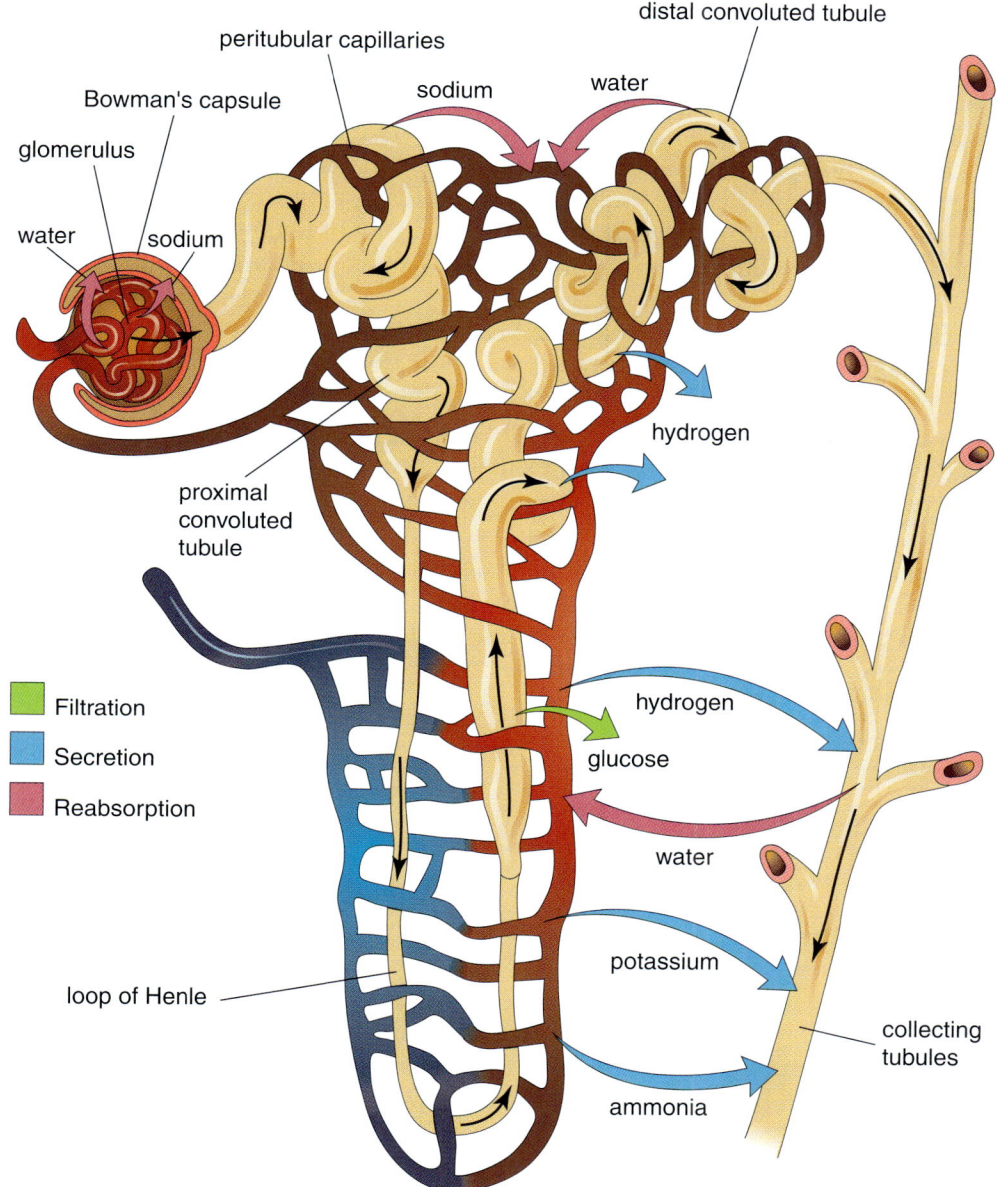

Figure 14.7 **Urine Formation**
Urine contains metabolic wastes filtered from the blood by the kidneys

URINALYSIS

Physicians are able to gain important knowledge about a patient's health by performing a urinalysis. The importance of studying urine in the diagnosis of human disease was known over 6000 years ago by several of the earliest civilizations in Africa and the Middle East. Most early approaches to studying urine involved merely looking at the urine for cloudiness or color. Sometimes the urine was tasted to detect the presence of sugar, a method for determining diabetes. The 15th century Swiss physician known as "Paracelsus," whose real name was Auroleus Phillipus Theostratus Bombastus von Hohenheim, was the first person to carry out chemical tests on urine as a way of diagnosing disease. Microscopic analysis of urine began in the 1800s when the microscope gained popularity as a tool for understanding disease.

Up until the 1950s, urine testing involved a laborious series of complex chemical tests and microscopic examinations that sometimes would take two or more days to complete. Urinalysis was made simpler and quicker with the invention of urine dipsticks, such as the kind used in home pregnancy tests, which carried out similar tests in one step and provided immediate results. Unfortunately, the dipsticks do not have the accuracy needed to diagnose complex medical conditions. This need led to the development of computerized urine analyzers that can instantaneously carry out over 20 chemical tests on urine. Urine analyzers can test many samples in a day and are popular in medical laboratories that serve large hospitals. A typical urinalysis determines the presence of bacteria, blood cells, epithelial cells, glucose, minerals, proteins, salts, and abnormal secretions of various organs. Urine is also tested for clarity, color, odor, and pH. Specific tests must be requested in order to analyze various other materials that show up in urine, such as drugs and hormones.

Filtration in the Glomerulus

Key Term: filtrate

Blood pressure is relatively high within the glomerulus of the renal corpuscle (Figure 14.8). This is partly because the diameter of the efferent arteriole is smaller than that of the afferent arteriole. It is similar to the way a hose sprayer with a very small opening attached to a hose with a larger opening increases the water pressure inside the hose. The glomerular capillary membrane is more permeable than the membrane of most tissue capillaries. This fact, and the high pressure, moves plasma, and many of the substances dissolved in it, out of the capillaries while retaining the blood's cellular components and almost all proteins. The resultant **filtrate** enters Bowman's capsule and is carried to the proximal convoluted tubule.

Filtrate The plasma and other substances that are removed from the blood in the renal corpuscle

Abnormal changes in arteriole constriction can lower glomerular blood pressure and lead to elevation or depression of normal filtration levels. Stress is one factor that can cause this by increasing sympathetic innervation of the arterioles. Increased permeability due to glomerular membrane damage can increase filtration levels and lead to leakage of cells and abnormal protein components into the filtrate. An amazing amount of filtration occurs in a kidney with normal function. Between 150 and 200 L of filtrate are produced daily. However, approximately only 1%, or 1 to 2 L, of this fluid becomes urine output. The rest is reabsorbed as it enters the renal tubular network.

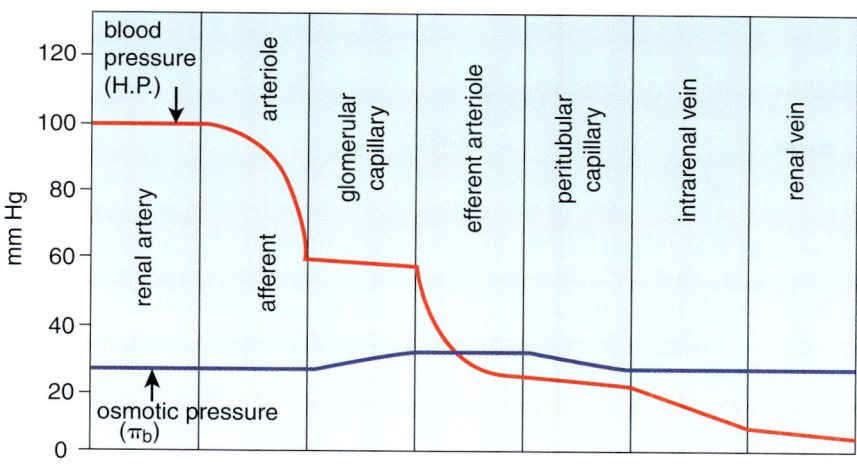

Figure 14.8 Pressures in Circulation
The tubular system of the nephron is responsible for the reabsorption of water and salts that would be lost in the urine.

- The afferent and efferent arterioles are the major resistance sites
- Thus these are the major sites for control of RBF.
- H.P. > π_b in glomerular capillaries; thus filtration occurs.
- π_b > H.P. in peritubular capillaries; thus absorption occurs.

WHAT GOES IN MAY COME OUT TOO QUICKLY

Urine testing for drug use is nothing new in our modern society. Many employers require applicants, and sometimes established employees, to submit urine samples for drug screening. Urine analysis for drugs can be used for several purposes, such as evidence in forensic investigations or proof that athletes are adhering to "no-drug-use" rules. Recently, legislation has been proposed to mandate urine analysis for public school students. Have you ever wondered how effective this method is at targeting drug abuse? The presence of drugs, or their metabolic products, in urine is certainly credible evidence that drugs have been introduced into the body. After all, what goes in comes out, in one form or another. However, is it possible that drugs can be eliminated from the body too quickly to detect? Unless random checks are done without prior notice, a person who habitually uses certain illicit drugs could abstain from use for a few days prior to urine testing, and the drug use could go undetected. Most stimulants (eg, cocaine and amphetamines), as well as many hallucinogenic drugs such as mescaline, phencyclidine (PCP), and methylenedioxymethamphetamine (Ecstasy), may be detectable in urine for only a couple of days after use. Marijuana, however, can usually be excreted at detectable levels for a few weeks. Anabolic steroid use, the topic of publicity in the professional and amateur sports arena, is also detectable for weeks or even months if administered by injection. Offenders often try to smuggle in substitute urine samples, but this is usually detected by collecting the urine in temperature-sensitive cups. There are also many adulteration tricks that are supposed to render a positive sample negative. A basic Internet search on the subject will reveal such products, the sale of which is legal is most states. Risk for adulteration, along with the short detection period, certainly limits urine collection as an effective tool to detect drug abuse. Hair analysis for drugs actually provides a more comprehensive history of drug use, as it can measure cumulative use over much longer periods of time. It is also more difficult to alter the results of tests conducted on hair.

Reabsorption in the Proximal Convoluted Tubule

  Key Terms: cotransported, symported

Most reabsorption occurs in the proximal convoluted tubule. The increased surface area of the microvilli of the epithelium provides optimal conditions for this process. Water, nutrients, and electrolytes travel back into the blood from the filtrate at this location through a combination of passive transport (osmosis and facilitated diffusion) and active transport (pumping). The series of transport mechanisms for the majority of these substances is as follows:

1. Sodium ions are actively pumped across the tubule membrane. This creates a gradient of the sodium ion concentration between the interstitial fluid and the blood in the peritubular capillaries, which are wrapped around the tubule. This concentration gradient causes sodium ions to move into the blood.
2. As more sodium ions enter the blood, its positive charge increases compared with that in the filtrate. The resultant electrical gradient attracts negative ions of the filtrate, such as chloride and phosphate, and they diffuse into the blood. As they continue to accumulate in the blood, the blood becomes hypertonic to the filtrate. This increased osmolarity causes osmosis to occur. In other words, water moves from the filtrate to the blood because the filtrate has become less concentrated (ie, contains more water) than the blood. Thus, water reabsorption occurs.
3. At the same time that reabsorption is taking place, facilitated diffusion occurs with the result of the reabsorption of nutrients, such as amino acids and glucose, which are too large to pass through membrane pores. The molecules "catch a ride" by binding to a membrane carrier molecule that simultaneously binds sodium ions on another portion of the membrane carrier molecule. As the sodium ions cross passively (due to the concentration gradient created earlier), nutrient molecules are **cotransported**, or **symported**, into the blood. The number of nutrient molecules in the blood that can be symported depends on the number of carrier molecules. When the body is healthy, it is able to clear the blood of normal levels of these molecules; so, very little to none remains in the filtrate to be excreted in the urine. However, if unusually high circulating levels of nutrient molecules are present in the blood, the carrier molecules become saturated and are unable to "keep up." This is the case with the elevated levels of blood glucose that diabetes mellitus causes. As you will see later in the discussion of pathology, detection of such molecules in the urine can indicate abnormal physiological activity.

During glomerular filtration, some proteins do get squeezed through the glomerular capillary membrane and into the filtrate. These large molecules cannot be reabsorbed by the processes just described. Instead, they move into the proximal convoluted tubules by endocytosis, where they are broken down into their amino-acid components. Facilitated diffusion reabsorption is responsible for their return to the blood.

Cotransported Moved across a membrane by attaching to carrier proteins that are attached to ions

Symported Synonym for cotransported

Good Choice Bad Choice

It does not take a college degree in physiology to realize that increasing the body's water loss will result in a lower reading when you step on the scale. Weight loss associated with the use of diuretics is certainly not a new phenomenon. For years, "water-pills" have been marketed to individuals looking for a quick fix in their weight management. Diuretics can produce a rapid loss of a few pounds, but this loss is transient and does not rid the body of fat. Continued use of diuretics can result in extreme dehydration, electrolyte imbalance, and dangerous pH levels. Women sometimes use diuretics to control the water retention that frequently accompanies hormonal fluctuations throughout their menstrual cycle. Is this a wise choice? Or, is there a danger associated with diuretic use on a short-term basis? Could the use of a diuretic for abnormal water retention mask the symptoms of an underlying urinary condition? What advice would you give an individual with little knowledge of physiology concerning the use of diuretics for weight management?

Reabsorption in the Loop of Henle

Reabsorption continues as the filtrate moves through the rest of the renal tubule's structure. In the descending side of the loop of Henle, about 20% of the water is reabsorbed. The ascending side of the loop, however, is impermeable to water, and it reabsorbs sodium ions and chloride ions by active transport. Removal of these solutes leaves the tubule fluid hypotonic (diluted), while the ions entering the interstitial fluid of the kidney medulla makes it hypertonic. A concentration gradient is created between the tubular fluid and the medulla's interstitial fluid. This gradient results from fluids in the tubular system, leaving the medulla to reenter the kidney cortex region where the distal convoluted tubules are located.

Reabsorption in the Distal Convoluted Tubule and Collecting Duct

Key Terms: dehydration, urine concentration, water conservation

The distal convoluted tubule continues to reabsorb ions, but is impermeable to water, so the tubule fluid becomes even more hypotonic. One would expect this hypotonic fluid to lose water molecules as it enters the collecting tube in the kidney medulla with its hypertonic interstitial fluid. However, the cell structure of the collecting-tube wall makes it impermeable to water. Without outside factors influencing this permeability, the tubule fluid would be carried through the collecting tube. It would eventually leave the body as urine containing a large amount of water that the body needs. **Dehydration** would quickly occur. Fortunately, the endocrine system controls ion and water reabsorption in the distal tubule and the collecting tube, so that they can perform the vital function of **water conservation**, or **urine concentration**.

Dehydration Abnormal loss of fluid from the body

Water Conservation The retention of water in the body

Urine Concentration The removal of water from the urine by reabsorption in the collecting tube before it leaves the body

Tubular Secretion

At the same time that reabsorption in the tubules is returning filtrate components to the blood, certain waste products and ions are being removed from the blood. These waste products and ions travel from the peritubular capillaries to the tubular fluid. This activity is known as tubular secretion. Urea and other waste products of protein metabolism enter the tubule fluid by this means. Many drugs are cleared from the blood stream by this method as well. The transfer of many ions from the blood stream through secretion helps to maintain a desirable body fluid pH level. Elevated hydrogen ion concentrations produce acidic conditions. Conversely, lowering the amount of hydrogen ions in body fluids makes the pH more basic (Figure 14.9). Potassium is an ion of particular interest; it is secreted in the distal convoluted tubule and collecting duct during hormonal control of urine production. Blood pressure regulation and cardiac function require particular concentrations of potassium ions to function normally.

Figure 14.9 Acid-Base Balance The pH of the blood is partly regulated by the transport of hydrogen ions and bicarbonate ions into or out of the blood by the kidneys.

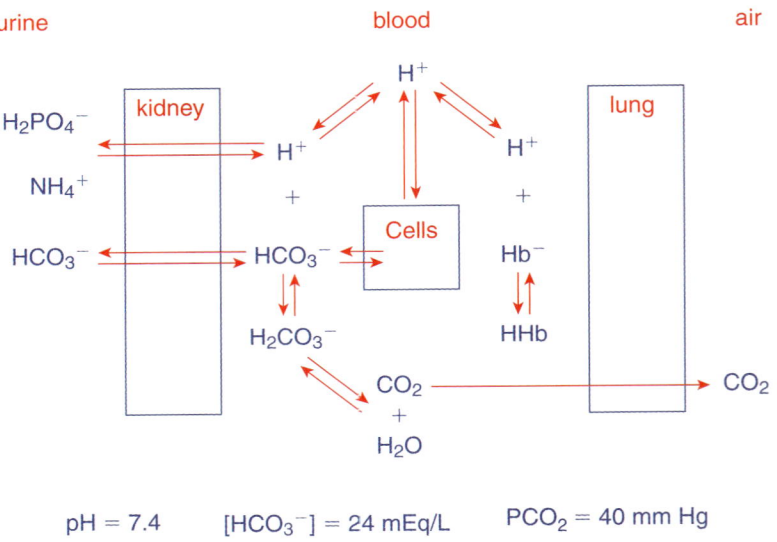

✓ Concept Check

1. Where does filtration occur, how is the filtrate formed, and what are some factors that can influence the filtration rate?
2. Describe what reabsorption in the urinary system means in general, and specify what substances are reabsorbed at specific locations of the nephron peritubular system.
3. What is tubular secretion, and what role does it play in waste removal and body pH maintenance?

DISCOVERY SCENE PLEASE ENTER DISCOVERY SCENE PLEASE ENTER

Could the women's problem in the CSI be explained using information about urine formation? How could it be related to her condition?

HORMONAL REGULATION OF URINE FORMATION

 Key Terms: atrial natriuretic factor (ANF), diuresis

Various hormones are involved in controlling the rate and volume of urine production (Figure 14.10). The release of each hormone is elicited by a specific change detected in the body. Some are stimulated by a change in blood pressure, while others are produced as a result of the body's ability to detect abnormal fluid composition. A discussion of some of the major hormones that affect the physiology of the urinary system is important for understanding how the filtration rate and urine concentration are controlled.

Antidiuretic hormone is produced by the pituitary gland in response to dehydration. It greatly influences **diuresis**, or excretion of water from the body. Its target is the collecting duct, and its action is to increase this structure's permeability to water. When water channels in the collecting duct membrane are open, which is due to ADH action, the higher water content inside the tubule moves to the medulla's hypertonic interstitial fluid. The water then moves to the blood stream to rehydrate the plasma.

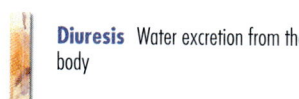 **Diuresis** Water excretion from the body

Figure 14.10 Control of Kidney Function
Hormones, like aldosterone, play an important role in controlling kidney function.

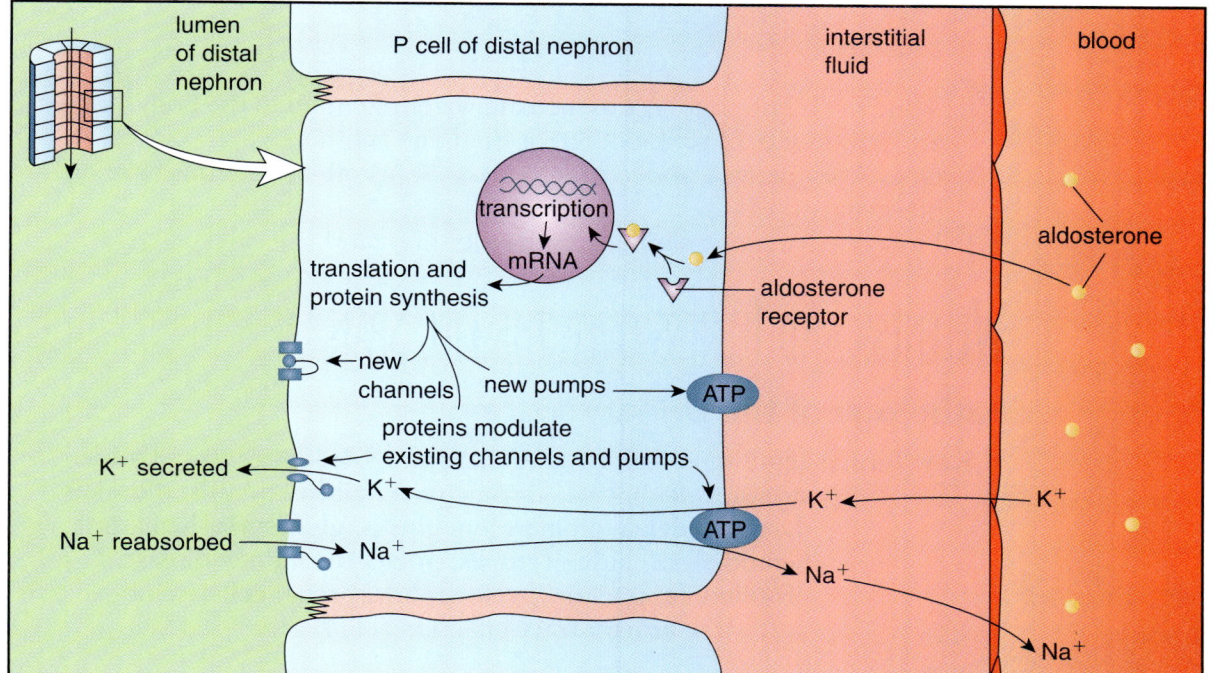

THE URINARY SYSTEM

The adrenal cortex steroid, aldosterone, is another hormone that plays a significant role in regulation of urine formation. In response to high levels of blood potassium ions, aldosterone is produced to increase water movement out of the distal convoluted tubule and the collecting duct in the kidney cortex region. It does this not by increasing permeability, but by stimulating reabsorption of sodium ions. The imbalance in the osmolarity of the tubule fluid and surrounding tissue is created by sodium ion reabsorption. This imbalance creates a concentration gradient for outward movement of water. However, the membrane must be permeable to water so the ADH action must coincide with aldosterone activity to aid water retention. As the blood becomes increasingly positive from reabsorbing sodium ions, it secretes potassium ions into the tubule to counterbalance the electrochemical gradient. Therefore, more potassium ions are excreted in the urine. Blood potassium levels fall, and aldosterone production ceases.

Special cells in the heart muscle secrete **atrial natriuretic factor (ANF)**, a hormone that works in a manner antagonistic to that of aldosterone. Its function is to lower blood volume and pressure. These decreases are induced by its ability to inhibit sodium ion reabsorption in the renal tubules. Lower sodium ion reabsorption limits water reabsorption into the blood, thus decreasing blood volume and pressure. Consequently, diuresis and sodium ion excretion in the urine are increased.

Atrial Natriuretic Factor (ANF) A hormone secreted by special cardiac cells; that lowers blood volume and blood pressure

The hormones just discussed all affect the renal tubules during reabsorption and secretion. One other hormone worth mentioning exerts its effects before this point in urine formation. It affects the glomerular filtration rate. As you will recall, angiotensin II is a hormone produced in response to the production of renin, an enzyme produced by the kidneys. Special cells, primarily in the afferent arterioles, produce renin when mechanoreceptors detect low blood pressure. Angiotensin II causes widespread constriction of blood vessels, but has an especially powerful vasoconstrictive effect on the efferent renal arteries. This constriction increases blood pressure within the glomerular capillaries and, consequently, increases the filtration rate.

✓ Concept Check

1. Which two hormones act to increase water retention in the body? How are their modes of action different from one another?
2. Does ANF increase or decrease diuresis? What effect does this have on blood pressure?
3. How does Angiotensin II affect the filtration rate?

WELLNESS AND ILLNESS OVER THE LIFE SPAN

Pathology of the Urinary System

Key Terms: acute renal failure, aminoaciduria, bladder cancer, bladder stones, calculi, cast, chronic renal failure, cystitis, diuretics, dysuria, edema, glomerulonephritis, glycosuria, hematuria, hemodialysis, kidney stones, polycystic kidney disease, proteinurea, pyelitis, pyelonephritis, renal cell carcinoma, urethritis, urinary tract infection (UTI)

CHAPTER 14

Many urinary system disorders have the potential to occur throughout the life of an individual. Most are renal function disorders, involving the kidneys, which can detrimentally affect the body's homeostasis. Other disorders are confined to the structures of the conducting system alone, but they can still have negative effects on renal function. Certain disorders may have a genetic link or may be the result of a congenital anomaly. Invasion by microorganisms can wreak havoc on the urinary system, as can some endocrine hormone imbalances and certain autoimmune disorders.

Disruption of the urinary system's ability to properly filter and reabsorb substances during urine formation in the nephron can disturb the balance of the body's acid base, electrolytes, and water. Imbalances can occur from physical damage to various renal tissues as a result of trauma, disease, and genetic and developmental errors. Inflammation of various tissues throughout the urinary system can also interfere with proper functioning of these vital processes or pose difficulty in the ability to transport urine out of the body. Most urinary system disorders fall under one or more of the following categories:

a. congenital disorders
b. infection and inflammation
c. immune disorders
d. hormonal disorders
e. degenerative disorders
f. tumors

It is beyond the scope of this text to discuss all urinary system disorders, but it is of clinical interest to present some of the more commonly encountered abnormalities, as well as to discuss some routine laboratory evaluations that can indicate certain disorders.

Congenital Disorders Congenital disorders are present at birth. There can be a genetic basis for their presence, or they may be due to some other factor that interferes with normal development. A common genetic disease of the kidneys is **polycystic kidney disease**. It is characterized by the growth of many fluid-filled cysts on the kidneys. These cysts damage renal tissue and disrupt the ability of the nephron to function properly. Over one-half million persons are affected by this disease, and about half of those are expected to require **hemodialysis** as the disease progresses. In this procedure, an individual's blood is shunted to a machine that artificially performs the functions of the nephrons. An artificial, semipermeable membrane and a dialysis fluid with a controlled composition are used to diffuse waste products out of the blood, while retaining the proper electrolyte and water levels in the blood. The blood is then returned intravenously to the body. Although, dialysis has certain drawbacks, such as the long, frequent sessions and the inherent risk for infection, it is certainly preferable to the possibility of complete renal failure and the eminent death that soon follows kidney failure. Another heritable factor is the inability to produce certain carrier proteins needed to transport glucose and specific amino acids back into the blood during tubular reabsorption. This can result in **glycosuria**, or **aminoaciduria**. Although the presence of these compounds in the urine does not usually indicate that the body is deficient in these nutrients, their presence greatly increases the solute concentration of the urine. Increased solute concentration in the filtrate can reduce the amount of osmotic water

Polycystic Kidney Disease A inherited disease that causes the growth of kidney cysts, which impairs kidney function

Hemodialysis A medical procedure that allows for artificial filtering of the blood

Glycosuria The presence of glucose in the urine

Aminoaciduria The presence of amino acids in the urine

Calculi Accumulation of amino acids or mineral crystals in the kidney or bladder

Nephrolith An alternate term for a calculus (singular for calculi)

reabsorption in the collecting ducts. Also, an accumulation of certain amino acids in the urine can result in their crystallization and subsequent formation of painful "stones," or **calculi**, in the kidney or bladder. A renal calculus is also known as a **nephrolith**.

HEMODIALYSIS

Fifty years ago kidney failure was a death sentence for the affected person. The most acute problem associated with kidney failure is the buildup of urea in the blood. Urea is toxic to highly metabolic cells and immediately affects nerve and muscle tissues. In the 1960s physicians began using dialysis machines to filter urea from the blood of people with kidney failure. This technique was called hemodialysis, meaning the filtering of blood. Early dialysis treatments were carried out for 6 to 8 hours every day. In effect, the patient had to live near the dialysis machine, had little time to work or even have a normal private life. Improvements in equipment reduced the dialysis time to 4 hours a day 3 times a week by the 1970s. People were still restricted to dialysis treatments in a clinic or hospital. Smaller and simpler dialysis machines were developed in the 1980s, making it possible for people to perform hemodialysis at home. This not only made it convenient for the patient, but it significantly reduced the medical costs of hemodialysis. The efficiency of hemodialysis is measured either as urine removal rate (URR) or dialysis clearance adequacy (Kt/V). Physicians use these measurements to calculate the time a patient must remain on dialysis machinery without developing urea poisoning. Currently, scientists are developing new types of dialysis membranes that are more effective at filtering urea from the blood. These smaller units may one day be small enough to carry around without impeding a person's daily routine. Thus, they can have a continuous hemodialysis system that works much like a kidney.

Urinary Tract Infection (UTI) Infection anywhere in the urinary tract

Urethritis Inflammation of the urethra

Cystitis Inflammation of the urinary bladder

Pyelitis Inflammation of the renal pelvis

Pyelonephritis Inflammation of the nephrons

Dysuria Painful urination

Pyuria The presence of white blood cells (WBCs) in the urine

Infection and Inflammation Bacterial infection can occur in any region of the urinary system. The medical condition commonly referred to as **urinary tract infection** (**UTI**) describes inflammation caused by bacteria in a particular structure or organ. Bacteria inhabiting the genital and rectal area of the body can enter the urinary tract through the urethra (due to its proximity to these areas), and cause infection and subsequent urethral inflammation, or **urethritis. Cystitis**, or inflammation of the urinary bladder, can easily occur as bacteria in the urethra move into the bladder. Bacteria invading these areas can travel upward into the ureters and cause **pyelitis**, or inflammation of the renal pelvis of the kidney. **Pyelonephritis** can occur if the bacteria reach the nephrons in the kidney cortex tissue. Painful urination, or **dysuria**, accompanies UTIs. Chemical analysis of urine collected from an infected urinary tract reveals high nitrate levels resulting from bacterial waste products and leukocytes. The presence of leukocytes in the urine is called **pyuria**. Microscopic observation of the urine reveals infective bacterial cells and the phagocytic WBCs combating them.

ROBOTIC REMOVAL

In the past 2 decades, minimally invasive surgical techniques have replaced many of the traditional surgical procedures to repair or remove damaged or diseased body parts. Through the use of an endoscope, a tube to which a tiny camera is attached, surgeons are able to view the body's interior on a video screen and perform "microsurgery" by manipulating surgical instruments inserted through the endoscope. This less-invasive approach decreases blood loss during surgery, minimizes the chance for bacterial invasion, reduces recovery pain and scarring, and dramatically reduces the length of hospital stays and recovery periods.

Endoscopic observation and surgery performed in the bladder is called cystoscopy. This technique has been commonly employed in urological surgery since the 1990s. Recent developments in robotic control give surgeons even more precise control over incisions and suturing. Since 2002, surgeons at the Glickman Urological Institute of the Cleveland Clinic Foundation in Cleveland, Ohio, have been forerunners in robotic surgery. Currently, they perform robotic surgery for two specific procedures: dismembered pyeloplasty and radical cystotomy with urinary diversion. In the first technique, robotic control of instruments allows a surgeon to remove an obstructed section of the urinary passageway at the junction of the ureter and renal pelvis with subsequent resection of the undamaged ureter to the renal pelvis. The latter involves robotic removal of the bladder (due to cancerous invasion of the muscle tissue in the bladder wall). The ureters are then joined together, and a section of the ileum (of the small intestine) that has been robotically excised is sutured to the joined ureters to serve as a passageway for urine to continue its exit from the body.

Immune disorders A common disorder, which can also be categorized as an inflammatory autoimmune disorder, is **glomerulonephritis**. This is a serious urinary system disorder that results from an abnormal response of the body's own immune system to glomerular cells. A streptococcal bacterial infection in the body is often a precursor to glomerulonephritis. Bacterial secretions deteriorate the integrity of the glomerular capillary membranes, making them more permeable to blood components, such as blood cells and proteins. As these substances leak out of the glomerulus, the elevated solute concentration that results in the filtrate leads to **edema**, or tissue fluid accumulation, and hypertension. The presence of red blood cells (RBCs) and abnormally high protein levels in the urine can indicate this condition; these conditions are known as **hematuria** and **proteinurea**, respectively. Blood cells and proteins in the filtrate can be caught in the renal tubules and collect into an aggregate. When this accumulated mass of cells dislodges and travels into the urine, it is known as a **cast**. Casts can be detected microscopically in urine sediment and indicate damage to the glomerulus. If left untreated, glomerulonephritis can become a chronic condition and lead to renal failure.

Glomerulonephritis An autoimmune disorder causing inflammation and deterioration of the glomerular membranes

Edema Accumulation of fluids in the body tissues

Hematuria Presence of red blood cells in the urine

Proteinurea The presence of abnormal protein levels in the urine

Cast An abnormal aggregate of cells or other substances that can be found in urine when they accumulate in renal tubules and dislodge into the tubular fluid

Diuretics Chemicals that increase the volume of water in urine

Hormonal Disorders Any condition causing a change in the levels of ADH, aldosterone, ANF, or angiotensin II can affect urinary production, composition, or output. Abnormally low aldosterone production causes Addison's disease. The effects of this disease on the urinary system involve an increase in sodium excretion, followed by excess water loss, and subsequent dehydration and hypotension. **Diuretics** also cause an increase in urine output because they decrease sodium absorption. They counteract the activity of ADH to remove water from the body. Diuretic drugs can be administered to treat

THE URINARY SYSTEM

hypertension because the increased water loss they cause decreases blood volume and lowers blood pressure. The excess fluid accumulation of edema can also be treated with diuretics.

Chronic Renal Failure Irreparable nephron damage and loss of kidney function

Acute Renal Failure Temporary loss of kidney function

Renal Cell Carcinoma Malignancy of the cells of the renal tubular lining

Bladder Cancer Malignancy of the tissue of the urinary bladder

Degenerative Disorders The categorization of urinary system disorders as degenerative includes any disorder that results from a deterioration of the cells or tissues of the structures responsible for a specific function. Tissue deterioration within the urinary system can result from a wide variety of factors: chronic infection, trauma, toxic environmental chemicals, tubule blockage, impaired vascular flow, and glomerulonephritis. These can all contribute to degenerative disorders. Degenerative disorders of the kidneys, which manifest in nephron impairment, can lead to the inability of the kidneys to form urine. This is known as **chronic renal failure**. Fortunately, not all nephron damage is irreparable, and less-severe damage may result in **acute renal failure**. Amazingly, proper nephron function in only one-third of a single kidney can keep a person alive. However, hemodialysis is necessary to maintain the normal homeostasis afforded by the kidneys when their physiological function falls below 25% of its normal operating capacity.

Tumors Cells of urinary system tissues can form malignant tumors when their growth occurs at an abnormally high rate. **Renal cell carcinoma** is a malignancy of the cells of the renal tubular lining. It is the most common form of kidney cancer, though **bladder cancer** occurs more often. Symptoms of these cancers may be similar to those of less-severe disorders, such as UTIs, so imaging techniques are necessary to confirm their diagnoses. Surgical tumor removal alone is often sufficient treatment in both renal and bladder cancers if detection occurs in the early stages of growth. Removal of a section of the affected organ, or even the complete organ, may be necessary in more advanced stages. Radiation and chemotherapy may be prescribed depending on the severity of progression. Renal cell carcinoma can be metastatic in progressive stages, while bladder cancer normally remains isolated in the organ of origin or in the organs adjacent to it.

Aging of the Urinary System

Key Terms: cystocele, nephroptosis

Although the urinary system is subject to the effects of aging, as are all body systems, it continues to adequately perform its function of waste removal. Reduction in the number of nephrons does occur, but this does not seem to have an extreme deleterious effect on the kidneys' ability to form urine. It is the increased incidence of diseases associated with aging that more directly contribute to most associated impairments. For instance, hypertension that commonly accompanies aging can reduce renal function. The increased number of mitotic divisions of cells, especially in the lining of the tubules and bladder, increases the likelihood of mutations due to telomere shortening. Consequently, the risk for malignancy increases with age. The accumulated exposure to environmental toxins increases this chance as well. Although it is not an extremely common condition, the forces of gravity and loss of body fat that often occur in the elderly population, increase the risk for **nephroptosis**, or floating kidney. Continual exposure to certain occupational or

Nephroptosis Movement of the kidney from its proper anatomical position to an inferior position

CHAPTER 14

Cutting Edge Research

DRUNK ON WATER?

Our society does not lack awareness of the need to consume adequate amounts of water. The marketing of bottled water and the media's bombardment of consumers about the health benefits associated with water consumption are rampant. However, there is another side to the story that is currently making sports headline news: overhydration, also known as water intoxication. The major result of an overabundance of water in the body is a dangerously low blood level of sodium, a condition known as hyponatremia, which has very detrimental effects on nerve conduction and cardiac function. The increased water retention that occurs can cause brain swelling, possibly leading to coma and even death. Symptoms include poor concentration, even delirium, impairment of muscle coordination, twitching, blurred vision, nausea, and vomiting, all of which mimic the symptoms of alcohol intoxication. This condition is not a new phenomenon in the medical field. Medical sources estimate that about 2% of hospital patients experience hyponatremia, and that it is the number one cause of patient electrolyte imbalance in hospitalized individuals. Administration of excess fluids and medications that influence water retention are major contributors. However, the death of Boston marathon runner Cindy Lucero in 2002 was due to hyponatremia. Her death was a major impetus in alerting health professionals and athletes to the potential risk for this condition created by excess water consumption during endurance athletic training or competition. In April of 2005, research physicians in Boston published a study in the *New England Journal of Medicine* in which the blood sodium levels of nearly 500 marathon runners were tested following competition. Hyponatremia was present in 13% of the participants, and blood sodium levels indicative of possible coma were detected in three of these individuals. Furthermore, sports drinks professing to maintain electrolyte balance were found to be ineffective in preventing this condition. Such research has lead to the routine monitoring of the blood sodium levels of participants at many endurance events who show symptoms of hyponatremia. In the past, it was not uncommon for health practitioners to assume that dehydration, not overhydration, was the cause of such symptoms. Ironically, fluid replacement therapy for treatment of dehydration could actually be fatal to victims of overhydration.

Sources: "McClean in the News," April 27, 2005: www.startribune.com
"Overhydration" at: http://www.healthatoz.com/healthatoz/Atoz/ency/overhydration.jsp)

recreational physical activities that jar the body can further contribute to the possibility of this condition. Herniation of the bladder, or **cystocele**, can occur as the continual pressure of the female bladder stresses the structural connections that hold it in place, which results in displacement downward into the vagina. The state of pregnancy and incidence of multiple pregnancies amplify this possibility. Degradation of the urethral sphincter muscles or the loss of their innervation due to injury or nervous tissue degeneration can cause incontinence (Figure 14.11). Males may experience urinary retention as early as age 40 years due to the normal hypertrophy of the prostate gland and subsequent restriction of the urethral passageway (Figure 14.12).

 **Cystocele** Herniation of the bladder into the vagina

THE URINARY SYSTEM

Figure 14.11 Control of Urination – Female
Degradation of urethral sphincter muscles due to age can cause incontinence

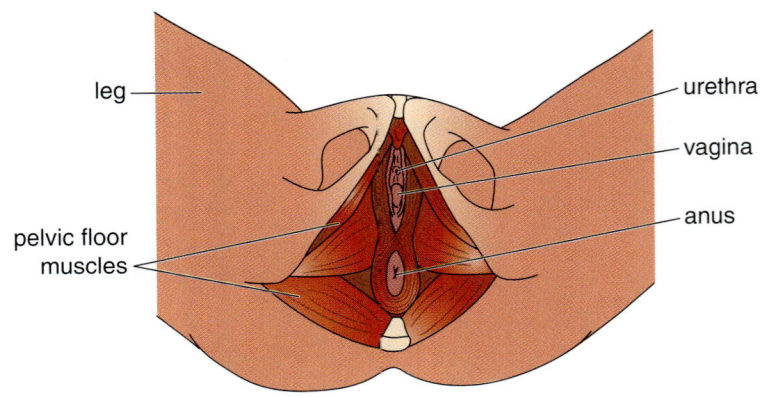

Figure 14.12 Control of Urination – Male
Hypertrophy of the prostate can cause urinary retention

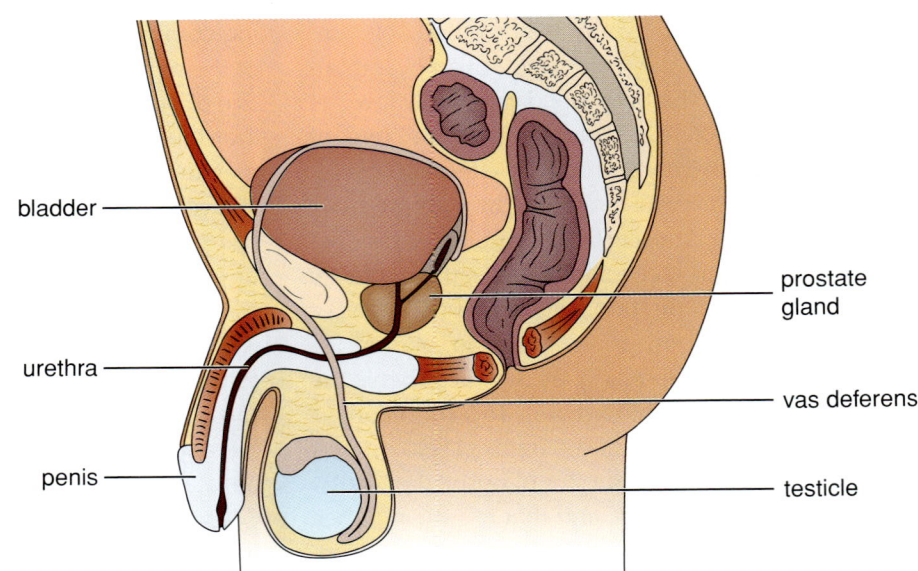

✓ Concept Check

1. Name one congenital urinary disease, and briefly describe its symptoms.
2. Differentiate the specific types of UTIs.
3. What urinary tract disease is usually categorized as an immune system disorder, but could also be classified as an inflammatory degenerative disease? Explain.

DISCOVERY SCENE PLEASE ENTER DISCOVERY SCENE PLEASE ENTER

Is the woman's condition in the CSI evident after reading about the pathology and aging of the urinary system? Could her condition be due to an injury or some type of disease? Is it possible that her age is a factor in the condition?

CSI – Case Study Investigation Conclusion

What information in this chapter gave you clues to identify the cause of the woman's condition? What about the fall or her surgery could have caused the problem? See how your analysis compares to the answer.

Answer:

The woman developed a condition called overactive bladder (OAB). It is primarily due to convulsive contractions of the smooth muscles surrounding the bladder (detrusor muscles). It causes high bladder pressure followed by the urgent need to urinate, or urgency. People with OAB often experience the need to urinate at unpredictable times and sometimes lose control of urination. It becomes an embarrassing problem that can interfere with a person's day-to-day routines. This condition has a variety of causes, including nerve damage caused by abdominal or pelvic trauma, or surgery. In this woman's case, it could not be determined if the fall or the surgery led to her condition. The inability to come up with the precise cause of a condition is unfortunately not an unusual situation in medicine.

This CSI was adapted from the following articles:

1. Fitzgerald MP, Brubaker L. The etiology of urinary retention after surgery for genuine stress incontinence. *Neurourol Urodynam*. 2001;20(1):13-21.

2. Sakakibara R, Hattori T, Uchiyama T, Yamanishi T, Ito H, Ito K. Neurogenic failures of the external urethral sphincter closure and relaxation; a videourodynamic study. *Auton Neurosci*. 2001;86(3):208-215.

Chapter Summary

The urinary system plays a major role in ridding the body of wastes. It works in concert with the respiratory system to maintain proper body pH. The endocrine system assists in controlling electrolyte and water levels. Blood filtration, and the subsequent formation of urine, occurs in the nephrons of the kidneys. These complex networks of capillaries and tubules function in a sequence of events that filter the blood and then reabsorb many of the salts and other small molecules back into the blood. Exchange of materials between these two nephron components occurs by a process of filtration and reabsorption. The remaining structures of the urinary system are composed of the conduction system, which carries urine out of the body. The conduction system has no effect on the urine's composition. Voiding urine involves both reflexive and voluntary control, but, neural damage or degeneration of body structures can lead to a loss of this control. All organs and structures of the urinary system can be invaded by bacteria, resulting in infection and inflammation. Certain hormonal imbalances can have grave effects on the ability of this system to function properly. The urinary system is subject to possible genetic and congenital anomalies, which, when present, play havoc with its ability to function properly. Renal and bladder malignancy can occur, the likelihood of which increases with age and prolonged exposure to environmental factors. The nephron degradation that accompanies aging does not seem to be significant enough to impair proper urine formation. Components of the urinary system involved in proper voiding of urine are more likely to be affected by the aging process.

Gross Anatomy of the Urinary System

- The kidneys are positioned on either side of the midline of the superior abdominal cavity.
- Each kidney is encased in a renal capsule and is retroperitoneally secured to the abdominal wall by renal fascia.
- A renal vein and artery exit or enter each kidney at its hilus.
- The inside of the kidneys have an outer cortex, an inner medulla, and a renal pelvis.
- Renal pyramids in the medulla are the site of urine collection. Urine drains into calyces that lead to the renal pelvis.
- A ureter extends from each hilus and runs inferiorly to transport urine to the urinary bladder for temporary storage. Detrusor muscles allow the bladder wall to expand and contract.
- Urine travels from the bladder to the body's exterior through the urethra.
- The urethral orifice is the terminal component of the urinary system.

Urine Voiding

- Micturition, or urination, involves both reflexive and voluntary control.
- Infants lack voluntary control of the external urethral sphincter.
- Incontinence is due to loss of bladder control in the mature urinary system.
- Conditions of abnormal urine voiding include anuria, urinary retention, oliguria, and polyuria.

The Nephrons

- Nephrons, which are located in the kidneys, perform all of the physiological functions of urine formation.

- Nephrons are composed of an integral complex of blood vessels and renal tubules.
- The nephrons' tubular system consists of Bowman's capsule, the proximal and distal convoluted tubules, the loop of Henle, and the collecting tubule.
- The blood vessel component of the nephrons begins with the afferent arteriole, narrows to form the glomerulus within Bowman's capsule, exits via the efferent arteriole, and ends with the peritubular capillary system.

Urine Formation

- Urine is formed in three stages: glomerular filtration, tubular reabsorption, and tubular secretion.
- Filtrate is formed in the glomerulus as plasma and many of its dissolved substances are pushed out of the capillaries into Bowman's capsule.
- Water, nutrients, and electrolytes travel back into the blood from the filtrate during reabsorption in the proximal tubules. A combination of passive transport (osmosis and facilitated diffusion) and active transport (pumping) are responsible.
- Absorption of water continues in the descending side of the loop of Henle.
- Water is not absorbed in the ascending side of the loop of Henle; however, the loop of Henle does absorb sodium and chloride ions.
- Ion absorption continues in the distal convoluted tubules and the collecting duct. Urine concentration is achieved through the hormonal control of water reabsorption at this location.
- The simultaneous secretion of waste products and ions from the peritubular capillaries into these terminal structures of the tubule network occurs as well.

Hormonal Regulation of Urine Formation

- Various types of receptors in the body activate the release of specific hormones that influence urine formation
- Antidiuretic hormone (ADH) reduces diuresis, or water excretion, and helps to prevent dehydration.
- High blood potassium ion levels stimulate the adrenal cortex to produce aldosterone, which decreases diuresis and reduces excretion of sodium.
- Atrial natriuretic factor (ANF) is produced by special cardiac cells. Its purpose is to lower blood pressure by reducing blood volume through an increase in diuresis.
- Angiotensin II elevates blood pressure through vasoconstriction. The consequent increase in pressure within the glomerular capillaries increases filtration and elevates urine output.

Pathology of the Urinary System

- Most urinary system disorders involve malfunction of the renal system, while others involve the structures of the conducting system.
- Urinary system disorder can be thought of as belonging to one or more of the following categories:
  - congenital disorders
  - infection and inflammation
  - immune disorders
  - hormonal disorders
  - degenerative disorders
  - tumors

Study Guide

- Congenital disorders include diseases that have a genetic basis or may result from an error in fetal development.
- Bacterial invasion is characterized by inflammation in specific portions of the urinary system's anatomy. Inflammation may also result from immune system disorders.
- Immune system disorders can disrupt the integrity of the filtration system, and allow nutrients and other blood components to be abnormally lost.
- Diseases that alter the normal hormone production involved in urine formation can result in deleterious alterations in urinary production, composition, or output.
- Renal diseases that result from degenerative disorders may lead to chronic renal failure.
- Treatment of malignancies depends on the location and progression of the disease.

Aging of the Urinary System

- Nephron reduction associated with aging has little effect on the ability of the kidneys to form urine.
- Health conditions, such as hypertension and increased incidence of other diseases that may accompany aging, are the major contributors to any age-related decrease in urinary function.

Key Terms

Gross Anatomy of the Urinary System
Kidneys
Umbilical cord
Ureters
Urethra
Urinary bladder

The Kidneys – External Anatomy
Adipose capsule
Hilus
Renal artery
Renal fascia
Renal vein
Retroperitoneal

The Kidney – Internal Anatomy
Calyces
Renal columns
Renal cortex
Renal medulla
Renal pelvis
Renal pyramids

The Urinary Bladder
Detrusor muscle
Internal urinary sphincter
Rugae
Transitional epithelium
Trigone

The Urethra
External urethral sphincter
Urethral orifice

Urine Voiding
Anuria
Catheter
Incontinence
Oliguria
Polyuria
Micturition
Nephron
Urinary retention

The Nephrons
Afferent arteriole
Bowman's capsule
Collecting tubule
Distal convoluted tubule
Efferent arteriole
Epithelial tubule

Study Guide

Glomerulus
Loop of Henle
Peritubular capillary system
Proximal convoluted tubule
Renal corpuscle
Renal tubules

Urine Formation
Glomerular filtration
Tubular reabsorption
Tubular secretion

Filtration in the Glomerulus
Filtrate

Reabsorption in the Proximal Convoluted Tubule
Cotransported
Symported

Reabsorption in the Distal Convoluted Tubule and Collecting Duct
Dehydration
Urine concentration
Water conservation

Hormonal Regulation of Urine Formation
Atrial natriuretic factor (ANF)
Diuresis

Congenital Disorders
Aminoaciduria
Bladder stones

Calculi
Glycosuria
Hemodialysis
Kidney stones
Polycystic kidney disease

Infection and Inflammation
Cystitis
Dysuria
Pyelitis
Pyelonephritis
Urethritis
Urinary tract infection (UTI)

Immune Disorders
Cast
Edema
Glomerulonephritis
Hematuria
Proteinurea

Hormonal Disorders
Diuretics

Degenerative Disorders
Acute renal failure
Chronic renal failure

Tumors
Bladder cancer
Renal cell carcinoma

Aging of the Urinary System
Cystocele
Nephroptosis

Check Your Understanding

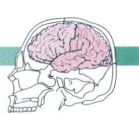

1. Urine formation occurs in:
    a. all organs of the urinary system
    b. the renal cortex
    c. nephrons in the kidneys
    d. the renal pelvis

2. An obstruction in the ureter would prevent the flow of urine to:
    a. the calyces
    b. the urinary bladder
    c. the renal pelvis
    d. the hilus

Study Guide

3. Females have a greater predisposition to UTI because of:
   a. their smaller bladder size
   b. the length of their urethra
   c. a difference in body pH
   d. their hormonal levels

4. When the stretch receptors of the micturition reflex arc are stimulated they elicit:
   a. contraction of the internal urinary sphincter muscles
   b. relaxation of the detrusor muscles
   c. relaxation of the external urinary sphincter muscles
   d. relaxation of the internal urinary sphincter muscles

5. Which of the following might be an indicator of diabetes mellitus?
   a. polyuria
   b. anuria
   c. dysuria
   d. oliguria

6. The renal corpuscle is composed of:
   a. renal tubules
   b. afferent capillaries and tubules
   c. Bowman's capsule and glomerulus
   d. the peritubular capillary system

7. Renal filtrate is initially formed in the:
   a. afferent arteriole
   b. renal pyramids
   c. peritubular capillaries
   d. renal corpuscle

8. Filtrate in Bowman's capsule normally contains:
   a. a high water content
   b. a high protein content
   c. a low water content
   d. no proteins

9. Sodium is reabsorbed in the proximal convoluted tubules due to:
   a. a concentration gradient between the blood and the tubule's interstitial fluid
   b. active transport mechanisms
   c. cotransportation
   d. the sodium/potassium pump

10. The loop of Henle:
    a. functions for filtration
    b. reabsorbs only water
    c. reabsorbs only ions
    d. reabsorbs both water and ions

11. Tubular secretion:
    a. occurs in Bowman's capsule
    b. returns filtrate components to the blood
    c. removes waste products from the blood
    d. increases the water content of the urine

12. Antidiuretic hormone:
    a. functions to retain water in the body
    b. increases urine volume
    c. is stimulated by blood potassium levels
    d. dehydrates the blood plasma

13. The inability to produce certain carrier proteins for cotransportation can result in:
    a. abnormally low levels of nutrients in the blood
    b. normal water volume in the urine
    c. normal nutrient levels, but high water content in the urine
    d. increased solute concentration of the urine

14. Hematuria, proteinuria, and the presence of casts in the urine are indicators of:
    a. bacterial infection
    b. glomerulonephritis
    c. hormonal irregularity
    d. polycystic kidney disease

15. The effects of aging on the urinary system are commonly attributed to:
    a. associated diseases of aging
    b. the inability of nephrons to function properly
    c. decreased production of hormones
    d. normal loss of neural innervation

A Case Study

DIURETIC USE IN HYPERTENSION TREATMENT

At her annual physical exam, your neighbor, who is a 60-year-old woman, is told that her blood pressure is 150/100. Previous blood pressure readings have always been less than 130/80, so she is alarmed to find that the pressure has risen so significantly. Her doctor has prescribed a blood pressure medication called lisinopril, which, he tells her, is an angiotensin-converting enzyme (ACE) inhibitor. She does not like the idea of having to be on a maintenance medication if there is any other alternative and, since her health insurance does not cover prescriptions, the cost of the medication is also a concern for her. After consulting with several friends whom she knows are also hypertensive, she learns that there is a variety of medications available to treat hypertension and that the cost of the prescriptions also varies widely. She remembers that she had previously read a magazine article that mentioned something about the use of a "water-pill" to control high blood pressure, but she does not really understand how that could aid in controlling hypertension.

Use the information provided in your textbook and the following Web sites to answer the following questions so that you can give your neighbor sound advice on how to pursue her situation.

- What therapy does current research show to be the most effective in the treatment of hypertension in persons over age 55 years? Is there necessarily unilateral agreement in these findings?
- What is the physiological basis for the use of a diuretic to treat hypertension? (How does increasing water loss in the urine help to reduce blood pressure?)

Study Guide

- What are some of the benefits of using diuretic treatment for hypertension versus alternative drug therapies? (eg, possible side effects, cost, etc.)

1. The ALLSTAT Study
   http://allhat.sph.uth.tmc.edu/

2. Treating Hypertension
   http://www.pbs.org/newshour/bb/health/july-dec02/hypertension.html

3. Water Pills Help All with Hypertension
   http://www.medicinenet.com/script/main/art.asp?articlekey=51967

4. Improving Medical Statistics – The ALLHAT Trial
   http://www.improvingmedicalstatistics.com/ALLHAT.htm

5. Diuretics in the Treatment of Hypertension: Current Status
   http://www.rxfiles.ca/acrobat/diuretic.pdf

Where Do We Go from Here?

People in health fields can use their knowledge of the urinary system to solve everyday problems. You may wish to use other resources, such as the Web sites below, in addition to your textbook to investigate the answers to each of the following situations:

1. A female friend of yours is embarrassed by the fact that lately she lacks control in urine voiding when she coughs, laughs, or jumps. Should she be concerned that she has a serious medical condition? What type of medical treatment might help her regain control?
   http://www.patient.co.uk/showdoc/23068767/
   http://www.vh.org/adult/provider/obgyn/urinaryincontinence/

2. Your child's daycare teacher has asked you for some advice on "potty training." She also would like to find information for parents of older toddlers who are experiencing frequent bedwetting. At what age can parents expect their child to gain voluntary control over urination? Is bedwetting merely under behavioral control, or could there be a physiological basis for its occurrence? What methods can be employed to help a child stop bedwetting?
   http://www.nlm.nih.gov/medlineplus/toilettrainingandbedwetting.html
   http://www.parentcenter.com/toilet-training

3. A man who has polycystic kidney disease is aware that his disease has a genetic basis. He and his wife are considering having a child, and he wants to know what the chances are that their child would be affected by the disease. Is this disease serious enough that if the chance is high, the two should consider the possibility of adoption rather than having their own child?
   http://www.pkdcure.org/
   http://kidney.niddk.nih.gov/kudiseases/pubs/polycystic/

4. You have been volunteering at a public health center. You notice that there seems to be a fairly high number of women seeking treatment for UTIs and fewer male patients with these symptoms. Is there a plausible

explanation for this observation, or is it merely coincidence? You also note that it is not uncommon for these patients to experience recurrent infections. Are certain individuals more susceptible to developing UTIs? What is the common treatment and prevention method for UTIs?
http://www.niddk.nih.gov/health/urolog/pubs/utiadult/utiadult.htm
http://womenshealth.about.com/cs/bladderhealth/a/UTI.htm

5. A friend of yours is training heavily in preparation for the Chicago marathon. It has been an exceptionally warm summer, and he is concerned about the potential for dehydration while he is running. Would you recommend that he drink exceptionally large volumes of water prior to running and continue hydrating himself throughout the course? Why or why not? What might be the danger of this practice? What is the recommended water intake regime for athletes participating in long-term physical exertion?
http://chemistry.about.com/cs/5/f/blwaterintox.htm
http://www.1960sports.com/features/features263.php

Skills Activities

1 Urine Chemical Analysis

Materials

- three artificial urine samples of the instructor's choice (one normal and two altered to indicate a specific pathology of choice) labeled A, B, and C
- protective gloves, only if biohazardous material, such as animal blood, has been added to a sample (it is strongly suggested that human blood not be used)
- beaker of disinfectant if necessary for biohazard disposal
- three urinalysis reagent chemical test strips with color chart/group
- clock with a second hand
- urine analysis handbook or computer with Internet access (suggested Web site:
http://www.texascollaborative.org/spencer_urinalysis/ds_sub2.htm)

Various urinalysis reagent strips for quick chemical analysis of urine samples are available commercially. These strips have replaced the laborious chemical analysis tests that were necessary in the past to obtain similar test results. The presence of certain metabolic products and/or blood components may indicate pathology or other abnormal body function. Measurements, such as pH and specific gravity, can also be ascertained. In the testing of real urine, it is imperative that samples are collected and analyzed in a prescribed manner to avoid contamination. In this activity, artificial urine will be utilized, so there is no need for these precautions. Each student group will record measurements for pH, glucose, protein, ketones, and blood on three artificial urine samples. These values will be collected using urinalysis reagent test strips. The results will then be compared with normal values obtained through the available resource material. A possible pathology for any abnormal samples will then be suggested.

Study Guide

Procedure

- Dip one test strip into urine sample A, being sure that all the test strip squares are submerged.
- Quickly remove the strip, running it along the top of the test tube to remove excess liquid.
- Turn the strip lengthwise on a paper towel to allow any excess urine to drain off without running from one test square to another. Begin timing for the appropriate number of seconds to wait before each test square color can be "read." (The color chart accompanying the test strips provides this information).
- After the appropriate time has passed for each test, compare the test square color for pH, protein, glucose, ketones, and blood with the corresponding color chart, and record the values in the data table below. Be sure to report units of measure as well as the numeric values.
- If the presence of blood is indicated, the test strip should be discarded in a beaker of disinfectant. If not, it can be discarded in a normal trash receptacle.
- Repeat the previous steps, and use a new test strip for each sample (B and C).

Data Table

Sample	pH	Protein	Glucose	Ketone	Blood
A					
B					
C					

- Using the resource material provided by your instructor (ie, a urinalysis handbook or suggested Web site) fill in the normal values for each of the chemical test results in the appropriate row of the following table: (Hint: If you use the suggested Web site, you will find a summary of normal values by accessing Activity 2.2 under the Chemical Testing section on the left-hand sidebar menu and viewing the work sheet in the directions section.)

Sample	pH	Protein	Glucose	Ketone	Blood
normal values					
A					
B					
C					

- For each abnormal value obtained in samples A, B, and C, place an arrow to indicate whether the value is high or low compared with the normal value.
- Complete the table by listing a possible pathology that might be indicated (if any) for each sample. The resource material provided should give you the necessary information to make this decision.

- Answer the following questions:
  1. If a "normal value" obtained for any of the above urine components was expressed as "weight unit/time" such as "mg/day," would a test strip reading allow you to make an accurate comparison? Why or why not?
  2. Is an abnormal pH reading always cause for alarm? Why or why not? What pH extremes might indicate concern, and what might this concern be?
  3. What conditions might cause a temporary elevation of glucose?
  4. Can a reagent test strip allow a health professional to detect whether or not intact RBCs are present? Why or why not?
  5. What would be a probable diagnosis for a patient whose urine test strip test showed a positive for nitrite levels and a positive for leukocyte esterase?

2 Microscopic Examination of Urine Sediment

Materials

- computer with Internet access
  (suggested Web site: http://www.texascollaborative.org/spencer_urinalysis/ds_act3-1.htm)

It takes a skilled technician to properly analyze urine sediment, but such analysis can provide medical professionals with valuable information about the health status and proper diagnosis of certain diseases. It is certainly beyond the scope of one lab activity to gain the adequate knowledge and skills to even attempt the microscopic analysis of a real urine sample. However, this Web site activity is proposed to provide students with information on the normal constituents of urine sediment and introduce them to a few components whose presence in certain levels may indicate pathology. Students will also have the opportunity to see representative photographs of actual microscopic specimens to gain a better understanding of the identification of the urine elements discussed.

- Access the suggested Web site by typing in the address provided in the Materials section above.
- Open activity 3.2 by choosing it on the left-hand sidebar menu. Be sure to look for it under the Activities section of subtopic 3: *Microscopic Examination of Urine Sediment*.
- As you read through the procedure steps for preparing and viewing a urine sediment slide, click on the underlined and highlighted terms: **casts**, **RBCs**, **WBCs**, **epithelial cells**, **microorganisms**, **crystals**, and **mucus** to view samples of each. As you will quickly realize, there is an overwhelming amount of information provided in each of these categories.
- Your instructor can offer you further assistance in choosing what information is most useful for his or her objective for this exercise and also provide questions or other assignments to assess your understanding of the specific desired objectives.

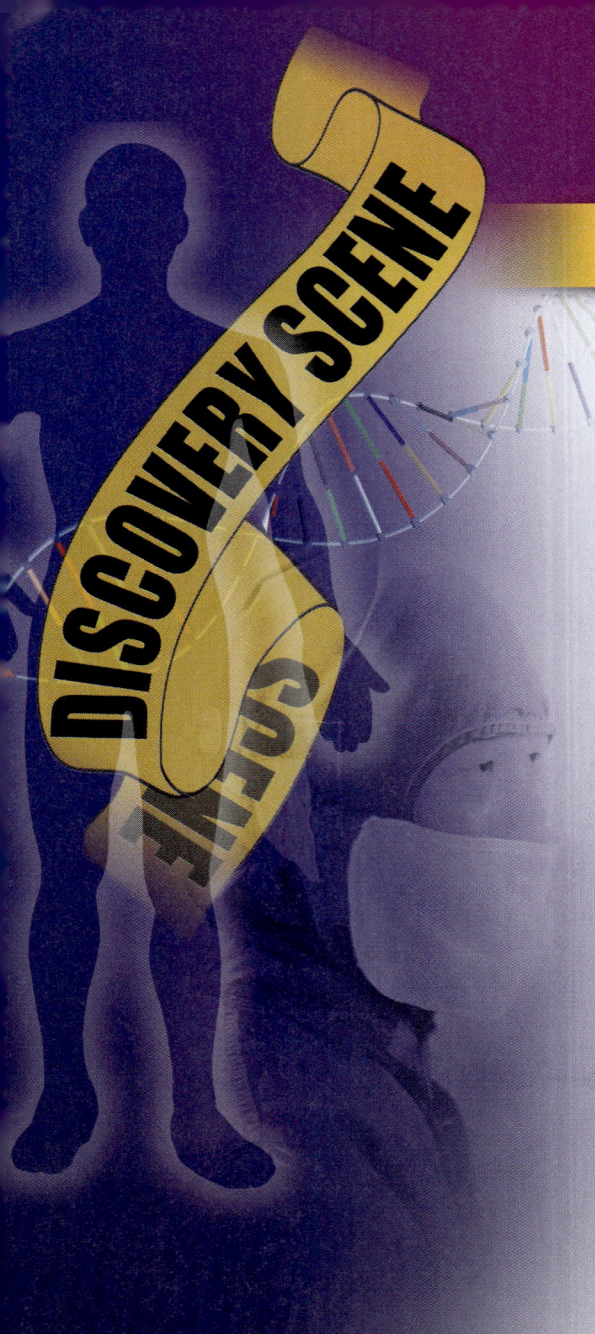

Case Study Investigation

Case Study Investigation #15

You are at a party where people are looking at family photographs. It seems strange to you that the grandmother showing the photographs keeps calling one of the young girls in the pictures "Edward." You think it is a nickname until you run into Edward at the party. Edward is a 6-foot-tall, 56-year-old male with a full gray beard. Edward describes a typical childhood as a girl until he turned about 12-years old. He then went through a confusing period during puberty that required psychiatric treatments and medical attention. It appeared that Edward was developing male secondary sex characteristics. He went on to explain that he had gender identity problems into early adulthood until he took on life as a male when he was 29 years old. How do you interpret Edward's situation?

CHAPTER 15

The Reproductive Systems and Human Development

Chapter Outline

Case Study Investigation (CSI)
Applied Learning Outcomes
Overview
Female Reproductive System
    Reproductive Tract
        Ovaries
        Fallopian Tubes
        Uterus
        Vagina
    Mammary Glands
Male Reproductive System
    Testes
    Seminal Vessels
    Penis
Basics of Sexual Reproduction
    Female Sexual Cycle
        Copulation
        Embryology and Pregnancy
Wellness and Illness over the Life Span
    Pathology of the Reproductive System
    Aging of the Reproductive System
CSI Conclusion
Study Guide

Applied Learning Outcomes

- Use the terminology associated with the reproductive system.
- Learn about the following:
  - organization of the female and male reproductive systems
  - structure of the female and male reproductive organs
  - function of the female and male reproductive organs
  - the female sexual response
  - human copulation, pregnancy, and development
- Understand the aging and pathology of the reproductive system.

Overview

Key Terms: gonad, ovary, puberty, secondary sex characteristics, sexual dimorphism, specialized germ cell (SGC), testicle, testis

Gonad Sexual reproductive organ

Specialized Germ Cell (SGC) A cell in the gonad involved in sexual reproduction

Ovary A female gonad

Testis or Testicle A male gonad

Sexual Dimorphism Developmental differences that distinguish the two genders

Secondary Sex Characteristics Anatomical features that distinguish males from females

Puberty The stage of development when sexual reproduction becomes possible

The main job of the reproductive system is to ensure the continuation of the human species. Humans, like all sexually reproducing organisms, contain specialized organs called **gonads** that aid in sexual reproduction. The gonad starts to develop in the embryo at 4.5 weeks and takes on its earliest functions at 11.5 weeks. At first the gonad is a mass of cells derived from the same mesoderm that forms the kidneys. Unique to the gonad are mobile cells called **specialized germ cells** (**SGCs**). The gonad, however, has not yet developed any gender differences. At 7 weeks, the SGCs develop the ability to undergo meiosis (Figure 15.1). After that point of development, the gonad becomes an **ovary** or a **testis**. Many people call the testis the testicle. The plural for testis is testes. It is not until much later that the gonads produce eggs or sperm.

Humans, like many animals, exhibit **sexual dimorphism**. In humans, sexual dimorphism is expressed as visible features called **secondary sex characteristics**. Secondary sex characteristics become evident after 14 weeks of embryological development. Further development of these sex characteristics takes place at **puberty**.

After puberty, the female body develops the physiology and structures to produce, store, and release eggs. The female body also has the structures to maintain the development and growth of the fetus until it can survive on its own. Like most mammals, human females feed their young the milk from specialized glands. The adult male body has the structures for producing, storing, and transporting sperm. Integrated into the reproductive systems of females and males is the urethra from the urinary system. This is the reason that disorders of one particular system sometimes carry over to the other organ system.

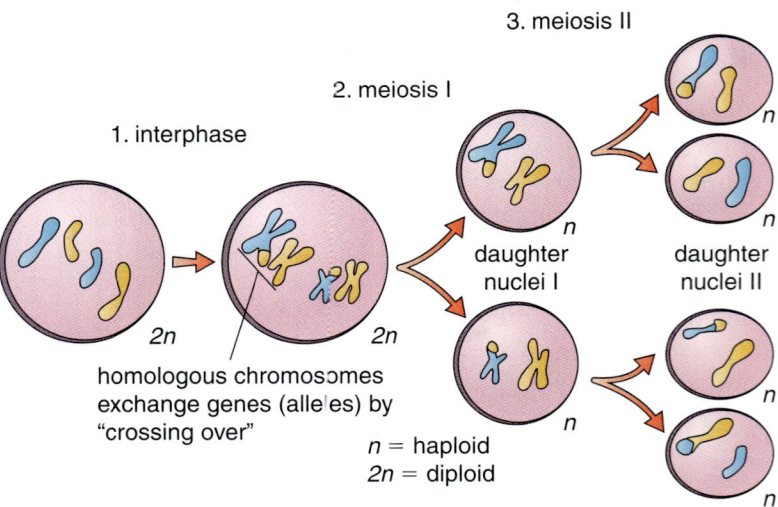

Figure 15.1 Secual Reproduction – Meiosis
Meiosis is the process of making eggs or sperm in the gonads

Good Choice Bad Choice

It is valuable to use what you have learned about the reproductive system to help others to better understand the world around them. It is very important to check your facts and seek further information about certain topics before discussing health and science issues. Here are some suggestions to foster a better public awareness of the reproductive system:

1. Speak to high school students about the risks for sexually transmitted diseases.
2. Assist at a clinic that provides prenatal health care to indigent women.
3. Volunteer at a hospice to help elderly people to better understand the facts behind the aging of the reproductive system.
4. Assist the school nurse at a school health day to teach children the basics of puberty and sexual development.

✓ Concept Check

1. What are the major functions of the reproductive system?
2. Explain the reproductive system process that produces eggs and sperm.
3. What is the relationship between the reproductive and urinary systems?

DISCOVERY SCENE PLEASE ENTER DISCOVERY SCENE PLEASE ENTER

CSI Break

How does information about sexual dimorphism help you to understand Edward's condition? Is it possible for a person's body to change gender after puberty? If so, how could such a developmental change be explained?

FEMALE REPRODUCTIVE SYSTEM

Key Terms: external genitalia, fallopian tube, mammary gland, reproductive tract, uterus, vagina

The female reproductive system is divided into the **reproductive tract** and the **mammary glands**. Most of the female reproductive system is internal (Figure 15.2). However, some components are external and often called the **external genitalia** (Figure 15.3). Organs of the female reproductive tract produce and transport the egg. Sexual reproduction and care of the developing fetus also takes place in the reproductive tract. It is composed of the ovaries, **fallopian tubes**, **uterus**, and **vagina**. Mammary glands are paired accessory organs of the female reproductive system. Both males and females possess mammary glands; however, the high estrogen levels present during puberty stimulate growth and development of the milk-producing glands only in females. Fat also accumulates in the female breasts after puberty.

Reproductive Tract Connected muscular tubes that are involved in female reproduction

Mammary Gland A milk-secreting organ of females

External Genitalia Sex organs on the outside of the body

Fallopian Tube A tube that extends from the uterus and ends proximal to the ovary

Uterus A pear-shaped organ that nourishes the growing embryo

Vagina A muscular canal running from the uterus to the exterior of the body

Figure 15.2 Female Reproductive System – Internal
The internal components of the female reproductive tract are located in the bottom of the pelvic cavity

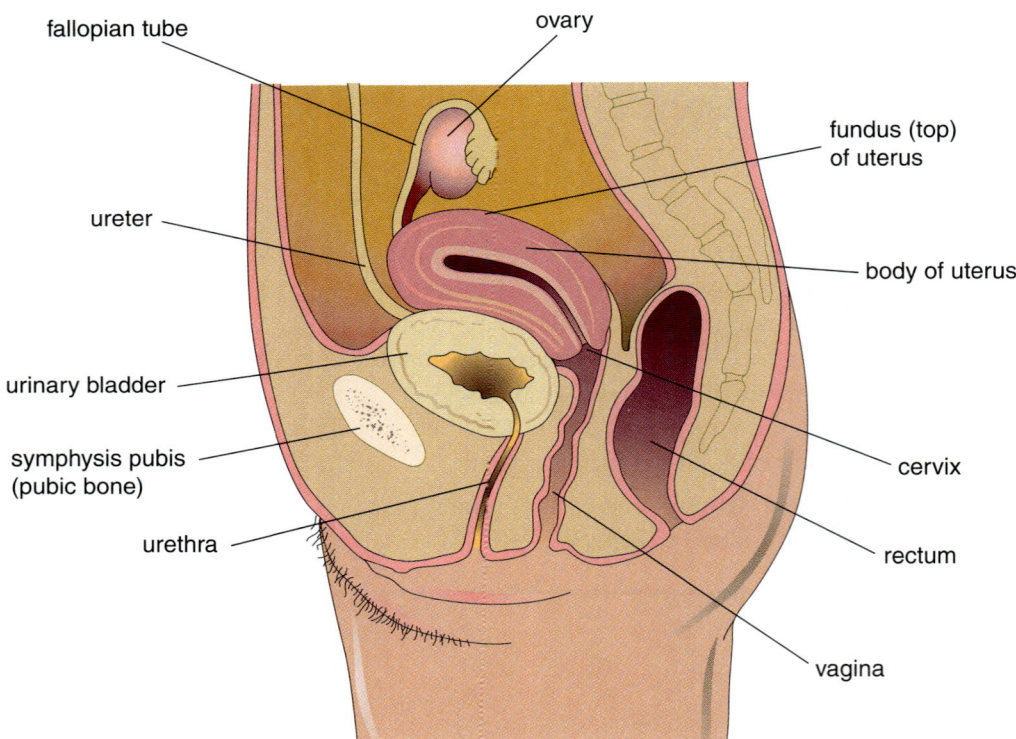

Figure 15.3 Female Reproductive System – External
The external female genetalia open into the internal components of the reproductive tract

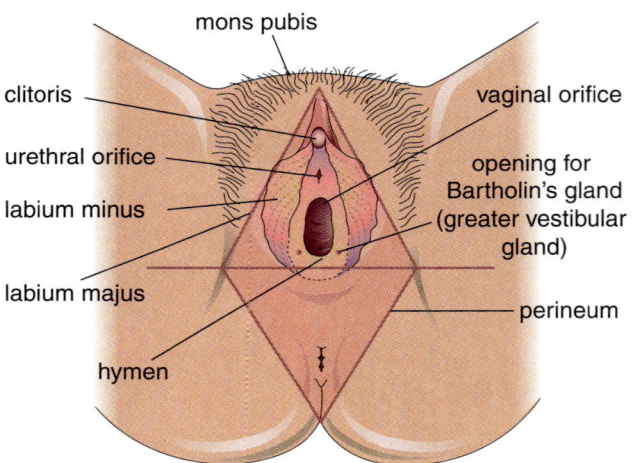

✓ Concept Check

1. Define the two parts of the female reproductive system.
2. Describe the organs of the female reproductive tract.
3. What is a mammary gland?

Reproductive Tract

Key Terms: androgen, aromatase, broad ligament, cervix, clitoris, clitoral hood, corpus luteum, desmolase, endometrium, erectile tissue, fertilization, fimbriae, graafian follicle, hymen, intersex, labia majora, labia minora, lower uterine segment, menstrual period, mons, myometrium, myosalpinx, oocyte, ova, ovarian follicle, ovarian ligament, oviduct, ovulation, perineum, Skene's gland, uterine fundus, vulva, womb

Ovaries

Ovaries are paired, oval-shaped organs responsible for egg formation and sex-hormone production (Figure 15.4). They are located on the lateral sides of the uterus, just below the opening to the fallopian tubes. The ovaries are attached to the uterus by the **ovarian ligament**. Each ovary is covered with a thin epithelium that is continuous with the peritoneum of the abdominal cavity. This explains why bacterial infections of the female reproductive tract can travel throughout the abdominal cavity. Underneath the epithelial covering is a rigid capsule made of fibrous connective tissue. Ovaries are composed of an outer cortex and an inner medulla layer. The cortex is composed of a cellular connective tissue where the eggs are located. An egg is also referred to as an **ovum** by many histologists. A loose connective tissue that contains numerous blood vessels and nerves makes up the medulla.

The cortex of the ovary does not produce eggs. Rather, it stores immature eggs in fluid-filled sacs called **ovarian follicles**. The eggs migrate into the ovarian follicles early in the development of the fetus. Ovarian follicles consist of an immature egg, or **oocyte**, which is surrounded by special follicle cells. An oocyte is a cell that has not yet undergone a complete meiosis. Many scientists believe that the ovary contains about 7 million oocytes by month 6 of embryological development. Approximately 40,000 to 60,000 oocytes are present by puberty. About 400 oocytes mature within a lifetime. One ovarian follicle at a time matures in an ovary. As a follicle matures, it enlarges and fills with fluids that nourish and protect the oocyte. The follicle migrates toward the surface of the ovary, facing the opening of the fallopian tube. It then becomes a mature, or **graafian follicle**. At this point, the oocyte reaches a further stage of meiosis where it is now called an ovum. The ovum ruptures out of the graafian follicle during **ovulation**.

> **Ovarian Ligament** A strip of connective tissue that attaches the ovary to the uterus
>
> **Ovum** A term for the female sex cell, or egg
>
> **Ovarian Follicle** A fluid-filled sac in which an egg matures
>
> **Oocyte** An immature egg
>
> **Graafian Follicle** A nearly mature egg or ovum
>
> **Ovulation** Release of an ovum from the ovary

Figure 15.4 **Ovaries**

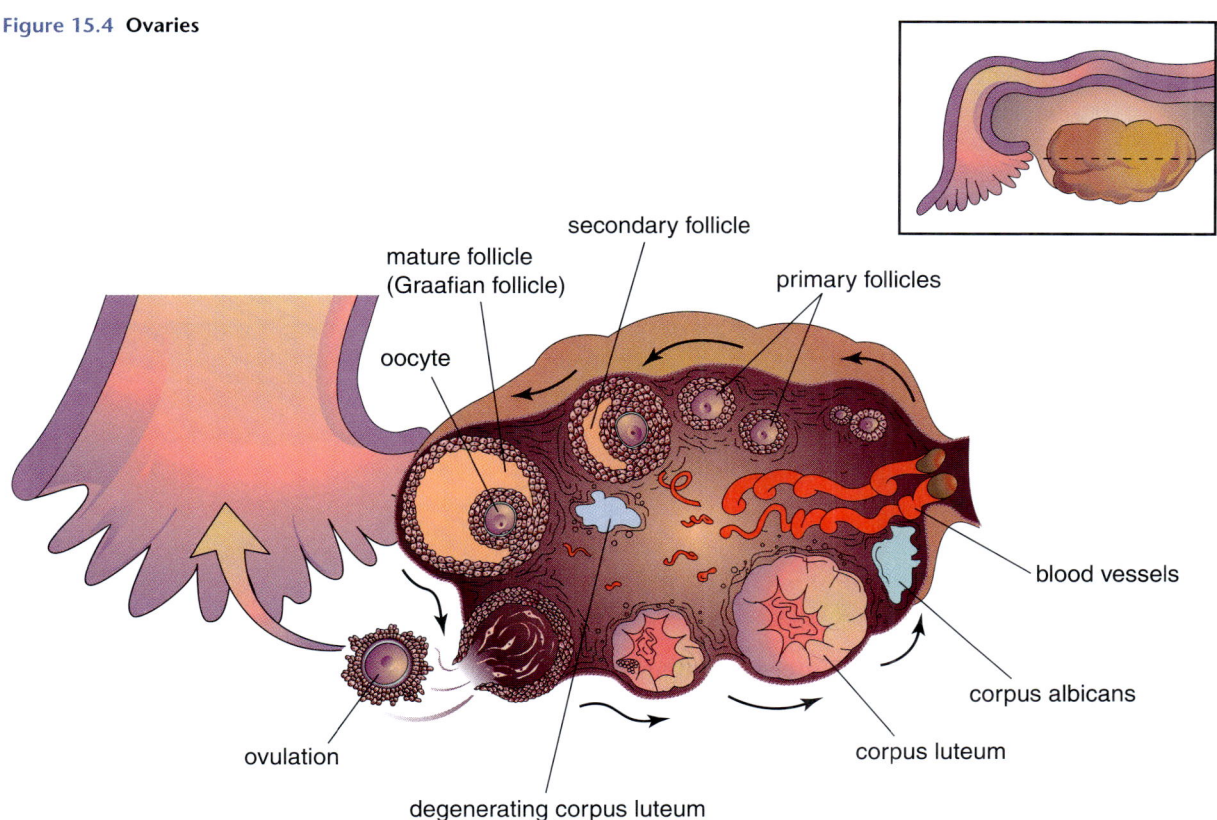

Aromatase An enzyme that converts androgens into estrogen

Androgen A hormone that maintains body structure and provides male sex characteristics

Corpus Luteum A structure formed in the follicle after the egg is released at ovulation

Desmolase An enzyme that helps to convert cholesterol into progesterone

Intersex A condition in which it is not clear at birth whether the individual is a male or a female

Oviduct Another name for fallopian tube

Fertilization The fusion of sperm and egg

Fimbriae Finger-like projections at the end of the fallopian tube

Myosalpinx The middle muscular layer of the fallopian tube

Broad Ligaments Two sheets of epithelium attached to the fallopian tube and uterus

Womb Another name for the uterus

Myometrium The muscular wall of the uterus

Ovulation is defined as the release of the ovum from the ovary. After release from the follicle, the egg enters the fallopian tube. Not all of the eggs reach the follicle tube; some accidentally fall into the abdominal cavity where they decay after approximately two days.

Follicles are also responsible for producing estrogen and secreting it into the blood stream. The enzyme **aromatase** converts hormones called **androgens** into estrogen. Androgens are secreted by the ovaries and adrenal glands. Small amounts of androgens are needed to maintain bone and muscle structure in females as well as males. However, large levels of androgens cause the male sex characteristics to develop. A structure called the **corpus luteum** makes progesterone. The corpus luteum forms in a follicle that has released an egg during ovulation. The enzyme **desmolase** produces progesterone from cholesterol molecules. The corpus luteum disintegrates if the egg is not fertilized. Some women have genetic defects that alter or diminish the function of aromatase and desmolase. This produces a wide variety of **intersex** conditions in which the female has different degrees of male secondary sex characteristics. The intersex condition usually does not possess both types of gametes. They have either a male or a female gonad which, in many cases, does not produce gametes due to abnormal levels of the appropriate sex hormones.

Fallopian Tubes The fallopian tubes, or **oviducts**, are 3-inch-long tubes that extend from each side of the uterus (Figure 15.5). One fallopian tube leads to each ovary on the respective side; however, they do not make contact with the ovary. They carry eggs and sperm to the site of **fertilization**. The end of the fallopian tube near the ovary has finger-like projections called **fimbriae**. Fimbriae contain a high density of ciliated cells whose wave-like movements sweep the ovulated egg into the opening of the fallopian tubes. The fimbriae are lined with a ciliated mucous membrane that secretes substances that sustain the egg and the sperm. These secretions also provide electrolytes and nutrients to the early stages of the embryo. Surrounding the mucous membrane is a smooth-muscle layer called the **myosalpinx**. Muscular contractions of the myosalpinx move the egg through the fallopian tube to the uterus. The serosa of the fallopian tube is composed of epithelial cells. The serosa is fused to two sheets of epithelial cells called the **broad ligaments.** The broad ligaments hold the fallopian tubes in place by securing them to the uterus.

Figure 15.5 Fallopian Tube

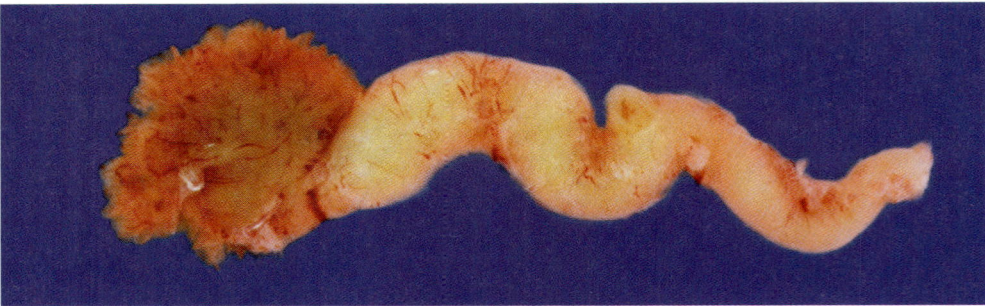

Uterus The uterus, or **womb**, is a hollow muscular organ where the embryo and fetus develop (Figure 15.6). It is composed of three layers, much like the fallopian tubes. The middle muscular layer, or **myometrium**, is very thick and is innervated by autonomic nerves. Muscles of the myometrium also contract in response to the hormone oxytocin. These contractions are particularly important during childbirth. The inner layer of the uterus is a thick mucosa called the **endometrium**. This layer is rich in blood vessels and varies in thickness with the **menstrual period**. The menstrual cycle is the periodic thickening and shedding of the endometrium. It prepares the lining of the uterus for development of the embryo. The outerlayer, or perimetrium, is a connective tissue covering that attaches laterally to the broad ligament.

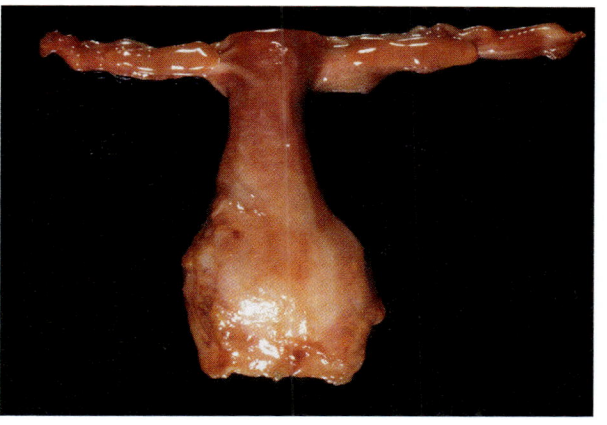

Figure 15.6 **Uterus**

Endometrium The mucous membrane lining of the inner surface of the uterus

Menstrual Period The cyclic shedding of the endometrium

GESTATION FACTS

It normally takes 253 to 303 days from the time of fertilization for a human fetus to fully develop. How do humans compare with other animals in their gestation time? Some animals gestate in the uterus just like humans. Others spend their time in an egg outside of the mother's body. The variation in gestation times for an animal depends on the particular species, and/or the climate in which it lives.

Animal Gestation Times

Animal	Gestation Time (Days)
cat	52–69
chimpanzee	227
cow	280
dog	53–71
elephant	510–730
gorilla	257
horse	329–345
kangaroo	32–39
pig	101–130
rabbit	30–35
rat	21
sheep	144–153
whale	365–547

Sources: New York Zoological Society; University of Michigan Museum of Zoology and Wildlife Conservation Society.

Uterine Fundus The top of the uterus

Lower Uterine Segment The lower third of the uterus

Cervix The lowermost part of the uterus, which opens into the vagina

Skene's Glands Mucus producing glands at the base of the female urethra

Perineum A diamond-shaped region making up the base of the pelvic region

Vulva The external female genitalia

Mons A pad of fat tissue that covers the pubic bone in females

Labia Majora Outer lips of the vulva

Labia Minora Inner lips of the vulva

Clitoris A small piece of erectile tissue within the labia minor

Erectile Tissue A tissue capable of filling with blood and swelling

Hymen A thin membrane partially covering the opening of the vagina

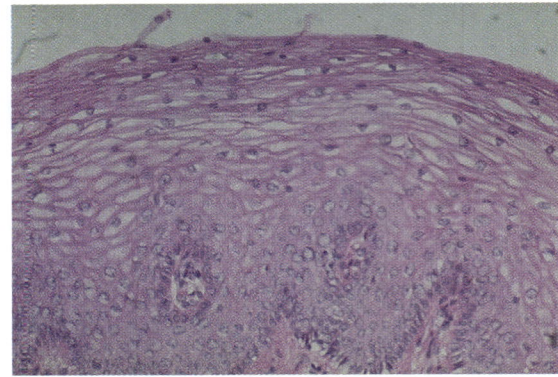

Figure 15.7 **Vaginal Tissue**

The uterus is divided into three regions. Connected to the fallopian tubes is the upper region of the uterus called the **uterine fundus**. The fundus has a thicker endometrium and myometrium than the rest of the uterus. It is in this region that the embryo normally develops. Below the fundus, making up the lower third of the uterus, is the **lower uterine segment** or body. The most inferior and final region is the **cervix**, making up the entrance to the uterus and slightly protruding into the vagina. It is composed of bands of muscles that contract during sexual activity and dilate during childbirth. The cervix has a rich supply of blood vessels and an abundance of lymphatic vessels.

Vagina The vagina is a muscular passage that connects the uterus to the external genitalia. It ranges in length from 2½ to 4 inches and is lined with a mucous membrane. The middle muscular layer has a large supply of blood vessels. These muscles are involved in the female sexual response and may assist the passage of sperm into the uterus. The walls of the vagina are highly flexible and lined with stratified, squamous, non-keratinized epithelium. This permits them to serve as a passageway for the baby during childbirth (Figure 15.7). Near the opening of the vagina is the urethra, which expels the urine past the external genitalia. At the opening of the urethra are mucous-producing glands called **Skene's glands**. The mucous is an important lubricant during sexual activity. Surrounding the vaginal opening and extending to the anus is a region of the pelvis called the **perineum**. The perineum is a diamond-shaped region making up the base of the pelvic region.

The external genitalia of the female are collectively called the **vulva** (Figure 15.3). Its components surround the vaginal opening. Forming the upper part of the vulva is the **mons**. It is a pad of fat tissue that covers the pubic bone above the vaginal opening. The mons is rich in nerves associated with sexual sensitivity. It is believed to protect the pubic bone from the forces of sexual intercourse. Located on both sides of the vaginal opening are the **labia majora** and **labia minora**. The labia majora are pads of fat tissue that wrap around the vulva from the mons to the perineum. They are covered with pubic hair after puberty, and have many sweat and sebaceous glands. Medial to the labia majora are the labia minora, which cover and protect the vaginal opening.

Within the upper region of the labia minor is the **clitoris**. The clitoris is a small piece of highly innervated **erectile tissue** involved in the sexual response. Covering the clitoris is a thin strip of labia minora called the clitoral hood. It protects the clitoris from abrasion. A membranous flap called the **hymen** partially covers the vaginal opening at birth. The hymen has a variety of sizes and shapes depending on how the reproductive tract formed during embryological development. Some hymens are solid and cover most of the vagina, while others have perforations and barely cover the vagina. It is typical for the hymen to erode with age or be broken during sexual activity.

Mammary Glands

Key Terms: areola, lactation, lactiferous duct, nipple

Mammary glands, or breasts, are specialized organs that secrete milk following pregnancy (Figure 15.8). They are composed of glandular tissue located in the subcutaneous tissue of the upper chest. The glandular tissue develops after puberty and contributes to the female secondary sex characteristics. Each mammary gland is divided into 15 to 20 lobes composed of loose connective tissue and glands. Many lymphatic vessels and lymph nodes are found within the lobes. Mammary glands also have a rich blood supply. The exocrine glands within the lobes secrete milk into **lactiferous ducts**. Lactiferous ducts carry milk to openings in the **nipples**. The nipple is a small raised area in the center of each mammary gland. It has openings through which milk flows from the lactiferous ducts. Around each nipple is a circular area of pigmented skin called the areola. The nipple and areola are heavily innervated with nerves associated with sexual stimulation and milk release.

Lactiferous Ducts Ducts of the mammary glands that carry milk to the nipples

Nipple A small raised area in the center of each mammary gland

Lactation The formation of milk by the mammary glands

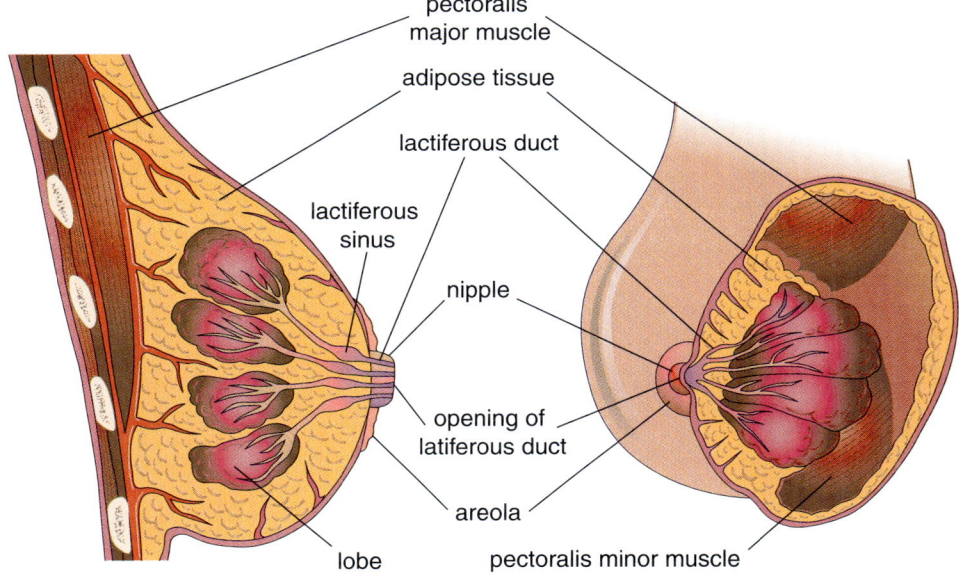

Figure 15.8 **Mammary Glands**

The production of milk by the mammary glands is called **lactation**. Lactation is stimulated by hormonal changes brought on at the end of pregnancy. Stimulation of the nipple by the baby's suckling encourages milk formation and release. Milk is a complete mixture of nutrients required for growth of the baby. Antibodies against general infections are deposited into the milk by the mammary gland's lymphatic system. The major components of human milk differ significantly from cow's milk (Table 15.1).

Table 15.1 Components of Human Milk and Cow Milk (grams per liter)

Milk Component	Human Milk (g/L)	Cow Milk (g/L)
Carbohydrates		
lactose	741	40
oligosaccharides	12	1
Lipids		
triglycerides	40	40
phospholipids	0.4	0.4
Proteins		
caseins	2	2.7
alpha-lactalbumin	2	0.1
lactoferrin	2	trace
Antibodies		
secretory immunoglobulin A (IgA)	2	0.03
beta-lactoglobulin	none	3.6
Minerals		
sodium ions	0.115	0.345
potassium ions	0.60	1.8
chloride ions	0.531	1.29
calcium ions	0.32	1.2
magnesium ions	0.017	0.16

Human milk has a higher carbohydrate and lower protein content than other animal milk. The amino-acid composition of human milk proteins differs significantly from the milk of other animals. Cow milk is lower in the amino acids needed for human CNS development. Baby formula is designed to provide all of the nutritional components of human milk; however, formula lacks the antibodies needed to assist the baby's ability to fight off infectious diseases.

THE RETURN OF WET NURSES?

It was not unusual for women throughout history to employ other women to breast feed their children. This was done by a woman hired as a wet nurse. A wet nurse would breast-feed a baby that was not her own. They were employed if a mother was unable to breast-feed her infant for a variety of reasons, which may have included illness or a reduction in milk production. Outbreaks of syphilis throughout 18th- and 19th-century Europe necessitated the use of disease-free wet nurses for fear that infected mothers would spread the disease to their babies by breast-feeding. The services of wet nurses were many times required by women who had multiple births. Wealthy women in 18th-century France hired wet nurses to give them freedom from breast-feeding. The services of wet nurses disappeared as people felt more comfortable feeding children cow milk and infant formulas. Today, it is not customary in Europe or North America for pregnant women to lend their services as a wet nurse. Wet nurses were looked down upon in society because many were from lower-income socio-economic groups.

Wet nurses may be on the return, however. It is now being looked upon as a noble profession. A wet-nurse business opened in Beverly Hills, California, in 2004. The services are provided to mothers who want their children raised on breast milk, but do not have the time to do it themselves. Women with breast implants may develop impaired milk production; they are also requesting wet-nurse services. Recent research shows that babies benefit emotionally and intellectually by suckling a breast. In addition, it is known that breast milk builds up the baby's immune system better than formula or the milk from other animals. Some psychologists are debating the value of wet nurses in child rearing. Many believe that the infants can develop separation anxiety when the wet nurse leaves upon weaning the baby. Nevertheless, the desire for wet-nurse services may reach the same level of demand as in 18th-century Europe.

Wolf JH. Mercenary hirelings or a great blessing?: doctors' and mothers' conflicted perceptions of wet nurses and the ramifications for infant feeding in Chicago, 1871-1961. J Soc Hist. 1999;33(1): 97-120.

✓ Concept Check

1. What are the functions of the ovaries?
2. Compare the structures of the uterus and the vagina.
3. Describe the structure of the mammary glands.

DISCOVERY SCENE PLEASE ENTER DISCOVERY SCENE PLEASE ENTER

What information about the female reproductive system is helpful in understanding the CSI? What features of Edward's body caused the physicians to identify him as a female? How could these features have changed to make him develop into a male?

MALE REPRODUCTIVE SYSTEM

 Key Terms: penis, phallus, scrotum, seminal vessel

The male reproductive system facilitates sexual reproduction and the elimination of wastes from the kidneys (Figure 15.9). As in the female, it is composed of internal components and external genitalia. The internal components include the **seminal vessels**. Seminal vessels are composed of a network of tubes and associated glands that assist with the survival and transport of sperm. The external components are the **penis** (**phallus**), testes, and **scrotum**. The ability for a male to reproduce requires that the reproductive system is able to deposit an adequate supply of sperm near the female's cervix. Accordingly, the testes can normally produce hundreds of millions of sperm that are placed near the cervix by the penis during sexual activity.

Seminal Vessels A network of tubes and glands that assist with the transport of sperm

Penis or Phallus An external part of the urinary and reproductive systems of the male

Scrotum A pouch of skin that encloses the testes

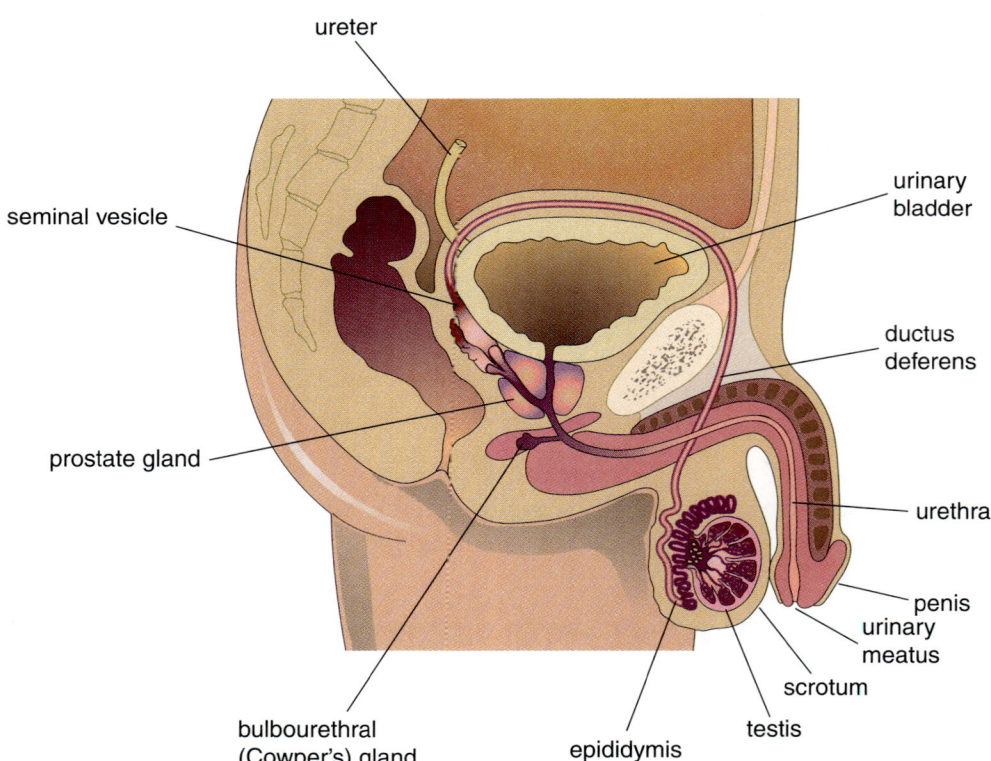

Figure 15.9 **Male Reproductive System**
The male reproductive system shares components with the urinary system

576

CHAPTER 15

Testes

Key Terms: cryptorchidism, epididymis, Leydig's cells, seminiferous tubules, undescended testis

Testes start out in the same location as the ovaries during fetal development, which is near the kidneys. They then descend through openings at the base of the pelvis and into the scrotum sometime just before birth. The scrotum is divided into two halves by a thin membrane. Each half holds one testis. Sometimes one or both testes do not descend. This condition, called **undescended testis**, or **cryptorchidism**, happens in about 4% of the males born in North America. In most of these cases, the testes usually descend by month 9. Surgical intervention is needed if they do not descend after that period. Compared with descended testes, undescended testes are more likely to produce less sperm and have a high probability of developing cancer. It is believed that testes require the cooler temperatures of the scrotum in order to produce viable sperm. Muscle contractions of the cremaster muscle then regulate the testes' temperature by adjusting the distance of the testes to the body. The testes become warmer when placed closer to the body and cooler when moved away.

The testes are covered with rigid connective tissue capsules. They protect the insides of the testes from being crushed by the physical forces associated with sexual activity and walking. A dense network of small, coiled tubes called the **seminiferous tubules** is surrounded by the capsules (Figure 15.10).

Each testis is believed to produce trillions of sperm in the lifetime of a male. **Leydig's cells** are interspersed between the seminiferous tubules and are responsible for producing testosterone. After the sperm are produced by meiosis, they move into a larger duct called the **epididymis**. The epididymis makes up the first portion of the seminal vessels. Abnormal sperm, and sperm that do not leave the epididymis, are broken down and absorbed.

Undescended Testis or Cryptorchidism A condition in which one or both testes do not pass into the scrotum

Seminiferous Tubules Tubes in the testes where sperm is produced

Leydig's Cells Cells that produce testosterone in the testis

Epididymis A tube where sperm are collected and stored after leaving the testis

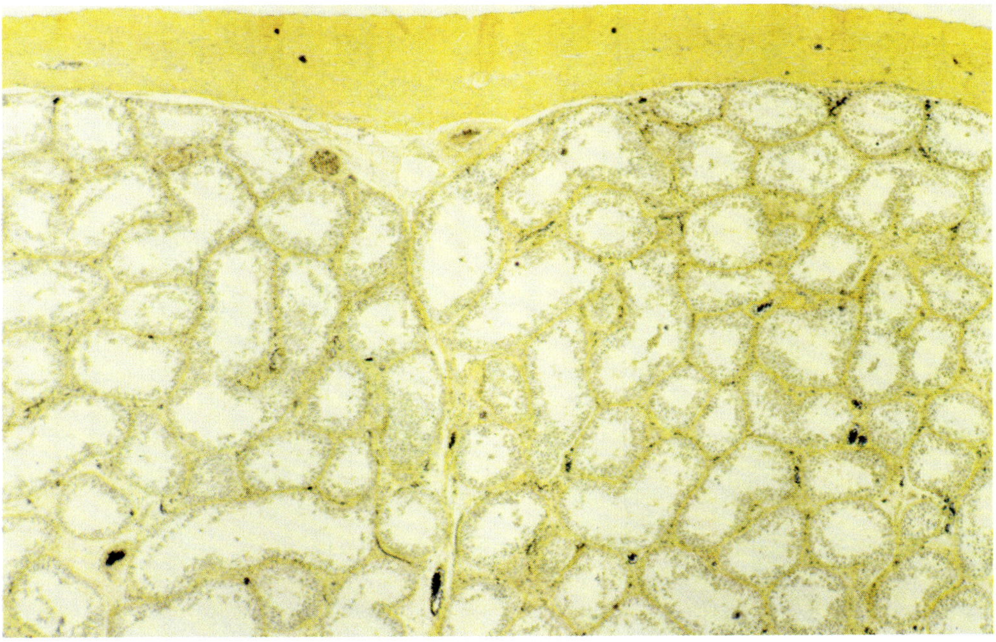

Figure 15.10 Testicular Tissue

Seminal Vessels

Key Terms: bulbourethral glands, Cowper's gland, ductus deferens, ejaculatory duct, prostate gland, semen, seminal vesicle, vas deferens

Vas Deferens or Ductus Deferens A thin tube that transports sperm from the testis to the urethra

Seminal Vesicles Glands that help produce semen

Semen A fluid containing sperm and seminal secretions

The epididymis covers the side portion of each testis (Figure 15.11). Fluids secreted by the epididymis nourish the sperm and permit them to mature until moving into the **vas deferens**, or **ductus deferens**. Sperm remain in the vas deferens until sexual stimulation causes them to be expelled. The vas deferens is a curved tube that runs up along the bladder and past glands called the **seminal vesicles**. Seminal vesicles are paired glands located posterior to the urinary bladder. They contribute to the production of a fluid called **semen**. Semen is composed of sperm and seminal secretions. The seminal vesicles produce the bulk of the semen. They secrete enzymes, fructose, hormones, lipids, and proteins that facilitate the survival and transport of sperm.

Figure 15.11 Seminal Vessels

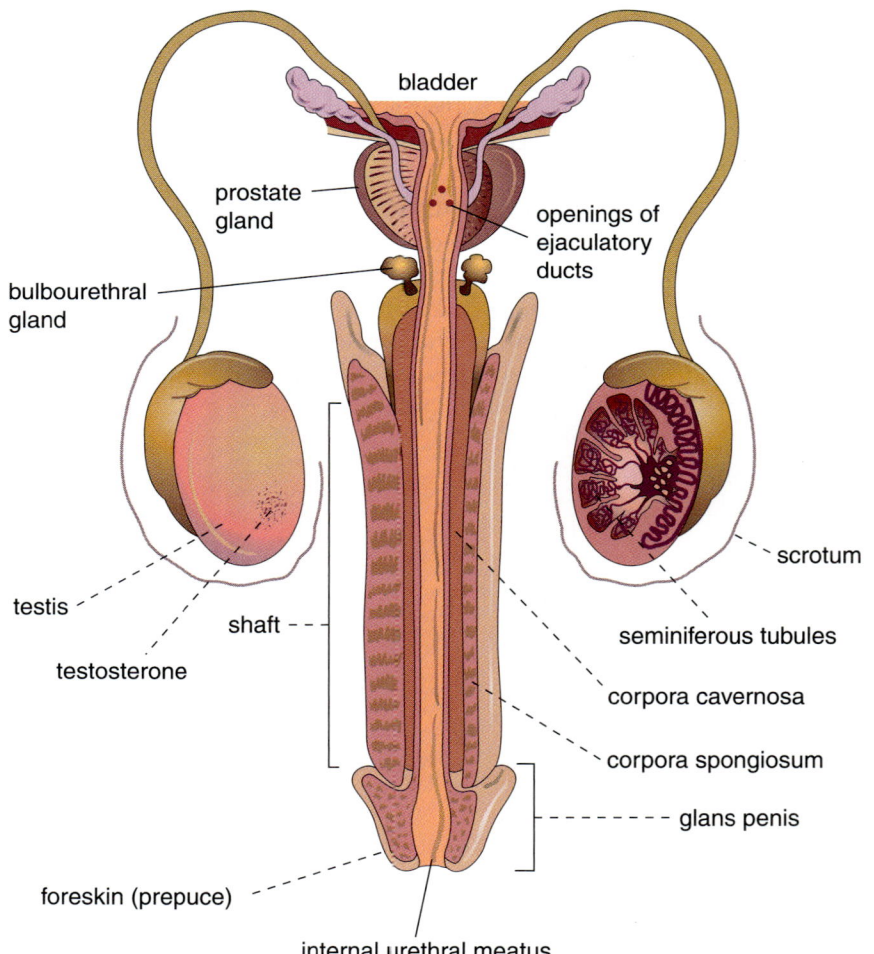

The semen passes into the **ejaculatory ducts**. These ducts come together and empty into the urethra. Surrounding the ejaculatory ducts and the base of the urethra is the **prostate gland**. The prostate gland secretes a muocus-like fluid into the semen. This secretion provides lubrication during sexual activity. Inflammation of the prostate due to various diseases can impede the passage of semen and urine, sometimes making it painful to urinate. The urethra passes by a pair of **Cowper's glands**, or **bulbourethral glands**, before entering the penis. Cowper's glands are a pair of pea-sized glands that lie beneath the prostate gland. They produce an alkaline fluid that neutralizes the acidic environment of the urethra. Sperm are disabled or killed by acidic conditions.

> **Ejaculatory Duct** A duct that opens into the urethra
> **Prostate Gland** A gland in the male that surrounds the base of the urethra
> **Cowper's or Bulbourethral Glands** A pair of glands that lie beneath the prostate gland

Penis

Key Terms: corpus cavernosum, corpus spongiosum, circumcision, dorsal vein, erection, foreskin, glans, prepuce

The penis is a tube of erectile tissue that serves as a passageway for semen and urine (Figure 15.12). A sheath of erectile connective tissue called the **corpus spongiosum** surrounds the urethra as it passes through the penis. The corpus spongiosum prevents the urethra from pinching closed and contributes to the male sexual response. At the distal portion of the corpus spongiosum is a large swelling called the **glans**. The glans is rich in nerves associated with sexual stimulation. Covering the glans is a cylinder of skin called the **prepuce**, or **foreskin**. Most scientists believe that the prepuce protects the glans from too much stimulation when the male is not sexually excited. However, many physicians are discovering that the prepuce can facilitate the spread of sexually transmitted diseases if not cleaned adequately. **Circumcision** is a surgical procedure in which the prepuce is removed.

> **Corpus Spongiosum** A sheath of erectile tissue in the penis that encloses the urethra
> **Glans** The swollen portion at the tip of the penis
> **Prepuce or Foreskin** A roll of skin that covers the glans of the penis
> **Circumcision** Surgical removal of the prepuce

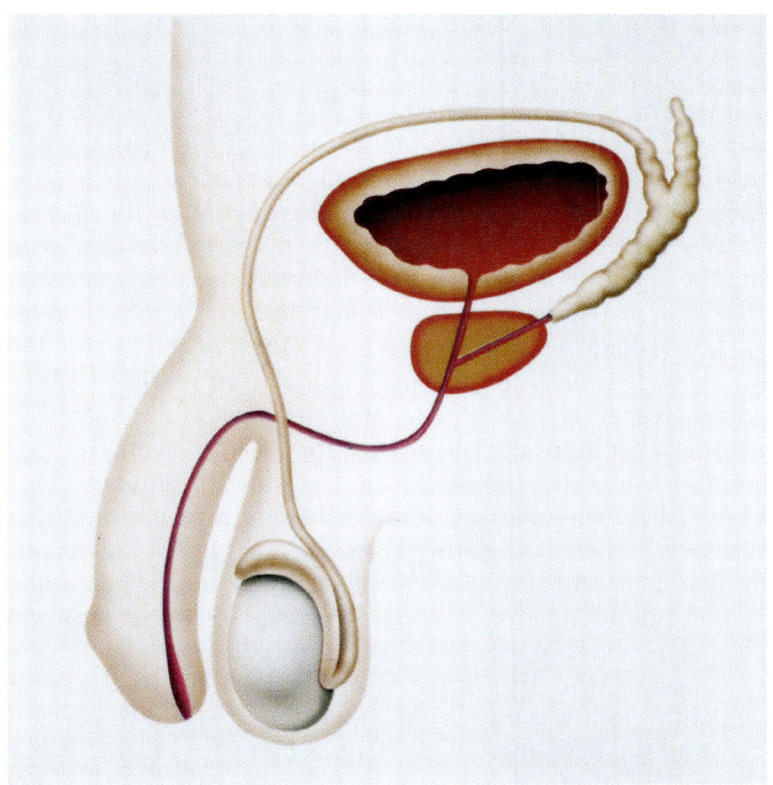

Figure 15.12 Penis

Corpus Cavernosum A large cylinder of erectile tissue in the penis

Dorsal Vein A large vein that runs along the dorsal length of the penis

Erection Enlargement and hardening of the penis during sexual excitement

Two large chambers of erectile tissue called the **corpus cavernosum** run along the dorsal surface of the penis. They originate at the base of the penis and end just behind the glans. The corpus cavernosum engorge with blood and make the penis erect during sexual excitement. Running between the corpus cavernosum, just underneath the skin, is the large **dorsal vein**. The male **erection** is achieved when blood fills the three cylinders of erectile tissue in the penis; the corpus spongiosum and the two cylinders of the corpus cavernosum.

✓ Concept Check

1. Describe the structure and function of the testes.
2. What are the components of the seminal vessels?
3. Explain the nature of the erectile tissue in the penis.

DISCOVERY SCENE PLEASE ENTER DISCOVERY SCENE PLEASE ENTER

How does an understanding of the male reproductive system contribute to the CSI? What male characteristics need to be present if Edward is truly a male? What changes could have occurred during puberty that would have produced male secondary sex characteristics in Edward?

Basics of Sexual Reproduction

Female Sexual Cycle

 Key Terms: follicular phase, luteal phase, menses, ovarian cycle, postovulation phase, preovulation phase, proliferation phase, uterine cycle

Ovarian Cycle The sequence of events that lead to ovulation

Uterine Cycle The sequence of events that prepare the uterus for pregnancy

Preovulation or Follicular Phase The ovarian cycle of events that take place before ovulation

Postovulation or Luteal Phase The ovarian cycle of events that take place after ovulation

The female sexual cycle, or menstrual cycle, is a series of events that prepare the body for pregnancy. This cycle is sometimes divided into the **ovarian** and **uterine cycles**. The ovarian cycle is the sequence of events that lead to ovulation. Events that prepare the uterus for pregnancy make up the uterine cycle. As discussed earlier in the chapter, each ovary carries all of the eggs a female would ever use in a lifetime. The ovarian cycle prepares one egg for pregnancy by maturing the egg and passing it out of the ovary. In some women, it is possible for both ovaries to undergo the cycle, simultaneously releasing an egg from each ovary. This situation produces fraternal twins if both eggs are fertilized.

The ovarian cycle is divided into two phases: **preovulation** (**follicular**) and **postovulation** (**luteal**) (Figure 15.13). In the preovulation phase, the follicle secretes estrogen that helps the oocyte mature. The estrogen also stimulates hormone production by the hypothalamus and anterior pituitary. Most noted is an increase in follicle-stimulating hormone and luteinizing hormone from the anterior pituitary. Follicle-stimulating hormone and luteinizing hormone stimulate the production of the corpus luteum. The levels of estrogen, follicle-stimulating hormone, and luteinizing hormone continue to elevate for approximately 7 to 9 days. The levels of these hormones then drop rapidly in response to rising progesterone levels. This progesterone is produced by the corpus luteum. Ovulation occurs at this point, about the day 14 of the cycle, causing the release of the egg from the follicle (Figure 15.14).

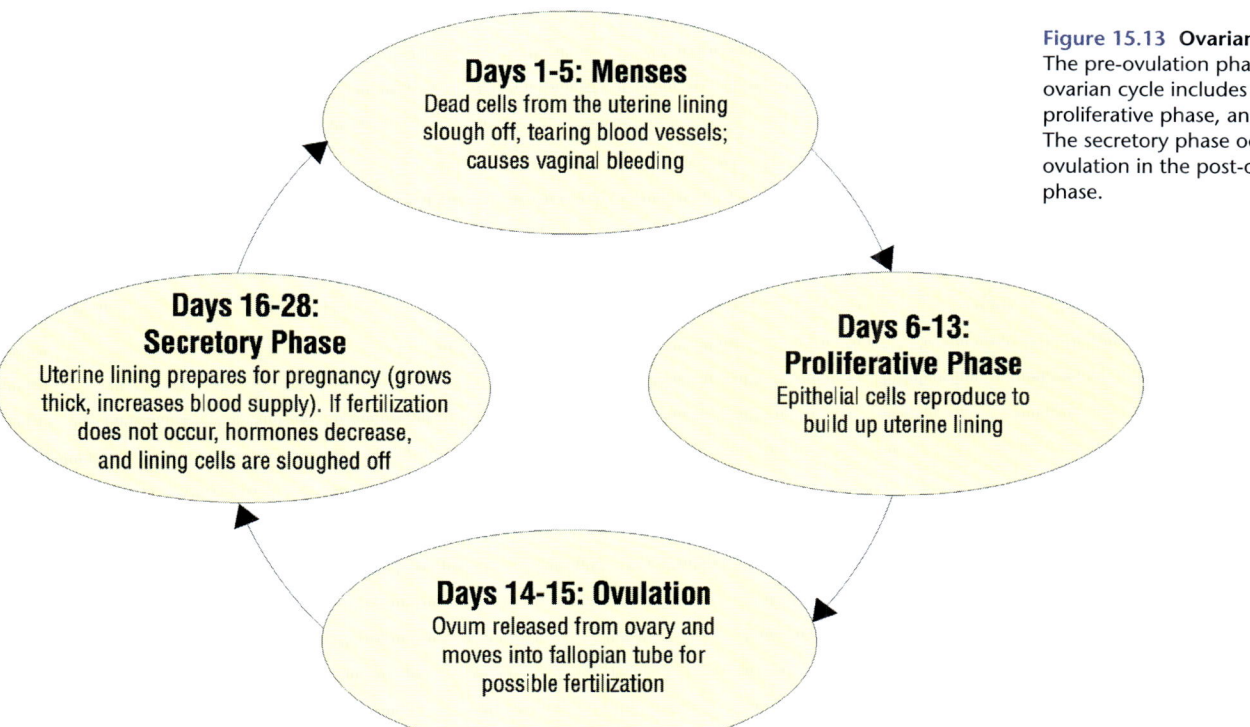

Figure 15.13 **Ovarian Cycle**
The pre-ovulation phase of the ovarian cycle includes the menses, proliferative phase, and ovulation. The secretory phase occurs after ovulation in the post-ovulation phase.

The corpus luteum now takes over in the postovulation phase. It secretes both estrogen and progesterone after ovulation. The corpus luteum will eventually degenerate if the female does not become pregnant. If pregnancy occurs, the corpus luteum is kept intact until the baby is born. The intact corpus luteum suspends the completion of the cycle, preventing any further menstrual cycles during pregnancy. Hormone fluctuations of the ovarian cycle control the events occurring in the uterine cycle. Most birth control pills interfere with the transition from the preovulation to postovulation phase. This, in turn, prevents ovulation.

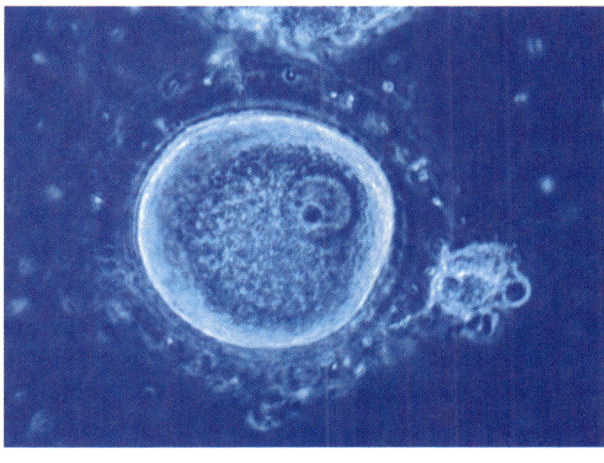

Figure 15.14 **Ovum**
The ovum is released into the fallopian tube from the ovarian follicle during ovulation

Elevated estrogen levels in the preovulation phase initiate the **proliferative phase** of the uterine cycle. During the proliferative phase, the cells of the endometrium undergo rapid divisions, thickening the uterine lining. Blood flow to the uterus also increases. Proliferation slows after ovulation. The endometrium slowly reaches its maximum thickness 8 to 10 days after ovulation. Estrogen and progesterone from the corpus luteum maintain the thickened endometrium. If pregnancy does not occur, estrogen and progesterone levels drop, causing the degeneration of the thickened endometrium. This begins a sequence of events called **menses**. Most physicians find that 28 days is the average length of a menstrual cycle. However, it can range from 23 to 35 days

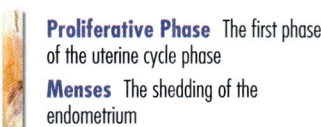

Proliferative Phase The first phase of the uterine cycle phase
Menses The shedding of the endometrium

THE REPRODUCTIVE SYSTEMS AND HUMAN DEVELOPMENT

in different women. The timing of the menstrual cycle is affected by nutrition, stress, and other health factors. Many studies show that the menstrual cycles of certain women will synchronize if they remain in proximity for long periods of time. This synchronization is due to pheromones secreted in the vaginal tract. The pheromones influence the timing of the cycle.

During menses, the proliferated cells die and shed from the lowest layer of the endometrium. Menses will generally overlap 1 to 2 days into the next pre-ovulatory phase. Blood vessels and cells of the endometrium break down into a red liquid resembling blood. This is what makes up the menstrual flow. A significant amount of iron is lost during menses, making women more susceptible to iron-deficiency disorders, such as anemia. Pregnancy will maintain the thickened endometrium throughout the development of the fetus. The "morning-after pill" and RU486 interfere with the proliferation and maintenance of the uterine lining. This cuts off the uterine cycle needed to carry out the pregnancy.

The blood-like appearance of the menstrual flow has led to many myths and superstitions about the menstrual cycle. Women in many cultures were considered unclean or unholy during their menses. In some cultures, women undergoing menses were not allowed in public until the flow stopped. The word taboo may come from the Polynesian word for unclean, which was used to describe menstruating women. Many cultures used fabric and leaves to keep the menstrual flow contained when in public. American military nurses during World War II developed the idea for tampons by using bandage gauze to reduce soiling of their uniforms by the menstrual flow.

Copulation

Key Terms: copulation, detumescence, ejaculation, erectile dysfunction, orgasm, sexual intercourse, urethral meatus

Copulation Or Sexual Intercourse
The act of mating

Erectile Dysfunction The inability to produce or maintain an erection

Urethral Meatus The external opening of the urethra

Copulation, or **sexual intercourse**, is the act of mating. The intent of copulation for reproductive purposes is to ensure fertilization of an egg. Fertilization is defined as the union of a sperm with an egg. Copulation begins with sexual arousal of the male and female reproductive systems. Sexual arousal in males stimulates the nerves of the autonomic nervous system, causing the veins to constrict and the arteries in the penis to dilate. This action engorges the erectile tissues of the penis with blood, causing an erection. An erection is required for the successful placement of sperm deep into the vagina. Inhibition of the parasympathetic nerve due to stress or neurological disorders inhibits the formation or maintenance of an erection. An inability to produce or maintain an erection is called **erectile dysfunction**. There are many causes of erectile dysfunction. Treatments vary greatly depending on the cause. Medications used to enhance erections are mistakenly thought to increase male arousal or sex drive. In actuality, these drugs encourage the vascular changes needed for an erection.

The act of rubbing the glans of the penis against the walls of the vagina produces spasms in the vas deferens. These spasms force the movement of sperm into the ejaculatory duct and stimulate Cowper's glands, prostate gland, and seminal vesicles to release their secretions. Fluids from Cowper's glands are released into the urethra before the sperm arrive to cleanse it of acid urine residues. During this process, sphincters at the base of the urinary bladder close tight to prevent urine from entering the urethra. The flowing semen is pushed into the urethra and out of the **urethral meatus**. This process is called

ejaculation. Autonomic neural impulses facilitate the contractions needed for ejaculation. It is estimated that 400 million sperm are ejaculated in an average intercourse. The sexual excitement associated with ejaculation is called an **orgasm**. Orgasm in males is associated with ejaculation. After ejaculation, the penis undergoes **detumescence**. During detumescence, the penis returns to its original size as blood flows out of the erectile tissue. A relaxation time of minutes to hours is often needed before a male can orgasm again.

Swelling of the erectile tissue of the clitoris initiates the female sexual response. Blood also rushes to the mammary glands causing them to swell and the nipples to become erect. Sometimes the skin of the face, chest, and neck will redden as blood flow to the skin increases. Nerves of the clitoris, nipples, skin, and vagina become more sensitive to touch as sexual excitement progresses in the female. At the same time, there is an increase in mucous production by Skene's glands and glands of the vagina. Stimulation of the clitoris induces muscle contractions of the vaginal tract. These contractions are one component of the female orgasm. Contractions of the vaginal tract are necessary to bring sperm to the egg. It is believed that contractions of the cervix pump the sperm into the base of the uterus. The completion of the female sexual response can follow one or more orgasms.

> **Ejaculation** The process of ejecting semen from the penis
> **Orgasm** An intense sensation that occurs at the height of sexual excitement
> **Detumescence** The loss of an erection

Embryology and Pregnancy

Key Terms: acrosome, amniotic fluid, amniotic sac, blastocyst, blastula, colostrum, conjoined twins, conception, embryogenesis, fraternal twins, gastrula, human chorionic gonadotropin (hCG), identical twins, implantation, labor, placenta

Fertilization of the egg occurs at the location where the egg enters the fallopian tube. Some people consider fertilization the point of human **conception**. Others interpret conception as a particular point of the embryo's development. Only a few hundred of the original 400 million sperm survive

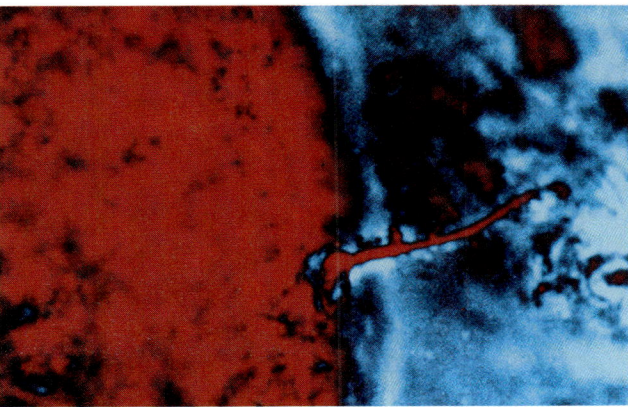

Figure 15.15 **Fertilization**
Fertilization occurs when the sperm penetrates the egg

the trip to the egg. The sperm can survive up to 48 hours in the female's reproductive tract. Sperm can swim several millimeters per second. Their progress is assisted by contractions of the uterus and fallopian tubes. Many sperm are destroyed in the female reproductive tract. The sperm die because of immune attack by the female reproductive tract and due to the acidic conditions of the mucous. Defective sperm end up decaying in the vagina or lower portions of the uterus.

Fertilization begins when the head of a sperm binds to the outer coating of the egg (Figure 15.15). The sperm then uses enzymes in a structure called the **acrosome** to create a small opening into the egg's covering. Fusion of the sperm and egg membranes permits the sperm's DNA to enter the egg. The remaining portions of the sperm are discarded. Fertilization stimulates the egg to finish meiosis, which is followed by rapid periods of mitosis and develop-

> **Conception** The point at which fertilization occurs
> **Acrosome** A packet of enzymes in a sperm's head

Embryogenesis The process of embryo formation

Fraternal Twins Twins produced by the simultaneous fertilization of two egg cells

Blastula or Blastocyst A hollow sphere of cells formed by repeated mitosis of the zygote

Implantation Attachment of the embryo to the endometrial lining

Placenta An organ that nourishes the developing fetus in the uterus

Pregnancy The condition in which an embryo is developing within the uterus

Human Chorionic Gonadotropin (hCG) A hormone produced by the placenta

ment called **embryogenesis**. The fertilization of two eggs by two sperm produces **fraternal twins**. Fraternal twins have the similarities other siblings share, though they are not usually identical in appearance.

Embryogenesis begins when the fertilized egg, now called a zygote, undergoes its first mitosis. Mitosis occurs rapidly in the zygote as it travels down the fallopian tube to the uterus (Figure 15.16). A zygote undergoes mitosis about every 7 hours. The typical body cell takes about 22 hours to complete mitosis. Each round of mitosis decreases the sizes of the cells by half. This occurs until the cells reach the average size of a typical human cell. As the embryo develops, it reaches a stage called the **blastula**, or **blastocyst**, just as it enters the fundus of the uterus 7 days after fertilization. The blastula is a hollow ball of cells that imbeds in the thickened uterine lining. Imbedding of the blastula into the endometrium is called **implantation**. It takes about 1 week for the fertilized egg to travel through the fallopian tube to the uterine lining. Implantation initiates the formation of the **placenta**, which is needed to nourish the remaining stages of embryogenesis (Figure 15.17). **Pregnancy** begins after successful formation of the placenta. The early placenta produces a variety of secretions that ensure that the mother's body will meet the needs of the fetus. **Human chorionic gonadotropin** (**hCG**) is a protein hormone that is produced by the placenta. The hormone maintains pregnancy by triggering the release of estrogen and progesterone. This hormone is also a useful indicator of pregnancy. It is present in the blood and urine in detectable amounts 10 days after fertilization.

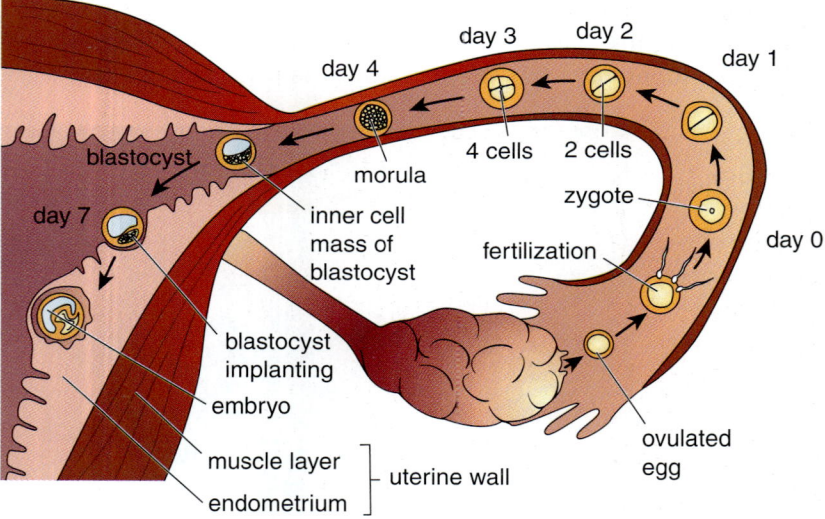

Figure 15.16 **Pregnancy**
The fertilized eggs travels down the fallopian tube for implantation in the uterus

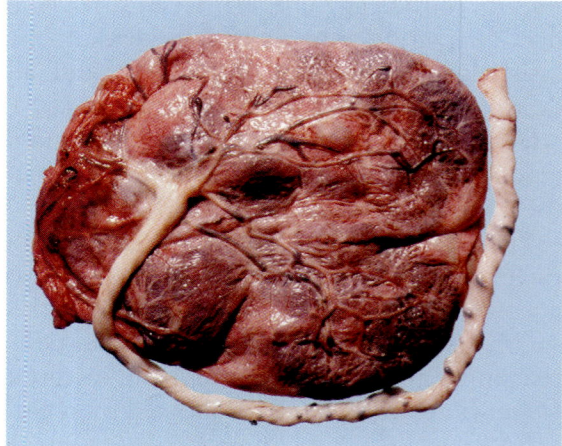

Figure 15.17 **Placenta**
The placenta is formed after implantation of the fertilized egg and nourishes the developing fetus

Once implanted, the blastula develops into a **gastrula**. It is at the gastrula stage that the ectoderm, mesoderm, and endoderm tissue germ layers form. **Identical twins** usually form at the gastrula stage. They are produced when the gastrula from one fertilized egg divides into two gastrulas. Identical twins share one placenta and start out with identical genetic material. **Conjoined twins** occur when the gastrula fails to split completely. This leaves the babies joined together in a variety of ways. Some are almost completely joined so that two heads share one body.

Further development of the gastrula produces the **fetus** (Figure 15.18). During fetal development, all of the major organ systems form from the three germ layers. The **amniotic sac**, or bag of waters, surrounds the fetus. Within the amniotic sac, the baby is cushioned by the **amniotic fluid**. The rate and degree of organ system development depends primarily on the mother's nutritional level. Alcohol, drugs, infectious diseases, and a variety of pollutants can alter the fate of fetal development. Genetic disorders can also produce a variety of alterations in fetal development, producing birth defects.

Gastrula An embryonic stage in which the three basic tissue germ layers form

Identical Twins Twins that develop from one fertilized egg

Conjoined Twins Twins whose bodies are joined together at birth

Fetus A stage of development before birth

Amniotic Sac A fluid-filled sac that surrounds the fetus

Amniotic Fluid Fluid within the amniotic sac

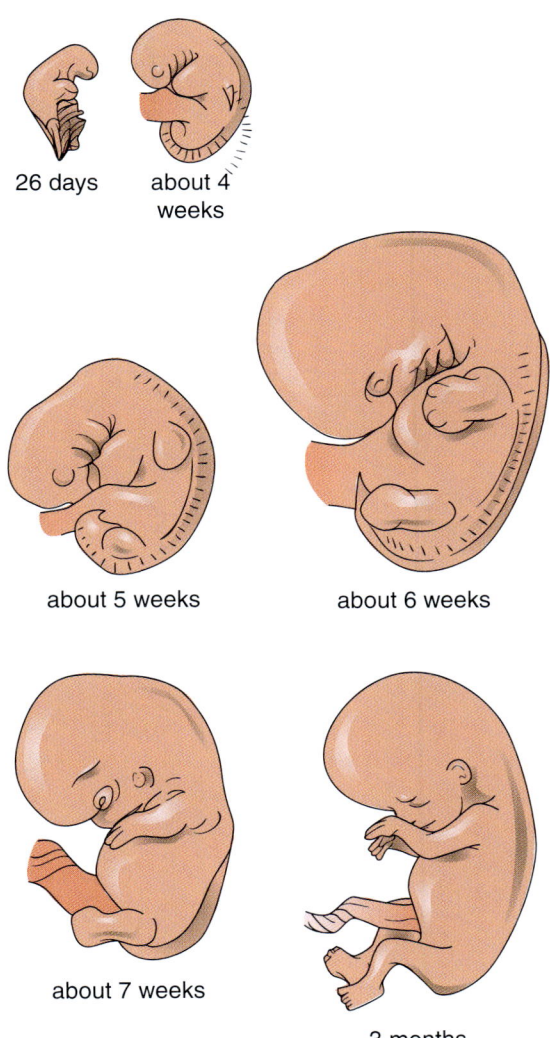

Figure 15.18 **Human Embryology** Further development of the gastrula leads to the formation of the fetus

THE REPRODUCTIVE SYSTEMS AND HUMAN DEVELOPMENT

The internal components of both genders are present by week 10 after fertilization (Figure 15.19). Development of sexual dimorphism begins 13 weeks after fertilization and continues until birth. The respective sex hormones secreted by the fetus' gonad maintain sexual dimorphism. Higher-than-usual levels of estrogen or testosterone in the mother's body could alter the development of sexual dimorphism and produce a variety of birth defects. Further sexual dimorphism changes occur after birth and during puberty in both males (Figure 15.20) and females (Figure 15.21).

Figure 15.19 Reproductive Development
The internal components of both genders are present at 10 weeks after fertilization

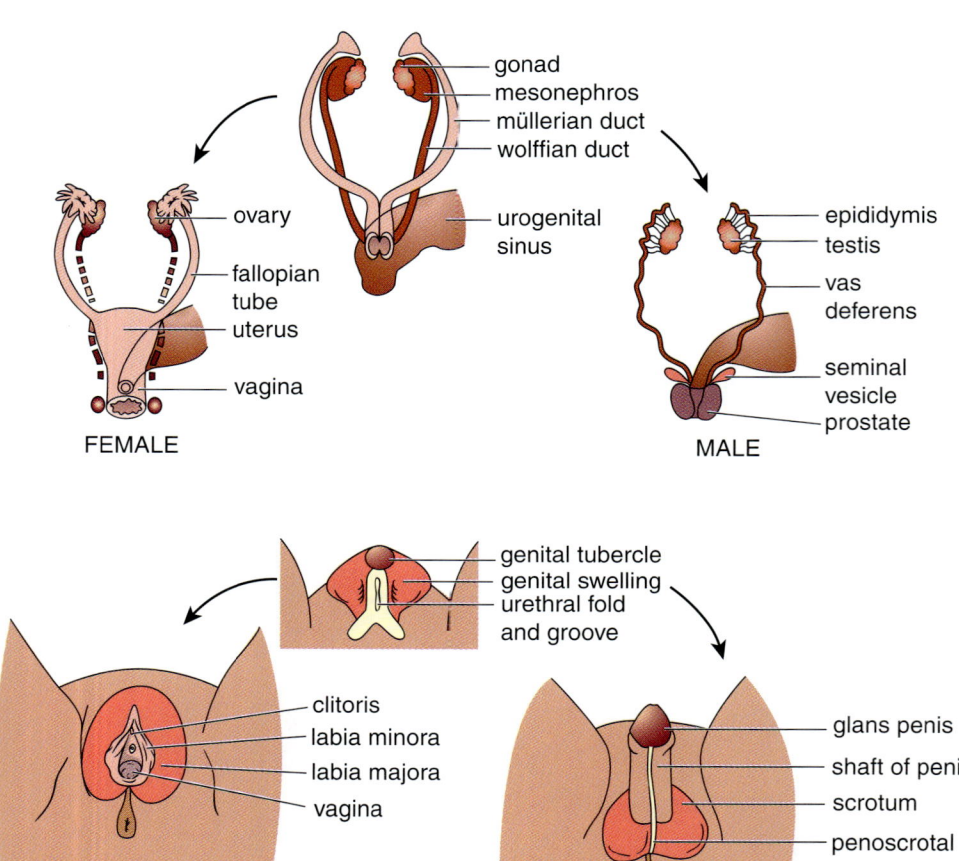

Genital Stage	Pubic Hair	Scrotum/Penis	Age at Onset (mean ±SD)
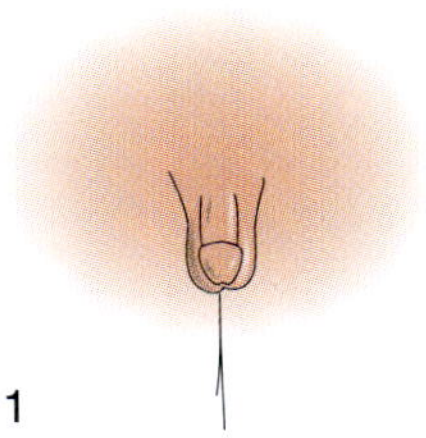 1	no coarse hair	prepubertal	
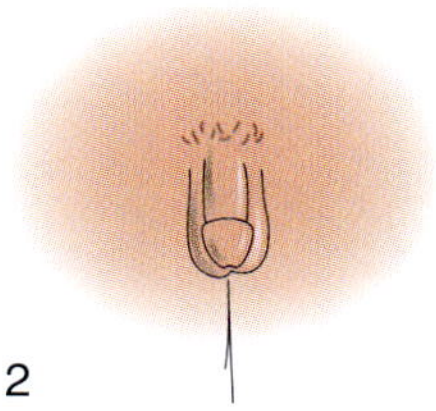 2	longer, silky hair appears at base of penis	scrotum and testes enlarge; skin of scrotum reddens and rugations appear	1.4 ± 1.1 yr
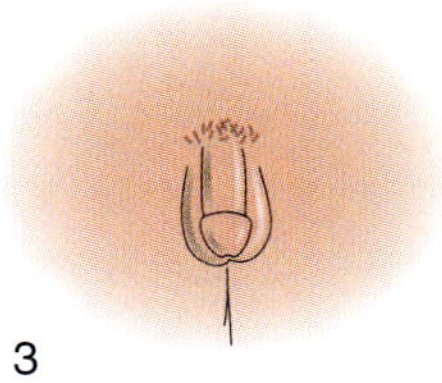 3	hair coarse, kinky, spreads over pubic bone	penis lengthens; testes enlarge further	12.9 ± 1 yr
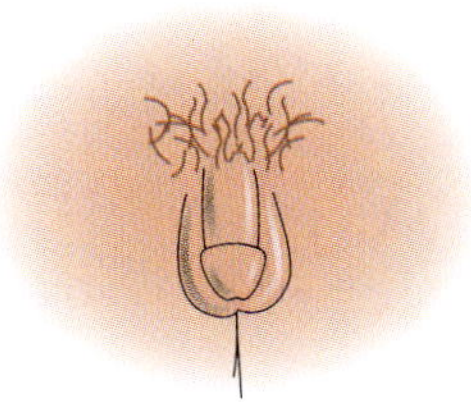 4	hair of adult quality but not spread to junction of medial thigh with perineum	penis growth continues in length and width; glans develops adult form	13.8 ± 1 yr
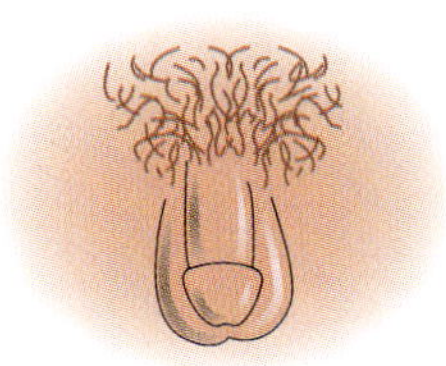 5	hair spreads to medial thigh	development completed; adult appearance	14.9 ± 1.1 yr

Figure 15.20 Tanner Staging – Male Development

THE REPRODUCTIVE SYSTEMS AND HUMAN DEVELOPMENT

Figure 15.21a Tanner Staging – Female Breast
The female breast develops slowly until puberty when it rapidly develops into the mature breast

Prepubertal
small nipple present

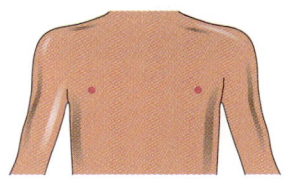

Breast bud stage
small amount (bud) of breast tissue and nipple develops

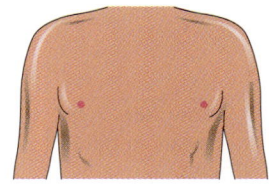

Elevation of breast contour
breast tissue growth continues; areolae enlarge

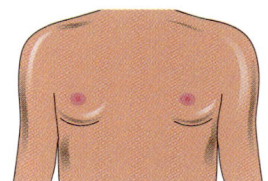

Development of secondary mound
areolae and papilla form a secondary mound

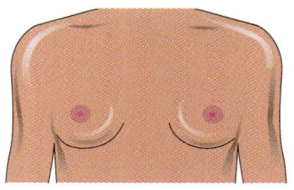

Adult form
development is complete; areola is flush with breast

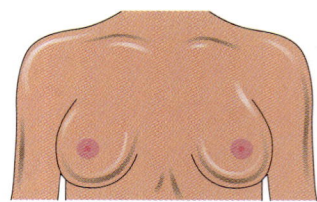

(a)

Figure 15.21b Tanner Staging – Pubic Hair Development

Prepubertal
no corse hair in pubic area

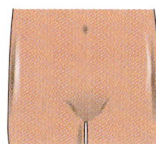

Development of silky hair
long, silky hair appears along labia

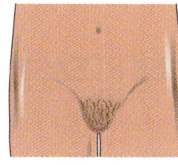

Spread of hair over mons pubis
coarse, kinky hair spreads over pubic bone

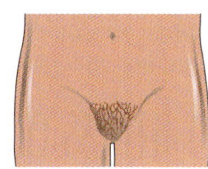

Devolopment of adult hair
hair of adult quality but has not spread to junction of medial thigh with perineum

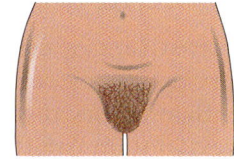

Spread of hair to thigh
hair development complete; hair also exists on medial thigh

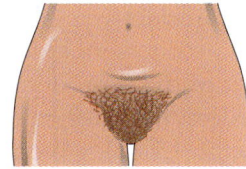

(b)

The fetus is almost fully formed by the end of month 8 of pregnancy. The lungs and parts of the digestive system usually remain incomplete. The beginning of childbirth is commonly called **labor**. Physicians are not completely sure about the factors that bring about the onset of labor. It is most likely due to hormone interactions between the fetus and the mother's pituitary gland. The first stage of labor is usually the dilation of the cervix. This is typically followed by powerful contractions of the myometrium. The amniotic sac ultimately ruptures, and the amniotic fluid flows out of the uterus through the vagina. Further contractions then push the baby past the cervix and through the vagina. The pituitary hormone oxytocin is important in maintaining the muscle contractions needed for labor. Shortly after the baby leaves the uterus, the placenta loosens its attachment to the endometrium. It can be expelled naturally after a few minutes to several hours after the baby is delivered. Right after birth, the mammary glands secrete a special fluid, called **colostrum**. Colostrum is higher in calories and nutrients than the regular milk produced a few days after labor.

Labor The beginning of childbirth
Colostrum A fluid produced by the mammary glands in late pregnancy and just after labor

THE REPRODUCTIVE SYSTEMS AND HUMAN DEVELOPMENT

✓ Concept Check

1. Describe the stages of the menstrual cycle.
2. Distinguish between the female and male sexual response.
3. Briefly describe the sequence of human embryogenesis and birth.

DISCOVERY SCENE PLEASE ENTER DISCOVERY SCENE PLEASE ENTER

Is it possible for Edward to have a menstrual cycle? What factors could prevent him from doing so? Is it possible that he could become pregnant? What effects would his outward male appearance play in the outcome of a pregnancy? Could something have happened during embryogenesis that caused him to have conflicting gender identity?

WELLNESS AND ILLNESS OVER THE LIFE SPAN

Pathology of the Reproductive Systems

Key Terms: breast cancer, cesarean section, cervical cancer, ectopic pregnancy, fibroids, genital warts, hypospadias, pelvic inflammatory disease (PID), placenta previa, prostate cancer, sexually transmitted diseases (STDs), testicular cancer

Hypospadias Abnormal development of the penis and male urethra

Sexually Transmitted Disease (STD) An infectious disease spread by sexual contact

Diseases of the reproductive system can be congenital, infectious, or degenerative. Congenital disorders include a variety of conditions that prevent the normal development of the reproductive system. Males and female have specific congenital disorders. In males, many of these conditions lead to abnormalities of the testes (Figure 15.22), which could result in the formation of defective sperm. Male fertility could be compromised if greater than 25% of the sperm are dead or defective. Other congenital conditions of the male affect the formation of the penis. There is evidence that certain pollutants that have a chemical structure similar to that of estrogen could impair penis formation in the fetus of a contaminated woman. **Hypospadias** is a birth defect in which the urethra and penis are not properly formed. Sometimes the urethra develops as an open slit in which urine and semen exit the base of the penis near the scrotum. This makes urination and sexual reproduction very difficult. In addition, it makes the man susceptible to urinary tract infections. Female are born with similar congenital disorders that alter the formation of the ovary or disfigure the shape of the reproductive tract. Other effects in females include deformation of the cervix and vagina.

Sexually transmitted diseases (**STDs**) are the most common infectious diseases of the reproductive system. They are predominantly spread through copulation; however, oral and anal sex are other means of spreading STDs. Arthropods, bacteria, fungi, protista, and viruses cause a variety of diseases spread through sexual contact. Health professionals are encouraged to report to public health officials any incidence of an STD diagnosed in a patient. Certain STDs must be reported to the Centers for Disease Control and Prevention (CDC) and the National Institutes of Health (NIH) The CDC's Web site has detailed information about the types and frequencies of STDs in the United States. They also track STDs that make their way into the United States from other countries. The United Nations World Health Organization

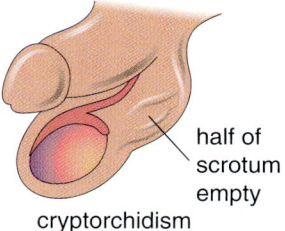
cryptorchidism — half of scrotum empty

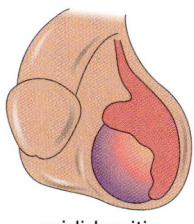
epididymitis

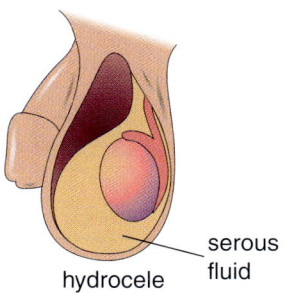
hydrocele — serous fluid

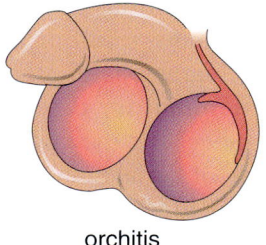
orchitis

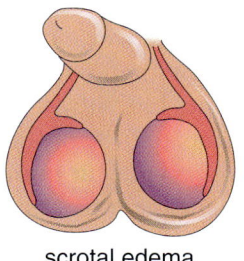
scrotal edema

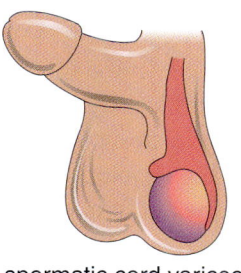
small testis

spermatic cord varicocele

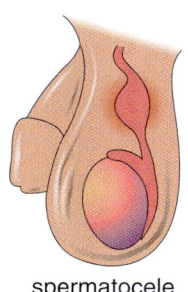
spermatocele

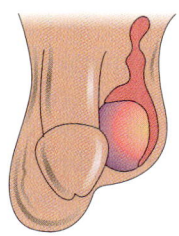
testicular torsion

Figure 15.22 **Pathology of the Reproductive System**
Various abnormalities of the testes occur during development of the male reproductive system

THE REPRODUCTIVE SYSTEMS AND HUMAN DEVELOPMENT

Cutting Edge Research
CONTROLLING BREAST CANCER WITH PREGNANCY

Many women would prefer to avoid the inconveniences and stress that pregnancy places on the body. However, recent studies show that pregnancy may have some benefits for the body. A protein called alpha-fetoprotein (AFP), which is produced during pregnancy, may inhibit the development of breast cancer. This finding came about after researchers discovered that the incidence of breast cancer is less in women who have had at least one full-term pregnancy. Women normally have a 13% chance of developing breast cancer; however, the chance is reduced to 7% in women who had children. Scientists surmised that AFP may play a role in reducing the incidence breast cancer. A team of researchers developed an artificial form of AFP called AFPep. The term AFPep is used to distinguish the artificial form of the protein from the naturally occurring AFP. Using mice in their studies, they compared AFPep with a drug therapy called tamoxifen, which is currently used to treat breast cancer. The results showed AFPep to be equally effective at reducing the incidence of breast cancer as tamoxifen. In addition, the cancer cells did not lose sensitivity to AFPep; cancer cells ultimately become resistant to tamoxifen after prolonged treatments. Another benefit of AFPep is that it is not nearly as toxic as other chemotherapy treatments. It is believed that during pregnancy, AFP plays a role in estrogen regulation. Most breast cancers are initiated by high levels of estrogen.

DeFreest LA, Mesfin FB, Joseph L, et al. Synthetic peptide derived from alpha-fetoprotein inhibits growth of human breast cancer: investigation of the pharmacophore and synthesis optimization. *J Pept Res.* 2004;63(5):409-419.

Pelvic Inflammatory Disease (PID) An infection of the upper female reproductive tract

Prostate Cancer Cancer of the prostate gland

Testicular Cancer Cancer of the testes

Breast Cancer Cancer of the mammary glands

Cervical Cancer Cancer of the cervix

(WHO) monitors STDs worldwide. Sexually transmitted diseases can range in severity from mild to fatal conditions. Mild STDs will cause slight inflammation or irritation of the reproductive system. Severe conditions can cause infertility or death. Many STDs spread to other organ systems, resulting in neural damage or immune system failure. Women with untreated bacterial STDs could develop a serious condition called **pelvic inflammatory disease** (**PID**). It causes inflammation of the uterus and fallopian tubes, and may even spread to the ovaries and peritoneum. It is a common cause of female infertility.

Degenerative disorders of the reproductive tract include cancers and structural changes to particular tissues or organs. **Prostate cancer** is a slow-growing, malignant growth in the prostate gland. Its cause is unknown, and it can spread to other tissues if left untreated. Most physicians believe it is stimulated by testosterone. The risk for developing prostate cancer increases with age. Approximately 70% of prostate cancers develop in men over age 65 years. Large cancers can restrict the flow of semen and urine. In contrast, **testicular cancer** is more common in younger males. It accounts for 1% of the cancers that occur in males. About 75% of testicular cancer cases occur in men between ages 20 and 49 years. If untreated, testicular cancer can be fatal if it spreads to other body organs. Modern methods of early detection make it possible to treat testicular cancer before it spreads.

Female reproductive system cancers include **breast** and **cervical cancer**. Breast cancers are usually due to a genetic condition that stimulates the growth of cancerous tumors in the glandular tissue of mammary glands. Estrogen stimulates these tumors, and they may be initiated by birth control pills or estrogen supplements. The tumors can be localized to one lobe or present in several lobes of one or both breasts. If untreated, they will spread into the blood and

CHAPTER 15

lymph nodes, producing a fatal condition. Most cases of cervical cancer are now accepted to be STDs caused by the human papillomavirus (HPV). This disease will damage the cervix and can invade the lower parts of the uterus. If untreated, cervical cancer can spread throughout the body and cause death. Current studies show that many males carry HPV on their penises. It causes an STD in males and females called **genital warts**. Females also develop genital warts in the reproductive tract. **Fibroids** are noncancerous tumors that form in the myometrium (Figure 15.23). They are categorized by their location within or upon the uterus. Some fibroids can grow as large as a grapefruit and cause severe cramping. Fibroids are known to prevent implantation of the embryo and are a cause of infertility in some females. Most fibroids do not have to be treated, though large ones must be surgically removed.

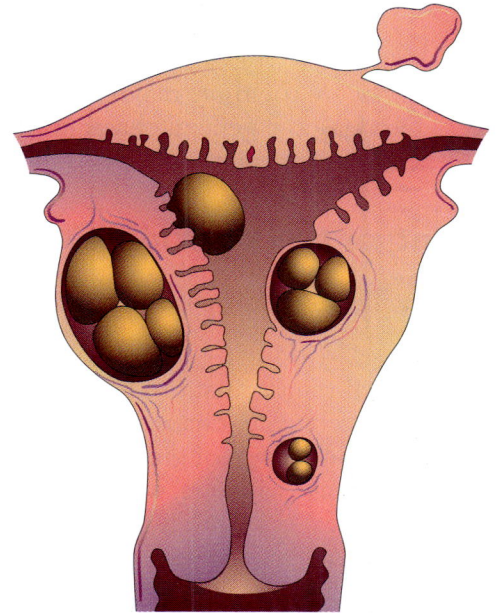

Figure 15.23 Fibroids
Fibroids are non-cancerous tumors that form in various areas of the uterus

Genital Warts Warts on the penis or female reproductive tract caused by the human papillomavirus

Fibroids Tumors of the myometrium

Ectopic Pregnancy Implantation that occurs outside the upper uterine cavity

Placenta Previa The location of the placenta in the lower part of the uterus

Cesarean Section Delivery of a fetus by surgical incision through the abdominal wall

Certain female reproductive disorders affect the placement of the placenta upon implantation of the embryo. **Ectopic pregnancies** occur when implantation takes place outside the upper uterine cavity. Abnormalities of the fallopian tubes can cause implantation within the abdominal cavity or on the walls of the fallopian tubes. A placenta will still form after implantation. However, a placenta that forms in the abdominal cavity may not provide the environment needed for fetal development. Fetal development in the fallopian tube will damage the fetus and rupture the walls of the fallopian tube. Sometimes the implantation occurs too low in the uterus, causing the placenta to cover the cervix. This produces a condition called **placenta previa** (Figure 15.24). Placenta previa does not affect the growth and development of the fetus; however, the placenta can block the passage of the baby out of the uterus. Placental detachment can also cause problems with placenta previa if blood vessels from the vagina feed into the placenta. Vaginal blood vessels can rupture and hemorrhage after the placenta is shed from the uterus after labor. Placenta previa babies are usually delivered surgically through the abdominal cavity using a procedure called a **cesarean section**.

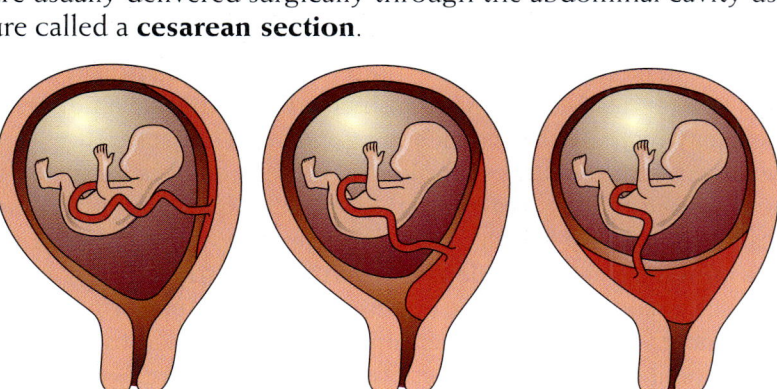

normal placenta   partial placenta previa   total placenta previa

Figure 15.24 Placenta Previa
Placentia previa can cause partial or total obstruction of the cervix

✓ Concept Check

1. What are the three major categories of reproductive system diseases?
2. Describe the general characteristics of sexually transmitted diseases.
3. Explain the common types of human reproductive system cancers.

Aging of the Reproductive Systems

Key Terms: andropause, menopause, prolapse, vesicoureteral reflux

Age-related transformations of the reproductive systems of men and women are mainly related to changes in hormone production and urinary system maintenance. Blood levels of growth hormone, thyroid hormones, and sex hormones decrease with age, which affects the maintenance of the secondary sex characteristics. In addition, it causes a slight atrophy of the reproductive organs. In females, the vaginal tract shrinks somewhat and reduces mucous production. This can make sexual intercourse more difficult. In males, the prostate gland enlarges, and the normal tissue develops into a scar-like tissue. This makes urination more difficult and restricts the flow of semen. As a result, urine can back up to the kidneys, producing a condition called **vesicoureteral reflux**.

Vesicoureteral Reflux A backup of urine to the kidneys
Andropause Age-related changes to the male reproductive system
Menopause The cessation of the menstrual periods
Prolapse Condition in which an organ becomes displaced

Specific aging events in males involve degenerative changes that affect sperm production and erectile function. This is most likely due to declining testosterone levels. Swelling of the prostate is not related to any subsequent decrease in fertility. These changes usually occur gradually after about age 50 years. They are due to the slow decline in fertility commonly called **andropause**.

It is typical for erections to occur less frequently as a male ages; however, this does not always lead to impotence. It is believed that 90% of impotency in older males is due to some type of medical condition unrelated to aging. Endocrine, neural, and vascular disorders can cause impotence. Recent studies show that male fertility also declines due to an increase in the number of abnormal sperm produced in males over age 45 years. Bladder and prostate cancers are more common in older men.

A progressive change called **menopause** typically defines aging of the female reproductive tract. Menopause is defined as the cessation of a woman's menstrual periods. It is caused by a decrease in sensitivity to follicle-stimulating hormone by the ovary. In turn this reduces estrogen production by the ovary. The adrenal glands and ovaries continue to produce testosterone, raising the relative levels of male hormones in the body. This leads to atrophy of the fallopian tubes and uterus. Many physicians believe that this elevation in testosterone levels may increase the female's sex drive at menopause, though some women lose their sex drive for psychological reasons or because sexual intercourse becomes painful. It is not unusual for muscles in the pelvis to lose tone. This can cause the urinary bladder, uterus, or vagina to drop out of position, producing a condition called **prolapse**. Breast tissue also atrophies and the breasts may sag as connective tissue undergoes a loss in supportive fibers.

There are many variables that determine the rate at which the reproductive system of a particular person ages. Genetics and lifestyle are two important aging factors. Many types of medications used to treat cardiovascular and psy-

> ### Good Choice Bad Choice
>
>
>
> A 60-year-old male neighbor wants to purchase a compound called hGH to improve sensuality with his spouse. What can you tell him about the benefits and risks of using hGH? What evidence is available supporting the effectiveness of hGH to reduce age-related changes to the male sexual response?

chiatric disorders worsen age-related changes to the reproductive system. Some physicians use hormone replacement therapy (HRT) to ward off degenerative changes of the reproductive system in women over the age of 50 years. However, there is little evidence that these therapies are effective in reducing reproductive tract degeneration in most people. Some women find benefit in taking HRT to reduce menopause symptoms, such as hot flashes. This therapy is usually most successful in people with underlying endocrine system disorders.

✓ Concept Check

1. What are the roles of hormones in the aging of the reproductive system?
2. Describe age-related changes in the male reproductive system.
3. Describe age-related changes in the female reproductive system.

DISCOVERY SCENE PLEASE ENTER DISCOVERY SCENE PLEASE ENTER

Have you solved the CSI yet? Did the information about aging of the reproductive system provide any more clues about Edward's condition? What aspects of aging could produce a change in gender?

CSI — Case Study Investigation Conclusion

What can you conclude about Edward's story? Does he suffer from a psychological disorder? Could his condition have been caused by aberrations in his sex hormones? Is it possible that he had a genetic condition that somehow caused him to change gender?

Answer:

Edward had a rare developmental condition called gonadal mosaicism. A person with gonadal mosaicism has tissues composed of genetically different cells. Sometimes this disorder is called true hermaphroditism because the person has both the female and male gonadal tissue. The term hermaphrodite was derived from the names of the Greek gods Hermes (a male god) and

THE REPRODUCTIVE SYSTEMS AND HUMAN DEVELOPMENT

Aphrodite (a female god, or goddess). This condition has a variety of causes and usually results in a variation in the number of chromosomes. It is found in approximately one in 25,000 births and can lead to intersex babies. Intersex is a condition in which physicians cannot determine a child's gender because of ambiguous external features. Edward was born with what appeared to be female sexual organs, except for an enlarged clitoris and labia. His gonadal mosaicism expressed both male and female characteristics in each gonad. During early development, the body takes on female characteristics until the gonads produce the appropriate hormones to form the secondary sex characteristics of the male. In gonadal mosaicism, the mixed gonads may not provide enough testosterone to direct the formation of the penis and scrotum. So, the child retains much of the female's external characteristics. Puberty elevates testosterone production and starts to bring out male features. Most people with gonadal mosaicism cannot produce gametes and have sexual identity problems.

The case study is based on Lynn Edward Harris, born Lynn Elizabeth Harris, in California in 1950. Harris was diagnosed with this rare condition in 1973 at age 23 years. How his disease developed was not fully understood until the publication of this phenomenon in the *New England Journal of Medicine* in 1998.*

This CSI was adapted from the following articles:

1. Houk CP, Lee PA, Rapaport R. Intersex classification scheme: a response to the call for a change. *J Pediatr Endocrinol Metab.* 2005;8(8):735-738.
2. MacGillivray MH, Mazur T. Intersex. *Adv Pediatr.* 2005;52:295-319.
3. Nihoul-Fekete C. Does surgical genitoplasty affect gender identity in the intersex infant? *Horm Res.* 2005;64(Suppl 2):23-26.
4. Strain L, Dean JC, Hamilton MP, Bonthron DT. A true hermaphrodite chimera resulting from embryo amalgamation after *in vitro* fertilization. *N Engl J Med.* 1989;15;338(3):166-169.*

Chapter Summary

The human reproductive system is designed to produce gametes to ensure perpetuation of the human species. Male sexual structures are designed to produce, store, and deliver sperm. External secondary sex characteristics ensure that the male's body is designed to successfully deliver sperm to the female reproductive tract. The female reproductive system not only produces eggs, but also must become a chamber for nurturing embryogenesis. A continuous series of events called the female menstrual cycle prepares the female's body for pregnancy. Human fertilization is possible upon ovulation. Human embryogenesis is a complex sequence of events that forms a fetus. As in most mammals, the fetus depends on nutrition from the placenta. The child is then dependent on breast milk as a source of nutrition after birth. Pathology of the reproductive system commonly reduces a person's ability to produce offspring. Some conditions lead to deadly infections or cancers. Aging of the reproductive system is usually associated with degeneration of the endocrine and urinary systems, though the gonads do lose function as a person ages.

Overview

- The reproductive system ensures the continuation of the human species.
- The reproductive system is designed to produce, store, and transport eggs or sperm.
- Gonads carry out meiosis to produce eggs or sperm.
- Humans exhibit sexual dimorphism that results in different reproductive systems.

Female Reproductive System

- The female reproductive system is composed of the reproductive tract and mammary glands.
- The female reproductive tract is composed of the ovaries, fallopian tubes, uterus, and vagina.
- Ovaries are responsible for storing and maturing eggs.
- The fallopian tubes transfer the egg to the uterus.
- The uterus is responsible for nurturing the developing embryo.
- The vagina makes up the external genitalia and assists with sexual reproduction.
- The clitoris is a piece of erectile tissue involved in sexual excitement.
- Mammary glands provide milk for the baby after childbirth.

Male Reproductive System

- The male reproductive system is composed of the testes, seminal vessels, and penis.
- The penis and scrotum form the external genitalia.
- Testes produce and mature sperm.
- Seminal vessel ducts include the epididymis, vas deferens, and ejaculatory duct.
- The seminal vessel glands include the seminal vesicles, prostate, and Cowper's glands.
- Semen is a combination of sperm and secretions from the seminal vessel glands.
- The penis is composed of erectile tissue and the urethra.

Study Guide

Basics of Sexual Reproduction
- Sexual reproduction requires sexual excitation of the male and female reproductive systems.
- The female has a menstrual cycle that prepares the body for pregnancy during ovulation.
- The menstrual cycle is divided into the ovarian and uterine cycles.
- Copulation is the transfer of sperm to the female reproductive tract.
- Fertilization takes place in the fallopian tubes.
- Implantation of the embryo takes place high up in the uterus.
- Embryogenesis is fueled by nutrients from the placenta.
- Fetal development continues until labor.
- Labor involves a series of events that expel the baby from the uterus.

Pathology of the Reproductive Systems
- Diseases of the reproductive tract are congenital, infectious, or degenerative.
- Congenital diseases affect the function of the gonads or the development of the other reproductive organs.
- Sexually transmitted diseases can be caused by arthropods, bacteria, protista, or viruses.
- Degenerative diseases include abnormal growths, including cancer.
- Reproductive disorders can include inappropriate implantation of the embryo.

Aging of the Reproductive Systems
- Most reproductive system aging is the result of the aging of the endocrine and urinary systems.
- Reproductive system aging reduces egg and sperm production.
- Reproductive system aging causes degeneration of the reproductive organs.
- The decrease in sex hormones is called andropause in males and menopause in females.

 Key Terms

Overview
Gonad
Ovary
Puberty
Secondary sex characteristics
Sexual dimorphism
Specialized germ cell (SGC)
Testis or testicle

Female Reproductive System
External genitalia
Fallopian tube
Mammary gland
Reproductive tract
Uterus
Vagina

Reproductive Tract
Androgen
Aromatase
Broad ligament
Cervix
Clitoris
Clitoral hood
Corpus luteum
Desmolase
Endometrium
Erectile tissue
Fertilization
Fimbriae

CHAPTER 15

Study Guide

Graafian follicle
Hymen
Intersex
Labia majora
Labia minora
Lower uterine segment
Menstrual period
Mons
Myometrium
Myosalpinx
Oocyte
Ova
Ovarian follicle
Ovarian ligament
Oviduct
Ovulation
Perineum
Skene's gland
Uterine fundus
Vulva
Womb

Mammary Glands
Areola
Lactation
Lactiferous duct
Nipple

Male Reproductive System
Penis or phallus
Scrotum
Seminal vessel

Testes
Cryptorchidism
Epididymis
Leydig's cells
Seminiferous tubules
Undescended testis

Seminal Vessels
Bulbourethral glands
Cowper's glands
Ductus deferens
Ejaculatory duct
Prostate gland
Semen
Seminal vesicle
Vas deferens or ductus deferens

Penis
Corpus cavernosum
Corpus spongiosum
Circumcision
Dorsal vein
Erection
Foreskin
Glans
Prepuce or foreskin

Female Sexual Cycle
Follicular phase
Luteal phase
Menses
Ovarian cycle
Postovulation phase
Preovulation phase
Proliferation phase
Uterine cycle

Copulation
Copulation
Detumescence
Ejaculation
Erectile dysfunction
Fertilization
Orgasm
Sexual intercourse
Urethral meatus

Embryology and Pregnancy
Acrosome
Amniotic fluid
Amniotic sac
Blastocyst
Blastula
Colostrum
Conjoined twins
Conception
Embryogenesis
Fraternal twins
Gastrula
Human chorionic gonadotropin (hCG)
Identical twins
Implantation
Labor
Placenta

Pathology of the Reproductive Systems
Andropause
Breast cancer
Cesarean section
Cervical cancer

THE REPRODUCTIVE SYSTEMS AND HUMAN DEVELOPMENT

Study Guide

Ectopic pregnancy
Fibroids
Genital warts
Hypospadias
Pelvic inflammatory disease (PID)
Placenta previa
Prostate cancer
Sexually transmitted diseases

Testicular cancer

Aging of the Reproductive Systems
Andropause
Menopause
Prolapse
Vesicoureteral reflux

Check Your Understanding

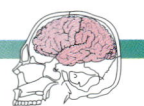

1. Gonads carry out _____ to produce eggs and sperm:
   a. mitosis
   b. binary fission
   c. embryogenesis
   d. meiosis

2. The reproductive tract shares this urinary tract structure:
   a. urethra
   b. ureters
   c. uterus
   d. utricle

3. Damage to the ovaries would affect the following functions:
   a. egg production
   b. estrogen production
   c. ovulation
   d. All of the above

4. Which is not part of the female reproductive system?
   a. Cowper's glands
   b. fallopian tubes
   c. uterus
   d. labia

5. Ovaries release eggs directly into the:
   a. vagina
   b. epididymis
   c. fallopian tube
   d. eustachian tube

6. The menstrual cycle affects the thickness of the:
   a. fallopian tubes
   b. vas deferens
   c. uterus
   d. ovary

7. The cervix is found at the proximal end of the:
   a. vagina
   b. uterus
   c. fallopian tubes
   d. urinary bladder

8. The average menstrual cycle is ___ days:
   a. 32
   b. 28
   c. 20
   d. 14

9. Ovulation usually occurs around day ____ of the menstrual cycle.
   a. 2
   b. 4
   c. 14
   d. 28

10. Sperm are stored in the following structure:
    a. Cowper's gland
    b. epididymis
    c. spermatorium
    d. testis

11. Sperm formation occurs in the:
    a. Cowper's gland
    b. epididymis
    c. prostate
    d. testis

12. Testosterone is produced by the:
    a. Cowper's gland
    b. epididymis
    c. prostate
    d. testis

13. The embryo normally develops in the:
    a. fallopian tubes
    b. epididymis
    c. uterus
    d. cervix

14. Most STDs today are spread by:
    a. drug use
    b. sexual contact
    c. kissing
    d. blood transfusions

15. Which is *not* a characteristic of reproductive system aging?
    a. estrogen level decline in females
    b. menopause
    c. an increase in a male's testosterone
    d. atrophy of the uterus

Study Guide

A Case Study

MANDATORY METHODS FOR CONTROLLING SEXUALLY TRANSMITTED DISEASES

Recent studies show that the incidence of STDs may be on the rise in North America. Over 30% of young, sexually active women tested positive for some type of STD; chlamydia and HPV were the most common. Studies of males show equally high incidences of HPV and other STDs. Other research shows that STDs in North America are likely to be unrecorded by public health agencies that are attempting to monitor the diseases. This lack of information reduces the effectiveness of controlling STDs. Therefore, government agencies are proposing a variety of strategies for controlling and monitoring STDs.

Circumcision is believed to be one possible strategy for controlling STDs. The procedure involves removal of the foreskin, leaving the distal end of the penis uncovered. Islamic and Jewish practices require circumcision of infant males. Many American physicians advise parents to have male children circumcised to facilitate cleanliness of the penis. Years of data indicate that circumcised males have lower rates of urinary tract infections, as well as lower rates of certain STDs. The evidence is so compelling that the American Academy of Pediatrics Task Force on Circumcision acknowledges that circumcision has potential medical benefits that outweigh the risks. Some governmental policies are giving incentives for physicians to encourage and perform circumcisions. One policy in New Mexico even requires insurance companies to pay for voluntary circumcisions for children and adults. Some legislators want to go as far as requiring circumcision of all newborn males to reduce the spread of STDs.

In 2004, the American Civil Liberties Union of Virginia found themselves at odds with a policy to monitor STDs. The pending legislation was called the State of Virginia House Bill 954. The bill permitted the courts to do mandatory testing of STDs for people convicted of drug-related crimes in Virginia because STDs are more prevalent in people who abuse illegal substances. Legislators felt that more measures had to be taken to reduce the spread of STDs. Currently, prisoners in several states and in all federal penitentiaries are tested for STDs. Prison officials are concerned about the spread of STDs among prisoners. Some public health officials are calling for mandatory STD testing of all individuals. They recommend that the testing be done during routine physical examinations and hospital visits. Other officials feel that all high school and college students should be tested before being permitted to attend classes. Again, the justification is to prevent a possible STD epidemic. Many agencies and companies require STD testing, including the Foreign Service and the military.

Use the information in this chapter and the following Web sites to answer the following questions about mandatory testing for STDs:

- What are the benefits and risks of mandatory circumcision?
- What are the benefits and risks of mandatory STD testing?
- What is the rationale for creating legislation to mandate circumcision and STD testing?
- How should physicians respond to privacy issues related to STD testing?
- Does the government have the right to implement mandatory circumcision?

CHAPTER 15

- Does the government have right to implement mandatory STD testing?
- What are your feelings about STD testing being required for your job?
- Should insurance companies have different medical coverage policies for people who are not circumcised?

1. Centers for Disease Control and Prevention (CDC)
   http://www.cdc.gov/std/

2. Circumcision Information and Resource Pages
   http://www.cirp.org/

3. Circumcision Resource Center
   http://www.circumcision.org/

4. WebMD
   http://www.Webmd.com/content/article/22/1728_55601.htm

5. National Coalition of STD Directors
   http://www.ncsddc.org/links.htm

References:

Buve A, Laga M, Remes P, Padian N, Morison L Ethics of mass STD treatment. *Lancet.* 2000;23;356(9235):1115-1116.

Datta, SD, Koutsky L, Douglas J, et al. Sentinel surveillance for human papillomavirus among women in the United States, 2003-2004. Program and abstracts presented of the 16th Biennial Meeting of the International Society for Sexually Transmitted Diseases Research; July 10-13, 2005; Amsterdam, The Netherlands. Abstract MO-306.

Donovan B. The repertoire of human efforts to avoid sexually transmissible diseases: past and present. Part 1: Strategies used before or instead of sex. *Sex Transm Infect.* 2000;76(1):7-12.

Schulte JM, Martich FA, Schmid GP. Chancroid in the United States, 1981-1990: Evidence for underreporting of cases. *MMWR.* 1992;41(No. SS-3):57-61.

Where Do We Go from Here?

1. A 40-year-old male classmate sees no need for getting a prostate exam. He claims that nobody in his family ever developed prostate cancer. What evidence would you present to convince him otherwise?
   EMedicine – Prostate Cancer:
   http://www.emedicinehealth.com/Articles/18465-1.asp

2. A friend read a blog story that male children in Japan are being born with defective sex organs. The story claims that the defects resulted from their mothers being exposed to certain pollutants during pregnancy. How would you explain the possible effects of the pollution on male development?
   e.Hormone: *http://e.hormone.tulane.edu/*

Study Guide

3. Some male classmates are joking around about going to an oyster bar before a big spring break party. They claim that oysters are an aphrodisiac. Is their claim based on any valid medical evidence?
   MedicineNet: http://www.medicinenet.com/

4. You are speaking to a class of middle school students about personal hygiene. A girl in the group asks why menstrual bleeding has to occur. How would you go about explaining this to the class?
   Kids Health: http://kidshealth.org/kid/

5. A friend asks you what the most common STD is in North America today. She also asks if condoms prevent the spread of all STDs. What would you explain to her?
   Medline Plus: http://www.nlm.nih.gov

Skills Activities

1 Predicting Birth Defects

Materials
- computer with internet access
- bookmark to Visembryo Web site: http://www.visembryo.com/baby/
- bookmark to Interactive Embryo Web site: http://www.anatomical-travel.com/CB_site/Conception_to_birth3.htm
- bookmark to Google to search chemicals and terms related to the activity
- spreadsheet software to record notes
- access to a printer to print comments

Physicians and scientists throughout the years have amassed detailed information about human embryological development. This information is stored in databases and image collections for use in understanding the causes of birth defects. The information is also a powerful tool for predicting the effects of chemicals on fetal development. Pathologists in particular use an understanding of embryology to better understand how certain drugs can affect a fetus. Today, it is common for attorneys and public health officials to use embryology databases to research incidents that could cause birth defects. This activity provides an embryology resource that could be used to predict the effects of particular drugs or pollutants on embryological development.

Go to the Interactive Embryo Web site and familiarize yourself with the progress of development from fertilization to the last week of pregnancy. Next, go to the Visembryo Web site and click on the different images representing the stages of human embryology. Again, become familiar with the different stages of development. Now, imagine that you are on a jury investigating the claims of four pregnant women who were exposed to a chemical-plant leak near their homes. Each woman is suing the chemical company for birth defects believed to be caused by exposure to the chemicals leaked into the air. They were not aware of the leak until public health officials noticed a rash of mysterious illnesses several months later. The chemical company manufactured various agricultural chemicals. An Environmental Protection

Agency (EPA) investigation discovered that the pesticide, aldrin, and the fungicide, dithiocarbamate, were released into the air. Your job is to determine whether the birth defects in these children were caused by either of the chemicals or by other factors not related to the chemical release. Use Google to research the nature of the pesticides, and the embryology Web sites to investigate the state of the embryo at the time of the mothers' exposure.

Woman 1
She said that she found out she was 1-week pregnant when the incident occurred. Her boy was born 1 month premature and had underdeveloped lungs. The physicians discovered a septal defect in the boy's heart that had to be surgically corrected. The women claims her two other children had no such problems.

Woman 2
Her physician confirms that she was 6 weeks pregnant during the time the chemicals were released. Her baby girl was born with facial defects that cause her eyes to be too widely spaced and her nose malformed. The girl also has fingers missing on one hand. This is her only child.

Woman 3
The woman says that she was 8 weeks pregnant at the time of the chemical release. Her baby girl was delivered prematurely with a condition called anencephaly. The child died 3 weeks after birth. The mother has one other child who was born with no problems.

Woman 4
A testimony from two physicians confirms the woman was 6 months pregnant at the time of the incident. Her baby boy was born with genital deformations that prevented the fusion of his penis and scrotum. The defects were surgically corrected. However, the boy may never have full use of the penis when he reaches reproductive age. The mother has a 2-year-old girl who was born with no problems.

On a spreadsheet, record the tissues or organ systems being developed in the fetus at the time during pregnancy that each women was exposed. Determine for each case whether the chemical exposure could have been responsible for each of the particular birth defects. Also, predict what probable developmental defects are possible given exposure to those chemicals at 1 week, 6 weeks, 8 weeks, and 6 months of pregnancy. In your analysis, think about how the chemical would travel to the baby to produce any damaging effects. Write a brief report explaining your conclusions as though you are using the information to determine whether the women have enough evidence to blame the chemical company for the birth defects.

Study Guide

Skills Activities

2 Modeling the Test for Human Chorionic Gonadotropin

Materials

- a bottle of red food coloring
- a supply of distilled water
- 3 g of calcium hydroxide in 200 mL of distilled water
- 16 g of ammonium oxalate in 200 mL of distilled water containing four drops of red food coloring
- 32 g of ammonium oxalate in 200 mL distilled water containing 4 drops of red food coloring
- 200 mL of distilled water containing four drops of red food coloring
- a classroom set of small vials labeled Negative Control containing distilled water
- a classroom set of small vials labeled Positive Control containing the 8 g of ammonium oxalate solution
- a classroom set of small vials labeled Subject 1 containing distilled water with the red food coloring
- a classroom set of small vials labeled Subject 2 containing the 8 g of ammonium oxalate solution.
- a classroom set of small vials labeled Subject 3 containing the 16 g of ammonium oxalate solution
- a classroom set of vials containing the calcium hydroxide solution. The vials should be labeled hCG Test Solution
- six droppers for each setup
- five small test tubes in a test-tube rack per setup
- a marker to label test tubes
- surgical gloves
- access to WebMd Web site at: http://www.Webmd.com/hw/being_pregnant/hw42062

Pregnancy testing detects (or does not detect if the test is negative) the presence of human chorionic gonadotropin (hCG). This protein is produced by the developing embryo. Human chorionic gonadotropin maintains the corpus luteum and causes it to secrete progesterone. Progesterone ensures that the uterine lining is thick and has ample blood vessels to sustain the growth of the fetus. Physicians can collect hCG from blood or urine samples. Home pregnancy test kits are designed to detect hCG from the urine. Pregnancy testing commonly uses a technique called an immunoassay. The test uses antibodies that specifically bind to hCG. If hCG is present, the antibodies stick to the hCG and stimulate a chemical reaction that indicates a positive test. Many physicians use a sensitive test in which they look for a clump of material that falls out of solution. This clump is formed when the antibodies bind to the hCG. The clump looks like a fuzzy, white residue on the bottom of a test tube. In this activity you will be asked to interpret the results of three samples being tested for hCG.

Chapter 15

Carry out the following steps to determine the presence of hCG in three urine samples:

1. Place five test tubes in the test tube rack.
2. Collect a set of vials labeled "test solution," "positive control," "negative control," "sample 1," "sample 2," and "sample 3."
3. Label the tubes No. 1 through 5.
4. Record the tube numbers on a sheet of paper.
5. Add 10 drops of "test solution" to each tube using a clean dropper.
6. Add five drops of "positive control" to test tube No. 1 using a clean dropper. Record the results on the sheet of paper.
7. Add five drops of "negative control" to test tube No. 1 using a clean dropper. Record the results on the sheet of paper.
8. Add five drops of "sample 1" to test tube No. 1 using a clean dropper. Record the results on the sheet of paper.
9. Add five drops of "sample 2" to test tube No. 1 using a clean dropper. Record the results on the sheet of paper.
10. Add five drops of "sample 3" to test tube No. 1 using a clean dropper. Record the results on the sheet of paper.
11. Record your conclusions for the observed results. Answer the following questions:

- What is the reason for performing the positive- and negative-control samples?
- What can you conclude about each sample?
- How do you explain the difference from the positive control seen in sample 3?
- You just discovered that sample 3 is a false-positive test result. Use the WebMD Web site link to determine the possible causes of the false-positive test result seen in this sample.

Note

It is recommended that you wear surgical gloves when handing the solutions. The samples should be handled as though they are actual human body fluids. All of the solutions can be stored in labeled bottles at room temperature for six months. Excess or waste solutions should be flushed down a drain.

GLOSSARY

A

Abdominal Refers to the stomach region of the abdomen

Abdominal Cavity The body cavity containing the liver, gall bladder, intestines, kidneys, spleen, and stomach

Abdominopelvic Cavity The body cavity containing the abdominal and pelvic cavities

Abdominopelvic Region Refers to the region of the body found below the breasts and above the groin

Abduction Movement of the arm or leg away from the midline of the body

Abductor Muscles Muscles that move a bone away from the body's midline

ABO Blood Group System A classification system for the proteins on human red blood cells

Accessory Digestive Organ The secretory organs associated with the digestive tract

Accumulated Cell Damage The sum effects of years of accumulated damage to a cell

Acetabulum A deep-socket depression on the pelvic girdle that forms the point of articulation with the femur

Acetyl Coenzyme A (Acetyl CoA) A metabolic compound related to acetic acid and vinegar

Acetylcholine A neurotransmitter that communicates with muscle cells

Acetylcholine Receptor A protein located on the sarcolemma that binds the neurotransmitter acetylcholine

Acid Water containing large amounts of hydrogen ions and fall within a pH range of 0 to 6

Acini Sac-like clusters of exocrine cells

Acne A common skin condition resulting in the overproduction of sebum

Acquired Immunity A specific set of events that the body uses to fight a particular infection

Acquired Immunodeficiency Syndrome (AIDS) An infectious disease caused by the human immunodeficiency virus (HIV)

Acromial Refers to the shoulder region of the body

Acrosome A packet of enzymes in a sperm's head

Actin One of the three proteins that form the thin myofilaments of muscle fibers; it forms the core of the fiber

Action Potential The electrical signal that rapidly travels along the axon of neurons

Active Immunity An immune system response gained by exposure to foreign antigens

Active Site An area of proteins called enzymes. It provides an attachment area for carrying out chemical reactions

Active Transport A means of moving particles across a membrane that requires cell energy

Active Transport Pumping The movement of particles across a membrane by protein "pumps"

Acute Occurring suddenly or severely over a short period of time

Acute Diarrhea A short-term, rapid-onset diarrhea

Acute Renal Failure Temporary loss of kidney function

Acute Respiratory Distress Syndrome (ARDS) The rapid onset of respiratory failure

Adaptive Refers to the ability of an organism's genes to respond to environmental changes

Adduction Movement of the arm or leg toward the midline of the body

Adductor Muscles Muscles that move a bone closer to the body's midline

Adenoids Glandular components of the immune system that are located in the nasopharynx

Adenomatous Polyp A polyp that can develop into cancer

Adenosine Triphosphate (ATP) A high-energy molecule formed during aerobic respiration. It is the most common energy transfer molecule and contains three phosphate functional groups

Adhesiveness A glue-like property, as with water

Adipose Capsule A cushioning of fat that encases each kidney

Adrenal Cortex The outer shell of the adrenal glands

Adrenal Glands Glands that produce glucocorticosteroids and mineralocorticosteroids

Adrenal Medulla The interior region of the adrenal glands

Adrenaline or Epinephrine A hormone produced by the adrenal glands in response to exercise, fear, or stress

Aerobic Respiration Cellular respiration that requires oxygen

Affector A neuron that transmits sensory information

Afferent Arteriole The small blood vessel that narrows to become the glomerulus

Afferent Nerve The part of the peripheral nervous system that carries information from the body to the central nervous system

Aging The gradual deterioration of a material or an object

Agonist A chemical that behaves like a hormone

Agranulocyte or Mononuclear White Blood Cell A WBC that lacks granules

Albinism A genetic disease characterized by the lack of melanin production in the eyes, hair, and skin

Aldosterone A steroid hormone that the adrenal cortex secretes to control sodium and potassium in the blood

Allergy An overreaction of the body's immune system

Allosteric site A region of an enzyme that binds to chemicals affecting enzyme activity

Allostery A change of enzyme activity in response to the binding of a molecule on the enzyme's allosteric site

Alveolar Bones Bones that develop from special cells found only in the jaw bones

Alveolus 1) A small sac-like structure at the end of a terminal bronchiole where gas exchange takes place; 2) Socket in the jawbone out of which a tooth grows

Amino Acids Structural units of proteins

Amino Group The functional group containing nitrogen, involved with the exchange of hydrogen ions and electrons with other molecules

Aminoaciduria The presence of amino acids in the urine

Amino-Sugars A mucopolysaccharide with an amino functional group

Amniotic Fluid Fluid within the amniotic sac of the fetus

Amniotic Sac A fluid-filled sac that surrounds the fetus

Amoebic Dysentery An inflammation of the intestines caused by a protistan called *Endamoeba histolytica*

Amphiarthrosis Joint A joint that permits only slight movement

Amyloid A protein-like material associated with cell injury

Amyloid Deposition or Amyloidosis The accumulation of amyloids in a cell

Amyotrophic Lateral Sclerosis (ALS) or Lou Gehrig's Disease A progressive neurological disease that causes the loss of neuron function

Anabolism A metabolic reaction that uses cell energy and results in the production of cell components

Anaerobic Respiration or Glycolysis Cellular respiration that does not require oxygen

Anal Canal A small portion of the rectum ending in the anus

Anal Sphincter The sphincter muscle of the anus

Anaphase I The third phase of meiosis I

Anatomy The structural makeup of an organism

Androgen A hormone that maintains body structure and provides male sex characteristics

Andropause Age-related changes to the male reproductive system

Anemia A deficiency of normal red blood cells in the bloodstream

Aneurysm An abnormal swelling of a blood vessel wall

Angina Pectoris Chest pain due to coronary heart disease

Angiogenic Factor A secretion that helps develop blood vessels

Angiogensis Factors Chemicals that stimulate the growth of new blood vessels

Angiotensin II A peptide produced from angiotensin I; it is involved in the maintenance of blood volume and pressure

Angulation A twisted change in the original shape of a bone as a result of damage

Anions Negatively charged ions

Ankylosing Spondylitis Arthritis affecting the spine

Anophase The third phase of mitosis

Antagonist A chemical that blocks the action of a hormone

Antagonistic Opposing movements

Anterior Meaning toward the front

Anterior Pituitary Gland The forward region of the pituitary gland

Antibiotic A chemical that harms or kills bacteria

Antibody or Immunoglobulin A protein produced by the immune system that binds to specific foreign antigens

Anticodon A unit of genetic code consisting of three consecutive nucleotides, complementary to the codon

Antigen A substances that can induce an immune response

Antioxidants Chemicals that protect the body by oxidizing before oxidizing agents can hurt the body

Antisense Refers to the DNA strand that does not carry the genetic code

Anuria The inability to produce urine

Anus The opening of the rectum to the outside of the body

Aorta The large artery that carries blood from the heart to the body

Aortic Valve The left semilunar valve

Apocrine Sweat Glands Sweat glands that secrete sweat into hair follicles

Apocrine Refers to secretions consisting of pinched-off cytoplasm, including some plasma membrane

Appendicitis Inflammation of the appendix

Appendicular Skeleton The part of the skeleton composed of the upper and lower appendages, and the bones that girdle them to the axial skeleton

Appendix A small pouch protruding from the cecum

Aqueous Humor A clear, watery fluid in the front of the eyeball

Arabinose A pentose monosaccharide found in vegetables and nutritional drinks

Arachnoid Mater The middle layer of the meninges, which nourishes and protects the brain and spinal cord

Areolar Connective Tissue A connective tissue that binds blood vessels, membranes, muscles, nerves, and skin to other structures

Aromatase An enzyme that converts androgens into estrogen; reduces the level of hormones needed for muscle mass and strength

Arrector Pili Muscle A smooth band of muscle that holds the hair erect

Arrhythmia Irregular rhythmic beating of the heart

Arterial Stiffness A progressive loss of flexibility of the arterial walls

Arteriole A small branch of an artery

Arteriosclerosis A gradual stiffening of the arterial walls due to aging

Arteriovenous Malformation An abnormal tangling of blood vessels in the brain that disrupts blood flow

Artery A blood vessel that carries blood from the heart to the body

Arthritis A condition causing swelling and stiffness in the joints

Arthropod A class of animals that have exoskeletons, jointed limbs, and segmented bodies

Articulation A junction between two or more bones

Artificial Immunization The therapeutic exposure to foreign antigens as a way of protecting a person from disease

Arytenoid Cartilage Two small cartilages at the back of the larynx

Ascending Colon The upward section of the large intestine on the right side of the body

Ascending Tract A longitudinal band of white matter in the spinal cord that carries sensory information to the brain

Asexual Division A type of division where two new cells develop from a single cell

Aspartate An excitatory amino acid neurotransmitter

Astrocyte A type of neuroglia providing organization and support for the nervous system

Atelectasis The complete or partial collapse of a lung

Atherosclerosis A progressive narrowing and hardening of arteries due to plaque formation

Athetosis A nervous system disorder that causes slow, involuntary movements of the hands and feet

Atlas The first cervical vertebra, which supports the head

Atom The smallest portion of an element that still retains its properties

Atomic Mass The sum of protons and neutrons in an atom's nucleus

Atomic Nucleus A core of dense material providing an element with its mass and physical properties

Atomic Number The number of protons in an atom

Atomic Orbitals (Shells) The regions around the nucleus where electrons are located

Atomic Structure The structure of atoms composing each element, consisting of an atomic nucleus and atomic orbitals

Atrial Natriuretic Factor (ANF) A hormone secreted by special cardiac cells that lowers blood volume and blood pressure

Atrioventricular (AV) Node A relay station between the atria and ventricles that coordinates atrial and ventricular connections

Atrioventricular (AV) Valves Relating to valves of the atria and ventricles of the heart; the mitral and tricuspid valves

Atrium One of the two thin upper chambers of the heart

Atrophy The wasting of a cell, tissue, or organ

Audition The ability to hear

Auricle or Pinna The external portion of the ear

Autoantibody An antibody that reacts against the person's own tissues

Autocrine A type of secretion that is self-governing and does not usually travel far in the blood

Autoimmunity A condition in which the body produces an immune response against its own organs or tissues

Autonomic The part of the peripheral nervous system responsible for such involuntary body functions as heartbeat, blood pressure, and digestion

Axial Skeleton The part of the skeleton composed of the spine, rib cage, and skull

Axis The second cervical vertebra; the atlas and head rotate upon it

Axoaxonic Synapse A synapse formed with an axon

Axodendritic Synapse A synapse formed with a dendrite

Axon Hillock A swelling at the cell body where the axon begins

Axon The fiber-like extension of a neuron that sends information to terminus

Axosomatic Synapse A synapse formed with a nerve cell body

B

Bacteria Very small, single-celled life-forms that is the most common microorganism on the body

Ball-and-Socket Joint A synovial joint that permits a variety of movements

Band or Stab An immature white blood cell

Basal Nuclei Four regions of gray matter found deep in the forebrain

Base or Alkaline Water with a low concentration of hydrogen ions

Basophil A white blood cell associated with allergies

Biceps A muscle with two origins

Bicuspid or Premolar Teeth Double-pointed teeth that are lateral to the canines

Bidirectional Communication Cell signals that pass in either direction between two cells

Bilateral Refers to body structures located laterally on both sides of the body

Bile A yellow-green fluid produced by the liver

Bilirubin A chemical breakdown product of hemoglobin

Bioactive Describes a substance that has an effect on living organisms

Bioactive Molecules Molecules that promote chemical reactions

Biochemicals Organic molecules produced by the chemical reactions of living organisms

Biochemistry or Molecular Biology The chemistry of the body's structures and functions

Biopsy The surgical removal of diseased cells for study

Bipolar Refers to a neuron that has two processes, one axon, and one dendrite

Bitter The taste that occurs when bases, such as soap, dissolve in the mouth

Bladder Cancer Malignancy of the tissue of the urinary bladder

Blastula or Blastocyst A hollow sphere of cells formed by repeated mitosis of the zygote

Blood Pressure The force of blood pushing against blood vessel walls

Blood Type Classification of a red blood cell according to the composition of proteins on its membrane

Blood Vessels A part of the cardiovascular system that carries blood throughout the body

Blood-Brain Barrier A barrier between brain blood vessels and brain tissues that restricts the passage of materials from the blood into the brain

B-lymphocyte A white blood cell that assists with the immune response

Body Mass Index (BMI) An indirect measure of body density

Bone A hard connective tissue

Bone Collar A layer of bone on the surface of the diaphysis

Bone Density Loss A loss of bone tissue that is less severe than osteoporosis

Bone Marrow Soft tissue within the medullary cavity of a bone

Botulism A serious illness caused by a toxin produced by certain bacteria

Bowman's Capsule An expanded portion of the renal tubule that contains the glomerulus

Brachial Refers to the arm

Brain Stem The part of the brain that is connected to the spinal cord

Breast Cancer Cancer of the mammary glands

Brevis Muscle The shortest muscle of a muscle group

Broad Ligaments Two sheets of epithelium attached to the fallopian tube and uterus

Bronchial Tree A network of passages that supplies the lungs with air

Bronchiectasis The abnormal stretching and dilation of the bronchi or bronchioles

Bronchiole The smaller subdivision of the bronchi

Bronchitis Inflammation of the bronchi

Bronchoconstriction The constriction of bands of smooth muscle in the terminal bronchioles

Bronchodilation The dilation of bands of smooth muscle in the terminal bronchioles

Bronchoneumonia A type of pneumonia scattered throughout the lung

Bronchospasm The tightening of bands of smooth muscle in the terminal bronchioles

Buccal Cavity The portion of the oropharynx associated with the lips and cheeks

Buffer Molecules that act as either hydrogen ion acceptors or donors

Bulk Active Transport The movement of large amounts of particles across a membrane due to membrane movement

Bulk Mechanical Transport The movement of large volumes of molecules and ions from one location to another

Bundle of His Specialized cells at the superior interventricular septum that receive nerve impulses from the atrioventricular node

Bursa A fibrous sack filled with synovial fluid

Bursitis The inflammation of a bursa

C

Cachexia Muscle loss associated with aging or disease

Calcaneus The largest of the tarsal bones, located at the lower and dorsal region of the foot forming the heel

Calcification The process by which bones harden

Calcitonin A hormone secreted by the thyroid gland that lowers blood calcium

Calculi A stone-like accumulation of amino acids or mineral crystals in the kidney or bladder

Calorie A standard unit of measurement equal to the amount of heat required to raise the temperature of one gram of water one degree Celsius

Calvaria The dome-shaped superior portion of the cranium

Calyces (singular is calyx) Branching extensions of the renal pelvis that transfer urine from the renal pyramids

Canaliculi Small channels that connect osteocytes

***Candida albicans* or Monilia** A yeast-like fungus responsible for causing infection of the skin, nails or vaginal area

Canine or Cuspid Teeth Sharp teeth located on each side of the mouth

Capillary A narrow thin-walled blood vessel

Capitate A large distal carpal bone

Carbo Loading Eating large amounts of glucose before an endurance event to maintain adequate glucose levels during activity

Carbohydrates Compound molecules that provide the body with energy

Carbon is an element found in all living organisms

Carbon Dioxide Intoxication A state resulting from excessive carbon dioxide in the environment or the blood

Carbon Skeleton Carbon atoms held together with covalent bonds

Carbonic Anhydrase An enzyme that converts carbon dioxide and water into bicarbonate ions

Carbonyl Group The functional group involved in the transfer of electrons between molecules

Carboxyl Group The functional group involved with the exchange of hydrogen ions and electrons with other molecules

Cardiac Cycle One complete contraction and relaxation of the heart

Cardiac Disorder Any disease of the heart

Cardiac Glands Pits of secretory cells in the upper region of the stomach

Cardiac Infarction Death of heart muscle due to lack of oxygen from the blood

Cardiac Ischemia A lack of sufficient oxygen for normal function of heart muscle

Cardiac Muscle The muscle of the heart

Cardiac Notch A region of the left lung where the heart sits

Cardiac Output The amount of blood the heart pumps each minute

Cardiac Region Upper region of the stomach

Cardiac Sphincter The upper sphincter of the stomach

Cardiovagal Baroreflex A reflex that adjusts the strength of heart contractions to body activity

Cardiovascular System Body system that regulates blood flow

Carpal Refers to the wrist

Carrier Protein or Channel A protein that moves molecules through a membrane as part of facilitated diffusion

Cartilage Flexible connective tissue

Cartilaginous Joint A joint formed of cartilage

Cast An abnormal aggregate of cells or other substances that can be found in urine when they accumulate in renal tubules and dislodge into the tubular fluid

Catabolism A metabolic reaction that breaks down molecules to provide the cell with energy and materials to perform anabolism

Catalyst Chemicals that start chemical reactions

Catecholamines A class of neurotransmitters related to amino acids

Catheter A tube inserted into the urethra to expel urine

Cations Positively charged ions

Caudal Meaning near the tail

Caudate Lobe A small subdivision of the liver

Cavity A hole in a tooth caused by decay

Cecum The first portion of the large intestine

Celiac Disease An inability to digest and absorb a certain protein found in wheat

Cell The basic structural and functional unit of the human body

Cell Cycle The events a cell goes through to carry out daily functions and the steps it takes to reproduce

Cell Theory or Cell Doctrine The assertion that all organisms are composed of cells

Cell Wall The covering of the cell membrane

Cell-Mediated Immunity The immune response coordinated by T-lymphocytes

Cellobiose An indigestible disaccharide; a combination of two glucose molecules

Cellular Aging The accumulation of molecular decay

Cellular Respiration The extraction of energy from the chemical breakdown of stored food molecules

Cellulose A fibrous plant polysaccharide that is indigestible to humans

Central Canal A centrally located cavity that runs the length of the spinal cord

Central Nervous System (CNS) A major division of the nervous system composed of the brain and spinal cord

Central Sulcus A groove separating the frontal lobe from the parietal lobe of the cerebrum

Centrifuge A machine that rapidly spins liquid samples and separates the materials by density

Centriole An organelle that assists the cell with reproduction

Centromere A region of the cell where doubled chromatids attach to one another

Cephalic Pertaining to the head

Cerebellum A large portion of the base of the brain that coordinates voluntary movements

Cerebral Hemispheres The left and right halves of the cerebrum

Cerebrospinal Fluid A nutrient-rich fluid that circulates around and through the brain and spinal cord

Cerebrovascular Disease Disorders of blood vessels in the brain

Cerebrum The part of the brain responsible for emotions, memory, motor movement, and thought

Cerumen A waxy secretion produced by the ceruminous gland

Ceruminous Gland The large gland found in the skin lining the ear canal

Cervical Refers to the neck region

Cervical Cancer Cancer of the cervix

Cervical Region The part of the spinal column comprising the neck

Cervical Vertebrae Seven vertebrae of the cervical region

Cervix The lowermost part of the uterus, which opens into the vagina

Cesarean Section Delivery of a fetus by surgical incision through the abdominal wall

Chambers Four sections of the heart through which blood is pumped

Chemical Bond The way atoms are attached to each other

Chemical Damage Damage caused by any chemical that breaks down cells or the connections between cells

Chemical Energy The conversion of a chemical into another form

Chemistry The branch of natural sciences dealing with

the composition of substances, their properties, and reactions

Chemoreceptor A sensory receptor that detects chemical stimuli

Chief Cells Cells of the stomach that secrete digestive enzymes

Chirality Describes isomers that form mirror image molecules

Cholecystokinin (CCK) A hormone the causes the pancreas to produce enzymes

Cholesterol The most common sterol, essential to body structure and brain development

Cholinergic Pertaining to the acetylcholine neurotransmitter

Chordae Tendineae Fibrous threads that attach papillary muscles to the atrioventricular valves

Chorea A nervous system disorder that causes muscular twitching of arms, legs, and face

Choroid A layer of blood vessels that lines the inner surface of the sclera

Choroid Plexus A collection of blood vessels in regions of pia mater in the brain

Chromatid One half of a doubled chromosome

Chromosome A thread-like collection of genes and other DNA found in the nucleus

Chronic Diarrhea A long-term, usually painful, diarrhea

Chronic Obstructive Pulmonary Disease (COPD) Any disorder that persistently obstructs bronchial airflow

Chronic Renal Failure Irreparable nephron damage and loss of kidney function

Chyme A liquid formed of partially digested food produced in the stomach

Cilia A hair-like organelle that helps move fluids over the surface of the cell

Ciliary Body Part of the eye that contains a focusing muscle and connective tissue

Ciliated Refers to cells with cilia

Circulating Monocyte A large phagocytic cell found in the blood

Circulatory System Pertains to blood circulation, blood vessels, and the heart

Circumcision Surgical removal of the prepuce

Cirrhosis A disease that causes the liver to become scarred and filled with fat

Clavicle or Collarbone The long bone that runs parallel to the first rib

Clavicular Refers to the region around the collar bone

Clitoris A small piece of erectile tissue within the labia minor

Clot A plug that forms in the blood to stop bleeding

Clotting Cascade A series of chemical reactions that cause clotting

Clotting Factor A chemical that initiates clotting

Coccygeal Vertebrae Three to five vertebrae of the coccyx region

Coccyx Fused coccygeal vertebrae

Coccyx Region The part of the spinal column comprising the tail bone

Cochlea A coiled organ in the inner ear that converts vibrations into neuron impulses that are sent to the brain

Codon A unit of genetic code consisting of a set of three consecutive nucleotides

Coenzymes Organic molecules that activate certain enzymes

Cofactor Elements or ions that facilitate enzyme activity

Cohesiveness A substance's tendency to stick to itself

Collagen A type of protein in connective tissue that provides strength

Collateral A branched axon

Collecting Tubule A straight tubule to which the distal tubules of several nephrons lead

Colon Another name for the large intestine

Colon Cancer A cancer that forms in the large intestine

Colon Polyp A small, fleshy outgrowth in the colon

Colostrum A fluid produced by the mammary glands in late pregnancy and just after labor

Columnar Cells shaped like a column

Commensals Beneficial microorganisms growing on the body

Comminuted or Compound Fracture A fracture in which one or more areas of bone are displaced

Common Bile Duct A tube through which secretions of the liver and pancreas pass into the duodenum

Compact or Cortical Bone Rigid outer shell of the bone

Complete Blood Cell Count (CBC) or Full Blood Cell Count (FCB) A series of tests that examine components of the blood

Compound Molecules of two or more different elements bonded together

Conception The point at which fertilization occurs

Condyloid or Ellipsoid Joint A synovial joint that permits a circular or oval pattern of movement

Cone A photoreceptor sensitive to bright light and color

Congenital Heart Disease A heart problem present before birth

Congenital A condition that appears at birth

Congestive Heart Failure (CHF) A condition in which the heart cannot pump out all of the blood entering the chambers

Conjoined Twins Twins whose bodies are joined together at birth

Conjunctiva A thin, transparent epithelium that covers the eye

Connective Tissue Tissue forming the supporting framework of the organs and the body

Constriction or Vasoconstriction The narrowing of the diameter of a blood vessel

Continuous Capillary A capillary that forms a continuous covering around the lumen

Contractile Cell A cell with a specialized membrane and cytoskeleton that permits it to change shape

Contractile Proteins The cytoskeleton proteins involved in muscle contraction

Contraction Muscle cell shortening

Contusion An injury caused by a direct hit or repeated battering of a muscle

Convolutions Folded ridges in the cortex of the brain

Copulation or Sexual Intercourse The act of mating

Cornea The clear covering at the front surface of the eye that permits light to enter

Coronal Suture The suture joining the frontal and parietal bones

Coronary Artery Vessels that supply oxygenated blood to the heart muscle

Corpus Callosum A band of white matter connecting the left and right hemispheres of the cerebrum

Corpus Cavernosum A large cylinder of erectile tissue in the penis

Corpus Luteum A structure formed in the follicle after the egg is released at ovulation

Corpus Spongiosum A sheath of erectile tissue in the penis that encloses the urethra

Cortex The gray matter covering the surface of the brain

Cortisol A type of glucocorticosteroid associated with stress-fighting and anti-inflammatory responses

Cotransported Moved across a membrane by attaching to carrier proteins that are attached to ions

Covalent Bond A bond between two elements sharing electrons

Cowper's or Bulbourethral Glands A pair of glands that lie beneath the prostate gland

Cramp The painful contraction of a muscle

Cranial Pertaining to the head

Cranial Base The portion of the cranium composed of the ethmoid and sphenoid bones

Cranial Bones Eight bones of the cranium that protect the brain

Cranial Cavity The body cavity containing the brain

Cranial Nerves Nerves of the PNS that arise from the brain

Cranium The skull

Craving The powerful and often uncontrollable desire for a substance

Creatine Phosphate A molecule that stores energy in muscle cells

Creutzfeldt-Jakob Disease A nervous system disease caused by a prion

Cricoid Cartilage A ring of cartilage that lies below the thyroid cartilage

Crossing Over The process of maternal and paternal chromosomes swapping DNA segments during meiosis

Crypts Pits of glandular cells found in the small intestine

Cubital Refers to the elbow

Cuboid A large tarsal bone that articulates with the calcaneus

Cuboidal Cells with a cubed shape

Cuneiforms Three of seven tarsal bones forming the middle region of the foot

Cuticle An outgrowth of the skin-nail fold

Cyst or Nodule A sac-like structure filled with a liquid or semisolid substance

Cystitis Inflammation of the urinary bladder

Cystocele Herniation of the bladder into the vagina

Cytokines Proteins that act as chemical messengers between cells

Cytokinesis The division of cytoplasm after karyokinesis

Cytoplasm The cell contents with the cell membrane

Cytoskeleton A meshwork of protein filaments in the cytoplasm giving the cell its shape and capacity for movement

Cytosol A gel-like fluid component of the cytoplasm

D

Deep Refers to any structure or region located away from the body's surface and toward the inside

Degenerative Refers to diseases that progressively deteriorate tissues

Dehydrated The state in which the body tissues are deprived of water

Dehydration Abnormal loss of fluid from the body

Deltoid or Triangular Muscles Muscles that have a broad origin, and focus to a narrow insertion point

Demyelination The loss of the myelin sheath of a neuron

Denaturation The process of altering a biochemical's structure

Dendrite Extensions of the nerve cell body that carry information to the nerve cell body

Dense Refers to connective Tissue that provides strength, storage, and flexibility

Dentin The bone-like substance of the tooth

Deoxyribonucleic Acid (DNA) The molecule inside cells that contains the genetic information and passes it from one generation to the next

Deoxyribose A pentose sugar involved in the composition of nucleotides

Depolarization The change in a neuron's membrane potential to a more positive potential

Depressor Muscles Muscles that produce a downward movement

Dermal (Intramembranous) Bones Bones that develop from embryonic connective tissue

Dermal Papilla A ridged layer of the dermis that is bound tightly to the stratum germinativum

Dermatitis A skin inflammation caused by an allergic reaction or contact with an irritant

Dermatomyositis Inflammation of the muscle and skin

Dermatophytes Fungi that eat hair, nails, and outer layers of the epidermis

Dermis The middle layer of skin, formed from mesenchyme cells

Descending Colon A portion of the large intestine that travels downward

Descending Tract A longitudinal band of white matter in the spinal cord that carries motor information to the body

Desmolase An enzyme that helps to convert cholesterol into progesterone

Desquamation The process of shedding dead skin cells

Detrusor Muscle The smooth muscle of the urinary bladder

Detumescence The loss of an erection

Developmental Anatomy The study of anatomical changes that occur during the growth of a human being

Dextrose The dietary form of glucose isomers

Diaphragm The muscle between the thorax and the abdomen that is used for breathing

Diaphysis The main body of a long bone

Diarrhea Frequent and watery bowel movements

Diarthrosis Joint A joint that permits a variety of movements

Diastole A part of the cardiac cycle during which the heart muscle relaxes and fills with blood

Diencephalon The part of the forebrain that contains the thalamus and hypothalamus

Differential White Blood Cell Count A measure of the percentage of each white blood cell type

Differentiation A process by which cells mature in order to carry out specific physiological tasks

Diffusion The mixing of liquid and gas molecules as a result of random thermal stirring

Diffusion Gradient The condition when a cell's internal and external environments have unequal quantities of particles

Digestive System Body system that regulates nutrition

Digestive Tract or Alimentary Canal The tubular organs through which food passes

Diglycerides A form of a glyceride lipid with two molecules of fatty acid

Dilation or Vasodilation The widening of the diameter of a blood vessel

Diploid Cells that contain the normal amount of chromosomes

Directional Orientation Refers to the anatomical view one has of a person

Directional Planes A series of terms that describe the way a body can be viewed and divided

Disaccharide Two covalently bonded monosaccharides

Dissociate When molecules of a compound break down into simpler molecules, atoms, or ions

Distal A body part located far from an attachment point

Distal Convoluted Tubule The third segment of the nephron's peritubular arrangement

Diuresis Water excretion from the body

Diuretics Chemicals that increase the volume of water in urine

Diverticulosis A condition in which pouch-like pockets develop in the large intestine

Division The initiation of cell replication

Dopamine A type of catecholamine neurotransmitter

Dormancy The suspension of active cell growth and development

Dorsal Toward the back

Dorsal Root Ganglion A bulge on a dorsal root that contains nerve cell bodies of afferent spinal nerve neurons

Dorsal Root The branch of a spinal nerve that arises from the dorsal part of the spinal cord

Dorsal Vein A large vein that runs along the dorsal length of the penis

Ductless Glands Glands without ducts to transport secretions

Ductus Arteriosus A blood vessel that connects the pulmonary artery to the aorta in the fetus

Duodenum The first section of the small intestine

Dura Mater The tough outermost layer of the meninges protecting the brain and spinal cord

Dysfunction Abnormal, impaired, or incomplete functioning of an organism, organ system, organ, tissue, or cell

Dysplasia A disorderly growth pattern in a tissue or organ

Dyspnea Shortness of breath; difficult or labored breathing

Dystrophy A progressive change in tissue due to a loss of nutrition or decreased blood flow

Dysuria Painful urination

E

Eccrine Sweat Glands Sweat glands that secrete mostly water and salts

Ectoderm The outer embryological germ layer forming the skin and brain

Ectopic Pregnancy Implantation that occurs outside the upper uterine cavity

Edema Accumulation of fluids in the body tissues

Effector A neuron that transmits motor information; another name for a target cell

Efferent Arteriole The blood vessel formed by the glomerulus where it exits the renal corpuscle

Efferent Nerve The part of the peripheral nervous system that carries information from the central nervous system to muscles and glands

Egg The female reproductive cell

Ejaculation The process of ejecting semen from the penis

Ejaculatory Duct A duct that opens into the urethra

Elastin A type of connective tissue protein that provides flexibility

Electrical Conduction System Specialized cardiac muscle cells that stimulate heart contraction

Electrical Energy The energy associated with the movement of electrons to produce a current

Electrocardiogram (ECG) a record of the heart's electrical activity

Electrocardiography A procedure that measures the electrical activity of the heart

Electrolytes Ions capable of conducting an electrical current in a solution

Electromagnetic Radiation A form of energy that travels in waves

Electromagnetic Spectrum The full breadth of wavelengths

Electron Transport Chain A chain of proteins on the mitochondrial membrane that transfers electrons for ATP production

Electrons Negatively charged particles that surround the atom's nucleus

Element A substance composed of atoms having identical numbers of subatomic parts that cannot be broken down into simpler substances by normal chemical means

Elephantiasis Blockage of the lymphatic system by infectious worms

Embryogenesis The process of embryo formation

Embryology The study of anatomical changes that occur during the growth of an embryo

Emphysema Enlargement and damage of the alveoli

Emulsion or Suspension A mixture of two or more liquids that do not readily combine

Enamel A hard layer that protects the dentin

Encephalitis Inflammation of the brain

Endergonic Chemical reactions that require cell energy

Endocarditis Inflammation of the inner lining of the heart

Endocardium The inner lining of the heart

Endochondral Ossification Bone formation that begins within a cartilage

Endochronal Bones Bones developed from embryonic cartilage

Endocrine Secretion A secretion sent directly into the bloodstream

Endocytosis A form of bulk active transport; the process of moving particles into a cell

Endoderm The inner embryological germ layer forming the digestive organs

Endometrium The mucous membrane lining of the inner surface of the uterus

Endomysium The connective tissue covering each muscle fiber

Endoneurium The connective-tissue sheath that surrounds each neuron and associated neuroglia, such as myelin

Endoplasmic Reticulum or ER An organelle responsible for the production of most of a cell's protein and lipid components

Endoskeleton An internal skeleton

Endotoxin A toxic substance found in certain bacteria

Energy The ability of chemical systems to do work or carry out change

Enlarged Heart A condition in which the heart is larger than normal

Enterocytes Absorptive cells of the small intestine

Enteroendocrine Cells Hormone-producing cells of the small intestine

Enterokinase An enzyme that activates intestinal zymogens

Entropy The measure of energy dispersal and disorganization in the universe

Envirome All of the environmental factors that affect the survival of an organism or society of organisms

Environmental Signals Signals that originate outside the body

Enzyme A functional protein

Eosinophil A blood cell associated with parasitic infections and allergies

Ependymal Cells A type of neuroglia that secretes fluids to protect the brain

Epicardium The outer layer of the heart formed by the visceral layer of the pericardium

Epidermis The outermost layer of skin

Epididymis A tube where sperm are collected and stored after leaving the testis

Epigastric The upper middle of the abdominopelvic region

Epilepsy A neurological disorder that produces recurrent seizures

Epimysium The fibrous connective tissue covering gross muscle

Epineurium The outermost layer of connective tissue covering a nerve

Epiphyseal Line A thin strip of bone marking the fusion of epiphyses to diaphysis

Epiphyseal Plate An actively growing area of bone

Epiphysis The end of a long bone that makes up a joint

Epithelial Tissue Tissue that covers external and internal body surfaces

Equatorial Plane The midline of the cell along which the cell will divide

Equilibrium A sensory system that detects the orientation of the body and head

Erectile Dysfunction The inability to produce or maintain an erection

Erectile Tissue A tissue capable of filling with blood and swelling

Erection Enlargement and hardening of the penis during sexual excitement

Erythroblast An immature red blood cell

Erythroblastosis Fetalis An immune response against a fetus' red blood cells

Erythrocytes or Red Blood Cells (RBCs) Blood cells responsible for transporting oxygen throughout the body

Erythropoiesis The formation of red blood cells

Erythropoietin (EPO) A protein secreted by the kidneys that stimulates red blood cell production

Esophagus A muscular tube through which food passes from the mouth to the stomach

Estrogen A sex hormone produced by the ovary, part of the sterol group

Ethmoid Bone The bone forming the roof of the nasal cavity and inner wall of the eye socket

Eukaryotes Cells characterized by containing their DNA inside a nucleus and having compartmentalized cytoplasm

Eustachian Tube The tube connecting the middle ear to the throat

Eversion Movement of the hand or foot so that the thumb or great toe moves away from the midline of the body

Excitable Membrane A membrane that responds to signals from other cells and the environment

Excitatory A stimulus that encourages an action potential

Excitatory Postsynaptic Potential (EPSP) A neural pathway that excites a postsynaptic neuron

Excretion The removal of waste from a cell using exocytosis

Exergonic Chemical reactions that release cell energy

Exocrine Secretion A secretion deposited into a body cavity through a duct

Exocytosis A form of bulk active transport; The process of moving particles out of a cell

Expiration or Exhalation The action of breathing out

Extension To straighten a joint

Extensor Muscles Muscles that increase the angle of a joint

External Auditory Meatus or Auditory Canal A structure that leads from the outer ear to the middle ear

External Ear The part of the ear that is visible

External Environment Conditions outside the cell

External Genitalia Sex organs on the outside of the body

External Respiration The process of inhalation and exhalation

External Stimuli A signal from outside the body that generates nervous system activity

External Urethral Sphincter A voluntary muscular ring located in the urethra

Extrapyramidal Tract A band of white matter on the ventral portion of the spinal cord that carries motor information

Extremities The upper and lower limbs

Extrinsic Aging Factor or External Aging Aging caused by environmental factors

Extrinsic Muscles 1) Muscles that originate outside of a structure and act as a unit; 2) The six muscles that move the eye

F

Facial Bones The 15 bones forming the face

Facilitated Diffusion A type of passive transport in which carrier proteins are used to move particles through a membrane

False or Fixed Vertebrae Vertebrae of the sacral and coccyx regions

Familial Periodic Paralysis Periodic weakness in the arms and legs; a genetic disorder

Fascia Fibrous tissue covering muscles, the skull, and some organs

Fascicle A bundle of muscle cells, or fibers

Fasciitis Inflammation of the hypodermis

Fat Soluble Able to dissolve in fat

Fatty Acid or Hydrocarbon A linear chain of carbons ending in a carboxyl functional group

Fatty Change The accumulation of lipids in the cell in response to cellular injury

Feces Waste eliminated from the large intestine

Femur The long bone of the thigh; the longest bone in the body

Fenestrated Capillary A capillary that has small openings around the endothelial covering

Fermentation A form of anaerobic respiration that produces lactic acid and alcohol

Fertilization The fusion of sperm and egg

Fetus A stage of development before birth

Fibrillation Rapid contractions of the heart muscles

Fibrin A protein necessary for blood clotting

Fibrinogen An inactive form of a protein necessary for blood clotting

Fibroblast A cell that secretes proteins that form collagen and elastin

Fibroids Tumors of the myometrium

Fibromyalgia A disorder of muscles and ligaments that causes widespread joint pain

Fibrous Joint A joint formed of fibrous connective tissue

Fibrous Pericardium The outermost layer of the pericardium

Fibula The lateral bone of the lower leg

Filamentous Refers to fungi characterized by masses of stringy cells

Filtrate The plasma and other substances that are removed from the blood in the renal corpuscle

Filtration A passive transport in which particles are removed from water by passage through a porous membrane

Fimbriae Finger-like projections at the end of the fallopian tube

Fine (Microscopic) Anatomy The study of anatomy concerned with microscopic features of the body

First-Degree Burn A burn involving superficial damage

Flaccid Paralysis Loss of muscle function due to a lack of muscle contraction

Flagella Swimming appendages found on some types of microorganisms

Flat Bone A bone that is thin and flattened, and often slightly curved

Flatulence Excessive gas production in the intestine

Flexion To bend a joint that decreases the angle between the bones of the limb

Flexor Muscles Muscles that decrease the angle of a joint

Flu or Influenza A contagious disease that is caused by the influenza virus

Fluid Mosaic Model Refers to the arrangement of proteins and lipids within the cell

Follicle Mite or Demodex An arthropod that causes an inflammation of the eyelash follicles

Folliculitis An inflammation of the hair follicles

Fontanelle A soft spot on an infant's skull where intramembranous development has not been completed

Food Intolerance The inability to digest or absorb a particular food

Foramen Magnum or Occipital Foramen The opening formed by the occipital bones through which the spinal cord enters the brain

Foramen Ovale An opening between the right and left atria

Forebrain The largest and anterior-most division of the embryonic brain

Fovea A depression in the retina that contains only cones

Fracture A crack or splinter in a bone

Fraternal Twins Twins produced by the simultaneous fertilization of two egg cells

Free Nerve Endings Pain-sensing nerves found in the lower part of the epidermis

Free Radicals Aggressive chemicals that readily react with biochemicals

Frontal Bone Bone forming the front part of the skull, shaping the forehead

Frontal Lobe The region of the brain situated directly behind the forehead

Fructose or Levulose A hexose monosaccharide that can be used as an energy source; fruit sugar

Functional Classification A system of joint classification based on the way joints move

Functional Group One or more elements attached to a carbon skeleton, responsible for the chemical activities of a biochemical

Functional Proteins Proteins that carry out functions that run the body

Fundic Glands Pits of secretory cells in lower region of the stomach

Fundic Region Middle region of the stomach

Fungi A microorganism found on the body, both single and multicellular

Furuncle or Boil An inflammation of a hair follicle

G

G_0 Differentiation stage of interphase

G_1 **or Gap 1** The first phase of interphase

G_2 **or Gap 2** The last phase of interphase

Galactose A hexose monosaccharide found in milk products and sugar beets

Galactosylceramide Beta-Galactosidase An enzyme that removes certain waste products from nerve cells

Gallbladder A muscular sac attached to the liver

Gallstone A solid material that forms in the gallbladder

Gamete or Haploid Cells that contain half the number of chromosomes of regular cells

Gamma-Aminobutyric Acid (GABA) An inhibitory amino acid neurotransmitter

Ganglia A collection of nerve cell bodies

Gastric Dysphagia A swallowing disorder

Gastric Reflux or Acid Reflux The backward flow of stomach contents into the esophagus

Gastric Stem Cells Cells that replace other glandular cells of the stomach

Gastrin A hormone that stimulates acid production

Gastrula An embryonic stage in which the three basic tissue germ layers form

Gastrulation A stage in embryonic development that forms the tissues of the digestive system

Gene A functional unit of heredity, one of many segments of DNA

Gene Regulatory Network or GRN Pieces of DNA that function as on and off switches for genes

Genetic Code The basis of DNA information, the specific order of DNA and RNA

Genetic Expression A process by which the genetic information in a cell is used to produce cell structures and carry out cell functions

Geniculate Refers to the knee region

Genital Warts Warts on the penis or female reproductive tract caused by the human papillomavirus

Genome The genetic material within the cell passed from generation to generation

Gladiolus The body of the sternum

Glandular Lobule Glandular clusters in an organ

Glans The swollen portion at the tip of the penis

Gliding Joint A synovial joint that permits a variety of side-to-side movements

Glioma A tumor that develops from neuroglia in the brain

Globulin A blood protein that acts like an antibody

Glomerular Filtration The process by which plasma and many dissolved substances are moved from the blood into Bowman's capsule

Glomerulonephritis An autoimmune disorder causing inflammation and deterioration of the glomerular membranes

Glomerulus A capillary loop within the nephron

Glottis The opening at the superior end of the larynx

Glucagon A peptide hormone produced by the pancreas that elevates blood glucose levels

Glucocorticosteroids Steroid hormones produced by the adrenal cortex that affect glucose metabolism in the body

Glucose A hexose monosaccharide that is the major energy source for most body functions

Glucosuria The presence of glucose in the urine

Glutamate An excitatory amino acid neurotransmitter

Glycemic Index A measurement indicating the amount of glucose available in a particular food

Glyceride The most abundant lipid in the body

Glycine An inhibitory amino acid neurotransmitter

Glycogen A highly branched polysaccharide of alpha-bonded glucose stored in the liver and muscle

Glycogen Storage Diseases Diseases that cause muscle weakness due to a diminished ability to use glucose

Glycosidic Linkage A type of covalent bond found in carbohydrates

Golgi Body or Golgi Apparatus An organelle responsible for modifying, storing, and shipping certain products from the endoplasmic reticulum

Gomphosis A synarthrodial joint formed by a conical process; it is held in a socket by a ligament

Gonad An organ of the reproductive system that produces eggs or sperm

Gout A metabolic disorder that causes severe inflammation of the joints

Graafian Follicle A nearly mature egg or ovum

Graded Effects Different pulling forces

Granulocyte or Polymorphonuclear White Blood Cell A WBC with conspicuous granules in the cytoplasm

Gray Matter The parts of the brain and spinal cord composed primarily of nerve cell bodies

Greenstick Fracture A fracture in which one side of the bone is broken and the other side is bent

Gross Anatomy The study of anatomy concerned with the features of the body visible to the naked eye

Growth Factor Chemicals that act as signals to initiate cell division and differentiation

Gustation The sense of taste

H

Hair Cortex The principle inner layer of the hair shaft

Hair Cuticle The outer layer of the hair shaft

Hair Cycle The mitosis cycle of the hair follicle

Hair Follicle or Hair Bulb An inward protrusion of the epidermis

Hair Medulla The inner layer of the hair shaft

Hair Papilla The base of the hair follicle

Hair Shaft The main part of the hair structure

Hamate A distal carpal bone

Hantavirus Pulmonary Syndrome A viral disease characterized by a sudden onset of fever, pain, and vomiting

Hapten A molecule that can cause an immune response when attached to blood proteins

Hard Palate The anterior part of the palate that covers the maxillary and palatine bones

Haversian Canal The internal structure of the osteon

Heart The hollow muscular organ that pumps blood throughout the body

Heart Rate The number of times the heart beats in one minute

Heat A form of energy that is transferred by a difference in temperature

Heimlich Maneuver First aid given to a choking victim

Helix A type of secondary structure of protein with a spring-like shape

Helper T-lymphocyte A type of T-lymphocyte that cooperates with B-lymphocytes

Hematocrit or Packed Cell Volume (PCV) The percentage of packed RBCs in a unit volume of whole blood

Hematopoietic or Multipotent Stem Cell A stem cell from which all RBCs and WBCs develop

Hematuria Presence of red blood cells in the urine

Hemidesmosome A specialized junction between an epithelial cell and the basement membrane

Hemodialysis A medical procedure that allows for artificial filtering of the blood

Hemoglobin A protein in red blood cells that carries oxygen

Hemogram Another name for complete blood cell count

Hemophilia A genetic disease that prevents normal blood clotting

Heparin A protein that helps prevent blood clotting

Hepatic Artery The artery that provides resources for the liver

Hepatic Flexure The last portion of the ascending colon

Hepatic Portal Vein A vein that carries blood from the small intestine to the liver

Hepatic Vein A vein that carries wastes from the liver

Hepatitis Inflammation of the liver

Hepatocyte A liver cell that carries out the major liver functions

Hernia The protrusion of an organ into surrounding tissues

Herpesvirus An inflammatory virus that grows in nerve tissue

Hexose Monosaccharide A monosaccharide with six carbon molecules

Hiatal Hernia A protrusion of the upper part of the stomach into the thorax

Hilum A depression of an organ where blood vessels enter the liver

Hilus The medial concave curve of each kidney

Hindbrain The lowermost portion of the embryonic brain, just above the spinal cord

Hinge Joint A synovial joint that permits an angular movement along a plane

Hirschsprung's Disease A congenital disorder in which the intestines do not have the normal network of nerves

Hodgkin's Lymphoma Cancers of the lymphatic system cells

Holocrine Gland Secretions composed of dead cells that swell with fat and then rupture

Homeostasis The natural tendency of a person to maintain psychological and physiological stability

Homologous Referring to chromosomes that are derived from a maternal egg and paternal sperm

Hormone Replacement Therapy (HRT) Treatment to replace hormones

Hormones Signals that originate inside the body

Human Chorionic Gonadotropin (hCG) A hormone produced by the placenta

Human Immunodeficiency Virus (HIV) An infectious virus that causes the immunodeficiency disease AIDS

Human Papilloma Virus A group of viruses that cause various types of warts in humans

Humerus The longest bone of the upper arm

Humoral Immunity The immunity that results from antibody production by B-lymphocytes

Hunger Center A region of the brain that signals hunger

Hyaline Cartilage A smooth cartilage covering the articular surfaces bones

Hydatid Lung Disease A lung infection caused by the inactive stage of a worm

Hydrocephalus Excess fluid in the ventricles of the brain

Hydrogen Bond A temporary weak bond between a partial positive hydrogen atom and a partial negative oxygen, nitrogen, or fluorine atom

Hydrogen Ion Acceptors Basic molecules that dissociate in water, absorbing hydrogen ions

Hydrogen Ion Donor Acidic molecules that dissociate in water, releasing hydrogen ions

Hydrogenate To add hydrogen to unsaturated fats

Hydrolysis A chemical process of decomposing a molecule with water

Hydrophilic A polar substance that dissolves in water

Hydrophobic or Lipophilic A nonpolar substance that does not mix with water

Hydrostatic Pressure The pressure of the water that is circulated in the blood and tissues

Hydroxyl Group The functional group containing oxygen; helps molecules to dissolve in water

Hydroxyl Ions Negatively charged oxygen particles bonded to hydrogen

Hymen A thin membrane partially covering the opening of the vagina

Hyoid Bone The U-shaped bone supporting the tongue and its muscles

Hyperosmotic Refers to a solution with a greater concentration of solute than exists in the cell

Hyperplasias The abnormal multiplication of normal cells

Hyperpolarization A condition in which a neuron's membrane potential becomes more negative than the resting potential

Hypertension Blood pressure above the normal range

Hypertonic Refers to an environment with a greater quantity of a particular molecule than exists in the cell

Hypertrophy The abnormal increase in cell size

Hypodermis or Subcutaneous Layer The innermost layer of skin

Hypogastric The middle section between the right and left inguinal sections of the abdominopelvic region

Hypoosmotic Refers to an environment with a greater quantity of a particular molecule than exists in the cell

Hypopigmentation The decrease in melanin production

Hypospadias Abnormal development of the penis and male urethra

Hypothalamus A region of the brain that controls endocrine activity

Hypotonic Refers to an environment with a lesser quantity of a particular molecule than exists in the cell

I

Identical Twins Twins that develop from one fertilized egg

IgA An antibody or immunoglobulin found in blood, tears, saliva, and mucous membranes

IgD An antibody or immunoglobulin present on the B-lymphocyte membrane

IgE The type of antibody or immunoglobulin most instrumental in allergic reactions

IgG The predominant type of antibody or immunoglobulin existing in the blood

IgM An antibody or immunoglobulin that gathers in clusters and is involved in killing bacteria

Ileocecal Valve A muscular valve separating the small intestine from the large intestine

Ileum The last section of the small intestine

Ilium The upper bone of the pelvic girdle

Immunization The process of inducing immunity

Immunodeficiency A decreased ability of the body to fight infection and disease

Impetigo An inflammatory bacterial disease of skin that produces rashes

Implantation Attachment of the embryo to the endometrial lining

Incisors The front (cutting) teeth

Incontinence The inability to hold urine

Incus The second bone in the series of three ossicles in the middle ear

Inferior A body part that is below another

Inferior Nasal Conchae Bones forming an inferior protrusion in the nasal cavity

Inferior Orbital Foramen A passageway to the eye for blood vessels and nerves

Inferior Vena Cava The large vein that carries blood from the lower body and organs to the right side of the heart

Inflammatory Bowel Disease (IBD) disease that causes irritation to the intestines

Inflammatory Response The swelling, redness, warmth, or pain in the area of an immune response

Ingestion Taking something into the body by the mouth

Inguinal Hernia A protrusion of the small intestine into pelvic muscles

Inherent Refers to qualities with which a person or animal is born

Inhibitory Postsynaptic Potential (IPSP) A neural pathway that inhibits a postsynaptic neuron

Inhibitory A stimulus that discourages an action potential

Innate Immunity or Nonspecific Immunity A general way in which the body responds to disease

Inner Ear The part of the ear that contains both the organs for hearing and balance

Innervate To supply nerves to an organ or other body part

Insertion (Muscle) The movable attachment point of a muscle to a bone

Inspiration or Inhalation The action of breathing in

Insula A region of the cerebrum associated with memories

Insulin A peptide hormone produced by the pancreas that lowers blood glucose levels

Insulin-Like Growth Factor-1 or IGF-1 A chemical needed for muscle cell growth, maintenance and repair

Insulin Receptor A membrane receptor activated by insulin that promotes glucose uptake, or absorption, by cells

Integument Skin

Integumentary System A system of tissues and organs associated with the skin

Intercalated Disks Structures that connect cardiac muscles cells to each other

Interferons A group of proteins that interfere with viruses

Intermediate Gray Matter A band of gray matter between the dorsal and ventral horns of the spinal cord

Internal Environment Conditions within the cell or the body

Internal Receptor A hormone receptor located within a cell

Internal Respiration The exchange of oxygen and carbon dioxide between the blood and body cells

Internal Stimuli A signal from inside the body that generates nervous system activity

Internal Urinary Sphincter An involuntary muscular ring at the exit of the urinary bladder

Interneuron A neuron that communicates only with other neurons

Interphase The non-divisional stage of the cell cycle the prepares the cell for division

Intersex A condition in which it is not clear at birth whether the individual is a male or a female

Intramembranous Ossification The formation of bone from connective-tissue membranes

Intrapleural Pressure The air pressure within the pleural membranes

Intrinsic Aging Factor Aging caused by the natural decline of cells

Intrinsic Beat A natural contraction cycle of cardiac muscles

Intrinsic Muscles The tongue muscles that attach to other tongue muscles and mucus membranes

Inulin A polysaccharide containing large amounts of fructose

Inversion Movement of the hand or foot so that the thumb or great toe moves toward the midline of the body

Involuntary Muscle Muscles that work without conscious effort

Ion An element that has gained or lost one or more electrons

Ionic Bond A bond between two electrically charged elements

Irregular Bone A bone having a unique, often complicated, shape that is not geometrically describable

Ischemic Attack A condition caused by insufficient blood flow to a body part

Ischium The lower bone of the pelvic girdle

Islets Clumps or islands of cells in an organ that secrete hormones

Islets of Langerhans Clumps of cells in the pancreas that secrete insulin and glucagons

Isomer One of two or more molecules with the same molecular formula, but different structural formulas

Isometric Muscle A muscle that is not lengthening or shortening

Isotonic Muscle A muscle that is actively shortening or lengthening

Isoosmotic Refers to a solution with an equal concentration of solute in the solution and the cell

Isotonic Refers to an environment with an equal quantity of a particular molecule in the solution and the cell

Isotope A variation of an element having the same number of protons, but a different number of neutrons

J

Jejunum The middle section of the small intestine

Juvenile Arthritis Arthritis that affects children

K

Karyokinesis The separation of the DNA into different nuclei

Keratin A yellow sulfur-rich protein that gives skin its strength

Keratocytes Cells that contain large amounts of the protein keratin

Kidneys The organs that form urine

Kilocalorie A standard unit of measurement, 1,000 times larger than a calorie

Kinetic Energy The energy associated with motion or action

Krabbe's Disease A degenerative disorder that affects the nervous system

Krause End Bulbs Nerves found in the mouth that respond to touch

Krebs Cycle or Tricarboxylic Acid Cycle (TCA) A series of chemical reactions whereby energy is obtained from the oxidation of certain molecules

Kupffer Cell A large phagocytic cell found in the liver

L

Labia Another name for lips

Labia Majora Outer lips of the vulva

Labia Minora Inner lips of the vulva

Labor The beginning of childbirth

Lacrimal Bones Bones forming the medial region of the orbits

Lacrimal Duct A tube connecting the orbit with the nasal cavity

Lacrimal Gland A gland that produces tears

Lactation The formation of milk by the mammary glands

Lacteal A collection of lymphatic vessels inside the villi

Lactic Acid A byproduct of fermentation that can cause soreness if built up in the muscles

Lactiferous Ducts Ducts of the mammary glands that carry milk to the nipples

Lactose A disaccharide; a combination of glucose and galactose

Lactose Intolerance The inability to digest lactose

Lacunae Cavities that store osteocytes

Lamellated or Pacinian Corpuscles Nerves that respond to hard pressure

Lamina Propria A layer of connective tissue underneath the epithelium of mucosa

Lanugo Temporary, fine body hair found on babies

Large Intestine A tube that extends from the small intestine to the rectum

Laryngeal Prominence The anterior region of the thyroid cartilage commonly called the Adam's apple

Laryngopharynx The lower part of the pharynx

Larynx The area of the throat that houses the vocal cords

Lateral Furthest from the midline of the body

Left Hypochondriac The upper left corner of the abdominopelvic region

Left Inguinal The lower left corner of the abdominopelvic region

Left Lobe A large subdivision of the liver

Left Lower Quadrant The quadrant containing the left hypochondriac, lumbar, epigastric, and umbilical regions

Left Lumbar The middle left corner of the abdominopelvic region

Left Upper Quadrant The quadrant containing the left hypochondriac, lumbar, epigastric, and umbilical regions

Lens The transparent structure inside the eye that focuses light rays for clear vision

Leukemia A cancer characterized by an increased number of abnormal leukocytes

Leukocytes or White Blood Cells (WBCs) Colorless blood cells that fight disease and maintain immune function

Levator Muscles Muscles that produce an upward movement

Levels of Organization or Hierarchies A series of ordered groupings within a system

Leydig's Cells Cells that produce testosterone in the testis

Lice Arthropods that irritate the skin and spread disease

Ligament Connective tissue that joins bone to bone

Ligand A chemical that attaches to a receptor

Limbic System A collection of nuclei at the base of the cerebrum associated with emotions

Linear Chains of polymers without branching

Lingual Membrane The covering of the tongue

Lingual Tonsils A mass of lymphatic tissue that covers the posterior region of the tongue

Lipid Hormones Hormones having a lipid structure

Lipids or Fats Simple molecules that provide the body with chemical signals, insulation, protective padding, and stored energy

Lipofuscin A brown waste material deposited in neurons

Lipomas Tumors formed in fat cells

Lobar Pneumonia A type of pneumonia limited to one lobe of the lung

Lobe A subdivision of the lung structure

Lobule The smallest subdivision of the lung lobes

Long Bone A bone having an elongated shape

Longitudinal Cerebral Fissure A crevice that separates the left and right hemispheres of the cerebrum

Longus Muscle The longest muscle of a muscle group

Loop of Henle The second segment of the nephron's peritubular arrangement

Loose Refers to connective tissue that provides attachment, stabilization, structure, and support for other tissues

Lower Appendages The hips, legs, knees, ankles, and feet

Lower Respiratory System The part of the respiratory system composed of the trachea, bronchial tree, and lungs

Lower Uterine Segment The lower third of the uterus

Lumbar Region The part of the spinal column comprising the dorsal section of the umbilical region

Lumbar Vertebrae Five vertebrae of the lumbar region

Lumen The space within the interior of a tubular body structure

Lung Cancer The uncontrolled growth of abnormal cells in lung tissue

Lung One of two large organs in which gas is exchanged between the blood and the environment

Lymph Node or Lymph Gland A small round gland that is part of the immune system

Lymph A colorless fluid that bathes cells and moves through lymphatic vessels

Lymphatic Sinuses A fluid-filled sack located in a lymph node

Lymphatic System Body system that regulates body fluids and helps fight disease

Lymphatic Trunk A network of lymph vessels that drains regions of the body

Lymphedema Swelling of the appendages due to blockage of the lymph vessels

Lymphocyte A white blood cell present in the blood and lymphatic system

Lymphoid A tissue composed of stem cells that gives rise to WBCs

Lymphoma A tumor that develops from cells of the immune system

Lysosome A vesicle responsible for recycling cell components

M

M Phase The mitosis phase of the cell cycles

Macrocytic Anemia Anemia characterized by RBCs that are larger than normal

Macrophage A large phagocytic white blood cell found in connective tissue

Mad Cow Disease or Bovine Spongiform Encephalopathy A neurotrophic virus that causes degeneration of the nervous system

Major Basic Protein A protein produced by eosinophils that harms or kills parasites

Malleus The first bone in the series of three ossicles in the middle ear

Malnutrition The condition resulting from a diet lacking a balanced molecular composition

Malpighian Layer The layer of epidermis containing melanocytes

Maltose A disaccharide; a combination of two glucose molecules

Mammary Gland A milk-secreting organ of females

Mandible The lower jawbone

Mannose A hexose monosaccharide that can be used as an energy source

Manubrium The upper region of the sternum

Mast Cell A type of basophil that causes inflammation

Mastication Chewing

Mastoid Process The attachment for neck muscles

Matrix Intercellular material found in connective tissue

Matter Material that has mass and occupies space

Maxillary Bones Bones forming the upper jawbone

Maximal Heart Rate The fastest heart rate possible under normal, maximal exercise conditions

Maximus Muscle The largest muscle of a muscle group

Mean Corpuscular Hemoglobin (MCH) The mass of hemoglobin molecules in each red blood cell

Mean Corpuscular Hemoglobin Concentration (MCHC) The average concentration of hemoglobin in red blood cells

Mean Corpuscular Volume (MCV) A measure of the average volume of red blood cells

Mechanical Damage Damage caused by any force that compresses, erodes, tears, or stretches the skin

Mechanical Energy The energy of motion or movement used to perform work

Meconium The intestinal contents of the fetus

Medial Nearest to the midline of the body

Mediastinum The body cavity between the lungs containing the pericardial cavity

Medulla Oblongata A section of the hindbrain connecting the pons to the spinal cord

Medullary or Marrow Cavity The hollow center of long bones

Megakaryocytes A bone marrow cell responsible for the production of blood platelets

Meiosis Sexual division

Meiosis I The first stage of meiosis

Meiosis II The second stage of meiosis

Melanin A black- or brown-colored chemical that gives color to skin

Melanoblast A cell that will develop into a melanocyte and produce pigment

Melanocyte An integumentary system cell that produces pigment

Melanosome The organelle in which melanin is produced

Melasma A disorder resulting in brown patches on the face

Melatonin A hormone responsible for regulating the body's daily rhythms

Membrane Diffusion A type of passive transport in which certain particles pass through a membrane

Membrane A sheet-like structure that surrounds a cell and keep its internal environment contained

Memory Cell A B-lymphocyte that stores the information to produce immune proteins

Meninges Membranes surrounding the brain and spinal cord

Meningioma A tumor that develops from the meninges

Meningitis Inflammation of the coverings around the brain

Menopause The cessation of the menstrual periods

Menses The shedding of the endometrium

Menstrual Period The cyclic shedding of the endometrium

Mental Foramen Passageways to the teeth for blood vessels and nerves

Merkel Cells Nerves sensitive to gentle physical sensations

Mesenchyme Embryonic connective tissue

Mesentery A membranous tissue that connects the small intestine to the peritoneum

Mesoderm The middle embryological germ layer forming bone and muscle

Messenger RNA or mRNA A nucleic acid derived from a copied segment of DNA during transcription

Metabolism The sum of all chemical reactions in the body that maintain homeostasis

Metal Any of several elements that conduct heat and electricity

Metaphase The second phase of mitosis

Metaphase I The second phase of meiosis I

Metaplasia An abnormal change in cell and tissue function

Metastasis The movement of diseased cells away from their original location to establish themselves in a new area of the body

Metastatic Pertaining to the ability of a cell or a group of cancerous cells to move throughout the body

Metatarsals The long bones of the foot

Microbe or Microorganism Any simple organism that can only be seen with a microscope

Microbial Damage Damage caused microorganisms on the skin

Microcirculation The circulation of blood flow through blood vessels

Microcytic Anemia Anemia characterized by RBCs that are smaller than normal

Microglia A type of neuroglia that maintains ion balances

Microvilli Small projections on the surface of epithelial cells of the villi in the small intestine

Micturition Elimination of urine from the bladder

Midbrain An arrangement of neurons that connects with the forebrain and organizes sensory information

Middle Ear The part of the ear inside the skull that contains the ear bones

Mineralocorticosteroids Steroid hormones produced by the adrenal cortex that affect salt and water balance in the body

Minerals Nutrients needed by the body

Minimus Muscle The smallest muscle of a muscle group

Ministroke A mild and temporary cutoff of the blood supply to the central nervous system

Mitochondria An organelle responsible for producing much of a cell's energy

Mitochondrial Myopathies Genetic mitochondrial abnormalities that prevent muscles from producing energy

Mitogen A factor that directs a cell to undergo division

Mitosis Asexual division

Mitral or Bicuspid Valve One of the two atrioventricular valves; it lies between the left atrium and left ventricle

mmHg The abbreviation for millimeters of mercury used to measure pressure

Molars The large teeth at the back of the mouth

Molecular Formula The number of atoms of an element present in a molecule

Molecular Level of Organization Groups of atoms making up molecules

Moles Heavily pigmented squamous cell tumors

Monoamine A neurotransmitter containing only one amino group

Monocyte A large white blood cell involved in immune defense

Monoglycerides A form of glyceride with one molecule of fatty acid

Monomer The simplest form of a biochemical

Monosaccharide A simple sugar consisting of a single sugar molecule that cannot be further decomposed. They are commonly used as a source of energy in organisms.

Mons A pad of fat tissue that covers the pubic bone in females

Morphology (Muscle) The shape of the muscle

Morphology The structural make-up of an organism, referring to differences and similarities in anatomy

Motor Cortex A strip of cortex involved in voluntary skeletal muscle control

Motor Nerve Cells Nerves that control skeletal muscle fibers

MSG The abbreviation for "monosodium glutamate," which is a seasoning often found in Chinese food

Mucosa The mucus membrane making up the inner layer of the digestive tract

Mucous Membrane Lubricated inner linings that secrete mucus

Mucous Neck Cells Cells of the stomach that secrete mucus

Multicellular Consisting of many cells

Multiple Sclerosis (MS) A potentially debilitating disease caused by demyelination

Multipolar Refers to a neuron that has one axon with many dendrites

Murmur An abnormal sound of the heart

Muscle Cell Contraction Stage The second stage of muscle cell contraction

Muscle Cell Relaxation Stage The last stage of muscle cell contraction

Muscle Fiber A muscle tissue cell

Muscle Sensitivity Continuous muscle pain due to tissue damage or disease

Muscle Tissue Tissue providing the body with movement and support

Muscular Dystrophies Diseases characterized by progressive weakness of voluntary muscles

Muscular System Body system that provides structure and movement

Muscularis Layer The smooth muscle layer of the digestive tract

Myelin A sheath protecting neurons

Myeloid Progenitor A stem cell that gives rise to platelets

Myeloma A cancer of the red bone marrow

Myoblast A stem cell that forms muscle tissue

Myocardium The muscle of the heart wall that contracts to pump blood

Myofibril Long cords of myofilaments that form parallel bundles that comprise most of a muscle cell's interior

Myofilaments Bands of proteins that compose the cytoskeleton of muscle cells

Myogenesis The process by which embryonic mesoderm cells become muscle tissue

Myoglobin A chemical that stores oxygen for muscle cells

Myoglobinurias Disorders that affect how myoglobin provides oxygen to muscles

Myometrium The muscular wall of the uterus

Myopathy or Neuromuscular Disorders Diseases characterized by the nervous system's inability to communicate with the muscular system

Myosalpinx The middle muscular layer of the fallopian tube

Myosin A protein of the thick myofilaments of muscle cells

Myositis Ossificans Bone growing within muscle tissue

Myotonia The slow relaxation of muscles after contraction

N

Nail Body or Nail Matrix The nail

Nail Plate The pink area underneath the nail body

Nail Root The area of the nail where growth takes place

Nasal Bones Bones forming the bridge of the nose

Nasal Cavity The body cavity behind the nose

Nasal Septum A bony plate that divides the nasal cavity

Nasopharynx The upper part of the throat behind the nose

Natural Immunization Natural exposure to foreign antigens

Natural Killer (NK) Cells T-lymphocytes that kill tumor cells and microorganisms

Navicular or Scaphoid Bone A proximal carpal bone

Navicular One of the tarsal bones located near the heel

Necrosis Localized tissue death

Negative Feedback A signal that inhibits an endocrine gland, preventing further secretion of a particular hormone

Nephrolith An alternate term for a calculus (singular for calculi)

Nephrons Tubular structures that filter the urine in the kidneys

Nephroptosis Movement of the kidney from its proper anatomical position to an inferior position

Nerve Cell Body The part of the neuron containing the nucleus

Nerve Tract A bundle of neurons following a path through the body or brain

Nervous System Body system that regulates the flow of information

Nervous Tissue Tissue that conducts and coordinates body information

Neural Crest Cells Fetal cells that are derived from the neural tube

Neural Pathways An arrangement of neurons connecting one part of the nervous system with another

Neural Stimulation Stage The first stage of muscle cell contraction

Neural Tube A structure in early fetal life that develops into the nervous system

Neuroblastoma A tumor that develops from immature nervous system cells

Neurodegenerative Disease A condition characterized by the deterioration of nervous tissue

Neurofascicle A bundle of neurons and associated neuroglia that form a part of a nerve

Neuroglia Nerve cells that assist neurons

Neuroma A tumor that develops from nervous system cells

Neuromuscular Junction The space between a nerve cell and a sarcolemma

Neuromyotonia A nerve disorder characterized by bouts of muscle twitching and stiffness

Neurons Nerve cells that transmit impulse throughout the body

Neurotransmitter A chemical used for cell communication

Neurotransmitter Receptor A cell membrane protein that binds to a signaling molecule to set off a cell response

Neurotrophic Refers to an organism that infects nerve tissue

Neurovascular Disease Disorders of blood vessels in the nerves

Neutral Neither basic nor acidic

Neutrophil The most common type of white blood cell

Nipple A small raised area in the center of each mammary gland

Nodes of Ranvier Gaps found between the Schwann cells

Non-Hodgkin's Lymphoma Cancers of the lymphocytes in the lymph nodes

Nonpolar Describes molecules lacking the electrically charged functional groups that aid in water solubility

Noradrenaline or Norepinephrine A hormone produced by the adrenal glands that has a stimulatory effect on the nervous system

Nose The entrance to the respiratory tract

Nostrils or Nares The openings in the nose through which air enters the nasal cavity

Nuclear Decay When the nucleus of an atom breaks down

Nuclear Envelope An organelle responsible for transmitting genetic information

Nucleic Acid Molecules involved in converting food to energy and an essential component of genetic material

Nucleoid A region in the cytoplasm of bacteria containing the genome

Nucleotide The building block of nucleic acids

Nucleus The central structure of a eukaryotic cell containing the DNA

O

Oblique Fracture Refers to a fracture occurring at angles on the bone

Obturator Foramen A passage for major blood vessels and nerves of the pelvis

Occipital Bone Bone forming the back part of the skull

Occipital Lobe The posterior region of the cerebrum

Ocular Refers to the eyes

Olecranon Process The proximal extremity of the ulna; "funny bone"

Olfaction The sense of smell

Olfactory Bulb An enlargement of the olfactory nerve that senses smell

Oligodendrite A type of neuroglia that produces myelin

Oligopeptides Small chains of amino acids

Oligosaccharide A polysaccharide consisting of three to 10 monosaccharides

Oliguria Decreased urine production

Oocyte An immature egg

Open Fracture A fracture in which the skin is pierced by the bone

Optic Cup A cuplike depression that develops into the sensory layer of the eyes

Orbital Ridge The thickened area of frontal bone above the orbits

Orbits Eye sockets

Organ A structure composed of more than one tissue that is specialized for a particular function

Organ Level of Organization Groups of tissues that perform similar specific functions

Organ System Level of Organization Groups of organs that perform similar specific functions

Organ Systems A collection of organs having related roles in the body's function

Organelle Specialized functional units of compartmentalized cytoplasm

Organic Chemistry The field of chemistry that studies matter composed of carbon

Organism An individual biological unit capable of reproduction

Orgasm An intense sensation that occurs at the height of sexual excitement

Origin The point of attachment of a muscle that remains fixed during contraction

Oropharynx The area of the throat at the back of the mouth

Osmolarity Refers to water's potential to move across a membrane

Osmosis Diffusion of water across a membrane

Osseus Bone tissue

Ossicles The three tiny bones in the middle ear

Ossification The process of bone formation

Osteoarthritis The deterioration of the articular cartilage covering the ends of bones

Osteoblasts Cells that build bone tissue

Osteocalcin Protein comprising bone matrix

Osteoclasts Cells that break down bone and cartilage

Osteocytes The mature cells of bone tissue

Osteomyelitis Inflammation of the bone caused by bacterial blood infections

Osteon The structural unit of compact bone tissue

Osteonecrosis A condition caused by osteocyte death due to the obstruction of blood flow

Osteoporosis A degenerative bone disorder that causes bone tissue loss

Oval Window A membrane separating the middle ear from the inner ear

Ovarian Cycle The sequence of events that lead to ovulation

Ovarian Follicle A fluid-filled sac in which an egg matures

Ovarian Ligament A strip of connective tissue that attaches the ovary to the uterus

Ovary A female gonad; produces eggs

Overhydration The state in which the body contains too much water causing dilution of the tissues

Oviduct Another name for fallopian tube

Ovulation Release of an ovum from the ovary

Ovum A term for the female sex cell, or egg

Oxidant A substance that oxidizes something

Oxidation The process of joining oxygen with another molecule; a chemical change in which an atom loses electrons

Oxidative Phosphorylation The process by which the electron transport chain produces ATP

P

P Wave Electrical activity of the sinoatrial node and atria

Palate The roof of the buccal cavity

Palatine Bone The bone forming the walls of the nasal cavity and the posterior roof of the mouth

Palmar Pertaining to the palm of the hand

Palpebrae The eyelids

Palsy A nervous system disorder that causes paralysis of a muscle or group of muscles

Pancreas A gland located near the stomach that produces digestive enzymes and hormones that regulate blood glucose levels

Pancreatic Duct A tube that collects pancreatic exocrine secretions

Pancreatitis Inflammation of the pancreas

Paneth Cells Cells of the small intestine that maintain beneficial microorganisms

Papillae Elevated areas of the tongue that contain taste buds

Papillary Muscles Muscles in the wall of the ventricles that control the atrioventricular valves

Paracrine A type of secretion that travels a short distance to target cells via the blood or other body fluids

Parallel Muscle Cells Muscle cells that run in the same direction

Paranasal Sinuses Air cavities within the facial bones

Parasympathetic Nerves The cranial and sacral divisions of the autonomic nervous system

Parathyroid Gland One of four small glands located behind the thyroid that increase calcium levels in the blood

Parathyroid Hormone (PTH) A hormone secreted by the parathyroid gland that regulates calcium levels in the blood

Parenteral Nutrition taken into the body by bypassing the digestive tract

Parietal Refers to the outer wall of a hollow body part, such as the stomach. It also refers to the thin linings covering body cavities

Parietal Bones The two bones on either side of the skull that form the sides and roof of the skull

Parietal Cells Acid-secreting cells in the stomach lining

Parietal Lobe A region of the cerebrum posterior to the frontal lobe

Parietal Pleura The outer membrane of the pleura

Parotid Glands The largest salivary glands, located in front of the ears

Partial Pressure The individual pressure exerted by a particular component of a gas mixture

Passive Immunity Immunity gained by the introduction of antibodies

Passive Transport Diffusion across a membrane which requires no cell energy

Patella The kneecap

Pathology The study of human diseases

Pedal Refers to the region around the foot

Pelvic Refers to the region around the hip bone

Pelvic Cavity The body cavity containing the rectum, reproductive system, and urinary bladder

Pelvic Inflammatory Disease (PID) An infection of the upper female reproductive tract

Penis or Phallus An external part of the urinary and reproductive systems of the male

Pentose A monosaccharide with five carbon molecules

Peptide Bond A bond created when the functional group of one amino acid covalently bonds to the carboxyl functional group of a neighboring amino acid

Peptide Hormones Hormones composed of amino acids

Peptides Linear polymers of amino acids

Percent Saturation The amount of oxygen that is dissolved in a solution of hemoglobin molecules

Pericardial Cavity The body cavity containing the heart

Pericarditis Inflammation of the membrane that surrounds the heart

Pericardium A membranous sac that encloses the heart

Perimysium The thin connective tissue covering each fascicle

Perineum A diamond-shaped region making up the base of the pelvic region

Perineurium The layer of connective tissue covering a nerve neurofascicle

Periodic Table A chart of all known elements arranged according to chemical properties

Periosteum A connective-tissue membrane that covers bones

Peripheral Nervous System (PNS) The part of the nervous system made up of neurons and neuroglia outside of the brain and spinal cord

Peripheral Neuropathy A condition of the nervous system that causes numbness, pain, tingling, or weakness in the peripheral nervous system

Peristalsis Weak, wave-like muscle contractions that push food, liquid, and waste through the digestive system

Peritubular Capillary System The network of renal tubules and capillaries distal to the glomerulus and Bowman's capsule of the renal corpuscle

Pernicious Anemia A type of anemia caused by vitamin B12 deficiency

Peyer's Patches Lymphoid tissues in the lining of the digestive system

pH "potential for hydrogen atoms," the measure of hydrogen ion concentration in water

pH Scale A scale measuring the concentration of hydrogen ions

Phalanges The bones of the fingers and toes

Pharynx The throat, or cavity, behind the mouth

Pheromones Secretions that leave the body and signal the cells of other organisms; chemicals secreted by apocrine sweat glands, believed to play a role in courtship and social behavior

Phosphate Group The functional group containing phosphorus, involved with the capture and release of energy

Phospholipid A lipid with a phosphate group attached to the glycerol of a diglyceride

Photoreceptor A sensory receptor that detects light stimuli

Phrenic Nerve A motor nerve to the diaphragm

Physical Properties Any characteristic that can be detected by the five human senses or devices that are extensions of the senses

Physiological Environment All the internal conditions that optimize individual cell function and body organization

Physiology The functions of an organism

Pia Mater The inner layer of the meninges, which nourishes and protects the brain and spinal cord

Pigmentation Skin coloration

Pineal Gland A gland responsible for producing melatonin and serotonin

Pinnate Muscle Cells Muscle cells that run in various directions

Pisiform Bone A small pea-shaped proximal carpal bone

Pituitary Gland or Hypophysis A gland that controls most of the other endocrine glands in the body

Pivot Joint A synovial joint that permits a rotation

Placenta Previa The location of the placenta in the lower part of the uterus

Placenta An organ that nourishes the developing fetus in the uterus

Plantar Pertaining to the sole of the foot

Plaque 1) A film of bacteria and bacterial waste that forms on teeth; 2) Hardened deposits that form on the inner walls of blood vessels or in a nerve cell

Plasma The fluid portion of the blood

Plasma Cell A B-lymphocyte that produces immune proteins

Plasmin An enzyme that dissolves the fibrin of blood clots

Plasminogen The inactive form of an enzyme that dissolves the fibrin of blood clots

Plasticity The ability of neurons to alter their function as a result of experience and usage

Pleura The membranes lining the lungs and thoracic cavity

Pleural Cavities The body cavities containing the left and right lungs

Pleuritis Inflammation of the pleura, also called pleurisy

Pneumonia Inflammation of the lungs caused by an infection

Pneumothorax The presence of air in the pleural membranes

Polar Having a stronger negative or positive charge concentrated on one side or region

Polycystic Kidney Disease A inherited disease that causes the growth of kidney cysts, which impairs kidney function

Polymer A complex biochemical made up of chains of molecules

Polypeptides or Proteins Large chains of amino acids

Polysaccharide A chain of monosaccharides

Polyuria The production of an increased amount of urine

Pons The topmost portion of the hindbrain

Posterior Toward the back

Posterior Pituitary Gland The rear region of the pituitary gland

Postganglionic Neuron A neuron situated after a synapse between two neurons

Postovulation or Luteal Phase The ovarian cycle of events that take place after ovulation

Postsynaptic Neuron A nerve cell that receives a signal across a synapse

Potential Energy The ability to do work

P-R (PQ) Interval The time interval between atrial depolarization and ventricular depolarization

Precancerous Genes Genes that have the potential to become cancerous under certain conditions

Preganglionic Neuron A neuron situated before a synapse between two neurons

Pregastric Factors Conditions that can dramatically affect food intake

Pregnancy The condition in which an embryo is developing within the uterus

Pre-mRNA Transcribed mRNA that is not yet prepared for translation

Preovulation or Follicular Phase The ovarian cycle of events that take place before ovulation

Prepuce or Foreskin A roll of skin that covers the glans of the penis

Presynaptic Neuron A nerve cell that sends a signal across a synapse

Primary Bronchi The first division of the respiratory tree following the trachea

Primary Response A response that occurs the first time an antigen enters the body

Primary Structure The initial arrangement of amino acids in a peptide bond, with the R group aligned on one side

Prion An infectious agent composed of a protein

Product The chemical produced as a result of an enzymatic reaction

Progesterone A female steroid sex hormone secreted by the ovary, which prepares the reproductive tract for pregnancy

Programmed Cell Death or Apoptosis The process by which cells program their own death

Prokaryote A cell with its genome located in the nucleoid and without specialized compartments in the cytoplasm

Prolapse 1) A condition in which an organ becomes displaced; 2) Stretching of the heart valves; the downward displacement or sinking of an organ or other part

Proliferative Phase The first phase of the uterine cycle phase

Pronator Muscles Muscles that turn the palm downward

Propagate The transmission of an action potential across a neuron cell membrane

Prophase The first phase of mitosis

Prophase I The first phase of meiosis I

Prostacyclin A lipid that prevents platelet activation

Prostate Cancer Cancer of the prostate gland

Prostate Gland A gland in the male that surrounds the base of the urethra

Protease An enzyme that digests proteins

Protein Synthesis or Gene Expression The process by which cells turn amino acids into proteins according to the genetic information contained within the genome

Protein Turnover The rate at which a cell replaces damaged proteins

Proteinurea The presence of abnormal protein levels in the urine

Prothrombin An inactive form of an enzyme that stimulates blood clotting

Protista A group of eukaryotes associated with disease

Proton A positively charged subatomic particle found in the nucleus

Proximal A body part located near an attachment point

Proximal Convoluted Tubule The first segment of the nephron's peritubular arrangement

Pseudostratified Appearing to be stratified but consisting of only one layer

Psoriasis An inflammation of the skin accompanied by increased skin cell production

Puberty The stage of development when sexual reproduction becomes possible

Pubic Refers to the groin region

Pubic Hair The hair around the genitals

Pubic Symphysis The articulation of the pubic bones

Pubis or Pubic Bones The lower front bones of the pelvic girdle

Pulmonary Artery The blood vessel that takes blood from the heart to the lungs

Pulmonary Circulation Circulation that supplies blood to the lungs

Pulmonary Valve The right semilunar valve

Pulmonary Veins The blood vessels that take blood from the lungs to the heart

Pulse A throbbing of the blood vessels produced by the heart beat

Pure Molecules Identical elements bonded together

Purines Nitrogen bases composed of adenine or guanine

Purkinje System Specialized muscle cells that carry the electric impulses through the ventricles of the heart

Pyelitis Inflammation of the renal pelvis

Pyelonephritis Inflammation of the nephrons

Pyloric Glands Mucus-secreting glands in the lower portion of the stomach

Pyloric Region Lower region of the stomach

Pyloric Sphincter The lower sphincter of the stomach

Pyrimidines Nitrogen bases composed of cytosine, thymine, or uracil

Pyruvic Acid A product of glucose produced during glycolysis

Pyuria The presence of white blood cells (WBCs) in the urine

Q

Q Wave The beginning of ventricular depolarization

QRS Complex A combined reading of the Q, R, and S waves of an electrocardiogram

Q-T Interval The time interval between ventricular depolarization and repolarization

Quadrangular Cartilage A vertical partition of cartilage that supports the nose

Quadrant Refers to the abdominopelvic regions as divided into four sections

Quadrate Lobe A small subdivision of the liver

Quadriceps Muscle A muscle with four origins

Quaternary Structure Two or more polypeptide chains combined to form a larger and more complex molecule

R

R Wave The electrical activity of ventricular contraction

Rabies A neurotrophic virus that is spread through the saliva of infected animals

Radial Glia Neuroglia that assist with the formation and maintenance of the developing nervous system

Radioactive Describes a substance that gives off energy due to the decay of its unstable atoms

Radiocarbon Dating A method for dating organic remains based on their content of carbon-14

Radius The shorter of the two forearm bones located laterally

Receptor A protein that permits a cell to detect stimuli

Receptor-Mediated Endocytosis A form of endocytosis using proteins to bind molecules

Recovery Phase A phase following the action potential that returns the cell to normal

Rectum The last section of large intestine

Red Marrow Produces red blood cells

Red Pulp An interior area of the spleen that contains RBCs

Reduced The process of a molecule gaining an electron or a hydrogen atom to its structure

Redundancy Two neural pathways that carry out equivalent functions

Reflex Arc A neural pathway that links a sensory receptor and an effector

Reflex An involuntary response to a stimulus

Reflux A backward flow of body fluids

Refractory Period The period immediately after an action potential in which the neuron cannot respond to another stimulus

Regulatory DNA Genes and DNA segments that regulate the expression of other genes

Regurgitation Backflow of blood through a heart valve

Releasers or Releasing Hormones Hormones released by the hypothalamus that control the anterior pituitary

Renal Artery A branch of the abdominal aorta that carries blood to the kidney

Renal Cell Carcinoma Malignancy of the cells of the renal tubular lining

Renal Column Renal cortex tissue that extends inward between the renal pyramids

Renal Corpuscle The renal structure formed by the glomerulus inside the Bowman's capsule

Renal Cortex The outer portion of the kidneys

Renal Fascia The connective tissue that secures each kidney to the posterior abdominal wall

Renal Medulla The soft, marrow-like structure in the center of each kidney

Renal Pelvis A cavity of the kidney where formed urine is collected before it enters the ureter

Renal Pyramid A triangular collection of tissue in the renal medulla

Renal Tubules Small tubes within the kidney that modify and transport urine

Renal Vein The vein that carries blood from the kidney to the inferior vena cava

Repolarize A change in a neuron's membrane potential that returns the potential to a negative value

Reproductive System Body system that regulates sexual function

Reproductive Tract Connected muscular tubes that are involved in female reproduction

Respiration The process of taking in oxygen and releasing carbon dioxide for cell metabolism

Respiratory Bronchiole The airways in the lung that branch off from the larger bronchi

Respiratory System Body system that regulates atmospheric gases and certain bodily wastes

Resting Potential The difference between the two sides of the neuron's membrane when the cell is not conducting an impulse

Restrictive Lung Disease A disease that restricts the expansion of the lungs

Reticulocyte An immature red blood cell

Reticulum A type of connective tissue that provides support

Retina The inside layer located at the back of the eye that contains photoreceptors

Retroperitoneal Behind the peritoneum

Reverberating Pathway A neuron pathway that stimulates itself continually

Reverse Peristalsis A peristalsis that forces food and liquid in the opposite direction of flow

Rh Factor The type of protein, also called D protein, on the surface of red blood cells in some blood types

Rheumatic Heart Disease Damage to heart valves caused by a bacterial infection

Rheumatoid Arthritis A condition in which the immune system attacks connective tissues

Rhodopsin A light-sensitive chemical contained in the rods of the retina

Rhomboideus Muscles Diamond-shaped muscles

Rib One of the 12 pairs of bones that form the bony wall of the chest or rib cage

Ribonucleic Acid or RNA Nucleotides containing ribose

Ribose A pentose sugar involved in the composition of nucleotides

Ribosome A structure found in the RER responsible for the manufacture of proteins

Right Hypochondriac The upper right corner of the abdominopelvic region

Right Inguinal The lower right corner of the abdominopelvic region

Right Lobe The largest subdivision of the liver

Right Lower Quadrant The quadrant containing the right inguinal, lumbar, hypogastric, and umbilical regions

Right Lumbar The middle right corner of the abdominopelvic region

Right Upper Quadrant The quadrant containing the right hypochondriac, lumbar, epigastric, and umbilical regions

Rigid Paralysis Loss of muscle function due to excessive muscle stiffness

Rigor Mortis Muscle stiffness due to calcium leakage after death

Ringworm or Tinea A type of dermatophyte or skin fungus infection

Rod A photoreceptor in the retina that is sensitive to dim light

Rotator Muscles Muscles that move a bone around its longitudinal axis in a circular direction

Rough Endoplasmic Reticulum or RER A region of the endoplasmic reticulum responsible for manufacturing proteins

Round Window A membrane in the cochlea that per-

mits pressure from sound waves to be released

Ruffini Receptors Nerves that respond to pressure or constant touch

Rugae Inward folds of an organ

S

S Wave The end of ventricular contraction

Sacral Region The part of the spinal column comprising the pelvic area

Sacral Vertebrae Five vertebrae of the sacral region

Sacrum Five fused sacral vertebrae

Saddle Joint A synovial joint that permits a variety of movements, primarily a rocking movement in two planes

Sagittal Suture The suture joining the parietal bones

Saliva A watery secretion from glands in the mouth

Salivary Amylase An enzyme in the saliva that digests starch

Salivary Gland A gland in the mouth that secretes saliva

Salmonella A type of bacteria that causes food poisoning

Salt The taste that occurs when salt dissolves in the mouth

Sarcolemma The membrane of muscle cells

Sarcomere The contractile unit of a muscle cell

Sarcoplasmic Reticulum A system of tubes that stores and transports the calcium needed for muscle contraction

Satellite Cells Neuroglia that cover the surface of neurons

Satiety Center A region of the central nervous system that signals that a person has eaten

Saturated Refers to fatty acids with one covalent bond per carbon

Scapula or Shoulder Blade The flat bone parallel to the vertebral column

Schwann Cells Neuroglia that surround axons

Sclera The tough outermost layer of the eye

Scleroderma A connective-tissue disorder that causes thickening of the skin

Scrotum A pouch of skin that encloses the testes

Sebaceous Gland A gland that secretes sebum into the hair follicle

Sebaceous Hyperplasia A disorder that affects oil glands

Seborrhoeic Keratosis A black or brown growth on the face or body

Sebum An oily secretion produced by the sebaceous glands

Secondary Bronchi A branch of the bronchi

Secondary Response A response that takes place when a subsequent encounter with an antigen occurs

Secondary Sex Characteristics Anatomical features that distinguish males from females

Secondary Structure The characteristic shape of peptide chains

Second-Degree Burn A burn involving the stratums spinosum and germinativum

Secretin A hormone that causes the pancreas to release digestive juices

Secretion The transport of molecules using exocytosis

Selectively Permeable Membrane A membrane allowing only certain molecules to through it

Semen A fluid containing sperm and seminal secretions

Semicircular Canals Structures of the inner ear that detect movement of the body in space

Semilunar Valve A heart valve that prevents blood from flowing back into the heart

Seminal Vesicles Glands that help produce semen

Seminal Vessels A network of tubes and glands that assist with the transport of sperm

Seminiferous Tubules Tubes in the testes where sperm is produced

Senescence The aging process in organisms

Sense Strand The strand of DNA carrying the genetic code

Sensory Cortex A strip of cortex involved in sensory interpretation

Sensory Receptor A nerve cell found in all skin layers

Septum A muscular wall that divides the heart chambers

Serosa or Adventitia The outer layer of the digestive tract

Serotonin A hormone involved with digestion, appetite, moods, and sleep

Serous Pericardium The innermost layer of the pericardium

Serratus Muscles Saw-tooth-shaped muscles

Serum Globulin A general antibody found in the blood serum

Sesamoid Bones Small bones that develop within tendons

Sexual Dimorphism Developmental differences that distinguish the two genders

Sexual Division A type of division where gametes are created

Sexually Transmitted Disease (STD) An infectious disease spread by sexual contact

Shaken-Baby Syndrome The severe injuries that result when a young child is violently shaken

Sheet A type of secondary structure shaped of protein similarly to a pleated sheet of paper

Shin Splint A painful condition of the anterior lower leg that develops from overuse of the ankle joint

Short Bone A bone having a square-like shape; one of four bone shapes

Sickle Cell Disease A genetic disorder of hemoglobin that affects the shape of the RBCs

Side Chain or R Group Used to characterize amino acids by the types of organic molecules they carry

Sigmoid Colon The last portion of the large intestine

Simple Refers to structures composed of a single layer of cells

Simple Fracture A crack in the bone structure

Sinoatrial (SA) Node Specialized cells of the heart that initiate the heart beat

Sinuses Small cavities found in bones surrounding the cranial cavity

Sinusoid A small cavity in the liver filled with blood

Skeletal Muscle Muscle attached to bone

Skeletal System Body system that provides support and movement

Skene's Glands Mucus producing glands at the base of the female urethra

Skin Appendages Complex structures that assist skin function

Skin Cancer A degenerative disorder with an underlying genetic component

Skin Needling An antiaging treatment in which needles are inserted into the skin to promote skin growth and swelling

Skin Tag Soft tumors

Skin-Nail Fold A fold of hardened skin overlapping the base a fingernail or toenail

Sleep Apnea Cessation of breathing during sleep

Small Intestine The part of the digestive tract located between the stomach and the large intestine

Smallpox A highly contagious, infectious, and often deadly viral disease

Smooth Endoplasmic Reticulum or SER A region of the endoplasmic reticulum responsible for carbohydrate and lipid production

Smooth Muscle Muscle found in the linings of blood vessels and tubular organs

Society Groups organisms interacting with each other

Sodium/Potassium Pump or Ion Channel A membrane protein channel that controls the ionic distribution of sodium and potassium inside and outside of the cell

Soft Palate The muscular, posterior part of the palate

Solar Lentigene A type of freckling

Solute Any particle that dissolves in a solvent

Solution A uniform mixture of substances dissolved in a liquid

Solvent A substance that dissolves other chemicals

Soma Another name for the nerve cell body

Somatic 1) Refers to any cell in the body other than reproductive cells; 2) The part of the peripheral nervous system associated with the voluntary control of body movements

Sour The taste that occurs when acids dissolve in the mouth

Spasm An involuntary, abnormal muscle contraction

Specialized Germ Cell (SGC) A cell in the gonad involved in sexual reproduction

Specific Heat The heat energy required to raise the temperature of a particular substance

Sperm The male reproductive cell

Sphenoid Bone The bone forming the anterior base of the cranium and the posterior orbit

Sphincter Muscles Muscles that decrease the size of an opening

Spider Veins A birthmark caused by enlarged veins, appears as a spider's web

Spinal Cavity The body cavity containing the spinal cord

Spinal Nerves Nerves that originate from the spinal cord and pass out of the vertebral column

Spindle Fibers Proteins attached to the centrioles during prophase

Spiral Fracture Refers to a fracture occurring in a twisting pattern on the bone

Spleen A large lymphatic organ in the abdominal cavity

Splenic Flexure The last portion of the transverse colon

Spongy Bone A synonym for cancellous bone; bone composed of a honeycomb-like network bony struts

Sprain An injury resulting from sudden or violent stress on a joint or muscle

Squamous Cells with a flat shape

Squamousal Suture The suture joining the parietal and temporal bones

S-T Interval The time interval between atrial depolarization and ventricular depolarization

Stapes The third bone in the series of three ossicles in the middle ear

Staphylococcal Scalded Skin Syndrome (SSSS) A potentially fatal disorder resulting in shedding and swelling of the skin

Staphylococcus aureus A species of bacteria that causes skin infections

Starch or Amylose An alpha-bonded glucose found in plants

Stem Cells Cells that retain their ability to undergo division

Sternum The breastbone

Steroids Long-chain fatty acids converted into ring shapes with a variety of functions

Sterols Complex lipids composed of carbon rings

Stiff-Man Syndrome Rigidity, and spasms of the spine and lower-extremity muscles

Stomach An enlarged, muscular sac-like organ of the digestive tract

Strain An injury due to overworking the muscle's force on the joints

Stratified Refers to multiple layers one on top of another

Stratum Basale or Stratum Germinativum The innermost layer of the epidermis

Stratum Compactum The second outermost layer of the epidermis

Stratum Corneum The outermost layer of the epidermis

Stratum Granulosum The middle layer of the epidermis

Stratum Lucidum A breakable layer of epidermis found in areas of thick skin

Stratum Spinosum The second innermost layer of the epidermis

Strawberry Hemangioma A birthmark caused by enlarged blood vessels

Stress Fracture Thin breaks in a bone that may be too small to detect and, generally, do not heal

Striations Muscle fibers that are grouped as visible bands

Stroke Volume The amount of blood pumped by the ventricle of the heart with each beat

Stroke A cerebrovascular disorder that occurs when the blood supply to part of the brain is suddenly interrupted

Structural Classification A system of joint classification based on tissue composition and structural complexity

Structural DNA Genes that carry the instructions for building structural components of a cell

Structural Formula The arrangement and bonding of elements comprising a compound

Structural Molecules Molecules that compose body parts

Structural Proteins Proteins that help build body structures

Styloid Process The slender process that projects from the temporal bone

Subacute A condition with a moderate duration and/or severity

Subarachnoid Space A cavity that lies between the arachnoid and pia mater of the meninges

Subatomic Particles The parts of an atom

Sublingual Glands A small salivary gland located under the tongue

Submandibular Glands The salivary gland near the lower edge of the mandible

Submucosa A connective tissue layer beneath the mucus membrane

Substrate A chemical that will be modified in an enzyme's active site

Sucrose A disaccharide; a combination of glucose and fructose

Sudden Cardiac Death An abrupt and unexpected death due to a loss of heart function

Sulfhydryl Group The functional group containing sulfur, involved with creating the structure of molecules

Superficial Refers to any body part or region close to the skin

Superior Nearest to the head

Superior Vena Cava The large vein that carries blood to the right side of the heart from the head and upper body

Supinator Muscles Muscles that turn the palms upward

Suppressor T-lymphocyte A T-lymphocyte that inhibits the immune response

Surface Feature Protrusions and edges on bone formed by the pull of ligaments and tendons

Surface Receptor A receptor located on the surface of a cell

Surfactant or Detergent Organic chemicals with a hydrophilic functional group at one end and a hydrophobic component at the other end

Surfactant Fluid in the lungs that helps to keep them open and expanded

Sweat Gland A gland that produces sweat

Sweet The taste that occurs when sugar dissolves in the mouth

Sympathetic Ganglion Chain A chain composed of the nerve cell bodies of sympathetic postganglionic neurons

Sympathetic Nerves The thoracic and lumbar divisions of the autonomic nervous system

Symported Synonym for cotransported

Synapse The junction where an impulse is transmitted from one neuron to another

Synaptic Cleft The gap at the junction where a signal is transmitted from one nerve cell to another

Synarthrosis Joint A joint that does not permit movement

Synchondrosis or Symphysis A synarthrodial joint formed by cartilage

Syndesmosis A synarthrodial joint formed by two or more ligaments

Synergistic Refers to muscles contracting together to produce a common effect

Synovial Capsule A fibrous, fluid-filled sack

Synovial Fluid Clear fluid that lubricates joint linings

Synovial Joint A joint formed by a synovial capsule

Synthesized Refers to the process of synthesis; composed

Syringomas Tumors that form in sweat glands

Systemic Circulation Circulation that supplies blood to all parts of the body, except the lungs

Systemic Lupus Erythematosus or Lupus An autoimmune disorder that causes inflammation of connective tissues throughout the body

Systole A part of the cardiac cycle during which the heart muscle contracts and forces blood out

T

T Cells A type of blood cell that protects the body from cancer cells, foreign materials, and viruses

T Wave The beginning of ventricular repolarization

Tactile or Meissner's Corpuscles Nerves found in the upper region of the dermis that responds to pressure

Talus One of seven tarsal bones

Tamponade Pressure on the heart that obstructs filling of the chambers

Tangle A conditon that causes neurons to lose their shape

Target Cells Cells with receptors sensitive to endocrine secretions

Tarsals The group of ankle bones

Taste Bud A small structure on the upper surface of the tongue that contains chemoreceptors

Telomere Shortening The process of a chromosome's telomeres becoming shorter after each round of mitosis

Telomere The end of a chromosome

Telophase The last phase of mitosis resulting in two similar cells

Telophase I The last phase of meiosis I

Temporal Bones The two bones that form the sides and base of the skull

Temporal Lobe A region of the cerebrum located inferior and lateral to the frontal and parietal lobes

Tendon Connective tissue that joins muscle to bone

Tensor Muscles Muscles that make a body part more rigid or tense

Terminal Bronchiole The smallest branches of the bronchioles that lead into the alveoli

Terminal Hair The hair of the head

Terminus The part of the neuron that transfers information to an adjacent neuron or muscle cell

Terpenoids Short-chain fatty acids that help fight disease

Tertiary Bronchi A branch of the secondary bronchi

Tertiary Structure The three-dimensional structure of a polypeptide chain

Testicular Cancer Cancer of the testes

Testis or Testicle A male gonad that produces sperm

Testosterone A sex hormone, part of the sterol group

Tetanus An infectious disease that causes rigid paralysis

Tetany Periods of arm and leg muscle spasms caused by calcium imbalances

Tetrodotoxin A potent nerve toxin derived from certain fish and frogs

Thalassemia A genetic disorder of hemoglobin in red blood cells

Thermal Energy The production of heat

Third-Degree Burn A burn that affects the entire epidermis

Thoracic Refers to the chest region

Thoracic Cavity The body cavity containing the esophagus, heart, lungs, and respiratory tree

Thoracic Region The part of the spinal column comprising the thorax, or chest

Thoracic Vertebrae Twelve vertebrae of the thoracic region

Thorax The rib cage and chest cavity

Threshold The level of stimulation needed to stimulate an action potential

Thrombin An enzyme that stimulates blood clotting

Thrombocytes or Platelets Blood cell fragments that control bleeding

Thrombosis A blood clot that forms in a blood vessel or in the heart

Thymosin A peptide hormone secreted by the thymus gland that causes the immune system to mature

Thymus Gland A gland located in the chest area that assists the immune response

Thyroid Cartilage The largest cartilage; it makes up the ventral and lateral part of the larynx

Thyroid Gland A gland located in the front of the neck that plays a role in regulating metabolism

Thyroxine A hormone secreted by the thyroid gland, which controls metabolism

Tibia The medial bone of the lower leg

Tissue An organized assembly of cells that have similar structures and perform a specific function

Tissue Level of Organization Groups of cells that perform similar specific functions

Tissue Monocyte Large phagocytic cells found in connective tissue

Tissue Plasminogen Activator (TPA) An enzyme that dissolves the fibrin in blood clots

Titin Filament A protein filament that supports the myofilaments

T-lymphocyte A white blood cell responsible for stimulating the immune system

Tonic Control Regular neural stimulation that maintains the health of glands and muscles

Tonsils Glandular components of the immune system located in the oropharynx

Tooth Decay Tooth destruction caused by bacteria

Toxicological Pertaining to the toxic effects of substances

Trabeculae Partitions that divide the lymph node into regions

Trachea The windpipe; a passage for the admission of air to the lungs

Tracheal Cartilage A ring of cartilage that supports the trachea

Trait A particular characteristic that distinguishes one person from another

Transcription The first phase of gene expression

Transducer A nerve cell that converts environmental stimuli into body signals

Transduction The conversion of a stimulus from one form to another

Transfer RNA or tRNA Carries anticodons

Transfusion The transfer of blood from one person to another

Transient Ischemic Attack (TIA) A "ministroke" caused by temporary loss of blood flow

Transitional Epithelium Epithelial tissue that can change shape with expansion and contraction of the underlying tissue

Translation The second stage of gene expression

Transport Vesicles Organelles responsible for transporting products within the cell

Transverse Colon A portion of the large intestine running horizontally below the diaphragm

Transverse Fracture Refers to a fracture occurring horizontally on the bone

Trapezium The lateral distal carpal bone

Trapezius or Trapezoid Muscles Muscles that have a broad origin, and focus to a narrow insertion point

Trapezoid A distal carpal bone

Traumatic Pertaining to an injury or wound

Tremor A nervous system disorder that causes uncontrollable, rhythmic, shaking movements

Triangular or Triquetral Bone A proximal carpal bone

Triceps Muscle A muscle with three origins

Tricuspid Valve One of the two atrioventricular valves; it lies between the right atrium and left ventricle

Triglycerides A form of glyceride with three molecules of fatty acid

Trigone The smooth triangular area of the urinary bladder floor

Trisaccharide Three bonded monosaccharides

Troclear Notch A cresent-shaped notch on the proximal end of the ulna that articulates with the humerus

Tropomyosin One of the three proteins that form the thin myofilaments of muscle fibers; it reinforces the actin core

Troponin One of the three proteins that form the thin myofilaments of muscle fibers

True or Movable Vertebrae Vertebrae of the cervical, thoracic, and lumbar regions

Tuberculosis (TB) A bacterial infection that usually affects the lungs

Tubular Reabsorption A process in the peritubular capillary system in which water, nutrients, and electrolytes travel back into the blood

Tubular Secretion The process by which certain waste products and ions are removed from the blood into the tubular fluid

Tunica Adventitia The outer layer of a blood vessel

Tunica Intima The inner layer of a blood vessel

Tunica Media The middle layer of a blood vessel

Turbinates Bony structures in the nasal cavity that clean and moisten air

Tympanic Membrane or Eardrum A thin epithelial layer at the end of the auditory canal

Tympanic Region The region containing the ear bones

U

Ulcer A condition caused by erosion of the dermis or the digestive tract mucosa

Ulna The longer of the two forearm bones

Ultraviolet (UV) Light The invisible rays of the electromagnetic spectrum

Umani The taste that occurs when the food ingredient monosodium glutamate (MSG) dissolves in the mouth

Umbilical The middle section between the right and left lumbar of the abdominopelvic region

Umbilical Cord The cord that transports blood, oxygen, nutrients, and wastes between the fetus and the mother during pregnancy

Undernutrition The condition due to a diet lacking sufficient molecules

Undescended Testis or Cryptorchidism A condition in which one or both testes do not pass into the scrotum

Unicellular Consisting of only one cell

Unilateral Refers to a single body part found in a lateral location

Unipolar Refers to a neuron that has only one process extending from the cell body

Unsaturated Refers to fatty acids that have double bonds and lack hydrogen

Upper Appendages The shoulders, arms, wrists, and hands

Upper Respiratory System The part of the respiratory system composed of the nose, nasal cavity, paranasal sinuses, eustachian tubes, and larynx

Urea A waste product consisting of amino functional groups

Ureters The tubes that send urine from the renal pelvis to the urinary bladder

Urethra The tube that transfers urine from the bladder to the body's exterior

Urethral Meatus The external opening of the urethra

Urethral Orifice The external opening through which urine leaves the body

Urethritis Inflammation of the urethra

Urinary Bladder A sac-like organ that serves as a reservoir for urine storage

Urinary Retention The inability to expel urine from the bladder

Urinary System Body system that regulates production, storage, and removal of urine

Urinary Tract Infection (UTI) Infection anywhere in the urinary tract

Urine Concentration The removal of water from the urine by reabsorption in the collecting tube before it leaves the body

Uterine Cycle The sequence of events that prepare the uterus for pregnancy

Uterine Fundus The top of the uterus

Uterus A pear-shaped organ that nourishes the growing embryo

Uvula A cone-shape projection posterior to the soft palate

V

Vaccination The therapeutic use of antigens to build up the immune system

Vaccine A drug preparation that contains an antigen

Vagina A muscular canal running from the uterus to the exterior of the body

Varicose Refers to a condition of being dilated or swollen

Vas Deferens or Ductus Deferens A thin tube that transports sperm from the testis to the urethra

Vascular Disorder A disease of the blood vessels

Vascularization The formation of blood vessels

Vasculature A network of blood vessels in an organ or body part

Vein A blood vessel that carries blood from the body to the heart

Vellus Hair Fine body hair

Venae Cavae Large veins that bring blood into the right side of the heart

Ventilation or Breathing The process of transporting air to the surface of the lungs

Ventral In humans, toward the front. In other animals, refers to the belly

Ventral Root The branch of a spinal nerve that arises from the ventral part of the spinal cord

Ventricle 1) One of the muscular lower chambers of the heart; 2) A collection of cavities within the forebrain containing cerebral spinal fluid

Venule A small branch of a vein

Vertebrae Bones that form the spinal column

Vertebral Canal A tube-like cavity that runs the length of the vertebral column

Vertebral Column A flexible column formed by the vertebrae

Vesicoureteral Reflux A backup of urine to the kidneys

Vestibule A portion of the inner ear between cochlear and semicircular canals that detects body position

Villi Finger-like projections on the small intestine mucosa

Viriod An infectious agent composed of pieces of RNA

Virus An infectious agent composed of a genome in a protein coat

Visceral Refers to the inner wall of an organ. It also refers to the coverings found directly on body parts

Visceral Pleura Two inner membranes of the pleura

Vitamin Bioactive molecules essential to the health of an organism

Vitiligo A skin disorder resulting in white patches on the skin

Vitreous Humor A gel-like fluid that fills the cavity behind the eye lens

Vocal Cords Two small bands of muscle within the larynx that produce vocal sounds

Voluntary Muscle Muscles that are under conscious control

Vomer The bone forming the inferior part of the nasal septum

Vomeronasal Gland A structure containing chemoreceptors, which are believed to sense chemicals associated with sexual behavior

Vulva The external female genitalia

W

Water Conservation The retention of water in the body

Water Excess or Water Intoxication The state in which the body contains too much water

Water Insoluble Not able to dissolve in water

Water Soluble Able to dissolve in water

Wavelength The length of a wave

Whiplash Nerve damage in the neck caused by an abrupt, forced movement of the head

White Blood Cell A large blood cell that helps fight infection

White Matter The parts of the brain and spinal cord composed primarily of axons and myelin

White Pulp A region of the spleen composed of lymphatic tissue

Wisdom Teeth The last set of molars at the back of the mouth

Womb Another name for the uterus

Wormian or Sutural Bones Bones that develop within flat bones of the skull

X

Xiphoid Process The tip of the sternum

Y

Yellow Marrow A food reserve for bone cells

Z

Z-Line A line across striated muscle fibers that marks the boundaries between each sarcomere

Zygomatic Bones Bones forming the cheeks

Zygomatic Process Articulates with the zygomatic bone to form the cheek

Zymogen An inactive form of an enzyme

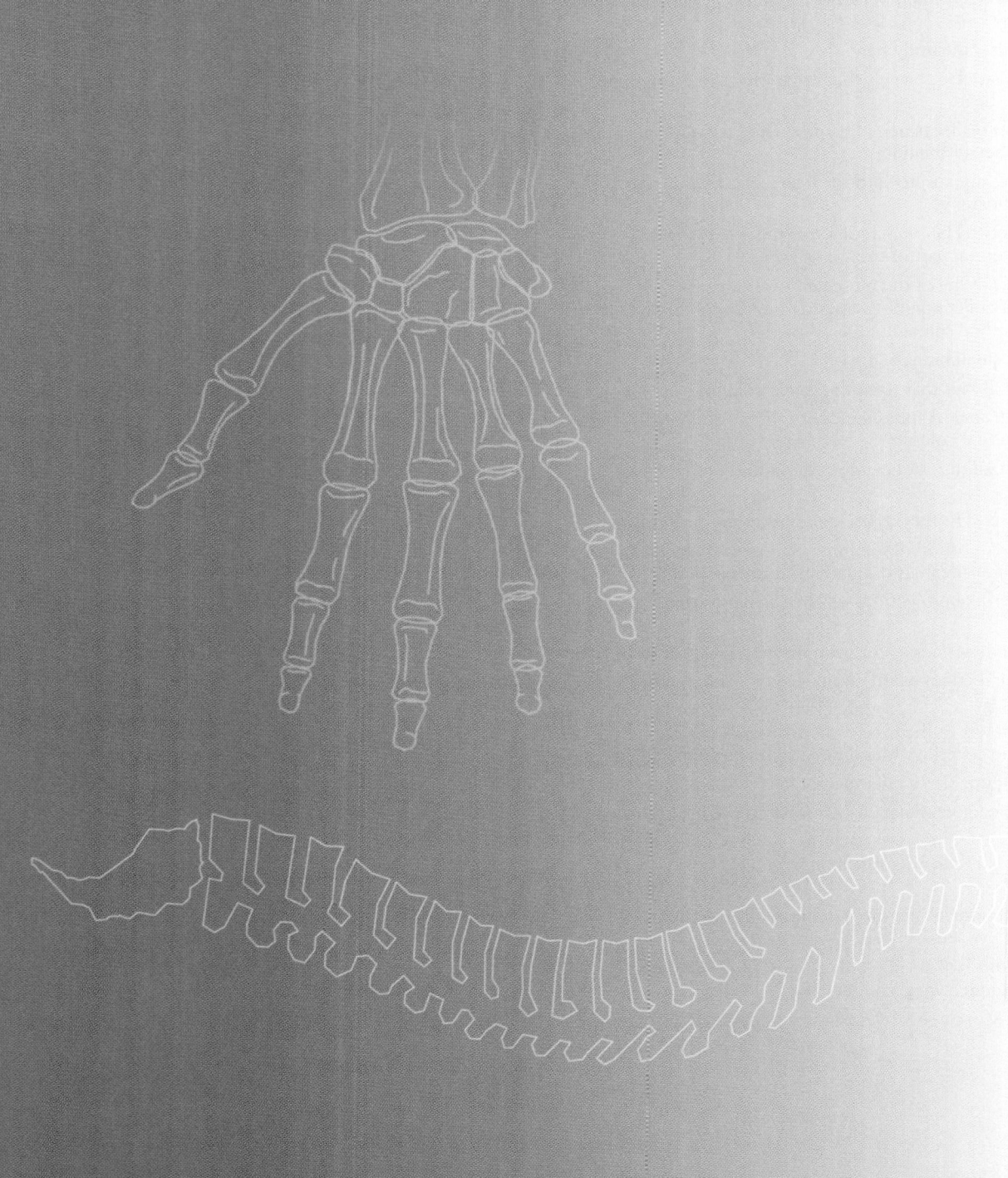

INDEX

A

Abdominal, 15, *15*
Abdominal cavity, 18, *19, 28*
Abdominopelvic cavity, 18, *19*
Abdominopelvic region(s), 17, *17*
  quadrants and, 16–18
Abducens nerve, 344*t*
Abduction, *13*, 14
Abductor muscles, 237
ABO blood group system, 453
Accessory digestive organs, 490
Accessory nerve, 344*t*
Accumulated cell damage, 120
Accutane, 144
Acetabulum, 184
Acetylcholine, 229
Acetylcholine receptors, 229
Acetyl coenzyme A (acetyl CoA), 98
Achilles tendon, *233*
Acid, 44
Acid-base balance, kidneys and, *544*
Acid reflux, 511, *511*
Acini, 504
Acne, 154
  pregnancy and, 144
Acquired immune response, 467
Acquired immunity, 465, 466, *467*, 468
Acquired immunodeficiency syndrome, 43, 474, *475*
Acromegaly, 279*t*
Acromial, 15, *15*
Acromion, *197*
Acrosome, 583
ACTH. *See* Adrenocorticotropic hormone
Actin, 227
Actin filament, *230*
Action potential, 301
Active immunity, 471
Active site, 84–85
Active transport, 86, 90–91
Active transport pumping, 90, *90*
Acupuncture, 312
Acute carbon dioxide intoxication, 456
Acute diarrhea, 511
Acute renal failure, 550
Acute respiratory distress syndrome, 389–390
Adam's apple, *378,* 379
ADAO. *See* Asbestos Disease Awareness Organization
Adaptive features, of skin, 136
Addison's disease, 279*t*, 549
Adduction, *13,* 14
Adductor muscles, 237, *239*
Adenine, 58
Adenoids, *377,* 378, *494*
Adenomatous polyps, 513
Adenosine triphosphate, 58, 99
ADH. *See* Antidiuretic hormone
ADHD. *See* Attention-deficit hyperactivity disorder
Adhesiveness, 40
Adipose capsule, 531
Adipose tissue, *109,* 141, 149
Adrenal cortex, 271, *271*
Adrenal glands, 28, *113,* 271, 271–272
  microscopic slide of, *289*

Adrenaline, 272, 470
  effects of, on daphnia, 290–291
Adrenal medulla, 271, *271*
Adrenocorticotropic hormone, *262, 269,* 270*t*
Adult teeth, *493*
Adventitia, 498
Aerobic respiration, 98
Affector, 304, 308
Afferent arteriole, 536
Afferent nerve paths, *332*
Afferent nerves, 331
AFP. *See* Alpha-fetoprotein
AFPep, 592
Aging
  of blood and lymphatic system, 475–476
  of body's chemistry, 61–62
  body structure and, 13
  of cardiovascular system, 436–438
  of cells, 120–121
  of digestive system, 514–516
  of endocrine system, 280–281
  genetics of, 92
  of integumentary system, *157*
  of muscular system, 244–245
  of nervous system, 314–316, 359–361
  of reproductive system, 594–595
  of respiratory system, 393–394
  of skeletal system, 207–208
  of skull, *208*
  of urinary system, 550–551
Agonists, 266
Agranulocytes, 454
AIDS. *See* Acquired immunodeficiency syndrome
Air, partial pressure and percent composition of, 387*t*
Air pollution, nervous cell aging and, 315
Albinism, 155
Alcmaeon, 4
Aldosterone, 271, 546, 549
Alimentary canal, 490
Alkaline solution, 44
Allergens, assessing, 486–487
Allergies, 475
Allosteric site, 85
Allostery, 85
Alpha cells, 504
Alpha-fetoprotein, 592
ALS. *See* Amyotrophic lateral sclerosis
Aluminum, case study on, 368
Alveolar bones, 188
Alveolar duct, *375, 382*
Alveolar sac, *375, 382*
Alveoli (alveolus), 179, *375, 380, 381,* 383
Alzheimer's disease, 119, 359, 360, 362
American Academy of Allergy and Immunization, 483
American Academy of Pediatrics Task Force on Circumcision, 602
American Civil Liberties Union, 602
American College of Physicians, 483
American Heart Association, 444
American Medical Association, 483
American Red Cross, 472
Amino acids, 54, 305
  arrangement of, 56

  standard, *54*
Aminoaciduria, 547
Amino group, 41
Amino-sugars, 53
Amniotic fluid, 585
Amniotic sac, 585
Amoebic dysentery, 512
Amphetamines, weight loss and, 306
Amphiarthrosis joints, 194
AMP kinase, aging and, 92
Ampulla of Vater, *505*
Amyloid deposition, 119
Amyloidosis, 119
Amyloids, 119
Amylose, 52
Amyotrophic lateral sclerosis, 311
Anabolic steroids, 278
Anabolism, 97
Anaerobic respiration, 98
Anal canal, *502,* 503
Anal sphincter, 503
Anaphase, 104, *104*
Anaphase I, 105, 106
Anatomical position, *9*
Anatomy, 4
  developmental, 6
  fine, 5
  gross, 5, 6
  roots of, 4–5
Anatoxin poisoning, 317
Anconeus muscle, *233*
Androgens, 271, 570
Andropause, 594
Anemia, 472
Aneurysm, 356, 430
ANF. *See* Atrial natriuretic factor
Angina pectoris, 430
Angiogenic (angiogenesis) factors, 137, 408
Angiotensin-converting enzyme (ACE) inhibitor, 559
Angiotensin II, 272
Angulation, 200
Animals, physical orientation terms for, 7
Anions, 81
Ankle joint, *172, 195*
Ankylosing spondylitis, 204
Anorexia, cachexia and, 244
Antacids, protein digestion and, 525–527
Antagonistic effect, in muscles, 236
Antagonistic movement, 13
Antagonists, 266
Anterior, 8, *9, 29*
Anterior chamber, *351*
Anterior cingulate, Stroop effect and, 370
Anterior pituitary gland, *262, 269, 269*
  major secretions of, 270*t*
Anterior serratus muscle, *232*
Anterior tibial muscle, *232*
Antiaging research, 92
Antibiotics, 457
Antibodies, 468, 469
Anticodons, 102
Antidiuretic hormone, *262, 269,* 545
Antigens, 465

639

Antioxidants, 62
Antisense, 101
Anuria, 536
Anus, 503
Aorta, *418, 419,* 421, *426*
Aortic semilunar valve, *417, 419*
Aortic valve, 421
Apex, *380, 419*
　of heart, *418*
Apocrine secretion, 143
Apocrine sweat gland, 143, *143*
APO4 gene, 360
Apolipoprotein E (APOE), 360
Apoptosis, 96, 120
Appendicitis, 502
Appendicular skeleton, 174, *183,* 183–186
Appendix, 502
Aqueous humor, 351
Arabinose, 52
Arachnoid mater, 335
ARDS. *See* Acute respiratory distress syndrome
Areolar connective tissue, 140
Aristotle, 4
Aromatase, 245, 570
Aromic orbitals (or shells), 35
Arrector pili muscle, *139, 142,* 146
Arrhythmia, 431
Arterial stiffness, 436
Arteries, *113, 412, 413*
　normal and blocked, *432*
Arterioles, 415
Arteriosclerosis, 431
Arteriovenous malformations, 356
Arthritis, 14, 204
Arthropods, 156
Articular cartilage, *190, 191*
Articulations, 175
Artificial bones, 202
Artificial immunization, 471
Artificial muscles, 227
Arytenoid cartilage, 379
Asbestos, 401
Asbestos Disease Awareness Organization, 401
Ascending colon, 502, *502*
Ascending tracts, 340
ASD. *See* Atrial septal defect
Asexual division, 103
Asexual reproduction, *104,* 104–105
Aspartate, 305
Aspirin, effects of, on cell function, 132–133
Asthma
　hot air balloon ride and, 377
　ozone and, 390
Astrocytes, 111, *112,* 298, *299*
Atelectasis, 390
Atherosclerosis, 431
Athetosis, 357, *357*
Athlete's foot, 156
Atlas, 181, *197*
Atomic mass, 36
Atomic nucleus, 35
Atomic number, 36
Atomic orbitals (or shells), 37
Atomic structure, 35, *35*
Atoms, 34
ATP. *See* Adenosine triphosphate
Atria, *422*
Atrial natriuretic factor, 546
Atrial septal defect, *433*
Atrial systole, *422*
Atrioventricular (AV) node, 423, *426*
Atrioventricular (AV) valves, 421, *422*
Atrium, 421
Atrophy, 119
　muscle, 236
Attention-deficit hyperactivity disorder, 306

Auditory canal, 352
Auditory tube, *494*
Auricle, 352, *352*
Autoantibodies, 475
Autocrine secretions, 265
Autoimmunity, 475
Autonomic nerves, 341
Autonomic nervous system, 344–346, *345*
　actions of, 346t
Autosomal-dominant porencephaly, 355
Avian influenza virus, 383
Axial skeleton, 174
Axillary, *15*
Axillary artery, *409*
Axillary lymph nodes, *463*
Axis, 181, *197*
Axoaxonic synapse, 306, *306*
Axodendritic synapse, 306, *306*
Axon hillock, 297
Axons, *111, 112,* 296, *296,* 297
Axosomatic synapse, 306, *306*

B

Bacteria, 93
　commensal, 149
　nervous system disease and, 310
Bacterial cell, *94*
Bacterial infection, in urinary system, 548
Balance, hearing and, 352–354
Baldness, 138
　male-pattern, 278
Ball-and-socket joint, 194, *195*
Bananas, nervous system and, 303
Bands, 462
Bartholin's gland, *568*
Basal nuclei, 336
Base, 44
Basement membrane, 140
Basilic vein, *414*
Basophils, 453, *461*
Benzene exposure, 477
ß-carotene, *49*
Beta cells, 504
Beta-galactosidase, 312
Bicarbonate (HCO3-), 81
Biceps brachii, *239*
Biceps muscle, 234
　of arm, *232*
　of thigh, *233*
Bicuspids, 497, *497*
Bicuspid valve, 421
Bidirectional communication, between nervous system cells, 295
Bilateral, 15, 16
Bile, 505, 507, 509
Bile duct, *505*
　in pancreas, *275*
Bilirubin, 460
Bioactive molecules, 40, 49
Biochemicals, 39
Biochemistry, 32
Biomechanics, 203
Biopsy, 119
Biopsy picture, *119*
Biotin, 50
Bipolar cells, 297
Bird flu, 383
Birth, 589
Birth defects, 129, 585, 586, 590
　predicting, 604–605
　thalidomide and, 43
Birthmarks, 154
Bitter taste, 347, *348*
Bladder, 530, *531,* 534, *534,* 578
Bladder cancer, 550, 594
Blastocyst, 584
Blastula, 584

Blepharoplasty, 241
Blood, *113*
　aging of, 475–476
　bone and, *461*
　common ailments in, 472–474
　components of, *451*
　fear of, 472
　flow of, through heart, *426*
　overview of, 450
　pathology of, 485
Blood-brain barrier, 298, 336–337, *338*
Blood cells
　formation of, 460–462, *461*
　function of, 455–460
　platelets, 454
　red, 451–453
　white, 453–454
Blood clotting, platelets and, 458
Blood donation, trivia about, 456–457
Blood gases, 450
Blood pressure, 408
Blood substitutes, 466
Blood sugar regulation, pancreas and, 275–276
Blood supply, to skin, *149*
Blood type, 453
Blood vessels, *190,* 408
B-lymphocytes, 454, *461, 463,* 468
BMI. *See* Body mass index
Body cavities, 18–20, *19*
　drawing, *29*
Body chemistry, aging of, 61–62
Body mass index, 224
Body orientation. *See* Human body orientation
Body quadrants, *17,* 18
Body regions, 14–18
　abdominopelvic regions and quadrants, 16–18
　general locations, 15–16
　surface feature coverage, *15*
Body structure, aging and, 13
Body system, *5, 33, 77*
Bonds
　covalent, 39, *40*
　hydrogen, 40, *40*
　ionic, 39, *40*
　types of, 39–40
Bone chart
　human, anterior view, *196*
　human, posterior view, *197*
Bone collar, 198
Bone density loss, 204
Bone fractures
　osteoporosis and, 204
　repair of, *201*
　types of, *200*
Bone marrow, *190, 191,* 192, 462
Bone(s), *115,* 172, 187–192
　artificial, 202
　blood and, *461*
　characteristics of, 175t
　cranial, 177
　development of, 197–200
　external features of, 188–189, *190*
　fractures of, 200
　growth of, *199*
　healing of, 201–202
　internal features of, 190–192, *191*
　structure of, 188–192, *189*
　study of, 219–220
　types of, *187,* 187–188
Botulism, 310, 317
Bovine spongiform encephalopathy, 311
Bowman's capsule, *537, 538, 539*
Brachial, 15, *15*
Brachial artery, *409*
Brachial muscle, *232*

Brachial plexus nerve, *342*
Brachiocephalic veins, *414*
Brachioradial muscle, *232, 233*
Brain, *116, 334, 335*
    aging and decrease in weight of, 359–360
    components of, 336–339, *337*
    hemispheres of, 337
    holes in, 355
    Stroop effect and, 369–370
Brain pathology, interpretation of, 371
Brain stem, 339
Brain tumors, 357
Breast cancer
    pregnancy and, 592
    robotic hands used for screening of, 155
Breast feeding, 69–70
    by wet nurses, 575
Breasts
    female, *573*, 573–574
        Tanner staging and, *588*
    reconstruction of, 241
Breathing, 374, 384–388
    gas exchange with, 387–388
    mechanics of, *384*, 385–386
Breathing capacity values, 395t
Brevis muscle, 235
Broad ligaments, 570
Bronchi, *375*
Bronchial tree, 375, 381, *381*
Bronchial tubes, *116*
Bronchiectasis, 390
Bronchioles, *380*, 381, *382*
Bronchitis, 388, *389*
Bronchoconstriction, 381
Bronchodilation, 381
Bronchopneumonia, 392
Bronchospasms, 381
Bronchus (bronchi), *379, 380*, 381
Bronze Age, 37
Buccal cavity, 492
Buffers, 44
Bulbourethral glands, *576, 578*, 579
Bulimia, cachexia and, 244
Bulk active transport, 90, 91
Bulk mechanical transport, 86
Bundle of His, 423, *426*
Bungee jumping, 12
Burns
    classification of, *151*, 151–152
    skin, 150
Bursa, 192

C

Cachexia, 244
Caffeine
    cell function and, 132
    effects of, on daphnia, 290–291
    nervous system and, 332–333
Calcanean tendon, *233*
Calcaneus, 186, *197*
Calcification, 198
Calcitonin, 274
Calcium (Ca2+), 81
Calcium chloride (CaCl$_2$), 39
Calculi, 548
Callus, 201
Calorie, 83
Calvaria, 177
Calyces, 531, *532*
Camphor, *49*
Canaliculi, 191
Cancellous bone tissue, 191
Cancer
    bladder, 550, 594
    breast, 592
    cervical, 592, 593
    colon, 513

    lung, 382, 391
    prostate, 592, 594
    skin, 153
    testicular, 592
Candida albicans (C. albicans), 155
Canine teeth, *493, 497*
Capillaries, *112, 113,* 141, 410
Capitate bones, 184
Carbohydrates, 45, 50–53, 60
"Carbo loading," 60
Carbon, 32, 39
Carbon dioxide
    in blood, 450
    intoxication, 456
    respiration and exchange of, 387–388
    transportation of, in blood, 455
Carbonic anhydrase, 456
Carbon isotopes, 36
Carbon skeleton, 40–41
Carbonyl group, 41
Carboxyl group, 41
Cardiac cycle, 425
Cardiac disorders, 430
Cardiac infarction, 420
Cardiac ischemia, 420
Cardiac muscle, 109, *110,* 225, 226
Cardiac notch, 382
Cardiac output, 427
Cardiac region, of stomach, 499
Cardiac sphincter, 498, 499, *499*
Cardiac vein, *418*
Cardiovagal baroreflex, 437
Cardiovascular (or circulatory) system, 112, *113,* 118, 407–447
    aging of, 436–438
    circulatory system vessels, 410–416
    function of heart, 425–427
    naming of, 408
    overview of, 408, 410
    pathology of, 430–436, *433*
    structure of heart, 417–425
    trivia about, 424–425
Caroll, James, 512
Carotid artery, *409*
Carpal bones, *172*
Carpals, 15, *15,* 183
Carpometacarpal joint, of thumb, *195*
Carpus, 196
Carrier protein or channel, 88
Cartilage, *115,* 172
Cartilaginous joint, 193
Case study investigations
    athlete's nervous system problems, 328, 361–362
    boy's bone breakages, 170, 209–210
    cardiovascular illness in runner, 406, 438
    classmate's intestinal ailment, 488, 516–517
    energy drink consumption and runner's collapse, 30, 63
    gender identity, 564, 595–596
    health club workout and illness, 372, 395–396
    hormone imbalance in young girl, 258, 281–282
    infant's condition, 74, 121
    knife injury and body damage, 2, 21
    muscle mass loss during space mission, 222, 246
    pesticides and uncle's illness, 292, 317
    skin condition in obese patient, 134, 159
    truck restoration hobbyist's ailment, 448, 477
    urinary problems in elderly neighbor, 528, 553
Cast, 549
Catabolism, 97

Catalysts, 83
Catch reflex, 325, 326–327
Catecholamines, 305
Catheter, 536
Cations, 81
Caudal, 8, *9,* 29
Caudate lobe, 506
Cavitational osteonecrosis, 217
Cavity(ies)
    body, 18–20, *19*
    in teeth, 206
CBC. *See* Complete blood count
CCK. *See* Cholecystokinin
CDC. *See* Centers for Disease Control and Prevention
Cecum, 502, *502*
Celgene Corporation, 43
Celiac disease, 510
Cell cycle, 97, 102–106
Cell doctrine, 93
Cell-mediated immunity, 470
Cellobiose, 52
Cells, *5, 33,* 76, *77*
    aspirin and function of, 132–133
    death of, 120
    description of, 94–96
    eukaryotic, 94
    functions of, 97–106
    human, *95*
    identifying structure and function of, 130–131
    membrane of, 93, *95*
    of microbes, 93–94
    osmosis and, 89, *90*
    pathology of, 119–120
    structure of, 92–96
    wall of, 93
Cell theory, 93
Cellular aging, 120–121
Cellular immunity, *467*
Cellular respiration, 98, *98*
Cellulose, 52
Centers for Disease Control and Prevention, 391, 444, 590
Central canal, 339
Central nervous system, 330
    afferent nerve paths and, *332*
    components of, *334,* 335–341
    infection of, 355
Central retinal artery and vein, *351*
Central sulcus, *337,* 339
Centrifuge, 450
Centrioles, 96
Centromere, 105
Cephalic, 8
Cerebellum, 339
Cerebral cortex, 336
Cerebral hemispheres, 337
Cerebrospinal fluid, 298
Cerebrovascular disease, 355
Cerebrum, 336
Cerumen, 143
Ceruminous glands, 143
Cervical, 15, *15*
Cervical cancer, 592, 593
Cervical lymph nodes, *463*
Cervical nerves, 334
Cervical region, of spinal column, 20, *20*
Cervical vertebrae, *173, 175,* 180, *181, 197*
Cervix, *568,* 572
Cesarean section, 593
Chambers, 420
Cheek bone, *172*
Chemical bonds, 38
Chemical damage, integumentary system and, 148

Chemical energy, 83
Chemistry, 32
Chemtrails, 323–324
Chewing tobacco, 420
CHF. *See* Congestive heart failure
Chicken pox, 316
Chief cells, 499
Childbirth, 589
Chiquita Brand bananas, 303
Chiral drugs, 43
Chirality, 42, *42*
Chlamydia, 602
Chloride (Cl-), 81
Chlorine (Cl), *82*
Cholecystokinin, 509
Cholesterol, 49, *49*, 505
    heart disease and, 432
Choline, 50
Cholinergics, 305
Chordae tendineae, 421
Chorea, 357, *357*
Choroid, 350, 351, *351*
Choroid plexus, 336
Chromatids, 103, 106
Chromosomes, 99
Chronic diarrhea, 512
Chronic obstructive pulmonary disease, 391, 456
Chronic renal failure, 550
Chyme, 509
Cigarettes, "safer," 386
Cigarette smoking, heart disease and, 444
Cilia, 96
Ciliary body, 351, *351*
Ciliated cells, 109
Circulating monocytes, 458
Circulation, fetal, *394*
Circulatory system, 408, *409*
    arteries, 410, 412–413, 415
    small vessels and capillaries, 415, *416*
    veins, 410, 412, 413, *414*, 415
    vessels of, 415–415
Circumcision, 579, 602
Circumduction, *13*
Cirrhosis, 513
Cisterna chyli, *463*
Citral, *49*
Civic responsibility: helping others with your knowledge
    biochemicals, 38
    blood and lymphatic system, 452
    body's organization, 80
    cardiovascular system, 410
    digestive system, 496
    endocrine system, 263
    integumentary system, 141
    muscular system, 224
    nervous system, 294
    public awareness of anatomy and physiology terminology, 6
    reproductive system, 567
    respiratory system, 374
    skeletal system, 176
    urinary system, 532
Clara cells, 383
Clavicle, *172*, *183*, *184*, *196*
Clavicular, 15, *15*
Clitoris, 568, *572*
Clostridium botulinum, 317
Clotting cascade, 458
Clotting factors, 458
CNS. *See* Central nervous system
Cobalamin (vitamin B-12), 50
Coccygeal nerves, *334*
Coccygeal vertebrae, 180
Coccyx region ("tail bone" region), of spinal column, 20, *20*
Coccyx (tailbone), *173*, 182, *196*
Cochlea, 353, *353*
Cochlear duct, *354*
Cochlear nerve, 352, *353*
Codons, 100, 102
Coenzymes, 85
Cofactors, 85
Cohesiveness, 40
Cold sores, 316
Collagen, 109, *142*
    aging and, 157
Collaterals, 297
Collecting duct, reabsorption in, 543
Collecting tubules, 538, *539*
    in nephron, *537*
Colon, 502
Colon cancer, 513
Colonic cleansing, weight loss and, 501
Colon polyp, 513, *513*
Colostrum, 589
Columnar cells, 108
Commensals, 149
Comminuted fractures, 200
Common bile duct, 505, *505*
Common peroneal nerve, *342*
Compact bones, *190*
Compact bone tissue, 190
Comparative anatomy, of primates, 221
Complements, 466
Complete blood count, 452
Compound fractures, 200
Compounds, 38
Conception, 583
Condyloid joint, 194, *195*
Cones, 351–352
Congenital disease, 310
Congenital disorders, of urinary system, 547
Congenital heart disease, 432
Congestive heart failure, 433
Conjoined twins, 585
Conjunctiva, 349, *351*
Connective tissue, 107, *108*, 109
    types of, *109*
Constipation, 515
Constriction, 412
Continuous capillaries, 415, *416*
Contractile cells, 224
Contractile proteins, 225
Contraction, 224
Contrails, 323
Contusions, 242
Convolutions, of brain, 336
COPD. *See* Chronic obstructive pulmonary disease
Copper (Cu), 81
Copulation, 582–583
Cornea, 350, *350*, *351*
Coronal suture, 178
Coronary arteries, *418*, 420
Coronary veins, 420
Coronary vessels, *419*
Corpus callosum, 337
Corpus cavernosum (corpora cavernosa), *578*, 580
Corpus luteum, 570, 581
Corpus spongiosum (corpora spongiosum), *578*, 579
Cortex
    of brain, 336
    renal, *531*, *532*
Cortical bone, *191*
Cortical bone tissue, 190
Corticosteroids, 271
Cortisol, 271
Cortisone, 470

Cosmetic surgery, 241
Costal cartilage, *172*, 182
Cotransported molecules, 542
Coughing, 379
Covalent bonds, 39, *40*
Cowper's gland, *576*, 579
Cramps, muscle, 242
Cranial, 8, *9*, 29
Cranial base, 177
Cranial bones, 177
Cranial cavity, 19
Cranial nerves, *334*, 342
    attachment of, *343*
    designations for, 344t
Cranium, 177, *183*
Cravings, 60
Creatinine, 251–252
Creatinine phosphate, 231
Creutzfeldt-Jakob disease, 57, 311, 371
Cricoid cartilage, *378*, 379
Crohn's disease, 516
Crossing over, 105
Cryptorchidism, 577
Crypts, 501
CSI. *See* Case study investigations
Cubital, 15, *15*
Cuboidal cells, 108
Cuboid bones, 186
Cushing's syndrome, 279t
Cuspid teeth, 497, *497*
Cuticle, 145, *145*
Cystitis, 548
Cystocele, 551
Cystoscopy, 549
Cysts, skin, 154
Cytokines, 201, 315
Cytokinesis, 105
Cytoplasm, 93, *95*, 128–129
Cytosine, 58
Cytoskeleton, 96
Cytosol, 95

D

Daphnia, effects of adrenalin and caffeine on, 290–291
Deciduous teeth, *493*
Deep, *9*, 15, 16, *29*
Deep muscles, 238
Deep peroneal nerve, *342*
Deep vein thrombosis, 435, *435*
Defibrillators, 434
Degenerative diseases, 310
    of skin, 152
    of urinary system, 550
Dehydrated, 79
Dehydration, 543
Deltoid (or triangular) muscles, 232, *232*, 233, *233*, 239, *240*
Dementia, genetics of, 360
Demodex, 156
Demyelination, 311, 312
Denaturation, 45
Dendrites, *111*, 296, *296*, 297, 301
    loss of, with age, 361
Denollet, Johan, 431
Dense connective tissue, 109, *109*
Dentin, 179, *179*
Deoxyribonucleic acid (DNA), 58, *95*
    cell damage and, 120
    developmental disorders and, 357
    skin damage and, 158
    structure of, *59*
Deoxyribose, 58
Depolarization, 302
Depressor muscles, 237
Dermal bones, 188
Dermal papillae, 140

Dermatitis, 153
Dermatomyositis, 242
Dermatophytes, 156
Dermis, 139, *139, 142*
Descending colon, 502, *502*
Descending tracts, 340
Desmolase, 570
Desquamation, 140
Detergents, 47
Detrusor muscle, 534
Detumescence, 583
Developmental anatomy, 6
Developmental disorders, of nervous system, 355
Dextrose, 51
Diabetes, 316
    obesity and, 159
    skin sore in patient with, 159
Diabetes insipidus, 279*t*
Diabetes mellitus, 279*t*
Dialysis clearance adequacy, 548
Diaphragm, 20, *28*, 116, *375*, 384, *384*, *418*
Diaphysis, 189, *190*
Diarrhea, 511
Diarthrosis joints, 194
Diastole, 425, *429*
Diencephalon, 336
Diet
    hair loss and, 138
    heart disease and, 432
    Krebs cycle and, 99
Dietary muscle loss, 244–245
Differential WBC count, 453
Differentiation, 76, 102
Diffuse endocrine structures, 260
Diffusion, 86, *86*, 88
Diffusion gradient, 87
Digestive organs, microscopic identification of, 524–525
Digestive process, 508–510
Digestive system, 112, *114*, 118, 489–527
    aging of, 514–516
    components of, *490*
    exercise and, 522–523
    fun facts about, 504
    glandular structures of, 504–507
    overview of, 490–491
    pathology of, 510–513
Digestive tract, 490
    components of, 491–503
Diglycerides, 48
Dilation, 413
Diploid cells, 103
Directional orientation, 8
Directional planes, 8
Directional terms, 13, 13–14, *28*
Disaccharides, 50, *52*
Distal, 8, *9, 29*
Distal convoluted tubule, 538, *539*
    in nephron, *537*
    reabsorption in, 543
Distal phalanx, *196, 197*
Diuresis, 545
Diuretics, 549
    hypertension treatment and, 559–560
    weight loss and, 543
Diverticulosis, 515, *515*
Division, 102
DNA. *See* Deoxyribonucleic acid
Dopamine, 305, 338, 339
Dormancy, 102
Dorsal, 8, *9, 29*
Dorsalis pedis, *409*
Dorsal recumbent position, 11–12, *12*
Dorsal root, 344
Dorsal root ganglion, 344

Dorsal vein, 580
Drug-breakdown residues, in hair, 146–147
Drug-related crimes, mandatory testing of STDs and, 602
Drug screening, urine samples for, 541
Ductless glands, 262
Ductus arteriosus, 423
Ductus deferens, *117, 576,* 578
Duodenum, *275,* 500, *500, 505*
Dura mater, 335
Dysfunction, 119
Dysphagia, 511
Dysplasia, 119
Dyspnea, 437
Dystrophy, 119
Dysuria, 548

E

Ear
    external, middle, and inner, *352*
    inner, *353*
    internal components of, 178
Eardrum, 353
Ear wax, 143
Eateral canthus, *350*
Eccrine sweat glands, *142,* 143, 150
ECGs. *See* Electrocardiograms
Ectoderm, 107, 137
Ectopic pregnancies, 593
Edema, 549
Effectors, 264, 304, 308
Efferent arteriole, 538
Efferent nerves, 331
Eggs, 103, 277, 583
Ejaculation, 583
Ejaculatory ducts, 579
Elastin, 109
    aging and, 157
Elbow
    joint, *172*
    superior radioulnar joint of, *195*
Electrical conduction system, 423
Electrical energy, 83
Electrocardiograms, 413, 428
Electrocardiography basics, 427–429, *428*
Electrolytes, 44, 224
Electromagnetic radiation, 62
Electromagnetic spectrum, 62
Electrons, 37
Electron transport chain, 99
Elements, 34, 37
    periodic table of, 34, *35*
Elephantiasis, 474, *474*
Ellipsoid joints, 194
Embryogenesis, 584
Embryology, 6, 583, *585*
Embryonic stem cell, cell cycle of, *315*
Emergency medical technician, 406
Emphysema, 388, *389*
Empodocles, 4
EMT. *See* Emergency medical technician
Emulsion, 47
Enamel, tooth, 179, *179*
Encephalitis, 311
Endergonic chemical reactions, 83
Endocarditis, 432
Endocardium, *419,* 420
Endochondral bones, 188
Endochondral ossification, 198
Endocrine disruptors, 266, 287
Endocrine glands, *261,* 268–278
    adrenal glands, 271–272
    gonads, 277–278
    hormones and, 259–291
    microscopic identification of, 289–290
    pancreas, 275–276
    pineal gland, 270

    pituitary gland, 268–270
    thymus gland, 276
    thyroid gland and parathyroid gland, 273–274
Endocrine secretions, 260, 265–266
Endocrine system, *113,* 118, *262*
    aging of, 280–281
    function of, 260
    pathology of, 278, 279*t*
Endocytosis, 91, *91*
Endoderm, 107
Endogenous induction, *314*
Endolymph, 353
Endometrium, 571, 582
Endomysium, 235
Endoneurium, 333
Endoplasmic reticulum, 95, 96
Endoscopes, 549
Endoskeleton, 172
Endosteum, *190*
Endotoxins, 310
Energy, 32, 83
Energy drink consumption, and runner's collapse, 30, 63
Enlarged heart, 433
Enterocytes, 501
Enteroendocrine cells, 501
Enterokinases, 509
Entropy, 61
Envirome, 78
Enviromics, 78
Environmental hormones, case study on, 287–288
Environmental Protection Agency, 401, 486
Environmental signals, 260
Enzymatic reactions, *84*
    energy and, 83–85
Enzymes, 55
Eosinophils, 453, 457
EPA. *See* Environmental Protection Agency
Ependymal cells, 111, *112,* 298, *299*
Epicanthic fold, *350*
Epicardium, 419, *419*
Epidermis, 138, *139,* 139–140, *142*
Epididymis, *576,* 577
Epigastric region, 17, *17*
Epiglottis, *273, 375, 377, 378,* 379, 492, *494*
Epimysium, 235, 236
Epinephrine, 272
Epineurium, 331
Epiphyseal line, 189, *191*
Epiphyseal plate, 189
Epiphysis, 189, *190*
Epithelial tissue, 107, *108,* 108–109
Epithelium, types of, *108*
EPSP. *See* Excitatory postsynaptic potential
Equatorial plane, 105
ER. *See* Endoplasmic reticulum
Erasistratus, 4, 5
Erectile dysfunction, 582
Erectile tissue, 572
Erection, 580
Erythroblastosis fetalis, 453
Erythrocytes, 450
Erythropoiesis, 460
Erythropoietin, 460
eSkeletons Project Web site, 221
Esophagus, *114,* 490, *490, 494, 498,* 498–499
    microscopic identification of, 524
Estrogen, 49, 277, 281, 580, 581
    environmental, 287
ETC. *See* Electron transport chain
Ethanol, 41
Ethmoid bones, 177
Ethmoid labyrinth, 178
Ethmoid sinuses, 376

Ethrythroblasts, 460, *461*
Eukaryote cell cycle, *103*
Eukaryotes, 94
　differences in gene expression between prokaryotes and, 102*t*
Eustachian tube, 352, 353, *354, 377, 494*
Eversion, *13*, 14
Excitable membrane, 227
Excitatory effect, 304
Excitatory postsynaptic potential, 307, *307*
Excretion, 91
Exercise
　digestive system and, 522–523
　stretching and, 174
Exhalation, 384, *384*
Exocrine glands, *261*
Exocrine secretions, 260
Exocytosis, 91, *91*
Expiration, 384, *384*, 386
Expiratory reserve volume, 394
Extension, *13*, 14
Extensor muscles, 237
Extensor retinaculum, *233*
External aging, 156
External auditory meatus, 352, *352*
External ear, 352
External environment, 79
External genitalia, female, 567, *568*
External oblique muscle of abdomen, *232*
External respiration, 384
External rotation, *13*
External stimuli, 294
External urethral sphincter, 535
Extrapyramidal tracts, 340
Extremities, 183
Extrinsic aging factors, 156
Extrinsic muscles, 350, 496
Eye, 349
　external features of, *350*
　internal features of, *351*
　sockets, 177
Eyelashes, 349
Eyelids, 349, *350*

F

"Face lifts," 241
Facial artery, *409*
Facial bones, 177, *183*
Facial nerve, 344*t*, *352*
Facial veins, *414*
Facilitated diffusion, 87
Falicform ligament, 506
Fallopian tubes, *117*, 567, *568*, 570, *570*, 583
False ribs, 182, *182, 197*
False vertebrae, 180
Familial periodic paralysis, 242
Fascia, 141
Fascicles, 235
Fasciitis, 141
Fatigue, effect of, on grip muscle action, 255, *256, 257*
Fats, in diet, 60
Fat soluble, 47
　vitamins, 49
Fatty acids, 47, *48*
　saturated, 48
　unsaturated, 48
Fatty change, 119
Fatty diets, Krebs cycle and, 99
FBC. *See* Full blood count
Feces, 509
Federal Trade Commission, 386
Females
　base femur length equation for, 220
　integumentary system of, *115*
　pelvic bones in, 185, *185*
　sexual arousal in, 583

　sexual cycle in, 580–582
　urethra in, 535
　urinary system in, *118*
　urination control in, *552*
Femoral, *15*
Femoral artery, *409*
Femoral nerve, *342*
Femoral veins, *414*
Femur, *28, 172, 183*, 184, 186, *196*
　head of, *197*
　height of person and length of, 220
　neck of, *197*
Fenestrated capillaries, 415, *416*
Fermentation, 98
Ferriar, John, 425
Fertilization, 570, 583, *583*
　reproductive development after, *586*
Fetal circulation, *394*
Fetal heart, *423,* 423–424
Fetal skull, *200*
Fetus, 585
Fevers, 466
Fibrillation, 433
Fibrin, 459
Fibrinogen, 459
Fibroblasts, 137, 140
Fibroids, uterine, 593, *593*
Fibromyalgia, 204
Fibrous joint, 193
Fibrous pericardium, 418
Fibula, *172, 183,* 186, *196*
"Fight or flight," 345
Filamentous fungi, 94
Filtrate, 540
Filtration, 87
Fimbriae, 570
Finasteride, 266
Fine anatomy, 5
Fine physiology investigations, 6
Fingernails, *115*, 145
First-degree burn, 151, *151*
Fixed vertebrae, 180
Flaccid paralysis, 242
Flagella, 93, 96
Flat bones, 188
"Flat-feet," 186
Flatulence, 509
Flexion, *13, 13*
Flexor muscles, 237
Floating kidney, 550
Floating ribs, *182, 196*
Flu, 391
Fluid mosaic model, 95
Folic acid, 50
Follicle mite, 156
Follicle-stimulating hormone, *262, 269,* 270*t,* 580
Folliculitis, 155
Fontanelles, 199
Food
　antioxidants in, 62
　intolerance to, 510
　meaning of organic label for, 34
Foramen magnum, 178
Foramen ovale, 423
Forearms, *183*
Forebrain, 336
Foreskin, *578,* 579
Formed elements, 450
Formulas
　molecular, 41
　structural, 41
Fovea, 351
Fowler's position, 11, *12*

Fractures
　bone, 200
　osteogenesis imperfecta and, 210
　osteoporosis and, 204
　repair of, *201*
　types of, *200*
Francis, George, 317
Fraternal twins, 584
Freckles, 153
Free nerve endings, 144
Free radicals, 62
Frontal bones, *172, 177, 196*
Frontal lobe, 337, *337,* 338
Frontal muscle, *232*
Frontal sinuses, 376, *377, 494*
Fructose, *50,* 51
FSH. *See* Follicle-stimulating hormone
Fugu (or pufferfish) poisoning, 89
Full blood count, 452
Functional classification, 192
Functional drinks, 180
Functional group, 40, *41*
Functional proteins, 54, 55
Functional residual capacity, 394
Fundic region, of stomach, 499
Fungus (fungi), 93, 94
　skin, 155–156

G

GABA. *See* Gamma-aminobutyric acid
Galactose, *50,* 51
Galactosidase, 495
Galactosylceramide, 312
Gallbladder, *114,* 490, 491, 506–507
　aging of, 516
Gallstones, 507
Gametes, 103, 277
Gamma-aminobutyric acid, 305
Ganglia, 333
Gas exchange (respiration), *385, 387,* 387–388
Gastin, 509
Gastric reflux, 511
Gastric stem cells, 499
Gastrocnemius muscles, *233, 239,* 240
Gastroesophageal reflux disease, 511
Gastrointestinal system, *114*
Gastrula, 585
Gastrulation, 491
Gender
　internal components of, post-fertilization, 586
　skeleton and determination of, 220
Gene expression, 99
　differences in, between eukaryotes and prokaryotes, 102*t*
　stages in, 100
Gene regulatory networks, 101
Genes, 99
　hair color and, 146
　precancerous, 153
　regulation of, *101*
Genetic code, 99
Genetic engineering, 78
Genetic expression, 96
Genetics
　of aging, 92
　DNA function and, *100*
Genetic skin disorders, 154–155
Geniculate, *15, 15*
Genitalia
　female, 567, *568*
　male, *579,* 579–580
Genital warts, 593
Genome, 93
GERD. *See* Gastroesophageal reflux disease
Gestation
　animal times of, 571*t*

facts about, 571
GH. *See* Growth hormone
Ginseng, pregnancy and, 84
Gladiolus, 183
Glands, 143, *143*
    endocrine, *261*
    exocrine, *261*
Glandular lobules, 504
Glandular secretions, 260
Glans, 579
Glaucoma, 351
Glial cells, types of, *112*
Gliding joint, 194, *195*
Glioma, 357
Globulin injections, 471
Glomerular filtration, 538
Glomerulonephritis, 549
Glomerulus, 536, *537, 539,* 540
Glossopharyngeal nerve, 344t
Glottis, *378,* 379
Glucagon, 275, 276
Glucocorticosteroids, 271
Glucose, 42, *50,* 51
    sources of, 71–72
Glutamate, 305
Gluteus maximus, *233,* 240
Gluteus medius, *233*
Glycemic index, 60
    of food, factors affecting, 72–73
Glycerides, 47, *48*
Glycerol, *48*
Glycine, 305
Glycogen, 52, 60
Glycogen storage diseases, 242
Glycolysis, 98
Glycosidic linkage, 50
Glycosuria, 547
Golgi apparatus, *95,* 96
Golgi bodies, 95, 297
Gomphosis, 194
Gonadal mosaicism, 595–596
Gonadotropic hormones, *262,* 269
Gonads, 277–278, 566
G_1 (gap 1) phase, of interphase, 103
G_0 stage, 103
Gout, 204
Graafian follicle, 569
Gracilis muscle, *232*
Graded effects, 236
Gradient, 87
Granulocytes, 453, *453*
Graves' disease, 279t
Gray matter, 336
Great adductor muscle, *233*
Greater pectoral muscle, *232*
Greater trochanter, 186, *197*
Great saphenous vein, *414*
Greenstick features, 200
Grip fatigue chart, *256*
Grip muscle action, fatigue and, 255, *256, 257*
GRNs. *See* Gene regulatory networks
Gross anatomy, 5, 6
Gross physiology studies, 6
Gross skeletal muscle types, 231–235
Growth factors, 226
Growth hormone, *262,* 269, 270t
Growth plate, 189
G_2 (gap 2) phase, of interphase, 103
Guanine, 58
Guillette, Louis, 287
Gulf War syndrome, 483
Gyri, of brain, 336

H

Hair, *115, 143, 146,* 146–147
    attachment of, to vein and artery, *147*
    cortex of, 146
    cuticle of, 146
    habits and loss of, 138
    new development of, *147*
    root of, *139*
    shaft of, *142,* 146
Hair bulb, *139, 143*
Hair color, genes and, 146
Hair cycle, 146
Hair follicles, *142, 143*
Hair follicle wall, *139*
Hair medulla, 146
Hair papilla, *139,* 146
Hamate bones, 184
Hamstring muscle group, *240*
Hantavirus pulmonary syndrome, 392
HAP. *See* Hazardous air pollutants
Haploid cells, 103
Hard palate, *492, 492*
Harris, Lynn Edward (born Lynn Elizabeth Harris), 596
Haversian canal, 190
Hazardous air pollutants, 323
hCG. *See* Human chorionic gonadotropin
Head of humerus, *197*
Hearing
    aging and loss of, 361
    balance and, 352–354
Heart, *28, 113,* 408, *417*
    adult, *419,* 420–423
    blood flow through, *426*
    electrical activity of, 427–429
    electrical conduction system in, *426*
    fetal, 423–424
    structure of, 417–424, *418*
Heart disease
    smoking and, 444
    Type D person and, 431
Heart murmurs, 434
Heart rate, 427
Heart sounds, stethoscopes and identification of, 445–446
Heat, 32
Heat regulation, integumentary system and, 149
Heimlich maneuver, 379
Helicobacter pylori (H. pylori), kissing and, 512
Helix secondary structure, 56
Helper T- lymphocytes, 468
Hematocrit, 450
Hematopoietic stem cell, 460
Hematuria, 549
Hemidesmosomes, 140
Hemodialysis, 547, 548
Hemoglobin, 56, 451, 455, 466
Hemogram, 452
Hemophilia, 473
Henle's loop, ascending and descending limbs of, *537*
Heparin, 507
Hepatic artery, 506
Hepatic ducts, *505*
Hepatic flexure, 502, *502*
Hepatic portal vein, 506
Hepatic vein, *414,* 506
Hepatitis, 513
Hepatocytes, 506, *507*
Heptens, 465
Heredity, 97
Hernias, 13, 513
Herophilus, 4, 5
Herpes, in nerve tract, *358*
Herpes infections, 316
Herpes virus, 311
Hexose monosaccharides, 51
H5N1 virus, 383
hGH, male sexual response and, 595

Hiatal hernia, 513
Hierarchies, 76
Hierarchy of human structure, 76–78
Hilum, 462, 506
Hilus, 531
Hindbrain, 336
Hinge joint, 194, *195*
Hip joint, *172*
Hippocrates, 4
Hip replacement, 208, *208*
Hirschsprung's disease, 312
Histamine, 457
Histology, of lung pathology, 403
HIV. *See* Human immunodeficiency virus
Hodgkin's lymphomas, 474
Holocrine glands, 143
Homeostasis, 59
    energy usage and, 83
    urinary system and, 530
Homologous pairs, 105
Hooke, Robert, 93
Hormonal disorders, of urinary system, 549–550
Hormone mimics, 266
Hormone replacement therapy, 281, 595
Hormones
    endocrine glands and, 259–291
    environmental, 287–288
    function of, 264
    lipid, 267
    origination of, 260
    peptide, 267
    pituitary, 269, *269*
    releasing, 269
    types of, 266–267
"Hot-tub lung," 395–396
HPV. *See* Human papillomavirus
HRT. *See* Hormone replacement therapy
Human body orientation, 7–14
    directional orientation, 8, 11–12
    movement, 13–14
Human cells, parts of, *95*
Human chorionic gonadotropin, 584, *584*
    modeling test for, 606–607
Human embryo, embryonic stages of, *585*
Human immunity, 467
Human immunodeficiency virus, 43, 474
Human molecules, 45–46, *46*
Human papillomavirus, 156, 593, 602
Human physiological environment, 79–92
    enzymatic reactions and energy, 83–85
    ions, 81–82
    molecular transport, 86–88
    osmosis, 89–91
Humeroulnar joint, *195*
Humerus, *172, 183,* 184, *196*
Humoral immunity, 470
Hunger center, 508
Hyaline cartilage, 189
Hydatid lung disease, 392, *392*
Hydrocarbons, 47
Hydrocephalus, 339
Hydrogel polymers, 227
Hydrogels, 165
Hydrogen, 39
Hydrogenates, 48
Hydrogen bonds, 40, *40,* 80, *80*
Hydrogen ion acceptors, 44
Hydrogen ion donors, 44
Hydrogen ions (H+), 44, 45
Hydrolysis, 83
Hydrophilic hormone, binding of, to membrane, *264*
Hydrophilic substances, 47
Hydrophobic hormone, crossing through membrane by, *265*

Hydrophobic substances, 47
Hydrostatic pressure, 410, *411*
Hydroxyl group, 42
Hydroxyl ions, 41, 44
Hydroxyl (or alcohol) group, 41
Hymen, *568*, 572
Hyoid bone, 179, *273*, *378*
Hyperosmotic solutions, 89
Hyperparathyroidism, 279*t*
Hyperplasia, 119
Hyperpolarization, 302
Hypersensitivities, 475
Hypertension, 434
    diuretics and, 559–560
Hypertonic environment, 87, 88
Hypertrophy, 120
Hypodermis, 139
Hypogastric region, 17, *17*
Hypoglossal nerve, 344*t*
Hypoglycemia, 276
Hypohphysis, 268
Hyponatremia, 551
Hypoosmotic solutions, 89
Hypopigmentation, 154
Hypospadias, 590
Hypothalamus, 268
    nerve connection to, *262*, *269*
Hypothyroidism, 279*t*
Hypotonic environment, 87, 88

I

IAQ. *See* Indoor air quality
IBD. *See* Inflammatory bowel disease
Identical twins, 585
IgA, 469
IgD, 469
IgE, 469
IgG, 469
IgM, 469
Ileocecal valve, 501, *502*
Ileum, 500, *500*, *502*
Iliac bone, *172*
Iliac veins, *414*
Iliohypogastric nerve, *342*
Iliotibial ligament, *233*
Ilium, 184, *196*
Immune disorders, of urinary system, 549
Immune response, 465–470
    acquired immunity, 466, 468
    inflammation, 470
    innate immunity, 465–466
    primary and secondary responses, 468–470
Immune system, *114*
    thymus gland and, 276
Immunization, 471
Immunodeficiency, 474
Immunoglobulins, 468
Impetigo, 155
Implantation, 584
Impotency, 594
Incisors, *493*, *497*, *497*
Incontinence, urinary, 536
Incus, *354*
Indoor air quality, 486
INDY, aging and, 92
Infectious diseases, 310
    of respiratory system, 391–392
Infectious skin disorders, 155–156
Inferior, 8, *9*, *29*
Inferior concha, *377*, *494*
Inferior lacrimal punctum, *350*
Inferior lobe, of lung, 382
Inferior nasal conchae, 177
Inferior orbital foramen, 178
Inferior palpebra, *350*
Inferior rectus, *351*
Inferior retinaculum of extensor muscle, *232*

Inferior vena cava, *414*, *417*, *418*, *419*, 421, *426*, 502
Infertility, fibroids and, 593
Inflammation, 469, 470
    in urinary system, 548
Inflammatory bowel disease, 512
Inflammatory response, 470
Influenza, 391
Infraspinous muscle, *233*
Ingestion, 508
Inguinal hernia, 513
Inguinal lymph nodes, *463*
Inhalation, 384, *384*
Inherent features, of skin, 136
Inhibitory effect, 304
Inhibitory postsynaptic potential, 307, *307*
Innate immunity, 465
Inner ear, 352, *353*
Innervate, 305
Inositol, 50
Insertion, 231
Inspiration, 384, *384*, 386
Inspiratory reserve volume, 394
Insula, 339
Insulin, 275
Insulin/insulin-like growth factor receptor, aging and, 92
Insulin-like growth factor -1 (IGF-1), 245
Insulin receptor, 275
Integument, 136
Integumentary system, 112, *115*, 118, 136–158, *137*
    aging of, 156–158, *157*
    description of, 136–136
    functions of, 148–152
    histology of, 167
    pathology of, 152–156
    skin appendages, 142–147
    skin structure, 138–141
Intercalated disks, 110, 226
Intercostal nerve, *342*
Interferons, 466
Intermediate gray matter, 340
Internal environment, 79
Internal receptors, 264
Internal respiration, 385
Internal rotation, *13*
Internal stimuli, 294
Internal urinary sphincter, 534
International Society of Regulatory Toxicology and Pharmacology, 483
International Wex Technologies, 89
Interneuron, *111*, *296*
Interphase, 103, *104*
    in meiosis, *106*
Intersex condition, 570, 596
Interventricular septum, *419*
Intervertebral joint, *195*
Intestines
    large, *502*, 502–503
    small, *500*, 500–501
Intra-atrial conduction system, *426*
Intramembranous bones, 188
Intramembranous ossification, 198
Intrapleural pressure, 386
Intrinsic aging factors, 156, 157
Intrinsic beat, 226
Intrinsic muscles, of tongue, 496
Inulin, 52
Inversion, *13*, 14
Involuntary muscle, 110
Involuntary muscle cells, 225
Iodine (I), 81
Ion channels, 229, 230
Ionic bonds, 39, *40*
Ions, 37, 81–82, *82*

IPSP. *See* Inhibitory postsynaptic potential
Iris, *350*, 351
Iron (Fe), 81
Irregular bones, 188
Irritable bowel syndrome, 516
Ischemic attacks, 356
Ischium, *173*, 184, *197*
Islets of Langerhans, 275, 504
Isomers, 41
    chirality and, 42, *42*
Isometric actions, 238
Isoosmotic solutions, 89
Isotonic actions, 238
Isotonic environment, 87, 88
Isotopes, 36
Isotretinoin, prenatal vitamins and, 144

J

Jackson, Michael, 154
Jaw, 497
    adult, *497*
    joint, *172*
Jejunum, 500, *500*
Jenner, Edward, 471
Jock itch, 156
Joint replacement, 208
Joints, *115*, 175, 192–195
    arthritis in, 14
    bouncing, physics of, 203
    functions of, 194, *195*
    structure of, 193, *193*
Jugular veins, *414*
Juvenile arthritis, 204

K

Karyokinesis, 105
Keratin, 140
Keratocytes, 140
Ketones, 276
Kidneys, *28*, *118*, *271*, 530, *531*
    acid-base balance and, *544*
    congenital disorders of, 547
    control of function of, 545
    external anatomy of, 531
    failure of, 548
    function of, *538*
    internal anatomy of, 533
    structure of, *532*
Kidney tubules, *262*, *269*
Kilocalorie, 83
Kinetic energy, 83
Kissing, ulcers and, 512
Knee cap, *183*
Knee-chest position, 12, *12*
Knee-jerk reflex, 325, 326
Knee joint, *172*, *195*
Krabbe's disease, 312
Krause end bulbs, *144*, 145
Krebs cycle, 98, 99
Kt/V. *See* Dialysis clearance adequacy
Kupffer cells, 458
Kuppfer cells, 506

L

Labia, 491
Labia majora, 572
Labia minora, 572
Labium majus, *568*
Labium minus, *568*
Labor, 589
Lacrimal bones, 177, 179
Lacrimal caruncle, *350*
Lacrimal ducts, 349, *350*
Lacrimal gland, 349, *350*
Lactase, 495
Lactation, 573
Lactic acid, 98
Lactiferous ducts, 573

Lactose, 52
Lactose intolerance, 52, 495
   in various populations, 495t
Lacunae, 190
Lally, David A., 174
Lamellated corpuscles, 145
Lamina propria, 498
Langerhans cells, 140
Lanugo, 137
Large intestine, *114,* 490, *490,* 502, 502–503
   disorders of, 515
   microscopic identification of, 525
Larval therapy, 165
Laryngeal prominence, 379
Laryngopharynx, *375,* 376, *377,* 494
Larynx (voice box), *116,* 273, 375, *375,* 378, *378*–379
Lateral, 8, *9, 29*
Lateral condyle, of femur, *197*
Lateral epicondyle, *197*
Lateral fissure, *337*
Latissimus dorsi, *233*
Leeuwenhoek, Antony van, 74, 93
Left atrium, *418, 422, 426*
   of heart, *417*
Left-handed chiral molecules, 42
Left hypochondriac region, 17, *17*
Left inguinal region, 17, *17*
Left lateral position, 12
Left lobe, of liver, 506
Left lower quadrant, 18
Left lumbar region, 17, *17*
Left pulmonary arteries, *418*
Left upper quadrant, 18
Left ventricle, *418, 419, 426*
Legend, 264
Lens, 351, *351*
Lesser trochanter, 186
Leukemia, 473
Leukocytes, 450
Leuteinizing hormone, *262, 269, 270t,* 580
Levator muscles, *232,* 237
Levels of organization, 76
Levodopa, 338
Levulose, 51
Leydig's cells, 577
LH. *See* Leuteinizing hormone
Lice, 156
Lifespan. *See* Wellness and illness over the lifespan
Ligaments, 172, 175
Limbic system, 339
Limbos, *350*
Limonite, *49*
Linear chains of polymers, 47
Lingual membrane, 496
Lingual tonsils, *377, 494,* 496
Lipid hormones, 267
Lipids, 45, 46–50, *48*
   diglycerides, 49
   fatty acids, 48
   glycerides, 47
   sterols, 49
   terpenoids, 49
Lipofuscin, 316
Lipomas, 153
Lipophilic substances, 47
Lips, *492*
Lisinopril, 559
Lithotomy position, 12, *12*
Liver, *114,* 490, 491, *505,* 506–507
   aging of, 516
   organization of, *507*
LLQ. *See* Left lower quadrant
Lobar pneumonia, 392
Lobes, of lung, 382

Lobules, 382
Long adductor muscle, *232*
Long bones, 187–188
Long extensor muscle of big toe, *232*
Long extensor muscle of digits, *232*
Longitudinal cerebral fissure, *337*
Long radial extensor muscle, of wrist, *233*
Longus muscle, 234
Loop of Henle, 538, *539*
   reabsorption in, 543
Loose connective tissue, 109, *109*
Lou Gehrig's disease, 311
Lower respiratory system, 375
Lower uterine segment, 572
Lucero, Cindy, 551
Lumbar nerves, *334*
Lumbar region, of spinal column, 20, *20*
Lumbar vertebrae, *173, 175,* 180
Lumen, 412
Lunate bones, 184
Lung cancer, 382, 391
Lung function models, lung capacity and, 403–405
Lungs, *116,* 374, 375, *375,* 382, 382–383, *418*
   collapse of, 390
Lunula, 145, *145*
LUQ. *See* Left upper quadrant
Lymph, 462
Lymphatic sinuses, 462
Lymphatic system, 112, 118, 141, *463*
   aging of, 475–476, *476*
   overview of, 450
   pathology of, 474–475, 485
Lymphatic trunks, 462
Lymphatic vessels, 410
Lymph capillaries, *411*
Lymphedema, 474, *474*
   managing, 476
Lymph glands, 462
Lymph nodes, *114,* 462, *464*
Lymphocytes (bone marrow), *114,* 453
Lymphoid progenitor cells, 460
Lymphoid stem cell, *461*
Lymphomas, 357, 474
Lymph vessels, *114,* 463
Lysosome, *95,* 96

M

Macrocytic anemia, 473
Macroglia, 298
Macrophages, 458, *461*
Macula lutea, *351*
Mad cow disease, 57, 311, 371
Magnesium (Mg^{2+}), 81
Major basic protein, 457
Male-pattern baldness, 138
Male reproductive system, pathology of, *591*
Males
   base femur length equation for, 220
   cardiovascular system in, *113*
   endocrine system in, *113*
   gastrointestinal system in, *114*
   immune system in, *114*
   musculoskeletal system of, *115*
   nervous system in, *116*
   pelvic bones in, 185, *185*
   reproductive system of, *117,* 576, 576–580
   respiratory system in, *116*
   sexual arousal in, 582
   Tanner staging and development of, *587*
   urethra in, 535
   urinary retention in, 551
   urination control in, *552*
Malleus, 353, *354*
Malnutrition, 60
   protein turnover and, 244
Malpighian layer, 140

Maltose, 52
Mammary glands, *28, 117, 262, 269,* 567, *573,* 573–574
Mandible, *172,* 196
Mandible bones, 177
Mannose, 51
Manubrium, 182, *182*
Marshall, Barry, 512
Mass, 32
Masseter muscle, *239*
Mast cells, 458
Mastication, 508
Mastoid process, 178
Mastoid sinuses, 178
Matrix, 109
Matter, 32
Maxilla, *196*
Maxillary bones, 177
Maxillary sinuses, 178, 376
Maximal heart rate, 436
Maximus muscle, 234
MCHC. *See* Mean corpuscular hemoglobin concentration
MCS syndrome. *See* Multiple chemical sensitivity syndrome
MCV. *See* Mean corpuscular volume
Mean corpuscular hemoglobin concentration, 452
Mean corpuscular volume, 452
Mechanical damage, integumentary system and, 148
Mechanical energy, 83
Meconium, 491
Medial, *9, 29*
Medial canthus, *350*
Medial condyle, of femur, *197*
Medial epicondyle, *197*
Medial orientation, 8
Median nerve, *342*
Mediastinum, 19
Medulla
   kidney, *531*
   renal, *532*
Medulla oblongata, 339
Medullary (marrow) cavity, *190,* 191
Megakaryocytes, 454
Meiosis, 103, *566*
   sexual reproduction by, 105–106, *106*
Meiosis I, 105, *106*
Meiosis II, 105, 106, *106*
Meissner's corpuscles, 144, *144*
Melanin, 168
Melanoblasts, 138
Melanocytes, 138
Melanocyte-stimulating hormone, *262, 269, 270t*
Melanosomes, 140
MELAS, 121
Melasma, 155
Melatonin, 270
Membrane, 86
Membrane diffusion, 87
Memory cells, 468, 470
Mendeleev, Dmitri I., 34
Meninges, 335
Meningioma, 357
Meningitis, 311
Menopause, 594
Menses, 581, *581, 582*
Menstrual cycle, 580, 582
Menstrual flow, 582
Menstrual period, 571
Mental foramina, 179
Menthol, *49*
Merkel cells, 144
Merocrine gland, *143*

INDEX

647

Mesenchyme, 137
Mesenteric veins, *414*
Mesenteries, 501
Mesoderm, 107, 137
Messenger RNA (mRNA), 100
Metabolism, 97, *98*
Metacarpal bones, *172*
Metacarpals, *183*
Metacarpus, *196*
Metals, 81
Metaphase, 104, *104*
Metaphase I, 105, *106*
Metaphase II, *106*
Metaplasia, 120
Metastasis, 120
Metastatic tumors, 357
Metatarsal bones, *172*
Metatarsals, *183*, 186, *196*
Microbes, 93
    cells of, 93–94
    cell structure of, *94*
Microbial damage, integumentary system and, 148
Microcirculation, 410
Microcytic anemia, 473
Microglia, 111, 298
Microglial cell, *112*, *299*
Microorganisms, 93
Microscopes, *6*
    urine analysis with, 540
Microvilli, 501
Micturition, 535
Midbrain, 336, 339
Middle concha, *377*, *494*
Middle ear, 352
Middle phalanx, *196*, *197*
Milk, components of human milk and cow milk, 574, 574t
Millimeters of mercury. *See* mmHg
Mineralocorticosteroids, 271
Minerals, 81
Minimus muscle, 234
Ministroke, 356
Minute respiratory volume, 394
Mitochondria, 96
Mitochondrial myopathies, 242
Mitochondrial myopathy, encephalopathy, lactic acidosis, and stroke syndrome. *See* MELAS
Mitochondria (mitochondrion), *95*, 105
Mitogens, 103
Mitosis, 103, *104*
    cellular aging and, 120
    in human zygote, 584
    in stratum germinativum, 140
Mitral stenosis, 438
Mitral valve, *417*, *421*, *422*, *426*
mmHg, 387
Modified Trendelenburg's position, 11, *12*
Molars, *493*, *497*, *497*
Molecular formulas, 41
Molecular level of organization, 76
Molecular transport, 86–91
    active transport, 90–91
    bulk mechanical transport, 90, *91*
    diffusion, 86–87
    osmosis, 89
    passive transport, 87–88
Molecules, 38
    bioactive, 40, 49
    chiral, 42, *42*
    human, 45–46, *46*
    nutrition and, 59–61
    parts of, 40–42
    properties of, 38–42
    structural, 40

Moles, 153
    ABCDs of, *153*
Monilla, 155
Monoamines, 305
Monocytes, 453, *461*
Monoglycerides, 48
Monomer, 46
Mononuclear WBCs, 454
Monosaccharides, 50, *50*
Mons, 572
Mons pubis, *568*
"Morning-after pill," 582
Morphology, 4, 231
Motor cortex, 338
Motor nerve cells, 226
Motor neuron, *111*, *296*
Mouth, *114*, *116*, *490*, 491–492, *492*, 496–497
Movable vertebrae, 180
Movement, terms for, 13–14
M phase (mitosis), 104
MS. *See* Multiple sclerosis
MSG, 347
MSH. *See* Melanocyte-stimulating hormone
Mucopolysaccharides, 53
Mucosa, 498
Mucous membranes, 136–137
Mucous neck cells, 499
Multiaxial ball-and-socket joint, *195*
Multicellular organisms, 76
Multiple chemical sensitivity syndrome, 483–484
Multiple sclerosis, 312
    thyroid hormone and, 280
Multipolar cells, 297
Multipotent stem cell, 460
Mummies, radiocarbon dating of, 37
Murmurs, heart, 434
Muscle cells
    contraction of, 229–230
    function of, 229–231
    relaxation of, 229, 230–231
    structure of, *228*
Muscle fatigue, grip muscle action and, 255, *256*, *257*
Muscle fiber, 226
Muscle loss, dietary, 244–245
Muscle mass loss, during space mission, 222, 246
Muscles
    ambient temperature and action of, 253–254
    artificial, 227
    atrophy of, 236
    contraction of, 230, *237*
    regeneration of, 238
    sensitivity of, 242
Muscles cells, structure of, 227–228
Muscle tissue, 107, *108*, 109–110
    types of, *110*, *225*, 226
Muscle upregulation, 244
Muscle wasting, progression of, *245*
Muscular dystrophies, 242
Muscularis layer, 498
Muscular system, 112, 118, 223–257
    aging of, 244–245
    muscles, 225–231
    musculature, 231–238
    musculature charts, 238–241
    overview of, 224
Musculature
    human, anterior view of, *232*
    human, posterior view of, *233*
    pathology of, 241–243, *243*
Musculature charts, 238–240
    human, anterior view, *239*
    human, posterior view, *240*

Musculoskeletal system, *16*, 115
Mycobacterium avium, 395
Mycobacterium tuberculosis, 392
Myelin, 111
    thyroid hormone and, 280
Myelin sheath, *111*, *296*, 303
Myeloid progenitor cells, 460
Myeloid stem cell, *461*
Myeloma, 206, *206*
Myocardium, 419, *419*
Myofibril contraction, *230*
Myofibrils, 228
Myofilaments, 227
Myogenesis, 226
Myoglobin, 231
Myoglobinurias, 242
Myometrium, 570
Myopathies, 242
Myosalpinx, 570
Myosin, 227
Myosin filament, *230*
Myositis ossificans, 242
Myotonia, 243

N

Nail fold, 145, *145*
Nail plate, 145
Nails, 145, *145*
    body of, 145, *145*
    root of, 145, *145*
Nares, 375, *376*
Nasal bones, 177, *494*
Nasal cavity, 19, *375*, *376*, *376*
Nasal septum, 180
Nasolacrimal duct, *350*
Nasolacrimal sac, *350*
Nasopharynx, 375, *376*, *377*, *378*, *494*
National Institutes of Health, 78, 204, 590
Natural drugs, synthetic drugs *versus*, 32
Natural immunization, 471
Natural killer (NK) cells, 466
Navicular bones, 184, 186
Necrosis, 120
Negative feedback, 265
Negative feedback loop, *274*
Nephrolith, 548
Nephrons, 536, *537*, 538
    pressures in circulation and, *541*
Nephroptosis, 550
Nerve cell body, 296
Nerve cell diseases, 310
Nerve damage, thyroid hormone and reversal of, 280
Nerves, *116*, *144*, 144–145
    structure of, 331–333
Nerve tracts, 331
    herpes in, *358*
Nervous cell replacement, targets for, *314*
Nervous system, 112, *116*, 118, 293–327
    aging of, 314–316, 359–361
    bananas and, 303
    caffeine and, 332–333
    central, 330
    components of, *334*, 334–346
    overview of, 294
    pathology of, 310–313, *311*, 354–359
    peripheral, 330
    poisoning of, 313
    public awareness of, 330
    structure of, 329–371
    traumatic damage to, 313
Nervous system cells
    neuroglia and stem cells, 298–300
    neurons, 296–297
    types of, *295*, 295–300
Nervous tissue, 107, *108*, 111, 112
Neural communication, types of, *307*

Neural crest cells, 295, 300
Neuralgia inducing cavitational osteonecrosis, 217
Neural injury, consequences of, *314*
Neural pathways, 306
Neural stimulation stage, 229
Neural tube, 294
Neural tube defects, 129
Neuroblastoma, 357
Neurodegenerative disease, 355, 358
Neurofascicles, 331
Neuroglia, 111, *112*, 294, *299*
    degradation of, 358–359, *359*
Neuroglia cells, 298–300
Neurolemmal node, *111*
Neuroma, 357
Neuromuscular disorders, 242
Neuromuscular junction, 229, *229*
Neuromyotonia, 243
Neuron communication, types of, *307*
Neuron excitability, 301
Neuron physiology, 301–305, *302*
Neurons, 111, *112*, *296*, 296–298, *299*
    trivia about, 309
    types of, *111*
    types of communication among, 305–307
Neurotransmitter receptors, 297
Neurotransmitters, 229, 303
    action of, *304*
    neurons and, *304*
    types of, 305
Neurovascular disease, 355
Neutral, 44
Neutrons, 36
Neutrophils, 453, 457, *461*
New England Journal of Medicine, 551, 596
Niacin (vitamin B-3), 50
NICO. *See* Neuralgia inducing cavitational osteonecrosis
NIH. *See* National Institutes of Health
Nipples, 573
Nitrogen, 39
    in blood, 450
Nodes of Ranvier, 299
Nodules, 154
Non-Hodgkin's lymphomas, 474
Nonpolar molecules, 46
Nonspecific immunity, 465
Noradrenaline, 272
Norepinephrine, 272
Nose and nasal cavity, *116*
Nostrils, 375, *376*
Nuclear decay, 36
Nuclear envelope, 95
Nuclear membrane, *95*
Nucleic acids, 45, 57–59, *58*
    oxidation damage and, 62
Nucleoids, 93
Nucleolus, *95*
Nucleotides, 57
Nutrition, molecules and, 59–61

O

OAB. *See* Overactive bladder
Obesity, diabetes and, 159
Oblique fractures, 200
Obturator foramen, 185
Occipital bones, *173*, 177, *197*
Occipital foramen, 178
Occipital lobe, *337*, 339
Occipital muscle, *233*
Ocular, 15
Oculomotor nerve, 344*t*
OI. *See* Osteogenesis imperfecta
Olecranon process, 184
Olfaction, *349*
    purposes of, 348

Olfactory bulb, 348, *349*, *376*
Olfactory nerve, 344*t*
Oligodendrites, 111, 298
Oligodendrocyte, *112*, *299*
Oligopeptides, 54
Oligosaccharides, 50, 53
Oliguria, 536
Oocyte, 569
Open fractures, 200
Optical rotation, 42
Optic cup, 349
Optic disc, *351*
Optic nerve, 344*t*, *351*
Oral cavity, 19, 492
Orbicularis oculi, *239*
Orbicular muscle of eye, *232*
Orbicular muscle of mouth, *232*
Orbital, 15
Orbital ridge, 177
Orbits, 177
Organ, 5, *33*, 77, *77*
Organelles, 94
Organic, meaning of, 34
Organic chemistry, 32, 33
Organism, 5, *33*, 77, *77*
Organization, levels of, *33*, *77*
Organ level of organization, 77
Organ of Corti, 353, *353*, *354*
Organs, systems and, 112, *113*, *114*, *115*, *116*, *117*, *118*
Organ system level of organization, 77
Organ systems, 77
Orgasm, 583
Orientation, directional, 8, 11–12
Origin, 231, 236
Oropharynx, *375*, *376*, *377*, 492, *494*
Osmolarity, 89
Osmosis, 86, 89, *90*
Ossicles, 353
Ossicular chain, *352*
Ossification, 198
Osteoarthtiris, 204
Osteoblasts, *189*, *191*, 198, 201, 202
Osteocalcin, 191
Osteoclasts, *191*, 198, 202
Osteocytes, *189*, 190, *191*
Osteogenesis imperfecta, 209–210
Osteomyelitis, 207
Osteon, 190
Osteonecrosis, 207
Osteoporosis, 204
    close up of, *205*
Osteoporotic vertebrae, normal vertebrae *versus*, *205*
OT. *See* Oxytocin
Oval window, 353
Ovarian cycle, 580, *581*
Ovarian follicles, 569
Ovarian ligament, 569
Ovary(ies), *113*, *117*, 277, *277*, 566, 567, *568*, *569*, 569–570
Overactive bladder, 53
Overhydration, 79
Oviducts, 570
Ovulation, 569, 570, 580
Ovum, 569, *581*
Oxalic acid, 204
Oxidants, 62
Oxidation, 62
Oxidative phosphorylation, 99
Oxygen, 39
    in blood, 450
    respiration and exchange of, 387
    transportation of, in blood, 455
Oxygen-enhancement products, 244
Oxytocin, 262, 269, 270*t*, 570, 589

Ozone, asthma and, 390

P

PABA. *See* Para-aminobenzoic acid
Pacinian corpuscles, *144*, 145
Packed cell volume, 450
Palate, 492
    hard and soft, *376*
Palatine bones, 177, 180
Palatine tonsil, *377*, *494*
Palmar, 16
Palpebrae, 349
Palpebral fissure, *350*
Palsy, 357
Pancreas, *114*, 275, 275–276, *490*, 491, 504, *505*
    aging of, 516
    microscopic slide of, *290*
Pancreatic duct, 275, 505, *505*
Pancreatic zymogens, 506*t*
Pancreatitis, 513
Paneth cells, 501
Pantothenic acid (vitamin B-5), 50
Papillae, 347
Papillary muscles, 421
Para-aminobenzoic acid, 50
"Paracelsus," 540
Paracrine secretions, 265
Parallel muscle cells, 233
Parallel muscles, 232, *234*
Paralysis, muscle, 242
Paranasal sinuses, 375, 376
Parasites, *156*
Parasitic worm, lifecycle of, *514*
Parasympathetic nerves, 345, *345*
Parathyroid gland, *113*, 273–274
    anterior and posterior views of, *273*
Parathyroid hormone, 274
Parenteral nutrition, 508
Parietal, 15, 16
Parietal bones, *173*, 177, *178*, *197*
Parietal cells, 499
Parietal lobe, *337*, 339
Parietal pericardium, *419*
Parietal peritoneum, *28*
Parietal pleura, 382
Parkinson, James, 338
Parkinson's disease, 339
    creative treatments for, 338
Parotid duct, *493*
Parotid glands, *493*, *496*
Partial pressure, 387
    percent composition of air and, 387*t*
Passive immunity, 471
Passive transport, 86, *87*, 87–88
Pasteur, Louis, 93
Patella, *172*, *183*, 186, *196*
Patellar ligament, *232*
Patent ductus arteriosus, *433*
Pathology, 6
Patient positioning, 11–12
PCV. *See* Packed cell volume
PDA. *See* Patent ductus arteriosus
Pearson, Mike Parker, 37
Pectineal muscle, *232*
Pectoral appendages, 183
Pectoral girdle, 183
Pectoralis major, *239*
Pedal, 15, *15*
Pelvic, 15, *15*
Pelvic bones, 184
    in males and females, *185*, 185–186
Pelvic cavity, 18, *19*
Pelvic girdle, 183
Pelvic inflammatory disease, 592
Pelvis, *532*
Penis, *117*, 576, *576*, *579*, 579–580

Pentose, 51
Peptic ulcers, 515
Peptide bonds, 55, *55*
Peptide hormones, 267
Peptides, 45, 53–56
Percent saturation, 455, *456*
Perfluorocarbons, 466
Pericardial cavity, 19
Pericardial space, *419*
Pericarditis, 434
Pericardium, 417, *418*
Perilymph space, *353*
Perimetrium, 571
Perimysium, 235
Perineum, *568,* 572
Periodic table, 34, *35*
Periosteum, 188, *190, 191*
Peripheral nervous system, 330, *342*
    afferent nerve paths and, *332*
    components of, *334,* 341–346
Peripheral neuropathy, 355
Peristalsis, 226, 508
Peritubular capillary system, 538
Pernicious anemia, 472
Peroneal vein, *414*
Pesticides, nervous system and, 317
Peyer's patches, 462, 501
pH, 44
Phagocytosis, 298, 501
Phalanges, *183,* 184
    of finger, *172*
    of toes, *172*
Phallus, 576
Pharyngeal tonsil, *377,* 494
Pharynx, *116,* 375, 376, *377,* 378, *490, 494*
Pheromones, 143, 265
Phosphate group, 41
Phosphate ($PO_4{}^{2-}$), 82
Phospholipids, 49
Phosphorus, 39
Photoreceptors, 351
Phrenic nerve, *342,* 385
pH scale, 44, *45*
Physical properties, 35
Physiological environment, 79
Physiology, 4, 6
Phytoestrogens, 287
Pia mater, 335
PID. *See* Pelvic inflammatory disease
Pigmentation, 138
Pineal gland, *113,* 270
Pinnate muscle cells, 233
Pinnate muscles, 232, *234*
Pisiform bones, 184
Pituitary gland, *113,* 260, 268–270
    aging of, 281
    major secretions of, 270*t*
Pituitary hormones, 269, *269*
Pivotal joint, 194, *195*
Placenta, 584, *584*
Placenta previa, 593, *593*
Plantar, 16
Plantar muscle, *233*
Plaque, 208
    arterial, 431
    in nerve cells, 316
Plasma, 450
Plasma cells, 468
Plasmin, 460
Plasminogen, 460
Plasticity, 360
Plastic surgery, 241
Platelets, 450, 454, *454,* 458–460, *459, 461*
Pleura, 382
Pleural cavities, 19
Pleuritis (pleurisy), 382

Plicae circulares, *500*
Pluripotential stem cell, *461*
Pneumonia, 392
Pneumothorax, 19, 390
PNS. *See* Peripheral nervous system
Poisoning
    anatoxin, 317
    of nervous system, 313
Polarity, 45
Polar molecules, 45
Pollution
    cytoplasmic function and, 128
    endocrine disruptor, 287
    ozone, 390
Polycystic kidney disease, 547
Polymers, 46
Polymorphonuclear WBCs, 453
Polypeptides, 54
Polyps, colon, 513, *513*
Polysaccharides, 50, 52–53, *53*
Polyunsaturated fats, 48
Polyuria, 536
Pons, 339
Popliteal, *409*
Popliteal lymph nodes, *463*
Popliteal vein, *414*
Port wine stains, 154
Posterior, 8, *9,* 29
Posterior pituitary gland, *262,* 269, *269*
    major secretions of, 270*t*
Postganglionic neurons, 345, 346
Postovulation phase, of female sexual cycle, 580–581
Postsynaptic neuron, 297
Postural changes, aging and, 208
Potassium (K+), 82
    in bananas, 303
    sodium balanced with, 272
Potential energy, 83
Precancerous genes, 153
Preganglionic neurons, 345, 346
Pregastric factors, 508
Pregnancy, 583–586, 589
    acne and, 144
    breast cancer and, 592
    ginseng and, 84
Premolars, 497
PremRNA, 101
Preovulation phase, of female sexual cycle, 580
Prepuce, *578, 579*
Presynaptic neuron, 297
Primary bronchi, 379
Primary immune response, 468, *468*
Primary structure, of amino acids, 56
Primates, comparative anatomy of, 221
Prions, 57, 93, 371
Prisons, controlling spread of sexually transmitted diseases in, 602
PRL. *See* Prolactin
Procaine hydrochloride, 341
Product, 85
Progesterone, 49, 277, 581
Programmed cell death, 96
Prohormones, recreational athletes and, 266
Prokaryotes, differences in gene expression between eukaryotes and, 102*t*
Prokaryotic cells, 93
Prolactin, *262, 269,* 270*t*
Prolapse, 435, 594
Proliferative phase, of uterine cycle, 581, *581*
Pronation, *13*
Pronator muscles, 237
Pronator teres muscle, *232*
Prone position, 11, *12*
Propanol isomer, *41,* 42
Prophase, 104, *104*

Prophase I, 105
Propogate, 301
Prostacyclin, 458
Prostaglandins, 267
Prostate cancer, 592, 594
Prostate gland, *117, 576, 578, 579*
    hypertrophy of, 551
Protease zymogens, 509
Protein, 54
    antacids and digestion of, 525–527
    contractile, 225
    dietary, 60–61
    functional, 54, 55
    Krebs cycle and, 99
    oxidation damage and, 62
    storage, 55
    structural, 54
    structure of, *56*
    synthesis of, 99
    turnover, 244
Proteinurea, 549
Prothrombin, 459
Protista, 93
Proton, 36
Proximal, 8, *9,* 29
Proximal convoluted tubule, 538
    in nephron, *537*
    reabsorption in, 542
Proximal phalanx, *196, 197*
P-R (PQ) interval, 428, *428*
Prusiner, Stanley B., 57
Pseudostratified epithelial cells, *108*
Pseudostratified epithelium, 108
Psoriasis, 154
PTH. *See* Parathyroid hormone
Puberty, 566
    sex hormones and, 280
Pubic, 15, *15*
Pubic arch, 185
Pubic bone, *173,* 184
    female, *568*
Pubic hair, 146
    Tanner staging for development of, *589*
Pubic symphysis, 185
Pubis, 184
Pulmonary arteries, *419*
Pulmonary arterioles, *382*
Pulmonary capillaries, *417*
Pulmonary circulation, 417, *417*
Pulmonary semilunar valve, *419*
Pulmonary valve, 421
Pulmonary veins, *414, 419,* 421, *426*
Pulmonary venules, *382*
Pulse, 408
Pupil, *350,* 351
Pupil reflex, 325
Pure molecules, 38
Purines, 58
Purkinje system, 423
P wave, in electrocardiogram, 428, *428*
Pyelitis, 548
Pyelonephritis, 548
Pyloric region, of stomach, 499
Pyloric sphincter, 499, *499,* 509
Pyridoxine (vitamin B-6), 50
Pyrimidines, 58
Pyruvic acid, 98
Pyuria, 548

Q

QRS complex, 428, *428*
Q-T interval, *428,* 429
Quadrangular cartilage, 375
Quadrant naming system, 18
Quadrate lobe, 506
Quadriceps muscle, *232, 234,* 239
Quaternary structure, 56

Q wave, in electrocardiogram, 428, *428*

R

Rabies, 311
Radial artery, *409*
Radial flexor muscle, of wrist, *232*
Radial glia, 299
Radial nerve, *342*
Radioactive isotopes, 36
Radiocarbon dating, 36
  of mummies, 37
Radius, *172, 183*, 184, *196*
Raffinose, 53
Ramon y Cajal, Santiago, 296
Rancho Bernardo Studies, 281
Randolph, Theron G., 483
Ranvier's nodes, 299
RBCs. *See* Red blood cells
Reabsorption
  in distal convoluted tubule and collecting duct, 543
  in loop of Henle, 543
  in proximal convoluted tubule, 542
Receptor cell, *111*
Receptor-mediated endocytosis, 91
Receptors, 262
  insulin, 275
  target cells and, *263*
Recovery phase, 302
Rectum, 490, *502*, 503, *503, 568*
  microscopic identification of, 525
Rectus abdominus, *232*
Red blood cell (RBC) count, 448
Red blood cells, 450, 451–453, *452, 461*
  disorders of, 473
  function of, *455*, 455–457
Red bone marrow, 192
Red pulp, 463
Reduced, 83
Redundancy, 360
Reflex arc, 308
Reflexes, *308*, 308–309
Reflux, 499
Refractory period, 302
Regulatory DNA, 99
Regurgitation, 435
Releasers (or releasing hormones), 269
Renal artery, 531, *531, 532*
Renal cell carcinoma, 550
Renal columns, 531
Renal corpuscle, *537*, 538
Renal cortex, 533
Renal fascia, 531
Renal function disorders, 546–550
Renal medulla, 531
Renal pelvis, 531
Renal pyramids, 533
Renal tubules, 536
Renal vein, 531, *531, 532*
Repolarized neurons, 302
Reproduction
  asexual, 104–105
  sexual, 105–106
Reproductive system, 112, 118, 565–607
  aging of, 594–595
  female, *117*, 567–574
  female, internal and external, *568*
  male, *117, 576*, 576–580
  male, pathology of, *591*
  overview of, 566
  pathology of, 590–593
Reproductive tract, 568–572
  female, 567
RER. *See* Rough endoplasmic reticulum
Respiration, 384
  external, 384
  internal, 385

Respiratory bronchiole, 381
Respiratory diseases, smoking and, 386
Respiratory system, 112, *116*, 373–405
  aging of, 393–394
  components of, *375*, 375–383
  lower, 375
  overview of, 374
  pathology of, 388–392
  upper, 375
Respiratory tract, lower and upper, *375*
Resting potential, 301
Restrictive lung disease, 391
"Rest to digest," 345
Reticulocytes, 451, *461*
Retina, 350, *350*, 351, *351*
Retroperitoneal position, 531
Reverberating pathway, 306
Reverse peristalsis, 508
R group, 55
Rheumatic heart disease, 434
Rheumatoid arthritis, 204
Rh factor, 453
Rhodopsin, 352
Rhomboideus muscles, 232, 233, *234*
Rib cage, *183*
  vertebral column and, 180–183
Riboflavin (vitamin B-2), 50
Ribonucleic acids (RNA), 58, *95*
Ribose, 58
Ribosomes, 95
Ribs, *172*, 182, *182, 196*
Right atrium, *417, 418, 422, 426*
Right-handed chiral molecules, 42
Right hypochondriac region, 17, *17*
Right inguinal region, 17, *17*
Right lobe, of liver, 506
Right lower quadrant, 18
Right lumbar region, 17, *17*
Right pulmonary arteries, *418*
Right upper quadrant, 18
Right ventricle, *418*, 419, *426*
Rigid paralysis, 242
Rigor mortis, 231
Ringworm, 156
RLQ. *See* Right lower quadrant
RNA. *See* Ribonucleic acids
Robotic hands, breast cancer screening and, 155
Robotic surgery, urinary system and, 549
Rods, 351, 352
Rotator muscles, 237
Rough endoplasmic reticulum, *95*, 96
Round window, *353*, 354
  of cochlea, 353
Ruffini receptors, *144*, 145
RU486, 582
Rugae, in urinary bladder, 534, *534*
Runner's collapse, energy drink consumption and, 30, 63
RUQ. *See* Right upper quadrant
R wave, in electrocardiogram, 428, *428*

S

Sacral nerves, *334*
Sacral plexus, *342*
Sacral region, of spinal column, 20, *20*
Sacral vertebrae, 180
Sacroiliac joint, *172*
Sacrum, *173*, 182, *196, 197*
Saddle joint, 194, *195*
"Safer" cigarettes, 386
Sagittal suture, 178
Saliva, 347, 496
Salivary amylase, 508
Salivary glands, *114, 490*, 491, 496
Salmonella, 511
Salt taste, 347, *348*

Sarcolemma, 227
Sarcomere, 228
Sarcomere shortening, 229
Sarcoplasmic reticulum, 228
SARS. *See* Severe acute respiratory syndrome
Sartorius muscle, *232*, 233
Satellite cells, 299
Satiety center, 508
Saturated fatty acids, 48
Scaphoid bones, 184
Scapula, *28, 173, 183*, 184, *196, 197*
Schleiden, Matthias, 93
Schwann, Theodor, 93
Schwann cells, 299, 331
Sciatic nerve, *342*
Sclera, 350, *350, 351*
Scleroderma, 204
Scrapie disease, 57
  in sheep brain, 371, *371*
Scrotum, 576, *578*
Sebaceous glands, *139, 142*, 143, *143*
Sebaceous hyperplasia, 153
Seborrhoeic keratosis, 153
Sebum, *142*, 143
Secondary bronchi, 381
Secondary sex characteristics, 566
Secondary structure, 56
Second-degree burn, 151, *151*
Second-hand smoke, 19
Secretary phase, of ovarian cycle, *581*
Secretin, 509
Secretion, 91
Selectively permeable membrane, 86
Sella turcica, 178
Semen, 578
Semicircular canals, 353, *353*, 354
Semilunar valves, 421, *422, 426*
Semimembranous muscle, *233*
Seminal vesicles, *117, 576, 578*
Seminal vessels, 576, *578*, 578–579
Seminiferous tubules, 577, *578*
Semitendinous muscle, *233*
Senescence, 61
Sensation, 150
Senses (human), 347–354
  hearing and balance, 352–354
  smell, 348
  taste, 347
  vision, 349–352
Sense strand, 101
Sensory cortex, 339
Sensory loss, with age, 361
Sensory nerves, 150
Sensory neurons, *111*, 296
Sensory receptors, 144
Septum, 420
  of heart, *426*
SER. *See* Smooth endoplasmic reticulum
Serosa, 18, 498
Serotonin, 270
Serous pericardium, 418
Serratus muscles, 234
Serum globulins, 507
Sesamoid bones, 188
Severe acute respiratory syndrome, 383
Sex hormones, 280
Sexual dimorphism, 566, 586
Sexual division, 103
Sexual intercourse, 582–583
Sexually transmitted diseases, 590
  mandatory methods in control of, 602–603
Sexual reproduction, 105–106
  copulation, 582–583
  embryology and pregnancy, 583–586, 589
  female sexual cycle, 580–582
  meiosis, 105–106, *106*, 566

SGCs. *See* Specialized germ cells
Shaft
   hair, *139*
   penile, *578*
Shaken-baby syndrome, 356
Sheep brain, with scrapie, 371, *371*
Sheet secondary structure, 56
Shin splint, 203
Shmaefsky, Nicole, 355
Shoichet, Molly, 202
Short bones, 188
Shoulder girdle, 183
Shoulder joint, *172, 195*
Sickle cell disease, 473
Side chain, 55
Sigmoid colon, 502, *502*
Simple columnar epithelial cells, *108*
Simple cuboidal epithelial cells, *108*
Simple epithelium, 108
Simple squamous epithelial cells, *108*
Sim's position, 12, *12*
Sinoatrial (SA) node, 423, *426, 427*
Sinuses, 19, 415
   paranasal, 376
Sinusoids, 506, *507*
SIR2/SIRT genes, aging and, 92
Sitostanol, *49*
Sitosterol, *49*
Sitting position, 11, *12*
Skeletal muscles, 109, *110, 115, 225,* 226
   action of, *236,* 236–238
   gross types of, 231–235
   structure of, *235,* 235–236
Skeletal system, 112, 118, 171–221
   aging of, 207–208
   anterior view of, *172*
   appendicular skeleton, 174, 183–186
   axial skeleton, 174, *175,* 176–183
   pathology of, 203–207
   posterior view of, *173*
Skeleton, determination of gender from, 220
Skene's glands, 572
Skin, *115,* 135–169, *262, 269*
   blood supply to, *149*
   functions of, *149*
   laser to, *150*
   trivia about, 148
Skin appendages, 142–147
   glands, 143
   hair, 146–147
   nails, 145
   nerves, 144–145
Skin cancer, 153
   signatures of, *153*
Skin disorders
   genetic, 154–155
   infectious, 155–156
Skinfold appendages, *142*
Skinfold calipers, *141*
Skin needling, 158, *158*
Skin structure, 138–141, *139*
   dermis, 140
   epidermis, 139–140
   subcutaneous layer, 141
Skin tags, 153
Skull, 176, 177–180, *183*
   aging of, *208*
   fetal, *200*
Sleep apnea, 391
Small intestine, *114,* 490, *490,* 500, 500–501
   microscopic identification of, 525
Smallpox, 471
Smell, 348
Smoking
   heart disease and, 444
   nervous cell aging and, 315
   respiratory diseases and, 386

skin aging and, 158
   thoracic cavity and, 19
Smooth endoplasmic reticulum, *95,* 96
Smooth muscle, 109, *110, 225,* 226
Society, 78
Sodium chloride (NaCl), 39
Sodium (Na+), 82, *82*
   potassium balanced with, 272
Sodium/potassium pumps, 229
Soft palate, 492, *492*
Solar lentigene, 153
Soleus muscle, *233*
Solute, 80
Solution, 80
Solvent, 80
Soma, 296
Somatic cells, 104
Somatic nerves, 341, 345
Sound waves, movement of, *354*
Sour taste, 347, *348*
Space mission, muscle mass loss during, 222, 246
Spasms, muscle, 242
Specialized germ cells, 566
Specific heat, 81
Sperm, 103, 277, 583
Sphenoidal air sinus, *494*
Sphenoid bones, 177
Sphenoid sinuses, 376, *377*
Sphincter muscles, 237, *505*
Sphincter of Oddi, *505*
Sphincters, urinary, 534
Spider veins, 154
Spinal accessory nerve, *342*
Spinal cavity, 20
Spinal column regions, 20
Spinal cord, *116,* 334, 335, *342,* 345
   components of, 339–341
   cross section of, *340*
Spinal nerves, 342, 344
Spindle fibers, 105
Spine of scapula, *197*
Spiral fracture, 200
Spleen, *114,* 462, *463*
Splenic flexure, 502, *502*
Splenius muscle of head, *233*
Spongy bone, *190,* 191, *191*
Spontaneous pneumothorax, 19
Sprains, 241
Squamousal suture, 178
Squamous cells, 108
SSSS. *See* Staphylococcal scalded skin syndrome
S (synthesis) phase, of interphase, 103
Stabs, 462
Stapes, 353, *354*
Staphylococcal scalded skin syndrome, 155
Staphylococcus aureus (S. aureus), 155
Starch, 52
STDs. *See* Sexually transmitted diseases
Stem cells, 107, *107,* 298, *300*
   cell cycle of, *315*
   muscle regeneration and, 238
   research on, 78
   training of, 78
Stent, *432*
Sternocleidomastoid muscle, *232,* 239, *240*
Sternum, *28, 172,* 182, *182, 183, 196,* 380
   regions of, 182
Steroids, 49
   abuse of, 251
   anabolic, 278
Sterols, 47, 49, *49*
Stethoscope, heart sounds identified with, 445–446
Stiff-man syndrome, 243
S-T interval, *428,* 429
Stomach, *114,* 490, *490,* 499, *499,* 500

microscopic identification of, 525
   peptic ulcers of, 515
   regions of, 499
Storage proteins, 55
Strains, muscle, 241
Stratified columnar epithelial cells, *108*
Stratified squamous epithelial cells, *108*
Stratum basale, 139
Stratum compactum, 140
Stratum corneum, 140, *142*
Stratum germinativum, 139, *142*
Stratum granulosum, 140
Stratum lucidum, 140
Stratum spinosum, 140, *142*
Strawberry hemangiomas, 154
Stress
   cell damage and, 120
   hair loss and, 138
Stress fracture, 204
Stretching, exercise and, 174
Striations, 110
Strokes, 356
Stroke volume, 427
Stroop, John Ridley, 369
Stroop effect, brain function and, 369–370
Structural classification, 192
Structural DNA, 99
Structural formulas, 41
Structural molecules, 40
Structural proteins, 54
Styloid process, 178
Subacute carbon dioxide intoxication, 456
Subarachnoid space, 335
Subatomic particles, 34
Subclavian veins, 414
Subcutaneous layer, 139, *139,* 141
Subcutaneous tissue, *142*
Sublingual glands, *492, 493,* 496
Submandibular duct, *492, 493*
Submandibular glands, *493,* 496
Submandibular lymph nodes, *463*
Submucosa, 498
Substrate, 85
Sucrose, 52
Sudden cardiac death, 434–435
Sulfate (SO$_4$$^{2-}$), 82
Sulfhydryl group, 41
Sulfur, 39
Sun exposure, skin aging and, 158
Sunscreen, effectiveness of, 168–169
Superficial, *9,* 15, 16, *29*
Superficial muscles, 238
Superficial peroneal nerve, *342*
Superior, 8, *9, 29*
Superior concha, *377, 494*
Superior lacrimal punctum, *350*
Superior lobe, of lung, 382
Superior longitudinal tongue muscle, 496
Superior palpebra, *350*
Superior radioulnar joint, of elbow, *195*
Superior rectus, *351*
Superior retinaculum of extensor muscle, *232*
Superior vena cava, *414, 417, 418, 419,* 421, *426*
Supination, *13*
Supinator muscles, 237
Supine position, 11, *12*
Suppressor T-lymphocyte, 470
Surface features, 175
Surface receptors, 264
Surfactants, 47, 383
Surfactant treatments, 393
Suspension, 47
Sutural bones, 188
Swallowing, 379
S wave, in electrocardiogram, *428, 428*
Sweat glands, *142,* 143

Sweat pore, *143*
Sweet taste, 347, *348*
Sympathetic ganglion chain, 345
Sympathetic nerves, 345, *345*
Sympathetic postganglionic neurons, 346
Sympathetic trunk, *334*
Symphysis, 194
Symported molecules, 542
Synapse, 297
Synaptic cleft, 297
Synarthrosis joints, 194
Synchondrosis, 194
Syndesmosis, 194
Synergistic effect, in muscles, 237
Synostosis, 194
Synovial capsule, 193
Synovial fluid, 192
Synovial joint, 193
Synthesized, 83
Synthetic drugs, natural drugs *versus*, 32
Syringomas, 153
Systemic capillaries, *417*
Systemic circulation, 417
Systemic lupus erythematosus, 204
Systole, 425, *429*

T

Tactile corpuscles, 144
Taenia coli, 503
Talus, *197*
Talus bones, 186
Tamoxifen, 592
Tamponade, 434
Tangles, 316
Tanner staging
　of female breast, *588*
　of male development, *587*
　of pubic hair development, *589*
Target cells, 262
　receptors and, *263*
Tarsal joint, *172*
Tarsals, *183*, 186
Tarsus, 196
Taste, 347, *348*
Taste buds, 347, *348*
TB. *See* Tuberculosis
TCA. *See* Tricarboxylic acid cycle
T cells, 276
Tears, flow of, *350*
Tectorial membrane, *353*
Teeth, *490*
　adult, *493*
　adult jaw and, *497*
　deciduous, *493*
Telomeres, 121
Telomere shortening, 121
Telophase, 104, *104*
Telophase I, 105
Telophase II, 106
Temperature
　body, endocrine secretions and, 265–266
　muscle action and, 253–254
Temporal artery, superficial, *409*
Temporal bones, *172, 177, 178, 196, 352*
Temporal lobe, *337, 339*
Temporal muscle, *232*
Tendons, *115, 172, 175*
Tensor muscle of fascia lata, *232*
Tensor muscles, 237, 238
Teres major muscle, *233*
Teres minor muscle, *233*
Terminal bronchioles, 381
Terminal hair, 146
Terminus, 296
Terpenoids, 49
Tertiary bronchi, 381
Tertiary structure, 56

Testes, *113, 117,* 566, *576, 577, 577, 578*
Testicular cancer, 592
Testis, 277, *277*
Testosterone, 49, 278, *578*
　virilism and, 281–282
Tetanus, 242
Tetanus bacterium, 310
Tetany, 243, 303
Tetraiodothyronine (T_4), 273
Tetralogy of Fallot, 433
Tetrodotoxin, 313, *313*
　in pufferfish, 89
Thalassemia, 473
Thalidomide®, 43
　myeloma and, 206
Thalidomide baby syndrome, 43
Thermal energy, 83
Thermal stirring, 86
Thiamin (vitamin B-1), 50
Third-degree burn, *151,* 151–152
Thoracic, *15,* 15
Thoracic cage, *182*
Thoracic cavity, 19, *19*
　smoking and, 19
Thoracic duct, *463*
Thoracic nerves, *334*
Thoracic region, of spinal column, 20, *20*
Thoracic vertebrae, *173,* 180, 181, *182,* 197
Thorax, 182
　skeletal aging in, *209*
Threshold, 236, 301
Throat, 116
Thrombin, 459
Thrombocytes, 450
Thrombosis, 435, *435*
Thumb, carpometacarpal joint of, *195*
Thymine, 58
Thymosin, 276
Thymus gland, *113,* 114, 276, *276, 463*
Thyroid cartilage, *377, 378,* 379, *494*
Thyroid gland, *113,* 273–274
　anterior and posterior views of, *273*
　microscopic slide of, *290*
Thyroid hormone, reversal of nerve damage
　and, 280
Thyroid-stimulating hormone, *262, 269, 270t, 273*
Thyrotropin, 273
Thyroxine, 265, 273
　role of, 266
TIAs. *See* Transient ischemic attacks
Tibia, *172, 183,* 186, *196*
Tibial nerve, *342*
Tibial vein, *414*
Tidal volume, 394
Tinea, 156
Tissue level of organization, 77
Tissue mast cell, *461*
Tissue monocytes, 458
Tissue plasminogen activator, 460
Tissues, *5, 33, 77,* 107–112
　connective, 109
　description of, 107
　epithelial, 108–109
　muscle, 109–110
　nervous, 111, *112*
　stem cell, *107*
　types of, *108*
Titin filament, 227
T-lymphocytes, 454, *461, 463, 464,* 468
Tobacco, heart disease and, 444
Toenails, *115,* 145
Tongue, *348, 377, 379, 492, 494*
　base of, *378*
　muscles of, *496*
Tonic control, 315

Tonsils, 114, *377,* 378, *463, 464, 494*
Tooth, 179, *179*
Tooth decay, 205, *206*
Total lung capacity, 394
Toxicological disease, 310
TPA. *See* Tissue plasminogen activator
Trabeculae, *191,* 463
Trabecular bone tissue, 191
Tracheal cartilages, 379
Trachea (windpipe), *116,* 273, 375, *375, 377, 378, 380, 418, 494*
　cartilage ring of, *381*
Traits, 97
Transcription, 100
Transducer, 150
Transduction, 309
Transfer RNA (tRNA), 101
Transfusions, 453
Transient ischemic attacks, 356
Transitional epithelial cells, *108*
Transitional epithelium, 109, 534
Translation, 100
Transmissible spongiform encephalopathy, 371
Transplantation, *314*
Transport vesicles, 95
Transverse colon, 502, *502*
Transverse fractures, 200
Transverse tongue muscle, 496
Trapezium bones, 184
Trapezius (or trapezoid) muscles, 232, *232, 233, 233, 234, 239, 240*
Trapezoid bones, 184
Trauma, nervous system pathology and, 355
Traumatic disease, 310
Tremors, 355, 357–358, *358*
Trendelenburg's position, 11, *12*
Triangular bones, 184
Triangular muscles, 232
Tricarboxylic acid cycle, 99
Triceps brachii, *240*
Triceps muscle, *233,* 234
Tricuspid valve, *417, 419,* 421, *422, 426*
Trigeminal nerve, 343, 344t
Triglycerides, 48, *48*
Trigone, 534
Triiodothyronine (T_3), 273
Trisaccharide, 53
Trochlear nerve, 344t
Trochlear notch, 184
Tropomyosin, 227
Troponin, 227
True ribs, 182, *182*
True vertebrae, 180
TSE. *See* Transmissible spongiform encephalopathy
TSH. *See* Thyroid-stimulating hormone
TTX. *See* Tetrodotoxin
Tuberculosis, 392
Tubular reabsorption, 538
Tubular secretion, 538, 544
Tumors
　brain, 357
　in urinary system, 550
Tunica adventitia, 410
Tunica intima, 412
Tunica media, 412
Turbinates, 376
T wave, in electrocardiogram, 428, *428*
Twins
　conjoined, 585
　fraternal, 584
　identical, 585
Tympanic membrane, *352,* 353, *354*
Tympanic region, 178
Type A persons, 431
Type B persons, 431

Type C persons, 431
Type D persons, heart disease and, 431

U

Ulcers, 513
  kissing and, 512
Ulna, *172, 183,* 184, *196*
Ulnar extensor muscle, of wrist, *233*
Ulnar flexor muscle, of wrist, *232, 233*
Ulnar nerve, *342*
Ultraviolet (UV) light, 62
  sunscreen and, 168–169
Umani taste, 347
Umbilical cord, 530
Umbilical region, 17, *17*
Undernutrition, 60
  protein turnover and, 244
Undescended testis, 577
Unicellular organisms, 76
Unilateral, 15, *16*
Unipolar cells, 297
United States Department of Agriculture, 34
United States Department of Energy, 227
United States Food and Drug Administration, 43
Unsaturated fatty acids, 48
Upper respiratory system, 375
Uracil, 58
Ureters, *118,* 530, 531, *531, 532,* 533, *534, 568, 576*
Urethra, *118,* 530, *531,* 535
  female, *568*
  male, *576*
  opening of, *534*
Urethral meatus, male, *578,* 583
Urethral orifice, 535
Urethritis, 548
Urinalysis, 540
Urinary bladder, *118,* 530, *531, 534, 534, 568, 576*
Urinary meatus, male, *576*
Urinary removal rate, 548
Urinary retention, 551
Urinary system, 112, 118, *118,* 529–563
  aging of, 550–551
  components of, *531*
  gross anatomy of, 530–535
  overview of, 530
  pathology of, 546–550
  robotic surgery and, 549
Urinary tract infection, 548
Urination, 535
Urine
  chemical analysis of, 561–563
  formation of, 538, *539*
  hormonal regulation of formation of, 545–546
  microscopic examination of sediment of, 563
Urine concentration, 543
Urine voiding, 535–536
Urothorax, 26
URR. *See* Urinary removal rate
Uterine cycle, 580
Uterine fundus, 572
Uterus, *117,* 567, 570–571, *571,* 583
  fundus and body of, *568*
  smooth muscle of, *262,* 269
UTI. *See* Urinary tract infection
Uvula, 492, *492*

V

Vaccination, 471
Vaccines, 471
Vacuoles, 96
Vagina, *117,* 567, *568,* 572, *572*

Vaginal orifice, *568*
Vagus nerve, *342,* 344t
Varicose veins, 436, *437*
Vascular dementia, 362
Vascular disorders, 430
Vascularization, 408
Vasculature, 420
Vas deferens, 578
Vasoconstriction, 412
Vasodilation, 413
Vegetables, antioxidants in, 62
Veins, *113, 412,* 415
  varicose, 436, *437*
Vellus hair, 146
Venae cavae, 421
Venous valves, identification of, 446–447
Ventilation, 374
Ventral, 8, *9, 29*
Ventral root, 344
Ventricles, 339, 421, *422*
Ventricular septal defect, 433
Venules, 415
Vermiform appendix, *502*
Vertebrae, 180
  osteoporotic *versus* normal, *205*
Vertebral canal, 335
Vertebral column, *28, 181, 183, 196*
  rib cage and, 180–183
Verticalis tongue muscle, 496
Vesicoureteral reflux, 594
Vestibular nerve, *352*
Vestibule, *353, 354*
Vestibulocochlear nerve, 344t, *354*
Veterinarians, 7
Villi, in small intestine, *500,* 501
Viral diseases, of the skin, 156
Virilism, 281–282
Viriods, 93
Virtual Hospital Medical Atlas Website, 167
"Virtual" patients, 413
Viruses, 93
  neurological disorders and, 358
Visceral, 16
Visceral peritoneum, *28*
Visceral pleura, 382
Vision, 349–352
  loss of, with age, 361
Vital capacity, 394
Vitamin A, *49*
Vitamin B complex, 50
Vitamin C, 50
Vitamin D, 49
  skin and production of, 150
Vitamin E, 49
Vitamin K, 49
  blood clotting and, 459
Vitamins, 49, *49*
  antioxidants in, 62
  fat soluble, 49
  prenatal, and isotretinoin, 144
  water-soluble, 50
Vitiligo, 154–155
Vitreous humor, 351
Vocal cords, *377,* 378, 379, *494*
Voiding, 535–536
Volkmann's canals, 191
Voluntary muscle, 110
Voluntary muscle cells, 225
Vomer bones, 177
Vomeronasal gland, 348, 376
VSD. *See* Ventricular septal defect
Vulva, *117,* 572

W

Wagers, Amy J., 238

Waldeyer, Heinrich, 296
Warren, Robin, 512
Warts, 156
Wasting, 99
Water, 38
  conservation of, 543
  molecular structure of, 80, *80*
  pH of, 44
  specific heat of, 81
Water excess, 79
Water insoluble, 46
Water intoxication, 79, 551
Water soluble, 45
Water-soluble vitamins, 50
Wavelength, 62
WBCs. *See* White blood cells
Weight loss
  amphetamines and, 306
  colonic cleansing and, 501
  diuretics and, 543
Weissman, Irving L., 238
Wellness and illness over the lifespan
  aging of body's chemistry, 61–62
  pathology of cells, 119–120
  pathology of musculature, 241–243
  pathology of nervous system function, 310–313
  pathology of skeletal system, 203–207
  pathology of the blood and lymphatic system, 472–475
  pathology of the cardiovascular system, 430–436
  pathology of the digestive system, 510–513
  pathology of the endocrine system, 278, 279t
  pathology of the integumentary system, 152–156
  pathology of the nervous system, 354–359
  pathology of the reproductive systems, 590–593
  pathology of the respiratory system, 388–392
  pathology of the urinary system, 546–550
Wet nurses, 575
Whiplash, 356
White blood cells, 201, 450, 453–454
  disorders of, 473–474
  lifetime of, 462
White blood cell (WBC) count, 448
White lung controversy, 401–402
White matter, 336
White pulp, 463
WHO. *See* World Health Organization
Wisdom teeth, *497, 497*
Womb, 570
World Health Organization, 590
Wormian bones, 188
Wound healing, larval therapy and, 165
Wrist joint, *172,* 195

X

Xanthines, 332
Xiphoid process, *172, 182,* 183

Y

Yellow bone marrow, 192

Z

Z-line, 228
Z membrane, *230*
Zygomatic arch, 179
Zygomatic bone, 177, *196*
Zygomatic process, 178
Zygote, 584
Zymogens, 505, 506t